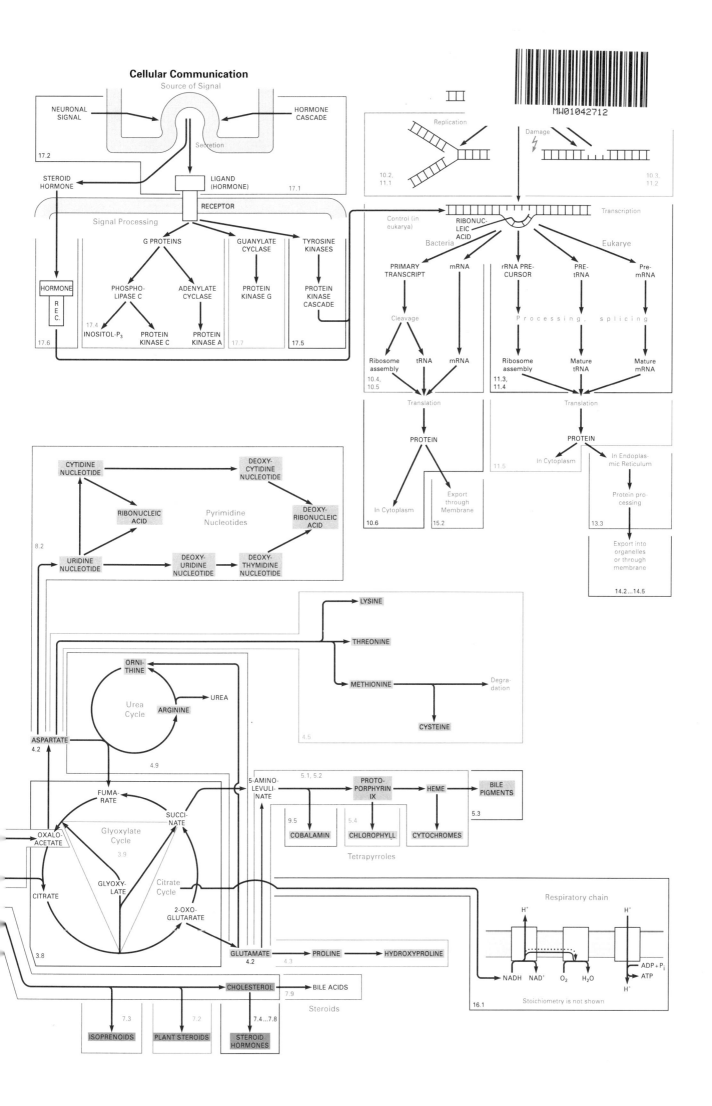

BIOCHEMICAL PATHWAYS:
AN ATLAS OF BIOCHEMISTRY AND MOLECULAR BIOLOGY

BIOCHEMICAL PATHWAYS:

An Atlas of Biochemistry and Molecular Biology

Edited by
Gerhard Michal

A JOHN WILEY & SONS, INC: and SPEKTRUM AKADEMISCHER VERLAG CO-PUBLICATION

 New York · Brisbane · Chichester · Toronto · Singapore · Weinheim

 Heidelberg · Berlin

Address of the editor:

Dr. Gerhard Michal
Kreuzeckstraße 19
D-82327 Tutzing
Germany

First published in Germany by
Spektrum Akademischer Verlag
Copyright © 1999 by Spektrum Akademischer Verlag GmbH, Heidelberg

English language edition published by
John Wiley & Sons, Inc.
Copyright © 1999 by John Wiley & Sons, Inc.

John Wiley & Sons, Inc.	Spektrum Akademischer Verlag
605 Third Avenue	Vangerowstraße 20
New York, NY 10158-0012	D-69115 Heidelberg
USA	Germany
Telephone: (212) 850-6000	Telephone: ++496221/91260

All rights reserved. This book protected by copyright. No part of it, except for brief excerpts for review, may be reproduced, stored in a retrieval system, or transmitted in any form or by any means, electronic, mechanical, photocopying, recording, or otherwise, without permission from the publisher. Requests for permission of further information should be addressed to the Permissions Department, John Wiley & Sons, Inc., 605 Third Avenue, New York, NY 10158-0012

Library of Congress Cataloging-in-Publication Data

Library of Congress Cataloging-in-Publication Data is available.
0-471-33130-9

The text of this book is printed on acid-free-paper.

Printed in Germany
10 9 8 7 6 5 4 3 2 1

Preface

This book is not intended to be a textbook of biochemistry in the conventional sense. There is no shortage of good biochemistry textbooks, which outline how biochemical knowledge has been gained, trace the logical and experimental developments in this field and present advances in their historical sequence.

In contrast, this book tries to condense important aspects of current knowledge. Its goal is to give concise information on the metabolic sequences in the pathways, the chemistry and enzymology of conversions, the regulation of turnover and the effect of disorders. This concentration on the sequence of facts has entailed the omission of researchers' names, experimental methods and the discussion of how results have been obtained. For information on these aspects, and for an introduction into the fundamentals of biological science, it is necessary to consult textbooks.

The scope of this book is general biochemistry, encompassing bacteria (and to some extent archaea), plants, yeasts and animals. Although a balanced representation is intended, personal interest naturally plays a role in the selection of topics. In a number of cases, the chemistry of the reactions is given in more detail, especially at metabolic key and branching points. Human metabolism, its regulation and disorders as a result of disease is a frequent topic. On the other hand, some chapters are especially devoted to bacterial metabolism.

This book grew out of my interest in metabolic interrelationships and regulation which was stimulated by my professional work at Boehringer Mannheim GmbH, Germany. Previously, this interest led me to compile the "Biochemical Pathways" wall chart, the first edition of which appeared 30 years ago. Two more editions followed, which have been widely distributed. As a result of this experience, I developed a preference for the graphic presentation of scientific facts. In contrast to texts, illustrations allow the simultaneous display of different aspects, such as structural formulas, enzyme catalysis and its regulation, the involvement of cofactors, the occurrence of enzymes in various kingdoms of biology, etc. This form of presentation facilitates a rapid overview. A standard set of conventions is used in all illustrations (representation of formulas, symbols for proteins, the use of colors, the shape of arrows, etc. – the rare exceptions are indicated), and this assists in finding the facts quickly.

Tables have been added to provide more detailed information. They list additional properties of the system components, homologies, etc. The text plays only a supportive role. It gives a concise description of the reactions and their regulation, and puts them into the general metabolic context. The arrangement of the text facilitates rapid finding of keywords (underlined), occurrence and location (in italics) and different or alternative metabolic states (listed). Details and discussion of special cases can be distinguished from the main text by the use of smaller print.

In many cases, current knowledge focuses on a limited number of species. A rough classification of the occurrence of pathways is given by the color or the reaction arrows in the illustrations, but both generalizations and specialization are expected to be found in the future, which will necessitate modification of the picture.

Clear representation of the multiple interconnections in metabolism poses a difficult task. In the wall chart, which was the precursor of this book, it was frequently necessary to cut off these interconnections in order to retain legibility. The illustrations in this book provide references to 'key compounds' (e.g., glucose, pyruvate, various amino acids) at the beginning and the end of the pathways in order to place the metabolic sequence into the general context. The interrelationships between these 'key compounds' are shown in the initial Figures 1.1-1 ... 1.1-3 (p. 1 and 2). Likewise, the text contains numerous cross-references to other chapters and sections. They are intended as the textual counterparts to the hyperlinks used in electronic representations. The pertinent decimal classification numbers are shown at the top of each page.

The literature references have been limited in number and they usually cite recent review articles and books, if possible, from readily accessible sources. They were selected to provide more detailed information on new developments and additional references for the interested reader. There are no references to long-established biochemical facts which can be found in any textbook. I hope that this restriction will be acceptable to readers, since a complete listing of all sources for the statements presented here would take up a major portion of this volume. To compensate for the omission of such general references, a special chapter on electronic data banks and major printed sources has been added at the end of the book.

This work could not have been compiled without the expert knowledge and the contributions of many coauthors, whose names are listed at the beginning of this book. They have written a considerable number of the sections. I wish to thank all of them most gratefully for their work and for their open, committed cooperation.

Likewise, I would like to express my best thanks to the scientists who have checked various chapters of the manuscript and have given their valuable advice on the selection of topics, the type of presentation and a large number of details. Besides the persons listed, I have obtained information from many other colleagues, to whom I am very grateful.

Further thanks are due to Spektrum Akademischer Verlag, Heidelberg and especially to Ms. Karin von der Saal who proposed publishing the 'Biochemical Pathways' chart in book form. She has helped me to solve many problems and has constantly furthered this project. I also want to express my thanks to the Universitätsdruckerei Stürtz, Würzburg for their efforts in converting my drafts of the illustrations into a printable form. The management of Boehringer Mannheim GmbH has often shown interest in the project, even after my retirement, and has supported it especially by providing library services. This is gratefully acknowledged.

Most of all I want to thank my wife Dea, who has often encouraged me during the long time required to finish this work. She has given me valuable advice and support in checking the text of the English edition. Without her understanding and her help this book would not have been brought to completion.

Tutzing, June 11, 1998 *Gerhard Michal*

Contributors to This Volume:

Dr. W. Ankenbauer, Boehringer Mannheim GmbH, D-82372 Penzberg (*Section 10.7*)

Doz. Dr. A. Brakhage, Institut für Genetik und Mikrobiologie der Universität, D-80638 München (*Section 15.8*)

Dr. H. Burtscher, Boehringer Mannheim GmbH, D-82372 Penzberg (*Sections 2.6, 11.1...11.5*)

Dr. B. Dohse, formerly Max-Planck-Institut für Biochemie, D-82152 Martinsried (*Section 16.2.1*)

Dr. E. Felber, Micromet GmbH, D-82152 Martinsried (*Section 19.2*)

Dr. A. Grossmann, Boehringer Mannheim GmbH, D-82372 Penzberg (*Sections 6.3.5, 9.6, 20.1...20.5*)

Dr. S. Hansen, Boehringer Mannheim GmbH, D-82372 Penzberg (*Section 11.6*)

Dr. A. Haselbeck, Boehringer Mannheim GmbH, D-82372 Penzberg (*Sections 13.3...13.5, 14.1...14.2, 19.3*)

Dr. C. Hergersberg, Boehringer Mannheim GmbH, D-82372 Penzberg (*Sections 14.3...14.5*)

Dr. W. Hösel, Boehringer Mannheim GmbH, D-82372 Penzberg (*Sections 13.3...13.5, 14.2, 19.3*)

Prof. Dr. M. Klingenberg, Institut für Physikalische Biochemie der Universität, D-80336 München (*Section 16.1*)

Dr. H. Koll, Boehringer Mannheim GmbH, D-82372 Penzberg (*Sections 14.3...14.5*)

Dr. B. König, Boehringer Mannheim GmbH, D-82372 Penzberg (*Sections 12.1...12.4*)

Prof. Dr. G.-B. Kreße, Boehringer Mannheim GmbH, D-82372 Penzberg (*Sections 2.3...2.5, 6.1...6.3, 14.6*)

Dr. M. Kromayer, Institut für Genetik und Mikrobiologie der Universität, D-80638 München (*Sections 10.2, 15.5...15.6*)

Dr. H. Lill, Boehringer Mannheim GmbH, D-82327 Tutzing (*Sections 20.1...20.5*)

Dr. B. Neuhierl, Institut für Genetik und Mikrobiologie der Universität, D-80638 München (*Sections 10.3...10.5, 15.4, 15.7*)

Prof. Dr. E. P. Rieber, Institut für Immunologie der Technischen Universität, D-01011 Dresden (*Section 19.1*)

Dr. L. Rüßmann, Boehringer Mannheim GmbH, D-82372 Penzberg (*Section 21.1*)

Dr. S. Schiefer †, formerly Boehringer Mannheim GmbH, D-82372 Penzberg (*Section 18.2*)

Dr. Y. Schmidt, Idea GmbH, D-81371 München, formerly Anatomische Anstalt der Universität, Lehrstuhl III, D-80336 München (*Section 17.2*)

Dr. R. Siegert, Max-Planck-Institut für Biochemie, D-82152 Martinsried (*Section 16.2.1*)

Dr. J. Wilde, formerly Institut für Physiologische Chemie der Technischen Universität, D-80802 München (*Sections 7.1...7.9, 9.11*)

Dr. R. Wilting, Novo Nordisk A/S, DK-2880 Bagsvaerd, formerly Institut für Genetik und Mikrobiologie der Universität, D-80638 München (*Sections 9.5, 10.6, 14.1, 15.1...15.3*)

The other chapters and sections were written by the editor.

Critical reviews and valuable information were obtained from the following scientists. Their efforts are gratefully acknowledged:

Prof. Dr. A. Böck, Institut für Genetik und Mikrobiologie der Universität, D-80638 München

Prof. Dr. J. Bode, Gesellschaft für Biotechnologische Forschung, D-38124 Braunschweig

Doz. Dr. T. Dondekar, European Molecular Biology Laboratory (EMBL), D-69012 Heidelberg

Prof. Dr. J. Gasteiger, Computer-Chemie-Centrum der Universität, D-91052 Erlangen

Doz. Dr. B. Kehrel, Med. Klinik und Poliklinik der Universität, D-48129 Münster

Prof. Dr. Ch. Lehner, Lehrstuhl für Genetik der Universität, D-95444 Bayreuth

Prof. Dr. H. K. Lichtenthaler, Botanisches Institut der Universität, D-76128 Karlsruhe

Prof. Dr. D. Oesterheldt, Max-Planck-Institut für Biochemie, D-82152 Martinsried

Prof. Dr. M. Savageau, Department of Microbiology and Immunology, University of Michigan, Ann Arbor 48109-0620, MI, USA

Dr. J. Sühnel, Institut für Molekulare Biotechnologie, D-07745 Jena

Doz. Dr. J. Unger, Anatomische Anstalt der Universität, D-80336 München

Dr. E. Wingender, Gesellschaft für Biotechnologische Forschung, D-38124 Braunschweig

Prof. Dr. H. G. Zachau, Institut für Physiologische Chemie der Universität, D-80336 München

Contents

Preface V

Contributors to This Volume VII

1 Introduction and General Aspects 1
1.1 Organization of This Book 1
1.1.1 Conventions Used in This Book 3
1.1.2 Common Abbreviations 3
1.2 Carbohydrate Chemistry and Structure 4
1.2.1 Structure and Classification 4
1.2.2 Glycosidic Bonds 4
1.3 Amino Acid Chemistry and Structure 5
1.3.1 Structure and Classification 5
1.3.2 Peptide Bonds 5
1.4 Lipid Chemistry and Structure 6
1.4.1 Fatty Acids 6
1.4.2 Acylglycerols and Derivatives 6
1.4.3 Waxes 6
1.4.4 Glycerophospholipids (Phosphoglycerides) 6
1.4.5 Plasmalogens 6
1.4.6 Sphingolipids 7
1.4.7 Steroids 7
1.4.8 Lipoproteins 7
1.4.9 Lipid Aggregates and Membranes 7
1.5 Physico-Chemical Aspects of Biochemical Processes 8
1.5.1 Energetics of Chemical Reactions 8
1.5.2 Redox Reactions 8
1.5.3 Transport Through Membranes 9
1.5.4 Enzyme Kinetics 9

2 The Cell and its Components 13
2.1 Classification of Living Organisms 13
2.2 Structure of Cells 13
2.2.1 Prokaryotic Cells 13
2.2.2 General Characteristics of Eukaryotic Cells 14
2.2.3 Special Structures of Plant Cells 16
2.2.4 Special Structures of Animal Cells 16
2.3 Protein Structure and Function 17
2.3.1 Levels of Organization 17
2.3.2 Protein Function 18
2.4 Enzymes 19
2.4.1 Catalytic Mechanism 19
2.4.2 Isoenzymes 20
2.4.3 Multienzyme Complexes 20
2.4.4 Reaction Rate 20
2.4.5 Classification of Enzymes 21
2.5 Regulation of the Enzyme Activity 21
2.5.1 Regulation of the Quantity of Enzymes 21
2.5.2 Regulation of the Activity of Enzymes 21
2.5.3 Site of Regulation 23
2.6 Nucleic Acid Structure 24
2.6.1 Components of Nucleic Acids 24
2.6.2 Properties of RNA Chains 24
2.6.3 Properties of DNA Chains 24
2.6.4 Compaction Levels of DNA Chains 25

3 Carbohydrates and Citrate Cycle 27
3.1 Glycolysis and Gluconeogenesis 27
3.1.1 Glycolysis 27
3.1.2 Regulation Steps in Glycolysis 28
3.1.3 Gluconeogenesis 29
3.1.4 Resorption of Glucose 30
3.1.5 Response of Animal Organs to High and Low Glucose Levels 31
3.2 Polysaccharide Metabolism 31
3.2.1 Structures 31
3.2.2 Biosynthesis of Polysaccharides 31
3.2.3 Catabolism of Polysaccharides 32
3.2.4 Regulation of Glycogen Metabolism in Mammals 32
3.2.5 Medical Aspects 34
3.3 Pyruvate Turnover and Acetyl-Coenzyme A 35
3.3.1 Pyruvate Oxidation 35
3.3.2 Regulation of Pyruvate Deydrogenase Activity 35
3.3.3 Acetyl-Coenzyme A (Acetyl-CoA) 35
3.3.4 Anaplerotic Reactions 36
3.3.5 Initiation of Gluconeogenesis 36
3.3.6 Alcoholic Fermentation 36
3.4 Di- and Oligosaccharides 37
3.4.1 Sucrose 37
3.4.2 Lactose 37
3.4.3 Other Glycosides 37
3.5 Metabolism of Hexose Derivatives 38
3.5.1 Uronic Acids 38
3.5.2 Aldonic Acids 38
3.5.3 Entner-Doudoroff-Pathway 38
3.5.4 Inositol 38
3.5.5 Hexitols 38
3.5.6 Mannose and Deoxy Hexoses 38
3.6 Pentose Metabolism 40
3.6.1 Pentose Phosphate Cycle 40
3.6.2 Other Decarboxylation Reactions 41
3.6.3 Plant Cell Walls 41
3.6.4 Pentose Metabolism in Humans 41
3.7 Amino Sugars 42
3.7.1 Biosynthesis 42
3.7.2 Catabolism 42
3.8 Citrate Cycle 43
3.8.1 Reaction Sequence 43
3.8.2 Regulatory Mechanisms 43
3.8.3 Energy Balance 43
3.9 Glyoxylate Metabolism 45
3.9.1 Glyoxylate Cycle 45
3.9.2 Other Glyoxylate Reactions 45

4 Amino Acids and Derivatives 46
4.1 Nitrogen Fixation and Metabolism 46
4.2 Glutamate, Glutamine, Alanine, Aspartate, Asparagine and Ammonia Turnover 46
4.2.1 Glutamine Metabolism 46
4.2.2 Glutamate Metabolism 47
4.2.3 Alanine Metabolism 47
4.2.4 Aspartate and Asparagine Metabolism 47
4.2.5 Transamination Reactions 48
4.3 Proline and Hydroxyproline 49
4.4 Serine and Glycine 50
4.4.1 Serine Metabolism 50
4.4.2 Glycine Metabolism 50
4.5 Lysine, Threonine, Methionine, Cysteine and Sulfur Metabolism 51
4.5.1 Common Steps of Biosynthesis and Their Regulation 51
4.5.2 Lysine Metabolism 51
4.5.3 Threonine Metabolism 52
4.5.4 Methionine Metabolism 52
4.5.5 Cysteine Metabolism 52
4.5.6 Sulfur Metabolism 52
4.5.7 Glutathione Metabolism 55
4.5.8 Reactive Oxygen Species, Damage and Protection Mechanisms 56
4.6 Leucine, Isoleucine and Valine 57
4.6.1 Biosynthetic Reactions 57
4.6.2 Degradation of Branched-Chain Amino Acids 57
4.7 Phenylalanine, Tyrosine, Tryptophan and Derivatives 59
4.7.1 Biosynthesis of Aromatic Amino Acids 59
4.7.2 Biosynthesis of Quinone Cofactors 60
4.7.3 Derivatives and Degradation of Aromatic Amino Acids 61
4.7.4 Catecholamines 63
4.7.5 Thyroid Hormones 63
4.7.6 Aromatic Compounds in Plants 64
4.8 Histidine 65
4.8.1 Biosynthesis 65
4.8.2 Interconversions and Degradation 65
4.9 Urea cycle, Arginine and Associated Reactions 66
4.9.1 Urea Cycle 66
4.9.2 Phosphagens (Phosphocreatine and Phosphoarginine) 67
4.9.3 Polyamines 67

5 Tetrapyrroles 68
5.1 Steps to Protoporphyrin IX 68
5.2 Hemoglobin, Myoglobin and Cytochromes 69
5.2.1 Heme Biosynthesis 69
5.2.2 Biosynthesis and Properties of Hemoglobin and Myoglobin 69
5.2.3 Oxygen Binding to Hemo- and Myoglobin 71
5.2.4 Cytochromes and Other Heme Derivatives 72
5.3 Bile Pigments and Bilins 72
5.3.1 Hemoglobin Oxidation and Bile Pigments 72
5.3.2 Bilins 73
5.4 Chlorophylls 74

6 Lipids 75

- 6.1 Fatty Acids and Acyl-CoA 75
  - 6.1.1 Biosynthesis of Fatty Acids 75
  - 6.1.2 Regulation of Fatty Acid Synthesis 76
  - 6.1.3 Fatty Acid Desaturation and Chain Elongation 77
  - 6.1.4 Transport and Activation of Fatty Acids 77
  - 6.1.5 Fatty Acid Oxidation 78
  - 6.1.6 Energy Yield of the Fatty Acid Oxidation 79
  - 6.1.7 Ketone Bodies 79
- 6.2 Triacylglycerols (Triglycerides) 79
  - 6.2.1 Biosynthesis of Triacylglycerols (Lipogenesis) 80
  - 6.2.2 Mobilization of Triacylglycerols (Lipolysis) 80
- 6.3 Phospholipids 81
  - 6.3.1 Occurrence of Phospholipids 81
  - 6.3.2 Glycerophospholipids 81
  - 6.3.3 Ether Lipids 83
  - 6.3.4 Sphingophospholipids 83
  - 6.3.5 Choline, Betaine, Sarcosine 84

7 Steroids 85

- 7.1 Cholesterol 85
  - 7.1.1 Biosynthesis 85
  - 7.1.2 Turnover of Cholesterol 86
  - 7.1.3 Function of Cholesterol in Membranes 86
  - 7.1.4 Regulation of Cholesterol Synthesis 86
  - 7.1.5 Cholesterol Homeostasis 86
- 7.2 Hopanoids, Steroids of Plants and Insects 88
  - 7.2.1 Hopanoids 88
  - 7.2.2 Phyto- and Mycosterols 88
  - 7.2.3 Ecdysone 88
- 7.3 Isoprenoids 89
  - 7.3.1 Terpenes 89
  - 7.3.2 *All-trans* Metabolites 89
  - 7.3.3 *Poly-cis* Metabolites 89
  - 7.3.4 Isoprenoid Side Chains 89
- 7.4 Steroid Hormones 91
  - 7.4.1 Biosynthesis 91
  - 7.4.2 Activation and Regulation of Steroid Hormones 91
  - 7.4.3 Transport of Steroid Hormones 91
  - 7.4.4 Degradation of Steroids 91
- 7.5 Gestagen 92
  - 7.5.1 Biosynthesis of Progesterone 92
  - 7.5.2 Gestagen Function, Transport and Degradation 92
- 7.6 Androgens 93
  - 7.6.1 Biosynthesis 93
  - 7.6.2 Transport and Degradation 93
  - 7.6.3 Biological Function of Androgens 93
  - 7.6.4 Medical Aspects 93
- 7.7 Estrogens 94
  - 7.7.1 Biosynthesis 94
  - 7.7.2 Transport and Degradation 94
  - 7.7.3 Biological Function of Estrogens 94
  - 7.7.4 Medical Aspects 94
- 7.8 Corticosteroids 95
  - 7.8.1 Biosynthesis 95
  - 7.8.2 Transport and Degradation 95
  - 7.8.3 Biological Function 95
  - 7.8.4 Medical Aspects 95
- 7.9 Bile Acids 97
  - 7.9.1 Occurence 97
  - 7.9.2 Biosynthesis 97
  - 7.9.3 Regulation of Biosynthesis 97
  - 7.9.4 Medical Aspects 97

8 Nucleotides and Nucleosides 99

- 8.1 Purine Nucleotides and Nucleosides 99
  - 8.1.1 Biosynthesis of Inosine 5′-Phosphate 99
  - 8.1.2 Interconversions of Purine Ribonucleotides 100
  - 8.1.3 ATP and Conservation of Energy 101
  - 8.1.4 Ribonucleotide Reduction to Deoxyribonucleotides 103
  - 8.1.5 Interconversions and Degradation of Purine Deoxyribonucleotides 104
  - 8.1.6 Catabolism of Bases 104
  - 8.1.7 Medical Aspects 104
- 8.2 Pyrimidine Nucleotides and Nucleosides 104
  - 8.2.1 Biosynthesis of Uridine 5′-Phosphate 104
  - 8.2.2 Interconversions of Pyrimidine Ribonucleotides 107
  - 8.2.3 Ribonucleotide Reduction and Interconversions of Pyrimidine Deoxyribonucleotides 107
  - 8.2.4 Catabolism of Bases 107
  - 8.2.5 Medical Aspects 107

9 Cofactors and Vitamins 108

- 9.1 Retinol (Vitamin A) 108
  - 9.1.1 Biosynthesis and Interconversions 108
  - 9.1.2 Biochemical Function 108
- 9.2 Thiamin (Vitamin B_1) 108
  - 9.2.1 Biosynthesis 108
  - 9.2.2 Biochemical Function 108
- 9.3 Riboflavin (Vitamin B_2), FMN and FAD 109
  - 9.3.1 Biosynthesis and Interconversions 109
  - 9.3.2 Biochemical Function 109
- 9.4 Pyridoxine (Vitamin B_6) 110
  - 9.4.1 Biosynthesis and Interconversions 110
  - 9.4.2 Biochemical Function 110
- 9.5 Cobalamin (Coenzyme B_{12}, Vitamin B_{12}) 111
  - 9.5.1 Biosynthesis of the Coenzyme and Reduction of the Vitamin 111
  - 9.5.2 Biochemical Function 111
  - 9.5.3 Siroheme and Coenzyme F_{430} 111
- 9.6 Folate and Pterines 113
  - 9.6.1 Tetrahydrofolate/Folylpolyglutamate 113
  - 9.6.2 General Reactions of the C_1 Metabolism 113
  - 9.6.3 Tetrahydrobiopterin 113
  - 9.6.4 Molybdenum and Tungsten Cofactors 113
  - 9.6.5 Methanopterin 113
- 9.7 Pantothenate, Coenzyme A and Acyl Carrier Protein (ACP) 115
  - 9.7.1 Biosynthesis and Interconversions 115
  - 9.7.2 Biochemical Function 115
- 9.8 Biotin 116
  - 9.8.1 Biosynthesis and Interconversions 116
  - 9.8.2 Biochemical Function 116
- 9.9 Nicotinate, NAD^+ and $NADP^+$ 117
  - 9.9.1 Biosynthesis and Degradation of NAD^+ and $NADP^+$ 117
  - 9.9.2 Mechanism of the Redox Reactions, Stereospecificity 117
  - 9.9.3 Biochemical Function of the Nicotinamide Coenzymes 118
- 9.10 Ascorbate (Vitamin C) 118
  - 9.10.1 Biosynthesis and Interconversions 118
  - 9.10.2 Biochemical Function 118
- 9.11 Calciferol (Vitamin D) 120
  - 9.11.1 Biosynthesis and Interconversions 120
  - 9.11.2 Biochemical Function 120
- 9.12 Tocopherol (Vitamin E) 120
- 9.13 Phylloquinone and Menaquinone (Vitamin K) 121
- 9.14 Other Compounds 121
  - 9.14.1 Lipoate 121
  - 9.14.2 Essential Fatty Acids ('Vitamin F') 121
  - 9.14.3 Essential Amino Acids 121

10 Nucleic Acid Metabolism and Protein Synthesis in Bacteria 122

- 10.1 Genetic Code and Information Transfer 122
  - 10.1.1 From DNA to RNA 122
  - 10.1.2 From Nucleic Acids to Proteins – The Genetic Code 122
  - 10.1.3 Influence of Errors 122
- 10.2 Bacterial DNA Replication 123
  - 10.2.1 Cell Cycle and Replication 123
  - 10.2.2 Initiation of Replication 123
  - 10.2.3 Elongation and Termination 123
  - 10.2.4 Fidelity of Replication 124
- 10.3 Bacterial DNA Repair 126
  - 10.3.1 DNA Damage 126
  - 10.3.2 Direct Reversal of Damage 126
  - 10.3.3 Excision Repair Systems 126
  - 10.3.4 Mismatch Repair 126
  - 10.3.5 Double-Strand Repair and Recombination 128
  - 10.3.6 SOS Response (Damage Tolerance Mechnism) 128
- 10.4 Bacterial Transcription 129
  - 10.4.1 RNA Polymerase 129
  - 10.4.2 Transcription 129
  - 10.4.3 Products of Transcription 129
  - 10.4.4 Fidelity of Transcription 130
  - 10.4.5 Inhibitors of Transcription 130
- 10.5 Regulation of Bacterial Transcription 131
  - 10.5.1 Regulation at the Initiation Step 131
  - 10.5.2 Regulation of Elongation 131
  - 10.5.3 Modification of Transcription Termination 132
  - 10.5.4 Integration of Metabolism by Stimulons 132
- 10.6 Bacterial Protein Synthesis 133
  - 10.6.1 Components of the Bacterial Translation System 133
  - 10.6.2 Aminoacylation of tRNAs 133

10.6.3 Polypeptide Synthesis 134
10.6.4 Fidelity of Translation 135
10.6.5 Selenocysteine 135

10.7 Degradation of Nucleic Acids 136
10.7.1 Exodeoxyribonucleases (*Exo*-DNases) 136
10.7.2 Endodeoxyribonucleases (*Endo*-DNases) 136
10.7.3 Ribonucleases (RNases) 136

11 Nucleic Acid Metabolism, Protein Synthesis and Cell Cycle in Eukarya 137

11.1 Eukaryotic DNA Replication 137
11.1.1 Cell Cycle and DNA Replication 137
11.1.2 Initiation of Replication 137
11.1.3 DNA Polymerases 138
11.1.4 Replication Forks 139
11.1.5 Telomeres 139
11.1.6 Fidelity of Replication 140

11.2 Eukaryotic DNA Repair 141
11.2.1 DNA Damage and Principles of Repair 141
11.2.2 Direct Reversal of Damage 141
11.2.3 Excision Repair System 141
11.2.4 Mismatch Repair 141
11.2.5 Double-Strand Repair and Recombination 141
11.2.6 DNA Repair and Human Diseases 142

11.3 Eukaryotic Transcription 143
11.3.1 RNA Polymerases 143
11.3.2 mRNA Transcription by RNA Pol II 143
11.3.3 Processing of mRNA 145
11.3.4 snRNA Transcription 146
11.3.5 rRNA Transcription by RNA Pol I 146
11.3.6 Processing of rRNA 146
11.3.7 tRNA Transcription by RNA Pol III 146
11.3.8 Modification/Processing of tRNAs 147
11.3.9 5S rRNA Transcription by RNA Pol III 147
11.3.10 Inhibitors of Transcription 148

11.4 Regulation of Eukaryotic Transcription 149
11.4.1 Structure of Core Promoter DNA Elements 149
11.4.2 Structure of Specific Transcription Factors 149
11.4.3 Modulation of the Transcription Rate 149

11.5 Eukaryotic Protein Synthesis 151
11.5.1 Components of the Translation System 151
11.5.2 Polypeptide Synthesis 151
11.5.3 Posttranslational Protein Processing 153
11.5.4 Translational Regulation 153
11.5.5 mRNA Degradation 153

11.6 Cell Cycle in Eukarya 154
11.6.1 Cyclins and Cyclin-Dependent Kinases 154
11.6.2 Regulation of G_1 to S Phase Transition in Yeast 154
11.6.3 Control of the Pre-Replication Complex Assembly in Yeast 154
11.6.4 Regulation of the G_1 to S Phase Transition in Mammals – The Role of the Rb Protein 155
11.6.5 Regulatory Mechanisms During M Phase (Mitosis) 156
11.6.6 Cell Cycle Checkpoints 157
11.6.7 Protein Degradation 157

12 Viruses 158

12.1 General Characteristics of Viruses 158
12.1.1 Genomic Characteristics 158
12.1.2 Structure 158

12.2 DNA Viruses 159
12.2.1 Bacteriophage λ 159

12.3 RNA Viruses 160
12.3.1 Tobacco Mosaic Virus 160

12.4 Retroviruses 161
12.4.1 Human Immunodeficiency Virus (HIV) 162

13 Glycosylated Proteins and Lipids 163

13.1 Glycosylated Proteins and Peptides 163
13.1.1 Glycoproteins 163
13.1.2 Proteoglycans 163
13.1.3 Peptidoglycans 164
13.1.4 Glycoprotein Degradation Diseases and Mucopolysaccharidoses 164
13.1.5 Repeating Units of Glycosaminoglycans as Components of Proteoglycans 164

13.2 Glycolipids 165
13.2.1 Glycosphingolipids 165
13.2.2 Glycoglycerolipids 166
13.2.3 Glycosylphosphopolyprenols 166

13.3 Protein Processing in the Endoplasmic Reticulum 167
13.3.1 Protein Synthesis and Import Into the Endoplasmic Reticulum (ER) 167
13.3.2 Location of the ER Proteins 167
13.3.3 Synthesis of Dolichol-Bound Oligosaccharides and N-Glycosylation 167
13.3.4 Formation of Lipid-Anchored Proteins in the ER 168
13.3.5 Acylation of Proteins 168

13.4 Glycosylation Reactions in the Golgi Apparatus 169
13.4.1 Formation of Glycoproteins 169
13.4.2 Formation of Proteoglycans 169
13.4.3 Formation of Glycolipids 169

13.5 Terminal Carbohydrate Structures of Glycoconjugates 171
13.5.1 Blood Groups 171

14 Protein Folding, Transport and Degradation 172

14.1 Folding of Proteins 172
14.1.1 Protein Folding in Bacteria 172
14.1.2 Protein Folding in the Eukaryotic Cytosol 172
14.1.3 Protein Folding in the Eukaryotic Endoplasmic Reticulum 172

14.2 Vesicular Transport and Secretion of Proteins 173
14.2.1 Pathways of Transport 173
14.2.2 Transport Vesicles 173

14.3 Protein Transport Into the Nucleus 174
14.3.1 Targeting Mechanism 174
14.3.2 Transport Mechanism 174

14.4 Protein Transport Into Mitochondria 175
14.4.1 Targeting Mechanism 175
14.4.2 Transport Mechanism 175

14.5 Protein Transport Into Chloroplasts 176
14.5.1 Targeting Mechanism 176
14.5.2 Transport Mechanism 176

14.6 Protein Degradation 177
14.6.1 Classification of Peptidases 177
14.6.2 Reaction Mechanism of Serine Peptidases 177
14.6.3 Reaction Mechanism of Cysteine Peptidases 177
14.6.4 Reaction Mechanism of Aspartate Peptidases 177
14.6.5 Reaction Mechanism of Metallopeptidases 177
14.6.6 Peptidase Inhibitors 177
14.6.7 Protein Degradation by the Ubiquitin (Ub) System 177

15 Special Bacterial Metabolism, Antibiotics 179

15.1 Bacterial Envelope 179

15.2 Bacterial Protein Export 180

15.3 Bacterial Transport Systems 181
15.3.1 Types of Active Transport 181

15.4 Bacterial Fermentations 182

15.5 Anaerobic Respiration 185
15.5.1 Redox Reactions and Electron Transport 185
15.5.2 Methanogenesis 186
15.5.3 Acetogenesis by CO_2 Fixation 186

15.6 Chemolithotrophy 187

15.7 Alkane and Methane Oxidation, Quinoenzymes 189

15.8 Antibiotics 190
15.8.1 Penicillin and Cephalosporin 191
15.8.2 Streptomycin 191
15.8.3 Erythromycin 192
15.8.4 Tetracycline 192

16 Oxidative Phosphorylation and Photosynthesis 193

16.1 Oxidative Phosphorylation 193
16.1.1 Energy Balance and Reaction Yield 193
16.1.2 Electron Transport System in Mitochondria 193
16.1.3 Bacterial Electron Transport Systems 195
16.1.4 H^+ Transporting ATP Synthase 195
16.1.5 Redox Potentials in the Respiratory Chain 196

16.2 Photosynthesis 196
16.2.1 Light Reaction 196
16.2.2 Dark Reactions 199

17 Cellular Communication 201

17.1 Intercellular Signal Transmission by Hormones 201
17.1.1 General Characteristics of Hormones 201
17.1.2 General Characteristics of Receptors 201
17.1.3 Insulin and Glucagon 202
17.1.4 Epinephrine and Norepinephrine (Catecholamines) 202
17.1.5 Hypothalamus-Anterior Pituitary Hormone System 202
17.1.6 Placental Hormones 206
17.1.7 Hormones Regulating the Extracellular Ca^{++}, Mg^{++} and Phosphate Concentrations 206

17.1.8	Hormones Regulating the Na$^+$ Concentration and the Water Balance 206	19		**Antimicrobial Defense Systems 235**
17.1.9	Hormones of the Gastrointestinal Tract 207	19.1		Immune System 235
17.2	Nerve Conduction and Synaptic Transmission 208		19.1.1	Cells of the Non Adaptive Immune Defense System 235
17.2.1	Membrane Potential 208		19.1.2	Development and Maturation of the Cellular Components 235
17.2.2	Conduction of the Action Potential Along the Axon 208		19.1.3	Antigen Recognition by B Lymphocytes 236
17.2.3	Transmitter Gated Signalling at the Synapse 208		19.1.4	Antigen Recognition by T Lymphocytes 239
17.2.4	Voltage Gated Signalling at the Synapse 210		19.1.5	Antigen Presentation by MHC Molecules 241
17.2.5	Postsynaptic Receptors 211		19.1.6	Cytokines and Cytokine Receptors 242
17.2.6	Axonal Transport 211		19.1.7	Regulation of the Immune Response 245
17.3	Principles of Intracellular Communication 212		19.1.8	IgE Mediated Hypersensitivity of the Immediate Type 246
17.4	Receptors Coupled to Heterotrimeric G-Proteins 213	19.2		Complement System 247
17.4.1	Mechanism of Heterotrimeric G-Protein Action 213		19.2.1	Activation of the Complement Pathways 247
17.4.2	cAMP Metabolism, Activation of Adenylate Cyclase and Protein Kinase A 214		19.2.2	Formation of the Membrane Attack Complex (MAC), Lysis of Pathogens and Cells 247
17.4.3	Activation of Phospholipase C and Protein Kinase C 215		19.2.3	Other Effects of the Complement System 248
17.4.4	Metabolic Role of Inositol Phosphates and Ca^{++} 216		19.2.4	Control Mechanism of the Complement System 248
17.4.5	Muscle Contraction 216		19.2.5	Medical Aspects 248
17.4.6	Visual Process 219	19.3		Adhesion of Leucocytes 249
17.4.7	Olfactory and Gustatory Processes 219			
17.4.8	Arachidonate Metabolism and Eicosanoids 220	20		**Blood Coagulation and Fibrinolysis 251**
17.5	Receptors Acting Through Tyrosine Kinases 222	20.1		Hemostasis 251
17.5.1	Regulatory Factors for Cell Growth and Function 222	20.2		Initial Reactions 252
17.5.2	Components of the Signal Cascades 222		20.2.1	Reactions Initiated at the Tissue Factor 252
17.5.3	Receptor Tyrosine Kinases 222		20.2.2	Contact Activation 252
17.5.4	Tyrosine Kinase Associated Receptors 224		20.2.3	Generation of Binding Surfaces 252
17.6	Receptors for Steroid and Thyroid Hormones, for Retinoids and Vitamin D 227	20.3		Coagulation Propagation and Control 253
			20.3.1	Requirements for Protease Activity 253
17.7	Cyclic GMP Dependent Reactions and Effects of Nitric Oxide (NO) 228		20.3.2	Pathways Leading to Thrombin 253
			20.3.3	Key Events 253
17.7.1	Membrane Bound Guanylate Cyclases 228		20.3.4	Controlled Propagation 254
17.7.2	Soluble Guanylate Cyclases and Their Activation by Nitric Oxide 228		20.3.5	Generation of Fibrin 254
		20.4		Platelets (Thrombocytes) 255
17.7.3	Protein Kinase G (PKG) 228	20.5		Fibrinolysis 257
			20.5.1	Pathways of Plasminogen Activation 257
18	**Eukaryotic Transport 229**		20.5.2	Control of Fibrinolysis 257
18.1	Systems of Eukaryotic Membrane Passage 229			
18.1.1	Channels and Transporters 229	21		**Further Information 258**
18.1.2	Import by Endocytosis and Pinocytosis 231	21.1		Electronic Storage of Biochemical Information 258
18.1.3	The Cytoskeleton as Means for Intracellular Transport and Cellular Movements in Eukarya 231	21.2		Printed Sources 260
18.2	Plasma Lipoproteins 232	22		**Index 262**
18.2.1	Apolipoproteins (Apo) 232			
18.2.2	Plasma Lipoprotein Metabolism 232			
18.2.3	Lipid Transport Proteins 233			
18.2.4	Lipoprotein Receptors 234			
18.2.5	Lipid Metabolic Disorders 234			

1 Introduction and General Aspects

1.1 Organization of This Book

This book deals with the chemistry of living organisms. However, this topic cannot be considered in an isolated way, but has to be placed into a more general context. In two introductory chapters, a short outline of interconnections with neighboring sciences is given. **Chapter 1** deals with the organic chemistry of important components present in living organisms and with the physical chemistry of reactions. **Chapter 2** describes the overall organization of cells and their organelles as well as the structure of proteins and nucleic acids. This is followed by a discussion of enzyme function, which depends on the protein structure and regulates almost all biological processes.

The biochemistry of living beings is a complicated network with multiple interconnections. **Figures 1.1-1...1.1-3** give a simplified survey of the main metabolic pathways in order to allow quick location of the detailed descriptions in this book. The decimal classification numbers in the various boxes refer to chapters and sections.

Chapters 3...8 are devoted to general metabolism, focusing on small molecules (carbohydrates, amino acids, lipids including steroids, nucleotides).

Figure 1.1-1, which abstracts these chapters, shows only biosynthetic pathways and sequences passed through in both directions (amphibolic pathways). This avoids a complicated presentation. (In the chapters, however, the degradation pathways of these compounds are usually dealt with immediately following the biosynthesis reactions.) Most of the compounds mentioned here are 'key compounds', which appear in the detailed figures later in this book either at the beginning or at the end of the reaction sequences. The classification of these compounds into chemical groups is indicated by the color background of the names.

Chapter 9 deals with vitamins and cofactors, which are involved in many reactions of general metabolism.

Chapters 10 and 11 describe the storage of information in DNA and its translation into proteins by bacteria and eukaryotes, respectively.

Figure 11.1-2 gives a short outline of these reactions, subdivided into bacterial reactions (left) and eukaryotic reactions (right).

Viruses, which utilize these mechanisms in hosts, are dealt with in **Chapter 12**.

Glycosylations of the formed proteins and related reactions with lipids are the subject of **Chapter 13**. **Chapter 14** deals with the folding and transport mechanisms of proteins.

Specialized bacterial reactions, including energy aspects, are described in **Chapter 15**. Aerobic respiration and its central role in energy turnover, as well as the photosynthetic reactions, which are the source of almost all compounds in living beings, are dealt with in **Chapter 16**.

Chapter 17 has the topic of cellular communication and of regulation mechanisms employed by multicellular organisms.

Figure 1.1-3 summarizes these multiple interconnections in a very short way. More details can be obtained from Figure 17.1-3.

Chapter 18 gives a survey of transport mechanisms, which transfer bulk compounds between cells.

Chapter 19 deals with the defense mechanisms of higher animals and **Chapter 20** with blood coagulation.

Every presentation can only contain a selection of the present knowledge. For this reason, the final **Chapter 21** is intended to assist in obtaining further information from electronic sources, which offer the most comprehensive collection of scientific results available today, as well as from printed sources.

Key to the Background Colors:
green = carbohydrates
blue   = amino acids
red   = lipids including steroids
orange = nucleotides
brown = tetrapyrroles
none   = compounds involved in general interconversions

The colors of the frames are for easy differentiation only.

Figure 1.1-1. Biosynthetic Reactions in General Metabolism

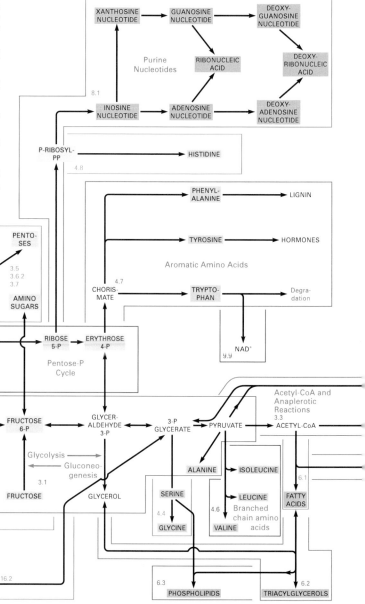

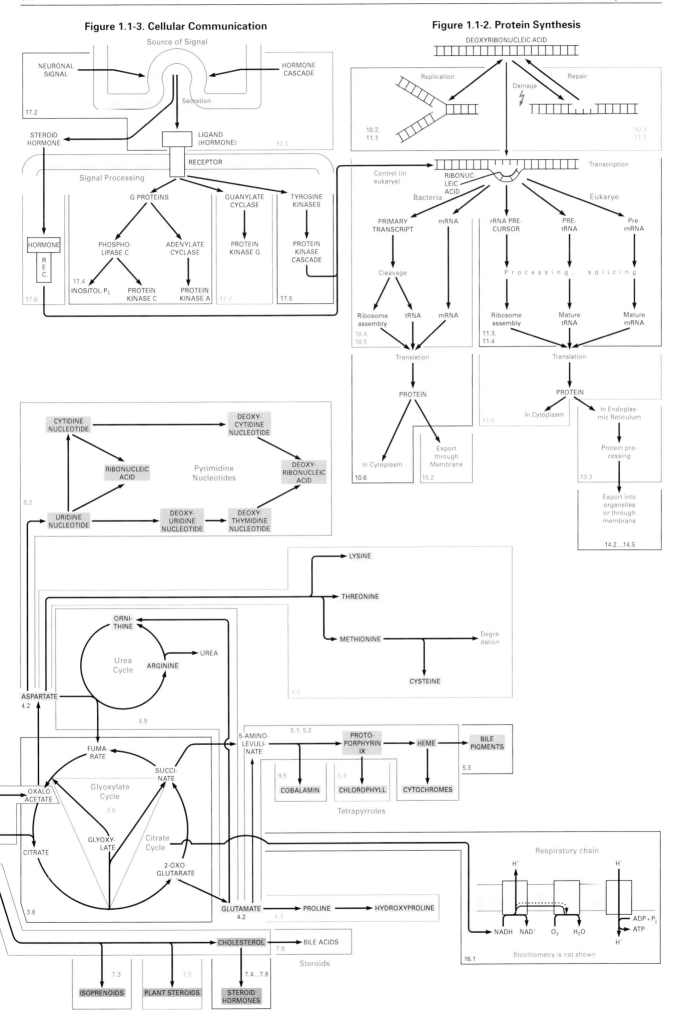

1.1.1 Conventions Used in This Book

1. A decimal classification system is used throughout with the subdivisions chapters-sections-subsections. Figures, tables and formulas are assigned to the sections, e.g. Fig. 9.6-1.

Reactions:

2. Whenever available, the 'Recommended Names' in the *Enzyme Nomenclature 1992* (Academic Press) are used for enzymes and substrates.

3. Substrates of enzymatic reactions are printed in black, enzymes in blue, coenzymes in red. Regulatory effects are shown in orange. This color is also used for pathway names and for information on the location of a reaction. For numbering systems, green is used.

4. The color of the reaction arrows shows where the reaction was observed (or at least where reasonable indications for its occurrence exist): black = general pathway, blue = in animals, green = in plants and yeasts, red = in prokarya (bacteria and archaea).

5. Bold arrows indicate main pathways of metabolism.

6. Points on both ends of an arrow ↔ indicate noticeable reversibility of this reaction under biological conditions. Unless expressly noted, this type of arrows does not indicate mesomeric (resonance stabilized) states of a compound, contrary to the use in organic chemistry.

7. Double arrows ⇌ are used, when the interconversion of two compounds proceeds via different reactions in both directions (e.g., for some steps of glycolysis).

8. Dashed reaction arrows show conversions with primarily catabolic (degradative) importance, full line arrows either mainly anabolic (biosynthetic) reactions or reactions in biological systems which are frequently passed through in both directions (amphibolic reactions).

Regulation:

9. Necessary cofactors and activating ions etc. are printed in orange next to a reaction arrow.

10. Full line orange arrows with an accompanying ⊕ or ⊖ indicate that the respective factor exerts 'fast' activation or inhibition of the reaction (by allosteric mechanisms or by product inhibition etc.). Dashed arrows are used, if the amount of enzyme protein is regulated, e.g., by varied expression or by changes in the degradation rate. If only one of multiple enzymes is regulated this way, it is indicated by Roman numbers.

Enzymes and Proteins:

11. When enzyme complexes are involved, the respective components are schematically drawn in blue-lined boxes with rounded edges. This does not express the spatial structure. If possible, interacting components are drawn next to each other.

12. When a sequence of domains occurs in a protein, special symbols are used for the individual domains. They are explained next to the drawing.

13. When the peptide chain has to be shown, helices are drawn as ⌇⌇⌇ (e.g., in transmembrane domains), otherwise they are symbolized as ∿.

Abbreviations and Notations:

14. Organic phosphate is generally abbreviated as –P, inorganic phosphate and pyrophosphate as P_i and PP_i, respectively. Only in drawings, where the reaction mechanism is emphasized, phosphate residues are shown as $-O-PO_3^{2-}$.

15. Braces { } are used for atoms or residues, which formally enter or leave during a reaction, if the molecular context is unknown.

16. While notations for genes are usually printed with small case letters (e.g., *raf*), the respective gene products (proteins) are written with a capitalized first letter (e.g., Raf). A number of proteins are defined by their molecular mass in kDa, e.g., p53.

17. When protein names are abbreviated, the notation frequently uses capitalized letters, e.g., cyclin dependent kinases = CDK in accordance with the literature.

18. A list of common abbreviations used throughout the book is given in 1.1.2. Less frequently used abbreviations are defined in the text.

Literature:

19. Only some recent references, primarily review articles and monographs are listed at the end of the various sections. For more details refer to the literature quoted in these references, to electronic data banks, to review books and journals and to textbooks of biochemistry. Chapter 21 contains a survey on electronic data banks and a list of printed sources, which have been used frequently during writing of this book.

1.1.2 Common Abbreviations
(Other abbreviations are defined in the text)

aa	Amino acid
Acc, AccH$_2$	Acceptor, reduced acceptor (unspecified)
ACP	Acyl carrier protein
ATP, ADP, AMP, A	Adenosine tri-, di-, monophosphate, adenosine
bp, kbp	base pair (in DNA), kilobase pairs
cAMP	Cyclic AMP = adenosine 3′,5′-monophosphate
cGMP	Cyclic GMP = guanosine 3′,5′-monophosphate
CoA-SH, CoA-S-	Coenzyme A
CTP, CDP, CMP, C	Cytidine tri-, di-, monophosphate, cytidine
Cyt	Cytochrome
Da, kDa	Dalton, kilodalton (unit of molecular mass)
dATP, dADP, dAMP, dA	Deoxyadenosine tri-, di-, monophosphate, deoxyadenosine
dCTP, dCDP, dCMP, dC	Deoxycytidine tri-, di-, monophosphate, deoxycytidine
dGTP, dGDP, dGMP, dG	Deoxyguanosine tri-, di-, monophosphate, deoxyguanosine
dTTP, dTDP, dTMP, dT	Deoxythymidine tri-, di-, monophosphate, deoxythymidine
DNA	Deoxyribonucleic acid
E	Enzyme
EC number	Enzyme classification according to the 'Enzyme Nomenclature' (1992)
ER	Endoplasmatic reticulum
ETF	Electron transferring flavoprotein
F_{430}	A corrinoid coenzme (Ni)
FAD, FADH$_2$	Flavin-adenine dinucleotide, reduced flavinadenine dinucleotide
Fd	Ferredoxin
FMN, FMNH$_2$	Flavin mononucleotide, reduced flavin mononucleotide
Fp	Flavoprotein
ΔG	Change of free energy (see 1.5.1)
G6P	Glucose 6-phosphate
GSH, GSSG	Glutathione, oxidized glutathione
GTP, GDP, GMP, G	Guanosine tri-, di-, monophosphate, guanosine
Ig	Immunoglobulin
ITP, IDP, IMP, I	Inosine tri-, di-, monophosphate, inosine
k	Velocity constant of a reaction (1.5.4)
K	Equilibrium constant of a reaction (see 1.5.1)
K_S, K_I, K_D	Dissociation constants (see 1.5.4, 17.1.2)
K_M	Michaelis constant (see 1.5.4)
kb	Kilobases (10^3 bases)
λ	Wavelength of light
Lip(SH)(SH), Lip(S)(S)	α-Lipoic acid, oxidized α-lipoic acid
NAD$^+$, NADH + H$^+$	Nicotinamide-adenine dinucleotide, reduced nicotinamide-adenine dinucleotide
NADP$^+$, NADPH + H$^+$	Nicotinamide-adenine dinucleotide phosphate, reduced nicotinamide-adenine dinucleotide phosphate
nt	Nucleotide
NTP, NDP, NMP, N	Any nucleotide tri-, di-, monophosphate or nucleotide
PAP	Adenosine 3′,5′-diphosphate
PAPS	3′-Phosphoadenylylsulfate
PEP	Phosphoenolpyruvate
P_i, PP_i	Inorganic phosphate, inorganic pyrophosphate
pK	Negative decadic logarithm of a dissociation constant (analogously: pH)
PQQ	Pyrroloquinoline quinone
PRPP	α-D-5-Phosphoribosylpyrophosphate
PyrP	Pyridoxal phosphate
RNA	Ribonucleic acid
mRNA, rRNA, tRNA	Messenger, ribosomal, transfer ribonucleic acid
R-S-S-R	Disulfide group of amino acids or peptides
S	Svedberg units (sedimentation coefficient)
SAH	S-Adenosylhomocysteine
SAM	S-Adenosylmethionine
THF	5,6,7,8-Tetrahydrofolate
THMPT	5,6,7,8-Tetrahydromethanopterin
ThPP	Thiamin pyrophosphate
UDPG	Uridine diphosphate glucoese
UQ, UQH$_2$	Ubiquinone, reduced ubiquinone
UTP, UDP, UMP, U	Uridine tri-, di-, monophosphate, uridine

Abbreviations for amino acids are listed in Figures 1.3.2 and 10.1-3, abbreviations for sugars in Fig.13.3-1.

1.2 Carbohydrate Chemistry and Structure

Carbohydrate monomers are of the general formula $(CH_2O)_n$. They have the chemical structure of aldehydes or ketones with multiple hydroxyl groups (<u>aldoses</u> and <u>ketoses</u>, respectively). A common name of mono- and dimers is '<u>sugar</u>'.

The large number of reactive groups, together with the stereoisomers causes a multiplicity of structures and reaction possibilities. Besides 'pure' carbohydrate monomers, oligo- (3.4) and polymers (3.2), there exist in nature carboxylic (3.5.1...2) and amino (3.7) derivatives, polyalcohols (3.5.5), deoxy sugars (3.5.6) etc.

Carbohydrates are the primary products of photosynthesis (16.2.2) and function as <u>energy storage forms</u> (e.g. starch, glycogen, 3.2), as part of <u>nucleic acid</u> and <u>nucleotide</u> molecules (8.1, 8.2), in <u>glycoproteins</u> (13.1) and <u>glycolipids</u> (13.2) and as <u>structural elements</u> in cell walls of *bacteria* (15.1), *plants* (3.6.3 and 13.2.2) and in the exoskeleton of *arthropods* (3.7.1). They are the most abundant chemical group in the biosphere.

1.2.1 Structure and Classification

The most simple carbohydrates are the <u>trioses</u> (C_3 compounds) glyceraldehyde (an aldose) and dihydroxyacetone (glycerone, a ketose). Larger molecules are <u>tetroses</u> (C_4), <u>pentoses</u> (C_5), <u>hexoses</u> (C_6), <u>heptoses</u> (C_7) etc.; the C_5 and C_6 molecules are most common.

Glyceraldehyde is the smallest aldose with an <u>asymmetric C-atom</u> (<u>chirality center</u>). Therefore there are 2 <u>stereoisomers</u> (<u>enantiomers</u>), which cause right and left rotation of polarized light. By the <u>Fischer convention</u>, they are named D- and L-form, respectively. For details, see textbooks of organic chemistry. Tetroses and larger carbohydrate monomers are classified (by comparison of the asymmetric center most distant to the aldehyde or keto group with D- or L-glyceraldehyde) as the D- and L-series of enantiomers (Fig. 1.2-1). With n-carbon aldoses, a total of 2^{n-2} stereoisomers exist, with n-carbon ketoses there are 2^{n-3} stereoisomers. <u>Epimers</u> are stereoisomers, which differ in the configuration at only one asymmetric C-atom. Most physiological sugars are of the D-configuration.

Figure 1.2-1. Nomenclature of D-Carbohydrates
The compounds printed in green are formally obtained by epimerization at the indicated positions. The L-enatiomers are the mirror images at a perpendicular mirror plane.

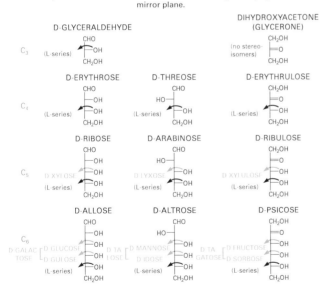

Aldopentoses, aldohexoses and ketohexoses (and higher sugars) can form cyclic structures (<u>hemiacetals</u> and <u>hemiketals</u>) by intramolecular reaction of their aldehyde or keto groups respectively with an alcohol group. This results in <u>pyranoses</u> (6-membered rings) and <u>furanoses</u> (5-membered rings, Fig. 1.2-2). The equilibrium is strongly in favor of the cyclic structures. The ring closure produces another asymmetric C-atom; the respective stereoisomers are named <u>anomers</u> (α- and β-forms).

The nonplanar <u>pyranose rings</u> can assume either <u>boat</u> (in 2 variants) or <u>chair conformation</u>. The substituents extend either parallel to the perpendicular axis (<u>axial</u>, in Fig. 1.2-3 printed in red) or at almost right angles to it (<u>equatorial</u>, printed in green). The preferred conformation depends on spatial interference or other interactions of the substituents.

Figure 1.2-2. Ring Closure of Carbohydrates

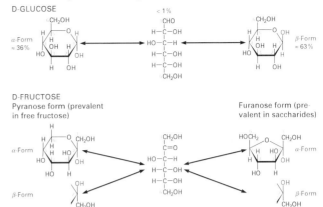

Although the bond angles of a <u>furanose ring</u> would permit an almost planar structure, the interference of substituents with each other causes a slight bending (<u>puckering</u>), e.g. to a <u>half-chair</u> (= <u>envelope</u>) structure in nucleotides and nucleic acids (Fig. 1.2-3).

Figure 1.2-3. Chair and Boat Conformations of Hexoses (Top) and Half-Chair (Envelope) Conformation of Pentoses (Bottom)

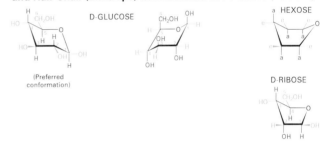

The linear form of carbohydrates usually shown as <u>Fischer projection</u> (ligands drawn horizontally are in front of the plane, ligands drawn vertically are behind the plane, e.g., in Fig. 1.2-1). The ring form is either drawn as <u>Haworth formula</u> (Fig. 1.2-2, disregarding the bent ring structure) or as <u>boat/chair</u> formula (Fig. 1.2-3).

1.2.2 Glycosidic Bonds (Fig. 1.2-4)

If the hemiacetal or hemiketal hydroxyl of a sugar is condensed with an alcoholic hydroxyl of another sugar molecule, a <u>glycosidic bond</u> is formed and water is eliminated. Since this reaction between free sugars is endergonic ($\Delta G'_0 = 16$ kJ/mol), the sugars usually have to be activated as <u>nucleotide derivatives</u> (3.2.2) in order to be noticeably converted. Depending on the configuration at the hemiacetal/hemiketal hydroxyl (1.2.1), there are α- or β-glycosides. Sugar derivatives, which contain a hemiacetal or a hemiketal group (e.g. uronic acids) are also able to form glycosidic bonds.

Since sugar molecules contain several alcoholic groups, various types of bonds are possible. Frequently, 1→4 or 1→6 bonds occur. With oligo- or polysaccharides, both linear or branched structures are found. Bond formation may also take place with alcoholic, phenolic or other groups of non-sugar molecules (aglycons).

Figure 1.2-4. Examples of Glycosidic Bonds

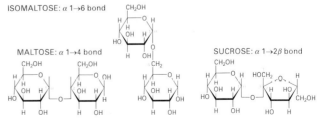

The metabolism of di- and oligosaccharides also involves many <u>isomerizations</u> (e.g. between aldoses and ketoses) and <u>epimerizations</u>. The conversions into <u>deoxy sugars</u> (rhamnose, fucose) are multistep reactions.

Literature:
Textbooks of organic chemistry.

1.3 Amino Acid Chemistry and Structure

All amino acids present in proteins carry a carboxyl- and an amino group, hydrogen and variable side chains (R) at a single (α-)carbon atom. Thus, this C-atom is asymmetric (1.2.1), with the only exception of glycine, where R = H. Almost all of the proteinogenic amino acids occurring in nature are of the L-configuration. (The 'L' is assigned by comparison with L- and D-glyceraldehyde, which are taken as standards, Fig. 1.3-1). A number of D-amino acids is found in *bacterial* envelopes (15.1) and in some antibiotics (15.8).

Unless otherwise stated, all amino acids discussed in the following sections are of the L-configuration.

Figure 1.3-1. Asymmetric Center of Amino Acids

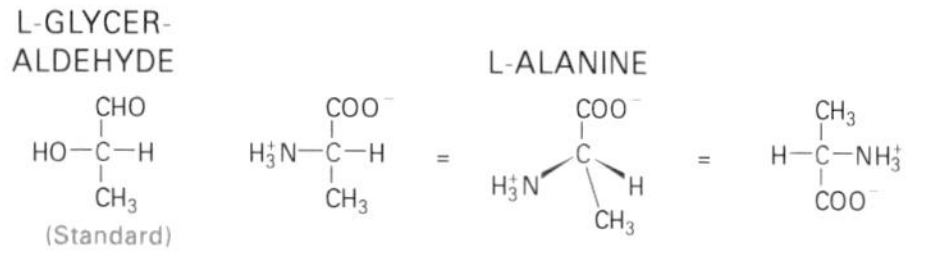

Chains of amino acids form the proteins and peptides, which, being enzymes, regulatory, mobility or structural compounds, are the central components in all living beings. Therefore they are the topic of most of this book. The protein synthesis is described in Chapters 10 and 11. The structure is dealt with in Section 2.3, which also gives a short listing of their functions.

1.3.1 Structure and Classification

The individual properties of the amino acids are determined by the side chain R. This is also the criterion for amino acid classification.

There are 20 standard ('classical') amino acids, which are incorporated as such into proteins, employing own codons (10.6, 11.5). As the 21st amino acid, selenocysteine is also incorporated directly by an unusual decoding procedure of mRNA (10.6.5). These amino acids are shown in Fig. 1.3-2. Nonstandard amino acids are produced by metabolic conversions of free amino acids (e.g. ornithine and citrulline) or by posttranslational modification of amino acids in proteins (e.g. by hydroxylation, methylation or carboxylation). Examples are given in Fig. 1.3-3.

At about neutral pH, the free amino acids are 'Zwitterion' dipols with charged carboxylate (dissociation constant $pK_1 = 1.82...2.35$) and amino groups ($pK_2 = 8.70...10.70$). In 7 cases, the side chains R also contain charged groups. Only the pK_R of histidine (6.04) is in the physiological range. In Fig. 1.3-2 and 1.3-3, the charged molecules are shown, while in the rest of the book, not ionized forms are presented for reasons of simplicity.

1.3.2 Peptide Bonds (Fig. 2.3-1)

Proteins and peptides are linear chains of amino acids connected by 'peptide' bonds between their α-amino and carboxylate groups. Since the formation of these bonds is endergonic, the reactants have to be activated as tRNA derivatives. Details are described in 10.6.2 and 10.6.3.

The peptide bonds are rigid and planar: The carboxylate-O and the amino-H are in *trans* conformation, the C-N bond shows partially double bond characteristics. Only peptide bonds followed by proline or hydroxyproline can alternatively be *cis* (6...10%). To some extent, both bonds in the backbone of the peptide chain extending from C_α can perform rotational movements (although there are still constraints on most conformations, which are shown in Ramachandran diagrams). Flexibility and constraints play a major role in the proper folding of the proteins (2.3.1).

Proteins and peptides carry charged amino- (N-) and carboxy-(C-)termini. Additional charges are contributed by the side chains. This allows the analytical separation by electrophoresis. It has to be considered, however, that the pK_R of amino acids in peptides differ from those in free amino acids due to the effects of neighboring groups.

Literature:
Meister, A.: *Biochemistry of the Amino Acids.* 2 Vols. Academic Press (1965).
Ramachandran, G.N., Sasisekharan, V.: Adv. Prot. Chem. 23 (1968) 326–367.
Rose, G.D. et al.: Adv. Prot. Chem. 37 (1985) 1–109.
Textbooks of organic chemistry.

Figure 1.3-2. Standard Amino Acids With Their 3- and 1-Letter Abbreviations

a) **Non-polar, aliphatic amino acids.** The non-polar side chains undergo hydrophobic interactions in protein structures. While the small glycine molecule allows high flexibility, the bulky proline confers enhanced rigidity to the structures.

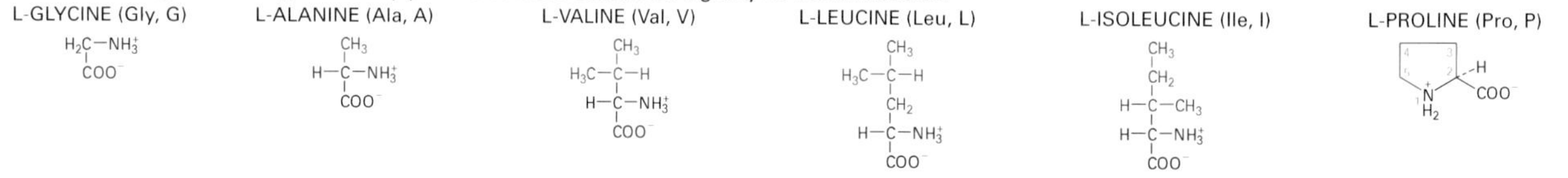

L-GLYCINE (Gly, G)   L-ALANINE (Ala, A)   L-VALINE (Val, V)   L-LEUCINE (Leu, L)   L-ISOLEUCINE (Ile, I)   L-PROLINE (Pro, P)

b) **Polar, uncharged residues R.** These functional groups are hydrophilic and can form hydrogen bonds with water or other polar compounds. Cysteine can easily be oxidized, resulting in intra- or intermolecular interconnections by disulfide bonds.

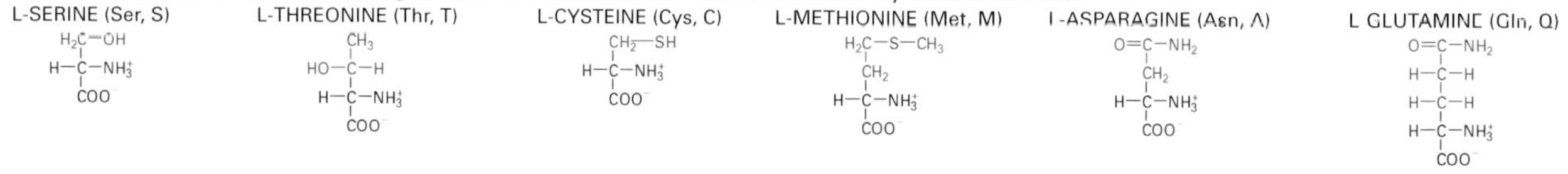

L-SERINE (Ser, S)   L-THREONINE (Thr, T)   L-CYSTEINE (Cys, C)   L-METHIONINE (Met, M)   L-ASPARAGINE (Asn, N)   L-GLUTAMINE (Gln, Q)

c) **Aromatic residues R.** The aromatic side chains are hydrophobic, while the hydroxyl group of tyrosine and the ring nitrogen of tryptophan form hydrogen bonds, which often play a role in enzyme catalysis.

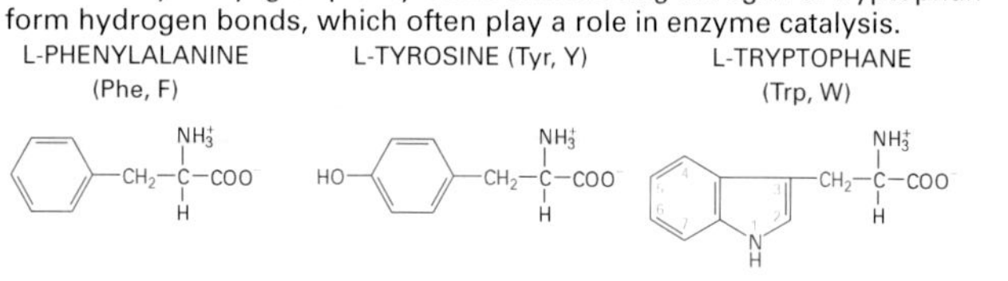

L-PHENYLALANINE (Phe, F)   L-TYROSINE (Tyr, Y)   L-TRYPTOPHANE (Trp, W)

d) **Positively charged side chains R.** The charged groups contribute in many cases to catalytic mechanisms and also influence the protein structure.

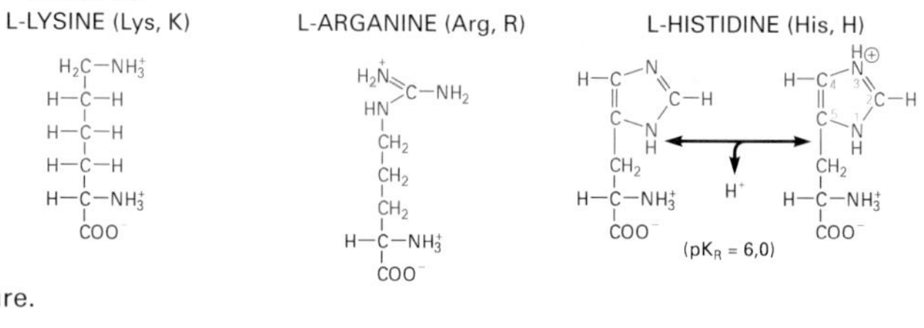

L-LYSINE (Lys, K)   L-ARGANINE (Arg, R)   L-HISTIDINE (His, H)   ($pK_R = 6.0$)

e) **Negatively charged side chains R.** The charged groups contribute in many cases to catalytic mechanisms and are also of influence to the protein structure.

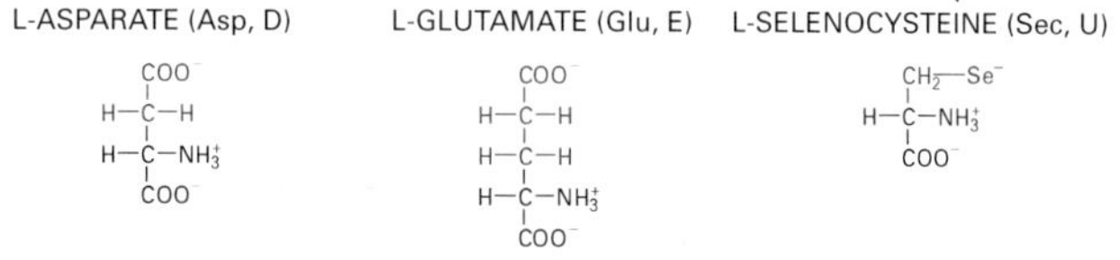

L-ASPARATE (Asp, D)   L-GLUTAMATE (Glu, E)   L-SELENOCYSTEINE (Sec, U)

Figure 1.3-3. Some Nonstandard Amino Acids

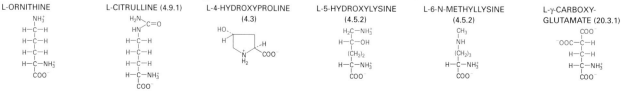

L-ORNITHINE   L-CITRULLINE (4.9.1)   L-4-HYDROXYPROLINE (4.3)   L-5-HYDROXYLYSINE (4.5.2)   L-6-N-METHYLLYSINE (4.5.2)   L-γ-CARBOXYGLUTAMATE (20.3.1)

1.4 Lipid Chemistry and Structure

The common denominator of lipids is their hydrophobic character and their solubility in organic solvents. Otherwise, they belong to different chemical classes. The biochemistry of most of them is described in Chapter 6, some other lipids are discussed in their metabolic context elsewhere (see cross-references below).

1.4.1 Fatty acids (Table 1.4-1, Fig. 1.4-1)

Fatty acids are characterized by a carboxylic group with a hydrocarbon 'tail'. The higher fatty acids are practically insoluble in water and show typical lipid properties. They serve in esterified form as triacylglycerols for energy storage or are, as glycerophospholipids, part of cellular membranes.

Contrary to them, the short-chain fatty acids are water soluble. They act as intermediates of metabolism and are dealt with in the respective chapters.

Higher fatty acids can also enter an amide bond (e.g., in ceramides). Some are precursors of other compounds (e.g., of prostaglandins, 17.4.8). Almost none of them occur in free form.

The predominant fatty acids in *higher plants* and *animals* have an even number of C atoms in the range of $C_{14}...C_{20}$ and are unbranched. Usually, more than half of all fatty acids are unsaturated. Monounsaturated fatty acids mostly contain a *cis*-double bond between C-9 and C-10. Often there exist additional double bonds in the direction towards the methyl terminus, usually with two saturated bonds in between (polyunsaturated fatty acids). Some of them cannot be synthesized in *animals* and have to be supplied by food intake (essential fatty acids). The notation of fatty acids is (number of C atoms) : (number of double bonds), e.g., for linoleic acid 18:2. The location of the double bonds is given as e.g., $\Delta^{9,12}$.

Polyunsaturated fatty acids are not usually present in *bacteria*, but there exist *cis*- and *trans*-monounsaturated, hydroxylated and branched fatty acids in many species.

While saturated fatty acids tend to assume an extended shape, unsaturated fatty acids show 30° bends at their double bonds (Fig. 1.4-1). This reduces van der Waals interactions between neighboring molecules and lowers the melting point (see textbooks of organic chemistry):

18:0 (70°)   18:1 (13°)   18:2 (–9°)   18:3 (–17°)

Table 1.4-1. Higher Fatty Acids Frequently Occurring in Nature

Number of C atoms	Saturated	Unsaturated, Number of Double Bonds. E = essential fatty acid for *humans*	
14	myristic acid		
16	palmitic acid		
18	stearic acid	1: oleic acid (Δ^9)	
		2: linoleic acid ($\Delta^{9,12}$)	E
		3: α-linolenic acid ($\Delta^{9,12,15}$)	E
20	arachidic acid	4: arachidonic acid ($\Delta^{5,8,11,14}$)	(E) *
24	lignoceric acid	1: nervonic acid (Δ^{15})	

* can be synthesized from the essential fatty acid linoleic acid

Figure 1.4-1. Structure of Saturated and Unsaturated Fatty Acids
(18:0 and 18:1, showing the bend)

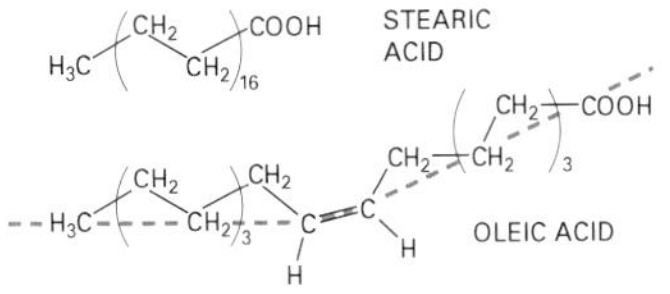

1.4.2 Acylglycerols and Derivatives (Fig. 1.4-2)

A major part of lipids occurring in *plants* and *animals* are triesters of glycerol (6.2) with higher fatty acids (triacylglycerols = triglycerides = neutral fat). In most of them, the fatty acids are different. Their type and the degree of their unsaturation determine the melting point.

Fats are solid, oils are liquid at room temperature. They are without influence on the osmotic situation in the aqueous phase due to their insolubility and do not bind water as, e.g., glycogen does. Thus, these compounds constitute an effective, convenient storage form of energy (≥ 10 kg in *adult humans*).

Their degree of oxidation is lower than that of carbohydrates or proteins, therefore they provide a higher energy during combustion: triolein yields 39.7 kJ/g. This is more than twice the value for anhydrous carbohydrates (17.5 kJ/g) or proteins, (18.6 kJ/g) and about 6 times the energy gained from degradation of these alternative compounds in their physiological state due to their water content.

Triacylglycerols do not contain any hydrophilic groups. If, however, only one or two of the hydroxyl groups of glycerol are esterified (mono- or diacylglycerols), the remaining polar hydroxyl groups allow the formation of ordered structures at water-lipid interfaces and of lipid bilayers (1.4.8). Therefore they can act as emulsifiers, e.g. during lipid resorption from the *intestine*.

The remaining hydroxyl groups of mono- and diacylglycerols can also carry sugar residues. These so-called glycoglycerolipids are constituents of *bacterial* cell envelopes (15.1), thylakoid membranes in *plants* and myelin sheaths of *neurons* in *animals*. They are dealt with in 13.2.2.

Figure 1.4-2. Structure of Acylglycerols, Glycoglycerolipids and Waxes

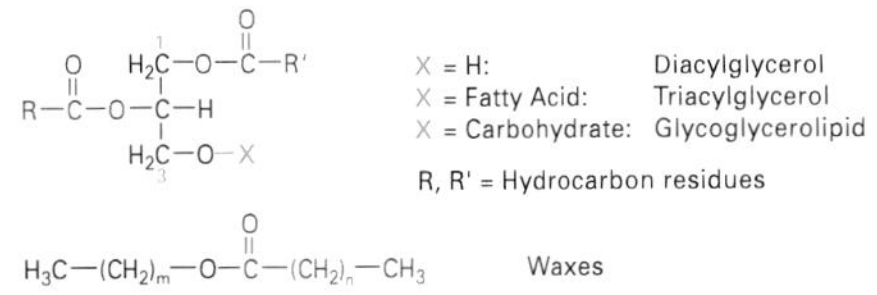

X = H: Diacylglycerol
X = Fatty Acid: Triacylglycerol
X = Carbohydrate: Glycoglycerolipid
R, R' = Hydrocarbon residues

Waxes

1.4.3 Waxes (Table 1.4-2, Fig. 1.4-2)

Waxes are esters of higher fatty acids with long-chain primary alcohols (wax alcohols) or sterols (Chapter 7), which are usually solid at room temperature.

They are more resistant than triacylglycerols towards oxidation, heat and hydrolysis (saponification). Frequently, they serve as protective layers, e.g. on leaves and fruits of *plants* or on skin, feathers and furs of *animals* (as secretions of specialized glands). They also are the material for the honeycombs of bees. In many *marine animals* they are the main component of lipids (for regulation of flotation and for energy storage). Fossil waxes occur in *lignite* and *bitumen*.

Table 1.4-2. Common Components of Waxes

Alcohol (primary, saturated)	Fatty Acid (saturated)
Cetyl alcohol (C_{16})	Lauric acid (C_{12})
Carnaubyl alcohol (C_{24})	Myristic Acid (C_{14})
Ceryl alcohol (C_{26})	Palmitic acid (C_{16})
Myricyl alcohol (C_{30})	Lignoceric Acid (C_{24})
	Cerotic Acid (C_{26})
	Montanic Acid (C_{28})
	Melissic Acid (C_{30})

1.4.4 Glycerophospholipids (Phosphoglycerides, Fig. 1.4-3)

In contrast to triacylglycerols, in glycerophospholipids only two of the hydroxyl groups of glycerol are esterified with long chain fatty acids, while the group at the 3-position (according to the *sn*-numbering system) forms an ester with phosphoric acid.

All glycerophospholipids have an asymmetric C-atom in the 2-position, they occur in nature in the L-form. Most common are saturated fatty acids (C_{16} or C_{18}) at the 1- and unsaturated ones ($C_{16}...C_{20}$) at the 2-position. Removal of one fatty acid yields lysoglycerophospholipids.

If the 3-position of glycerol carries only phosphoric acid, the compound is named phosphatidic acid. However, in most cases the phosphate group is diesterified. This extra residue ('head group', Y in Fig. 1.4-3) determines the class of the compound.

Figure 1.4-3. Classes of Glycerophospholipids

Basic Structure: Diacyl-glycerophospholipid   Inositol, a ligand Y

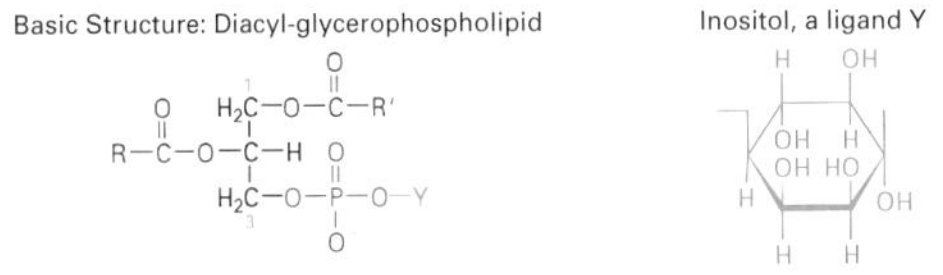

Class	Y =	Formula of Y (F = Fatty acid)
Phosphatidic acid	H	-H
Phosphatidylethanolamine	ethanolamine	$-CH_2-CH_2-NH_2$
Phosphatidylcholine (lecithin)	choline	$-CH_2-CH_2-N(CH_3)_3^+$
Phosphatidylserine	serine	$-CH_2-CH(NH_2)-COOH$
Phosphatidylinositol	inositol	see above
Diphosphatidylglycerol (cardiolipin)	glycerol-phosphatidic acid	$-CH_2-CHOH-CH_2O-(PO_2)-$ $-O-CH_2-CHOF-CH_2OF$

These compounds are more polar than mono- or diacylglycerols and form the major part of biological membranes (1.4.8).

1.4.5 Plasmalogens (Fig. 1.4-4)

This group of compounds is related to diacylglycerophospholipids (1.4.4). Also, the head groups (Y) are similar. However, the 1-position of glycerol is not esterified, but carries an α,β-unsaturated alcohol in an ether linkage. They are major components of the *CNS*, *brain* (> 10 %), *heart* and *skeletal muscles*, but little is known about their physiological role.

Figure 1.4-4. Structure of Plasmalogens

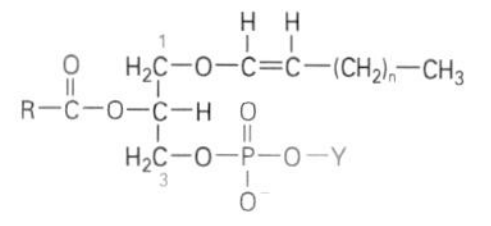

1.4.6 Sphingolipids (Fig. 1.4-5)

Sphingolipids are important membrane components. They are derivatives of the aminoalcohols dihydrosphingosine (C_{18}), sphingosine (C_{18} with a *trans* double bond) or their C_{16}, C_{17}, C_{19} and C_{20} homologues.

Ceramides are N-acylated sphingosines. If the hydroxyl group at C-1 is esterified with phosphocholine, phosphoethanolamine etc., sphingomyelins (sphingophospholipids) are obtained. If, alternatively, the hydroxyl group is glycosylated, glycosphingolipids (cerebrosides) result. This latter group of compounds is described in 13.2.1.

Figure 1.4-5. Basic Structures of Sphingolipids

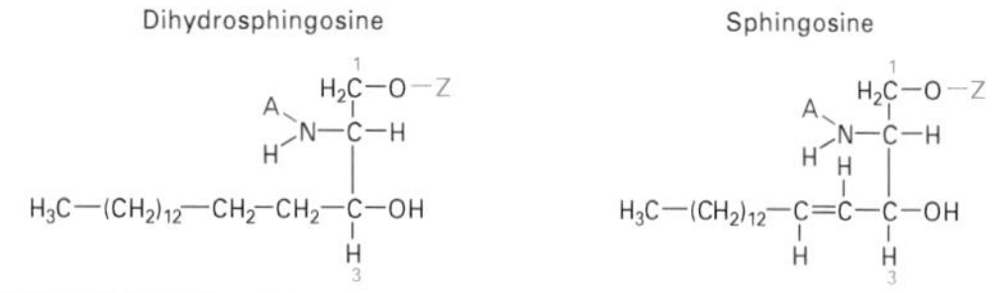

Class	A (= N-Substituent)	Z	Formula of Z
Sphingosine	H	H	-H
Ceramides	higher fatty acid	H	-H
Sphingo-myelins	higher fatty acid	phosphoethanolamine or phosphocholine	$-PO_2-O-CH_2-CH_2-OH$ $-PO_2-O-CH_2-CH_2-N(CH_3)_3^+$
Glyco-sphingolipids	higher fatty acid	carbohydrate chains	-carbohydrates

1.4.7 Steroids

Steroids are derivatives of the hydrocarbon cyclopentanoperhydrophenanthrene (Fig. 1.4-6).

Biologically important steroids carry many substituents: generally there is a hydroxy or oxo group at C-3. In addition, several methyl, hydroxy and oxo, in some cases also carboxy groups occur. In many cases, there is a larger residue bound to C-17. Frequently, some double bonds are present. In a few cases, ring A is aromatic. Substituents below the ring system are designated α and above the ring system β (see Fig. 7.1-5).

Steroids are membrane components and participants as well as regulators of metabolism. A detailed description is given in Chapter 7.

Figure 1.4-6. Structure of Cyclopentanoperhydrophenanthrene

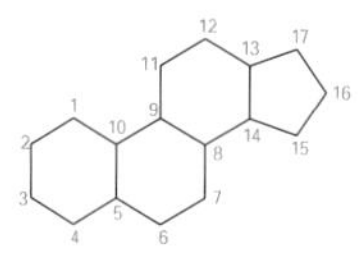

1.4.8 Lipoproteins

The major function of lipoproteins is the transport of lipids. They contain nonpolar lipids (triacylglycerols, cholesterol esters) in their core, surrounded by a layer of polar compounds (glycerophospholipids, cholesterol, proteins, Fig. 18.2-1). This group of compounds is discussed in context with their transport function in 18.2.

1.4.9 Lipid Aggregates and Membranes (Fig. 1.4-7)

Compounds, which contain hydrophobic (aliphatic or aromatic hydrocarbons) and hydrophilic regions (charged or polar alcoholic or phenolic groups) are called amphiphilic molecules. Among the above listed compounds, glycerophospholipids, sphingolipids, glycoglycerolipids, mono- and diacylglycerols, but also alkali salts of fatty acids (soaps) show this property. Molecules of this kind arrange in unique ways when being in contact with an aqueous phase:

- They can associate to spherical micelles with only hydrophilic groups located on the outside of a monomolecular layer. The interior of larger micelles can be filled with 'neutral' lipids lacking polar groups (e.g. triacylglycerols and cholesterol esters), yielding mixed micelles (detergent effect).

  This arrangement is mostly made by 'single-tailed' molecules, e.g. monoacylglycerol, soaps and artificial detergents and occurs only above a critical micelle concentration (cmc, <1 μmol/l for biological lipids), otherwise the molecules cannot shield their hydrophobic tails from water.

- A feature of enormous importance in biology is the formation of lipid bilayers of about 6 nm thickness, which face aqueous phases on both sides. This is the basic arrangement of all cellular membranes including intracellular ones.

  Best suited for membrane formation are compounds with space filling hydrophobic areas (otherwise they would form micelles), such as glycerphospholipids and sphingolipids. In addition, the membranes usually contain many other compounds, primarily proteins (which contribute to the membrane function as well as to transport and metabolism) and cholesterol. The membrane constituents can move laterally within the membrane (fluid mosaic model).

Characteristics of membranes: The fluidity of the membranes depends on the lipid interior of the bilayer: it increases, when the order of arrangement decreases (e.g. caused by bent unsaturated fatty acids). Biological membranes undergo a phase change at a transition temperature (usually between 10 ... 40° C), above which lateral movements of membrane components can take place more easily, although the basic structure is still kept up. Cholesterol (which by itself does not form bilayers) widens the transition range.

The lipids of the membranes are asymmetrically distributed. The 'flip-flop' exchange rate to the other membrane side is very low ($t_{1/2}$ = days, much higher in *bacteria*). In *human* erythrocytes, the distribution is as follows:

mol-% of total lipids	total	outer leaflet	inner leaflet
Phosphatidylcholine	30	22	8
Phosphatidylethanolamine	31	7	24
Phosphatidylserine	9	0	9
Sphingomyelin	25	21	4

Transmembrane proteins (integral proteins) span the membrane with α-helices of about 19 amino acids. These domains contain mostly hydrophobic amino acids, which interact with the membrane lipids. If the protein contains several transmembrane domains, they can assume a ring-like structure, which is hydrophobic on the outside and forms a hydrophilic channel at the inside. This enables the passage of hydrophlic compounds through the membrane, e.g., for the import of metabolites or ions (17.2.3, 18.1.1). Peripheral proteins are attached by hydrophobic interactions with the membrane lipids or by electrostatic or hydrogen bonds with integral proteins (e.g., *liver* cytochrome b_5). They can be removed under relative mild experimental conditions. Membrane-associated proteins are bound to the membrane by lipid anchors, such as phosphatidylinositol (13.3.4) or isoprenoids (7.3.4).

Literature:

Dawidowicz, E.A.: Ann. Rev. of Biochem. 56 (1987) 43–61.

Vance, D.E., Vance, J.E. (Eds.): *Biochmistry of Lipids, Lipoproteins and Membranes.* *New Comprehensive Biochemistry* Vol. 31. Elsevier (1996).

Figure 1.4-7. Structure of Lipid Aggregates

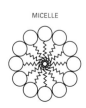
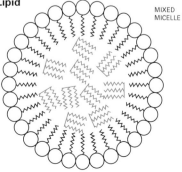
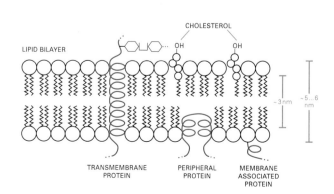

1.5 Physico-Chemical Aspects of Biochemical Processes

There are surely a number of readers, which are less inclined to deal with a fairly large number of mathematical formulas. However, formulas are necessary to describe biochemical processes quantitatively. Considering this, the mathematical part of this book has been concentrated into this section, while usually other chapters refer to it.

Only the most important equations required for discussion of biochemical reactions are presented. In order to facilitate their use, companion equations are given, which show the numerical values of the factors and the dimensions of the terms. For derivation of the equations, refer to textbooks of physical chemistry. The units and constants used in the following paragraphs are listed in Table 1.5-1.

Table 1.5-1. Measures and Constants (Selection)

Measure:	Unit:	Equivalents / Value of Constant	Equivalents in SI Basic Units
Length	meter (m)	1 mm = 10^{-3} m, 1 µm = 10^{-6} m, 1 nm = 10^{-9} m, 1 Å (Ångström) = 10^{-10} m	SI basic unit
Volume	cubic meter (m^3)	1 l (liter) = 10^{-3} m^3; 1 ml = 10^{-3} l, 1 µl = 10^{-6} l	Derived SI unit
Mass	kilogram (kg)	1 g (gram) = 10^{-3} kg; 1 mg = 10^{-3} g, 1 µg = 10^{-6} g	SI basic unit
Time (t)	second (sec)	1 msec = 10^{-3} sec, 1 µsec = 10^{-6} sec, 1 nsec = 10^{-9} sec, 1 psec = 10^{-12} sec	SI basic unit
Temperature (T)	Kelvin (K)	0 K = -273.16 °C	SI basic unit
Quantity of matter	mol	1 mol = $6.0221 * 10^{23}$ [moleculs or ions] 1 mmol = 10^{-3} mol, 1 µmol = 10^{-6} mol, 1 nmol = 10^{-9} mol, 1 pmol = 10^{-12} mol This unit is also applied to photons. 1 Einstein = 1 mol photons.	SI basic unit
Electric current	Ampere (A)		SI basic unit
Force	Newton (N)	1 N = 1 [m * kg * sec^{-2}]	1 N = 1 [m * kg * sec^{-2}]
Pressure	Pascal (Pa)	1 Pa = 1 N * m^{-2}, 1 kilopascal (kPa) = 10^3 Pa, 1 atm = 101.325 kPa	1 Pa = 1 [m^{-1} * kg * sec^{-2}]
Energy	Joule (J)	1 J = 1 [N * m], 1 kilojoule (kJ) = 1000 J 1 cal (calorie) = 4.181 J, 1 kcal = 4.181 kJ (Non-SI unit)	1 J = 1 [m^2 * kg * sec^{-2}]
Electric charge	Coulomb (C)	1 C = $6.241 * 10^{18}$ electron charges	1 C = 1 [A * sec]
Electric potential	Volt (V)	1 V = 1 [J * C^{-1}], 1 mV = 10^{-3} V	1 V = 1 [m^2 * kg * sec^{-3} * A^{-1}]
Constants:	**Abbreviation:**		
Avogadro's number	N	N = $6.0220 * 10^{23}$ [mol^{-1}] (see 'quantity of matter', above)	N = $6.0220 * 10^{23}$ [mol^{-1}]
Boltzmann's constant	k_B	k_B = $1.3807 * 10^{-23}$ [J * K^{-1}]	k_B = $1.3807 * 10^{-23}$ [m^2 * kg * sec^{-2} * K^{-1}]
Molar gas constant	R	R = N * k_B = 8.31441 [J * mol^{-1} * K^{-1}]	R = 8.31441 [m^2 * kg * sec^{-2} * mol^{-1} * K^{-1}]
Faraday's constant	F	F = 1 N electron charges = 96484.5 [C * mol^{-1}] = 96484.5 [J * V^{-1} * mol^{-1}]	F = 96484.5 [A * sec * mol^{-1}]
Planck's constant	h	h = $6.6262 * 10^{-34}$ [J * sec]	h = $6.6262 * 10^{-34}$ [m^2 * kg * sec^{-1}]

In calculations in this book, usually **l, g, kJ** and **mV** are used. Since the constants have then to be expressed in these units, their numerical value changes by the factor 10^{-3} or 10^{-6} in the respective formulas.

1.5.1 Energetics of Chemical Reactions

To each component of a system, an amount of free energy G is assigned, which is composed of the enthalpy H (internal energy + pressure * volume) and of the entropy S (measure of disorder). While the absolute values are not of importance, the change of G (ΔG) is decisive for chemical reactions:

$$\Delta G = \Delta H - T * \Delta S. \qquad [1.5\text{-}1]$$

or

$$\Delta G \, [kJ * mol^{-1}] = \Delta H \, [kJ * mol^{-1}] - T \, [K] * \Delta S \, [kJ * mol^{-1} * K^{-1}]. \qquad [1.5\text{-}1a]$$

A reaction proceeds spontaneously only, if ΔG is negative.

In biochemistry, ΔG of reactions are usually listed as $\Delta G_0'$, which is obtained at standard conditions of 298 K (25 °C), pH 7.0 and a reactant concentration of 1 mol/l each except for water, where the normal concentration of 55.55 mol/l and gases, where a pressure of 101.3 kPa (= 1 atm) are taken as unity and thus do not appear in the formula.

If the reactant concentrations (henceforth written as [X]) of a reaction A + B + ... = Z + Y + ... differ from 1 mol/l each, ΔG can be calculated by:

$$\Delta G = \Delta G_0' + R * T * 2.303 * \log \frac{[Z] * [Y] * ...}{[A] * [B] * ...} \quad \text{(end products)} \atop \text{(starting comp.)} \qquad [1.5\text{-}2]$$

or

$$\Delta G \, [kJ * mol^{-1}] = \Delta G_0' + 0.00831 * T * 2.303 * \log \frac{[Z] * [Y] * ...}{[A] * [B] * ...} \qquad [1.5\text{-}2a]$$

Reaction sequences can be calculated by addition of ΔG's of the individual reactions.

A reaction is at equilibrium, if $\Delta G = 0$. Then the equilibrium constant

$$K = \frac{[Z] * [Y] * ...}{[A] * [B] * ...} \quad \text{(end products)} \atop \text{(starting compounds)} \qquad [1.5\text{-}3]$$

can be calculated as follows:

$$\Delta G_0' = -R * T * 2.303 * \log K; \quad K = 10^{(-\Delta G / R * T * 2.303)} \qquad [1.5\text{-}4]$$

or

$$\Delta G_0' \, [kJ * mol^{-1}] = -0.00831 * T * 2.303 * \log K;$$
$$K = 10^{(-\Delta G / 0.00831 * T * 2.303)}. \qquad [1.5\text{-}4a]$$

Enzymes cannot shift the equilibrium, they only increase the reaction velocity. The kinetics of enzyme catalyzed reactions are dealt with in 1.5.4.

1.5.2 Redox Reactions

Redox reactions are reactions, where one compound is reduced (electron acceptor A), while its reaction partner is oxidized (electron donor B) by transfer of n electrons:

$$A_{ox}^{n+} + B_{red} = A_{red} + B_{ox}^{n+}.$$

The change of free energy during such a reaction is described by a formula, which is analogous to Eq. [1.5-2]:

$$\Delta G = \Delta G_0' + R * T * 2.303 * \log \frac{[A_{red}] * [B_{ox}^{n+}]}{[A_{ox}^{n+}] * [B_{red}]} \quad \text{(end products)} \atop \text{(starting comp.)} \qquad [1.5\text{-}5]$$

or

$$\Delta G \, [kJ * mol^{-1}] = \Delta G_0' + 0.00831 * T * 2.303 * \log \frac{[A_{red}] * [B_{ox}^{n+}]}{[A_{ox}^{n+}] * [B_{red}]} \qquad [1.5\text{-}5a]$$

w expresses the work gained by transferring n mol charges (= n Faraday, F) across a potential difference of $\Delta E = E_{end} - E_{begin}$

$$w = -n * F * \Delta E. \qquad [1.5\text{-}6]$$

Since a positive amount of work diminishes the free energy of the system

$$w = -n * F * \Delta E = -\Delta G \qquad [1.5\text{-}6a]$$

or

$$\Delta G \, [kJ * mol^{-1}] = +n * 0.0965 * \Delta E \, [mV], \qquad [1.5\text{-}6b]$$

equation [1.5-5] can also be written as:

$$\Delta E = \Delta E_0' + \frac{R*T}{n*F} * 2.303 * \log \frac{[A_{red}]*[B_{ox}^{n+}]}{[A_{ox}^{n+}]*[B_{red}]} \quad \text{(end products)} \atop \text{(starting comp.)} \qquad [1.5\text{-}7]$$

or

$$\Delta E \,[mV] = \Delta E_0' + \frac{0.00831*T}{n*0.0965} * 2.303 * \log \frac{[A_{red}]*[B_{ox}^{n+}]}{[A_{ox}^{n+}]*[B_{red}]} \qquad [1.5\text{-}7a]$$

$\Delta E_0'$ is the difference of the <u>redox potentials</u> of this reaction (or the <u>electromotive force</u> across membranes, 1.5.3) under biochemical standard conditions (298 K = 25 °C, pH 7.0 and a reactant concentration of 1 mol/l each). Only water, which is present in a concentration of 55.55 mol/l and gases, with a pressure of 1 atm are taken as unity.

Redox potentials: The reaction can be divided into two half reactions (e^- = electrons):

$$A_{red} = A_{ox}^{n+} + n\,e^- \quad \text{and} \quad B_{red} = B_{ox}^{n+} + n\,e^-$$

The zero value of the redox potential is by convention assigned to the potential of the half reaction $2\,H^+ + 2\,e^- = H_2$ at a platinum electrode at pH = 0, 298 K (25 °C) and a hydrogen pressure of 101.3 kPa (= 1 atm). Thus, under the standard conditions used in biochemistry (pH = 7.0), E_0' ($2H^+/H_2$) = –410 mV.

Correspondingly, the half reactions can be expressed as:

$$E_A = (E_0')_A + \frac{R*T}{n*F} * 2.303 * \log \frac{[A_{ox}^{n+}]}{[A_{red}]} \qquad [1.5\text{-}8]$$

or

$$E_A\,[mV] = (E_0')_A + \frac{0.00831*T}{n*0.0965} * 2.303 * \log \frac{[A_{ox}^{n+}]}{[A_{red}]} \qquad [1.5\text{-}8a]$$

and analogously for B.

Various redox potentials can be combined this way: $\Delta E = E_B - E_A$ (A being the electron acceptor and B being the electron donor). The reactions proceed spontaneously only, if ΔE is <u>negative</u>, i.e. when the potential changes to a more negative value.

Redox potentials are usually plotted with the minus values on top. A spontaneous reaction proceeds in such a plot from top to bottom (e.g., Fig. 16.2-4).

In the literature, the definition of ΔE is not uniform. In a number of textbooks it is defined in opposite order as above: $\Delta E = E_{begin} - E_{end}$. Therefore, ΔE and $\Delta E_0'$ have to be replaced by $-\Delta E$ and $-\Delta E_0'$, respectively. This effects Eqs. [1.5-6] ... [1.5-8a] and has to be considered when making comparisons.

1.5.3 Transport Through Membranes

Uncharged molecules: If an uncharged compound A is present on both sides of a permeable membrane in different concentrations, its passage through the membrane is accompanied by a change of free energy. In biochemistry, this situation occurs mostly at <u>cellular membranes</u> (or membranes of organelles). For <u>import</u> into cells, the following equation applies:

$$\Delta G = R*T*2.303*\log \frac{[A_{inside}]}{[A_{outside}]} \qquad [1.5\text{-}9]$$

or

$$\Delta G\,[kJ*mol^{-1}] = 0.00831*T*2.303*\log \frac{[A_{inside}]}{[A_{outside}]} \qquad [1.5\text{-}9a]$$

Thus, the transport ocurs spontaneously only at <u>negative</u> ΔG, (when $[A_{inside}] < [A_{outside}]$), thus from higher to lower concentrations.

Correspondingly, for <u>export</u> from cells, the quotient is reversed

$$\Delta G = R*T*2.303*\log \frac{[A_{outside}]}{[A_{inside}]} \qquad [1.5\text{-}9b]$$

Charged molecules: The situation is more complicated, if there exists a <u>potential difference</u> $\Delta\Psi$ across the membrane (e.g., by non-penetrable ions)

$$\Delta\Psi = \Psi_{inside} - \Psi_{outside} \qquad [1.5\text{-}10]$$

and the compounds passing through the membrane carry <u>Z positive charges/molecule</u> (or –Z negative charges/molecule). The contribution of the charges to ΔG (with the prefix of Z corresponding to the + or – charge of the ions) is expressed by:

$$\Delta G_{charge\,transport} = Z*F*\Delta\Psi \qquad [1.5\text{-}11]$$

or

$$\Delta G_{charge\,transport}\,[kJ*mol^{-1}] = Z*0.0965*\Delta\Psi\,[mV]. \qquad [1.5\text{-}11a]$$

Thus, for an <u>import process</u>, Eq. [1.5-9] and Eq. [1.5-11] have to be combined:

$$\Delta G = R*T*2.303*\log \frac{[A_{inside}]}{[A_{outside}]} + Z*F*\Delta\Psi \qquad [1.5\text{-}12]$$

or

$$\Delta G\,[kJ*mol^{-1}] = 0.00831*T*2.303*\log \frac{[A_{ins.}]}{[A_{outs.}]} + Z*0.0965*\Delta\Psi\,[mV] \qquad [1.5\text{-}12a]$$

Correspondingly, for an <u>export</u> process,

$$\Delta G = R*T*2.303*\log \frac{[A_{outside}]}{[A_{inside}]} - Z*F*\Delta\Psi \qquad [1.5\text{-}12b]$$

The prefix of the last term in this equation is the opposite one of Eq. [1.5-12], since the membrane potential (Eq. 1.5-10) has the opposite effect on the energy situation.

An equilibrium exists, if $\Delta G = 0$. Then the <u>equilibrium potential</u> $\Delta\Psi_0$ [mV] can be obtained by the <u>Nernst equation:</u>

$$\Delta\Psi_0 = -\frac{R*T}{Z*F}*2.303*\log \frac{[A_{inside}]}{[A_{outside}]} \qquad [1.5\text{-}13]$$

or

$$\Delta\Psi_0\,[mV] = -\frac{0.00831*T}{Z*0.0965}*2.303*\log \frac{[A_{inside}]}{[A_{outside}]} \qquad [1.5\text{-}13a]$$

An extension of this formula to the equilibrium potential of several ions is the <u>Goldman equation</u> (see 17.2.1).

Literature:
Textbooks of physical chemistry.

1.5.4. Enzyme Kinetics

The biochemical base of enzyme catalysis is dealt with in 2.4. In the following, the mathematical treatment of the kinetics is given in some more detail.

Velocity of reactions: The <u>reaction rate</u> v for conversion of a single compound A → product(s) (<u>first order reaction</u>) is proportional to the concentration of this compound [A], while for a two-compound reaction A + B → product(s) (<u>second order reaction</u>) it depends on the number of contacts and thus on the concentration of both components (Eq. [1.5-14] and [1.5-15]). The proportionality factor k is termed <u>rate constant</u>.

Eq. [1.5-15] can also be applied for the formation of a <u>complex</u> and Eq. [1.5-14] for the decomposition of this complex. This includes substrate-enzyme complexes (see below), ligand-receptor complexes (17.1-2), antigen-antibody complexes (19.1.7) etc.

$$v = -\frac{d\,[A]}{dt} = k*[A] \qquad [1.5\text{-}14]$$

$$v = -\frac{d\,[A]}{dt} = -\frac{d\,[B]}{dt} = k*[A]*[B] \qquad [1.5\text{-}15]$$

Enzyme catalyzed one-substrate reaction: The theory of the enzyme-catalyzed conversion of a <u>single reactant</u> (the <u>substrate</u>, S) is based on the assumption that the enzyme (the catalyst, E) and this substrate form a complex (ES) by a reversible reaction. This step is kinetically treated like a two-compound reaction (rate constants k_1 and k_{-1} for formation and decomposition, respectively). The complex is then converted into the product (P) with the rate constant k_2. The conversion into P is considered to be irreversible at the beginning of the reaction, when practically no product is present.

$$E + S \underset{k_{-1}}{\overset{k_1}{\leftrightarrow}} ES \overset{k_2}{\to} EP \to E + P \qquad [1.5\text{-}16]$$

Therefore, for the formation of the enzyme-substrate complex, Eq. [1.5-15] has to be applied, while for its decomposition into its components, as well as for its conversion to the products, Eq. [1.5-14] is valid. There is actually an intermediate step ES → EP before the product is released. Its rate constant is not treated as a separate entity in most discussions of kinetic behavior, but is combined with the dissociation step to k_2. This is also done in the following considerations.

Usually, the substrate is in large excess over the enzyme. In this case, after a short 'transient phase', [ES] can be considered to be sufficiently constant (steady-state assumption). Disregarding the reverse reaction by using the situation immediately after the transient phase (see above) one obtains

$$\frac{d[ES]}{dt} = 0 = k_1 * [E] * [S] - k_{-1} * [ES] - k_2 * [ES] \qquad [1.5\text{-}17]$$

If one assumes that the rate determining process is the reaction ES → E + P, the initial reaction rate v_0 can be written as a function of [ES], which is analogous to Eq. [15.1-14]

$$v_0 = k_2 * [ES]. \qquad [1.5\text{-}14a]$$

By using a term for the total concentration of enzyme $[E_t] = [E] + [ES]$, by expressing the maximum reaction rate V_{max}, which is obtained, when all of the enzyme is saturated with substrate ($[ES] = [E_t]$) as

$$V_{max} = k_2 * [E_t], \qquad [1.5\text{-}14b]$$

and by introducing the Michaelis constant K_M

$$K_M = \frac{k_{-1} + k_2}{k_1}, \qquad [1.5\text{-}18]$$

one obtains the so-called Michaelis-Menten equation

$$v_0 = \frac{V_{max} * [S]}{K_M + [S]}, \qquad [1.5\text{-}19]$$

which shows the dependency of the reaction rate on the substrate concentration (first-order reaction). The plot of reaction rate vs. substrate concentration is a rectangular hyperbola (Fig. 1.5-1).

These formulas describe only the forward reaction. If the reverse reaction is included, the equivalent to Eq. [15.1-19] is

$$v = \frac{\dfrac{(V_{max})_f * [S]}{(K_M)_f} - \dfrac{(V_{max})_r * [P]}{(K_M)_r}}{1 \dfrac{[S]}{(K_M)_f} + \dfrac{[P]}{(K_M)_r}} \qquad [1.5\text{-}20]$$

where $(V_{max})_f$ and $(K_M)_f$ are identical to V_{max} and K_M in Eq. [1.5-19], while the terms $(V_{max})_r$ and $(K_M)_r = (k_{-1} + k_2)/k_{-2}$ are formed analogously for the reverse reaction.

Michaelis Constant: As can be derived from Eq. [1.5-19], the Michaelis constant K_M equals the substrate concentration at half the maximal reaction rate. Instead of obtaining this value from a plot according to Fig. 1.5-1, it is more convenient to use the reciprocal of the Michaelis-Menten equation, which yields a linear plot (at least in the ideal case, Lineweaver-Burk plot, Fig. 1.5-2a):

$$\frac{1}{v_0} = \frac{K_M}{V_{max} * [S]} + \frac{1}{V_{max}} \qquad [1.5\text{-}21]$$

If $1/v_0$ is plotted vs. $1/[S]$, then the intersections of this line with abscissa and ordinate allow the determination of K_M and V_{max}.

A disadvantage of the Lineweaver-Burk plot is the accumulation of measuring points near the ordinate (see the markings on the abscissa of Fig. 1.5-2a). Therefore other ways of plotting have been proposed. Hanes used another transformation of the Michaelis-Menten equation

$$\frac{[S]}{v_0} = \frac{K_M}{V_{max}} + \frac{[S]}{V_{max}} \qquad [1.5\text{-}21a]$$

The plot of $[S]/v_0$ vs. $[S]$ yields a line with the abscissa intersection $-K_M$ and the ordinate intersection K_M/V_{max}. The slope equals $1/V_{max}$ (Fig. 1.5-2b).

Still another method, the so-called 'direct plot', has been proposed by Eisenthal and Cornish-Bowden. The Michaelis-Menten equation is rearranged as follows:

$$V_{max} = v_0 + \frac{v_0}{[S]} * K_M \qquad [1.5\text{-}21b]$$

For each individual measurement, $-[S]$ is marked on the abscissa and v_0 on the ordinate and a line is drawn through both points. The intersection of these lines has the abscissa value K_M and the ordinate value V_{max} (Fig. 1.5-2c).

Figure 1.5-2. Linear Plots of an Enzyme Catalyzed Reaction

a) Lineweaver-Burk Plot

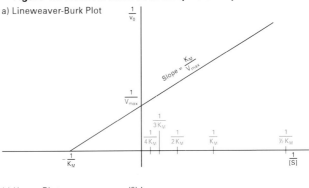

b) Hanes Plot

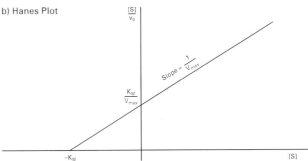

c) Eisenthal and Cornish-Bowden Plot

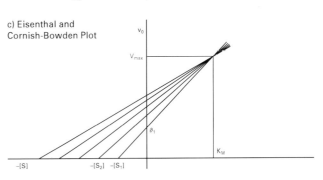

Figure 1.5-1. Reaction Velocity of an Enzyme Catalyzed Reaction
The velocity at $[S] = n * K_M$ is shown

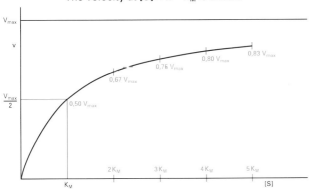

However, the most accurate method is the statistical evaluation of the measurements. In spite of this, the Lineweaver-Burk plot will be used in the following graphical representations, since it is the best known one.

Characterization of enzyme activities: The <u>enzyme activity</u> is defined as the quantity of substrate turned over per time unit in the presence of a given amount of enzyme. Thus the standard dimension would be [mol * sec^{-1}] = <u>katal</u>. For practical reasons, usually the activity is expressed as [μmol * min^{-1}]. This term is named <u>international unit</u> (<u>U</u>), if the measurement is performed under standard conditions (with isolated enzymes at conditions, which are optimized as much as possible). The <u>specific activity</u> is the enzyme activity per unit of weight, e.g., per mg and is frequently used to characterize the degree of purification of isolated enzymes.

The <u>turnover number</u> of an enzyme is defined as the number of molecules converted by one molecule of enzyme per unit of time, if the enzyme is saturated with substrate ([E_t] = [ES]). It is identical to the rate constant k_2 and can be calculated from Eq. [1.5-14b] as $k_2 = V_{max}/[E_t]$. Most turnover numbers are in the range of 1...10^4, the value for catalase is 4 * 10^7.

Most reactions *in vivo* proceed below the saturation limit of the enzyme, frequently at [S] = 0.01 ... 1 K_M. By the combination of Eq. [1.5-17], Eq. [1.5-18] and Eq. [1.5-14a] one obtains

$$v_0 = \frac{k_2}{K_M} * [E] * [S] \qquad [1.5\text{-}22]$$

At low substrate concentration, only a small portion of the enzyme forms an enzyme-substrate complex and [E] ≈ [E_t] = constant. The term k_2/K_M indicates how often a contact of enzyme and substrate leads to a reaction and is therefore a measure of the <u>catalytic efficiency</u>. It has an upper limit of ca. 10^9 [l * mol^{-1} * sec^{-1}], when practically every contact leads to a reaction and the reaction rate is determined by the diffusion speed. The value for catalase (4 * 10^8) is one of the highest observed.

Inhibitions: The mathematical treatment of an inhibited reaction depends on the mechanism of the inhibition. The general principles of inhibition are described in 2.5.2.

<u>Competitive inhibition</u>: The inhibitor competes with the substrate for reversible binding to the active site of the enzyme. The enzyme-substrate and the enzyme-inhibitor complexes are formed with the <u>dissociation constants</u> K_S and K_I, respectively.

$$K_S = \frac{[E] * [S]}{[ES]} \qquad [1.5\text{-}23]$$

$$K_I = \frac{[E] * [I]}{[EI]} \qquad [1.5\text{-}23a]$$

This results in the equation

$$\frac{1}{v_0} = \frac{K_M}{V_{max} * [S]} * \left(1 + \frac{[I]}{K_I}\right) + \frac{1}{V_{max}} \qquad [1.5\text{-}24]$$

In the Lineweaver-Burk plot, lines obtained at different inhibitor concentration intersect at the ordinate (Fig. 1.5-3a).

<u>Uncompetitive inhibition</u>: The inhibitor reacts reversibly only with the enzyme substrate-complex, but does not affect its formation. The dissociation constant is K'_I.

$$K'_I = \frac{[ES] * [I]}{[ESI]} \qquad [1.5\text{-}25]$$

This yields the equation

$$\frac{1}{v_0} = \frac{K_M}{V_{max} * [S]} + \frac{1}{V_{max}} \left(1 + \frac{[I]}{K'_I}\right) \qquad [1.5\text{-}26]$$

In the Lineweaver-Burk plot, parallel lines are obtained at different inhibitor concentrations (Fig. 1.5-3b).

<u>Noncompetitive and mixed inhibition</u>: If the inhibitor binds both to the enzyme and to the enzyme-substrate complex according to Eqs. [15.1-23a] and [15.1-25] and prevents the product formation, the following equation results

$$\frac{1}{v_0} = \frac{K_M}{V_{max} * [S]} * \left(1 + \frac{[I]}{K_I}\right) + \frac{1}{V_{max}} * \left(1 + \frac{[I]}{K'_I}\right) \qquad [1.5\text{-}27]$$

If the affinities of the inhibitor to the enzyme and to the enzyme-substrate complex are equal ($K_I = K'_I$), then the lines obtained at different inhibitor concentrations intersect in the Linewaever-Burk plot at the negative abscissa (K_M remains unchanged, <u>noncompetitive inhibition</u>, Fig. 1.5-3c). Otherwise, they intersect in the second quadrant (left of the ordinate, <u>mixed inhibition</u>, Fig. 1.5-3d).

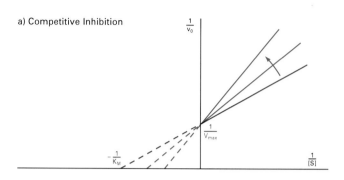

Figure 1.5-3. Lineweaver-Burk Plots of Inhibited Reactions
Red = uninhibited reaction, blue = inhibited reaction,
arrow = shift of the plot at increasing inhibitor concentrations

a) Competitive Inhibition

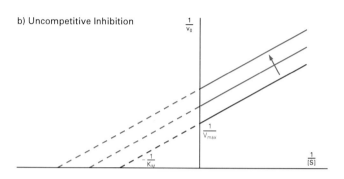

b) Uncompetitive Inhibition

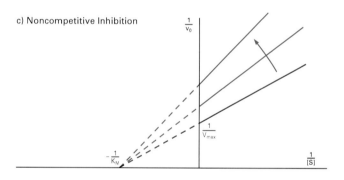

c) Noncompetitive Inhibition

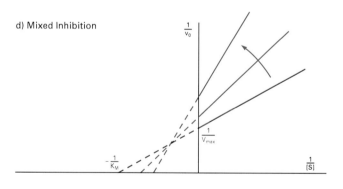

d) Mixed Inhibition

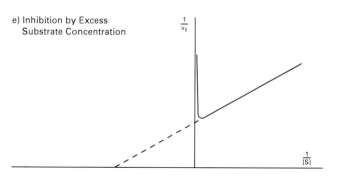

e) Inhibition by Excess Substrate Concentration

Inhibition by excessive substrate concentrations: If the reaction velocity decreases at very high substrate concentrations, this results in a Lineweaver-Burk curve bent upwards near the ordinate This situation is mostly observed in *in vitro*-experiments (Fig. 1.5-3e).

Two-substrate reactions: The formulas describing the kinetics are derived from the same assumptions as for one-substrate reactions. Their form depends on the reaction sequence. They involve separate Michaelis constants for the turnover of each substrate.

The Cleland nomenclature system uses the following expressions for the number of the substrates and products of the reaction: 1 – Uni, 2 – Bi, 3 – Ter, 4 – Quad. The substrates are named A, B, C ..., the products P, Q, R ... and the enzyme species (original state, intermediates and final state) E, F, G ... If all components have to combine before the reaction takes place, this is called a sequential reaction. This may take place in an ordered way or at random. If, however, one component leaves the enzyme before the other enters, it is a ping-pong reaction. The mechanisms are schematically drawn in Fig. 1.5-4.

Figure 1.5-4. Types of Two-Substrate-Two-Product (Bi-Bi) Reactions
The enzyme is represented by the horizontal line

a) Ordered Sequential Bi-Bi

b) Random Sequential Bi-Bi

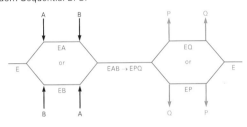

c) Ping-Pong Bi-Bi

The formula for an ordered sequential Bi-Bi reaction is

$$\frac{1}{v_0} = \frac{1}{V_{max}} + \frac{(K_M)_A}{V_{max} * [A]} + \frac{(K_M)_B}{V_{max} * [B]} + \frac{(K_M)_{AB}}{V_{max} * [A] * [B]} \quad [1.5\text{-}28]$$

The general formula for random sequential reactions is very complicated.
A ping-pong Bi-Bi reaction is described by

$$\frac{1}{v_0} = \frac{1}{V_{max}} + \frac{(K_M)_A}{V_{max} * [A]} * \frac{(K_M)_B}{V_{max} * [B]} \quad [1.5\text{-}29]$$

If in second order reactions the concentration of one of the substrates is very much above the respective Michaelis constant, then the terms containing this concentration in Eqs. [1.5-28] and [1.5-29] are practically zero and the equations become identical with Eq. [15.1-21], allowing the same evaluation as with a first order reaction.

If a series of measurements are made, in which one substrate is varied, while the other is kept constant, then one obtains Lineweaver-Burk plots, which formally resemble those obtained with inhibited reactions. However, increasing concentrations of the second substrate shift the lines the other direction (Fig. 1.5-5). Ordered sequential mechanisms yield a series of lines, which intersect left of the ordinate (above or below the abscissa), while ping-pong mechanisms yield parallel lines.

Dependence of reactions on temperature and activation energy: A more refined consideration of the reaction sequence Eq. [1.5-16] shows, that only collisions of the reactants above a certain energy level will lead to the formation of complexes, e.g., ES and EP. Also, the reaction ES → EP requires an initial energy input. Thus, the reaction has to cross 'energy hills', which represent metastable states (Fig. 2.4-1). They are called transition complexes X^+ and can either return to the original components or progress towards the products of the reaction, quickly achieving equilibrium in both cases. For obvious reasons, the highest 'hill' is the rate determining one and has to be the one considered further. (It takes the place of [ES] in the previous equations.) Thus, the equilibrium for formation of this complex can be described analogously to Eq. [1.5-3] by

$$K = \frac{[X^+]}{[A] * [B]} \quad [1.5\text{-}30]$$

Figure 1.5-5. Lineweaver-Burk Plots of Two-Substrate Reactions
Arrow = shift of the plot when the concentration of the other substrate is raised

a) Ordered Sequential Bi-Bi Reaction

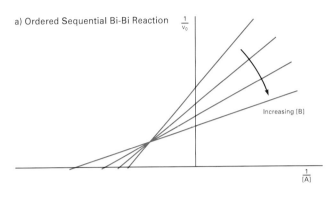

b) Ping-Pong Bi-Bi Reaction

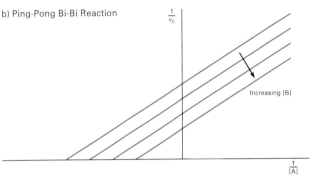

The energy required for its formation is called activation energy ΔG^+, which can be calculated from this equilibrium by applying Eq. [1.5-4] as

$$\Delta G^+ = -R * T * 2.303 * \log \frac{[X^+]}{[A] * [B]} \quad [1.5\text{-}31]$$

According to Eq. [1.5-14] the reaction rate for formation of the product(s) from this complex is expressed by $v_0 = k * [X^+]$. When combining this equation with Eq. [1.5-31], one obtains the following formula for the temperature and ΔG^+ dependence:

$$v_0 = \text{const.} * [A] * [B] * 10^{-\Delta G^+ / 2.303 * R * T} \quad [1.5\text{-}32]$$

The increase of the reaction rate with rising temperature is limited, however. When the enzyme becomes thermally denatured, the rate drops (Fig. 2.4-4).

Fractal enzyme kinetics: The above considerations assume 'ideal' conditions: purified enzymes, low concentrations, free movement of the reactants. However, *in vivo* the situation is different. Based on a power-law derivation it has been shown, that, e.g., restrictions in space require the introduction of non-integer powers >1 to the concentration terms in Eq. [1.5-17]:

$$\frac{d[ES]}{dt} = \alpha_1 * [E]^g * [S]^h - \alpha_{-1} * [ES] - \alpha_2 * [ES] \quad [1.5\text{-}33]$$

The consecutive equations change analogously. This system, which is under development at present, is called fractal kinetics. Its main implications are:
- K_M is dependent on the enzyme concentration, it decreases with increasing enzyme concentration
- The plot of enzyme activity vs. substrate concentration has a tendency towards a sigmoid shape even with monomeric enzymes
- The velocity of the reaction increases if the movements are, e.g., restricted to surface interfaces (e.g., 6.3.2) or to one dimension (e.g., by sliding along nucleic acid strands, 10.2.2, 11.5.2 or by 'substrate channeling', 4.7.1)
- In sequences of reactions, the flux responses are faster and the accumulation of intermediates is lower as compared to the Michealis-Menten assumption.

In some respects, fractal kinetics resemble allosteric situations (2.5.2). Velocity calculations according to this theory have a tendency to yield higher values as according to the Michaelis-Menten theory, which represents a borderline case of a more general treatment, but is still of value for understanding the basic principles of enzyme catalysis.

Literature:
Cornish-Bowden, A., Wharton, C.W.: *Enzyme Kinetics*. IRL Press (1988).
Dixon, M., Webb, E.C.: *Enzymes*. 3rd Ed. Academic Press (1979).
Fersht, A.: *Enzyme Structure and Mechanism* (2nd Ed.). Freeman (1985).
Savageau, M.A., J. theor. Biol. 176 (1995) 115–124.
Savageau, M.A.: BioSystems, in press (1998).
Sigman, D.S., Boyer, P.D. (Eds.): *The Enzymes*. 3rd Ed. Vols. 19 and 20. Academic Press (1990 and 1992).

2 The Cell and Its Contents

This chapter presents selected information on the structure and organization of living organisms and their major components to serve as as a background for the biochemical text of this book. For more details, refer to textbooks of biology.

2.1 Classification of Living Organisms

Life is associated with a number of characteristics, such as propagation, metabolism, response to environmental influences, evolution etc. For all living beings, cells are the basic unit of organisation. While unicellular organisms exist as separate entities, the various cells of multicellular organisms fulfill different functions and the organism depends on mutual cellular interaction.

There are several systems of classification of living organisms. From a phylogenetic viewpoint, the classification into the 3 domains *bacteria*, *archaea* and *eukarya* (which are further subdivided) appears most justified (Table 2.1-1). When common aspects of *eubacteria* and *archaea* are discussed, the term *prokarya* is used.

The metabolic reactions in this book are indicated by colored arrows. Since frequently the occurrence of the reactions is known only for a few species and also in order to prevent an 'overloading' of the figures by information details, the arrow colors have been combined into (black) general metabolism, (red) *bacteria* and/or *archaea*, (green) *plants* and/or *fungi* and/or *protists*, (blue) *animals*.

Living organisms exhibit a high degree of order. The sum of all endogeneous life processes results in a steady decrease of free energy (1.5.1). Therefore, life can only be kept up by an energy input from the environment, either as light energy or by uptake of oxidizable compounds. Another essential requirement of life is the availability of an adequate carbon source. Living beings can be classified according to the mode of energy uptake and the carbon source used (Table 2.1-2).

During the oxidation of compounds, electrons are released, which have to be taken up by a terminal electron acceptor. Energy-wise oxygen is most favorable (16.1). Previous to its appearance in the primeval atmosphere and even now in oxygen-free habitats, living organisms must use other acceptors (Table 2.1-3).

Literature:
Fox, G.E. *et al.*: Science 209 (1980) 457–463.
Holt, J.G. *et al.*: *Bergey's Manual of Descriptive Bacteriology* (9th Ed.). Williams and Wilkins (1994).
Margulis, L., Schwartz, K.V.: *Five Kingdoms*. 2nd. Ed. Freeman (1987).
Woese, C.R. *et al.*: Proc. Natl. Acad. Sci USA 87 (1990) 4576–4579.

2.2 Structure of Cells

2.2.1 Prokaryotic Cells (Fig. 2.1-1)

The genetic information is stored in a single, circular double helix of deoxyribonucleic acid (DNA, 2.6.4). It is located in the central portion of the cell in a densely packed form (nucleoid), but without a special separation from the rest of the cell. Its replication and the translation of the information into protein structures is described in Chapter 10.

In prokarya, frequently a number of plasmids may occur, which also consist of circular DNA and replicate independently of the main DNA. They carry only a few genes. Although plasmids are usually not essential for survival, they are involved in DNA transfer during conjugation, provide resistance to antibiotics etc. Some plasmids can be reversibly integrated into the main DNA (episomes). Similar properties are exhibited by DNA viruses and retroviruses (12.2, 12.4). The translocation of genetic material is not discussed here.

The cytoplasm is a semifluid, concentrated solution of proteins, metabolites, nucleotides, salts etc. It also contains several thousand ribosome particles involved in translation (10.6). It is the site of most metabolic reactions and exchanges material in a controlled way with the environment (15.1 ... 15.3).

Prokaryotic cells are surrounded by an envelope (Fig. 15.1-1). It has not only an enclosing and protective function. Rather, metabolic reactions take place at transmembrane proteins (e.g., respira-

Table 2.1-1. Some Typiceal Properties of Living Organisms (Exceptions exist)

Domains	**Bacteria**	**Archaea**	**Eukarya**			
Kingdoms	Bacteria	Archaea	Protists *	Plants	Fungi	Animals
Nucleus	no (common term **prokarya**)		yes			
Genome	circular, ca. $10^6 ... 5 * 10^7$ kb, extra plasmids		linear, $10^7 ... > 10^{11}$ kb, organized in several chromosomes			
RNA polymerase	one type		several types			
Starting amino acid for translation	formylmethionine		methionine			
Reproduction	binary scission		asexual/sexual	asexual/sexual	asexual/sexual	asexual/sexual
Cellular organization	unicellular (some are aggregated)		mostly unicellular	multicellular	uni-/multicellular	multicellular
Nutrition (Table 2.1-2)	chemoorganotrophic, photoautotrophic or photoheterotrophic	chemolithotrophic, photoautotrophic or chemoorganotrophic	chemoorganotrophic or photoautotrophic	photoautotrophic	chemoheterotrophic including saprobiontic	chemoheterotrophic
Size of cells	average 1 ... 5 μm, wide variation		average 10 ... 100 μm, wide variation			
Cell membranes	rigid, contain peptidoglycans	rigid, without peptidoglycans	rigid or soft	rigid, contain cellulose and lignin	rigid, contain chitin	soft, lipid bilayer only
Internal membranes	no		yes, they enclose organelles /vesicles			

* Algae, protozoa, fungi-related. The exact demarcation is under discussion.

Table 2.1-2. Sources of Carbon and of Energy

	Phototrophy (Energy input by light)	Chemotrophy (Energy provided by oxidizable compounds from the environment)
Autotrophy (only CO_2 needed as carbon source)	*green plants*, some *protists*, photosynthesizing *bacteria* (16.2)	*Prokarya* (mainly *archaea*). Oxidation of inorganic material (chemolithotrophy, 15.6)
Heterotrophy (organic compounds needed as carbon source)	some *prokarya*	All *animals* and *fungi*, non-green *plants*, many *protists* and *prokarya*. Oxidation of organic material (chemoorganothrophy). Included are *saprobiontes* (use decaying organic material) and *parasites* (feed from living beings)

Table 2.1-3. Terminal Electron Acceptors for Oxidation Reactions

	Atmospheric Oxygen		
	Not required		Required
Energy obtained by	anaerobic respiration (15.5)	fermentation (15.4)	aerobic respiration (16.1)
Electron acceptors	oxidized external compounds (mostly inorganic)	internally generated compounds	atmospheric O_2
Organisms	anaerobes: part of *archaea* and *bacteria*		All other organisms
	facultative anaerobes (*bacteria*)		

tion and ATP synthesis) or at membrane associated proteins. In *bacteria*, the sequence of membrane components from the interior outwards are:
- the plasma membrane, a lipid bilayer with embedded proteins (1.4.8)
- the rigid cell wall, which in the case of *bacteria* consists of either multiple layers (*Gram positive bacteria*) or a single layer (*Gram negative bacteria*) of peptidoglycans (murein).
- (only in *Gram negative bacteria*): an additional outer membrane
- (frequently): an additional gelatineous capsule superimposed on the cell wall. It consists mainly of polysaccharides (polymerized glucose, rhamnose, uronic acids etc.). There also may be mucus layers.

Extensions of the cell envelope are pili and flagella, which provide for cellular contact, conjugation, propulsion etc.

The composition of an *E.coli* cell by weight is H_2O 70%, protein 15%, DNA 1%, RNA 6%, polysaccharides 3%, lipids 2% (both are mainly present in the envelope), small organic molecules 1%, inorganic molecules 1%.

Mycoplasms are a group of *bacteria*, which lack a cell wall. Among them are the smallest self-reproducing organisms (0.10...0.25 μm diameter).

Archaea differ from *bacteria* by
- another composition and arrangement of rRNAs
- differences in the RNA polymerase and in the translation mechanism (Table 2.1-1)
- different composition of the cellular envelope. E.g., murein (15.1) is absent, acylglycerols are replaced by branched chain glycerol ethers (6.3.3)
- unusual pathways of metabolism and habitats (methanogens, 15.5.2, halobacteria, 16.2.1, thermophiles etc.)

2.2.2 General Characteristics of Eukaryotic Cells (Figures 2.2-1, 2.2-2)

As compared to *prokaryotic* cells, *eukaryotic* cells exhibit a much more complicated structure. Inside the plasma membrane there are the nucleus and the cytoplasm, which encompass the fluid cytosol and many organelles. These are compartments enclosed by individual membranes, which are devoted to specific functions.

Nucleus: The common denominator of eukaryotic cells is the presence of a separate nucleus, which contains the major portion of the genetic material of the cell. (The rest is present in *mitochondria* and *chloroplasts*, see below.) The nuclear DNA is organized in a number of chromosomes. Each double helix of chromosomal DNA (2.6.3) can be present once (in haplont organisms) or twice (in diplont organisms). During cell division (11.6), the condensed chromosomes separatedly arrange themselves. Otherwise they are combined with proteins as a ball of chromatin with an elaborate fine structure (2.6.4).

The number of chromosomes present in the various species differ widely (from 4 to > 500; humans 46 in the diploid set). While bacterial genomes contain $< 10^6 ... 5 * 10^7$ bp, the diploid set of *eukaryotic* DNA varies between ca. 10^7 bp (some *fungi*) and $> 10^{11}$ bp (*lungfish*, some *algae*). The diploid human genome contains $5.8 * 10^9$ bp ($2 * 2,900,000$ bp).

The nucleus contains in addition to the DNA also the nuclear matrix, which is composed of the enzymes and factors required for DNA replication, DNA repair, transcription and processing of the transcription products (11.1...11.4).

The nucleus is surrounded by a double membrane of lipid bilayers with integrated proteins. Nuclear pores (14.3.2, ca. 125 nm diameter) span the nuclear membrane and enable the transport of proteins, rRNA etc. The inner surface of the nuclear membrane is covered by nuclear lamina, a net of protein fibers, which stabilizes the structure and provides attachment points for the chromatin (2.6.4). During cell division, the nuclear membrane dissolves.

Cytosol: Although in *eukarya* many cytosolic functions have been taken over by specific organelles (see below), a large number of metabolic reactions still take place here. This includes glycolysis (3.1.1) and the synthesis of cytosolic proteins (11.5) and fatty acids (6.1.1).

Endoplasmic reticulum (ER, 13.3): This is a labyrinth of membranes, which frequently encompasses half of the total amount of membranes. It consists of a system of sac- and tubelike structures, which locally expand into cisterns. Its internal lumen is connected with the intermembrane space of the *nuclear membrane*.

Part of the ER is studded on the outside with ribosomes (rough ER), which take part in protein synthesis (11.5). The proteins thus formed enter the ER lumen and are processed mainly by glycosylation (13.3.1). They leave the ER via vesicles (14.2.2). The

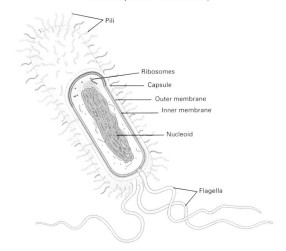

Figure 2.2-1. General Structure of a Bacterial Cell
After Campbell, N.A.: *Biology* 4/e. Benjamin/Cummings 1996. The colors are for easy differentiation only.

Figure 2.2-2. General Structure of a Plant Cell (Top) and an Animal Cell (Below)
After Campbell, N.A.: *Biology* 4/e. Benjamin/Cummings 1996. The colors are for easy differentiation only.

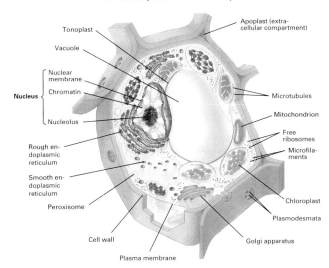

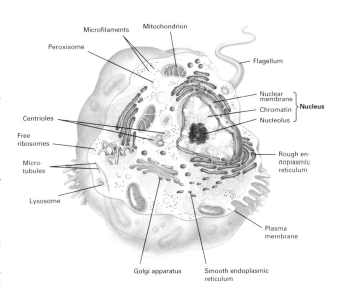

rough ER is also the site of membrane phospholipid synthesis, thus providing membrane material for the departing vesicles (6.3.2, 13.3.4).

The other part of the ER is free of ribosomes (smooth ER). Enzymes of the smooth ER are involved in the synthesis of fatty acids (desaturation, 6.1.3), phospholipids (6.3.2), steroids, especially steroid hormones (7.5.1) and other lipids. They are located on the outside of this organelle. The smooth ER plays a role in detoxification by hydroxylation reactions. Its equivalent in *muscles* (sarcoplasmic reticulum) stores Ca^{++} ions and is thus involved in the many reactions regulated by this ion (17.4.4). In liver, the smooth ER is also the storage site for glycogen (3.2).

Golgi apparatus (13.4): This organelle consists of stacks of flattened membrane sacs, which are especially numerous in *plants* (dictyosomes). The properties on both sides of an individual sac, as well as throughout the whole stack differ (details in 13.4). Their main function is the further processing and sorting of proteins and their export to the final targets. In most cases, these are membrane or secreted proteins. In addition, the Golgi apparatus also produces polysaccharides (e.g., hyaluronic acid, 13.1.5), glycosphingolipids (13.2.1) etc.

Lysosomes: Lysosomes are vesicles of 0.1...0.8 μm diameter, which are enclosed by a lipid bilayer. They are formed by budding from the *trans* side of the Golgi apparatus (13.4, 14.2.1). These organelles are filled with many enzymes for polysaccharide, lipid, protein and nucleic acid degradation. They fuse with endosomes containing internalized lipoproteins (18.2.4) or phagocytosed nutrients (18.1.2) and hydrolyze these compounds (Fig. 2.2-3). They act also on intracellular material to be removed and even contribute to the apoptosis (programmed cell death, 11.6.6) of their own cell. Lysosomes of special cells (e.g., *macrophages*) destroy, in the same way, *bacteria* or *viruses* as a defense mechanism (19.1.7).

The degradative enzymes of lysosomes exhibit an activity optimum at pH = 5, which is also the pH of the lysosome lumen. It is kept constant by continuous proton pumping into the lysosome lumen. The pH difference to the usual cytosolic pH of 7.0 is a safety measure, since after accidental leakage of some lysosomes, the released enzymes are almost inactive at the cytosolic pH. Only a cumulated release from many lysosomes is deleterious to the cell.

Insufficient activity of lysosomal enzymes is the reason for many diseases. Examples are gangliosidoses (13.2.1) and glycogen storage diseases (e.g., Pompe's disease, 3.2.5).

Peroxisomes: This is another example, how enclosure into an organelle allows reactions to take place, which would otherwise be deleterious to the rest of the cell. Peroxisomes (ca. 0.5 μm diameter) are surrounded by a single membrane. They are generated from components of the cytosol and do not bud from other membranes. The main task of these organelles is the performance of monooxygenase (hydroxylase) or oxidase reactions, which produce hydrogen peroxide (H_2O_2). This dangerous compound is immediately destroyed by catalase within this enclosed space:

$X\text{-}H + O_2 + H_2O = X\text{-}OH + H_2O_2$ or

$X\text{-}CH_2OH + O_2 = X\text{-}CHO + H_2O_2$, followed by

$2\ H_2O_2 = 2\ H_2O + O_2$

Reactions of this kind are the oxidative degradation of fatty acids (6.1.6), of alcohols etc. In glyoxisomes of plants, the reactions of the glyoxylate cycle (3.9.1) take place.

Mitochondria: In a typical *eukaryotic* cell, there are in the order of 2000 of these organelles, which are often of ellipsoidal shape (length ca. 1...10 μm). They have a smooth outer membrane and a highly folded inner membrane with numerous invaginations (cristae), which contain most of the membrane bound enzymes of mitochondrial metabolism. The internal area contains the mitochondrial matrix, while the intermembrane space is the narrow area between both membranes. Since protons are permanently pumped out through the inner membrane, the matrix is more alkaline than the intermembrane space and the cytoplasm (pH 8 vs. ≈ pH 7).

Mitochondria are the site of respiration and ATP synthesis (16.1.2, 16.1.4), but also of many other central reactions of metabolism, e.g., citrate cycle (3.8.1), fatty acid oxidation (6.1.5), glutamine formation (4.2.1) and part of the pathway leading to steroid hormones (7.5.1, 7.8.1). The latter sequence, as well as the initiation of gluconeogenesis (3.3.4) are examples of how the site of reactions frequently changes from one organelle to another one within a single pathway.

Besides *chloroplasts* (2.2.3), *mitochondria* are the only organelles which are equipped with own (circular) DNA, RNA and ribosomes and thus can perform their own protein synthesis. The components and the mechanism resembles more the bacterial than the eukaryotic system. However, less than 10 % of the mitochondrial proteins are generated by this means, the rest is encoded by nuclear DNA and imported (14.4). *Mitochondria* reproduce by a binary scission similar to *bacteria*. Their membranes do not exchange material with the rest of the cellular membrane system; the membrane proteins are produced on internal or *cytosolic* ribosomes. The membranes contain many transport systems resembling those of *bacteria*. These and other arguments are the base of the endosymbiont theory, which assumes that mitochondria have originated from ingested *bacteria* (e.g., aerobically living heterotrophs). They provide an effective energy production, but are dependent on the host in many aspects.

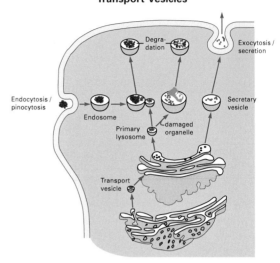

Figure 2.2-3. Interrelationship of Rough Endoplasmic Reticulum, Golgi Apparatus, Lysosomes, Endosomes and Transport Vesicles

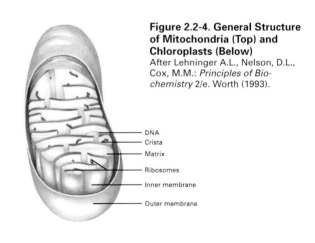

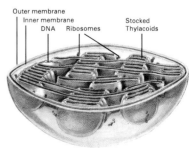

Figure 2.2-4. General Structure of Mitochondria (Top) and Chloroplasts (Below)
After Lehninger A.L., Nelson, D.L., Cox, M.M.: *Principles of Biochemistry* 2/e. Worth (1993).

Cytoskeleton: The internal cytoskeleton is a general component of *eukaryotic* cells, but is of special importance in *animal* cells. Here it is the primary factor for maintenance of the external structure, which, in *plant* cells, is achieved by the rigid cell wall (2.2.3). The cytoskeleton also provides anchoring points for the organelles and for some enzymes and even contributes to movements either by its own formation and degradation or by motor proteins moving along the filaments.

Components of the cytoskeleton are microtubules, actin filaments and intermediary filaments (Table 2.2-1). More details on their composition, formation and degradation are presented in 18.1.3.

Table 2.2-1. Components of the Cytoskeleton

	Microtubules	Microfilaments	Intermediary Filaments
Components	α/β-tubulin	actin	keratin, vimetin, desmin etc.
Shape	hollow tube of 13 tubulin sequences	2 strings of actin monomers, coiled	variable, frequently coiled coil of α helices
Diameter	25 nm	7 nm	8 ... 12 nm
Examples of function	chromosome separation (11.6.5), axonal transport (17.2.6), other movements	muscle contraction (17.4.5), movement of the cytoplasm, cell division (furrow ingression, 11.6.5)	structure stabilization, anchoring of organelles

The hollow microtubules (Fig. 18.1-3) frequently originate from the centrosome, which is located close to the nucleus (11.6.5). The moving part of cilia and flagella are circular arrangements of tubulin. Attached dynein molecules cause movements of the tubules relative to each other and thus curvature of these cellular annexes (18.1.3).

Actin filaments associate with myosin. The heads of the myosin molecules can 'walk along' the string of actin monomers. The best known action of actin filaments is the muscle contraction (17.4.5), but they are also involved in many other, mostly movement functions. E.g., the cytoplasmic movement in some *algae* is caused by the transport of organelles along actin filaments by myosin.

Intermediary filaments are composed of variable units, depending on the particular cell type. In general they have a structural function. They are more long-lasting than the other components of the cytoskeleton, but are also subject to rearrangements. Vimetin occurs in, e.g., *endothelial cells* and *adipocytes* and anchors the nucleus and the fat droplets. Desmin filaments keep the Z disks of muscle cells in place (17.4.5). Neurofilaments reinforce the long axons of *neuronal cells*. The nuclear lamina (see above) also consist of intermediary filaments.

2.2.3 Special Structures of Plant Cells

Plant Cell Wall: Cell walls are an essential factor distinguishing *plant* and *animal* cells. They provide stability and prevent an expansion of the cell beyond its fixed size. These secondary walls consist mainly of cellulose (3.2.2), hemicelluloses (3.6.3) and lignin (4.7.6). During an early phase of formation, the primary walls contain considerable amounts of pectate (3.5.6) and glycoproteins (13.1.1). On the inside, the cell contents are enclosed by the usual lipid bilayer membrane. Plasmodesmata interconnect neighboring cells and enable the transport of water and metabolites.

Chloroplasts: These are lens-shaped organelles of about 1 ... 5 µm diameter, which occur only in photosynthesizing green plants. They are enclosed by a double membrane with a thin intermembrane space in between. The interior contains the fluid stroma and a third membrane system surrounding the thylakoid space. This system has the shape of interconnected flat disks, which, in most cases, are stacked on top of each other (grana). The thylakoid membranes are the site of photosynthesis (16.2.1). In photosynthesizing bacteria, their role is taken over by the cytoplasmic membrane.

Due to the permanent pumping of protons into the thylakoid space, its interior is much more acidic than the stroma (pH = 5 vs. pH = 8). While some chloroplast proteins originate from the own protein synthesis system, the majority are nuclear encoded and are imported by a special mechanism into the stroma and the thylakoid space (14.5.2).

Analogous to *mitochondria* (2.2.2), *chloroplasts* resemble *bacteria* in several aspects (protein synthesis, membrane structure, reproduction etc.). Likewise the endosymbiont theory refers to them. Possible precursors of chloroplasts could be photoautotrophic cyanobacteria.

Vacuoles: Although vacuoles also occur in other kingdoms (e.g., protists), they are most prominent in *plants*. Young plant cells contain several vacuoles, which originate in vesicles released from the endoplasmic reticulum and the Golgi apparatus. In mature plant cells they combine to form a single central vacuole, which is enclosed by a membrane (tonoplast).

The central vacuole can occupy up to 90% of the total cell volume and is a storage space for inorganic salts, saccharose, proteins, pigments and waste. The accumulation of these compounds causes the inflow of water by osmosis and keeps up the internal pressure (turgor).

2.2.4 Special Structures of Animal Cells

Animal cell membrane and extracellular matrix (Fig. 2.2-5): The actual cell membrane consists only of the lipid bilayer. However, it is covered by a complex, gelatinous extracellular matrix, which is formed by glycoproteins secreted from the cell. Its main component is collagen (2.3.1). The collagen fibers are enclosed by a network of carbohydrate rich proteoglycans (13.1.2). Tasks of the extracellular matrix are the interconnection and anchoring of the cells, the support of the structure and to a certain extent also the localization of cells (e.g., during embryonic development).

Depending on their organ specific structure, they allow the passage of fluids and dissolved material at varying degrees (e.g. filtration effects in the *glomerulus* of the *kidney*). The components of the extracellular matrix are connected via linker proteins (mostly fibronectin) with membrane spanning proteins (integrins). The integrins, in turn, are associated at the interior side of the membrane with microfilaments of the cytoskeleton.

Figure 2.2-5. Structure Elements of the Extracellular Matrix

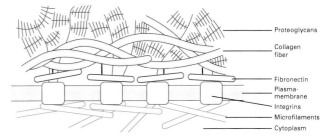

Interconnection between cells (Fig. 2.2-6): Neighboring cells (especially *epithelial cells*) are interconnected in several different ways. Tight junctions firmly attach the cells to each other and prevent any fluid passage between them. Point desmosomes (also frequently ocurring in *plants*) are firm interconnections of neighboring cells via a network of glycoprotein filaments. They are reinforced on the inside of the cells by filaments of the cytoskeleton. Gap junctions are small channels passing through both membranes of neighboring cells. They allow the exchange of small molecules and ions (17.2.3) and represent an equivalent of plant plasmodesmata (2.2.3). Pores, channels, transporters and receptors involved in material and signal passage through individual membranes are described elsewhere (17.1.2, 17.2.3, 18.1).

Figure 2.2-6. Types of Contact Between Cells

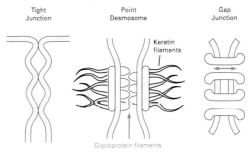

Literature:
de Duve, C.: *A Guided Tour of the Living Cell*. Scientific American Books (1984).
Textbooks of biology, e.g. Campbell, N.A.: *Biology* 4th Ed. The Benjamin/Cummings Publ. Co (1996).

2.3 Protein Structure and Function

While the two previous sections dealt with the cellular structures, the rest of this chapter describes properties of the two most important classes of macromolecules: proteins and nucleic acids.

Proteins and peptides are linear polymers which are built from the 21 naturally occuring L-amino acids (1.3 and Chapter 4, including selenocysteine, 10.6.5) and are linked by peptide bonds (1.3.2, Fig. 2.3-1). Peptides containing D-amino acids are less frequent. They occur mostly in *microorganisms*. Chains of up to ≈100 amino acid residues are usually named 'oligo-(or poly-)peptides', larger ones are termed 'proteins'. However, the expression 'peptides' is frequently also used for the whole class of compounds containing peptide bonds.

All polypeptides and proteins are synthesized by ribosomal synthesis, using mRNA as the source of information for the sequence of amino acids (10.6, 11.5). However, a number of small peptides are formed by a non-ribosomal sequence of specific enzyme reactions, e.g., glutathione (4.5.7), the penicillin precursor peptide (15.8.1) and many peptide antibiotics.

After formation of the peptide chain, additional post-ribosomal modifications of the amino acid components may take place, e.g., hydroxylation (4.3, 4.5.2), carboxylation (4.2.2, 9.13), methylation (e.g., of glutamate during chemotactic mechanisms), acylation (acetylation, myristoylation, palmitoylation, 13.3.5), phosphorylation (2.5.2, 17.5), glycosylation (13.3.3, 13.4.1) and formation of disulfide bonds between the -SH groups of cysteine (14.1.3).

In a number of cases, a long peptide is cleaved after synthesis, e.g., to be activated for a special function. Examples are hormones (17.1.3, 17.1.5), digestive enzymes (2.5.1) and blood coagulation factors (20.2...5).

2.3.1 Levels of Organization

The amino acid sequence contains all the information necessary to determine the three-dimensional structure of the protein which is assumed to be the thermodynamically most stable one among the kinetically possible ways of folding (14.1)

Several levels of structural organization can be distinguished:

Primary structure: This is the amino acid sequence (Fig. 2.3-1).

The CO-NH bond is fairly rigid and usually assumes the *trans* configuration. To a limited extent, rotations are possible around both other bonds of the peptide backbone (torsion angles Φ and Ψ). Steric restrictions are caused by the side chains of the amino acids.

Figure 2.3-1. Primary Structure of a Polypeptide Chain in Extended Conformation
The distances (in nm = 10 Å) and the bond angles are shown. The green quadrangles indicate the rigid structure of the peptide bonds. Ψ and Φ are the angles characterizing the rotation around the C_α–CO and N–C_α bonds (in this figure $\Psi = \Phi = 180°$). The formation of hydrogen bonds leading to α helices is indicated.

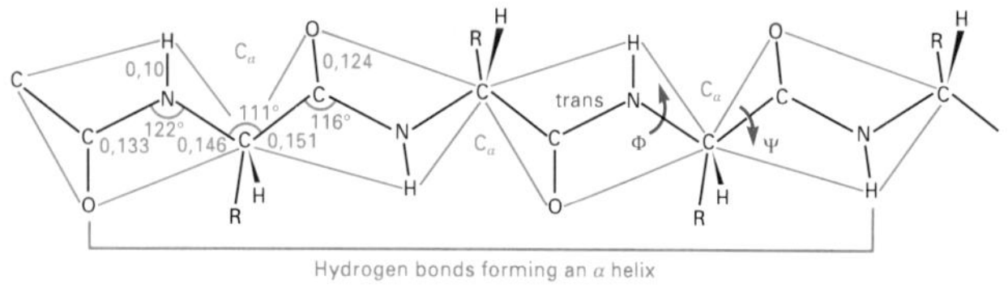

Secondary structure: These are regular arrangements of the backbone of the polypeptide chain, which are stabilized by hydrogen bonds between amide and carboxyl groups of the peptide. Frequently occurring secondary structures are the α-helix and the β-pleated sheet (Fig. 2.3-2). In globular proteins, a variable portion (usually below 40%) is arranged in less regular loop structures, which are frequently found on the outside of water-soluble proteins.

The α helix is a rod-like structure, which is formed by hydrogen bonds between a CO group and the fourth NH group of the sequence (shown in Fig. 2.3-1). Helices formed by hydrogen bonds between other CO-NH pairs are rare. There are also van der Waals bonds across the helix. The side chains of the amino acids point to the outside. Almost all α helices occurring in nature are right-handed (i.e. when looking along the axis of the helix from the N to the C terminus, the chain turns clockwise). One turn is formed by 3.6 residues, the pitch amounts to 0.54 nm. α-Helices occur both in globular and in fibrous proteins including transmembrane sections.

The β-pleated sheet forms slightly folded planes. This structure is also stabilized by hydrogen bonds, however, between different strands. 2...15 strands may arrange themselves in parallel or antiparallel directions. The side chains of consecutive amino acids extend sideways in opposite directions.

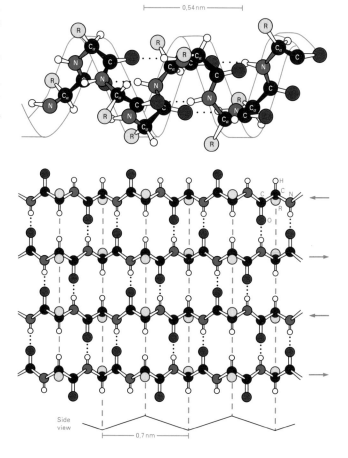

Figure 2.3-2. α Helix (Top) and Antiparallel β Pleated Sheet Structure (Bottom)
C = black, O = red, N = blue, H = white, side chains = grey

Tertiary structure is a term which refers to larger arrangements, but it cannot be clearly separated from secondary structure elements. In general, it encompasses the following:

Supersecondary structures (superfolds) are combinations of secondary structure elements, such as the coiled-coil α helices (e.g., in many fibrillar proteins) or barrels (in globular proteins). A number of structural motifs are occurring in these superfolds.

Two or more α helices can associate into an antiparallel arrangement, which allows interactions of their side chains (αα motif). This is an essential element of fibrous proteins (e.g., in α keratin, Fig. 2.3-5 and in the segments of spectrin, which is present in *erythrocytes*), but does also occur in globular proteins (e.g., in myoglobin, Fig. 5.2-3).

A number of common motifs occurring in globular proteins are caused by the turns of the peptide backbone at the end of β pleated sheets (Fig. 2.3-3): Antiparallel β pleated sheets frequently form β loops (also called hairpin loops), which are stabilized by hydrogen bonds between a CO group and the third NH group of the sequence. A series of such turns can form a β-meander. Another arrangement of turns is the so-called 'Greek key'. Parallel β pleated sheets require crossover connections, which often assume right-handed helix structures (βαβ motif). These loops and crossover connections frequently occur at the outside of globular proteins.

Figure 2.3-3. Elements of Supersecondary Protein Structures

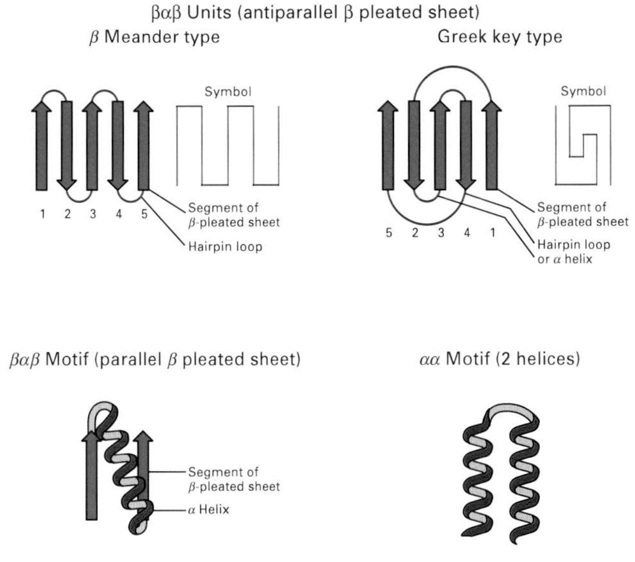

βαβ Units (antiparallel β pleated sheet)

β Meander type   Greek key type

βαβ Motif (parallel β pleated sheet)   αα Motif (2 helices)

Figure 2.3-4. Examples for Supersecondary Structures

β Barrel (Triosephosphate isomerase)   Saddle (Carboxypeptidase A)

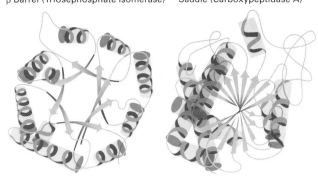

Figure 2.3-5. Structure of α Keratin

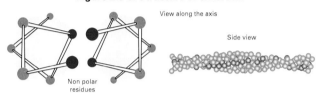

Figure 2.3-6. Structure of Collagen

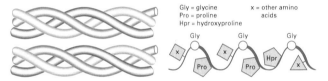

Gly = glycine   x = other amino acids
Pro = proline
Hpr = hydroxyproline

The 'plane' of β-pleated sheets in globular proteins is always slightly twisted in a right handed direction. A large β-pleated sheet can roll up to a β barrel structure (e.g., triosephosphate isomerase). Another possibility is a saddle-shaped arrangement (e.g., bovine carboxypeptidase A, Fig. 2.3-4).

Structural domains are structurally independent globular arrangements of ca. 100...200 amino acid residues. They are connected with each other by flexible polypeptide segments (e.g. in immunoglobulins, Fig. 19.1-2).

Globular proteins are the result of the folding of complete single polypeptide chains into quite compact structures.

Globular proteins are usually arranged with hydrophobic amino acid residues in the interior and hydrophilic residues facing the outside. There are exceptions: Internal polar groups can be required for the protein function (e.g. in the active center of enzymes, 2.4.1). Transmembrane protein sections are usually structured as an α helix composed of 19...20 amino acids with nonpolar side chains. A number of transmembrane passes can be circularly arranged with hydrophobic residues facing the lipid membrane and polar residues forming a channel, which permits the passage of polar or charged molecules (18.1.1). The size of the pore and the electric charges of its lining determine its specificity. Often the passage is regulated by controlled movements of the protein. Still more complicated are transporters, which interact with the cargo and often require an input of energy (18.1.1).

Fibrous proteins frequently consist of several helices turned around each other as a 'coiled coil'.

α Keratin is the principal component of the epidermal layer and also of hair, feathers etc. Regularly spaced nonpolar residues in both α helices interact noncovalently, form a hydrophobic contact strip and cause both helices to wind in a left handed mode around each other (Fig. 2.3-5) as the primary keratin element.

Collagen is the major component of the connective tissue of the skin and of blood vessels. In collagen type I, three individual helices [left handed(!), 3.3 amino acids/turn, pitch 1 nm] wind around each other, forming a right handed superhelix (pitch 8.6 nm total length ca. 300 nm, Fig. 2.3-6). Its formation is also caused by a regular amino acid arrangement within the individual helices: Every third amino acid is glycine, which points towards the center of the superhelix. Its amino group exhibits strong hydrogen bonds with a CO group of another chain. The other amino acids are mostly proline and hydroxyproline, which contribute to the rigidity due to their inflexible bond structure (1.3.2). Individual superhelices associate to fibers.

Generally, protein domains having more than 30 % of their sequence in common will adopt the same overall folding structures. However, even proteins with lower sequence similarities can belong to the same structural family.

Quaternary structure: This term is used for the association of two or more polypeptide chains into a multi-subunit aggregate. This arrangement can contribute to

- fast and effective transfer of the substrate from one active center of enzymes to the next, e.g., in
  - fatty acid synthase (6.1.1)
  - biotin dependent CO_2 transfer (Fig. 6.1-2)
- allosteric regulation (2.5.2), e.g., of
  - hemoglobin (Fig. 5.2-3).

2.3.2 Protein Function

Proteins and peptides play crucial roles in virtually all processes occurring in living cells and organisms. Their functional diversity is represented by the examples in Table 2.3-1.

Table 2.3-1. Biological Function of Proteins

Function	Section*	Examples
Enzyme catalysis	2.4, 2.5	lactate dehydrogenase, trypsin, DNA polymerases
Transport	5.2.3, 18.1...2	hemoglobin, serum albumin, membrane transporters
Storage	5.2.1	ovalbumin, egg-white protein, ferritin
Motion	17.4.5, 18.1.3	myosin, actin, tubulin, flagellar proteins
Structural and mechanical support	2.2.2, 12.1.2	collagen, elastin, keratin, viral coat proteins
Defense	19.1...2	antibodies, complement factors, blood clotting factors, protease inhibitors
Signal transduction	17.1...7	receptors, ion channels, rhodopsin, G proteins, signalling cascade proteins
Control of growth, differentiation and metabolism	10.5, 11.4, 11.6, 17.1...7, 19.1.6	repressor proteins, growth factors, cytokines, bone morphogenic proteins, peptide hormones, cell adhesion proteins
Toxins	6.3.2, 17.4.1	snake venoms, cholera toxin

* Only the sections are named, where the principle of action is presented. Additional information is given in the sections dealing with the individual proteins.

Literature:

Kleinkauf, H., von Döhren, H.: Eur. J. Biochem. 236 (1996) 335–351.
Kyte, J.: *Structure in Protein Chemistry*. Garland (1994).
Matthews, B.W.: Ann. Rev. Biochem. 62 (1993) 139–160.
(various authors): Methods in Enzymology, especially Vols. 82, 85, 91, 117, 130, 131, 144, 145. Academic Press.

2.4 Enzymes

Enzymes are proteins which catalyze the turnover of other compounds (substrates) without being consumed themselves and without changing the equilibrium of the reaction.

- They accelerate the rate of reactions of their substrate by factors up to ca. 10^{15} as compared to the uncatalyzed reaction.
- They show substrate specificity by limiting their action to one substance or a small number of structurally related substances.
- They also show reaction specificity by catalyzing only one of several possible ways of substrate conversion.

Whereas reaction specificity usually is absolute, the degree of substrate specificity may vary among different enzymes.

2.4.1 Catalytic Mechanism (For quantitiative aspects, see 1.5.4)

Every conversion of a molecule into another one requires at first an input of energy (activation energy ΔG^+), which leads to a transition state. From this state, either the conversion to the product or a return to the original molecule takes place.

Single-substrate reaction: In an enzyme-catalyzed reaction involving a single substrate (S), this molecule combines with the enzyme (E) to form an enzyme-substrate complex (ES), followed by conversion to the product (P) which is then released. The enzyme is then free to enter the next catalytic cycle:

$$E + S \leftrightarrow ES \leftrightarrow EP \leftrightarrow E + P$$

The interaction of the enzyme and the substrate facilitates the formation of the transition complex. Therefore, a lower amount of free energy ΔG^+ is required for its formation and thus for the consecutive conversion of the substrate to the product (Fig. 2.4-1). This results in an increase of the reaction velocity. The same is valid for the reverse reaction. Since the equilibrium constant K is determined by the overall reaction rate constants k_f and k_r for the forward and reverse reactions, respectively

$$K = \frac{k_f}{k_r} \qquad [2.4\text{-}1]$$

and enzymatic catalysis increases both rate constants by the same factor, their ratio remains constant and the equilibrium does not change.

There is an exponential relationship between lowering of the activation energy and increase of the reaction rate (Eq. [1.5-32]). 'Good' enzymes, which noticeably decrease of the activation energy, elevate the substrate turnover tremendously.

If the activation energy ΔG^+ [kJ * mol^{-1}] necessary for an uncatalyzed reaction is lowered by $\Delta\Delta G^+$ [kJ * mol^{-1}] in the presence of the enzyme, an increase of the reaction rate by the factor $e^{\Delta\Delta G^+/RT}$ takes place. At ambient temperature, a lowering by $\Delta\Delta G^+ = 5.7$ [kJ/mol] increases the rate constant tenfold, a lowering by $\Delta\Delta G^+ = 34$ [kJ/mol] increases the rate millionfold.

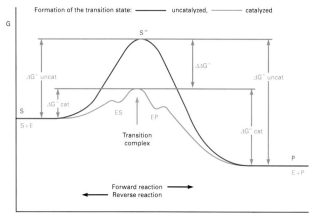

Figure 2.4-1. Changes of the Free Energy ΔG^+ During the Course of the Reaction S → P (For Details, see 1.5.4)

Two-substrate reactions: These reactions proceed analogously via enzyme-substrate complex formation. There are several possible ways of interaction between enzyme E and the substrates A and B (Details in 1.5.4):

- They may form a ternary intermediate complex (EAB). After the catalyzed reaction has taken place (in the case of two products EAB → EPQ), the products leave the complex. Formation and decomposition of the complex can take place either in definite order (ordered sequential mechanism) or at random (random sequential mechanism). This is the usual way when a group or an ion is directly transferred from one substrate to another.

  Examples for the ordered mechanism are most NAD$^+$, NADH, NADP$^+$ or NADPH dependent dehydrogenase reactions (9.9.2) where the first contact to the enzyme is made by the coenzyme. A random mechanism is performed by some kinases.

- The substrate A may contact the enzyme E (or its prosthetic group) and modify it in some way (designated as F). Then it leaves as product P, before the second substrate B reacts with the modified enzyme, reestablishes its primary state and leaves as product Q. Thus, this ping-pong-mechanism is a sequence of binary intermediates.

  $E + A \leftrightarrow EA, \quad EA \leftrightarrow FP, \quad FP \leftrightarrow F + P$
  $F + B \leftrightarrow FB, \quad FB \leftrightarrow EQ, \quad EQ \leftrightarrow E + Q$

  Examples for this reaction mechanism are chymotrypsin (Fig. 14.6-1) and transaminases (Fig. 4.2-3). In the latter case, the cofactor pyridoxal phosphate, which exchanges amino and oxo groups, represents the temporarily modified part of the enzyme.

Catalytic center: The part of the enzyme structure involved in the binding of the substrate and in catalysis is called the catalytic center (active site). It usually encompasses only a small part of the enzyme and consists of a few amino acids, which are often present on non-contagious locations of the protein chain (and in some cases even on different protein chains of multimeric enzymes). Prosthetic groups or cofactors (see below) may also be present in the catalytic center.

The catalytic center is a three-dimensional unit, which performs interactions with definite structures of the substrate

- for recognition, providing the specificity and
- for catalysis.

Usually, the catalytic center has a cave-like structure which permits the exclusion of water (unless it is a reaction partner), but contains some polar amino acids involved in the catalytic mechanism. In a number of cases (e.g., carboxypeptidase A, 14.6.5) a movement of amino acid residues seals the center after the substrate has entered, while in other cases, structural movements are part of the catalytic process. Movements during allosteric regulation play a special role (2.5.2).

Typical catalytic mechanisms: The enzymes may facilitate the formation of the transition complex in various ways, e.g., by

- general acid and base catalysis (proton donation and abstraction)
- covalent bond formation (the bonds must not be too strong, therefore the reacting group must be highly polarizable, e.g. imidazoles or thiols)
- redox effects (by metal ions, enzyme residues or cofactors)
- electrostatic effects (by metal ions or amino acid residues, enhanced by a mainly non-polar environment)
- orientation and proximity effects (moving the reactants into favorable positions and diminishing their relative motions)
- straining the bound substrate(s) into a transition-state like configuration

Examples for such mechanisms are found in various places in this book.

Cofactors: While in a number of reactions the enzyme protein exerts the catalytic activity on its own, in other cases, cofactors are needed for catalysis (Table 2.4-1). They can be

- Coenzymes, which are transiently associated with the enzyme and frequently serve as a carrier for atoms or effect the molecule transfer to other enzymes (e.g. NAD$^+$, coenzyme A).
- Prosthetic groups, which are firmly bound to the enzyme (e.g., biotin). The inactive enzyme protein is then termed apoenzyme which, upon combining with the prosthetic group, forms the enzymatically active holoenzyme.

Effects of pH: In most cases, positively or negatively charged amino acid side chains are involved in substrate binding and catalysis. Therefore, most enzymes

Table 2.4-1. Examples of Cofactors in Enzyme-Catalyzed Reactions

Coenzymes or Prosthetic Groups	Function (Examples)	Section
Metal ions (e.g., Mg^{++}, Mn^{++}, Zn^{++}, Cu^{++}, Co^{++}, Fe^{++})	participation in many enzyme-catalyzed reactions	14.6.5
Nicotinamide nucleotides (NAD^+, NADH, $NADP^+$, NADPH)	redox reactions, hydrogen transfer	9.9.2
Flavin nucleotides (FAD, FMN)	redox reactions, electron and hydrogen transfer	9.3.2
Quinones (e.g., ubiquinone, phylloquinone, pyrroloquinoline quinone)	redox reactions, electron and hydrogen transfer	4.7.2, 15.7
Coenzyme A (CoA)	acyl group transfer	9.7.2
Pyridoxal phosphate (PyrP)	aminoacyl group interconversions, decarboxylation	9.4.2
Thiamine pyrophosphate (ThPP)	group transfer and decarboxylation	9.2.2
Biotin	CO_2 transfer	9.8.2
S-Adenosyl methionine (SAM)	methyl group transfer	4.5.4
Deoxyadenosyl cobalamine (Coenzyme B_{12})	methyl group transfer	9.5.2
Lipoate (Lip)	transfer and oxidation of carbonyl compounds	9.14.1
Nucleoside diphosphates (UDP, CDP, ADP, GDP, dTDP)	glycosyl group transfer	3.2.2, 3.4.1, etc.

Examples of the contribution of enzyme amino acid residues to the catalytic mechanism are shown in Fig. 14.6-1 and 14.6-2 (chymotrypsin and pepsin), for the involvement of metal atoms in Fig. 14.6-3 (carboxypeptidase A) and for a non-covalently bound prosthetic group in Figs. 4.2-3, 4.4-2 and 4.4-3 (pyridoxal phosphate). The structure of the intermediate complex of L-lactate and NAD^+ with the catalytic center of L-lactate dehydrogenase is shown in Fig. 2.4-2.

have a characteristic pH value at which their activity is maximal (Fig. 2.4-3). Above and below this pH the acitivity declines.

Effects of temperature: The elevation of the temperature increases greatly the reaction rate (Eq. [1.5-32]). Beyond an optimum temperature, however, denaturation of the enzyme protein results in a quick decrease of the enzyme activity (Fig. 2.4-4). For many *mammalian* enzymes, this optimum temperature is about 40...50 °C. There are enzymes of thermophilic *archaea*, which are still active at >100 °C.

2.4.2 Isoenzymes

Within a single given species, variants of an enzyme may be present. This may be caused by the occurrence of multiple gene loci coding for distinct versions of the enzyme protein, or by the existence of multiple alleles at a single gene locus. Such variations are designated isoenzymes if they represent multiple forms of the same enzyme activity present in the same species, but are encoded by independent genes. Multiple forms of enzymes resulting from different post-translational modifications are not true isoenzymes.

Well-known examples of isoenzymes are *animal* lactate dehydrogenases (e.g., *bovine, porcine,* or *rabbit*) which are formed from random association of two types of subunits (H- and M-types) into tetramers, resulting in five discernible species HHHH, HHHM, HHMM, HMMM, MMMM. In *mammalian* creatine kinases, two different subunit types (B- and M-type) can associate to give three forms of the dimeric enzyme. Isoenzymes may have different kinetic or regulatory properties.

2.4.3 Multienzyme Complexes

Inside the cell, enzymes frequently catalyze multistep anabolic or catabolic sequences where the product of one enzyme becomes the substrate of the next. In many cases, the individual enzymes are dissolved in the *cytosol* without physical association, and the substrates which are turned over diffuse more or less rapidly from one enzyme molecule to the next. However, there are also higher levels of organization in multi-enzyme systems:

- Two or more enzyme activities may reside within one polypeptide chain which, therefore, comprises more than one active site (multifunctional enzymes), e.g., some enzymes of the purine or pyrimidine synthesis in *vertebrates* (8.1.1, 8.2.1).

- Several individual enzymes may be associated non-covalently or even covalently to give a multienzyme complex, such as the pyruvate (3.3.1) and the 2-oxoglutarate dehydrogenase complexes (3.8.1), the biotin-containing carboxylases (9.8) or the polyketide synthases (15.8.3...4). The fatty acid synthase of *mammals* (6.1.1) is a homodimer of two cooperating multifunctional enzyme chains.

- The most highly organized enzyme systems are associated with supramolecular structures, such as ribosomes (10.6, 11.5), the proteasome (14.6.7) or the components of the respiratory chain which spans the inner membrane of *mitochondria* (16.1.2).

2.4.4 Reaction Rate (For a Detailed Discussion, see 1.5.4)

The dependence of the initial reaction rate v_0 of an enzyme reaction on the concentration of a single substrate [S] is given by the Michaelis-Menten equation (Eq. [1.5-19]):

$$v_0 = \frac{V_{max} * [S]}{K_M + [S]}$$

Figure 2.4-3. pH Dependence of an Enzyme-Catalyzed Reaction (Schematically)

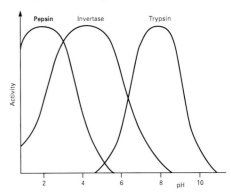

Figure 2.4-4. Temperature Dependence of an Enzyme-Catalyzed Reaction (Schematically)

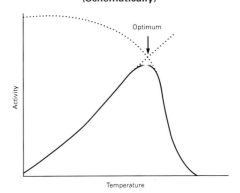

Figure 2.4-2. Catalytic Center of L-Lactate Dehydrogenase
(Blue: Enzyme, Red: NAD^+, Black: L-Lactate)

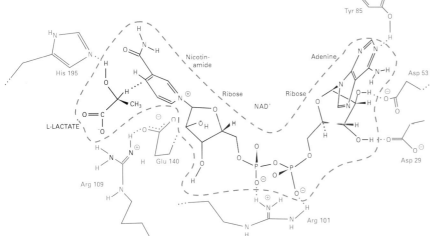

which plots as a hyperbolic curve (Fig. 1.5-1). V_{max} is the maximum velocity obtained with a fixed amount of enzyme, when the enzyme is saturated with substrate. K_M, the Michaelis constant, equals the substrate concentration at which $v = 0.5\ V_{max}$. A similar plot is obtained for a two-substrate reaction, when the concentration of one substrate is kept constant and only the other is varied. Since *in vivo* the substrate concentrations are frequently in the range of the K_M value, this can serve regulatory effects (Fig. 3.1-4).

2.4.5 Classification of Enzymes

A Nomenclature Committee set up by the International Union of Biochemistry and Molecular Biology (IUB) established a system for the classification of enzymes, which was updated several times. The 1992 edition lists 3196 enzymes. This system has the goal to provide unambiguous enzyme names based on the nature of the catalyzed reaction. Although there are some problems of applying the name to the actual situation (e.g., to the direction of the reaction taking place in nature) and some names are unwieldy, the application of this 'EC list' has improved greatly the terminology of the scientific literature. As far as possible, this system has been used for the enzymes dealt with in this book.

The EC list is based on a decimal classification system with classes, subclasses, sub-subclasses and individual enzyme names. Besides the systematic names, often 'trivial names' are admitted, which are easier to use. The main arrangement is as follows:

1 Oxidoreductases

    1.1 Acting on the CH-OH group of donors
        1.1.1 With NAD^+ or $NADP^+$ as acceptor
            1.1.1.1 Alcohol dehydrogenase
            1.1.1.2 ...

1.2 Acting on the aldehyde and oxo group of donors, 1.3 acting on the CH-CH group of donors, 1.4 acting on the CH-NH$_2$ group of donors, 1.5 acting on the CH-NH group of donors, 1.6 acting on NADH and NADPH, 1.7 acting on other nitrogeneous groups as donors, 1.8 acting on a sulfur group of donors, 1.9 acting on a heme group of donors, 1.10 acting on diphenols and related substances as donors, 1.11 acting on a peroxide as acceptor, 1.12 acting on hydrogen as a donor, 1.13 acting on single donors with incorporation of molecular oxygen (oxygenases), 1.14 acting on paired donors with incorporation of molecular oxygen, 1.15 acting on superoxide radicals as acceptors, 1.16 oxidizing metal ions, 1.17 acting on CH_2 groups, 1.18 acting on reduced ferredoxin as a donor, 1.19 acting on reduced flavodoxin as a donor, 1.97 other oxidoreductases

2 Transferases

2.1 Transferring one-carbon groups, 2.2 transferring aldehyde or ketone residues, 2.3 acyltransferases, 2.4 glycosyltransferases, 2.5 transferring alkyl or aryl groups other than methyl groups, 2.6 transferring nitrogeneous groups, 2.7 transferring phosphorus-containing groups, 2.8 transferring sulfur-containing groups

3 Hydrolases

3.1 Acting on ester bonds, 3.2 glycosidases, 3.3 acting on ether bonds, 3.4 acting on peptide bods (peptidases), 3.5 acting on carbon-nitrogen bonds other than peptide bonds, 3.6 acting on acid anhydrides, 3.7 acting on carbon-carbon bonds, 3.8 acting on halide bonds, 3.9 acting on phosphorus-nitrogen bonds, 3.10 acting on sulfur-nitrogen bonds, 3.11 acting on carbon-phosphorus bonds, 3.12 acting on sulfur-sulfur bonds

4 Lyases

4.1 Carbon-carbon lyases, 4.2 carbon-oxygen lyases, 4.3 carbon-nitrogen lyases, 4.4 carbon-sulfur lyases, 4.5 carbon-halide lyases, 4.6 phosphorus-oxygen lyases, 4.99 other lyases

5 Isomerases

5.1 Racemases and epimerases, 5.2 *cis-trans* isomerases, 5.3 intramolecular oxidoreductases, 5.4 intramolecular transferases (mutases), 5.5 intramolecular lyases, 5.99 other isomerases

6 Ligases

6.1 Forming carbon-oxygen bonds, 6.2 forming carbon-sulfur bonds, 6.3 forming carbon-nitrogen bonds, 6.4 forming carbon-carbon bonds, 6.5 forming phosphoric ester bonds

Literature:

Fersht, A.: *Enzyme Structure and Mechanism.* 2nd Ed. Freeman (1983).
International Union of Biochemistry and Molecular Biology. Nomenclature Committee: *Enzyme Nomenclature 1992.* Academic Press.
Sigman, D.S., Boyer, P.D. (Eds.): *The Enzymes* Third Ed. Vols. XIX and XX Academic Press (1990, 1992).
(Various authors): *Methods in Enzymology.* More than 250 volumes, especially Vols. 63, 64, 87, 249. Academic Press.

2.5 Regulation of the Enzyme Activity

Since enzymes catalyze almost all metabolic reactions of living beings, it is of utmost importance that their activities be strictly regulated according to the needs of the cell or organism. Regulation of enzyme activity occurs at several levels. Characteristically, control mechanisms regulating the enzyme synthesis at the transcriptional or translational level are slow (often with response times of hours or even days), whereas rapid regulatory mechanisms act directly on the enzyme molecules.

2.5.1 Regulation of the Quantity of Enzymes

Regulation of enzyme synthesis and degradation: The biosynthesis of some enzymes is constitutive, i.e., they are formed independently of the environmental or metabolic conditions of the cell. However, for most enzymes (as well as for other proteins) the production is regulated:

- The rate of gene expression (and thus, the amount of enzyme protein synthesized) can be modulated at the transcription level.

    A well-known example is the repression of the *lac* operon in *E. coli* by the *lac* repressor (10.5.1), and the induction of the biosynthesis of β-galactosidase, galactose permease, and thiogalactoside transacetylase by lactose or isopropyl-thiogalactoside acting as inducers.

- Regulation of the rate of protein synthesis at the translation level, e.g., by modification of the activity of initiation factors, occurs mostly on *eukarya* (11.5.4).

- The length of the mRNA half-life is another parameter which influences the amount of the synthesized protein (11.5.5). It is usually very short in *bacteria*, but varies greatly in *eukarya*.

- In addition to the regulation of protein synthesis, the amount of enzyme present may also be influenced by the rate of degradation (or turnover) of the enzyme protein.

Zymogen activation: Some enzymes are synthesized as inactive precursors (zymogens or proenzymes) which have to be processed into their active form by limited proteolysis. Examples are digestive enzymes such as chymotrypsinogen or trypsinogen (Fig. 2.5-1). They are synthesized in *mammalian pancreas* as zymogens. After hormone-controlled release into the *small intestine* they are irreversibly processed by trypsin or enteropeptidase, respectively, to become active proteases (Fig. 17.1-11).

Figure 2.5-1. Conversion of Chymotrypsinogen Into Chymotrypsin

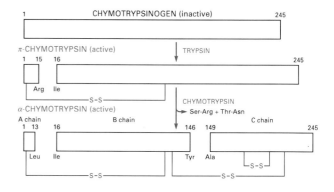

2.5.2 Regulation of the Activity of Enzymes (For a Mathematical Treatment see 1.5.4)

Regulation depending on substrate concentration: In the most simple case, the dependency of the reaction rate on the substrate concentration according to the Michaelis-Menten equation (Eq. [1.5-19], Fig. 1.5-1) is a means of controlling the substrate throughput.

Since *in vivo*, the substrate concentrations frequently are in the range of the K_M value, any variation in substrate concentration will result in a corresponding change of the reaction rate (e.g., if the substrate concentration rises, the conversion will be faster). An example is glucokinase (Fig. 3.1-4).

Competitive inhibition (Fig. 2.5-2a): Competitive inhibitors react reversibly with the normal substrate binding site of the enzyme to form an enzyme-inhibitor complex which prevent substrate binding and conversion to the reaction products. Therefore, competitive inhibitors are usually substances with some structural similarity to the substrate. Competitive inhibition can be reversed by increasing the substrate concentration. Formally, the inhibitor increases the apparent K_M value of the substrate, while V_{max} remains constant (Fig. 1.5-3a).

Noncompetitive inhibition (Fig. 2.5-2b): This reaction type is found when the inhibition cannot be overcome by increased substrate concentration. Presumably, the inhibitor binds to an enzyme locus different from the substrate binding site, both on the free enzyme and on the enzyme-substrate complex and prevents the turnover to the products. This is equivalent to a deactivation of a portion of the enzyme. Formally, the inhibitor decreases V_{max}, while the K_M value of the substrate remains constant (Fig. 1.5-3c). If, however, the inhibitor binds to the free enzyme and to the enzyme-substrate complex with different affinities, this is called a **mixed inhibition:** V_{max} is decreased, while K_M is elevated (Fig. 1.5-3d).

Uncompetitive inhibition (Fig. 2.5-2c): There are also cases where both K_M and V_{max} are affected in a different way: both V_{max} and K_M decrease. One possible reason is that the inhibitor binds only to the enzyme-substrate complex, but does not affect its formation.

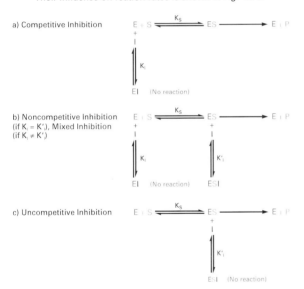

Figure 2.5-2. Different Inhibition Mechanisms of Enzyme-Catalyzed Reactions
Their influence on reation rates is shown in Fig. 1.5-3.

Allosteric regulation: Many oligomeric enzymes are regulated by inhibition or activation due to interaction with an allosteric effector (an inhibitor or activator) at a binding site separate from the active center of the enzyme (negative or positive cooperative regulation). The regulator substance changes activity through the induction of conformational changes in the enzyme which may be concerted (Monod-Changeux-Wyman symmetry model) or sequential due to 'induced fit' (Koshland-Nemethy-Filmer sequential model). Allosteric regulation may also be exerted by the substrates or products of the reaction themselves (homotropic regulation). Allosteric enzymes usually do not obey Michaelis-Menten kinetics, but show sigmoid (S-shaped) substrate-saturation curves (blue curve in Fig 2.5-3).

The theory of allosteric mechanisms actually deals with the binding of ligands (S) to oligomeric proteins, no matter whether there is a consecutive reaction or not. Therefore, this discussion applies to the oxygen binding to hemoglobin (5.2.3) as well as to the substrate binding to an enzyme. Since enzymatic reactions require the previous formation of the enzyme-substrate complex (2.4.1), allosterism is a means of regulating the reaction rate.

Figure 2.5-3. Saturation Curves of Allosteric Proteins, Influence of Activators and Inhibitors
The degree of saturation Y is proportional to the reaction rate in case of allosterically controlled enzymes.

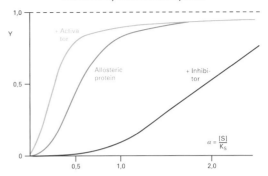

The protein subunits may exist in two possible states: R (relaxed) or T (tight) with different binding affinities to the ligand. In the following discussion, tetrameric proteins are assumed, which are very common among allosteric proteins.

Symmetry model: In the case of homotropic regulation (the regulation is performed by the ligand S itself), the following assumptions are made:
- T as well as R forms can bind the ligand, however, with different binding affinities. The T form has the lower affinity. Thus, low or high binding affinity only depends on the fact if the subunit is in the T or R state.
- All 4 subunits of the tetramer have to be either in the T or in the R state (T_4 or R_4, respectively).
- The T state tetramers are in equilibrium with those R state tetramers, which have bound the same number of ligands.
- Due to close coupling between the subunits, all of them shift simultaneously from the T to the R state (and vice versa, concerted allosteric transition (Fig. 2.5-4).
- The more of the 4 subunits have bound a ligand, the more likely is the common transition from the T to the R state.

Therefore, the binding of a ligand facilitates the binding of the next ligand. This leads to a steep increase of the saturation curve at medium and high ligand concentrations, causing a sigmoid shape. The amount of cooperativity (interaction between the subunits) is expressed by the degree of upward curvature, that is, by the deviation from a simple hyperbolic curve.

If the allosteric regulation is performed by activators or inhibitors (heterotropic regulation), it is assumed, that activators bind exclusively or at least preferentially to the R state. Their action can be considered similarly to the action of an additional substrate molecule: They promote the transition from the T to the R state and thus enhance substrate binding. Inhibitors, however, bind to the T state. They decrease the tendency of transition from the T to the R state and thus antagonize the effect of bound substrate on the binding of additional substrate.

Therefore, activators shift the saturation curve to the left (to lower substrate concentrations) and inhibitors to the right (to higher substrate concentrations, Fig.2.5-3). The same effect can be seen with the reaction velocity plots of allosterically controlled enzyme reactions.

The efficiencies of allosteric inhibitors (I) or activators (A) depend (among other factors) on their concentrations in relation to their dissociation constants, $\beta = [I]/K_I$ and $\gamma = [A]/K_A$, while the concentration of the ligand is expressed by $\alpha = [S]/K_S$. If L is the ratio of T and R states in absence of ligand, L = $[T_0]/[R_0]$, then the degree of saturation Y of an n-meric protein (for a tetrameric protein n = 4) by the ligand is expressed by the formula

$$Y_S = \frac{\alpha * (1 + \alpha)^{n-1}}{(1 + \alpha)^n + \frac{L * (1 + \beta)^n}{(1 + \gamma)^n}} \qquad [2.5\text{-}1]$$

In Fig. 2.5-3, the respective data are: n = 4, L = 1000, for the inhibited curve $\beta = 2.5$, $\gamma = 0$, for the activated curve $\gamma = 100$, $\beta = 0$.

Sequential model: This model also assumes T and R states of the allosteric protein, but with important differences:
- There exist only complexes of the ligand L with the R state (RL).
- Hybrids of T and R states are possible [e.g., $T_3(RL)$, $T_2(RL)_2$, $T(RL)_3$].
- Ligand binding induces the conformational change of the particular subunit ('induced fit') from the T to the R state.
- The transition of a subunit from the T to the R state leads to a cooperative interaction with the neighboring subunit, which facilitates its ligand binding and thus its transition from the T to the R state (Fig. 2.5-4).
- The more of the subunits are already in the R state, the more interactions with the remaining T state subunits take place. This gradually increases the ease of their transition into the R state.

The strength of the interaction between the subunits determines the degree of cooperativity. In case of very strong coupling, the transition from T to R states for all 4 subunits takes place almost simultaneously, which corresponds to the assumptions of the symmetry model.

Although both models can be adapted fairly well to many actual situations, they represent two contrasting assumptions. In general, the reactions of allosteric proteins contain elements of both models.

If the metabolic situation changes, the regulation system is reversed. The phosphorylated glycogen synthase is dephosphorylated and thus activated by protein phosphatase 1. The phosphorylated phosphorylase a is dephosphorylated and deactivated by the same enzyme protein phosphatase 1, which is controlled by the hormone insulin in a multistep mechanism.

Thus, each of the regulating enzymes, phosphorylase kinase and protein phosphatase 1 influence both glycogen synthesis and degradation, however, in opposite directions.

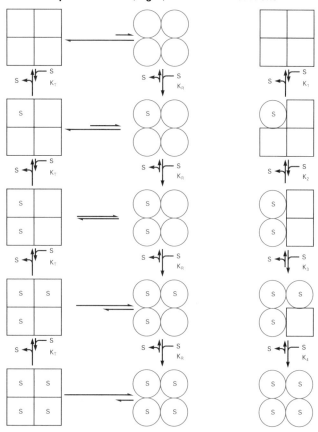

Figure 2.5-4. Interconversions between T and R States According to the Symmetry Model (Left) and to the Sequential Model (Right) of Allosteric Reactions

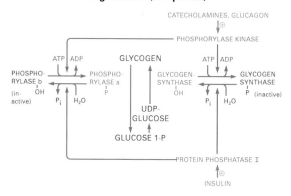

Figure 2.5-5. Regulation of Glycogen Synthesis and Degradation (Simplified)

Cascade mechanisms: The effect of a regulated change in the activity of an enzyme is sometimes amplified through a cascade mechanism: the first enzyme activates a second enzyme, this, in turn, a third one and so on. This causes a potentiation of the regulatory effect and very rapidly produces a large amount of the active form of the last enzyme. Activation may proceed via proteolysis (e.g., in blood coagulation, 20.3) or by phosphorylation/dephosphorylation mechanisms (e.g. in the regulation of glycogen synthesis and degradation mentioned above, regarding the intermediate steps between hormone action and activation of phosphorylase kinase and protein phosphatase, Fig. 3.2-5).

2.5.3 Site of Regulation

In multistep metabolic pathways, the product of a late (or the last) step frequently acts as an inhibitor of the first committed step in this pathway ('feedback inhibition'). This way, the end product of a pathway controls its own synthesis and prevents useless accumulation of intermediates (Fig. 2.5-5). All of the regulation mechanisms mentioned above can be applied.

Examples are the regulation of glycolysis (Fig. 3.1-3), pyruvate dehydrogenase (Fig. 3.3-2), citrate cycle (Fig. 3.8-3), fatty acid synthesis (Fig. 6.1-5), cholesterol synthesis (Fig. 7.1-7), bile acid synthesis (Fig. 7.9-3), the repression of genes coding for the enzymes of lactose and tryptophan synthesis (Figs. 10.5-2 and 10.5-5) etc.

Since many metabolic pathways are branched, complex regulatory patterns such as multiple inhibition, sequential inhibition, or the presence of isoenzymes with different regulatory properties can be observed.

Examples for such mechanisms are, e.g., the regulation of the threonine, methionine and lysine biosynthesis (Fig. 4.5-1), aromatic amino acid biosynthesis (Fig. 4.7-2), purine nucleotide biosynthesis (Fig. 8.1-5), ribonucleoside-diphosphate reductase (Fig. 8.1-7) etc.

Literature
Lipscomb, W.N.: Chemtracts-Biochemistry and Mol. Biol. 2 (1990) 1–15.
(various authors): *Methods in Enzymology*. More than 250 volumes, especially Vol. 63, p. 383–515, Vol. 64, p. 139–192. Academic Press (1979/1980).

Covalent modification: Another method of reversible regulation of enzyme activities involves covalent modifications by attachment or removal of special groups. There exist two possibilities:

- The enzyme is normally inactive, but is interconverted to an active form by the addition of a group. Examples are
  - phosphorylation (e.g., glycogen phosphorylase, 3.2.4)
  - adenylylation (e.g., *E. coli* glutamine synthetase, 4.2.1)
  - NAD ribosylation (17.4.1).
- The enzyme is normally active, but is inactivated by the addition of a group.
  An example is the phosphorylation of glycogen synthase (3.2.4).

These modification reactions are catalyzed by more or less specific, regulatory enzymes which themselves may be subject to various control mechanisms.

There are cases where enzymes which catalyze the interconversion of two compounds in different directions are both regulated by such covalent modifications.

An example is the formation and degradation of glycogen in *animals*. (Fig. 2.5-5. This figure is simplified, more details are given in Fig. 3.2-5.) Glycogen degradation is effected by phosphorylase. This enzyme is activated by phosphorylation in a reaction catalyzed by phosphorylase kinase (change from phosphorylase b into phosphorylase a). Glycogen synthase, which effects glycogen formation, is deactivated by phosphorylation. This reaction is also catalyzed by phosphorylase kinase. Phosphorylase kinase, in turn, is activated in a multistep reaction by hormones (glucagon, epinephrine).

2.6 Nucleic Acid Structure

2.6.1 Components of Nucleic Acids

Nucleic acids are sequences of nucleotides (Chapter 8) interconnected by phosphate diester bonds. Thus they contain two structural features: the information-carrying bases and a sugar-phosphate 'backbone'. In ribonucleic acid (RNA), the sugar is ribose, in deoxyribonucleic acid (DNA), it is deoxyribose (Fig. 2.6-1).

The sugar residues of the backbone are linked at their 5' and 3' positions by a phosphate diester bridge. This gives the nucleic acid strand a distinct polarity (5' and 3' end). Nucleic acids form unbranched chains of greatly varying length (from a few building blocks up to millions).

The bases are purines (adenine, guanine, 8.1) or pyrimidines (cytidine, thymine in DNA, uracil in RNA, 8.2). They can interact via hydrogen bonds with each other according to distinct rules (Watson-Crick pairing rules, Fig. 2.6-2):

- adenine only pairs with thymine (in DNA) or uracil (in RNA), interacting via 2 hydrogen bonds
- guanine only pairs with cytidine, interacting via 3 hydrogen bonds.

Since the distances of the bases and the angles of the base-sugar bonds are exactly equal for both Watson-Crick pairs, an exchange of bases does not affect the backbone structure of the double helix (see below).

In some cases, non-Watson-Crick base pairs occur, e.g. of the Hoogsteen type (Fig. 2.6-2). They are found, e.g., in tRNA (10.6.1) or in 'wobble base pairing' during transcription (10.1.2).

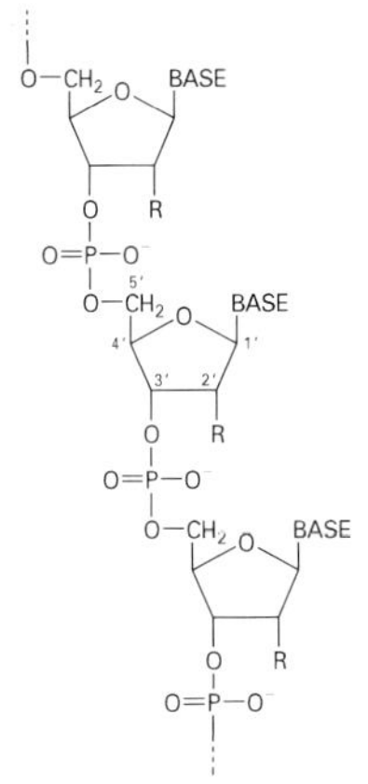

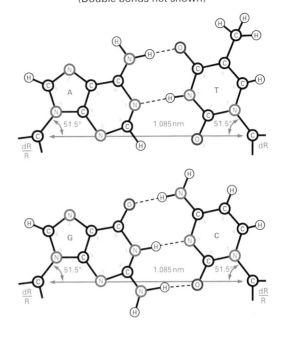

Figure 2.6-2. Watson-Crick Base Pairing Rules
(Double bonds not shown)

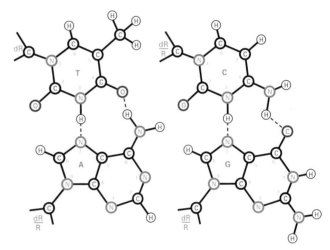

Figure 2.6-3. Non-Watson-Crick Base Pairs (Hoogsteen Pairs; double bonds not shown)

Figure 2.6-1. General Structure of Nucleic Acids
(R= -OH in RNA, R = H in DNA)

The bases are the elements of the information carried by nucleic acids (analogously to 'bits' in computer science), but they are also indispensable for the structural and catalytic features of nucleic acids.

DNA serves almost exclusively for information storage. RNA molecules have many different functions (10.4, 10.6, 11.3, 11.5):

- structural and catalytic roles in ribosomes (rRNA)
- information transfer (mRNA)
- translation of the nucleic acid information into the protein information (tRNA)
- catalytic roles during posttranscriptional processing of other RNAs (as 'ribozymes', e.g., in spliceosomes)
- information storage in RNA viruses (12.3, 12.4)

2.6.2 Properties of RNA Chains

Likely due to the presence of the 2'-OH group, steric inhibition prevents RNAs from assuming a DNA-like B-type double helix conformation. They are mainly single-stranded with intrastrand base-pairing forming stems, loops etc., which resemble the less favored A-type DNA helices (see below). The same A-type helix structure is assumed by the short DNA-RNA hybrids, which are formed during transcription (10.4.2, 11.3), initiation of DNA replication (10.2.2, 11.2) and reverse transcription of retrovirus RNA (12.4.1).

Depending on their different tasks, the properties of RNA molecules vary greatly:

- tRNAs have a cloverleaf arrangement of loops, which are folded into an 'L' shape (Fig. 10.6-1). They consist of 60...100 nucleotides.
- mRNAs are mainly linear, but hairpin loops have functional properties. Examples are the termination of transcription (10.5.3) or the regulation of gene translation, e.g. by binding of regulatory proteins (11.5.4) or in polycistronic mRNA. mRNAs vary greatly in size.
- rRNAs of several sizes occur in ribosomes (Tables 10.6-1 and 11.5-1). They contain a large number of stem loops. Although variations in the base sequences have occurred, many aspects of the secondary structure have been conserved during evolution.

2.6.3 Properties of DNA Chains

DNA occurs mainly as a double helix of two antiparallel strands with complementarily paired bases. The polar sugar-phosphate backbone is located on the outside and the hydrophobic bases are stacked in the central area with their planes parallel to each other, almost perpendicular to the long axis of the helix. This stacking allows hydrophobic interactions, which are responsible for the stability of the double helix in the first place. The helix is usually 'right handed' (i.e. when looking along the axis of the helix, the strands are coiled clockwise) with a major and a minor groove. Both grooves are wide enough to allow proteins to come into contact with the bases (Fig. 2.6-4).

Figure 2.6-4. Structure of the B-Type DNA Double Helix

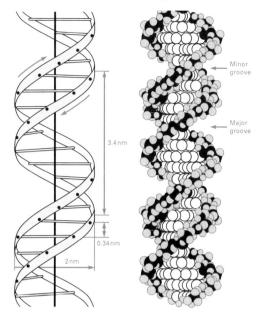

Table 2.6-1. Different Forms of the DNA Double Helix

	B Type	A Type	Z Type
Diameter (nm)	ca. 2	ca. 2.6	ca. 1.8
Turning mode	right handed	right handed	left handed
Base pairs/turn	10.4...10.65	11	12
Pitch/base pair (nm)	0.34	0.26	0.37
Pitch/turn (nm)	3.4	2.8	4.5
Occurrence	most common	bacterial spores, DNA-RNA hybrids	in alternating purine-pyrimidine sequences, during torsional stress or when stabilized by supercoiling, proteins, methylation etc.

The DNA double helix is plectonemically coiled, i.e. the helices can only be separated by unwinding the coils.

Several helix forms have been described (Table 2.6-1). Some interconversions of these helix forms are possible depending on the concentration and types of salts present. The natural form of DNA is the B type helix, which is also the most prevalent form in solution. In this helix, the major and minor grooves are most pronounced (2.2 nm and 1.2 nm wide, respectively). Adjacent base pairs are rotated by 34.5 to 35.5°. This twist angle can change depending on the sequence which may result in kinking of the double helix. This kinking can also be caused by other properties of the DNA or by proteins. The twist can be superimposed on the double helix, resulting in supercoiling. A helix can become positively (twist in the direction of winding) or negatively supercoiled. Supercoiling results in a more compact structure of DNA. This is very important in DNA packaging.

Torsional stress due to supercoiling can be overcome by the formation of DNA structures other than the B-form. E.g., negative supercoiling is a strong driving force for the stabilization of Z-DNA. During transcription, positive supercoils are formed in front of the transcription apparatus and negative supercoils behind it. These supercoils are controlled by enzymes (11.1.4).

2.6.4 Compaction Levels of DNA Chains

The genetic material of all organisms and viruses exists in the form of tightly packaged nucleoprotein. A high degree of compaction is necessary in order to store the long DNA molecules within cells.

The circular DNA of *E.coli* is ca. 1600 µm long and has to be placed into a cell of 1...3 µm length. In *humans*, the diploid DNA has a linear equivalent (contour length, sum of all 46 chromosomes) of $2 * 1$ m, which has to be packed into a nucleus of 10 µm diameter. The DNA of the individual chromosomes has contour lengths of 16...82 mm.

Bacterial genomes: They contain several DNA binding proteins (20% by mass), some of which are small and highly basic. They condense the DNA and wrap it into a bead-like structure. This nucleoid is kept together by RNA and protein which forms the core of the condensed chromosome. The rest of the DNA exists as a series of highly twisted or supercoiled loops.

Eukaryotic genomes: They are arranged in several chromosomes which contain chromatin composed of

- a double helix of DNA (e.g., $48 * 10^6 ... 240 * 10^6$ bp in the various *human* chromosomes)
- protein (about the same quantitiy by weight as DNA):
  – histones (small basic proteins, most abundant, see below)
  – nuclear scaffold proteins (acidic non-histone proteins, see below)
  – some enzymes (DNA polymerases, DNA topoisomerase II etc.),
- nuclear RNA. Much of the RNA in chromatin (<10% of DNA mass) are nascent chains still associated with the template DNA.
- some lipids.

Since DNA cannot be directly packed into the final dense structure of chromatin, a hierarchy of organization levels is necessary (Fig. 2.6-6, Table 2.6-2).

Specific proteins contribute to these steps:

- Histones are small basic proteins (11...21 kDa), which are involved in the first packaging level of DNA.

  In most *eukarya* there are 5 types: H1, H2A, H2B, H3 and H4. They are rich in lysine and arginine. The amino acid sequences of histones have been extremely conserved during evolution (e.g., in histone H4 only 2 out of 102 amino acids differ between *pea seedlings* and *calf thymus*, and even these exchanges are conserved, Val→Ile and Lys→Arg). There are many posttranslational modifications (methylations, acetylations, phosphorylations). Part of them are reversible and seem to be connected to the cell cycle.

- For the higher compaction mechanisms, a set of highly conserved non-histone (ribonucleo-)proteins, the so-called nuclear scaffold proteins (or nuclear matrix) are extremely important. Their organization and function are still not fully understood.

  The nuclear scaffold proteins provide a three-dimensional structural framework for DNA, and even seem to contribute to tissue-specific gene expression. They also contain topoisomerases, which allow unwinding of DNA supercoils. Eukaryotic DNA has many scaffold-attached regions (SARs) or matrix-associated regions (MARs) of 200 bp length, mostly A/T rich and containing special sequences like topoisomerase cleavage sites.

Organization levels: The fundamental organizational unit of the eukaryotic chromosome is a histone-DNA complex, the nucleosome (Fig. 2.6-5). Folding of chromosomal DNA into core nucleosomes results in a 7-fold compaction in length. The nucleosomes are linked by a short stretch of 'linker' DNA (normally 30...60 nt). Histone H1 is bound to this region. The binding of histone H1 increases the supercoiling of DNA and plays a major role in higher order structure and in chromatin condensation. A continuous string of nucleosomes forms a 10 nm filament.

At the center of this nucleosome core particle, there is an octameric complex of histone proteins: a central tetramer composed of two molecules each of H3 and H4 is flanked on either side by a dimer of H2A/H2B. This core self-assembles in the presence of the protein nucleoplasmin. It resembles a short cylinder. It binds 146 bp of DNA into 1.65 left handed turns around the outside, followed by linker DNA. Thus, up to 240 bp DNA are organized per nucleosome. In the nucleated *erythrocytes* of *birds, fish* and *amphibians* the histone H1 variant H5 can take the role of H1, when the chromatin is inactive. This seems to be necessary for very dense packaging.

Figure 2.6-5. Structure of the Nucleosome Core

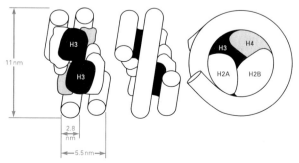

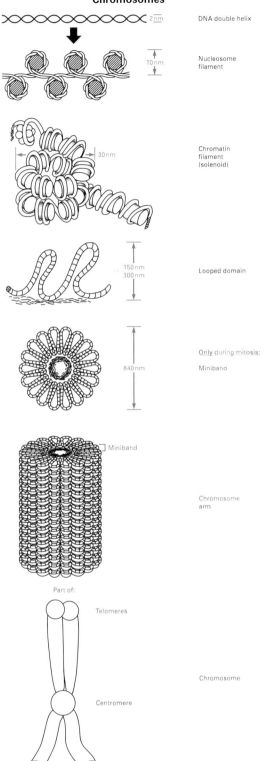

Figure 2.6-6. Organization Levels of *Eukaryotic* Chromosomes

The next higher level of compaction is reached when the 10 nm fiber of nucleosomes forms a left-handed hollow helix of 6 nucleosomes per turn, the 30 nm chromatin filaments. They have the shape of a solenoid with an 11 nm pitch per turn. This aggregation is apparently effected by the histone H1 molecules, which polymerize and form a band in the center of the helix. The degree of compaction is between 35 and 50.

The solenoids are compacted further to form looped DNA domains. Each loop is 150...300 nm long and contains about 50 solenoid turns. Loop domains are thought to be the basic unit of higher order DNA structure in all *eukaryotic* cells. In the interphase nucleus (11.6), this is the maximum compaction level of euchromatin, which represents the transcriptionally active part of DNA. The loops are attached to the nuclear matrix (netlike nuclear lamina) at sites where origins of replication exist. Active genes seem to be located close to these regions. An interphase nucleus contains ca. 50 000 loop domains.

The most highly condensed eukaryotic DNA known is in the *mammalian sperm nucleus*, which is about sixfold more condensed than the mitotic chromosome (see below). The packaging system differs from normal cells (e.g., histones are replaced by protamines).

Mitosis: The degree of compaction varies throughout the cell cycle. In the course of mitosis the loop domains of euchromatin are further packaged to form a 250 nm fiber, which coils to form the arms of a metaphase chromosome. One layer of the coil (diameter ca. 840 nm) consists of 18 loops and is also called a miniband. The packaging ratio (length of DNA/length of the unit containing it) is now up to 10 000...12 000.

Centromeres and Telomeres are two special regions of eukaryotic chromosomes. Centromeres are *cis*-acting genetic loci, which are essential for proper segregation of chromosomes during mitosis and meiosis and are made up of very large regions of repeated sequences. Telomeres are specialized DNA-protein complexes that form the ends of linear chromosomes (11.1.5). They are important for keeping the integrity of individual chromosomes.

All of the vital processes of DNA metabolism must deal with the topological complexity of the chromosome. The chromatin structure is involved in the regulation of replication, transcription, repair and recombination. The topological organization of DNA is both cell and tissue-specific and DNA can take many forms of higher order structure necessary for the expression of only the appropriate tissue-specific genes.

It should be kept in mind that many of the compaction mechanisms described above take place at the same time in different parts of the same DNA molecule. Simultaneously, this DNA molecule is involved in many processes, such as transcription, replication and repair (10.2...4, 11.1...3). It is very surprising how such widely divergent actions can result in almost perfectly ordered processes.

Literature:
Auffinger, L. *et al.*: J. Mol. Biol. 273 (1997) 54–62.
Benbow, R. M. : Sci. Progress Oxford 76 (1992) 425–450.
Dickerson, R.: Nucl. Acids Res. 26 (1998) 1906.
Getzenberg, R. H. *et al.*: J. Cell. Biochem. 47 (1991) 289–299.
Gutell, R.R. *et al.*: Progr. Nucl. Acid Res. Mol. Biol. 32 (1985) 155–216.
Luger, K. *et al.*: Nature 389 (1997) 251–260.
Smith, M. M.: Curr. Opinion in Cell Biology 3 (1991) 429–437.

Table 2.6-2. Condensation Levels of Eukaryotic DNA

Level	Total base pairs/unit	Size (nm)	Condensation ratio
DNA	10.5/turn	2.6	1
Nucleosome filament	150...240/nucleosome	10 (diameter)	7
Chromatin filament (solenoid)	900...1500/turn	30 (diameter)	35...50
Looped Domains	20,000...100,000/loop	150...300 (length of loop)	1,700...2,500
Only during mitosis:			
Miniband	360,000...1,800,000	840	10,000...12,000
Chromosome arm		840 (diameter)	

3 Carbohydrate Metabolism and Citrate Cycle

3.1 Glycolysis and Gluconeogenesis

3.1.1 Glycolysis (Fig. 3.1-1)

Glycolysis is the conversion of glucose (or, in a wider sense, of other hexoses) to pyruvate. In consecutive metabolic steps, pyruvate is oxidized in the citrate cycle (3.8) or, with insufficient oxygen supply, converted to lactate (e.g. in *animals* and some *microorganisms*) or to ethanol (e.g. in *yeast*) in order to reconstitute NAD$^+$, which is required for further progress of glycolysis.

Glycolysis is a key reaction of metabolism. It takes place in almost all living cells and
- supplies energy (in the form of 2 ATP/1 glucose metabolized)
- supplies reducing equivalents (in the form of NADH), which yield additional ATP under aerobic conditions (16.1) or are consumed by reductions (e.g. under anaerobic conditions)
- converts carbohydrates into compounds which undergo terminal oxidation (acetyl-CoA, 3.8) or are used for biosynthesis (e.g. glycerol, acetyl-CoA).

Glucose + 2 ADP + 2 P$_i$ + 2 NAD$^+$ = 2 pyruvate$^-$ + 2 ATP +
    + 2 H$_2$O + 2 NADH + 2 H$^+$ $\Delta G'_0 = -84$ kJ/mol

Glucose + 2 ADP + 2 P$_i$ = 2 lactate$^-$ + 2 ATP + 2 H$_2$O + 2 H$^+$
  $\Delta G'_0 = -136$ kJ/mol

Glucose + 2 ADP + 2 P$_i$ = 2 ethanol + 2 ATP + 2 H$_2$O + 2 CO$_2$
  $\Delta G'_0 = -174$ kJ/mol

The enzymes are present in the *cytosol*, but are partially bound to structures (*cellular membrane*, in *eukarya* also to the *mitochondrial outer membrane* and the *cytoskeleton*). Many enzymes are associated with each other or are interconnected by a common product/substrate (substrate channeling).

Glucose directly enters the main pathway of glycolysis (Fig. 3.1-2) by phosphorylation (3.1.2). Mannose is first phosphorylated and then isomerized to fructose 6-P. (Its biosynthesis is dealt with in 3.5.6). The interconversion of galactose and glucose (via 1-phosphates and UDP-derivatives) by epimerization and transferase reactions takes place during both galactose catabolism and anabolism (3.4.2).

Fructose is converted in *liver, kidney* and *intestinal mucosa* by fructokinase to fructose 1-P (Fig. 3.1-2). This reaction is independent of hormones (therefore diabetics can tolerate fructose). In *liver* and *kidney*, this compound is cleaved by fructose bisphosphate aldolase B to glycerone-P (dihydroxyacetone-P, which is a member of the main path of glycolysis) and to glyceraldehyde (which enters the main path by phosphorylation to its 3-phosphate). In other tissues (e.g. *muscle*), phosphorylation of fructose to fructose 6-P takes place at a low rate. In hereditary fructose intolerance, fructose bisphosphate aldolase B is lacking in *liver*. The accumulated fructose 1-P inhibits fructose 1,6-bisphosphatase and fructose bisphosphate aldolase and therefore disturbs glycolysis and gluconeogenesis.

Sorbitol is oxidized to fructose in various tissues. Therefore its metabolism is also hormone independent.

Figure 3.1-2. Fructose Metabolism

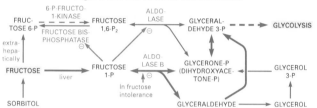

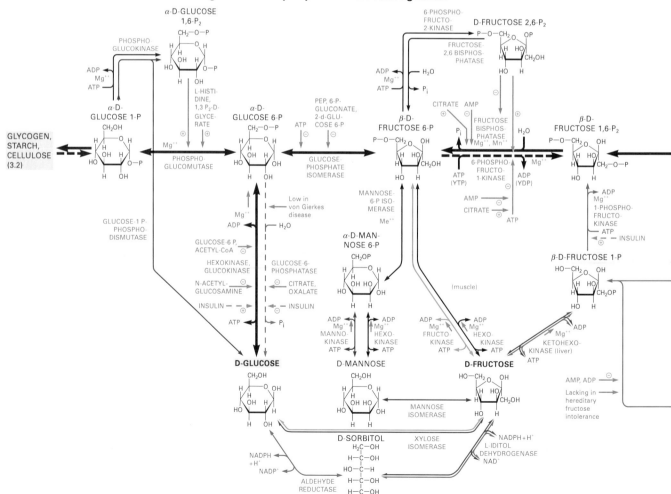

Figure 3.1-1. Glycolysis and Gluconeogenesis

The sequence of the glycolysis reactions and the formation of some compounds of importance for biosynthesis (glycerol, alanine etc.) are shown in Fig. 3.1-1.

Cofactors for the phosphoglucomutase and phosphoglyceromutase reactions are glucose-1,6-bisphosphate and 2,3-bisphosphoglycerate, respectively, which are formed by kinase reactions from glucose 1-phosphate and 3-phosphoglycerate. In the mutase reaction, they confer a phosphate moiety (via phosphorylation of the enzyme) to the substrate. Thus, the cofactor is turned into the product, while the substrate is converted into the cofactor.

Isomerase and epimerase reactions usually involve the formation of enediolate intermediates by removal of a proton by a basic group at the enzyme. The reprotonation yields a different product.

Glyceraldehyde-3-P dehydrogenase uses an oxidation reaction to obtain a high-energy phosphate bond (which in consecutive reactions enables ATP formation):

D-Glyceraldehyde 3-P + NAD$^+$ + P$_i$ = 1,3-P$_2$ D-glycerate + NADH
$\Delta G_0' = + 7.5$ kJ/mol (actual ΔG_{phys} almost 0).

3.1.2 Regulation Steps in Glycolysis

It is obvious, that this key metabolic pathway has to be strictly regulated. In *animals*, this occurs at many levels (Fig. 3.1-3). Under physiological conditions, the major regulated enzymes glucokinase (or hexokinase), phosphofructokinase and pyruvate kinase have restricted flow rates, the substrates pile up to some extent and

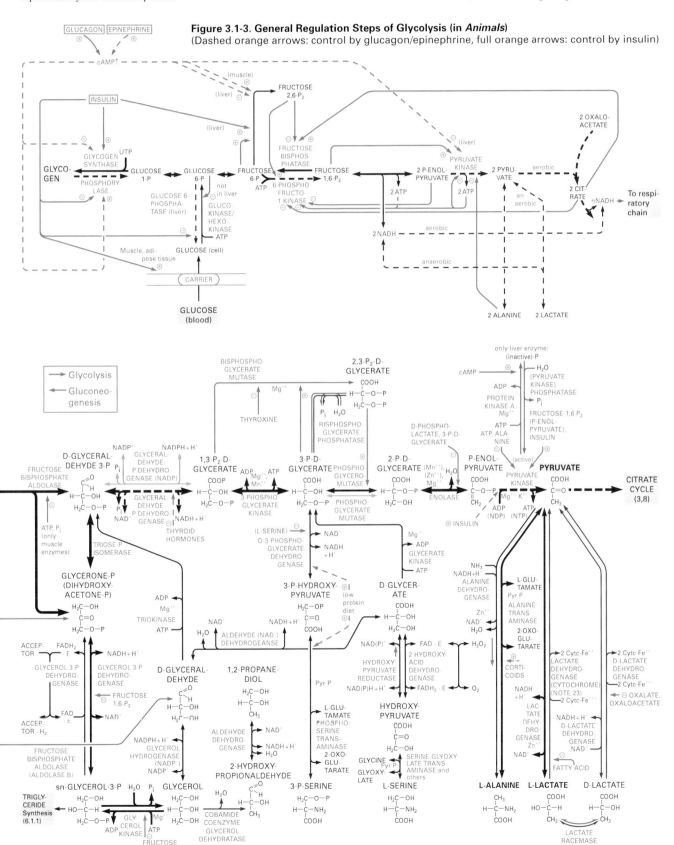

Figure 3.1-3. General Regulation Steps of Glycolysis (in *Animals*)
(Dashed orange arrows: control by glucagon/epinephrine, full orange arrows: control by insulin)

the reactions are far from equilibrium (ΔG_{phys} strongly negative), while the other reactions are almost at equilibrium (ΔG_{phys} close to zero) and thus reversible.

Phosphorylation of glucose takes place by the reaction

Glucose + ATP = glucose 6-P + ADP

The reaction is catalyzed in *liver* by glucokinase (K_M ca. 8 mmol/l). The enzyme is not saturated by substrate under physiological conditions. Thus, a higher glucose concentration increases the rate of phosphorylation (Fig. 3.1-4, glucose utilization as part of glucose homeostasis). The reaction is not inhibited by its end products ADP and glucose 6-P. The phosphorylation is practically irreversible, but is counteracted *in vivo* by glucose 6-phosphatase (see below).

In *other organs*, the reaction is catalyzed by hexokinase (K_M ca. 0.1 mmol/l). The enzyme is saturated by substrate under physiological conditions. It is inhibited by its end products ADP and glucose 6-P. This achieves a steady supply rate for glycolysis intermediates. The enzyme is bound in a regulated way to the *mitochondrial* outer membrane (e.g. in *brain* at the receptor porin → direct coupling to the energy state of the cell).

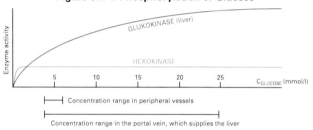

Figure 3.1-4. Phosphorylation of Glucose

Dephosphorylation: Glucose 6-phosphate is transferred to the *endoplasmic reticulum* of *liver, kidney* and *intestine* in a carrier-dependent way and is hydrolyzed there. The glucose formed returns to the *cytosol* and is secreted into the *bloodstream*. This causes elevation of low blood glucose levels. The enzyme is inhibited by insulin.

Since the reaction of 6-phosphofructo 1-kinase (PFK, tetrameric, 4*85 kDa) is the committed step in glycolysis (= first unambigous step, since the earlier reactions can also lead to other pathways), it is the major regulation point of this pathway. A plethora of mechanisms exist (Fig. 3.1-5). They encompass also fructose bisphosphatase (FBPase, the corresponding reaction of gluconeogenesis) and prevent in this way the futile simultaneous occurence of glycolysis and gluconeogenesis.

- PFK is allosterically inhibited by high ATP levels, this is counteracted by elevation of AMP levels (sensing of the energy supply, Pasteur-effect)
- PFK is allosterically inhibited by low pH (sensing of acidification due to lactate formation, 3.1.5)
- PFK is inhibited by high citrate concentrations (by increasing the ATP effect, indicating sufficient material for biosynthesis or for oxidation in the citrate cycle)
- The concentration of fructose 2,6-P_2, an activator of PFK and inhibitor of FBPase activities in *liver*, is regulated by fructose 6-P (indicating sufficient substrate) and by hormones (sensing the general metabolic situation). This effector is synthesized and degraded by the enzymes 6-phosphofructo-2-kinase and fructose-2,6-bisphosphatase, respectively, which are located on the same polypeptide chain.
  - Fructose 6-P activates 6-phosphofructo-2-kinase and inhibits fructose-2,6-bisphosphatase, thus increasing the fructose 2,6-P_2 level and promoting PFK activity (feed-forward activation).
  - The bifunctional peptide itself is regulated by phosphorylation and dephosphorylation (Fig. 3.1-5). In *liver*, the dephosphorylated state leads to an elevated level of fructose 2,6-P_2, thus to PFK activation and to enhanced glycolysis. This situation is achieved by insulin, which allosterically inhibits phosphorylation. Phosphorylation (which is activated by glucagon) reverses the situation and decreases the fructose 2,6-P_2 level. PFK is inhibited and FBPase is activated, the reaction is shifted from glycolysis to gluconeogenesis. In *muscle* there exists a different isoenzyme, which responds the opposite way to phosphorylation and dephosphorylation.

Pyruvate kinase (tetrameric, 4*57 kDa) is inhibited by its reaction products ATP and alanine (formed from pyruvate by transamination, Fig. 3.1-5). It is feed-forward activated by fructose 1,6-P_2.

In addition, the *liver* isoenzyme (but not the *muscle* isoenzyme) is regulated via its phosphorylation state: Glucagon promotes the inactivation by phosphorylation (Fig. 3.1-5). This reduces specifically the glucose consumption in *liver* and leaves the consumption in other organs unaffected.

Induction of enzymes: Besides the 'fast' regulation by allosteric mechanisms, phosphorylation and dephosphorylation, the quantity of enzymes is also under hormonal control by induction and repression of the enzyme synthesis. The counteracting effects of hormones on various enzymes is shown in Table 3.1-1.

3.1.3 Gluconeogenesis

Gluconeogenesis is the formation of glucose (or its phosphates, which are further converted to poly- or oligosaccharides, 3.2 and 3.4) from non-carbohydrate sources. Examples are amino acids (alanine is first converted into pyruvate, other amino acids are converted into citrate cycle intermediates, 3.8), glycerol, in *plants* also fatty acids via the glyoxylate cycle (3.9.1). In *animals*, this takes place in the *liver* and to a minor extent in the *kidney cortex*. Its major purpose is to keep up a sufficient glucose supply to *brain* and *muscles*.

Most reaction steps are a reversal of the respective glycolysis reactions, with the exception of those reactions which are highly regulated in glycolysis (3.1.2) and thus usually far from equilibrium. They are replaced by alternative reactions, which are energetically more favorable in the direction of glucose synthesis (glucose-6-phosphatase instead of glucokinase/hexokinase; hexose diphosphatase instead of phosphofructokinase). The complicated reactions for the circumvention of pyruvate kinase and their regulation are discussed in 3.3.5.

Figure 3.1-5. Detailed Regulation Mechanisms of 6-Phosphofructo 1-Kinase, Fructose Bisphosphatase and Pyruvate Kinase (in *Animals*)

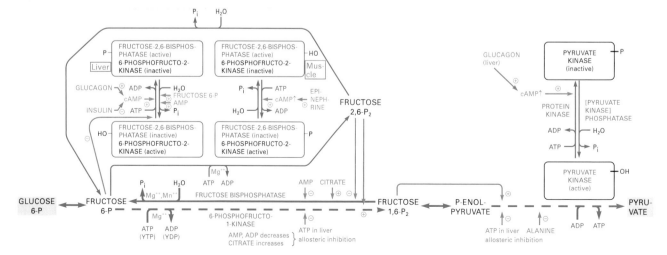

Table 3.1-1. Modulation of Enzyme Expression by Insulin, Catecholamines/cAMP and Glucocorticoids (L = in liver, A = in adipose tissue)

	Insulin induces (# only in presence of glucose)				Insulin represses	
	Glycolysis enzymes:		**Other enzymes:**			
Catecholamines repress (via cAMP)	glucokinase	(3.1.2) L	GLUT4 transporter	(3.1.4) A		
	phosphofructokinase	(3.1.2) # L A	acetyl-CoA carboxylase	(6.1.1) L A		
	fructose-6-P 2-kinase	(3.1.2) # L	fatty acid synthase	(6.1.1) L A		
	pyruvate kinase	(3.1.2) L	lipoprotein lipase	(18.2.1) A		
Catecholamines induce (via cAMP)					**Gluconeogenesis enzymes:**	
					pyruvate carboxylase	(3.3.4) L
					PEP carboxykinase	(3.3.4) L
Glucocorticoids induce					glucose 6-phosphatase	(3.1.2) L
					fructose-1,6-bisphosphatase	(3.1.2) L
					pyruvate carboxylase	(3.3.4) L
					PEP carboxykinase	(3.3.4) L
Glucagon induces (via cAMP)					fructose-1,6-bisphosphatase	(3.1.2) L
					PEP carboxykinase	(3.3.4) L

While the pathway glucose → pyruvate yields 2 ATP (3.1.1), the reverse conversion consumes 6 ATP in order to make it thermodynamically feasible.

2 Pyruvate$^-$ + 4 ATP + 2 GTP + 2 NADH + 6 H$_2$O =
glucose + 4 ADP + 2 GDP + 6 P$_i$ + 2 NAD$^+$ + 2 H$^+$ $\Delta G'_0 = -37.7$ kJ/mol

Regulation of gluconeogenesis: Besides the inhibition of the fructose bisphosphatase reaction by fructose 2,6-P$_2$ (3.1.2), this enzyme is also inhibited by AMP and activated by citrate. Further, the expression of enzymes is regulated by hormones (Table 3.1-1).

3.1.4 Resorption of Glucose (Fig. 3.1-6)
The passage of glucose through most cell membranes proceeds via the regulated transport proteins GLUT1...5 with 12 transmembrane domains each. Glucose transport is provided by conformation changes of the protein.

• The uptake of glucose from the *intestine* into the *epithelial cells* takes place through the Na$^+$-glucose symporter (Table 18.1-3). Glucose passes from these cells into the *bloodstream* via GLUT5.

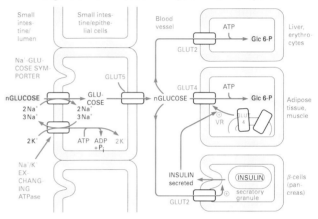

Figure 3.1-6. Glucose Resorption and Transport

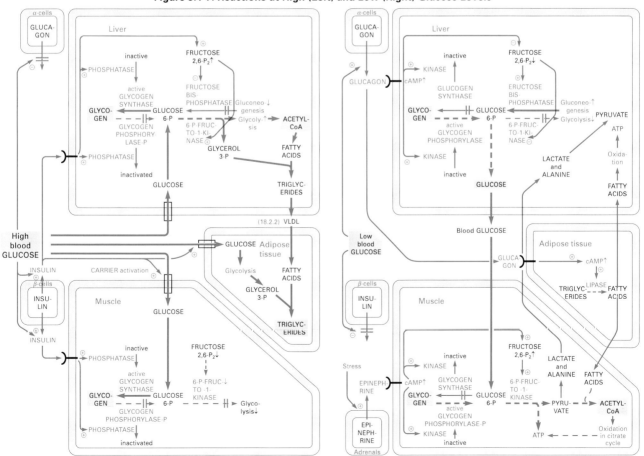

Figure 3.1-7. Reactions at High (Left) and Low (Right) Glucose Levels

- GLUT1 and 3 are almost ubiquitous in *mammalian* cells. With a low K_M (ca. 1 mmol/l) they provide the basic glucose supply.
- *Liver* and *pancreatic β-cells* possess also the GLUT2 protein with the high K_M of ca. 20 mmol/l, well above the physiological glucose blood levels (4...7 mmol/l). Thus glucose uptake is proportional to the blood level and high only at elevated glucose concentrations.
- GLUT4 with an intermediate K_M of 8 mmol/l provides glucose uptake into *muscles* and *adipose tissue*. The number of transport proteins at the cellular membrane increases greatly in presence of insulin, indicating sufficient glucose supply. This takes place by reversible transfer of GLUT4 from *internal vesicles* to the *cytoplasmic membrane*.

3.1.5 Response of Animal Organs to High and Low Glucose Levels (Fig. 3.1-7)

The various organs respond differently to varying blood glucose concentrations, depending on their function. Both glycolysis and storage as glycogen play a role. The mechanisms of glycogen formation, degradation and their hormonal control are described in detail in 3.2.

- *Blood* contains a small share of glucose and functions only as a transport medium between organs.
- *Liver* is the central storage organ for glycogen and acts as a buffer during the intervals of food uptake. It can release glucose for use by other organs.
- *Muscles* store a considerable amount of glycogen, but can use it only for their own purposes.
- All other organs consume glucose taken up from blood. Some of them can also use other compounds for their metabolism.

Effects After Glucose Intake:

At high glucose levels, storage reactions and conversions for biosynthetic purposes prevail.

- In *liver*, both the glucose uptake by GLUT2 (3.1.4) and its phosphorylation by glucokinase (3.1.2) respond to the high glucose concentration and remove a great portion of glucose from the bloodstream by forming glucose 6-P. The antagonizing reaction by glucose 6-phosphatase is inhibited.
- In the *β cells* of *Langerhans islets* in the *pancreas* increased glucose 6-P formation is the primary signal for insulin release (17.1.3).
- Glucose uptake in *muscles* and *adipose tissue* is increased by insulin, which acts on the GLUT4 transporter (see above). In these and in other organs, glucose phosphorylation is performed by hexokinase.
- Glycogen synthesis from glucose 6-P is activated in *liver* and *muscle*, while glycogen degradation is inhibited.
- Glycolysis from glucose 6-P is activated in *liver* and *adipose tissue*. This enables the formation of triglycerides as energy storage forms and of amino acids for biosynthetic reactions.
- Gluconeogenesis is generally inhibited.

Effects of Starvation or in Sudden Energy Demand:

At low blood glucose levels, glycogen synthesis is inhibited, while glycogen degradation and glucose release from *liver* is activated.

- In *liver*, the glucose 6-P formed from glycogen is dephosphorylated and glucose is released into the *blood* in order to keep up the physiological glucose level. This is essential especially for *brain/nerves* (daily consumption of an *adult human* 150 g), *adrenal medulla* and *erythrocytes*. The glycolysis steps beyond glucose 6-P are inhibited in *liver*.
- In *muscle* during exercise, glucose uptake and endogeneous glycogenolysis take place to meet the ATP demand. The glucose 6-P formed from glycogen is not dephosphorylated to glucose, but rather passes through the glycolysis sequence to pyruvate and further on to lactate. The lactate formed is carried via the bloodstream to the *liver*, where it is reconverted to glucose ('Cori cycle'). Besides being reduced to lactate, pyruvate can also be transaminated to alanine (4.2.3), which likewise is transported to the *liver* and serves as source for gluconeogenesis.
- In *muscle* and other organs, proteins are degraded to amino acids (Chapter 4), which are used for gluconeogenesis or are oxidized directly.
- In *adipose tissue*, triglycerides are degraded (6.2.2) and supply fatty acids and glycerol to *liver and muscle*. There they can be used for gluconeogensis or are directly oxidized.

Literature:

Hers, H.G., Van Schaftingen, E.: Biochem. J. 206 (1982) 1–12.
Lemaigre, F.P., Rousseau, G.G.: Biochem. J. 303 (1994) 1–14.
Mueckler, M.: Eur. J. Biochem. 219 (1994) 713–725.
Pilkis, S.J. et al.: Ann. Rev. of Biochem. 64 (1995) 799–835.
Pilkis, S.J., Granner, D.K.: Ann. Rev. of Physiol. 52 (1992) 885–909.
Silverman, M.: Ann. Rev. of Biochem. 60 (1991) 757–794.
(various authors): The Enzymes of Glycolysis. Phil. Trans. R. Soc. London 293 (1981) 1–214.

3.2 Polysaccharide Metabolism

Glycogen in heterotrophic organisms (*animals, fungi, bacteria*) and starch (in *plants*) are polymeric storage forms for carbohydrates, which decrease the osmotic pressure. Glycogen is essential for glucose homeostasis as a readily available source (providing more than half of the glucose consumed in *human adults*). Cellulose is a structure-forming component of *plant cell walls*, but also occurs in *marine invertebrates*. The amount synthesized in *plants* is estimated to be 10^{12} t/year. It is the compound present in largest quantities in the biosphere. Dextrans are highly branched storage polymers of glucose, produced by *bacteria* and *yeasts*. Fructans are soluble storage polymers of fructose in *plants*. In these polymers, the sugar residues (hexoses) are interlinked with glycosidic bonds (For details, see 1.2.2).

3.2.1. Structures (Fig. 3.2-1)

Starch is a mixture of amylose and amylopectin. Amylose has a linear structure of ca. $1...5 * 10^3$ glucose units, partially forming a lefthanded helix. The bonds are of $\alpha 1 \rightarrow 4$ configuration. Amylopectin (ca. $10^4...10^5$ glucose units) and glycogen (similar size) also contain branched structures with $\alpha 1 \rightarrow 6$ bonds at the branch points. Between the branch points there are 24...30 glucose units in amylopectin and 8...12 glucose units in glycogen.

Cellulose has a linear structure of 2000...8000 glucose units, forming parallel microfibrils. Since the bonds between the glucose residues are of $\beta 1 \rightarrow 4$ configuration, every other glucose is 'inverted'. This allows the formation of additional hydrogen bonds between consecutive glucose units within the chain, as well as with neighboring chains of the microfibril and hemicellulose and protein molecules of the matrix.

Dextrans consist of several thousand mostly $\alpha 1 \rightarrow 6$ linked glucose units. Also $\alpha 1 \rightarrow 2$, $\alpha 1 \rightarrow 3$ and $\alpha 1 \rightarrow 4$ bonds occur. Fructans are polyfructose chains, which are attached to a single sucrose molecule at various positions (in kestoses to the fructose moiety; levan type: $\beta 2 \rightarrow 6$, inulin type: $\beta 2 \rightarrow 1$).

All of these compounds have only 1 reducing residue per molecule (free C-1 of the terminal glucose).

3.2.2. Biosynthesis of Polysaccharides

The formation of the glycosidic bonds in polysaccharides proceeds in most cases via an activated incoming sugar:

Glucose 6-P = glucose 1-P

Glucose 1-P + UTP (ATP, GTP) = UDP (ADP, GDP)-D-glucose + PP_i
  $\Delta G'_0$ = ca. 0 kJ/mol

$PP_i + H_2O = 2\ P_i$ $\Delta G'_0$ = –33.5 kJ/mol (drives the previous reaction to the right)

UDP (ADP, GDP)-D-glucose + polysaccharide$_n$ =
  = polysaccharide$_{n+1}$ + UDP (ADP, GDP)

UTP takes part in glycogen and in *bacterial* cellulose synthesis, ATP in starch synthesis, GTP and UTP in cellulose synthesis.

Glycogen synthesis (Fig. 3.2-2) in *vertebrate liver* and *muscles* starts, when the protein glycogenin (37 kDa) catalyzes the glucosylation of its own tyrosine residue and the extension of the chain to 8 glucose units. Then glycogen synthase takes over and enlarges the molecule as long as the synthase remains in firm contact with glycogenin (which remains in the center of the glycogen). The branching mechanism during glycogen synthesis is catalyzed by 1,4-α-glucan branching enzyme, which transfers a terminal fragment of ca. 7 glucose residues to the C-6 hydroxyl of a glucose in the same or in another chain (Fig. 3.2-2). This enlarges the number of non-reducing ends, which are the attack points for degradation by phosphorylase.

Starch synthesis takes place in *plant chloroplasts* (16.2.2) from fructose 6-P, which is converted to glucose 6-P and further on to glucose 1-P and to ADP-glucose (Fig. 3.2-3). The latter reaction is the regulated one: low P_i and high 3-phosphoglycerate in *chloroplasts* promotes it. This situation occurs, if there is a high concentration of sucrose in the *cytosol*. This way, the synthesis rate of both photosynthesis products starch and sucrose is coordinated.

The action of starch synthase results in the linear product amylose, which is thereafter branched to amylopectin similarly as in glycogen synthesis, but at different chain lengths. The product is stored in *chloroplast* or in *leukoplast* granules (in *heterotrophic plants*).

Cellulose synthesis from UDP-D-glucose or GDP-D-glucose is catalyzed by cellulose synthases, which are bound to the *plasma membrane* (Fig. 3.2-4).

Frequently, the product of photosynthesis sucrose (16.2.2) is converted at the *membrane* by sucrose synthase (3.4.1) and provides the activated glucose for cellulose synthesis at the location of use.

Sucrose + UDP = UDP-glucose + fructose

The synthesis of dextrans and fructans also starts from sucrose, however by direct transfer of hexose units without intermediate nucleotide derivatives. Dental plaques consist of dextrans.

3.2.3. Catabolism of Polysaccharides

Glycogen degradation in *animals* (Fig. 3.2-2) starts at the non-reducing ends and is initiated by the action of phosphorylase (dimeric, 2 * 97 kDa). This reaction determines the hydrolysis rate and is strictly regulated (for details see 3.2.4).

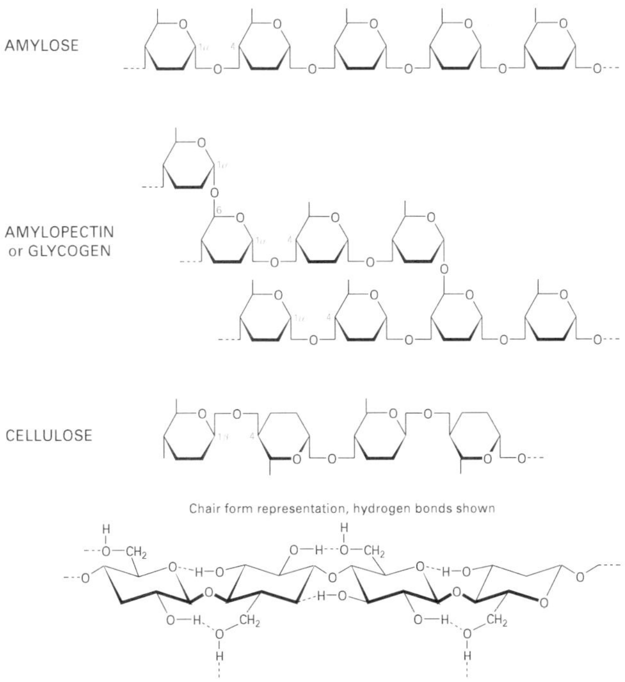

Glycogen$_n$ + P$_i$ = glycogen$_{n-1}$ + glucose 1-P $\Delta G'_0$ = + 3.1 kJ/mol

Since, however, the physiological concentration ratio P$_i$/glucose 1-P is 30...>100, the reaction is driven to the right (ΔG_{phys}, ca. – 6 kJ/mol). It yields a phosphorylated product without requiring ATP. The reaction requires pyridoxal phosphate, which acts first as a proton donor and then as a proton acceptor (acid-base catalysis).

When the phosphorylase reaction has shortened the outer glycogen chains to a length of about 4 glucose units, a transfer reaction by 4-α-glucanotransferase takes place. The remaining single α1→6 bound glucose is hydrolytically removed by amylo-1,6-glucosidase, which is another function of the same peptide chain.

The degradation of starch can take place hydrolytically (generally) or by the starch phosphorylase reaction (analogously to glucagon, but only in *plants*). See Fig. 3.2-3.

α-amylase (in *animal saliva* and *pancreatic secretions*, in *plants, fungi, bacteria*) is an *endo*-enzyme (randomly acting on inner bonds) and produces besides α-maltose and maltotriose also α-dextrin, which contain many α1→6 bonds. The latter compound is debranched by hydrolytic removal of the α1→6 bound glucoses. The linear oligosaccharides are thereafter degraded to glucose. Alternatively, starch phosphorylase converts them to glucose 1-P, similarly to above.

β-amylase (in *germinating plant seeds/malt*) is an *exo*-enzyme, removing β-maltose units from the non-reducing ends (The α-bond in starch is inverted to the β-configuration). The reaction is interrupted when 1→6 bonds are reached; a limit dextrin remains. Debranching (similar to above) has to take place.

Starch$_n$ + H$_2$O = starch$_{n-2}$ + β-maltose

Cellulose is mostly cleaved to cellobiose by cellulase (Fig. 3.2-4). This is an *endo*-enzyme occuring in *bacteria* (including the *bacteria* in the intestinal tracts of *ruminants*), *protozoa, fungi* and *insects*, e.g., termites.

Cellulose + n H$_2$O = cellobiose.

3.2.4. Regulation of Glycogen Metabolism in Mammals (Fig. 3.2-5)

The central importance of glycogen for glucose metabolism requires a tight control of its synthesis and degradation. Both phosphorylase and glycogen synthase are hormonally regulated via phosphorylation and dephosphorylation cascades, for obvious reasons in opposite directions (Details of such cascade mechanisms are described in 17.3 and 17.4). Generally,

- phosphorylation of the various enzymes is initiated by, e.g, epinephrine (in *muscle*) and glucagon (in *liver*) and favors glycogen degradation;
- dephosphoryation is intitiated by insulin and favors glycogen synthesis.
- Furthermore, allosteric mechanisms provide another level of regulation.

Figure 3.2-1. Structures of Amylose, Amylopectin, Glycogen and Cellulose

Figure 3.2-2. Glycogen Synthesis and Degradation

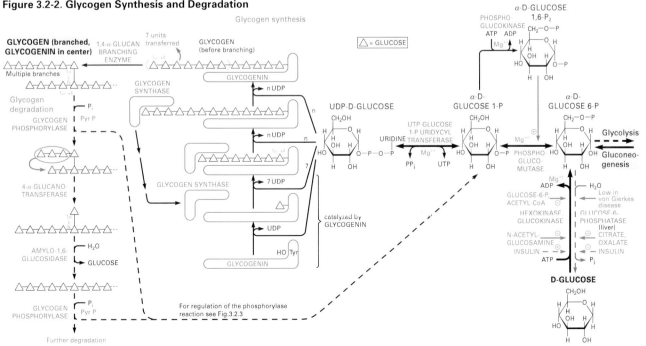

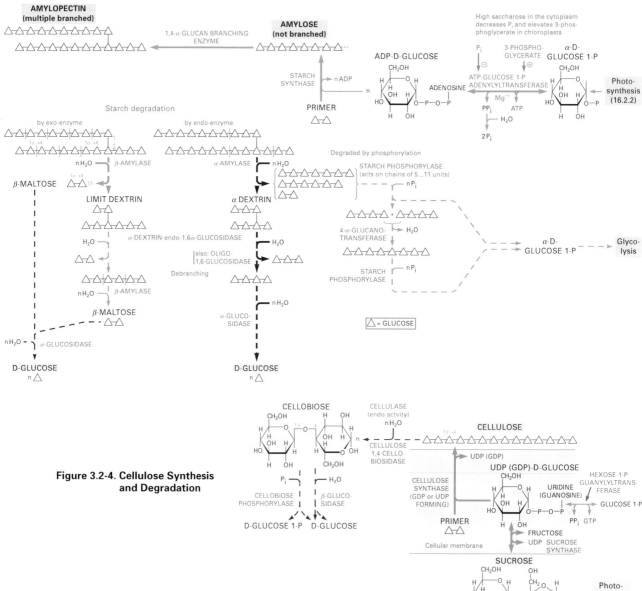

Figure 3.2-3. Starch Synthesis and Degradation

Figure 3.2-4. Cellulose Synthesis and Degradation

Phosphorylation: Phosphorylase b, the usually inactive, phosphate free form is activated by phosphorylation at Ser 14 to phosphorylase a (for allosteric interconversions see below). The activating phosphorylase kinase itself is activated by phosphorylation, catalyzed by protein kinase A, which is cAMP dependent and thus hormone controlled (17.4.2, e.g., by epinephrine).

Phosphorylase kinase also requires Ca^{++} for activation. In the phosphorylated state, already a moderate Ca^{++} elevation is sufficient. The elevation is sensed by calmodulin (17.4.4), which constitutes the δ subunit of the phosphorylase kinase $(αβγδ)_4$. This way, phosphorylase kinase integrates stimulatory effects by hormones (via phosphorylation) and neuronal impulses (via Ca^{++} response).

Glycogen synthase exists in dephosphorylated and (9-fold) phosphorylated forms. In contrast to phosphorylase, the dephosphorylated form (glycogen synthase a) is generally active, while the phosphorylated form (glycogen synthase b) requires a high level of glucose 6-P for activity (operates therefore only at high glucose supply). The phosphorylation is performed by the cAMP dependent protein kinase A and some other kinases.

Dephosphorylation: The phosphorylase system is deactivated by dephosphorylation either of phosphorylase a or of phosphorylase kinase. In both cases, the reaction is catalyzed by protein phosphatase 1 (PP1). The same enzyme also removes the phosphates from glycogen synthase and activates it.

The catalytic subunit of PP1 (37 kDa) obtains the affinity to glycogen particles (and thus to its protein substrates, which are associated with glycogen) by combination with the G-subunit ('glycogen binding', 160 kDa). Phosphorylation of the G-subunit by an insulin-dependent kinase in a way not completely known enables the association and promotes the phosphatase activity. If, however, the G-subunit becomes phosphorylated at another site by the cAMP dependent protein kinase A, the association of the catalytic and the G-subunits is prevented and the catalytic subunit remains inactive.

Additionally, the activity of PP1 is prevented by the inhibitor 1. This, however, only takes place if the inhibitor was phosphorylated by the cAMP dependent protein kinase A.

Allosteric mechanisms: Both phosphorylase a and phosphorylase b exist in active R and inactive T-forms (2.5.2).

In *liver*, glucose binding causes a shift of phosphorylase a from the R- to the T-form. This exposes the bound phosphate and enables inactivating dephosphorylation to yield phosphorylase b. Thus phosphorylase acts as a glucose sensor and prevents glycogenolysis if abundant glucose is available. Thereafter, the formed phosphorylase b releases the phosphatase, which is now free to act on glycogen synthase and activate it for formation of glycogen.

In the resting *muscle*, phosphorylase b prevails in the inactive T-form. There are two ways of activation when work requires energy supply:

- Hormones cause phosphorylation of phosphorylase b to a as described above.

Figure 3.2-5. Regulation of Glycogen Synthesis and Degradation in *Animals* (Contrary to the arrow colors in other Figures, red arrows indicate here reactions and regulation mechanisms leading to glycogen synthesis, green arrows leading to glycogen degradation)

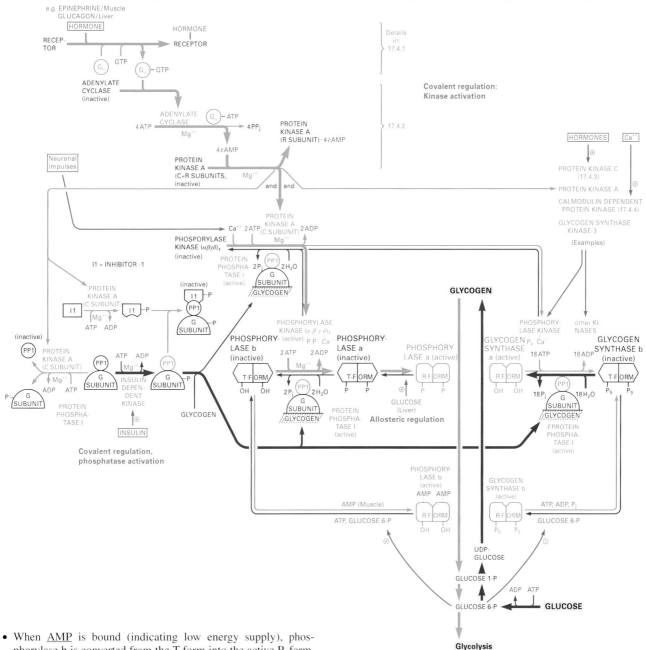

- When AMP is bound (indicating low energy supply), phosphorylase b is converted from the T-form into the active R-form. This is counteracted by ATP and glucose 6-P (indicating sufficient energy and glucose supply).

3.2.5 Medical aspects

Many diseases are caused by inheritable defects of the enzymes involved in glycogen metabolism (Table 3.2-1). Except in disease IX, glycogen is either elevated or of abnormal structure.

Table 3.2-1. Hereditary Glycogen Storage Diseases

Type	Name	Enzyme Deficiency	Tissue
I	von Gierke's disease	glucose 6-phosphatase	*liver, kidney*
II	Pompe's disease	α-1,4 glucosidase	*lysosomes*
III	Cori's disease	amylo-1,6-glucosidase	*general*
IV	Andersen's disease	1,4-α-glucan branching enzyme	*liver, general?*
V	MacArdle's disease	glycogen phosphorylase	*muscle*
VI	Hers' disease	glycogen phosphorylase	*liver*
VII	Tarui's disease	phosphofructokinase	*muscle*
VIII		phosphorylase kinase	*liver*
IX		glycogen synthase	*liver*

Literature:

Alonso, M.D. et al.: FASEB J. 9 (1995) 1126–1137.
Browner, M.F., Fletterick, R.J.: Trends in Biochem. Sci. 17 (1992) 66–71.
Hers, H. G. *et al.* in Scriver, C.R. *et al.* (Eds.): *The Metabolic Basis of Inherited Disease* 6th Ed. McGraw-Hill (1989) 425–452.
Krebs, E.G.: Angew. Chem. 105 (1993) 1173–1180.
Larner, J.: Adv. in Enzymol. 63 (1990) 173–231.
Lomako, J. *et al.*: FASEB J. 7 (1993) 1386–1393.
Roach, P.J. *et al.*: Adv. Enzyme Regul. 31 (1991) 101–120.

3.3 Pyruvate Turnover and Acetyl-Coenzyme A

3.3.1 Pyruvate Oxidation (Fig. 3.3-1)

Pyruvate oxidation is common to all *aerobic organisms*. By action of the pyruvate dehydrogenase enzyme complex, pyruvate is converted to acetyl-CoA (3.3.3), the activated form of acetate. In *eukarya*, pyruvate is at first transported into the *mitochondria*, where this reaction takes place.

If anaerobic conditions exist, pyruvate is converted instead into reduced compounds, such as lactate (e.g. in muscle of *animals*) or ethanol (e.g. by *yeast*).

The oxidative decarboxylation is catalyzed by the multi-enzyme complex pyruvate dehydrogenase (lipoamide) consisting of the subunits pyruvate dehydrogenase (E_1), dihydrolipoamide acetyltransferase (E_2) and dihydrolipoamide dehydrogenase (E_3). They catalyze the following reactions (ThPP = thiamine pyrophosphate, Lip = lipoamide, for mechanism see 9.2.2 and Fig. 3.3-4):

E_1: Pyruvate⁻ + ThPP-E_1 + H⁺ = hydroxyethyl-ThPP-E_1 + CO_2
  Hydroxyethyl-ThPP-E_1 + lipoamide-E_2 =
     acetyl-dihydrolipoamide-E_2 + ThPP-E_1
E_2: Acetyl-dihydrolipoamide-E_2 + CoA-SH = acetyl-CoA + dihydrolipoamide-E_2
E_3: Dihydrolipoamide-E_2 + NAD⁺ = lipoamide-E_2 + NADH + H⁺

In the enzyme complex from *E. coli* (4600 kDa), 8 trimers of E_2, 12 dimers of E_1 and 6 dimers of E_3 subunits are arranged in highly symmetrical cubic order. The multienzyme complex from *eukarya* (8400 kDa) contains a 'nucleus' of 60 E_2 monomeric subunits surrounded by 30 E_1 dimers and 6 E_3 dimers, as well as 1...3 copies of pyruvate dehydrogenase kinase and -phosphatase. In both cases, lipoic acid (9.14.1) is bound to the ε-amino group of an E_2-lysine residue (hence 'lipoamide'). This 'arm' moves the attached acetyl group from E_1 to E_3. This enhances the reaction speed, coordinates the regulation of the reactions and avoids side reactions.

3.3.2 Regulation of Pyruvate Dehydrogenase Activity

In *eukarya*, E_2 and E_3 are competitively inhibited by the products of their reaction, acetyl-CoA and NADH, respectively. The *eukaryotic* E_1-subunit is inactivated by covalent phosphorylation at a serine moiety and activated by dephosphorylation. Acetyl-CoA and NADH promote the inactivating reaction. The modifying enzymes are attached to the E_2-'nucleus' of the multienzyme complex.

At high product concentrations, the reaction course of the E_2 and E_3 enzymes can be reversed. The E_1-catalyzed reaction, however, is irreversible.

Figure 3.3-2. Regulation of the Reactions Catalyzed by Pyruvate Dehydrogenase Subunits

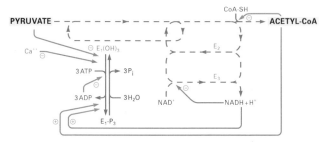

3.3.3 Acetyl-Coenzyme A (Acetyl-CoA)

Acetyl-CoA is an example of an 'energy-rich' thioester bond ($\Delta G_0'$ for hydrolysis = –31.5 kJ/mol). This is similar to the change of free energy during hydrolysis of high energy phosphate bonds of ATP ($\Delta G_0'$ = –30.5 kJ/mol for the γ-bond, –32.2 kJ/mol for the β-bond). Thus, acetyl-CoA can be used for ATP formation in acetate fermentations (15.4).

Acetyl-CoA plays a central role in metabolism. It is the common degradation product not only of carbohydrates, but also of fatty acids and of ketogenic amino acids (lysine and leucine, as well as of parts of the carbon skeleton of isoleucine, phenylalanine, tyrosine, tryptophan and threonine). For details, see the respective pages.

Acetyl-CoA may enter the citrate cycle (3.10) for degradation, but it can also be the origin of synthesis of fatty acids (6.1) and of cholesterol (7.1).

Figure 3.3-1. Reactions of Pyruvate and Phosphoenolpyruvate

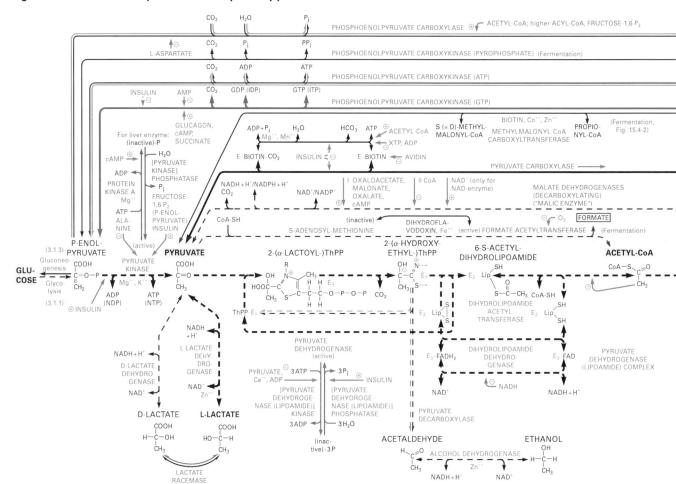

3.3.4 Anaplerotic Reactions

If members of the citrate cycle are used for biosyntheses, insufficient oxaloacetate is available for the reaction with acetyl-CoA. Anaplerotic ('filling up') reactions are required. The carboxylation reactions starting from pyruvate require an energy source, while this is not necessary in conversions of the 'energy-rich' phosphoenolpyruvate.

- Pyruvate carboxylase reaction (in *liver* and *kidney*):

    Pyruvate + ATP + HCO_3^- = oxaloacetate + ADP + P_i.

    The enzyme uses biotin (9.8) as a prosthetic group, which is attached by its valerate side chain to an ε-amino group of lysine, forming a mobile 'arm'. The enzyme is strongly activated allosterically by acetyl-CoA (for restarting the citrate cycle) and inhibited by high levels of nucleoside triphosphates (indicating sufficient energy supply) or insulin.

- 'Malic enzyme' reaction (frequent in *eukarya* and *prokarya*):

    Pyruvate + NADPH + CO_2 = malate + $NADP^+$.

    The energy for carboxylation is provided by NADPH oxidation. The reverse reaction in the *cytosol* is used to supply NADPH (Fig. 6.1-1).

- Phosphoenolpyruvate carboxykinase reaction (in *heart* and *skeletal muscle*):

    Phosphoenolpyruvate + CO_2 + GDP = oxaloacetate + GTP.

    The reverse reaction is used in gluconeogenesis (3.3.5). In *plants*, the guanosine nucleotides are replaced by adenosine nucleotides. *Bacteria* use similar reactions (with phosphate, see Fig. 3.3-1) mainly for oxaloacetate formation during fermentations (15.4).

- Phosphoenolpyruvate carboxylase reaction (in *higher plants, yeast, bacteria*). This enzyme is also part of the CO_2 pumping mechanism (16.2.2):

    Phosphoenolpyruvate + CO_2 + H_2O = oxaloacetate + P_i.

- In *bacterial* fermentations, pyruvate carboxylation can also be achieved by CO_2 transfer from other compounds via carboxyltransferases (15.4).

3.3.5 Initiation of Gluconeogenesis

The pyruvate kinase reaction is highly exergonic in the direction from phosphoenolpyruvate (PEP) to pyruvate ($\Delta G_0' = -23$ kJ/mol) and under *in-vivo* conditions irreversible. Therefore, pyruvate cannot be used directly for gluconeogenesis. The pyruvate carboxylase reaction, energized by ATP hydrolysis initiates a bypass reaction and forms the 'energy-rich' compound oxaloacetate. In the PEP carboxykinase reaction, it is decarboxylated and accepts concomitantly a phosphate group from GTP (in *animals*) or ATP (in *plants*), yielding PEP. Insulin inhibits, glucagon activates both *animal* enzymes. The standard $\Delta G_0'$ of the overall reaction is 0.9 kJ/mol, but under physiological conditions ΔG_{phys} amounts to ca. –25 kJ/mol, making it irreversible.

The carboxylation of pyruvate takes place only in *mitochondria*, while the conversion to PEP can occur either in *mitochondria* or in the *cytosol* (species dependent, in *humans* in both compartments). The further steps of gluconeogenesis (3.1.3) generally take place in the *cytosol*. Therefore, either oxaloacetate or PEP have to leave the mitochondria in order to enter this pathway. While PEP can be transported across the mitochondrial membrane, in *animal tissues* oxaloacetate has to be exported via the malate shuttle (Fig. 3.3-3) or alternatively via the aspartate shuttle (Fig. 16.1-2). The selected route depends on the cytosolic NADH requirements.

Figure 3.3-3. Transfer of Compounds Through the *Mitochondrial* Membrane by the Malate Shuttle

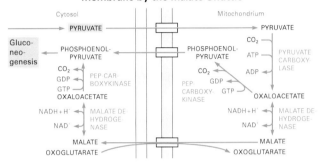

3.3.6 Alcoholic Fermentation

Under anaerobic conditions, *yeast* and a number of *bacteria*, but also some *higher plants* convert pyruvate to acetaldehyde by the action of pyruvate decarboxylase (Fig. 15.4-2). The first step is identical with the E_1 reaction of pyruvate dehydrogenase (3.3.1). Instead of transferring the acyl group of hydroxyethyl-ThPP to lipoamide, however, the group is eliminated. (In Figure 3.3-4, the actually existing ionized forms are drawn, while in the main figure the non-ionized forms are shown.) Acetaldehyde is afterwards reduced to ethanol, thus regenerating NAD^+. This is similar to lactate formation in *animals* under anaerobic conditions.

In *other bacterial species*, different mechanisms exist for fermentative pyruvate turnover (e.g. pyruvate: ferredoxin oxidoreductase or pyruvate formate-lyase; the acetyl-CoA formed is converted via acetaldehyde to ethanol, 15.4).

Literature:
Attwood, P.V.: Int. J. Biochem. Cell Biol. 27 (1995) 231–249.
Mallevi, A. *et al.*: Curr. Opin. Struct. Biol. 2 (1992) 877–887.
Patel, M.S., Roche, T.E.: FASEB J. 4 (1990) 3224–3233.

Figure 3.3-4. Reaction Mechanisms of Pyruvate Dehydrogenase and Decarboxylase

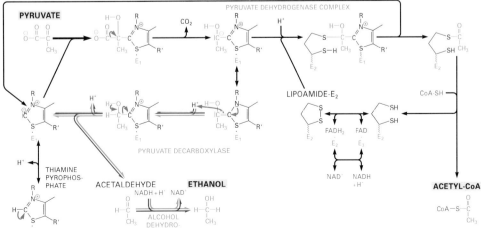

3.4 Di- and Oligosaccharides

3.4.1. Sucrose (Fig. 3.4-1)

Besides starch, sucrose is an important product of photosynthesis in *plants* (16.2.2). The primarily formed triosephosphates are exported from the *chloroplasts* to the *cytosol*, converted to UDP-glucose and fructose 6-P and condensed to sucrose 6-P by sucrose-P synthase (Fig. 16.2-8). The phosphate is then removed by sucrose-P phosphatase.

The regulation of sucrose synthesis takes place at the fructose 1,6-bisphosphatase step by the concentration of the inhibitory fructose 2,6-P_2 (compare gluconeogenesis, 3.1.3). An additional regulation point is the sucrose-P synthase, which is activated by the substrate precursor glucose 1-P and inhibited by the product P_i. Additionally, this enzyme activity is decreased by phosphorylation and enhanced by dephosphorylation (similarly to glycogen synthase, Fig. 3.2-5). The sucrose concentration, in turn, regulates the starch synthesis (3.2.2).

Sucrose is a transport form of carbohydrates, as well as the precursor of starch (in cells distant from the site of photosynthesis) and of cellulose. The nucleotide sugars, which are necessary for their synthesis, are formed by the reversible enzyme sucrose synthase:

Sucrose + UDP = UDP-D-glucose + fructose.

Other synthesis reactions take place by direct transglycosylations, e.g. the formation of fructans (in *plants*, 3.2.2) and of dextrans (in *bacteria* and *yeast*). Synthesis of the raffinose family starts with an isomerization of UDP-D-glucose to UDP-D-galactose, its condensation with *myo*-inositol to galactinol, followed by a transglycosylation reaction with sucrose.

Sucrose in food is cleaved in the *intestine* by α-glucosidase or by β-fructofuranosidase (invertase).

3.4.2 Lactose (Fig. 3.4-2)

Lactose synthesis takes place in the *mammary gland* of *mammals*. UDP-D-glucose is epimerized to UDP-D-galactose. Then, it is condensed with glucose.

The enzyme lactose synthase consists of the subunits galactosyl transferase (which preferably condenses UDP-D-galactose with N-acetylglucosamine, e.g. in synthesis of complex glycoproteins, 13.4) and α-lactalbumin (which changes the specificity, so that the transferase accepts glucose). During pregnancy, the transferase biosynthesis is induced by insulin, cortisol and prolactin, while the lactalbumin biosynthesis is inhibited by progesterone. Shortly before birth, the progesterone concentration decreases and this inhibition ceases. Lactose synthesis starts.

For catabolism, lactose is hydrolyzed in the *intestine* by β-galactosidase (lactase). After resorption, it is phosphorylated in the *liver* to galactose 1-phosphate. Galactose 1-phosphate reacts with UDP-D-glucose to yield glucose 1-phosphate and UDP-D-galactose. This reaction is, catalyzed by UDP-D-glucose-hexose-1-P uridylyltransferase. UDP-D-galactose is epimerized afterwards to UDP-D-glucose.

While in *infants*, β-galactosidase is generally present, it exists only at a low level in *adult* black and oriental population (lactose intolerance). In hereditary galactosemia, the uridylyltransferase has low activity in *liver* and *erythrocytes*, causing an increase of galactose 1-P. This compound inhibits phosphoglucomutase, glucose 6-phosphatase and glucose-6-P dehydrogenase, causing serious disturbances in glucose metabolism.

In *bacteria*, β-galactosidase is an inducible enzyme (for regulation see 10.5.1).

3.4.3 Other Glycosides

Many other α- and β-saccharides are synthesized by reaction of nucleotide sugars with other sugars or aglycons. A large number has been found in *plants*, *yeast* etc. Also, the synthesis of homo- or heteropolymeric glycosides usually starts from nucleotide sugars (e.g. 3.2, 13.3, 13.4).

Literature:
Heldt, H.W.: *Plant Biochemistry and Molecular Biology*. Oxford University Press (1998).
Huber, S.C. et al.: Int. Reviews of Cytology 149 (1994) 47–98.
Kretchmer, M.: Scientific American 227 (4) (1972) 74–78.
Smith, H.S. et al.: Plant Physiology 107 (1995) 673–677.

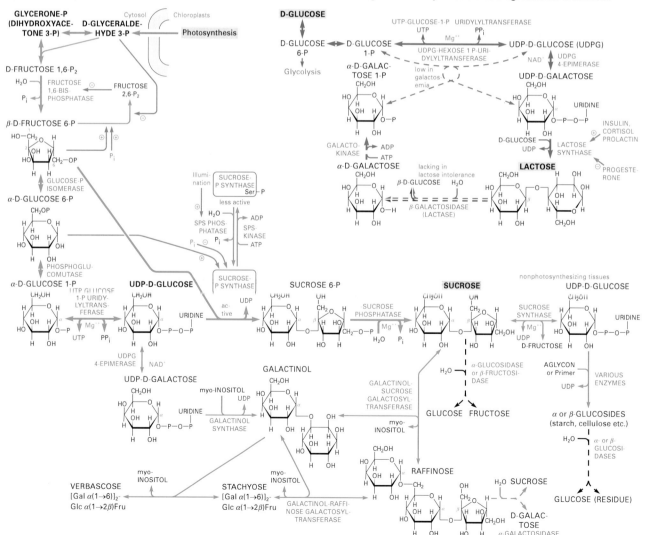

Figure 3.4-1. Synthesis and Metabolism of Sucrose

Figure 3.4-2. Synthesis and Degradation of Lactose

3.5 Metabolism of Hexose Derivatives

3.5.1 Uronic acids (Fig. 3.5-1/next page)

Uronic acids are derivatives of hexoses, in which the hydroxyl group at C-6 is oxidized to a carboxyl group. This primary oxidation takes place with UDP-glucose or GDP-mannose (not with the free sugar). By 4' or 5' epimerizations (1.2.1), UDP-D-glucuronate is converted into UDP-D-galacturonate or into UDP-L-iduronate, respectively. Uronic acids (as well as aldonic acids, 3.5.2) and their derivatives have a strong tendency to form internal esters (lactones).

UDP-glucuronate reacts with many aglycons (primary amines, alcohols, carbonic acids) to glucuronides. Analogous reactions take place with other UDP-uronates. In *animals*, glucuronidation is of importance for excretion in urine. Uronates are important components of proteoglycans (13.1.2). Alginate (the polymerization product of D-mannuronate and L-guluronate) occurs in the *cell walls* of *brown algae* and is used in food industry.

Additional oxidation of D-glucuronate at C-1 to the carboxyl level yields the dicarboxylic D-glucaric acid.

3.5.2 Aldonic Acids (Fig. 3.5-1)

If the oxidation of D-glucose to the carboxylic function occurs at C-1, D-gluconate results. This reaction may take place either with free glucose or with its 6-phosphate derivative. The product of the latter reaction, 6-P-gluconate, may enter the pentose phosphate cycle (3.6.1). In *bacteria*, it is an intermediate of the Entner-Doudoroff pathway (3.5.3).

Another pathway leading to aldonic acids is the reduction of the C-1 hemiacetal group of D-glucuronate to the hydroxyl function, which yields L-gulonate. [The change to the L-form is not a conversion to an enantiomer (1.2.1), but the conventional numbering of this compound starts at the opposite end.] Oxidation of L-gulonate or its lactone results in 3-dehydro-L-gulonate or ascorbate (vitamin C, 9.10), respectively.

3.5.3 Entner-Doudoroff Pathway (Fig. 3.5-1, see also 15.4)

This pathway [also named 2-dehydro-(or keto-)3-deoxy-6-P-gluconate pathway] is frequently used by *bacteria* (e.g. *Zymomonas*) for degradation of gluconate and glucose. After phosphorylation, 6-P-gluconate is dehydrated to 2-dehydro-3-deoxy-6-P-gluconate and cleaved into pyruvate and glyceraldehyde 3-P (which is then also converted to pyruvate); further reactions lead to ethanol.

Via several intermediate steps, glucuronate and galacturonate can also yield 2-dehydro-3-deoxy-6-P-gluconate and are degraded the same way afterwards.

3.5.4 Inositol (Fig. 3.5-1)

myo-Inositol is formed from glucose 1-phosphate in a NAD^+-catalyzed oxidation/reduction reaction. It is a cyclic alcohol with 6 hydroxyl groups, one of many stereoisomers. In the phosphorylated form, it plays an important role in intracellular signal transfer (17.4.4) and is also present in phospholipids (6.3) and in glycolipid anchors of proteins (13.3.4). In *plants*, inositol phosphates are present in large quantities; part of them may have a storage function. For degradation, *myo*-inositol can be oxidized to glucuronate.

3.5.5 Hexitols

Many aldoses and ketoses can be reduced by NAD^+ or $NADP^+$ dependent reactions to the corresponding linear alcohols. Hexitols (C_6, e.g. sorbitol, Fig. 3.1-1), as well as pentitols (C_5, 3.6) frequently occur in *plants*. In human nutrition, they are used as food additives due to their sweet taste; their metabolism in *humans* starts with reconversion to the respective sugars.

3.5.6 Mannose and Deoxy Hexoses (Fig. 3.5-2)

Isomerization of fructose 6-P by the respective enzymes results in either glucose 6-P (3.1.1) or in the other epimer, mannose 6-P. Conversion to mannose 1-P and further on to GDP-mannose yields the activated sugar as a precursor of glycoproteins and glycolipids (13.3; 13.4) as well as of mannuronates (e.g. alginate, 3.5.1).

The biosyntheses of L-rhamnose (6-deoxy-L-mannose) and L-fucose (6-deoxy-L-galactose) proceed via dehydration, epimerization and consecutive reduction of dTDP-glucose and of GDP-mannose, respectively (Fig. 3.5-2). L-Rhamnose combines with polymeric galacturonate (pectate) to form rhamnogalacturonan, which is essential for the formation of *primary plant cell walls* (3.6.3). L-Rhamnose is also present in *bacterial cell walls* (15.1). L-Fucose is an important component of many glycoproteins (13.3 and 13.4).

Literature:
Ruoslahti, E.: Ann. Rev. of Cell Biol. 4 (1988) 229–255.
Varner, J.E., Lin, L.S.: Cell 56 (1989) 231–239.
(General): Textbooks of organic chemistry.

Figure 3.5-2. Mannose and Deoxy Hexoses

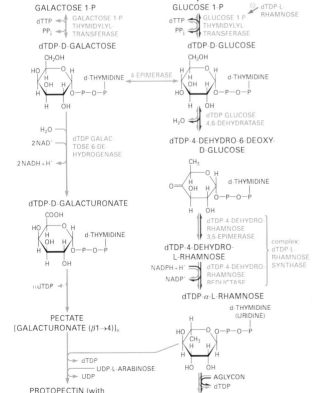
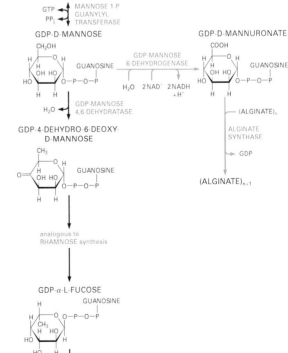

Figure 3.5-1. Acidic Hexose Derivatives and Inositol

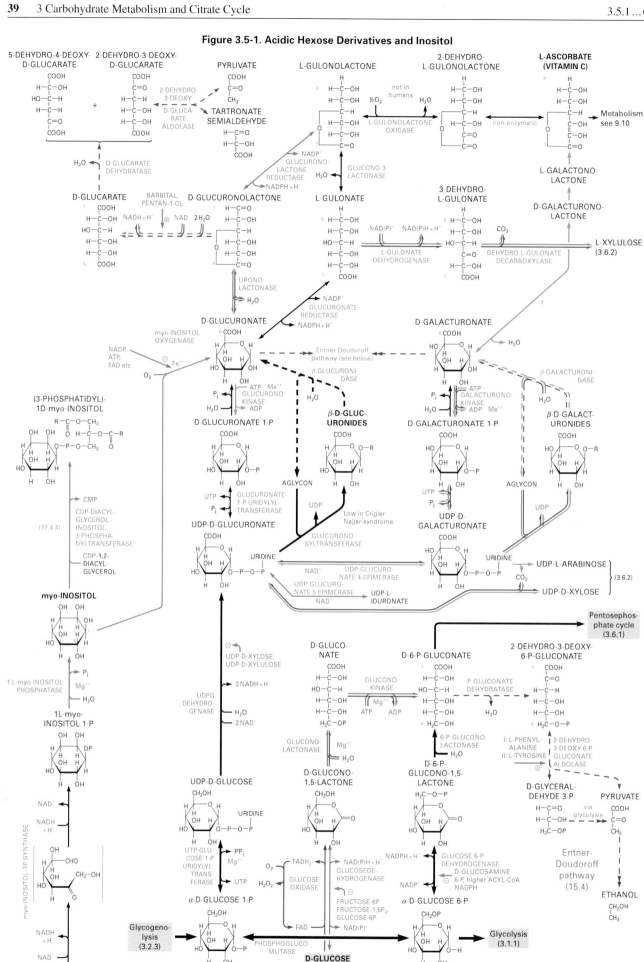

3.6 Pentose Metabolism

Pentoses are essential parts of nucleic acids and nucleotides, glycoproteins, plant cell walls etc. They are usually generated via hexose intermediates. The formation of deoxyribose (present in DNA) occurs by reduction of ribonucleotides and is described there (8.1.4).

3.6.1 Pentose Phosphate Cycle (Fig. 3.6-1)

The pentose phosphate cycle is a pathway of glucose turnover alternative to glycolysis and occurs in most species. Its major function is the production of reducing equivalents (in form of NADPH) and of pentoses and tetroses for biosynthetic reactions (nucleoside and amino acid syntheses) in variable ratios (see below).

The enzymes are present in the *cytosol*. In *humans*, major activities are found in *liver, adipose tissue, lactating mammary glands, adrenal cortex, testis, thyroid gland* and *erythrocytes*.

Glucose 6-P is converted into ribulose 5-P by two dehydrogenase reactions (yielding 2 NADPH) and a decarboxylation step. The initial glucose-6-P dehydrogenase is the regulated enzyme. Its activity depends on the $NADP^+$ concentration, NADPH inhibits. Isomerization and epimerization reactions of ribulose 5-P yield ribose 5-P and xylulose 5-P, respectively.

Ribose 5-P can be used for biosynthetic purposes. Otherwise, C_2-units ('active glycolaldehyde') are moved by transketolase (TK) and a C_3-unit by transaldolase (TA), resulting in C_4 and C_7 intermediates and finally in fructose 6-P (glucose 6-P) and glyceraldehyde 3-P. The latter compound can either enter glycolysis or be reconverted into glucose 6-P. The stoichiometry for this reconversion is

6 Glucose 6-P + 12 $NADP^+$ + 7 H_2O =
  12 NADPH + 12 H^+ + 6 CO_2 + 5 glucose 6-P + P_i.

Thus, the pentose cycle can meet different requirements of metabolism. If there is excessive demand for pentoses, the TK and TA reactions of the pentose cycle can run backwards.

The TK reaction requires thiamine pyrophosphate (ThPP) as a coenzyme. It reacts with the keto moiety of the substrate (xylulose 5-P or sedoheptulose 7-P) similarly to the mechanism of the pyruvate decarboxylase reaction (Fig. 3.3-4). Following cleavage, the remaining activated intermediate (in this case the 1,2-dihydroxyethyl residue) is transferred. For details, see 9.2.2.

TA does not require a coenzyme. It starts the reaction by forming a Schiff base between a ε-lysine group of the enzyme with the keto moiety of the substrate (sedoheptulose 7-P). This leads to an aldol cleavage, releasing erythrose 4-P. The remaining activated residue is then transferred to the acceptor glyceraldehyde 3-P, resulting in fructose 1,6-P_2. In the otherwise analogous fructose-bisphosphate aldolase reaction (3.1.1) the activated residue (dihydroxyacetone-P) is released after protonation (Fig. 3.6-2).

Figure 3.6-2. Aldol Cleavage Reactions of Carbohydrates

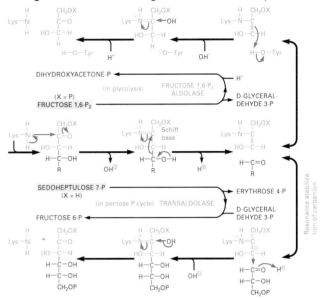

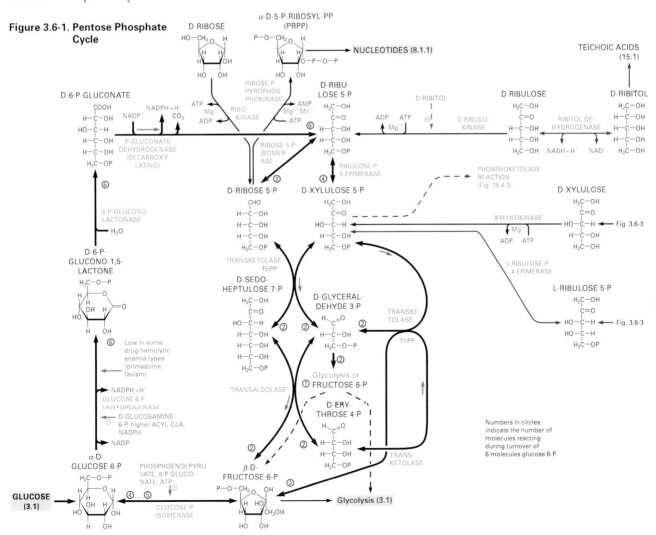

Figure 3.6-1. Pentose Phosphate Cycle

Numbers in circles indicate the number of molecules reacting during turnover of 6 molecules glucose 6-P

In *humans,* deficiency of glucose-6-P dehydrogenase (and thus of NADPH which is essential for glutathione regeneration, 4.5.7) causes hemolytic anemia after administration of some drugs (e.g. primaquine), which produce elevated H_2O_2 levels.

In *bacteria,* the phosphoketolase cleavage of xylulose 5-P leads to lactate and ethanol (Fig. 15.4-3). The reduction of D-ribulose yields ribitol, which is an essential component of teichoic acids (15.1).

3.6.2 Other Decarboxylation Reactions (Fig. 3.6-3)

Pentoses can be also formed from UDP-derivatives of uronic acids (UDP-glucuronate, UDP-galacturonate) by decarboxylation. This pathway is especially prominent in *plants*, the products (e.g. L-arabinose, D-xylose) occur in their *cell walls* (3.6.3).

The oxidation products of L-gulonate and its lactone (3-dehydro-L-gulonate and of L-ascorbate = vitamin C, respectively) yield upon decarboxylation C_5 compounds (L-xylulose, L-xylonate and L-lyxonate) or are metabolized differently (9.10).

3.6.3. Plant Cell Walls

Primary *plant* cell walls show a wide variation of composition. In a typical example (*Sycamore maple*), they consist of roughly similar quantities of rhamnogalacturonan (3.5.6), arabinogalactan, xyloglucan, cellulose (3.2.2) and glycoproteins (13.1). In *mosses* and in the animal kingdom (*arthropods*), cellulose is replaced by chitin (linear polymer of N-acetylglucosamine, 3.7.1).

Cell wall synthesis (Reactions are schematically shown in Fig. 3.6-3, structures in Figs. 3.6-4 and 3.6-5): After cell division, the synthesis of primary cell walls starts at a middle lamella consisting of protopectin (rhamnogalacturonan, Fig. 3.5-2). This structure contains many negative charges due to its galacturonic acid content. It binds Ca^{++} and Mg^{++} and is highly hydrated. Hemicelluloses (xylans, xyloglucan, derivatives of mannose, galactose, fucose etc.) and arabinogalactan are produced by *dictyosomes* (*plant Golgi apparatus*) and become attached by covalent bonds. Glycoproteins (with up to 30% hydroxyproline) are located between the hemicellulose molecules. Then the cellulose fibers are synthesized by enzymes located at the plasma membrane (3.2.2). The single straight β-glucose chains of cellulose are combined by hydrogen bonds in overlapping fashion, forming a fiber bundle of about 70...100 neighbouring chains with a partially crystalline structure. This bundle becomes attached to the hemicelluloses by many hydrogen bonds and provide the tensile strength. Plasmodesmata (plasma-membrane lined channels of about 5 nm diameter) cross the cell wall and allow movement of fluids and metabolites (12.3).

The secondary cell wall is formed from the primary one by thickening, by deposition of additional cellulose layers and of lignin (4.7.4), which fills the spaces between the fibers (analogously to reinforced concrete).

Degradation of wood: The complex structure requires many enzymes. They are mostly of *bacterial* or *fungal* origin, among them cellulase (EC 3.2.1.4) for cellulose; polygalacturonase (EC 3.2.1.15) and α- and β-rhamnosidases (EC 3.2.1.40 and 43) for pectins; arabinogalactan-endo-galactosidases (EC 3.2.1.89 and 90) and arabinan-endo-arabinosidase (EC 3.2.1.99) for arabinosides; xylan xylosidase (EC 3.2.1.37 and 72) for xylans. A lignin degrading enzyme is lignostilbene α,β-dioxygenase (EC 1.13.11.43).

3.6.4 Pentose Metabolism in Humans

Dietary L-arabinose in *humans* is metabolized by *intestinal bacteria* via L-ribulose to D-xylulose 5-phosphate, which is part of the pentose phosphate cycle (3.6.1) D-xylose is also converted to D-xylulose 5-phosphate. High pentose content of food, as well as a low inherited enzyme activity lead to pentosurias.

Literature:

Heldt, H.W.: *Plant Biochemistry and Molecular Biology.* Oxford University Press (1998).
Lewis, N.G., Yamamoto, L.: Ann. Rev. of Plant Physiol. and Plant Mol. Biol. 41 (1990) 455–496.
Raven, P.H. et al.: *Biology of plants.* 5th ed. Worth (1992).

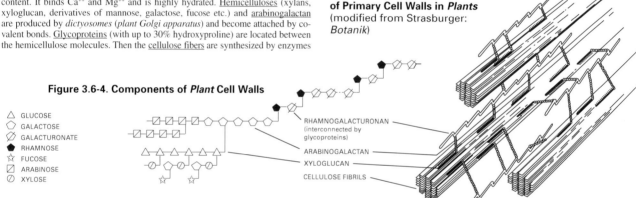

Figure 3.6-5. A Structure Model of Primary Cell Walls in *Plants* (modified from Strasburger: *Botanik*)

Figure 3.6-4. Components of *Plant* Cell Walls

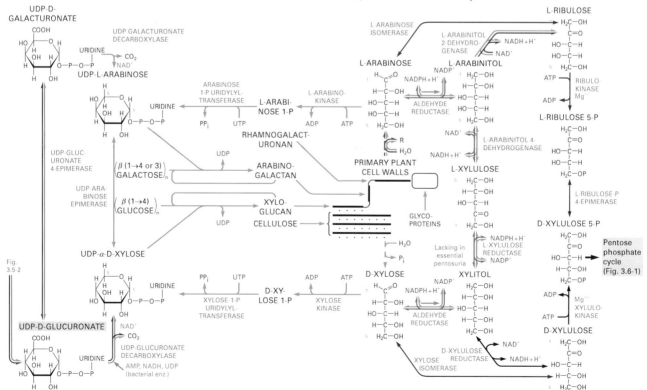

Figure 3.6-3. Formation of C_5 Compounds by Decarboxylation and Cell Wall Synthesis in *Plants*

3.7 Amino Sugars

Amino sugars are sugar derivatives, in which a hydroxyl group has been replaced by an amino group. Frequently, they are acetylated. Other derivatives are, e.g., N-acetylneuraminic acid (sialic acid) and N-acetylmuramic acid. They play a great role in glycoproteins and glycolipids, as components of cell walls and of exoskeletons (see below).

3.7.1 Biosynthesis (Fig. 3.7-1)

The biosynthesis of amino sugars starts from D-fructose 6-P by a transamination reaction with glutamine, yielding D-glucosamine 6-P. Apparently, a Schiff base is the intermediate. Thereafter, acetylation by acetyl-CoA takes place, followed by a mutase reaction to yield N-acetylglucosamine 1-P and by the formation of the UDP-derivative. UDP-N-acetyl-D-glucosamine is the substrate for epimerization reactions at the 4-position (yielding UDP-N-acetylgalactosamine) or at the 2-position (yielding UDP-N-acetyl-mannosamine, which, however, is immediately hydrolyzed to N-acetyl-D-mannosamine). Acetylation reactions can also take place with unphosphorylated amino sugars.

Both UDP-N-acetylglucosamine and UDP-N-acetylgalactosamine take part in the synthesis of many glycoproteins and glycolipids in *animals* and *plants* (13.3; 13.4).

The linear polymerization of UDP-N-acetylglucosamine by chitin synthase yields chitin [($\beta 1 \rightarrow 4$-glucosamine)$_n$], which is structurally very similar to cellulose (3.2.2). The interaction of neighboring molecules and therefore the strength of the structure even surpasses cellulose. Chitin is a major component of the exoskeleton of *invertebrates* (*crustaceans*, *insects*), it also occurs in the cell walls of many *fungi* and *algae*. The chitin fibers provide strength to the elastic framework of glycoproteins or glycans. Deacetylation of chitin yields chitosan, which also occurs in cell walls of *fungi*.

The condensation of UDP-N-acetylglucosamine with phosphoenolpyruvate and the consecutive reduction results in N-acetylmuramate, the starting compound for the formation of murein (15.1), which is an essential component of *bacterial* cell envelopes.

The synthesis takes place at undecaprenyl anchors by adding amino acids (part of them in the D-configuration, which renders them resistant to common proteases) and N-acetylglucosamine. Both amino sugar derivatives alternate in the backbone chain, while the amino acids interconnect these chains (15.1).

Either N-acetyl-D-mannosamine or its 6-phosphate can be condensed with phosphoenolpyruvate to yield N-acetylneuraminate (sialate) or its 6-phosphate, respectively, which is thereafter dephosphorylated. The ring form of this compound includes the 3 added C-atoms. Sialate is an essential component of gangliosides (13.2.1). For introduction into these compounds, it is activated as CMP-derivative.

3.7.2 Catabolism

Murein is split by lysozyme between the N-acetylmuramate and N-acetylglucosamine residues, yielding disaccharides. Lysozyme is present in *tears*, *sneeze mucus* and other body secretions as well as in *bird eggs* (protective function).

The degradation of chitin can take place by chitinase, an *endo*($\beta 1 \rightarrow 4$) glucosaminidase, which performs random hydrolysis, but also by lysozyme. Remaining chitobiose is further cleaved by β-N-acyl-hexosaminidase.

Terminal neuraminic acid in glycoproteins and glycolipids is removed by neuraminidase (13.1.1). This is frequently the first step of degradation.

Glucosamine 6-P can be reconverted into fructose 6-P by an isomerase reaction, NH_3 is liberated.

Literature:
Kumagai, I. *et al.*: J. Biol. Chem. 267 (1992) 4608–4612.
Muzzarelli, R.A.: *Chitin*. Pergamon Press (1977).

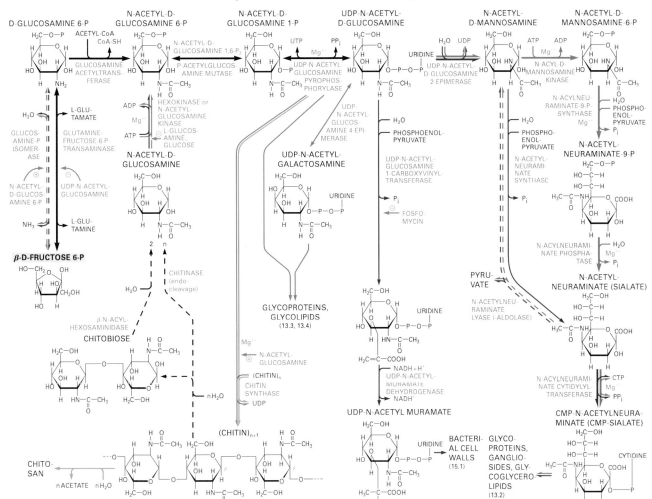

Figure 3.7-1. Metabolism of Amino Sugars

3.8 Citrate Cycle

The citrate cycle (Krebs cycle, tricarboxylic acid cycle) plays the central role in metabolism both of *eukaryotes* and most *prokaryotes* (Fig. 3.8-1). It is the major site of oxidation of carbon chains from carbohydrates, fatty acids (both entering via acetyl-CoA) and many amino acids to CO_2 and water. It supplies NADH as essential substrate for the oxidative phosphorylation in the respiratory chain (16.1) and thus has a major role in energy metabolism. It also provides intermediates for the synthesis of amino acids (4.2.2, 4.2.4) and porphyrins (5.1). For the anaplerotic reactions, which are required to replace cycle compounds removed for biosynthesis, see 3.3.4.

The cycle operates only under aerobic conditions, since it requires the quick reconstitution of NAD^+ from NADH by the respiratory chain (16.1.2).

Acetyl-CoA + 3 NAD^+ + FAD + GDP (or ADP) + P_i + 2 H_2O =
2 CO_2 + 3 NADH + $FADH_2$ + GTP (or ATP) + 2 H^+ + CoA
$\Delta G_0' = -41$ kJ/mol

In *green sulfur bacteria* (*Chlorobiaceae*), some other *bacteria* and *archaea*, the 'reductive citrate cycle' is operative, which yields acetyl-CoA by autotrophic CO_2 fixation under anaerobic conditions. It resembles the citrate cycle running backwards. It is speculated that inorganic catalysis of analogous reactions might have played a role in origin of life on earth.

mate). The 2-oxoglutarate dehydrogenase enzyme complex is very similar in its structure to the pyruvate dehydrogenase complex (3.3.1) and uses the same coenzymes. However, no regulation of its activity by phosphorylation and dephosphorylation takes place. Succinyl-CoA ligase (-synthetase) uses the energy of succinyl-CoA hydrolysis for the formation of a high-energy phosphate bond (GTP in *mammals*, ATP in *plants* and *bacteria*). The membrane-bound succinate dehydrogenase is a member of the electron transport chain and feeds protons and electrons into the quinone pool (16.1.2, 16.1.3). The resulting fumarate is hydrated and oxidized, yielding oxaloacetate, whose condensation with acetyl CoA yields citrate and closes the cycle.

The interconversion of malate and oxaloacetate is also used for the initiation of gluconeogenesis to circumvent the irreversible pyruvate kinase reaction (3.3.5). It is a means for transport through membranes (Fig. 3.3-3, Fig. 6.1-1, Fig. 16.1-2, Fig. 16.2-10).

3.8.2 Regulatory Mechanisms (Fig. 3.8-3)

The citrate cycle in *muscle* is regulated in a relatively simple way by substrates and products. The regulated enzymes are citrate synthase, isocitrate dehydrogenase and oxoglutarate dehydrogenase. (For the regulation of pyruvate dehydrogenase, which provides the 'fuel' acetyl-CoA, see 3.3.2.)

Figure 3.8-1. Interrelations of the Citrate Cycle With Other Metabolic Pathways (blue: amino acids, red: carbohydrates, green: lipids)

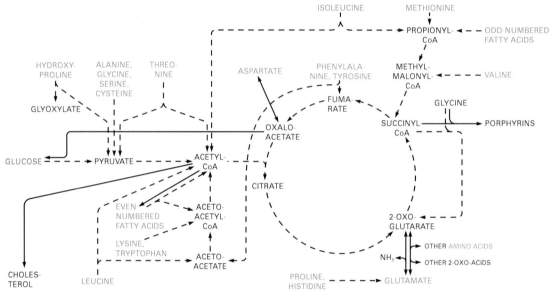

3.8.1 Reaction Sequence (Fig. 3.8-2)

In *eukarya*, the enzymes of the cycle are located in the *mitochondrial matrix*, only succinate dehydrogenase is part of the *inner mitochondrial membrane*. Some of these enzymes also exist in the *cytosol*, fulfilling other tasks, e.g., as members of the urea cycle, 4.9.1.

Citrate is formed by a condensation reaction of acetyl-CoA and oxaloacetate, catalyzed by citrate synthase.

The structure of this enzyme favors the initial condensation reaction to citryl-CoA by an ordered sequential mechanism (1.5.4): an 'open form' binds oxaloacetate, the conformation change to the 'closed' form generates the acetyl-CoA binding site. After condensation, hydrolytic removal of coenzyme A takes place.

This reaction also plays a role in the export of acetyl-CoA to the *cytosol* (e.g. for fatty acid synthesis, Fig. 6.1-1). Since acetyl-CoA cannot pass the mitochondrial membrane, it is converted by citrate synthase into citrate, which is exported via an antiport carrier and is cleaved in the *cytosol* by the ATP-energized citrate lyase. While the acetyl-CoA is metabolized, oxaloacetate is reduced to malate or is converted into pyruvate, which returns to the *mitochondrion* by a carrier.

Another citrate lyase reaction occurs in citrate fermentation (Fig. 15.4-3).

Cis-aconitate and oxalosuccinate are labile intermediates of the aconitate hydratase (aconitase) and the isocitrate dehydrogenase reactions, respectively. They remain bound to the enzymes. The NAD^+-dependent *mitochondrial* form of isocitrate dehydrogenase is a member of the citrate cycle (in *mammals*), while the $NADP^+$-dependent form also occurs in *cytosol* and is responsible for NADPH and oxoglutarate supply (e.g. for conversion to gluta-

As described for glycolysis (3.1.2), the ΔG_{phys} of the regulated enzymes is strongly negative. Major mechanisms are the inhibitions by NADH (control by the redox state) and by ATP, the activations by ADP (control by the energy state of the cell) and by Ca^{++} (in muscles, as a result of muscular activity). Some feedback inhibitions by products also exist. Since usually the *in vivo* concentrations of acetyl-CoA and oxaloacetate do not saturate the citrate synthase, this enzyme is additionally regulated by the availability of the substrates.

The quantity of circulating compounds changes, when compounds are removed for biosynthesis or when citric cycle intermediates are generated by catabolic reactions (e.g. by transaminations of glutamate or aspartate (4.2.2, 4.2.4), formation of succinyl-CoA by degradation of odd-numbered fatty acids, 6.1.5 or of branched-chain amino acids, 4.6.2). This is taken care of by control of anaplerotic reactions (3.3.4), prevalently in *liver*.

In *bacteria* (*E. coli*), the isocitrate dehydrogenase becomes inactivated by phosphorylation of the active site serine-113.

3.8.3 Energy Balance

The chemical combustion of pyruvate yields $\Delta G_0' = -1145$ kJ/mol, while the conversion of pyruvate to acetyl-CoA ($\Delta G_0' = -34$ kJ/mol) and the oxidation of acetyl-CoA in the citrate cycle (see above) combined, decrease the $\Delta G_0'$ only by -75 kJ/mol. Thus, the major part of the free energy of the oxidation reaction (ca. 94%) is stored in NADH, $FADH_2$ and GTP (or ATP). 4 NADH are formed in the pyruvate-, isocitrate-, 2-oxoglutarate- and malate dehydrogenase reactions, 1 $FADH_2$ in the succinate dehydrogenase reaction, 1 GTP (in *animals*) or ATP (in *plants* or *microorganisms*) in the succinate-CoA ligase (= succinyl-CoA synthetase) reaction.

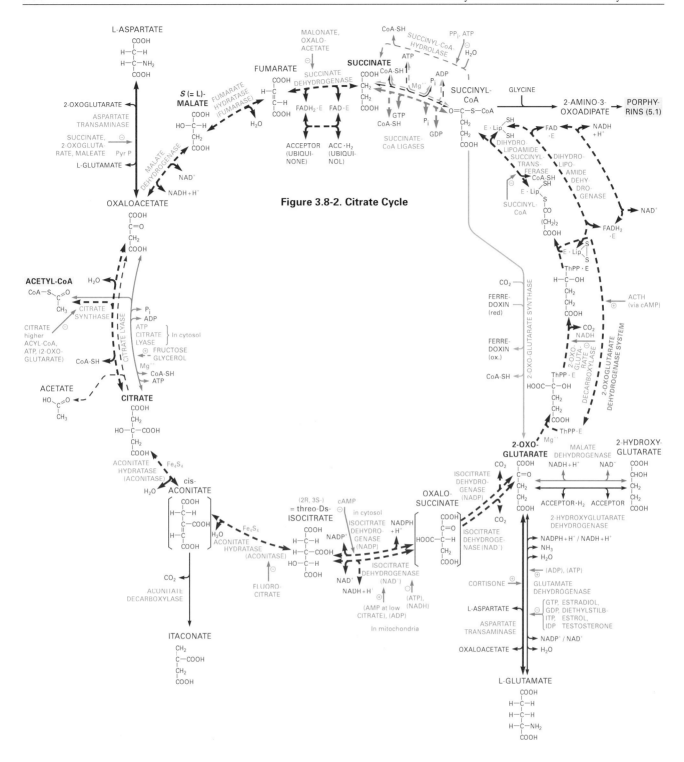

Figure 3.8-2. Citrate Cycle

In the respiratory chain (16.1), additional ATP is produced by oxidation of NADH (2.5 *cytosolic* ATP/1 NADPH) and FADH$_2$ (1.5 *cytosolic* ATP/1 FADH$_2$), yielding a maximum of $4*2.5 + 1.5 + 1 =$ 12,5 mol ATP/mol pyruvate. The free energy conserved in these bonds are $\Delta G_0' = 12{,}5 * 30.5$ kJ $= 381$ kJ, resulting in an energy yield of about 33%. For conversion of glucose, the ATP and NADH yield of glycolysis has to be added (3.1.1).

Frequently in the literature, a maximum yield of 3 ATP/1 NADH and a total yield of 15 mol ATP/mol pyruvate (or 38 mol ATP/mol glucose) are stated. This figure refers to the ATP obtained within the *mitochondria*, while the figures above also consider the energy requirement of ATP export into the *cytosol* (16.1.1). All of these values are only approximate, since in the living cell the actual conditions differ from the standard conditions and also the efficiency of the oxidative phosphorylation varies according to the metabolic situation.

Literature:
Perham, R.N.: Biochemistry 30 (1991) 8501–8512.
Remington, S.J.: Curr. Top. Cell Reg. 33 (1992) 202–229.
Wächtershäuser, G.: Proc. Natl. Acad. Sci (USA) 87 (1990) 200–204.
Walsh, C.: *Enzymatic Reaction Mechanisms*. Freeman (1979).

Figure 3.8-3. Regulation of the Citrate Cycle

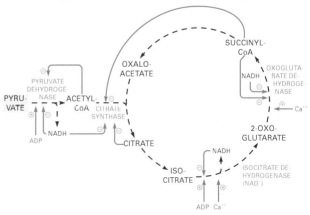

3.9 Glyoxylate Metabolism

3.9.1 Glyoxylate Cycle (Fig. 3.9-1)

The glyoxylate cycle occurs in *plants and bacteria*. The major purpose of this cycle in *plants* is the conversion of acetyl-CoA (from fat degradation) into malate (for gluconeogenesis, 3.3). The specific enzymes of this cycle occur in *glyoxisomes*, a special type of peroxisomes mostly present in tissues, where fat conversion to carbohydrates is important (e.g. in seed during germination).

Acetyl-CoA combines with oxaloacetate and is converted into isocitrate analogously to the citrate cycle reactions. The consecutive cleavage reaction by isocitrate lyase yields succinate and glyoxylate. The latter compound condenses with another acetyl-CoA to malate. This reaction is catalyzed by malate synthase, which resembles citrate synthase. Malate is oxidized to oxaloacetate, which can start another round of the glyoxylate cycle. Succinate, however, passes over to the *mitochondria* and is converted into oxaloacetate (or malate?) by the citric acid cycle enzymes. These compounds leave the *mitochondria* via translocator mechanisms and initiate the gluconeogenesis pathway in the *cytosol* (3.3.5). The overall reaction is

2 Acetyl-CoA + 2 NAD^+ + FAD = oxaloacetate + 2 NADH + $FADH_2$ + 2 H^+

Some authors assume, that malate from the *glyoxisomes* enters the *cytosol*, while the succinate, which is imported into the *mitochondria*, is converted into oxaloacetate, transaminated into aspartate, transported via a shuttle mechanism to the *glyoxisomes* and reconverted into oxaloacetate.

Similar reactions also take place in the *cytoplasm* of *bacteria*. Their purpose is mainly the metabolism of acetate, which is taken up from the medium and converted into acetyl-CoA before entering this pathway. In *Escherichia coli*, isocitrate dehydrogenase (which competes with isocitrate lyase for the substrate in the same compartment) is inactivated by phosphorylation and activated by dephosphorylation of the enzyme (3.8.2). The modifying bifunctional kinase/ phosphatase is allosterically regulated by intermediates of the citrate cycle, which also allosterically regulate the isocitrate lyase in the opposite direction. These mechanisms determine the share of isocitrate being converted by the citrate and glyoxylate cycles.

3.9.2 Other Glyoxylate Reactions

In *photosynthesizing plants*, the oxygenase side activity of rubisco yields 2-phosphoglycolate (Fig. 16.2-8), which after dephosphorylation leaves the *chloroplasts* via a translocator mechanism and enters the *peroxisomes* (Fig. 3.9-2). There it is oxidized to glyoxylate, the H_2O_2 formed is decomposed by catalase. A transamination reaction yields glycine. The complicated condensation, oxidation and partial decarboxylation to serine takes place in *mitochondria* by the multienzyme complex glycine cleavage enzyme and glycine hydroxymethyltransferase (4.4.2). Serine is converted via further steps into phosphoglycerate, which reenters the Calvin cycle (16.2.2). However, this 'photorespiration' detour causes a loss of energy (1 decarboxylation, 1 dephosphorylation).

The oxidation of glyoxylate to oxalate and further on to formate and CO_2 occurs in *plants* (as a side activity of the *peroxisomes*) and in *bacteria* (Fig. 3.9-3).

The location of the complete glyoxylate and H_2O_2 metabolism in *glyoxisomes/peroxisomes* protects the rest of the cell from these highly toxic compounds.

Literature:
Douce, R.: Biochem. Soc. Transactions 22 (1994) 184–188.
Gerhardt, B.: Progress Lipid Res. 31 (1992) 417–446.
Husic, D.W. *et al.*: CRC Crit. Revs. in Plant Sciences Vol. 1 (1987) 45–100.
Holms, W.H.: Curr. Top. Cell. Regul. 28 (1986) 69–106.
Moore, T.S.: Lipid Metabolism in Plants. CRC Press (1993).

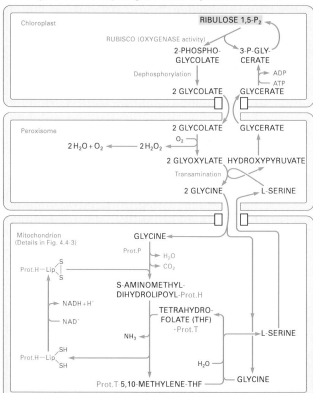

Figure 3.9-2. Recycling of Photorespiration Products

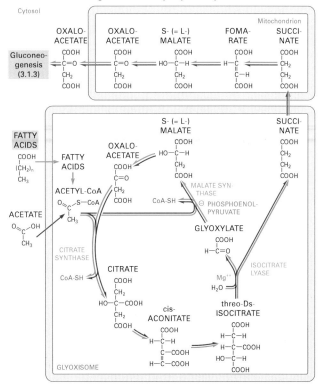

Figure 3.9-1. Glyoxylate Cycle

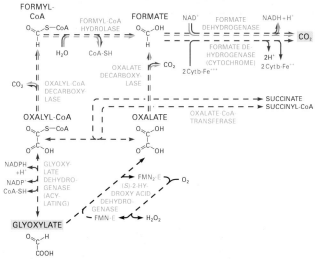

Figure 3.9-3. Oxidation of Glyoxylate

4 Amino Acids and Derivatives

4.1 Nitrogen Fixation and Metabolism

Most of the nitrogen in the biosphere is bound in amino acids and nucleotides (which are formed from amino acids). Since only a small number of *bacteria* are able to convert atmospheric nitrogen into utilizable compounds (primarily NH_3/NH_4^+), especially *plants* had to adapt themselves to a limited nitrogen supply.

Nitrogen fixation is performed by *bacteria* living as symbionts with *plants* (e.g. *Rhizobium*) as well as by some free-living *bacteria* (e.g. *Azotobacter, Cyanobacteria*). About $7 ... 11 * 10^7$ t/year are converted in this way (as compared with $3 * 10^7$ t/year by industrial means). Cleavage of the excessively stable triple bond of N_2 ($\Delta G_0' = 945$ kJ/mol) by nitrogenase requires a high energy input in the form of ATP and reducing equivalents:

$N_2 + 8\ H^+ + 8\ e^- + 16\ ATP + 16\ H_2O = 2\ NH_3 + H_2 + 16\ ADP + 16\ P_i$.

Nitrogenase is composed of 2 parts:

- a homodimeric Fe protein ($2 * 32$ kDa) with 2 ATP binding sites and a Fe_4S_4 cluster located between both subunits (dinitrogenase-reductase)
- a MoFe protein [220 kDa, $(\alpha\beta)_2$, actual dinitrogenase]. Each $\alpha\beta$ heterodimer contains two Fe_4S_4 'P'-clusters linked with each other (located between the α and β subunits) and a Fe_4S_3-$MoFe_3S_3$ complex of cubane structure linked by 3 S bridges (located in a cavity of the α subunit). In some organisms, Mo can be replaced by V or Fe.

An electron liberated by oxidative or photosynthetic reactions (16.2.1) is transferred via ferredoxin to the Fe-protein and reduces it. (In some species, the reductant ferredoxin is replaced by flavodoxin. After a conformation change, which is energized by the hydrolysis of 2 ATP, the electron passes on to the Mo-Fe protein. After 8 such rounds, the Fe-Mo complex of the MoFe protein is reduced. It is then able to reduce N_2 to $2\ NH_3$ (Fig. 4.1-1) and in a side reaction $2\ H^+$ to H_2, which then partially counteracts the first step of N_2 reduction.

Figure 4.1-1. Mechanism of the Nitrogenase Reaction
(Dotted connections indicate electron transfer reactions)

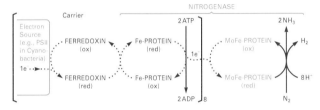

Anaerobic or at least microaerobic conditions are required. The protection against O_2 in *Rhizobium* takes place through the symbiontic synthesis of leghemoglobin by the host plant, which binds O_2 with high affinity and releases it in limited quantities to the *bacterial membrane* (location of the respiratory chain), thus avoiding interference with the nitrogenase activity, which is located in the *cytoplasm*.

Circulation of nitrogen: Ammonia obtained by the nitrogenase reaction or by degradation procedures is introduced into amino acids by glutamate ammonia ligase (glutamine synthase), in the second place by glutamate dehydrogenase reactions (4.2.2), which take place in all living beings. Other amino acids are obtained from glutamine and glutamate mostly by transamination reactions (4.2). The nucleotide-N is contributed by amino acids (8.1.1, 8.2.1).

Many *soil bacteria* derive metabolic energy from the oxidation of ammonia to NO_2^- and NO_3^- (nitrification, Fig. 15.6-1), while *facultative anaerobic bacteria* and *plants* can reduce them again (nitrate ammoniafication, Fig. 15.5-1). Alternatively, the reduction leads to nitrogen (denitrification, 15.5).

Essential amino acids: While *plants* and *bacteria* are able to synthesize all amino acids (4.2...4.9), *mammals* are unable to synthesize some of them and have to obtain them by food intake (Table 4.1-1).

Table 4.1-1. Essential and Non-essential Amino Acids for *Humans*

Essential:	histidine, isoleucine, leucine, lysine, methionine, phenylalanine, threonine, tryptophan, valine
Non-essential:	alanine, arginine, asparagine, aspartate, cysteine, glutamate, glutamine, glycine, proline, serine, tyrosine

Surplus amino acids cannot be stored, they are degraded. In most *terrestrial vertebrates*, amino-N is converted to urea and excreted (4.9.1). Urea can be easily cleaved by *bacteria*, resulting in ammonia. Ammonia is also the terminal product of amino acid degradation in other species.

Literature:
Deng, H., Hoffmann, R.: Angew. Chem Int. Ed. 32 (1993) 1062–1065.
Kim, J., Rees, D.C.: Biochemistry 33 (1994) 389–397.
Mylona, P. et al.: The Plant Cell 7 (1995) 869–885.

4.2 Glutamate, Glutamine, Alanine, Aspartate, Asparagine and Ammonia Turnover

All amino acids of this group are connected via transaminations to the citrate cycle or to pyruvate. Since these compounds can be used for gluconeogenesis (3.1.3), these amino acids are termed glucogenic. Glutamine is the primary entrance gate for ammonia into *bacterial* and *plant* metabolism. In *animals*, which obtain amino acids by food intake, glutamate plays the central role in amino acid interconversions.

4.2.1 Glutamine Metabolism (Fig. 4.2-2)

Glutamine synthesis: The ubiquitous glutamate-ammonia ligase (glutamine synthetase) performs the ATP-energized ammonia binding reaction:

Glutamate + ATP + NH_3 = glutamine + ADP + P_i

The *E.coli* enzyme consists of 12 identical subunits. Due to its central role in nitrogen metabolism, it is regulated at three levels (Fig. 4.2-2):

① Many end products of biosynthesis allosterically inhibit the ligase in a cumulative feedback fashion (each end product causes partial inhibition).

② The sensitivity to allosteric inhibitors is increased by reversible adenylylation of tyrosine-397. This effect becomes more pronounced, the more the subunits are adenylylated.

③ The adenylylation level, in turn, is regulated by controlled uridylylation of an auxiliary protein P_{II}, which associates with the adenylyl transferase.

Figure 4.2-2. Regulation of Glutamate-Ammonia Ligase in *E. coli*

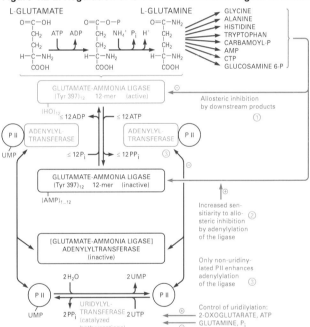

In *mammals*, the enzyme is present in the *mitochondria* of all organs; in *liver*, however, only in the small portion of paravenous cells. Its regulation is much simpler: it is only activated by oxoglutarate. By action of this enzyme, free ammonia from degradation procedures is bound. The resulting glutamine is the major transport form of amino groups between organs. Free ammonia is toxic beyond moderate concentrations, therefore its blood concentration is kept low (< ca. 60 µmol/l, except in portal vein blood).

In *plant chloroplasts*, the enzyme mostly acts to recover ammonia, which is liberated at the glycine oxidation step of photorespiration (3.9.2) in the neighboring mitochondria. It also occurs in *leukoplasts* of *roots*, where it binds the ammonia taken up from the soil or obtained by reduction of nitrate.

Glutamine conversions and degradation: Glutamine can be directly incorporated into proteins (10.6, 11.5). The amino group of glutamine can be transferred by transaminase reactions (4.2.5) to other moieties for synthesis of amino acids (4.2...4.9), amino sugars (3.7.1), nucleotides (8.1, 8.2), NAD (9.9.1) etc. The glutamate formation by glutamate synthase is described below. In *animals*, liberation of ammonia by glutaminase occurs in several organs. The resulting glutamate is oxidized there for energy supply or enters biosynthetic reactions.

4.2.2 Glutamate Metabolism (Fig. 4.2-1)

Glutamate synthesis: In *bacteria* and in *plant chloroplasts*, the main pathway to glutamate starts from 2-oxoglutarate and glutamine (4.2.1) as the amino source. It is catalyzed by glutamate synthase in a reduction-transamination reaction.

The major glutamate production in *animals*, however, takes place by transamination between 2-oxoglutarate and amino acids to be catabolized (Fig. 4.2-3). This proceeds mainly in the *liver*, which is the central organ for amino acid interconversions and acts also as a 'buffer' after resorption.

For removal of free ammonia and simultaneously for glutamate biosynthesis, *liver mitochondria* use the glutamate dehydrogenase reaction (operating in the direction of glutamate formation).

Glutamate conversions: Besides being incorporated into proteins and peptides (e.g. glutathione, 4.5.7) and taking part in glutamine formation, glutamate acts as donor of amino groups for biosynthetic reactions in numerous transamination reactions.

Additional glutamate reactions are:
- Glutamate, as well as its decarboxylation product 4-aminobutyrate (GABA) act in the *CNS* as neuronal transmitters (Table 17.2-2). The 4-aminobutyrate degradation leads finally to succinate, it also occurs in *bacteria*.
- Condensation of glutamate with acetyl-CoA yields N-acetyl-glutamate, the initial compound for ornithine and arginine synthesis and an activator of carbamoyl synthesis (urea cycle, 4.9).
- Phosphorylation and consecutive reduction leads to glutamate 5-semialdehyde, which is the precursor of both pathways to proline (4.3) and to ornithine. Ornithine, in turn, can be converted into arginine (4.9.1).
- Posttranslational γ-carboxylation of glutamate in coagulation factors is essential for their activity (20.3.1). This reaction requires the presence of vitamin K (phylloquinone, 9.13).

Glutamate degradation: Depending on the metabolic state, glutamate dehydrogenase in *liver mitochondria* (see above) also may operate in the catabolic direction by converting glutamate into 2-oxoglutarate.

Ammonia, which is liberated this way, either enters the urea cycle (4.3.1) or is directly excreted (species-dependent). The other product, 2-oxoglutarate,
- is oxidized in the citrate cycle (3.8.1) or
- enables by transamination the conversion of many other amino acids into their respective oxo acids.
  - The oxo acids are then oxidized (Fig. 3.8-1).
  - Alternatively, these oxo acids can enter the gluconeogenetic pathway. Glucogenic amino acids are: Ala, Arg, Asn, Asp, Cys, Gln, Glu, Gly, His, Ile*, Met, Phe*, Pro, Ser, Thr*, Trp*, Tyr*, Val. The amino acids marked with * are also ketogenic; they yield upon degradation ketone bodies (acetyl-CoA, acetoacetate, 6.1.7). Strictly ketogenic are only Lys and Leu.

Thus, glutamate dehydrogenase plays a central role in amino acid metabolism. In *vertebrates* it is allosterically regulated by the energy situation: It is inhibited by GTP (some of which is formed in the citrate cycle, 3.8.1) and ATP, but activated by GDP and ADP. In a number of organisms, the enzyme uses both NAD$^+$ and NADP$^+$.

Glutamate, as well as a number of other amino acids, can also be degraded by amino acid oxidases. Some of them are fairly unspecific. In *humans*, they occur in the endoplasmic reticulum of *liver* and *kidney*.

4.2.3 Alanine Metabolism (Fig. 4.2-1)

Alanine is an essential component of *bacterial* murein walls (15.1). In *animals*, alanine is an important transport metabolite for amino groups besides glutamine. Pyruvate, which is abundantly generated in *muscles* during exercise, accepts the amino groups from glutamate by transamination and passes them on to the *liver*, where the reverse reaction reconstitutes pyruvate (used for gluconeogenesis, Fig. 3.1-7) and glutamate (glucose-alanine cycle).

(muscle)
Pyruvate + glutamate ⇌ alanine + oxoglutarate
(liver)

4.2.4 Aspartate and Asparagine Metabolism (Fig. 4.2-1)

Aspartate is connected by a transaminase reaction with the citrate cycle component, oxaloacetate. This reaction is used for aspartate biosynthesis as well as for degradation.

Oxaloacetate + glutamate = aspartate + oxoglutarate

The *mitochondrial* and *cytosolic* isoenzymes of aspartate (ASAT, GOT) and alanine transaminases (ALAT, GPT, see above 4.2.3) are of importance for diagnosis of liver damage.

Aspartate is the starting point of important biosynthetic pathways:
- The condensation with carbamoyl-P is the initiation reaction for pyrimidine biosynthesis (8.2.1).
- The pathway to inosine monophosphate and later to adenosine monophosphate involves 2 condensation steps with aspartate (8.1.1, 8.1.2).
- The condensation with citrulline leads to the member of the urea cycle, argininosuccinate, which is a precursor of arginine (4.3).
- Aspartate phosphorylation, yielding aspartyl-P leads in *bacteria* and *plants* to the biosynthesis of the essential amino acids methionine, threonine, lysine and isoleucine (4.5.2...4, 4.6.1).
- Decarboxylation (in *bacteria*) yields β-alanine, which is used in pantothenate (9.7.1) and carnosine biosynthesis (4.8).

Asparagine results from the asparagine-ammonia ligase reaction with glutamine as an amino group donor. Its degradation takes place in the *cytosol* by the asparaginase reaction (which proceeds analogously to the glutaminase reaction) or in *mitochondria* by transamidation and oxidation.

Figure 4.2-1. Metabolism of Glutamate, Aspartate and Related Compounds

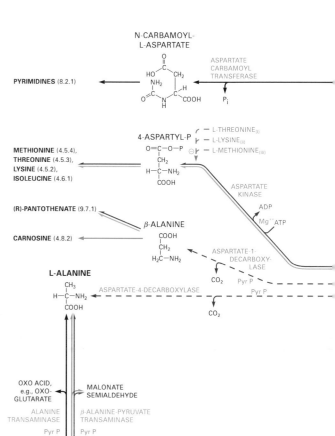

4.2.5 Transamination Reactions (Fig. 4.2-3, next page)

Transamination is an important step for synthesis and degradation of many amino acids, as well as of other amino compounds (e.g. D-glucosamine, 3.7.1). It takes place by <u>pyridoxal-P</u> dependent transfer between amino (in some cases amido, e.g. glutamine) and oxo compounds ($\supset C-NH_2 + \supset C=O \longleftrightarrow \supset C=O + \supset C-NH_2$).

The catalyzed reaction starts by nucleophilic attack of the substrate amino group on the Schiff base structure, which exists between pyridoxal phosphate and an ε-lysine group of the enzyme and replaces it by a '<u>aldimine</u>' Schiff base between the substrate and pyridoxal phosphate. Abstraction of a proton by the Lys-NH$_2$ group of the enzyme yields a resonance-stabilized intermediate. The consecutive protonation results in a '<u>ketimine</u>' Schiff base (tautomerization). Hydrolysis leads to release of the oxo acid. After the binding of another oxo acid, the reaction proceeds in the reverse direction to yield the respective amino acid (Ping Pong Bi Bi-mechanism, 1.5.4).

In the aldimine structure, not only the C_α–H bond, but also the C_α–COOH bond, the C_α–R bond and the next intra-R bond are labilized, since cleavage of each of them leads to the resonance-stabilized carbanion. The pyridine ring acts as an electron sink. This allows several reaction variants (for details see 9.4.2):

- Cleavage of C_α–H bond: removal of amino group, yielding an oxo group; the consecutive reversal reaction results in <u>transamination</u>
- Cleavage of C_α–COOH bond: <u>decarboxylation</u> (e.g. aspartate 1- or 4-decarboxylase, glutamate and histidine decarboxylases)
- Cleavages regarding R (α,β and β,γ eliminations): e.g. serine dehydratase (Fig. 4.4-2) and kynureninase (Fig. 4.7-4).

Literature:

Bender, D.A.: *Amino Acid Metabolism*. 2nd Ed. Wiley (1985).
Hayashi, H. *et al.*: Ann. Rev. of Biochem. 59 (1990) 87–110.
Liaw, S.-H., Eisenberg, D.: Biochemistry 33 (1994) 675–681.
Meister, A.: *Biochem. of the Amino Acids*. 2nd. Ed. Academic Press (1965).
Umbarger, H.E.: Ann. Rev. of Biochem. 47 (1978) 522–606.

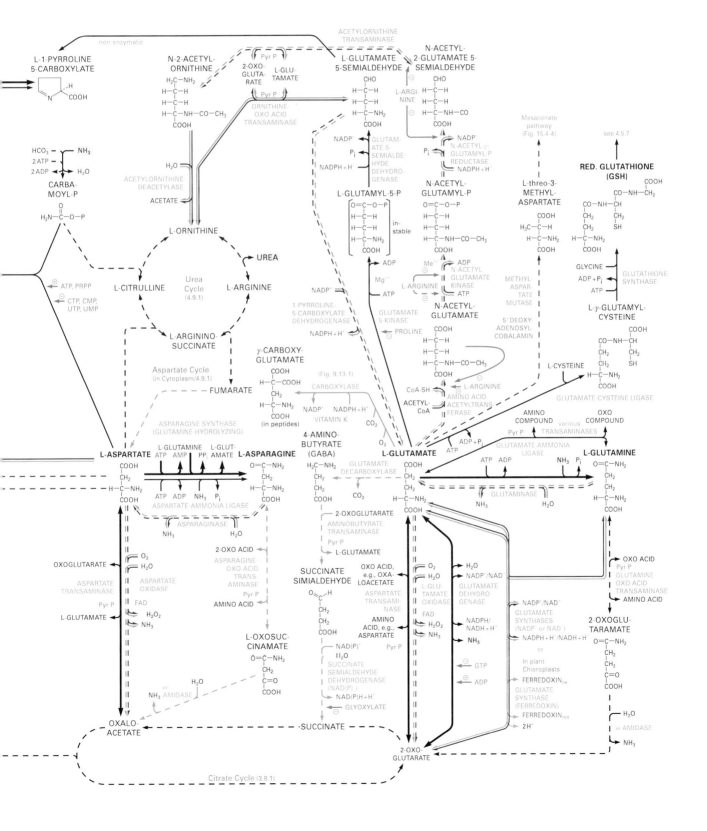

Figure 4.2-3. Mechanism of Transamination Reactions

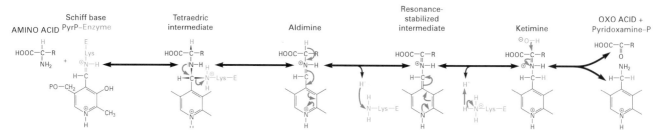

4.3 Proline and Hydroxyproline (Fig. 4.3-1)

Biosynthesis: Glutamate is phosphorylated and consecutively reduced to glutamate semialdehyde (Fig. 4.2-1). This compound cyclizes nonenzymatically to L-1-pyrroline-5-carboxylate. An NADH or NADPH dependent reduction leads to the nonessential amino acid proline.

In proteins, the rigid ring structure of proline does not allow rotation at the carboxylate-C–N bond and impedes the formation of α-helices. Therefore, proline puts many constraints on the protein structure. On the other hand, amino acid-proline bonds can assume both *trans-* and *cis*-configurations (1.3.2). The isomerization by peptidyl-proline-*cis-trans*-isomerases (PPI) is frequently the rate determining step in protein folding procedures (Table 14.1-1). This enzyme is an essential subunit of proline-4-hydroxylase in *eukarya* (see below).

4-Hydroxyproline (Hyp) is produced posttranslationally by procollagen-proline 4-dioxygenase (proline 4-hydroxylase), which is an intermolecular dioxygenase containing Fe^{++}. Besides proline, Hyp is a major component of collagen (2.3.1) and stabilizes it by formation of hydrogen bonds. Small amounts of 3-hydroxyproline and 5-hydroxylysine (4.5.2) are also present.

Proline-4-dioxygenase binds in ordered sequence, Fe^{++}, the cosubstrate 2-oxoglutarate, O_2 and a peptide containing the sequence X-Pro-Gly (preferably with X = Pro). O_2 is activated by interaction with the bound Fe^{++} and performs a nucleophilic addition to oxoglutarate. The complex hydroxylates proline; then the products leave the enzyme. Ascorbate is oxidized in substoichiometric amounts and apparently prevents Fe from being in the oxidized state after the reaction cycle has ended. A similar situation exists with procollagen-lysine 5-dioxygenase (4.5.2). Diminished hydroxylation of proline due to a lack of ascorbate prevents proper the formation of collagen fibers, their melting temperature is diminished (scurvy).

Degradation of proline starts with the oxidation of 1-pyrroline-5-carboxylate, finally yielding glutamate and oxoglutarate. Apparently, glutamate 5-semialdehyde is an intermediate. Degradation of hydroxyproline takes place analogously, leading to hydroxyglutamate. After cleavage, transamination results in pyruvate and glycine.

Bacterial enzymes can perform racemization of L-proline and L-hydroxyproline to D-proline and D-*allo*-4-hydroxyproline, respectively. Their degradation takes place by D-amino acid oxidase.

Literature:

Adams, E., Frank, L.: Ann. Rev. of Biochem. 49 (1980) 1005–1061.
Counts, D.F. *et al.*: Proc. Natl. Acad. Sci. (USA) 75 (1978) 2145–2149.
Myllylä, R. *et al.*: Biochim. Biophys. Res. Commun. 83 (1978) 441–448.

Figure 4.3-1. Proline and Hydroxyproline Metabolism

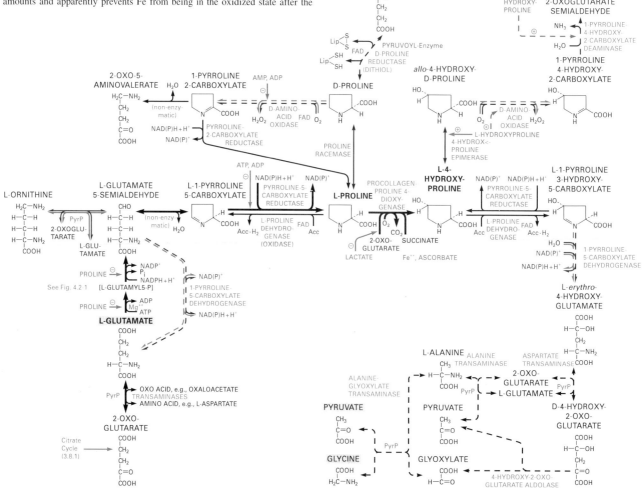

4.4 Serine and Glycine

Serine and glycine are nonessential amino acids. They are derived from 3-phosphoglycerate. In *plants* and *bacteria*, serine can be directly converted into cysteine, while in *animals*, the essential amino acid methionine is needed as reaction partner for cysteine biosynthesis (4.5.5). The interconversion of serine and glycine requires the participation of the tetrahydrofolate C_1-transfer system (9.6.2). Glycine is an inhibitory neurotransmitter (Table 17.2-2).

4.4.1 Serine Metabolism (Fig. 4.4-1)

The biosynthesis of serine starts from 3-phosphoglycerate and proceeds by a sequence of dehydrogenase (to 3-P-hydroxypyruvate), transaminase (to 3-P-serine) and phosphatase reactions. In *animals*, the controlled enzyme is the third one of this pathway (phosphoserine phosphatase) instead of the usual initial one.

Besides its occurrence in proteins, serine is a component of glycerophospholipids (6.3.2) and the origin of sphingosine and ceramide biosynthesis (6.3.4). The interconversion with glycine proceeds in both directions (4.4.2). In *plants* and *bacteria*, serine can be acetylated and thereafter the acetyl group be exchanged with sulfide, resulting in cysteine. (*Animals*, however, synthesize cysteine from the essential amino acid methionine, 4.5.4.) via the intermediate cystathionine.

Serine degradation: The major route for serine catabolism is the serine dehydratase reaction, leading to pyruvate (Fig. 4.4-2). This way, serine (and glycine after interconversion to serine) can enter the gluconeogenesis pathway (3.3.5).

The dehydratase and transaminase (4.2.5) mechanisms are related, both require pyridoxal phosphate. In both cases, an aldimine structure is formed. In the transaminase reaction, after removal of the α-hydrogen of the amino acid, there is a protonation of the C-4' atom of pyridoxal phosphate (Fig. 4.2-3). In the dehydratase reaction, however, β-elimination of the hydroxyl group from serine takes place.

Reconversion of serine to 3-phosphoglycerate via the non-phosphorylated compounds hydroxypyruvate and glycerate is another way of serine utilization. In *liver* and *kidney* of *animals*, it is used for gluconeogenesis. In *plants*, it is part of the photorespiration sequence (3.9.2, 16.2.2).

4.4.2 Glycine Metabolism (Fig. 4.4-1)

Both glycine synthesis and catabolism proceed mainly by interconversion with serine. This involves a C_1 group transfer by tetrahydrofolate (THF, 9.6.2).

For glycine formation, a methylene group is moved from serine to THF by the pyridoxal dependent enzyme glycine hydroxymethyltransferase (also named serine hydroxymethyltransferase), yielding 5,10-methylene-THF and releasing glycine. The folate coenzyme passes on this C_1 moiety to various acceptors for biosynthetic purposes.

In the opposite direction, a molecule of glycine is at first converted by the *mitochondrial* glycine cleavage system to 5,10-methylene-THF and CO_2 (Fig. 4.4-3). This system is a multi-enzyme complex, which resembles the pyruvate dehydrogenase complex (3.3.1) and consists of the components

- glycine dehydrogenase (decarboxylating) (P protein, contains pyridoxal-P)
- aminomethyltransferase (H protein, contains an 'arm' of lipoic acid for transport between P and T proteins)
- aminomethyltransferase (T protein, contains tetrahydrofolate)
- dihydrolipoyl dehydrogenase (L protein)

The initial decarboxylation reaction is due to the labilization of the C_α–COOH bond in the aldimine structure (cf. transaminase mechanism, 4.2.5).

One of the various possible reactions of the 5,10-methylene THF product is the C_1 transfer to a second glycine molecule by glycine hydroxymethyltransferase, resulting in serine formation. Thus, this enzyme catalyzes, together with the glycine cleavage system, the reversible reaction:

2 Glycine + H_2O + NAD^+ = serine + CO_2 + NH_4^+ + NADH.

Glycine hydroxymethyltransferase can also react in absence of tetrahydrofolate. This way it catalyzes the cleavage of threonine, resulting in glycine and acetaldehyde ('threonine aldolase', Fig. 4.5-2).

Another way of glycine synthesis is by transamination of glyoxylate, which originates from glycolate. It is part of the photorespiration sequence in *plants* (3.9.2, 16.2.2), but also occurs in non-photosynthetic organisms, e.g. *yeast*, for gluconeogenetic purposes.

Literature:
Rapoport, S. *et al.*: Eur. J. Biochem. 108 (1980) 449–455.
Snell, K.: Adv. Enzyme Regul. 22 (1984) 325–400 and 30 (1990) 13–32.
Umbarger, H.E.: Ann. Rev. of Biochem. 47 (1978) 533–606.

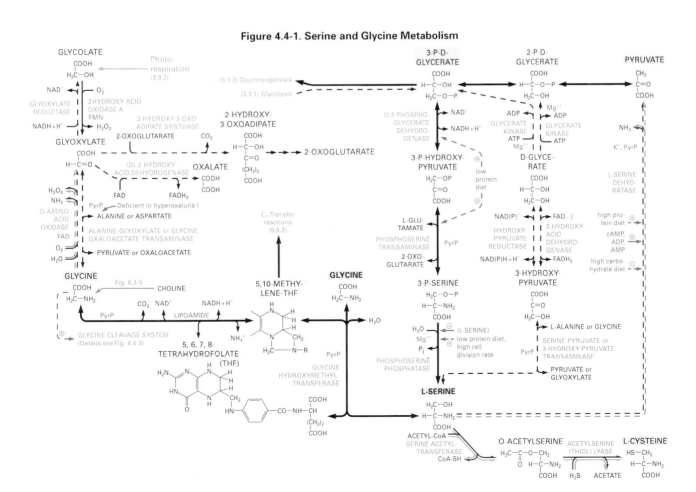

Figure 4.4-1. Serine and Glycine Metabolism

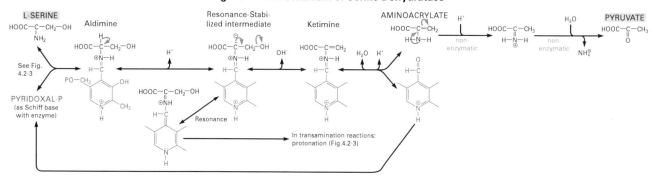

Figure 4.4-2. Mechanism of Serine Dehydratase

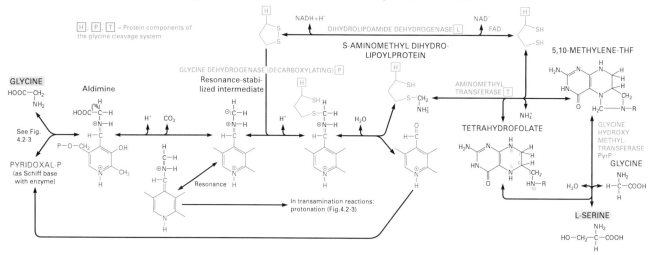

Figure 4.4-3. Synthesis of Serine from Glycine (Reversible)

4.5 Lysine, Threonine, Methionine, Cysteine and Sulfur Metabolism

This group of amino acids is formed from aspartate. Only in *fungi*, lysine originates from 2-oxoglutarate. Although cysteine is the only one out of this group which is not considered essential for *mammals*, it still requires the essential amino acid methionine for its synthesis. Precursors of isoleucine are both threonine and pyruvate. Its biosynthesis is discussed in 4.6.1.

4.5.1 Common Steps of Biosynthesis and Their Regulation (Fig. 4.5-1)

The biosynthesis of lysine (in *bacteria and most plants*), threonine and methionine starts with the phosphorylation of aspartate. As usual, the first committed step of the pathway is the point of action for regulatory mechanisms. Since the pathways for synthesis of the individual amino acids branch later, each amino acid has to perform its own control. In various organisms, different systems have been realized. They are excellent examples of the various possibilites for control of metabolism.

In *E. coli*, there are 3 aspartate kinases. Each of them is feedback inhibited by one of the amino acids (multiple enzyme control). Additionally, feedback inhibition of the respective enzymes after branch points takes place. Each of the aspartate kinases I and II (which are regulated by threonine and methionine) exists together with the respective homoserine dehydrogenase as a bifunctional enzyme on a single peptide chain.

The operon encompassing the 3 structural genes coding for aspartate kinase I/ homoserine dehydrogenase I, homoserine kinase and threonine synthase is controlled by a single promoter-operator locus. Bivalent repression takes place by threonine and isoleucine. The repression of aspartate kinase II activity by methionine is also enhanced by isoleucine.

In *Rhodopseudomonas spheroides*, however, only aspartate semialdehyde (the last common intermediate) inhibits the single aspartate kinase, while in *R.capsulata* and others, a synergistic inhibition of the aspartokinase by lysine and threonine takes place (cooperative feedback control). Still other variants have been found in other organisms.

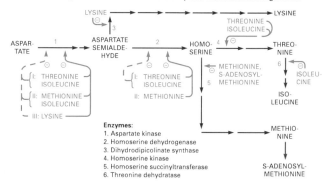

Figure 4.5-1. Regulation of Threonine, Methionine and Lysine Biosynthesis in *E. coli*

Feedback inhibition is indicated by solid arrows, repression of enzyme synthesis by dashed arrows. I, II and III indicate different enzymes with individual regulation.

Enzymes:
1. Aspartate kinase
2. Homoserine dehydrogenase
3. Dihydrodipicolinate synthase
4. Homoserine kinase
5. Homoserine succinyltransferase
6. Threonine dehydratase

4.5.2 Lysine Metabolism (Fig. 4.5-2)

Lysine biosynthesis in *bacteria*: The individual pathway begins with a condensation reaction of aspartate semialdehyde with pyruvate. The product cyclizes immediately. After reduction, the non-cyclic form is stabilized by succinylation or acetylation. Further reactions lead to diaminopimelate, which is also a component of *bacterial* cell walls (15.1). By decarboxylation, lysine is obtained. In some *bacteria*, a reductive amination leads directly from Δ^1-piperidine 2,6-dicarboxyate to diaminopimelate.

Lysine biosynthesis in *fungi*: These organisms use a completely different pathway for biosynthesis, which originates from 2-oxoglutarate and acetyl-CoA. The first steps resemble the beginning of the citric acid cycle, employing homologous (= homo) compounds. The oxoadipate formed undergoes a transamination reaction, followed by the reduction to the semialdehyde (possibly via an adenylylated intermediate). The ε-amino group is not introduced by transamination, but rather in a sequence involving reduction, formation of the covalently linked intermediate saccharopine and oxidative cleavage to yield L-lysine. This pathway is regulated by feedback inhibition of the initial condensation reaction.

Biological role of lysine: Lysine is involved in various mechanisms of enzyme cytalysis (e.g. transaminases, 4.2.5). Biotin (9.8) and lipoic acid (9.14.1) are bound to lysine residues of enzymes. In procollagen, lysine is 5-hydroxylated in the *endoplasmic reticulum* (analogous to proline, 4.3). After its glycosylation, final folding to the collagen triple helix take place (2.3.1). A similar hydroxylation of trimethylated lysine is an intermediate step in the synthesis of carnitine, which is important for uptake of fatty acids (6.1.4).

Lysine degradation: There are several degradation mechanisms. The one prevalent in *mammalian liver* starts with reactions, which are essentially the reversal of the biosynthesis reactions in *fungi* (see above). Likely, the roles of NAD^+ and $NADP^+$ are exchanged. Oxidative decarboxylation of 2-oxoadipate results in glutaryl-CoA (the homologue of succinyl-CoA), which after a second decarboxylation yields crotonyl-CoA. This is a member of the fatty acid degradation pathway (6.1.5), which leads to acetyl-CoA. Thus, lysine is a ketogenic amino acid.

4.5.3 Threonine Metabolism (Fig. 4.5-2)

Threonine biosynthesis: The individual part of the pathway is the isomerization of homoserine to threonine.

This reaction takes place by the formation of a phosphate ester with the hydroxyl group and a consecutive β,γ-elimination reaction requiring pyridoxal phosphate. (For other examples of pyridoxal-P catalyzed reactions, see Figures 4.2-3, 4.4-2 and 4.4-3.) The kinase reaction is competitively inhibited by threonine.

Threonine conversions and degradation: By action of threonine dehydratase, threonine is converted to 2-oxobutyrate by a reaction analogous to the serine dehydratase reaction (Fig. 4.4-2). 2-Oxobutyrate acts as a precursor of isoleucine (4.6.1) or is degraded (4.5.4).

For degradation, threonine is cleaved directly or after oxidation, yielding glycine (which is converted via serine to pyruvate, 4.4.2) and acetaldehyde or acetate, which consecutively are converted into acetyl-CoA.

4.5.4 Methionine Metabolism (Fig. 4.5-2)

Methionine biosynthesis: The γ-hydroxyl group of homoserine is activated by succinyl- or acetyl- or by phosphate groups (in *higher plants*). In *bacteria*, this reaction is synergistically inhibited by methionine and S-adenosylmethionine. Then a transsulfuration with cysteine takes place. The resulting cystathionine is cleaved to yield homocysteine. Several variants of this sequence exist, see footnote to Fig. 4.5-2. Homocysteine is then methylated by methyltetrahydrofolate (9.6.2, in most cases), resulting in methionine. Various diseases are caused by defects of these enzymes.

Biological roles: Methionine (in *bacteria*: formylmethionine) is the starting amino acid in protein biosynthesis (10.6.3, 11.5.2). Homocysteine is likely a risk factor for arteriosclerosis.

S-Adenylylation of methionine leads to S-adenosyl-L-methionine (SAM) with a positively charged sulfur atom, which activates the neighboring methyl group. This compound is most important as methyl group donor in transfer reactions (e.g., 4.9.1, 6.3.2, 10.3.4).

Thus, by the formation of methionine and this activation reaction, the moderate methylation ability of methyltetrahydrofolate (9.6.2) is converted into a more reactive mode. The adenylylation reaction is feedback-inihibited by its products S-adenosyl-L-methionine, PP_i and the cleavage product P_i.

Decarboxylation of SAM yields S-adenosylmethioninamine, which enters some biosynthetic reactions (4.9.3). A conversion to ethane occurs in *fruits*.

After transfer of the methyl group, the resulting S-adenosyl-homocysteine (SAH) is deadenylated to homocysteine, which enters another methylation-demethylation cycle or is degraded.

Methionine degradation: The demethylation product homocysteine undergoes a condensation reaction with serine, yielding cystathionine. After releasing cysteine (which is the biosynthesis reaction of this amino acid in *animals*), 2-oxobutyrate is oxidatively decarboxylated, resulting in propionyl-CoA. Then, carboxylation by the biotin dependent enzyme propionyl-CoA carboxylase takes place, which resembles the acetyl-CoA carboxylase reaction. After epimerization, a mutase reaction converts L-methylmalonyl-CoA into the citric acid cycle component succinyl-CoA.

The same last steps of this sequence take place during catabolism of threonine, valine and isoleucine (4.6.2) and of odd-numbered fatty acids (6.1.5). The mechanism of biotin-dependent carboxylations is dealt with in 9.8.2. The mutase reaction is one of the 2 *mammalian* reactions employing a vitamin B_{12} derivative (9.5.2). Deficiencies of propionyl-CoA-carboxylase lead to an increase in propionate in blood. An increase of methylmalonate takes place, if the mutase enzyme or the coenzyme B_{12} biosynthesis is defective.

In *bacteria*, propionate fermentation proceeds via essentially the same sequence in the opposite direction. The CO_2 released from methylmalonyl-CoA is transferred by a biotinyl protein to pyruvate, yielding propionyl-CoA and oxaloacetate (Fig. 15.4-2).

Possibly there exists in *mammals* an additional catabolic pathway, which proceeds via transamination of methionine, decarboxylation and liberation of methanethiol. Its oxidation leads to CO_2 and sulfate.

Methionine → oxo acid → 3-methylthiopropionate → methanethiol →
$$CO_2 + SO_4^{--}$$

4.5.5 Cysteine Metabolism

Cysteine biosynthesis (Fig. 4.5-2): *Bacteria* and *plants* are able to convert L-serine into L-cysteine via acetylserine. The acetyl group is directly exchanged with H_2S, which is provided by reduction of sulfate (Fig. 4.5-4). *Animals*, however, require homocysteine (from methionine degradation) as source of sulfur and produce cysteine via the condensation product cystathionine.

Biological role of cysteine: The thiol group of cysteine takes part in a number of enzyme reaction mechanisms (e.g. glutathione reductase, 4.5.7). It also forms FeS centers in electron transfer proteins involved in, e.g., respiration and photosynthesis (16.1, 16.2). The oxidation of cysteine to the disulfide cystine (Cys-S-S-Cys) plays an essential role in formation and maintenance of the secondary structure of proteins (2.3.1, 14.1.3). On the other hand, cysteine oxidation by atmospheric oxygen is a frequent cause of protein inactivation.

Cysteine conversions and degradation (Fig. 4.5-3): Oxidation of the -SH group and decarboxylation result in taurine, which is a conjugation partner of bile acids (7.9) and is present in *retina, brain, lymphocytes* etc. It may have a detoxifying and membrane protecting effect. The final product of other cysteine degradation pathways is pyruvate. The release of sulfur can proceed in several ways (as H_2S, SO_3^{2-} or SCN^-).

4.5.6 Sulfur Metabolism (Fig. 4.5-4)

In the biosphere, sulfur plays a role

- in oxidized form, primarily as $-OSO_3H$ in sulfated glycoproteins (13.1) and glycolipids (13.2) and as a conjugation partner for excretion (e.g., 7.6...7.8)
- in reduced form as -SH and -S- in cysteine and methionine (also in proteins), glutathione and redox centers (16.1.2, 16.2.1)
- *Anaerobic bacteria* use reduction and oxidation of sulfur compounds for anaerobic respiration and for chemolithotrophy. The energy aspect of these reactions is dealt with in 15.5 and 15.6. Slow reduction to the -SH level can be performed by *plants*.

Metabolism of sulfate: For conjugation as well as for reduction reactions, sulfate has to be activated in an ATP-dependent reaction to adenylylsulfate (APS). Since $\Delta G_0'$ for hydrolysis of the sulfate-phosphate bond (–71 kJ/mol) is much larger than that for the α-β pyrophosphate bond in ATP (–32.2 kJ/mol), the reaction has to be 'pulled' by hydrolysis of the liberated pyrophosphate and by additional phosphorylation of APS to 3'-phosphoadenylylsulfate (PAPS). PAPS introduces sulfate residues in many compounds; the resulting adenosine 3',5'-diphosphate (PAP) is then hydrolyzed to yield 5'-AMP.

For conversion to sulfite in *photosynthesizing plants*, reduced ferredoxin (from photosynthesis, 16.2 or regenerated by NADPH) reduces thioredoxin (ca. 100 amino acids, containing the sequence Cys-Gly-Pro-Cys) from the -S-S- to the $(-SH)_2$ state. Catalyzed by PAPS reductase, PAPS is then converted to free sulfite and PAP. The further reduction to sulfide is catalyzed by sulfite reductase, an enzyme containing siroheme (9.5.3), which closely resembles the nitrite reductase. The reduction equivalents are supplied by ferredoxin. Also *yeast* and a number of *bacteria* use PAPS as an intermediate. *Animals* are unable to reduce sulfate.

Continuation on p. 55

Figure 4.5-2. Lysine, Threonine, Methionine and Cysteine Metabolism

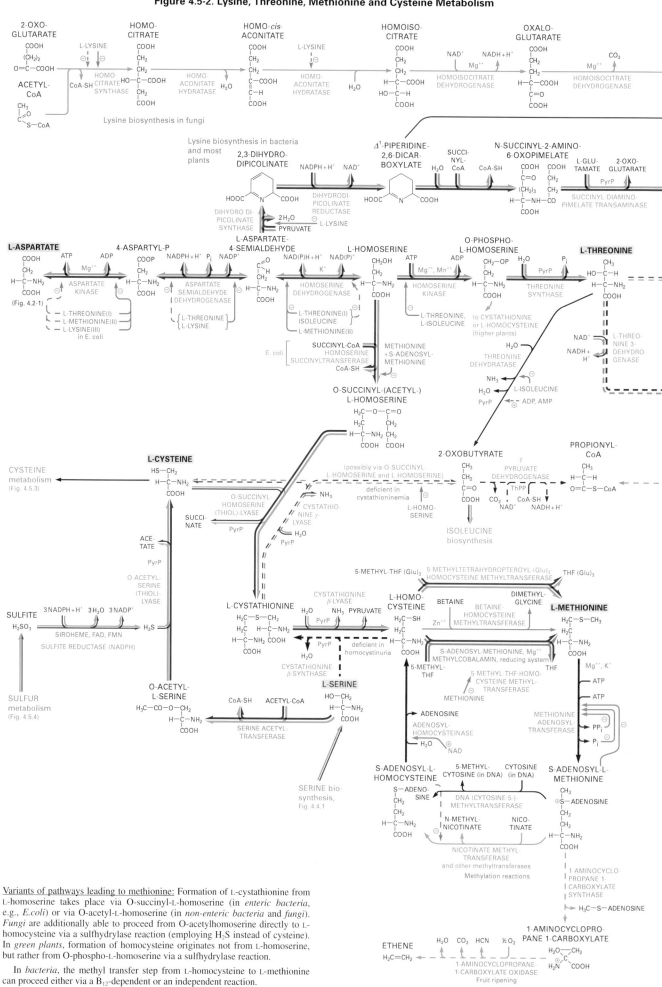

Variants of pathways leading to methionine: Formation of L-cystathionine from L-homoserine takes place via O-succinyl-L-homoserine (in *enteric bacteria*, e.g., *E.coli*) or via O-acetyl-L-homoserine (in *non-enteric bacteria* and *fungi*). *Fungi* are additionally able to proceed from O-acetylhomoserine directly to L-homocysteine via a sulfhydrylase reaction (employing H_2S instead of cysteine). In *green plants*, formation of homocysteine originates not from L-homoserine, but rather from O-phospho-L-homoserine via a sulfhydrylase reaction.

In *bacteria*, the methyl transfer step from L-homocysteine to L-methionine can proceed either via a B_{12}-dependent or an independent reaction.

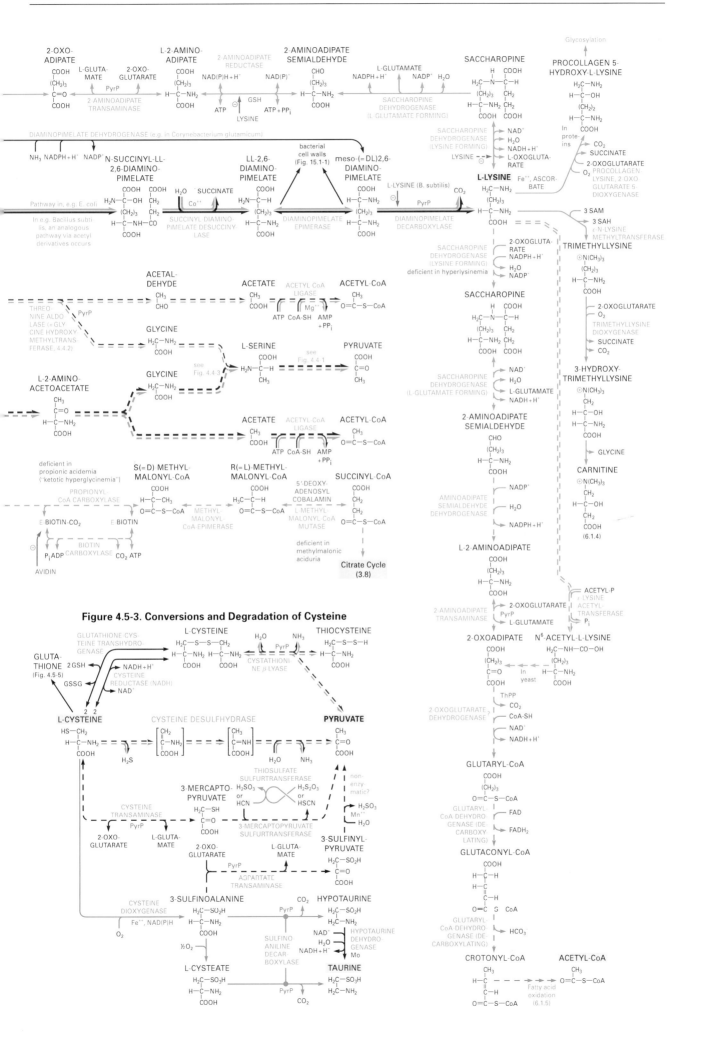

Figure 4.5-3. Conversions and Degradation of Cysteine

Another pathway in *plants*, which is under discussion, starts directly from APS (15.5) and proceeds via reduction of a glutathione thiosulfate intermediate to the disulfide level (not shown).

Metabolism of sulfide: Fixation of sulfide occurs in *plants* and *bacteria* by reaction with acetylserine, yielding cysteine (4.5.5). The further reactions of sulfur in amino acids are dealt with in 4.5.4, 4.5.5 and 4.5.7.

4.5.7 Glutathione Metabolism (Fig. 4.5-5)

The tripeptide glutathione (GSH, γ-Glu-Cys-Gly) is synthesized by specific enzymes and is *intracellularly* present in high concentrations (in *animals* ca. 2...5 mmol/l, more than 99% in reduced form). Glutathione is an important reductant ($E_0' = -230$ mV). The oxidized form (GSSG) resulting from these reactions is reduced again in a NADPH-dependent reaction by glutathione reductase (see below). GSH shows many detoxifying and cytoprotective effects. Major reaction types are:

- Removal of H_2O_2 (directly or indirectly) by peroxidase reactions (4.5.8)

- Stabilization of the redox state of peptides (e.g. insulin) and proteins by the protein-disulfide reductase reaction

    $2\ GSH + R_1-Cys-S-S-Cys-R_2 = GSSG + R_1-Cys-SH + R_2-Cys-SH$

    In some cases, the reduction of proteins is performed by thioredoxin instead, e.g. for activation of Calvin cycle enzymes (16.2.2).

- Reduction of ribonucleotides (e.g. in some bacteria, 8.1.4)

- Cellular import of amino acids by the γ-glutamyl cycle.

    A noticeable portion of the intracellularly synthesized glutathione is exported, in *animals* mostly from *liver* and *kidney*. At the outside of the cellular membrane of *animals* and *yeast*, GSH transfers its glutamate moiety to amino acids to be imported, catalyzed by the membrane bound enzyme γ-glutamyl-transpeptidase (γ-GT). After membrane passage, the dipeptide is hydrolyzed and the temporarily formed 5-oxyproline is reconverted to glutamate. This reaction is energized by ATP hydrolysis.

- Conjugation reactions, mainly for neutralizing toxic compounds (xenobiotics), but also for biosynthesis (Fig. 17.4-15).

    The thiol group of GSH reacts with C=C bonds, carbonyl groups, sulfates etc. This is catalyzed by various, mostly unspecific glutathione transferases. In *plants*, the conjugates are then actively imported into *vacuoles*.

- Formation of phytochelatins in *plants*.

    As a detoxification reaction, phytochelatins form complexes with heavy metals, which are then transported into *vacuoles*.

The reduction of GSSG in order to regenerate GSH takes place by glutathione reductase. The reaction mechanism involves the initial reduction by NADPH of a Cys-S-S-Cys bond in the enzyme, formation of a charge-transfer complex of one Cys-S- with FAD, nucleophilic attack of the other Cys-SH on GSSG and cleavage of the mixed disulfide. Other disulfide oxidoreductases (e.g. dihydrolipoamide dehydrogenase, 3.3.1) react in an analogous mode.

Lack of NADPH leads to insufficient reduction of GSSG. This is especially manifest in *erythrocytes*, where NADPH formation can take place only by the glucose-6-phosphate dehydrogenase reaction (3.6.1), since mitochondria are absent. An inherited diminished activity of this enzyme (X-chromosomal, occurring in more than 100 million persons) is further decreased by some antimalarial drugs and by fava beans (vicia faba). This leads to hemolytic anemia.

Literature for metabolism of the amino acids:
Cooper, A.J.L.: Ann. Rev. of Biochem. 52 (1983) 187–222.
Meister, A., Anderson, M.E.: Ann. Rev. of Biochem. 52 (1983) 711–760.
Meister, A.: J. Biol. Chem. 263 (1988) 17205–17208.
Schwenn, J.D.: Z. Naturforsch. 49c (1994) 531–539.
Umbarger, H.E.: Ann. Rev. of Biochem. 47 (1978) 533–606.

Figure 4.5-4. Sulfur Metabolism

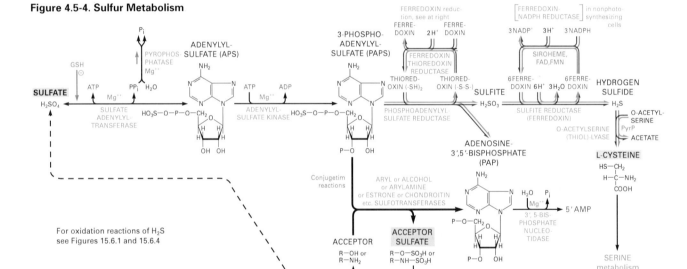

Figure 4.5-5. Glutathione Metabolism

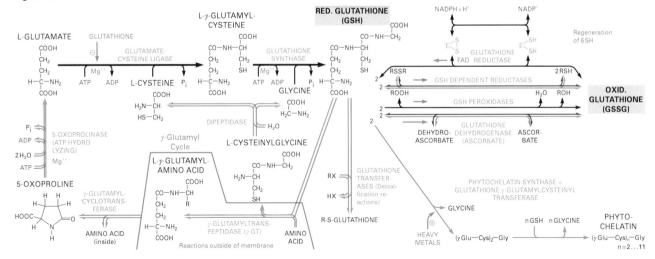

4.5.8 Reactive Oxygen Species, Damage, and Protection Mechanisms (Fig. 4.5-6)

Molecular oxygen, the most important element of aerobic life, can enter into some side reactions, which release dangerous reactive oxygen species (ROS) or reactive oxygen intermediates (ROI). Therefore, appropriate protection mechanisms are essential. On the other hand, ROS can be used for defense and for regulatory reactions. In the following, selected reactions are shown.

Superoxide radical ($O_2^{\bullet-}$): This radical is formed by one-electron transfer to O_2 ($\Delta E_0' = -330$ mV).

$O_2^{\bullet-}$ is generated by side reactions of some oxidases (e.g., xanthine oxidase, 8.1.6), of the photosynthesis complex I, of electron carriers in the respiratory chain (mainly ubiquinone), or by transfer from semiquinone intermediates (①, e.g., during monooxygenase reactions in the *ER*). $O_2^{\bullet-}$ is also produced by high-energy irradiation. The aggressive compound acts not only as a reductant, but also sometimes as an oxidant (e.g., on sulfite). Superoxide dismutase (SOD) converts it into the likewise toxic hydrogen peroxide H_2O_2 (see below). The same reaction also takes place spontaneously.

The superoxide radical seems to be the most toxic ROS. It shows a certain chemical selectivity, e.g., for reactions with unsaturated phospholipids in biomembranes leading to lipid peroxidation. It is also involved in reactions causing, e.g., ischemia-reperfusion injury.

Activated *macrophages* generate $O_2^{\bullet-}$ during inflammation by the cytochrome b_{558} containing NADPH oxidase ② and convert it by SOD to H_2O_2. This compound is used by myeloperoxidase to generate the aggressive hypochlorite (HOCl) for defense purposes ③.

$NADPH + 2\ O_2 = 2\ O_2^{\bullet-} + NADP^+ + H^+$
$2\ O_2^{\bullet-} + 2\ H^+ = H_2O_2 + O_2$
$H_2O_2 + Cl^- = HOCl + OH^-$

In addition, $O_2^{\bullet-}$ reacts rapidly with NO (17.7.2) yielding peroxynitrite ④

$O_2^{\bullet-} + NO^{\bullet} = O=N-O-O^-$

This compound is highly reactive and produces, e.g., peroxides. At present, it is not exactly known if this reaction plays a role in NO metabolism.

Hydrogen peroxide (H_2O_2) is a reaction product of various oxidases (e.g., in *peroxisomes*, 2.2.2 and in the *ER*, 13.3) and a product of the SOD reaction ⑤.

H_2O_2 toxicity may be partly caused by direct inactivation of enzymes and (mostly reversible) of hemoproteins. H_2O_2 also can give rise to the very deleterious hydroxyl radical $HO^{\bullet}$ by reaction with semiquinones (⑥, see below). It is questionable, whether a metal ion dependent cleavage of H_2O_2 to form hydroxyl radicals (Fenton reaction) takes place *in vivo*, since free heavy metal ions are practically absent under these conditions.

The **hydroxyl radical ($HO^{\bullet}$)** is the most reactive, shortlived and unspecific ROS. It reacts with nearly all biomolecules and may thereby start radical chain reactions ⑦.

$HO^{\bullet} + ...-HCH-... = ...-HC^{\bullet}-... + H_2O$

Its main source is probably the reaction of H_2O_2 with semiquinones and other reductants. This may explain the toxicity of quinone compounds of foreign (xenobiotic) origin.

Peroxide radicals ($ROO^{\bullet}$) arise by the very fast, spontaneous reaction of oxygen with organic radicals (...-$HC^{\bullet}$-..., ⑧) which have been previously formed by hydroxyl or superoxide radicals in an attack reaction (see above). The peroxide radicals are intermediates in the chain reactions of lipid peroxidation (of membranes, LDL, etc.).

$...-HC^{\bullet}-... + O_2 = ...-HCOO^{\bullet}-...$

Singlet oxygen ($^1O_2$): This is the only physiologically relevant 'excited state' of molecular oxygen. Singlet oxygen may be produced during photosynthesis or nonenzymatically by reactions of hypochlorite ⑨ and peroxynitrite (see above).

$HO-X + H_2O_2 = H^+ + X^- + H_2O + {}^1O_2$ (e.g., X = Cl)

In oxidation reactions, singlet oxygen directly accepts 2 electrons with the normal antiparallel spin, while 'regular' oxygen would require 2 electrons with the uncommon parallel spin. Thus it is usually restricted to consecutive 1-electron transfer reactions (e.g., those catalyzed by flavin coenzymes, 9.3.2), while $^1O_2$ can react either way. This oxygen species is, e.g., involved in photosensitization processes (illumination of FMN, retinal, porphyrins etc.). This can be the cause of dermatoses.

Regulatory functions of reactive oxygen species: Some ROS appear to be involved in signal transduction, e.g., by oxidative activation of NF-κB or by modulation of tyrosine phosphorylation cascades (17.5).

Antioxidant Defenses: In all aerobic living organisms, both $O_2^{\bullet-}$ and H_2O_2 must be removed for protection.)

Various superoxide dismutases (SOD) disproportionate $O_2^{\bullet-}$ but still yield H_2O_2 as in the spontaneous reaction ⑤. This compound is destroyed by the ubiquitously occurring catalase ⑩ and by various peroxidases ⑪ including glutathione peroxidase ⑫ (in *animals*, it also removes organic peroxides) and ascorbate peroxidase ⑬ (in *plants*). The resulting monodehydroascorbate can be recycled by ferredoxin or by NADPH, while dehydroascorbate is recycled by GSH. Since NADPH is required for the reconversion of GSSG into GSH, an adequate supply is essential. The critical situation in *erythrocytes* has been described in 4.5.7.

Molecular scavengers, such as α-tocopherol (vitamin E, 9.12), interrupt the sequence of radical transfers in peroxidative chain reactions by accepting the radical function themselves. These long-living radicals then react with themselves or with physiological antioxidants (ascorbate, uric acid). The singlet oxygen radical is preferably trapped by β-carotene (7.3.2), ascorbate, GSH and somewhat less by α-tocopherol.

Literature:

Finkel, T.: Current Opinion in Cell Biology 10 (1998) 248–253.
Fridovich, I.: Ann. Rev. of Biochem. 64 (1995) 97–112.
Rice-Evans, C.A., Burdon, R.H. (Eds.): *Free Radical Damage and its Control. New Comprehensive Biochemistry* Vol. 28: Elsevier (1994).

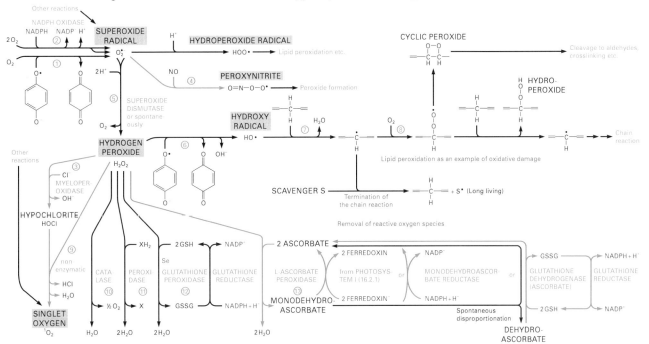

Figure 4.5-6. Generation of Reactive Oxygen Species (Red Background) and Their Removal

4.6 Leucine, Isoleucine and Valine

The essential amino acids L-leucine, L-isoleucine and L-valine are commonly named 'branched chain amino acids'. Their carbon skeleton originate from pyruvate (in case of isoleucine additionally from threonine). Their biosynthesis pathways are closely interrelated.

Leucine is a strictly ketogenic amino acid (the degradation products are acetoacetate and acetyl-CoA), isoleucine is ketogenic (acetyl-CoA) as well as glucogenic (succinyl-CoA), while valine is completely glucogenic.

4.6.1 Biosynthetic Reactions (Fig. 4.6-1)

The synthesis pathways for valine and isoleucine in *plants* and *bacteria* begin with the condensation of a 2-oxoacid with hydroxyethyl-ThPP, an intermediate of the pyruvate dehydrogenase reaction (3.3.1). This is followed by a reduction-alkyl group migration reaction which results in dihydroxy acids. Dehydration leads to the oxo acid precursors, which are converted to the amino acid by transamination. The leucine biosynthesis starts from the oxo acid precursor of valine and resembles the initial 3 steps of the citrate cycle (3.8.1).

In *green plants*, the biosyntheses of valine and isoleucine occur in the *chloroplasts*, while in *yeasts*, they are located in the *mitochondria*. The specific part of leucine biosynthesis in *yeasts* takes place either in the *mitochondria* or in the *cytosol*.

In some *bacteria*, L-3-*threo*-methylaspartate, an intermediate of the mesaconate pathway (Fig. 15.4-4), can also be converted into 2-oxobutyrate and act as a precursor of isoleucine. Acetolactate is also an intermediate during mixed acid fermentation (Fig. 15.4-2), while 2-oxoisovalerate is the starting compound for coenzyme A synthesis (9.7.1).

Regulation: There are many antagonisms of the three amino acids in feedback regulation, control of expression and active transport.

Isoleucine acts as a negative effector on threonine dehydratase, while valine antagonizes its effect. Control of the following biosynthesis reactions is complicated, since the pathways leading to valine and isoleucine (and the beginning of leucine biosynthesis) proceed through identical steps. Frequently multiple enzyme control takes place by individual end-product inhibition of isoenzymes (e.g., up to 6 acetolactate synthase isoenzymes in *Enterobacteria*). The regulation of gene expression varies between species. In *E.coli*, the genes for threonine dehydratase, dihydroxyacid dehydratase and branched chain amino acid transaminase are located in a single operon, which is multivalently repressed by the 3 branched chain amino acids. The genes for the initial acetolactate synthase isoenzymes are individually regulated.

The first specific enzyme of leucine biosynthesis, 2-isopropylmalate synthase, is feedback-inhibited by leucine, which generally occurs after branch points. The enzymes for the individual part of the leucine biosynthesis one encoded by a single operon, which is coordinately repressed or derepressed over a 1000-fold range.

In *animals*, these amino acids are preferably taken up by peripheral organs (*skeletal* and *cardiac muscle, kidney*) instead of by the *liver*. This is effected by the tissue distribution of the specific L-amino acid transporter.

4.6.2 Degradation of Branched-Chain Amino Acids (Fig. 4.6-1)

The intial catabolic steps for all 3 amino acids are catalyzed by the same *mitochondrial* enzymes. The sequence begins with reconversion to the respective oxo acids. Then a decarboxylation-dehydrogenation reaction similar to the pyruvate dehydrogenase reaction

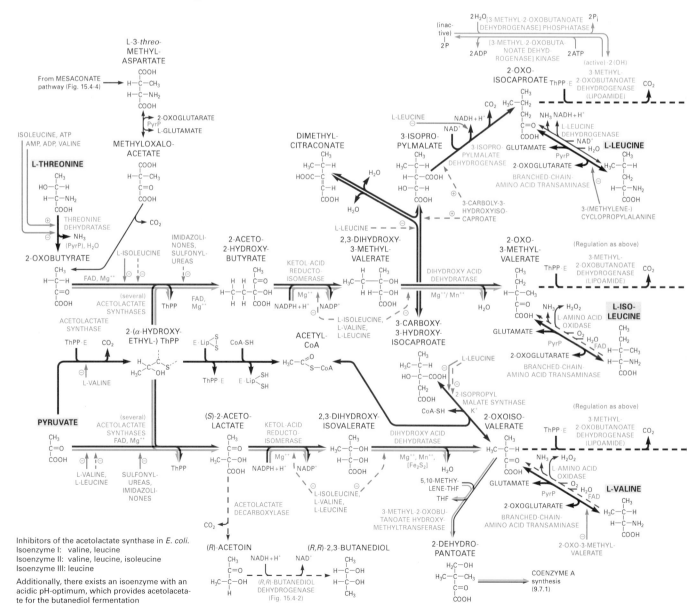

Figure 4.6-1. Metabolism of the Branched Chain Amino Acids

(3.3.1) follows. In isoleucine and valine catabolism, the resulting CoA derivatives undergo β-oxidation reactions analogously to the respective compounds during fatty acid degradation.

In isoleucine degradation, the resulting 2-methylacetoacetyl-CoA is cleaved to yield acetyl-CoA and propionyl-CoA. The latter compound is converted via methylmalonyl-CoA into succinyl-CoA as described in methionine catabolism (Fig. 4.5-2).

3-Hydroxyisobutyryl-CoA formed from valine, however, loses its CoA-group, is oxidized to methylmalonate semialdehyde and then further on to methylmalonate, which is converted into the CoA derivative. The degradation steps end with succinyl-CoA.

The 'CoA detour' in valine degradation is required, since the carbon chain is too short for regular β-oxidation. Some *bacteria* convert methylmalonate semialdehyde into propionyl-CoA instead. The intermediates of catabolism, isobutyryl-CoA and 2-methylbutyryl-CoA can also be used for the synthesis of branched-chain fatty acids (6.1.1).

While the first steps of leucine catabolism are analogous to both other amino acids, a biotin-dependent carboxylation step (9.8.2, analogous to acetyl-CoA carboxylase, 6.1.1) precedes the hydratase reaction. The final compound, 3-hydroxy-3-methylglutaryl-CoA is cleaved in *mitochondria* to acetyl-CoA and acetoacetate. It can be also used for cholesterol biosynthesis (7.1.1).

In *mammals*, the first specific enzyme for catabolism, 3-methyl-2-oxobutanoate dehydrogenase (branched-chain α-keto acid dehydrogenase), is enzymatically inactivated by phosphorylation and activated by dephosphorylation. In resting *muscle*, the enzyme is usually in the inactivated state. Thus, liberated branched chain amino acids are not catabolized, but released into the bloodstream. During exercise, the enzyme is activated and the amino acids are degraded and oxidized for energy production in the organ itself.

Defects in the initial decarboxylation-dehydrogenation enzyme system cause maple-sugar disease. This can occur with any of the 3 enzymes involved. There is an elevation in blood of all 3 branched chain amino acids and of the respective 2-oxo acids. Also 2-hydroxy acids are found, which arise by reduction of the oxo acids. Serious disturbances of the CNS and early death can be the consequences, if not treated by an appropriate diet.

Literature:

Harris, R.A. *et al.*: J. Nutr. 124 (1994) 1499S–1502S.
Kishore, G.M., Shah, D.M.: Ann. Rev. of Biochem. 57 (1978) 627–663.
Patel, M.S., Harris, R.A.: FASEB J. 9 (1995) 1164–1172.
Umbarger, H.E.: Ann. Rev. of Biochem. 47 (1978) 533–606.

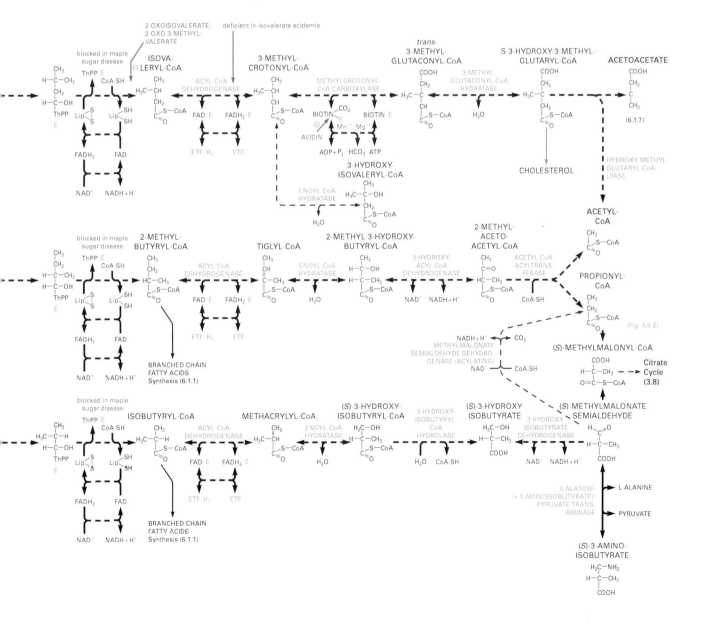

4 Amino Acids and Derivatives

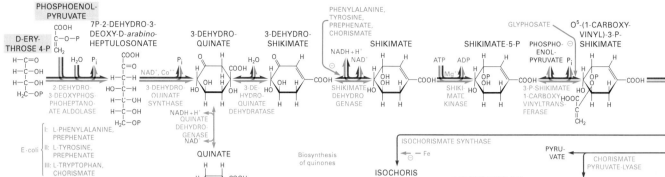

Figure 4.7-1. Biosynthesis of Aromatic Amino Acids and of Quinone Cofactors

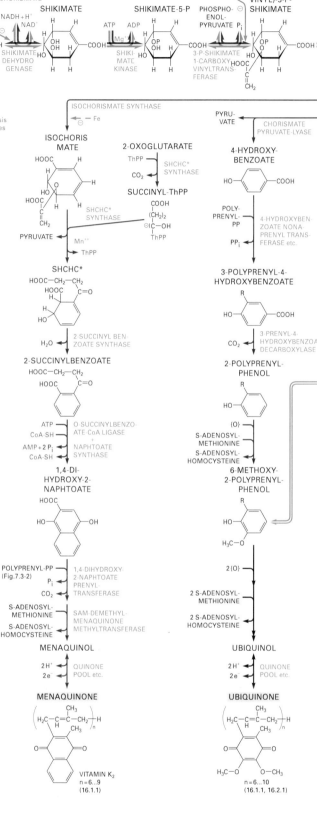

4.7 Phenylalanine, Tyrosine, Tryptophan and Derivatives

The amino acids of this group are commonly named 'aromatic amino acids'. Tryptophan and phenylalanine are essential for *mammals*. This is indirectly also valid for tyrosine, since it is formed from phenylalanine (*bacteria*, however, can bypass the phenylalanine step). The first part of biosynthesis is common to all three compounds.

NAD, NADP and a number of important biogenic amines and quinone compounds are derived from these amino acids or from their precursors. Tryptophan is frequently the limiting amino acid for protein biosynthesis.

The degradation of all these amino acids yields ketogenic acetoacetate. Tyrosine and phenylalanine also produce glucogenic fumarate. Tryptophan yields alanine, which can be transaminated to glucogenic pyruvate.

4.7.1 Biosynthesis of Aromatic Amino Acids (Fig. 4.7-1)

Common ('shikimate') pathway: The biosynthesis starts with phosphoenolpyruvate (PEP) and erythrose 4-P (from the pentose-P cycle, 3.6.1). Their condensation product cyclizes in a reaction requiring only catalytic amounts of NAD. Formation of two double bonds and introduction of another molecule phosphoenolpyruvate yield chorismate.

The latter reaction is inhibited competitively to PEP by glyphosate (P-O-CH$_2$-NH-CH$_2$-COOH), a widely used weed killer (Round-up®). Interestingly, other PEP-reactions are not inhibited.

Chorismate is a branch point for different pathways leading to phenylalanine and tyrosine, to tryptophan, to ubiquinone, phylloquinone and menaquinone (4.7.2) and to 4-aminobenzoate (9.6.1).

Phenylalanine and tyrosine biosynthesis: A mutase reaction yields prephenate, which undergoes either a dehydratase/decarboxylase or a reduction/decarboxylase reaction (in some *bacteria*) resulting in aromatization. The oxo analogues of phenylalanine and tyrosine obtained in this way are then converted into the respective amino acids by transamination.

In *chloroplasts* of *higher plants*, prephenate is first transaminated to arogenate, while the dehydratase/decarboxylase or reduction/decarboxylase reactions take place afterwards (arogenate pathway, not shown).

Mammals obtain tyrosine from phenylalanine by a hydroxylation reaction catalyzed by phenylalanine 4-monooxygenase, which uses tetrahydrobiopterin as a hydrogen donor (9.6.3). The reduced tetrahydrobiopterin is regenerated by NAD(P)H, catalyzed by dihydrofolate reductase.

The conversion into tyrosine is also the initial step of phenylalanine degradation. Disturbances of this reaction lead to phenylketonuria (4.7.3).

Tryptophan biosynthesis: Removal of the phosphoenolpyruvate moiety from chorismate and amination yields the aromatic compound anthranilate. Condensation with 5-P-ribosyl-PP (compare 8.1.1, an activated form of ribose 5-P) supplies two carbons to the 5-membered tryptophan ring, the rest of the ribose moiety is then cleaved off and replaced by L-serine.

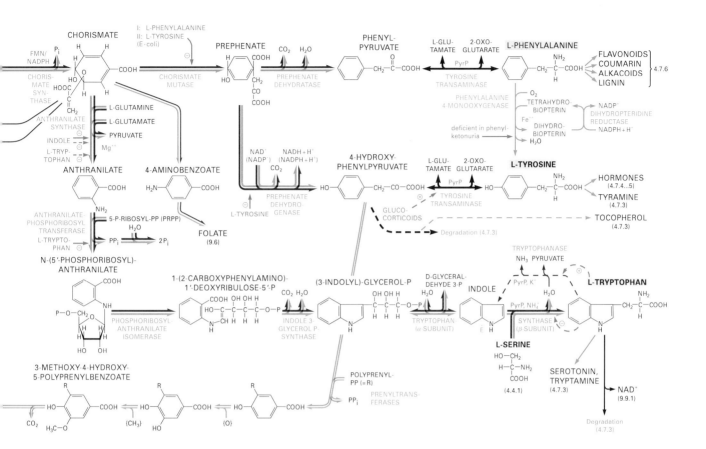

In *E. coli*, these two final reactions are performed by the α (29 kDa) and β (43 kDa) subunits of the tetrameric enzyme tryptophan synthase ($α_2β_2$). The combination of the (individually active) subunits greatly increases the reaction rate. This is due to direct transfer of the indole intermediate between the α and β subunits through a 'tunnel' in the enzyme ('channeling effect'), which prevents loss of this lipophilic compound through membranes. Serine forms a Schiff base with pyridoxal phosphate present in the β subunit, which is then dehydrated and nucleophilically attacked by indole.

Regulation: The regulatory systems differ greatly among species. Two examples are schematically shown in Fig. 4.7-2.

In *Bacillus subtilis*, each amino acid inhibits the first enzyme of its individual pathway after the respective branch point. Prephenate and chorismate, which are the end products of the common part of the pathway, inhibit its initial reaction (sequential control). Some of these enzymes are bifunctional.

A different system exists in *E.coli*: The initial reaction of the common pathway is catalyzed by 3 isoenzymes; each of which is inhibited by an amino acid end product (multiple enzyme control). After the first branch point, anthranilate synthase is inhibited by tryptophan, while the 2 bifunctional isoenzymes catalyzing the reactions to phenylalanine and tyrosine are correspondingly inhibited by these amino acids. After branching of these pathways, there is another inhibition by the respective amino acids (compare Fig. 4.5-1).

In *Neurospora crassa*, the inhibition pattern of the three isoenzymes catalyzing the first enzyme of the common pathway is analogous to *E.coli*. Anthranilate synthase is an individual enzyme. The other enzymes of the common pathway and of the anthranilate pathway (after the branch point) form multi-enzyme complexes, which allow catalytic facilitation by direct transfer to the next enzymatic function. Enzyme associations or multifunctional enzymes occur also in other organisms.

4.7.2 Biosynthesis of Quinone Cofactors

The quinone cofactors play an essential role in hydrogen transfer reactions, e.g. during aerobic and anaerobic respiration (16.1.1, 15.5) and photosynthesis (16.2.1). They form the 'quinone pool' in membranes. Phylloquinone is the cofactor for glutamate carboxylation, which is necessary for coagulation reactions (20.3.1).

Menaquinone and phylloquinone biosynthesis in *bacteria* (Fig. 7.7-1): By action of a mutase, chorismate is converted into isochorismate. A thiamine-PP requiring reaction effects decarboxylation of oxoglutarate. Then a condensation with isochorismate and the removal of pyruvate, takes place. After dehydratation (causing aromatization), the dihydrosynaphtoate structure is formed. It reacts with un-

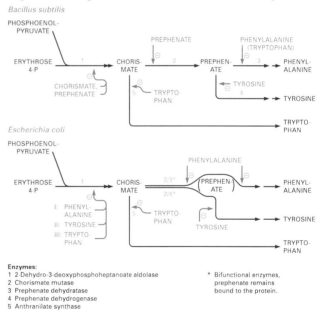

Figure 4.7-2. Regulation of Aromatic Amino Acid Biosynthesis

Enzymes:
1 2-Dehydro-3-deoxyphosphoheptanoate aldolase
2 Chorismate mutase
3 Prephenate dehydratase
4 Prephenate dehydrogenase
5 Anthranilate synthase

* Bifunctional enzymes, prephenate remains bound to the protein.

saturated polyprenyl-PP of various lengths (up to n = 15 isoprene units, Fig. 7.3-2), followed by methylation of the quinol and oxidation to menaquinone (vitamin K_2). The *plant* product phylloquinone is synthesized similarly. It differs in having a phytyl side chain of 4 isoprene units with a single double bond (vitamin K_1).

Ubiquinone biosynthesis (Fig. 7.7-1): In *bacteria*, the reaction starts from chorismate. After the removal of pyruvate, the 4-hydroxybenzoate formed is prenylated and decarboxylated. This is followed by hydroxylation and methylation reactions. In *yeast* and in *animals*, the starting compound for biosynthesis is 4-hydroxyphenylpyruvate (which can be obtained from tyrosine by transamination). The attachment of a polyprenyl chain (6...10 isoprene units, Fig. 7.3-2) is followed by hydroxylation, methylation and decarboxylation. The last steps are identical to the pathway in *bacteria*.

4.7.3 Derivatives and Degradation of Aromatic Amino Acids

Phenylalanine derivatives: The most important compounds directly derived from phenylalanine are found in *plants*. They are described in 4.7.5.

Tyrosine derivatives (Fig. 4.7-3): The product of tyrosine hydroxylation by tyrosine 3-monooxygenase is dihydroxyphenylalanine (dopa). This reaction resembles the conversion of phenylalanine to tyrosine and also employs tetrahydrobiopterin.

The result of dopa decarboxylation is dopamine. This compound is taken up by *neurons* and acts as a neurotransmitter (Table 17.2-2). An insufficient supply leads to Parkinson's syndrome. By a second hydroxylation at the 3-position with the participation of ascorbate, the important hormones norepinephrine (4.7.4, noradrenaline) and further on epinephrine (adrenaline) are generated. Iodination of tyrosine leads to thyroid hormones (4.7.5).

In *mammalian melanocytes*, tyrosine is oxidized to dopa and further on to dopaquinone, which after cyclization and oxidative condensation reactions yield melanins (dark pigments of the skin). This reaction is activated by irradiation. If the enzyme is deficient, albinism occurs. Similar dopa oxidations in *plants* (by phenoloxidases) lead to darkening of cut fruits or branches.

Decarboxylation of tyrosine without previous hydroxylation (e.g. by intestinal *bacteria*) yields tyramine, which elevates blood pressure.

Phenylalanine and tyrosine degradation (Fig. 4.7-3): Fermenting *bacteria* usually degrade both compounds by deamination and shortening of the side chain. In *animals*, however, the degradation of phenylalanine starts with its conversion to tyrosine (4.7.1).

Disturbances of the monooxygenase system lead to phenylketonuria, a very frequent hereditary disease in *humans* (frequency ca. 1 : 10000). Transamination to phenylpyruvate and consecutive reduction to phenyllactate or other reactions take place. The metabolites disturb neuronal development. The disease is treated by a low-phenylalanine diet.

For degradation, tyrosine undergoes transamination, followed by a decarboxylating dioxygenase reaction, yielding homogentisate. This enzyme requires the precence of ascorbate (or of another reducing agent). A second dioxygenation opens up the aromatic ring. The final products are fumarate and acetoacetate.

In *plants*, homogentisate is the precursor of plastoquinone and of tocopherol (vitamin E, 9.12). Their formation begins by the attachment of a polyprenyl-PP residue (for plastoquinone) or of a phytyl-PP residue (for tocopherol), followed by methylations and in case of tocopherol by a cyclization reaction.

Tryptophan derivatives (Fig. 4.7-4): This amino acid is an important source of biogenic amines. Via 5-hydroxylation and decarboxylation, serotonin is generated, which is ubiquitously occuring in *animals* (as a neurotransmitter, Table 17.2-2) and in *plants*. Its formation in *brain* is regulated by the tryptophan concentration (due to a high K_M of tryptophan hydroxylase, 2.5.2). The first step of serotonin catabolism is an oxidation by monoamine oxidase type A. The end product of serotonin degradation is 5-hydroxyindoleacetate, which is secreted in urine.

After its synthesis in the *CNS* and in *gastrointestinal cells* of *higher animals*, serotonin is interneuronally transported to vesicles at the nerve endings or it is released into the bloodstream, where it is transported by *platelets* to receptors, which are located on *smooth muscle, endothelial* and *epithelial cells, neurons* and *platelets*. See Tab. 17.2-2 for neuronal effects. In addition, serotonin relaxes the smooth muscles of *blood vessels* and of the *gastrointestinal tract*.

In the *pineal gland,* serotonin is is converted by acetylation and methylation into melatonin.

Figure 4.7-3. Derivatives and Degradation of Phenylalanine and Tyrosine

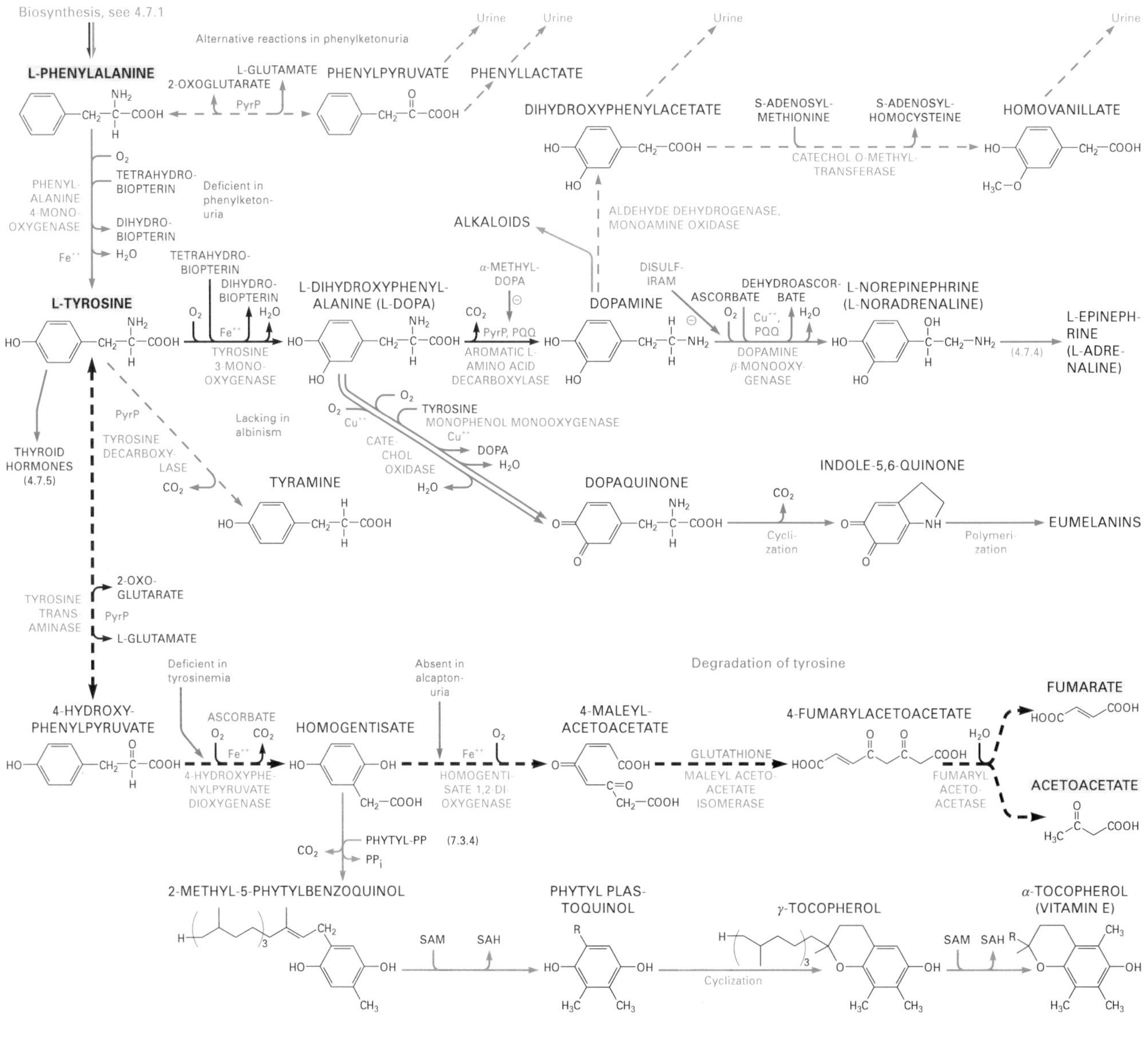

Figure 4.7-4. Derivatives and Degradation of Tryptophan

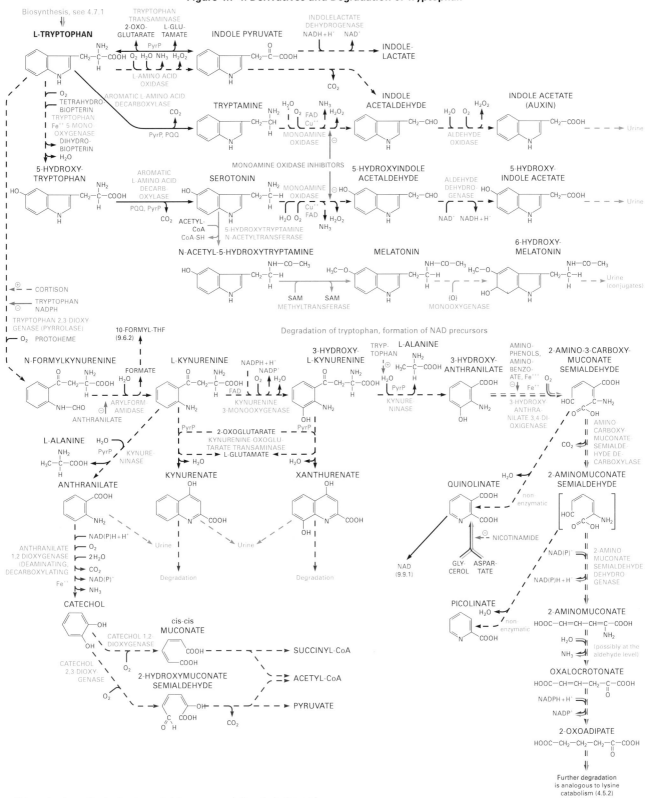

This amine is synthesized and secreted in pronounced diurnal rhythm and acts as an antagonist to melanotropin (Fig. 17.1-4). In *mammals* it antagonizes the function of the thyroid gland and the secretion of luteinizing hormone (17.1.5) and generally lowers metabolism ('sleep hormone'). Catabolism of melatonin takes place by hydroxylation and excretion in urine.

By direct decarboxylation of tryptophan without previous hydroxylation, tryptamine is obtained. Further oxidation of the side chain leads to indole acetate (auxin), which is a growth factor for *plants*. Alternatively, its biosynthesis can proceed via indolepyruvate. Besides in *plants*, tryptamine formation takes place in *bacteria*, in *kidney* and *liver*.

Tryptophan is the precursor of NAD and NADP (9.9.1). The pathway for synthesis of these coenzymes branches off from tryptophan catabolism.

Tryptophan degradation (Fig. 4.7-4): The major catabolic pathway in *mammalian liver* begins with the oxidative cleavage of the 5-membered ring and the consecutive liberation of formate (which is taken up by tetrahydrofolate, 9.6.2), resulting in kynurenine. After hydroxylation, a shortening of the side chain and release of alanine by action of kynureninase takes place.

The *cytoplasmic* kynureninase reaction is a β,γ-elimination requiring pyridoxal phosphate. If this cofactor is present only at low levels, the substrates are rather transaminated to kynurenate and xanthurenate (mostly by a *mitochondrial* transaminase which is less affected by pyridoxal phosphate deficiencies) and excreted. *Bacteria* can further degrade both acids.

Kynureninase can also act before hydroxylation takes place, resulting in anthranilate. Its conversion to catechol, ring opening and further degradation is a common way of tryptophan catabolism in *bacteria*. *Animals*, however, excrete anthranilate in conjugated form in *urine*.

After the kynureninase step in *mammalian* metabolism, the action of 3-hydroxyanthranilate 3,4-dioxygenase causes release of alanine and ring opening. The resulting semialdehyde can spontaneously cyclize to quinolinate, which is the initial compound for NAD and NADP biosynthesis (9.9.1). The same cyclization reaction can also take place after the β-carboxyl group has been removed, resulting in picolinate. The main degradation pathway leads to crotonyl-CoA (which is a member of the fatty acid degradation pathway) and further on to acetyl-CoA. The steps after 2-oxoadipate are also identical with lysine catabolism (4.5.2).

Literature:
Barker, H.A.: Ann. Rev. of Biochem. 50 (1981) 23–40.
Bender, D.A.: *Amino Acid Metabolism.* 2nd Ed. Wiley (1985).
Crawford, I.P., Stauffer, G.V.: Ann. Rev. of Biochem. 49 (1980) 163–195.
Eisensmith, R.C., Woo, S.L.C.: Mol. Biol. Med. 8 (1991) 3–18.
Kaufman, S.: Adv. in Enzymology 67 (1993) 77–264.
Kishore, G.M., Shah, D.M.: Ann. Rev. of Biochem. 57 (1988) 527–663.
Pau, P. *et al.*: Trends in Biochem. Sci. 22 (1997) 22–27.
Umbarger, H.E.: Ann. Rev. of Biochem. 47 (1978) 533–606.

4.7.4 Catecholamines

Catecholamines is the common term for the important hormones norepinephrine (noradrenaline) and epinephrine (adrenaline).

These compounds are produced by the *adrenal medulla* and by *postganglionic nerve terminals*, which are members of the *adrenergic nervous system* of *vertebrates*. While in some species norepinephrine is the major product, *humans, dogs* and some other species produce mostly epinephrine. These compounds are also found in *lower animals* and *plants*.

Biosynthesis (Fig. 4.7-5): Hydroxylation of tyrosine and consecutive decarboxylation yields dopamine (4.7.3). This compound is transferred by a specific carrier from the *cytosol* into *chromaffin granules*. Dopamine β-monooxygenase, a PQQ (15.7) and Cu^{++} dependent enzyme, catalyzes the hydroxylation to norepinephrine. A consecutive methyl transfer from S-adenosylmethionine, catalyzed by phenylethanolamine N-methyltransferase, leads to epinephrine. The catecholamines are stored in the granules and are released upon arrival of signals from the sympathetic nervous system.

Regulation of the biosynthesis (Details in Fig. 17.1-3): Norepinephrine inhibits the tyrosine monooxygenase, epinephrine inhibits the methyltransferase by allosteric mechanisms. Neuronal signals induce the expression of both biosynthetic hydroxylases and glucocorticoids the expression of the methyltransferase.

Hormone effects: The hormones act on α and β receptors, which are present in different organs. Although stimulation of both receptor types often cause opposite effects, they cooperate in response to physical or emotional stress situations by enhancing the energy supply and directing it to the sites where it is primarily needed (by influencing the circulation, increasing the heart output etc.). For details, see 17.1.4 and Table 17.2-2.

The antihypertensive drugs α-methyltyrosine and α-methyldopa are metabolized to the α-methyl anlogue of epinephrine (side chain methylated), inhibit the hormone synthesis and antagonize the hormone action at the receptor.

Degradation (Fig. 4.7-5): The degradation starts by methylation and oxidation reactions, catalyzed by catechol-O-methyltransferase (COMT) and monoamine oxidase (MAO), respectively. The products are excreted in urine, some of them in conjugated form.

4.7.5 Thyroid Hormones

Thyroid hormones (T_3 = 3,5,3′-triiodothyronine and T_4 = thyroxine = 3,5,3′,5′-tetraiodothyronine) are the other important hormone group in *vertebrates* derived from tyrosine.

Biosynthesis (Fig. 4.7-6): The formation of the thyroid hormones takes place in the *follicles* of the *thyroid gland*. The epithelial cells synthesize thyreoglobulin (dimeric, 2 * 330 kDa, 2 * 72 tyrosyl residues). By action of membrane bound thyreoperoxidase and H_2O_2, 10…20 % of the tyrosine residues get mono- or diiodinated. Coupling of the aromatic rings yields T_4 or T_3 and some inactive reverse T_3 (rT_3, 3,3′,5′-triiodothyronine). The prohormone is stored in the *follicular lumen*.

For secretion, thyreoglobulin reenters via pinocytosis (18.1.2) the epithelial cells. The vesicles fuse with *lysosomes*; the thyreoglobulin is then degraded. The iodinated amino acids are liberated and enter the *bloodstream*. Most of them are bound to thyroxin binding globulin (TBG, Table 4.7-1).

Table 4.7-1. Concentration of Thyroid Hormones in Blood

	total	free	% of total
T_3	1.5…3.5 nmol/l	4…9 pmol/l	0.1…0.6 %
T_4	60…140 nmol/l	10…25 pmol/l	0.04…0.07 %

Regulation: All steps of the biosynthesis are regulated by TSH (thyreoglobulin synthesis, iodination, coupling and vacuole formation, 17.1.5).

Figure 4.7-5. Biosynthesis and Degradation of Catecholamines

Figure 4.7-6. Biosynthesis and Degradation of Thyroid Hormones

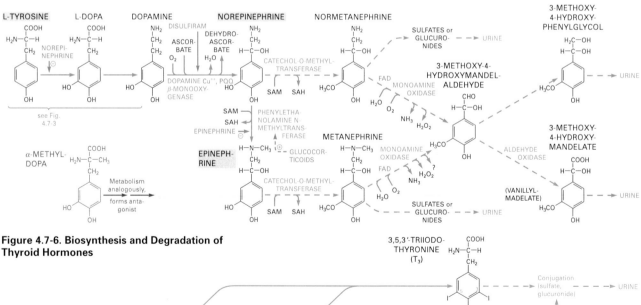

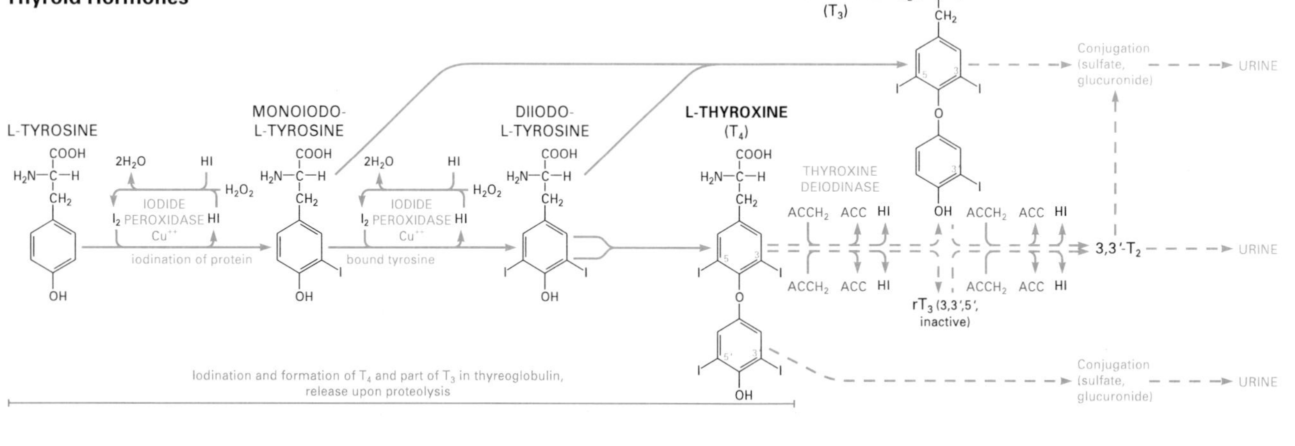

Hormone effects: Only the free hormones are biologically active. The lipophilic hormones pass through the membrane and meet their respective receptors intracellularly. In the *nucleus*, they modulate transcription (17.6). Generally, they enhance metabolism and development. Cretinism and goiter are consequences of defects in hormone synthesis or action (Details in 17.1.5)

Interconversions and degradation: After release from thyreoglobulin, part of T_4 is converted into T_3 in *kidney* and *liver*. Its hormonal activity is about three times higher than T_4. On the other hand, also inactive rT_3 is formed. Degradation of the hormones primarily involves deiodation by thyroxine deiodinase, the liberated iodide is reused. Further steps are conjugation with sulfate or glucuronate and excretion in bile and urine.

Literature on Catecholamines and Thyroid Hormones:
Brent, G.A.: New Engl. J. Med. 331 (1994) 847–853.
Cooke, B.A. *et al.* (Eds.): *Hormones and Their Actions. Part I. New Comprehensive Biochemistry* Vol. 18A. Elsevier (1988).
Nagatsu, T.: *Biochemistry of Catecholamines*. Univ. of Tokyo (1973).

4.7.6 Aromatic Compounds in Plants (Fig. 4.7-7)

Phenylpropanoids is the common term for a large number of compounds in *plants*, which are derivatives of phenylalanine. The removal of ammonia by action of phenylalanine-ammonia lyase (PAL) yields *trans*-cinnamic acid, which contains an unsaturated side chain. The enzyme is induced e.g. by illumination or by infections and is feedback-inhibited by its product. Then hydroxylations and also frequently methylations of the newly generated hydroxyl groups take place. Reduction of the CoA-activated carboxyl groups to the alcohol level results in 4-coumarol, ferulol and sinapol. These so-called monolignols are the starting compounds for synthesis of lignin.

The extracellular process of lignin synthesis is initiated by the formation of a radical, presumably by H_2O_2 action (4.5.8) and progresses via chain reaction mechanisms. The result is a close-meshed, irregular network. Its overall composition depends on the ratio of the originating alcohols and the reaction conditions and varies among different species. Lignin is the second most frequent compound in the biosphere (after cellulose, annual synthesis rate ca. $2 * 10^{10}$ t). It effects the pressure resistance of *plant* cell walls (3.6.3). Only a few organisms, mostly *fungi*, can degrade lignin. Suberine has a similar structure with alcoholic groups esterified by (mostly) long-chain fatty acids.

Flavonoids are compounds, in which another aromatic ring is bound to the C-9' atom of cumarate. In stilbenes, the aromatic ring is bound to the C-8' atom after removal of the C-9' atom. The C-atoms of the newly formed ring are contributed by 3 malonyl-CoA molecules (malonate pathway of aromatization). Frequently, these compounds are glycosylated afterwards.

The synthesizing enzymes, especially chalcone synthase and chalcone isomerase are closely controlled. E.g., elicitors (signal compounds indicating an infection, injury etc.) induce their expression. The synthesized flavonoids play a role in defense mechanisms (as phytoalexins). Other flavonoids serve protection purposes (e.g., tannin).

2-Hydroxylation of 4-coumarate and ring closure yields umbelliferone. Further steps lead to psoralene, a compound which, after illumination, reacts with DNA and blocks transcription. Oxidative shortening of the phenylpropanoid side chain results in phenolic acids, e.g. salicylic acids.

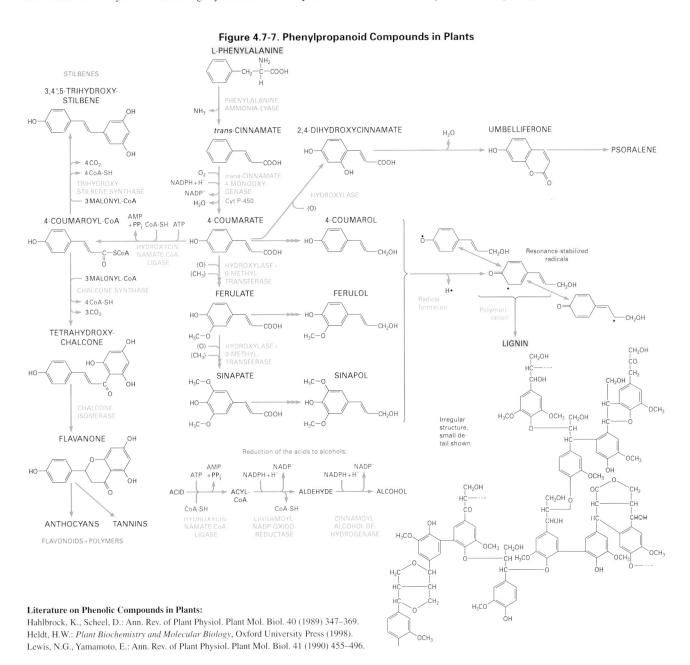

Figure 4.7-7. Phenylpropanoid Compounds in Plants

Literature on Phenolic Compounds in Plants:
Hahlbrock, K., Scheel, D.: Ann. Rev. of Plant Physiol. Plant Mol. Biol. 40 (1989) 347–369.
Heldt, H.W.: *Plant Biochemistry and Molecular Biology*, Oxford University Press (1998).
Lewis, N.G., Yamamoto, E.: Ann. Rev. of Plant Physiol. Plant Mol. Biol. 41 (1990) 455–496.

4 Amino Acids and Derivatives

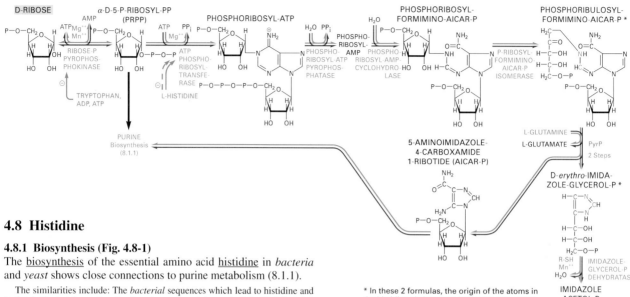

Fig. 4.8-1. Biosynthesis and Degradation of Histidine

* In these 2 formulas, the origin of the atoms in the histidine skeleton are indicated: red = from ribose, blue = from ATP, green = from glutamine

4.8 Histidine

4.8.1 Biosynthesis (Fig. 4.8-1)

The biosynthesis of the essential amino acid histidine in *bacteria* and *yeast* shows close connections to purine metabolism (8.1.1).

The similarities include: The *bacterial* sequences which lead to histidine and purine both start from phosphoribosyl pyrophosphate; part of the histidine ring is derived from purine; AICAR (see below) is released during histidine biosynthesis and is utilized in purine formation. This connection and the presence of the catalytically active imidazole group in these compounds might be a hint for a phylogenetic development from the 'RNA world' to the 'protein world' and from ribozyme (e.g., 10.4.3) to enzyme catalysis. The pathway in higher *plants* has not been elucidated in all details.

The activated compound phosphoribosyl-PP reacts with the purine ring of ATP, which is thereafter opened. After isomerization, cleavage of this molecule releases 5'P-ribosyl-5-amino-4-imidazole carboxamide (AICAR). This is also an intermediate of purine formation (8.1.1) and can be converted with a few steps into ATP, thus recovering this compound, which has been previously utilized for histidine biosynthesis.

During the same reaction step, glutamine adds an amino group to the rest of the molecule, then closure of the imidazole ring takes place. After 4 more steps, histidine is obtained.

Histidine inhibits the first individual step of the pathway, ATP-phosphoribosyltransferase.

4.8.2 Interconversions and Degradation (Fig. 4.8-1)

An important product obtained by histidine decarboxylation in *animals* is the biogenic amine histamine.

The reaction proceeds in *mast cells* and in many *tissue cells*. Histamine is released from *mast cells* in IgE-mediated hypersensitivity of the immediate type (19.8.1, allergic reactions). It acts on specific receptors (Table 17.2.2). This leads to bronchoconstriction and to NO release, which in turn results in vasodilatation (17.7.2). Histamine also stimulates the secretion of HCl into the *stomach*. Histamine formation as well as degradation is hormone or cytokine controlled. Histamine is also present in *stinging nettles, bee venom*, secret of *mosquitoes* etc. The development of antihistamines is an important issue in pharmacology.

The dipeptide carnosine (β-alanylhistidine) occurs in large quantities in *muscles*. It is synthesized in an ATP-dependent reaction from histidine and β-alanine. It provides histidine for metabolism if its supply by food is insufficient. Another *muscle* component is 3-methylhistidine, which is present in actin and myosin (17.4.5). The methylation is a post-transcriptional reaction.

Histidine degradation starts with a non-oxidative deamination resulting in urocanate. This is followed by hydratation (which saturates two double bonds!) and hydrolytic opening of the ring. The resulting formiminoglutamate transfers its formimino group to tetrahydrofolate (9.6.2), glutamate is obtained.

Histidinemia results from deficiency of histidine ammonia-lyase. In the case of folate deficiency, formiminogluatamate appears in urine.

Literature:
Umbarger, H.E.: Ann. Rev. of Biochem. 47 (1978) 533–606.

THF = tetrahydrofolate
SAM = S-adenosylmethionine
SAH = S-adenosylhomocysteine

4.9 Urea Cycle, Arginine and Associated Reactions

In *animals*, ammonia, which is liberated by the degradation of amino acids and which is not incorporated into other compounds, has to be excreted.

Animals living in water can easily do this with the free compound (ammonotelic animals). This requires the simultaneous, constant excretion of water, which is not possible in terrestrial animals. Due to the toxicity of the free compound, these animals convert ammonia into nontoxic urea in the *liver* (urotelic animals). *Birds*, terrestrial *reptiles* and many *insects*, however, excrete uric acid (8.1.6, uricotelic animals).

4.9.1 Urea Cycle (Krebs-Henseleit Cycle, Fig. 4.9-1)

The urea cycle is not only a means to remove ammonia from the body, but CO_2 as well. In *animals*, *plants* and *bacteria*, it is also the source of the proteinogenic amino acid arginine, of polyamines and guanidino compounds (4.9.2).

In *animals*, the first steps of the urea cycle proceed in the *mitochondria* of the *liver*. While ammonia release due to degradation reactions already takes place there, a major part of CO_2 has to pass from the cytosol into the mitochondria, where it is converted into bicarbonate by carbonate dehydratase (carboanhydrase). The ATP-driven carbamoyl-P synthesis by carbamoyl synthase I (ammonia) from NH_3 and HCO_3^- requires N-acetylglutamate (4.2.2) as an allosteric activator.

The concentration of N-acetylglutamate parallels the glutamate concentration, thus it apparently acts as a sensor of glutamate increase resulting from elevated protein degradation or supply from food.

The glutamine hydrolyzing carbamoyl synthase II occurs in the *cytosol* and is involved in pyrimidine biosynthesis (8.2.1). In *bacteria*, this is the only type of carbamoyl synthase.

The condensation reaction with ornithine (biosynthesis see 4.2.2) yields citrulline. (Both amino acids are not proteinogenic.) Citrulline leaves the *mitochondria* and condenses with aspartate in an ATP-dependent reaction to argininosuccinate. Cleavage of this compound yields the proteinogenic amino acid arginine and fumarate.

The latter compound is converted via malate to oxaloacetate. Although these reactions are identical to citrate cycle reactions, they take place in the *cytosol*. Oxaloacetate is reconverted by transamination into aspartate (aspartate cycle). Finally, arginase cleaves arginine, liberates urea and reconstitutes ornithine, which returns to the *mitochondria*.

While carbamoyl-P synthase and ornithine carbamoyl transferase are associated with the *inner membrane* of liver mitochondria, the other enzymes of the cycle are associated with the outside of the *outer membrane*. Thus they are in close proximity (forming a metabolon), which allows direct substrate passage from one enzyme to the next (substrate channeling).

The urea cycle is energized by the hydrolysis of 4 energy-rich phosphate bonds (3 ATP → 2 ADP + AMP). However, they are reconstituted during oxidative phosphorylation, using the NADHs formed by the glutamate dehydrogenase (which releases ammonia) and malate dehydrogenase reactions.

The expression of all enzymes of the urea cycle in *animals* is increased during protein rich diets. Also in *bacteria*, the expression of all enzymes of arginine synthesis is co-regulated by a common mechanism (regulon, 10.5.4).

Figure 4.9-1. Urea and Aspartate Cycles, Polyamine and Creatine Synthesis

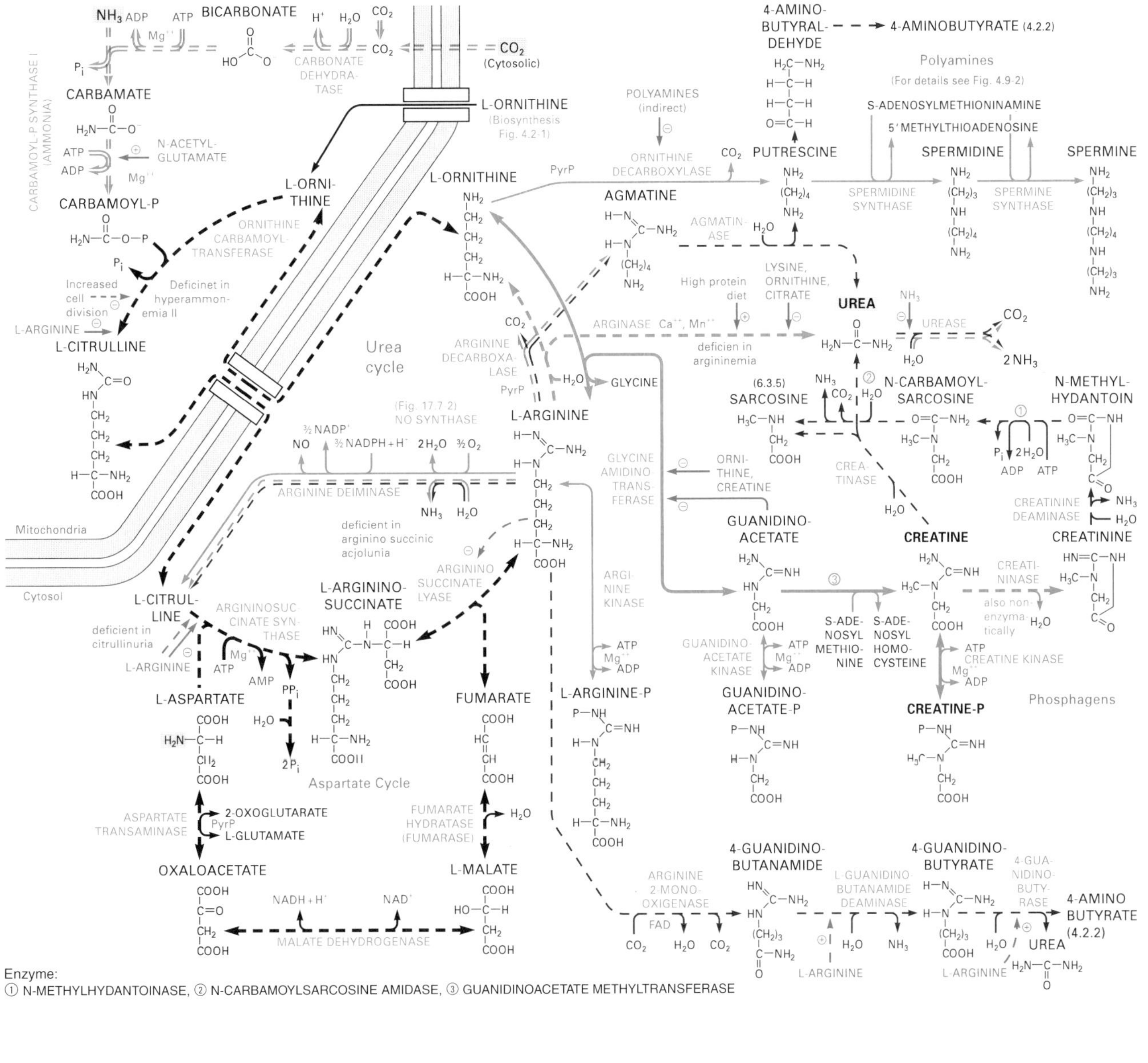

Enzyme:
① N-METHYLHYDANTOINASE, ② N-CARBAMOYLSARCOSINE AMIDASE, ③ GUANIDINOACETATE METHYLTRANSFERASE

Minor activities of the urea cycle enzymes are also found in other organs besides the liver. Apparently their major task is the biosynthesis of arginine or the provision of ornithine for synthesis of polyamines (4.9.3).

Deficiencies of all enzymes of the urea cycle have been observed (Table 4.9-1, combined homozygotic frequencies of these diseases ca. 1/25000). They lead to increased blood concentration of their particular substrates, but also of their precursors in the cycle as far back as ammonia. The primary effect of hyperammonemia is brain damage.

Table 4.9-1. Diseases Caused by Deficiencies in Urea Cycle Enzymes

Deficient Enzyme	Disease
Carbamoyl-P synthase (ammonia)	hyperammonemia I
Ornithine carbamoyltransferase	hyperammonemia II
Argininosuccinate synthase	citrullinuria
Argininosuccinate lyase	argininosuccinic aciduria
Arginase	argininemia
N-acetylglutamate synthase (forms activator)	N-acetylglutamate synthase deficiency

Degradation of arginine, as well as of the other urea cycle intermediates to glutamate occurs via glutamate semialdehyde by reversal of their biosynthesis (except the phosphorylation step).

Literature:
Batshaw, M.L.: Ann. Neurol. 35 (1994) 133–141.
Grisolia, S. et al.: The urea cycle. Wiley (1976).
Meijer, A.J. et al.: Physiol. Rev. 70 (1990) 701–748.

4.9.2 Phosphagens (Phosphocreatine and Phosphoarginine, Fig. 4.9-1)

In *vertebrates*, the guanidino residue of arginine can be transferred to glycine by action of glycine amidinotransferase. This reaction yields guanidinoacetate and ornithine. Guanidinoacetate is further methylated to creatine. These reactions take place in *liver, kidney* and *brain*. Via the bloodstream, a considerable amount of creatine is transported to the *skeletal muscles*.

Phosphorylation of creatine results in the high energy compound (8.1.3) phosphocreatine. The formation of phosphocreatine is endergonic:

Creatine + ATP = phosphocreatine + ADP $\Delta G'_0 = +12.6$ kJ/mol

$$K = \frac{[\text{phosphocreatine}] * [\text{ADP}]}{[\text{creatine}] * [\text{ATP}]} = 6.18 * 10^{-3}$$

Due to this equilibrium, phosphorylation of creatine can only take place when there is ample supply of ATP. On the other hand, quick reconstitution of ATP is possible when high energy demand causes the dephosphorylation of ATP to ADP.

Correspondingly, creatine phosphorylation takes place inside the *mitochondria*, phosphate transfer to ADP proceeds in the *cytosol*. This way, phosphocreatine acts as a 'buffer' and enlarges the amount of quickly available energy (especially in *skeletal muscle*, but also in *heart and brain*, less in other tissues). The same function in *invertebrates* is exerted by phosphoarginine.

The lability of phosphocreatine is caused by the competition of the electron withdrawing imino and phosphoryl groups, which decrease the resonance stabilization of the C-NH-P bridge. This effect surpasses the analogous one in pyrophosphate bonds (e.g. in ATP), as can be seen by the $\Delta G'_0$ for hydroysis

phosphocreatine → creatine + P_i $\Delta G'_0 = -43.1$ kJ/mol
ATP → ADP + P_i $\Delta G'_0 = -30.5$ kJ/mol

For excretion in urine, creatine is converted to creatinine.

Literature:
Bessman, S.P., Carpenter, C.L.: Ann. Rev. of Biochem. 54 (1985) 831–862.

4.9.3 Polyamines (Fig. 4.9-2)

Decarboxylation of ornithine results in the diamine putrescine. In *bacteria*, this compound can also be formed from arginine by a two-step reaction.

Inhibition of ornithine decarboxylase takes place by tight, noncovalent binding of specific proteins (antizymes) and indirectly by polyamines. In *mammals* (not in *yeast* or *bacteria*), ornithine decarboxylase has a high turnover rate ($t_{1/2} = 10...30$ min). Its expression is enhanced and the enzyme degradation is delayed by growth factors (NGF, EGF, PDGF), insulin, corticosteroids and testosterone.

The addition of one and two propylamino residues leads to spermidine and spermine, respectively. These polyamines (except spermine) occur ubiquitously; especially in quickly dividing cells. They are essential growth factors. Their highest concentration is found at the end of the G_1 phase.

Possibly their biological function is the stabilization of DNA by association of their amino groups with the phosphate residues of DNA. They also have an enhancing effect on RNA synthesis and on tRNA and ribosome stability.

S-adenosylmethioninamine is the donor of the aminopropyl groups. This compound is generated by decarboxylation of S-adenosylmethionine; the decarboxylase contains the unusual cofactor pyruvate instead of pyridoxal phosphate. While in the transfer reactions from S-adenosylmethionine the methyl group is transferred (4.5.4), the spermidine and spermine synthase reactions cleave the other sulfonium-carbon bond and transfer the aminopropyl group (Fig. 4.9-2), yielding 5'-methylthioadenosine instead of S-adenosylhomocysteine.

Degradation of polyamines: In *animals* and *yeast*, spermine and spermidine are acetylated to their N^1 derivatives and then oxidized at the secondary amino group by polyamine oxidase, releasing aminopropionaldehyde and finally putrescine. This compound is converted into 4-aminobutyrate and further on via succinate semialdehyde to succinyl-CoA, ending up in the citrate cycle. In some *bacteria*, instead of aminopropionaldehyde, 1,3-diaminopropane is released.

Literature:
Canellakis, E.S. et al.: Biosci. Rep. 5 (1985) 189–204.
Large, P.J.: FEMS Microbiol. Rev. 8 (1992) 249–262.
Tabor, C.W., Tabor, H.: Ann. Rev. of Biochemistry 53 (1984) 749–790.
Verdine, G.L.: Cell 76 (1994) 198–206.

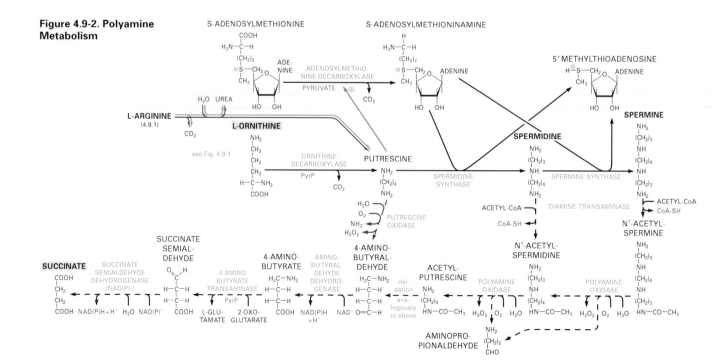

Figure 4.9-2. Polyamine Metabolism

5 Tetrapyrroles

In tetrapyrroles, 4 pyrrole residues are interconnected by methine (–CH=) bridges. Most important are circular compounds with a system of conjugated double bonds: porphyrins and chlorins. (In chlorins, one or two of the pyrroles are partially reduced, but the uninterrupted conjugated ring system is kept up.) Usually, the pyrrole moieties are substituted by methyl, vinyl, acetate or propionate residues. Members of this group are present in all living beings. The most important representatives are hemes (present in hemoglobin and cytochromes) and chlorophylls, which contain metal ions (Fe^{++} and Mg^{++}, respectively) in the center of the ring system. The biosynthetic pathways branch at protoporphyrin IX.

A related ring system is the corrin ring (present in cobalamin, 9.5), where two of the pyrrole moieties are directly interconnected without a methine bridge. This compound and the porphyrins siroheme and coenzyme F_{430}, share the initial portion of their biosynthetic pathways with hemes and chlorophylls.

5.1 Steps to Protoporphyrin IX

The formation of the key intermediate 5-aminolevulinate proceeds via 2 different pathways (Fig. 5.1-2):

- The succinyl-CoA pathway is used by *mammals*, *birds* and some bacteria (e.g. *Rhodobacter*)
- The glutamate (= C_5) pathway is used by *higher plants*, *algae* and many *bacteria* and *archaea*.

Succinyl-CoA pathway: Succinyl-CoA and glycine are the starting compounds. Their condensation in *mitochondria* is followed by decarboxylation, resulting in 5-aminolevulinate. This compound is then transported into the *cytoplasm*.

The 5-aminolevulinate synthase reaction is pyridoxal-P dependent and involves the labilization of the C_α-H bond of glycine by formation of a Schiff base (9.4.2). It is only step requiring an activated substrate (succinyl-CoA) and is also the attack point for regulatory mechanisms.

- In the *liver* of *higher animals*, heme inhibits the enzyme directly and also represses its synthesis. Additionally, it blocks the transport of the synthesized enzyme from the *cytoplasm* into the *mitochondria*. Thus, compounds which diminish the heme pool increase the enzyme activity. Most important are compounds which have to be hydroxylated for detoxification. They stimulate the heme-requiring synthesis of cytochrome P-450 (Fig. 7.4-1) and thus reduce the amount of free heme. Since 5-aminolevulinate synthase has a short half-life (1...3 hours), there is a quick response to regulation mechanisms.
- Contrary to this, heme in *reticulocytes* (maturating *erythrocytes*, Fig. 19.1-2) stimulates the globin synthesis by controlling the translation of mRNA (5.2.2). Thus, both components of hemoglobin are obtained in a balanced way.

5-Aminolevulinate synthase also catalyzes the condensation of acetyl-CoA and glycine, resulting in L-2-aminoacetoacetate. Further steps lead either to 1-amino-2-propanol, which is required for coenzyme B_{12} synthesis (9.5) or via methylglyoxal to D-lactate.

Glutamate pathway: In *plants* and many *bacteria*, glutamate is converted to 2-aminolevulinate with the carbon skeleton remaining intact. The pathway involves an unusual activation of the carboxylate group for the consecutive reduction by formation of Glu-tRNA^{Glu}. Then follows an 'internal transaminase reaction', which moves the amino group from the C-2 to the C-1 position.

In *plants*, this reaction takes place in the *chloroplasts*. It is activated by light. The first 2 steps are inhibited by heme (which also occurs in chloroplasts). The last reaction requires PyrP in the pyridoxamine form, which forms a Schiff base with the semialdehyde. Apparently, all the consecutive steps up to chlorophyllide are also located in the chloroplasts. In addition, some later steps of the heme synthesis may occur in *mitochondria*.

Further steps: The condensation of 2 molecules 2-aminolevulinate yields porphobilinogen (PBG) and is catalyzed by porphobilinogen synthase (5-aminolevulinate dehydratase).

This reaction also proceeds via Schiff bases. They are formed with amino groups of the enzyme, which is an octamer. The metal ion required varies between species.

The condensation of 4 molecules porphobilinogen results in uroporphyrinogen III (Details in Fig. 5.1-1). The first step of this reaction is catalyzed by hydroxymethylbilane synthase (porphobilinogen deaminase) and involves the formation of the linear tetrapyrrole hydroxymethylbilane. This labile compound will spontaneously cyclize into the symmetric uroporphyrinogen I, unless a second enzyme, uroporphrinogen-III (co)synthase is present. This enzyme effects an 'inversion' of pyrrole ring D before the closure to the asymmetric uroporphyrinogen III ring takes place.

Hydroxymethylbilane synthase contains an unique dipyrromethane cofactor (which is the condensation product of 2 PBG residues) covalently bound to an enzyme cysteine. Base-catalyzed removal of an amino group from free PBG enables its condensation with the cofactor. This is repeated 3 more times. The resulting linear hexapyrrole is then hydrolytically cleaved and the tetrapyrrole hydroxymethylbilane is released. The reaction of the cosynthase likely involves a spiro intermediate. The pathway leading to coenzyme B_{12} branches off from uroporphyrin III (9.5).

Decarboxylation of all 4 acetate side chains of uroporphyrin III by a single enzyme yields coproporphyrinogen III. In *eukarya*, this compound is transferred back into the *mitochondria*. By oxidative decarboxylation, 2 propionate side chains are converted into vinyl groups, resulting in protoporphyrinogen IX. Finally, by oxidation of the methylene bridges connecting the pyrrole residues to methenyl bridges, protoporphyrin IX is obtained. This is a macrocycle with conjugated double bonds. In anaerobic conditions, other compounds act as electron acceptors in the place of oxygen.

Protoporphyrin IX is the branching point for the pathways leading to heme and to chlorophyll.

Figure 5.1-1. Mechanism of Uroporphyrinogen III Synthesis

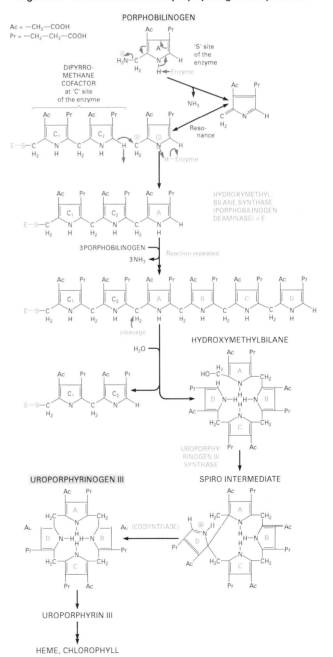

Figure 5.1-2. Biosynthesis of Hemoglobin and Cytochromes

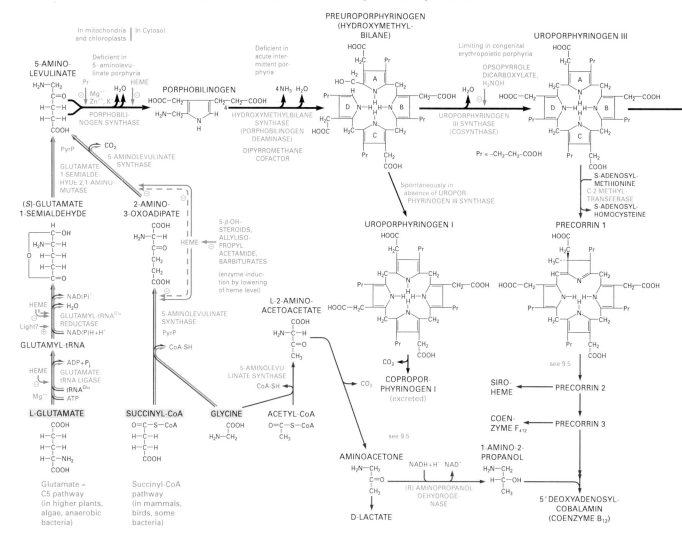

Medical aspects: Defects of almost all of the heme synthesis enzymes are known (see Fig. 5.1-2). The diseases are collectively named porphyrias and are characterized by accumulation of the respective substrates.

In general, acute porphyrias are caused by enzyme defects previous to uroporphyrinogen III and non acute porphyrias by deficiencies at later steps. Most porphyrias are associated with skin lesions due to photosensitisation by light-absorbing porphyrins (strongest at $\lambda \approx 400$ nm = Soret band) in addition to other disturbances. Frequent congenital diseases are the acute intermittent porphyria (affects hydroxymethylbilane synthase) and the porphyria cutanea tarda (affects uroporphyrinogen decarboxylase). The congenital erythropoietic porphyria (deficiency of uroporphyrinogen III synthase) prevents formation of uroporphyrinogen III, therefore the non enzymatic conversion to uroporphyrinogen I and consecutive decarboxylation to coproporphyrinogen I takes place (which causes red urine color).

Porphobilinogen synthase is inactivated by lead, which displaces the activating zinc. The resulting increase of 5-aminolevulinate causes psychoses (This compound resembles the neurotransmitter 4-aminobutyrate, Table 17.2-2).

Literature:
Jordan, P.M.: *Biosynthesis of Tetrapyrroles. New Comprehensive Biochemistry* Vol. 19. Elsevier (1991).
Louie, G.V. *et al.*: Nature 359 (1992) 33–39.
Warren, M.J., Scott, A.I.: Trends in Biochem. Sci. 15 (1990) 486–491.

5.2 Hemoglobin, Myoglobin and Cytochromes

5.2.1 Heme Biosynthesis

Heme (protoheme) is the first compound in the biosynthetic pathway with an inserted iron atom. This reaction is catalyzed by ferrochelatase (heme synthase, protoheme ferro-lyase), which is located at the inner mitochondrial membrane or at the *chloroplast* membrane (in *plants*).

In higher *animals*, ca. 85% of the heme production takes place in *reticulocytes* (immature *erythrocytes*, Fig. 19.1-3) to provide hemoglobin. Most of the rest is produced in the liver and is then converted into cytochrome P-450 (Fig. 7.4-2). This is the prosthetic group of hydroxylases, which are involved in detoxifying and in biosynthetic reactions. Correspondingly, heme biosynthesis is regulated in liver and reticulocytes in different ways (5.1).

Iron metabolism in higher *animals* (Fig. 5.2-1): In the *intestine*, food iron is reduced from the Fe^{+++} to the more soluble Fe^{++} state. During resorption by the *mucosal cells* of the *intestine*, Fe attaches to mucin. Inside the cells, it is transported by mobilferrin to ferritin, which is the main Fe storing protein of the body (450 kDa, binds about 5000 Fe^{+++} ions in the internal cavity). Ferritin responds to low iron levels in *blood plasma* by releasing iron. This is taken up by to transferrin (88 kDa, binds 2 Fe^{+++}) and is transported by the bloodstream to other cells. The expression of these iron binding proteins is regulated by the iron-sensing protein, which binds to DNA when the blood iron level is low.

Fe storage also occurs in cells of the *reticuloendothelial system* and of *liver* by binding to ferritin and to hemosiderin. It is released from the bound forms by reduction to the Fe^{++} level by ferritin reductase (transferrin reductase) and is then reoxidized by ferrioxidase I (coeruloplasmin).

The permanent binding of Fe to proteins greatly reduces the generation of reactive oxygen species by free Fe (4.5.8).

5.2.2 Biosynthesis and Properties of Hemoglobin and Myoglobin

Hemoglobin (Hb) has the task of delivering oxygen to the organs of *vertebrates*. *Invertebrates* use different compounds: hemocyanin (contains Cu^{++}) or hemerythrin (contains nonheme-Fe). The O_2 transport by proteins is necessary since the O_2 solubility in blood plasma is too low to satisfy the oxygen demand of the tissues. Myoglobin takes care of the oxygen transport inside of rapidly respiring *muscles*. *Mammals* living in water also use myoglobin for oxygen storage. Hemoglobin and Myoglobin contain 4 and 1 heme groups, respectively, surrounded by globulin.

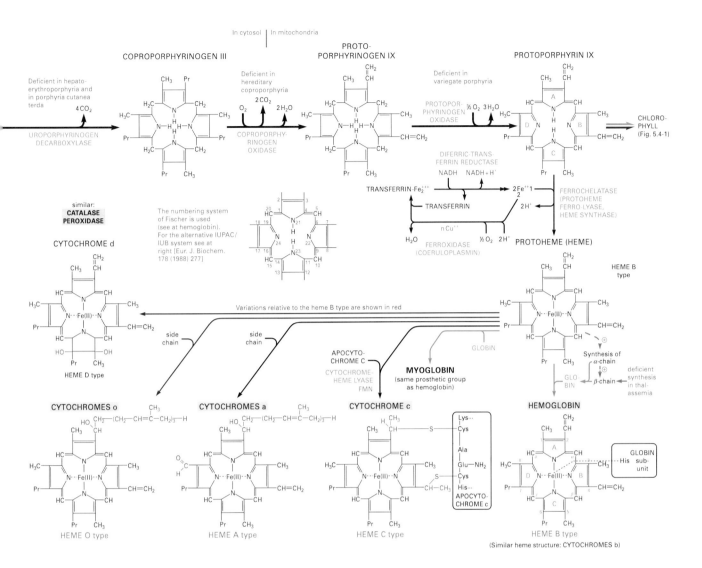

Figure 5.2-1. Transport and Storage of Iron in Higher Animals

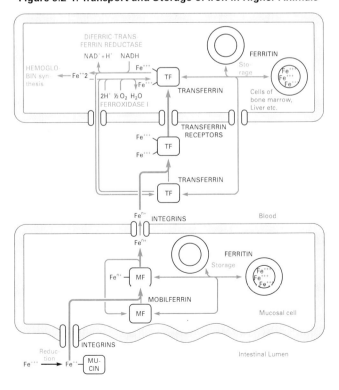

Biosynthesis: In *reticulocytes* (immature *erythrocytes*), heme enhances the rate of globin synthesis. The regulation takes place at the translational level (Fig. 5.2-2).

In absence of heme, the heme-controlled repressor (HCR) is activated by a mechanism, which is not known in detail so far. It catalyzes the phosphorylation of the translation initiation factor eIF-2 (see Fig. 11.5-2). The phosphorylated factor firmly binds the GDP-GTP exchange factor eIF-2B and thus prevents any further GDP-GTP exchange at eIF-2, which is necessary for its function. The initiation of transcription stops. Heme prevents the activation of HCR and thus enables continuous globin production.

In addition, the synthesis of the β-chain is controlled by the α-chain.

Structure (Fig. 5.2-3): Hemoglobins are tetramers of roughly globular shape, containing 2 * 2 identical subunits ($\alpha_2\beta_2$, 4 * 17 kDa), which are kept together by many noncovalent interactions. Each subunit contains one heme molecule, which is partially hidden in a 'pocket'. The interior of the protein is filled with nonpolar amino acids except for two histidine residues on both sides of the heme ring ('proximal' and 'distal' histidine).

Figure 5.2-2. Control of Globin Synthesis by Heme-Controlled Repressor HCR

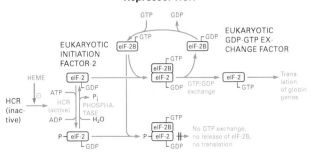

In humans, the globin subunits of hemoglobin change during life: embryonic state $\zeta_2\varepsilon_2$ or $\alpha_2\varepsilon_2$, fetal state $\alpha_2\gamma_2 = HbF$, adult state 97.5% $\alpha_2\beta_2 = HbA_1$ and 2.5% $\alpha_2\delta_2 = HbA_2$.

Myoglobin consists of a single chain with one heme molecule, which closely resembles a hemoglobin subunit (16.7 kDa).

In hemoglobin and in myoglobin, the coordination sites of Fe are occupied by the 4 porphyrin-N and the proximal histidine residue. The occupation of the sixth site and the oxidation state of Fe differs (Table 5.2-1):

Table 5.2-1. Environment of the Heme Group

Hemo-/myoglobin	Sixth Ligand	Oxidation state of Fe
Deoxy-	none	II
Oxy-	O_2 (loosely bound)	II
CO-	CO (firmly bound)	II
Met- (= Ferri-)	H_2O (firmly bound)	III

Both with free hemes and with hemo-/myoglobin, O_2 binds at an angle to the axial direction, while CO binds to free heme in axial direction. In hemo-/myoglobin, it is forced by the distal histidine to obtain a more unfavorable tilted position. This reduces the binding affinity of CO considerably (from 25 000 times the affinity of O_2 to 200 times) and provides partial protection against the CO produced by heme degradation (5.3.1).

Figure 5.2-4. Structural Shifts During Oxygenation of Hemoglobin

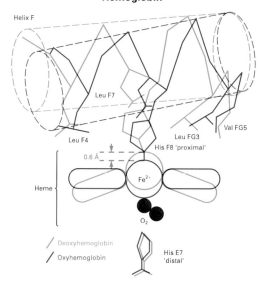

Figure 5.2-3. Structure of Myoglobin and Hemoglobin

5.2.3 Oxygen Binding to Hemo- and Myoglobin

Under normal conditions, the transition from deoxy- to oxyhemo-/oxymyoglobin and back does not affect the Fe^{++} oxidation state.

The protein environment greatly reduces the oxidation rate of the heme iron to the methemoglobin (Fe^{+++}) state. In addition, the NADH-dependent methemoglobin reductase reconstitutes the Fe^{++} state in *erythrocytes*. The formation of methemoglobin is increased in mutations, which change amino acids in the neighborhood of the heme-Fe and thus stabilize the Fe^{+++} state (methemoglobinemia).

Oxygen binding by the monomeric myoglobin takes place without major structural modifications. This is different in hemoglobin: Each of its 4 subunits can bind one oxygen molecule. This causes considerable modifications in the tertiary (folding state) and quaternary structure (association of subunits, 2.3.1).

Hemoglobin oxigenation involves a chain of events (Fig. 5.2-4):

- Fe which was located ca. 0.06 nm outside the plane of the heme ring, moves to its center. This enables the contact of Fe with O_2.
- The proximal histidine is pulled closer to the heme plane.
- The 'F' helix, which contains the proximal histidine, is tilted and laterally moved about 0.1 nm across the heme plane.
- This structural change effects a movement of the subunits about 0.6 nm relative to each other along the $\alpha_1\beta_2$ and $\alpha_2\beta_1$ interfaces. This is a transition from one stable state to another (from T to R state). This way, the oxygen binding to one subunit of hemoglobin affects the configuration of the other subunits and gives rise to cooperative behavior (compare 2.5.2).

Dissociation curves: The O_2 dissociation curve for the monomeric myoglobin has the usual hyperbolic shape, while the curve for the tetrameric hemoglobin is sigmoid (Fig. 5.2-5).

For myoglobin, the following formulae can be derived analogously to the formulae for enzyme-substrate binding (1.5.4).

The dissociation constant K_D for the equilibrium state

$$Mb + O_2 \underset{k_{-1}}{\overset{k_1}{\rightleftharpoons}} Mb\text{-}O_2 \quad \text{is calculated as} \quad K_D = \frac{k_{-1}}{k_1} = \frac{[Mb] * [O_2]}{[Mb\text{-}O_2]}. \quad [5.2\text{-}1]$$

Figure 5.2-5. Oxygen Dissociation Curves for Hemoglobin —— (black) and Myoglobin —— (red)

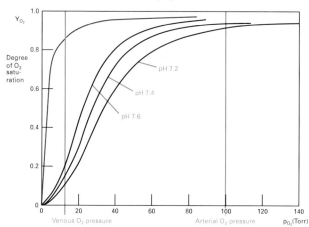

Using $[Mb_t]$ for the total myoglobin concentration $[Mb] + [Mb\text{-}O_2]$, a derivation analogous to the Michaelis-Menten equation [Eq. 1.5-19], but without further turnover to products yields

$$[Mb\text{-}O_2] = \frac{[Mb_t] * [O_2]}{K_D + [O_2]} \quad \text{or} \quad \frac{[Mb\text{-}O_2]}{[Mb_t]} = Y = \frac{[O_2]}{K_D + [O_2]} \quad [5.2\text{-}2]$$

where Y is the degree of myoglobin saturation. This results in a hyperbolic plot of Y vs. $[O_2]$. The term $[O_2]$ can be expressed as partial pressure pO_2.

For the hemoglobin tetramer, a derivation has to consider the positive cooperativity (compare 2.5.2). The binding of the second, third and fourth oxygen molecules proceeds gradually easier than that of the first one. The most radical assumption is the 'all or none'-principle, disregarding the intermediates $Hb\text{-}O_2$, $Hb\text{-}(O_2)_2$ and $Hb\text{-}(O_2)_3$. Then the dissociation constant K_D for the equation

$$Hb + 4\,O_2 \underset{k_{-1}}{\overset{k_1}{\rightleftharpoons}} Hb\text{-}(O_2)_4 \quad \text{is calculated as} \quad K_D = \frac{k_{-1}}{k_1} = \frac{[Hb] * [O_2]^4}{[Hb\text{-}(O_2)_4]}. \quad [5.2\text{-}3]$$

Using the substitution $[Hb_t]$ for the total hemoglobin concentration $[Hb] + [Hb\text{-}(O_2)_4]$ analogously to above, the following equation results:

$$[Hb\text{-}(O_2)_4] = \frac{[Hb_t] * [O_2]^4}{K_D + [O_2]^4} \quad \text{or} \quad \frac{[Hb\text{-}(O_2)_4]}{[Hb_t]} = Y = \frac{[O_2]^4}{K_D + [O_2]^4}. \quad [5.2\text{-}4]$$

The generalized case with the exponent n is called the Hill equation.

Since the partially saturated intermediates also have to be considered, the observed Hill coefficient n is less than the number of binding sites. It can be taken as a measure of cooperativity. For hemoglobin with 4 binding sites, Hill coefficients of 2.8...3.0 have been observed, indicating a fairly high cooperativity. Correspondingly, the share of partially oxygenated species is relatively small (Fig. 5.2-6).

A more refined treatment of cooperativity is given by the symmetry and the sequential models (2.5.2).

Figure 5.2-6. Share of the Hemoglobin Species With Different Degrees of Oxygen Saturation

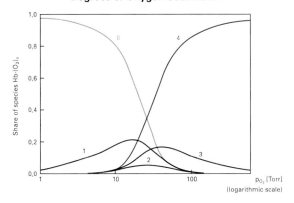

Table 5.2-2. Types of Cytochromes

Heme Type	Substitution at Position			Axial Fe Ligands	
	2	4	8		
B	vinyl	vinyl	methyl	His	His
A	2-farnesyl-1-hydroxyethyl	vinyl	formyl	His	His
C	1-mercapto-ethyl	1-mercapto-ethyl	methyl	His	Met
O	2-farnesyl-1-hydroxyethyl	vinyl	methyl		
	2	5	6		
D	vinyl	methyl + hydroxyl	propionyl + hydroxyl		
(Pos. 4 and 8 as in cytochrome b)					

Table 5.2-3. Examples of Cytochromes

Heme Type	Cytochromes (Occurrence)	Section
A	a, a_3 (respiratory chain in *mitochondria*)	16.1.2
	a_1 (terminal oxydase in *bacteria*)	
B	$b_L = b_{566}$, $b_H = b_{560(562)}$ (respir.chain in *mitochondria*)	16.1.2
	b_{556}, b_{558}, b_{595} (bd complex), b_{562} (bo complex) (respiratory chain in *bacteria*)	16.1.3
	b_L, b_H (photosynthesis in *plants* + *cyanobacteria*)	16.2.1
	b_L, b_H (photosynthesis in *purple bacteria*)	16.2.1
	b_2 (*yeast* lactate dehydrogenase)	3.1.1
	b_5 (microsomes in *animals*)	6.3.2
	P-450 (monooxygenases)	7.4.1
C	c, c_1 (respiratory chain in *mitochondria*)	16.1.2
	f (photosynthesis in *plants* + *cyanobacteria*)	16.2.1
	c_1, c_2 (photosynthesis in *purple bacteria*)	16.2.1
	c_5 (*bacterial* nitrate assimilation)	15.5
O	o (*bacteria*)	16.1.3
D	d (chlorine structure, *bacteria*)	16.1.3
	Similar: bacterial catalase	

Dondekar, T. *et al.*: EMBO J. (1991) 1903–1909.
Hershey, J.W.B.: Ann. Rev. of Biochem. 60 (1991) 717–755.
Jordan, P.M.: *Biosynthesis of Tetrapyrroles. New Comprehensive Biochemistry* Vol. 19. Elsevier (1991).
Perutz, M.F.: Ann. Rev. of Biochem. 48 (1979) 327–368.
Warren, M.J., Scott, A.I.: Trends in Biochem. Sci. 15 (1990) 486–491.
Watowich, S.J. *et al.*: J. Mol. Biol. 209 (1989) 821–828.

5.3 Bile Pigments and Bilins

5.3.1 Hemoglobin Oxidation and Bile Pigments

Erythrocytes have a limited lifetime (about 120 days in *humans*), then they are degraded in the *spleen* and *bone marrow*. The amino acids originating from globin are reutilized, while the degradation products of heme (except iron) are excreted.

The heme ring is cleaved by membrane-bound heme oxygenase (decyclizing) between the pyrroles A and B and the carbon of the methenyl bridge is liberated as CO. The enzyme is a mixed function oxygenase, related to the cytochrome P-450 enzymes (7.4.1). The resulting linear product biliverdin is excreted by *reptiles* and *birds*, while in *mammals* a reduction to bilirubin takes place. This lipophilic compound forms a complex with albumin for the transport in blood ('indirect bilirubin', named because of its behavior in analysis). After entering the *liver*, the propionate side groups are esterified with glucuronic acid in order to increase the solubility ('direct bilirubin'). Finally, bilirubin mono- and diglucuronides are excreted via *bile* into the *intestine*. There, anaerobic intestinal *microorganisms* take over, removing the glucuronide groups and reducing bilirubin to urobilinogen and stercobilinogen. Some of the former is reabsorbed from the *intestine*, converted in the *kidney* to urobilin and excreted in urine. The major portion of the heme degradation products (bile pigments) appears as stercobilin in stool and causes its dark color.

Bilirubin shows an effective antioxidant activity and removes peroxide radicals (4.5.8) in *blood* and in membranes.

Medical aspects: If a high amount of bilirubin is present in blood, deposition of this compound in the *skin* and in the *eye* leads to jaundice. This may be caused by elevated red cell destruction, obstruction of the *bile duct* or diminished glucuronidation in the *liver* (especially in premature babies, where the enzyme activity has not been fully developed).

Influence of CO_2 and 2,3-bisphosphoglycerate: The conformation change of hemoglobin, which is caused by oxygen binding, also effects a release of protons.

$$Hb\text{-}(O_2)_z + O_2 = Hb\text{-}(O_2)_{z+1} + y\ H^+ \quad (y \approx 0.6)$$

Therefore, a more alkaline pH (with lower proton concentration) 'pulls' the reaction towards higher oxygen binding, while more acid pH faciliates the oxygen release (Bohr effect, Fig. 5.2-5). Since *capillary blood* is more acid due to the CO_2 production of the tissues and its conversion to $HCO_3^- + H^+$ by carbonic anhydrase, the oxygen transfer from the capillaries into the tissues is enhanced. On the other hand, CO_2 removal in the *lung* facilitates the saturation of hemoglobin with O_2.

In addition, CO_2 interacts directly with hemoglobin and causes conformation changes, which influence the O_2 binding in the same direction as the pH effect.

2,3-Bisphosphoglycerate (3.1.1), which is present in *mammalian erythrocytes*, binds more strongly to deoxygenated hemoglobin than to the oxygenated form. As a result, the hemoglobin saturation curve is shifted to higher O_2 concentrations. This allows easier dissociation of O_2 from hemoglobin. The same effect is achieved in *birds* by inositol hexaphosphate (17.4.4) and in *fishes* and most *amphibians* by ATP.

Diseases in *humans*: More than 300 genetic abnormalities of hemoglobin are known. Sickle cell anemia is caused by a mutation in the β-chains of hemoglobin ('Hb S', Glu 6 → Val). This does not affect the properties of oxygenated hemoglobin. However, in the shifted structure of deoxygenated HbS, hydrophobic interactions of the mutated $β_2$-Val 6 with $β_1$-Phe 85 and Leu 88 of another Hb molecule is possible. This decreases the solubility and leads to the formation of fibrous HbS strands, which change the erythrocyte shape into a 'sickle' form. The shorter lifetime of the affected erythrocytes apparently provides some protection against malaria. Thalassemias are disorders of hemoglobin synthesis. The α or the β globin chains are either absent or are synthesized only at a reduced rate. They also have some protective effect against malaria. Methemoglobinemias have been mentioned above.

5.2.4 Cytochromes and Other Heme Derivatives (Fig. 5.1-2)

Cytochromes are hemoproteins taking part in redox reactions by reversible transitions between the Fe^{++} and Fe^{+++} states. Cytochromes are members of electron transfer chains (respiratory chain, photosynthesis) and participate in enzymatic reactions (e.g. cytochrome P-450). They occur in many variants in almost all organisms (except in a number of anaerobic bacteria). Hemes are also present in other enzymes, e.g. catalase and peroxidases (4.5.8).

Cytochromes are grouped by the substitutions at the porphyrin ring and by the fifth and sixth ligands of the Fe atom (in addition to the 4 ligands provided by the porphyrin ring, Fig. 5.1-2, Table 5.2-2). The ring substitutions of cytochromes b are identical with protoporphyrin IX and with heme in hemoglobin. Cytochrome d is a chlorin (5.4) with a reduced ring c. The individual cytochromes are named by the peak wavelength of the light absorption band α.

Many cytochromes are transmembrane proteins with several membrane-spanning helices. In some cases, 2 hemes are connected with a single protein chain (e.g. cytochrome b_H, b_L, Fig. 16.1-1). Other cytochromes are peripheral proteins, which act as 'electron shuttles'. A number of cytochromes are shown in Table 5.2-3.

Literature:
Baldwin, J.: Trends in Biochem. Sci. 5 (1980) 224–228.
Beale, S. in Neidhardt, F.C. et al. (Eds.): *Escherichia coli and Salmonella*. ASM Press (1996) 731–748.

5.3.2 Bilins

Bilins are open-chain tetrapyrroles, which function as photosynthetic light absorbing chromophores (phycocyanobilin, phycoerythrobilin, 16.2.1) when bound to special proteins (blue phycocyanins, red phycoerythrins). They are present in *cyanobacteria, red algae* and *cryptophytes*. In *higher plants* and in some *algae*, a related bilin is the chromophore of the photosynthetic pigment phytochrome. The structure of some of these compounds is shown in Fig. 5.3-2, their absorption spectra in Fig. 16.2-4.

Biosynthesis: Apparently, the reactions are similar to the heme biosynthesis steps described in 5.1 and 5.2. Also, the cleavage of protoheme to biliverdin proceeds analogously. Then a NADPH-dependent reduction to phycocyanobilin follows, which is different from the reduction to bilirubin in animals. Simultaneously, a cystein-SH of the apoprotein binds to the vinyl group of Ring B. In *cyanobacteria*, phycoerythrins and phycocyanobilins are stacked on top of the photosynthetic reaction centers (16.2.1).

Literature:

Jordan, P.M. (Ed.): *Biosynthesis of Tetrapyrroles. New Comprehensive Biochemistry Vol. 19.* Elsevier (1991).
Stocker, R. *et al.*: Science 235 (1987) 1043–1046.

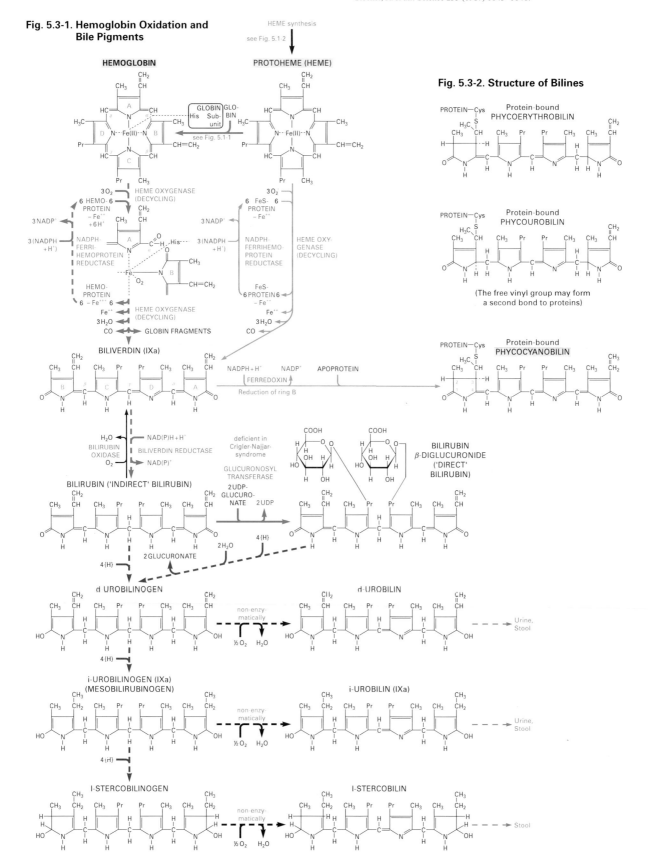

Fig. 5.3-1. Hemoglobin Oxidation and Bile Pigments

Fig. 5.3-2. Structure of Bilines

5.4 Chlorophylls

Chlorophylls are important photosynthetic pigments of *plants* and *bacteria*. As far as is known, the steps of their biosynthesis are similar in both kingdoms, only the last reactions differ. In *plants*, the synthesis of chlorophylls takes place in *chloroplasts*. Chlorophylls are characterized by a central Mg^{++} atom and by reduction of ring D (chlorin ring system). The propionate group at ring D is esterified with an isoprenoid residue (in many cases phytol). The function of these compounds is discussed in 16.2 and their absorption spectra are shown in Fig. 16.2-4.

Biosynthesis: Analogous to heme synthesis (Fig. 5.1-2), the individual branch of chlorophyll synthesis starts from protoporphyrin IX with the introduction of a metal atom (in this case magnesium) by magnesium chelatase. This reaction is ATP dependent and is feedback inhibited by chlorophyllide a, a later product of the biosynthesis pathway. Then the propionate group at ring C is esterified, followed by β-oxidation of this residue, resulting in a β-oxo acid structure. This is the prerequisite of the cyclization step, which leads to ring E of the compound Mg-2,4-divinylpheoporphyrin a_5 monomethylester. Apparently, the esterification prevents an interfering spontaneous decarboxylation. Then a reduction step involves the vinyl group at ring B. The consecutive reduction of ring D marks the transition from the porphyrin to the chlorin ring system. In many (but not all) cases, this is a light requiring reaction, resulting in chlorophyllide a. Finally, an isoprenoid side chain is introduced by an esterification reaction. If phytyl-PP (7.3.4) is the reaction partner, chlorophyll a results. It is the most abundant chlorophyll in *green plants*.

Higher plants and *green algae* also contain chlorophyll b, where the methyl group in position 3 has been converted into a formyl group (–CHO) by an O_2 and $NADP^+$ dependent reaction. It is still uncertain, whether the oxidation takes place with chlorophyll a or with an earlier intermediate of the biosynthesis. Chlorophyll c occurs in many *algae*. It is not a chlorin, but a porphyrin with an unreduced ring D. Likely, it is synthesized directly from Mg-divinylphaeoporphyrin a_5.

A fast regulation of chlorophyll synthesis in *angiosperms* by light takes place at the protochlorophyllide reductase step. In addition, the enzymes for the aminolevulinate synthesis (5.1) are induced by light. Little is known about chlorophyll degradation.

Bacteriochlorophylls occur in several variants. In the most common bacteriochlorophylls a and b, ring B is also reduced (in addition to ring D, 'bacteriochlorin ring systems', Fig. 5.4-1). Pheophytins are Mg-free chlorophyll-like compounds. They occur in the photosystem II of *plants* (pheophytin a) or in the photosynthetic reaction center of *bacteria* (pheophytins a, b or g). It is assumed that they are generated from the respective chlorophylls by loss of the metal atom.

Light absorption mechanism: If a photon of suitable wavelength is absorbed by a chromophore molecule, an electron is elevated to an excited 'singlet' state. This is only possible if the energy difference of the electron orbitals equals the energy of the photon. The energy required to reach an excited state is lower, if a long conjugated system of double bonds is present in the light absorbing molecule. When the electron returns to the original state, the energy is released as heat, as chemical energy, by fluorescence or by phosphorescence. If there is an excitable molecule nearby, an exciton transfer of energy can take place. In photosynthesizing organisms, the collected energy is funneled in this way to the reaction centers (16.2.1).

Literature:
Jordan, P.M. (Ed.): *Biosynthesis of Tetrapyrroles. New Comprehensive Biochemistry Vol. 19.* Elsevier (1991).
Walker, C.J., Willows, R.D.: Biochem. J. 327 (1997) 321–333.

Fig. 5.4-1. Biosynthesis of Chlorophylls

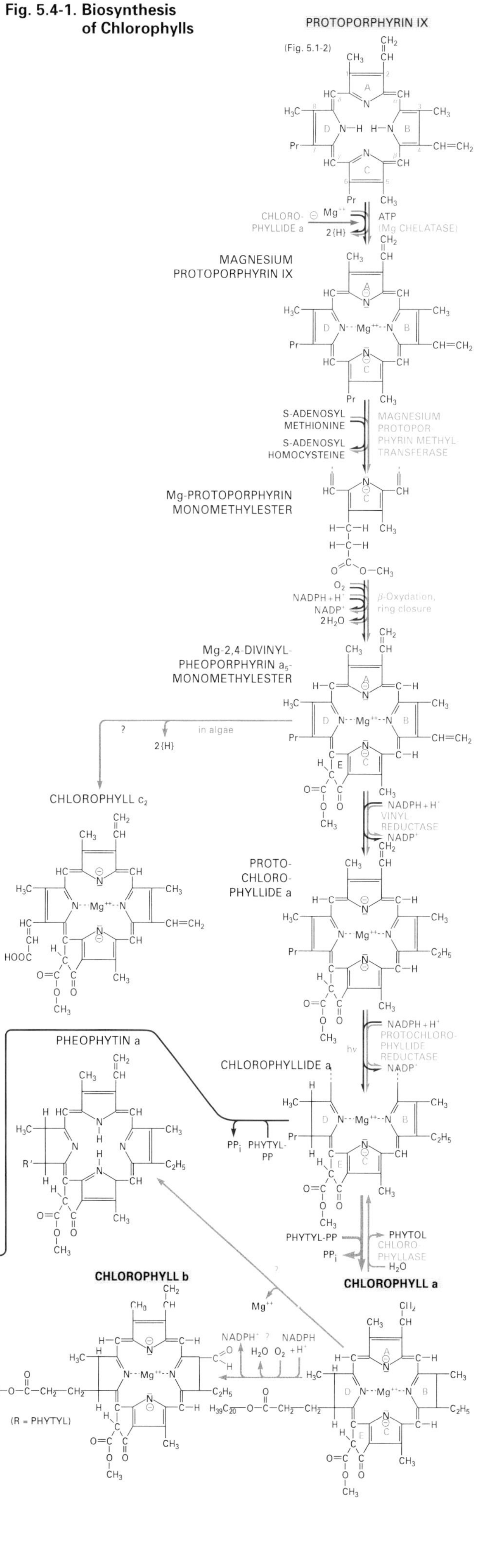

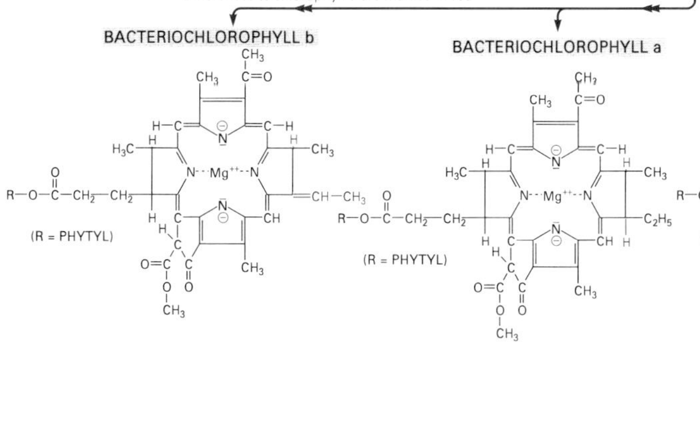

6 Lipids

The chemistry and structure of major lipids is presented in Section 1.4.

6.1 Fatty Acids and Acyl-CoA

Most fatty acids of biological importance have a chain length in the range of $C_{14} \ldots C_{20}$. Only a small portion of them occurs as free fatty acids in nature. In esterified form they are essential constituents of membranes (e.g., as phospholipids, 6.3 and glycolipids, 13.2). A smaller portion of amide derivatives are also formed (e.g., ceramides). Biosynthesis and degradation of fatty acids proceed by addition or removal of C_2 units, however in different ways (Table 6.1-1).

Table 6.1-1. Differences Between Fatty Acid Biosynthesis and Degradation Mechanisms in Eukarya

	Biosynthesis	Degradation
Location	cytosol (in *plants*: stroma of chloroplasts)	mitochondrial matrix (in *plants*: glyoxisomes)
Enzyme	multienzyme complex	individual enzymes, partly associated
Fatty acid chain is bound to	acyl carrier protein	coenzyme A
Added / removed C_2 unit	malonyl-CoA – CO_2	acetyl-CoA
Configuration of the hydroxyacyl derivative	D	L
Redox coenzyme	NADPH (*yeast:* + FADH$_2$)	NAD$^+$ and FAD

6.1.1 Biosynthesis of Fatty Acids

The primary source for the biosynthesis of fatty acids in *animals* are carbohydrates, followed by a number of amino acids. Their degradation yields acetyl-coenzyme A, which is present in *mitochondria* (3.3.1, 4.2...4.7). This is the starting compound of fatty acid biosynthesis, which proceeds in the *cytosol* of many organs (with preference in the *liver*) according to the formula

  8 acetyl-CoA + 14 NADPH + 7 ATP =
   = palmitate + 14 NADP$^+$ + 8 CoA + 6 H$_2$O + 7 ADP + 7 P$_i$.

Since acetyl-CoA cannot pass through the *mitochondrial* membrane into the *cytosol*, it has to be transferred there via the citrate/pyruvate shuttle (Fig. 6.1-1). Its carboxylation to malonyl-CoA is performed by the multienzyme complex acetyl-CoA carboxylase. Biotin functions as a CO_2 carrier (9.8.2).

In *E. coli*, the carboxylase complex is composed of 3 subunits. The biotin-carboxyl carrier protein (BCCP, 22 kDa, Fig. 6.1-2) uses a flexible arm to move the biotin group back and forth between the biotin carboxylase (49 kDa) and transcarboxylase (35 + 33 kDa). In *eukarya*, the three components are combined in a single multifunctional polypeptide chain (210...265 kDa).

The consecutive reactions of fatty acid synthesis are catalyzed in *fungi* and *animals* by the cytoplasmic multienzyme complex fatty acid synthase (240...275 kDa), which consists of two (*yeast*: α,β) or one (*vertebrates*) multifunctional polypeptide chains. The yeast enzyme has the structure $\alpha_6\beta_6$, the *animal* enzyme is a homodimer (Fig. 6.1-3). A 10 kDa acyl carrier protein (ACP) component acts as the central carrier of the growing acyl chain, using a 4′-phosphopantetheine moiety as a 'swinging arm' (2 nm length, 9.7.2) to move the acyl intermediates to the various active sites.

In *plant plastides* and *bacteria,* the reaction steps are catalyzed by individual enzymes, which are closely associated.

Here, a preferred alternative is the use of acetyl-CoA instead of acetyl-S-enzyme for condensation with malonyl-ACP in the synthase reaction. Out of the 3 *plant* oxoacyl synthases, enzyme III catalyzes the condensation of acetyl-CoA with malonyl-ACP, enzyme I the elongation up to C_{16} and enzyme II the final step to C_{18}.

Figure 6.1-1. Citrate-Pyruvate Shuttle for Acetyl-CoA Export From *Mitochondria*

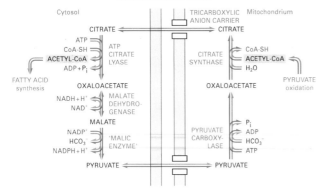

Figure 6.1-2. Carboxylation of Acetyl-CoA in *E. coli*

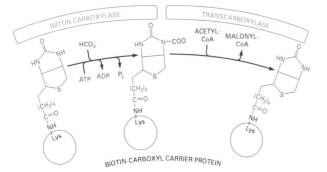

Figure 6.1-3. Schematic Drawing of the Animal Fatty Acid Synthase Dimer
(modified after Joshi and Smith).

Green and blue arcs: Functional interaction variants of the acyl-ACP moiety with the enzyme domains.
X = domain without enzymatic function.
Z = NADPH binding sites.
The numbers correspond to Fig. 6.1-4.

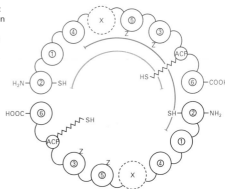

Reaction sequence (Fig. 6.1-4): Specific acyl transferases (one in *animals*, two in *yeast*) transfer an acetyl residue from acetyl-CoA to a cysteine-SH of the 3-oxoacyl synthase component ('peripheral SH group') ① and a malonyl residue to the 'central' SH group of the ACP component ①ₐ of the multienzyme system. Condensation with an acetoacetyl chain (bound to the central SH group) occurs with simultaneous decarboxylation, thereby energetically favoring chain elongation ②. This step is irreversible. Then the resulting C_4 residue is reduced by NADPH at the oxoacyl reductase component ③, dehydrated to a 2,3-desaturated intermediate ④ and reduced again by the enoyl reductase ⑤ (also using NADPH as a reductant; NADH in *E.coli*). The *yeast* enzyme contains flavin. Finally, the butyryl chain is transferred to the 'peripheral' SH group ①. Condensation with another malonyl residue initiates another round of chain elongation.

About half of the NADPH required for the reduction steps is provided by the 'malic enzyme' reaction (Fig. 6.1-1), the rest is supplied by the initial reactions of the pentose phosphate cycle (3.6.1).

The fatty acid synthesis is terminated in *vertebrates* by hydrolysis of the long-chain (mainly C_{16}) acyl residue by palmitoyl thioesterase ⑥ to give free palmitic acid.

A similar reaction takes place in *plants*. The enzyme prefers oleyl-ACP. In *yeast*, a specific palmitoyl transferase transfers the product to CoA, yielding palmitoyl-(or stearoyl-)CoA. In *bacteria*, long-chain acyl residues are directly transferred to glycerophosphate for the synthesis of membrane phosphatidates.

Figure 6.1-4. Steps of Fatty Acid Synthesis

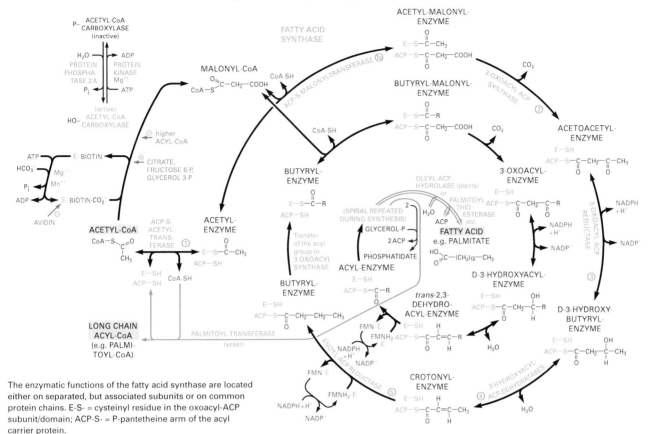

The enzymatic functions of the fatty acid synthase are located either on separated, but associated subunits or on common protein chains. E-S- = cysteinyl residue in the oxoacyl-ACP subunit/domain; ACP-S- = P-pantetheine arm of the acyl carrier protein.

Biosynthesis of fatty acids of medium chain length: Fatty acids of $C_6...C_{10}$ length are present especially in milk fat. In lactating mammary glands of *ruminants*, their presence is due to the broad specificity of acyl transferase which already removes medium chain-length fatty acids from the complex. In contrast, the mammary glands of *non-ruminants* and the *preen glands* of *birds* contain a second 29-kDa thioesterase which releases $C_8 ... C_{14}$ fatty acids.

Biosynthesis of odd-numbered and branched fatty acids: Due to the addition of C_2 units during synthesis, fatty acids with an even number of carbon atoms are most abundant in nature. However, fatty acid synthase can also utilize propionyl-CoA instead of acetyl-CoA as a primer, generating saturated fatty acids with odd carbon-numbers (mainly C_{15} and C_{17}). If malonyl-CoA is replaced by methylmalonyl-CoA, methyl-branched fatty acids are formed (e.g. in s*ebaceous glands*, where they can be the major products).

6.1.2 Regulation of Fatty Acid Synthesis (Fig. 6.1-5)

The carboxylation of acetyl-CoA is the committed step of fatty acid biosynthesis (for definition, see 3.1.1) and the point of action for regulatory mechanisms.

In *vertebrates*, acetyl-CoA carboxylase exists as an enzymatically inactive homodimer (450 kDa) and as an active, filamentous polymer (up to 10^4 kDa). The interconversion is allosterically stimulated by citrate (feed-forward activation, indicating sufficient supply of the precursor in the *cytosol*) and inhibited by palmitoyl-CoA (which is elevated during starvation due to degradation of fats). Furthermore, acetyl-CoA carboxylase is inactivated by phosphorylation and reactivated by dephosphorylation. Citrate can partially counteract the loss of activity by phosphorylation, while palmitoyl-CoA enhances the phosphorylation effect.

Inactivating phosphorylations at different serine residues are catalyzed by an AMP-dependent protein kinase (which, in turn, is regulated by activating phosphorylation and by inactivating dephosphorylation and appears to be the most important kinase), by acetyl-CoA carboxylase kinase type 2 (in *mammary glands*) and by cAMP dependent protein kinase A (17.4.2). The reactivating dephosphorylation of the carboxylase is catalyzed by protein phosphatase 2A. The interconversions are under the control of hormones, although the mechanisms are not known in detail.

Figure 6.1-5. Control Mechanisms for Acetyl-CoA Carboxylase in Liver

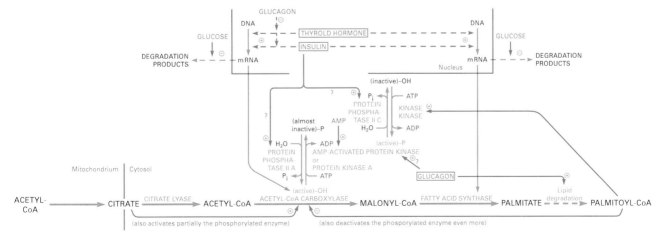

- In well-fed states with ample supply of glucose, insulin is secreted (Fig. 3.1-7). This effects a low degree of phosphorylation of the enzyme. The activity of the carboxylase and the rate of fatty acid synthesis is high.
- During fasting and in sudden energy demand, glucagon and epinephrine, respectively, are secreted (Fig. 3.1-7). In response to these hormones, the degree of inactivating phosphorylation is high (in *liver* and in *adipocytes*, respectively) and the rate of fatty acid synthesis is low or nil, since at first, the glucose requirements of various cells (primarily *brain* and *erythrocytes*) have to be satisfied.
- Fatty acid synthesis is also low in diabetes, where there is a low insulin level.

The major portion of fatty acid synthesis and consecutive esterification takes place in the *liver* (for transport of lipids to other organs, see 18.2). There are interconnections with the regulation of cholesterol synthesis (7.1.4).

Long-term adaptive control of fatty acid synthesis occurs through changes in the rates of synthesis (at the transcription level) and degradation of the enzymes participating (acetyl-CoA carboxylase and fatty acid synthase).

The formation of fatty acid synthase- and malic enzyme-mRNAs is enhanced by insulin and thyroid hormones and decreased by glucagon and cAMP. Polyunsaturated (but not saturated) fatty acids decrease the rate of transcription. High glucose levels stabilize the mRNAs.

In *E. coli*, citrate has no effect on regulation. The activity of the carboxylase is controlled by guanine nucleotides.

6.1.3 Fatty Acid Desaturation and Chain Elongation (Fig. 6.1-7)
There are preferred sites of desaturation of fatty acids (Fig. 6.1-6). The numbering system can start at either end.

Figure 6.1-6. Common Sites of Desaturation (Example: C_{18} Acid)

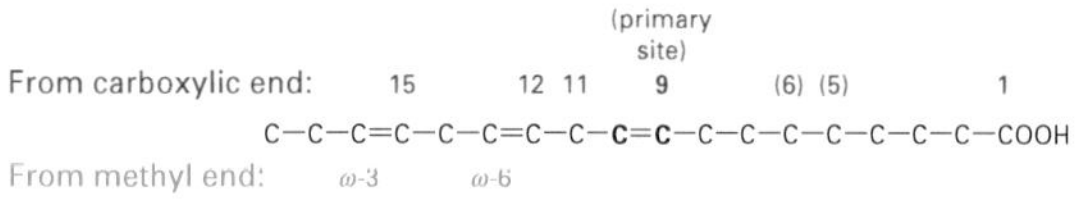

```
                             (primary
                               site)
From carboxylic end:   15    12 11  9    (6) (5)        1
                     C–C–C=C–C–C=C–C=C–C–C–C–C–C–C–C–COOH
From methyl end:            ω-3      ω-6
```

- In *mammals*, desaturation takes place with the CoA derivatives of preformed long chain fatty acids (C_{14} ... C_{18}), preferably at the 9-10 bond of stearoyl-CoA, yielding the *cis*-compound oleyl-CoA.

  This is an oxidative reaction in the *endoplasmic reticulum* of the *liver*, catalyzed by stearoyl-CoA desaturase (Δ^9-desaturase, 53 kDa, a non-heme iron protein, highly sensitive to cyanide). The enzyme complex contains also cytochrome b_5 (16.7 kDa) and cytochrome-b_5 reductase (43 kDa). The expression of the enzymes depends greatly on the nutritional state.

  Mammals are unable to introduce double bonds beyond the Δ^9 position and have to obtain polyunsaturated fatty acids by food intake from *plant* sources (essential fatty acids: linoleic and linolenic acids, 18:2 and 18:3, respectively).

  However, they can form double bonds closer to the carboxyl end, e.g. at 5-6 or 6-7 bonds of linoleic and linolenic acids. The enzymatic mechanism is similar to introduction of the first double bond (see above). Additionally, chain elongation may take place. One of the products is arachidonic acid (17.4.8).

- In *plant plastids*, the most common initial desaturation takes place at the 9-10 bond by stearoyl-ACP desaturase, which is a soluble homodimer. A similar reaction is performed by *bacteria* under aerobic conditions.

  After reduction by ferredoxin, the desaturase binds oxygen by a diiron-cluster, which then attacks the fatty acid, removes hydrogen and forms the double bond. Thereafter, other enzymes desaturate at definite distances from the methyl end, e.g. at ω-6 and ω-3, resulting in *cis*-polyunsaturated fatty acids.

  Starting from 9,10-unsaturated fatty acids, *bacteria* can form 9,10-cyclopropane fatty acids by introduction of a methylene residue from S-adenosylmethionine.

- In *anaerobic bacteria*, unsaturated fatty acids are produced by a variation of the fatty acid synthase system (6.1.1). A different 3-hydroxydecanoyl-ACP-dehydratase catalyzes an alternative dehydratase reaction at the C_{10} level of fatty acid synthesis.

  The reaction product is not the *trans*-2,3-, but rather the *cis*-3,4-unsaturated acid. Immediately thereafter the synthase reaction takes place, which is catalyzed by a special 3-oxoacyl-ACP synthase. The final result is *cis*-vaccenate (18:1, *cis*-Δ^{11}), a major fatty acid of microorganisms.

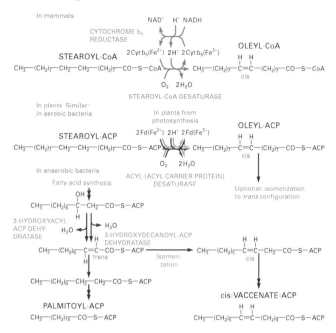

Figure 6.1-7. Mechanisms of Desaturation

- *Bacteria* can also produce *trans*-unsaturated acids by isomerization of preformed *cis*-unsaturated fatty acids.

  This is apparently a measure to control membrane fluidity. This proceeds e.g., in the gastrointestinal tract of *ruminants*. The *trans*-acids are consecutively found in milk and in beef fat. (They also occur in chemically hydrogenated plant oils.) Their influence on metabolic processes in *mammals* has not been unambiguously resolved.

Importance of unsaturated fatty acids: Unsaturated fatty acids are essential for membrane fluidity. Polyunsaturated fatty acids are precursors of biologically active molecules, e.g. eicosanoids (17.4.8). Their controlled release from membrane phospholipids may even initiate control functions by modulation of transcription via specific receptors and response elements, of protein kinase C isoforms (17.4.3), of guanylate and adenylate cyclases etc. In *humans*, the lack of polyunsaturated acids causes severe disturbances (weight loss, change in lipid composition and in organ function, death). If saturated fatty acids prevail in food over polyunsaturated ones, atherosclerosis and coronary heart disease may result.

Elongation: In *animals*, the *de novo*-synthesis in the *cytosol* usually ends with palmitate. In the *endoplasmic reticulum*, further condensations of malonyl-CoA with long chain acyl-CoA take place, followed by reactions analogous to the reaction sequence above (6.1.1). This leads to C_{18} and higher acids. There are two different condensation enzymes for saturated and unsaturated fatty acids.

6.1.4 Transport and Activation of Fatty Acids (Fig. 6.1-8)
Fatty acids are transported between *animal* organs either as unesterified fatty acids complexed with serum albumin or in the form of triacylglycerols associated with lipoproteins (18.2.2). These are hydrolyzed outside of the cells by membrane-bound lipoprotein lipase to yield free fatty acids. Once long chain fatty acids have entered the cells, they either diffuse or are transported to *mitochondria*, *peroxisomes* or the *endoplasmic reticulum* where they are activated by formation of their CoA thioesters for consecutive conversions. Fatty acid binding proteins (FASBPs, 14 ... 15 kDa) in the *cytosol* of various animal tissues may function as carriers.

Activation of fatty acids by ATP dependent fatty-acid-CoA ligases (acyl-CoA synthetases, Fig. 6.1-8), a group of enzymes with different chain length specificities, occurs via the intermediate acyl adenylate. Hydrolysis of the released diphosphate by pyrophosphatase drives the reaction. The acyl residue is then transferred to the SH group of coenzyme A to give an acyl-CoA thioester which serves as a substrate for β-oxidation (6.1.5) or for triglyceride biosynthesis (6.2.1). In *mammals*, the long-chain-fatty-acid-CoA ligase is a transmembrane protein of the outer *mitochondrial* membrane and is also present in the *endoplasmic reticulum* and in *peroxisomes*.

For degradation via the *mitochondrial* β-oxidation pathway, the acyl group of *cytosolic* acyl-CoA compounds passes the mitochondrial membrane bound to carnitine (4.5.2) via a carnitine shuttle operated by carnitine-acylcarnitine translocase (Fig. 6.1-8) and is afterwards transferred to CoA again.

Mammalian carnitine acyltransferases exhibit different chain length specificities: two isoenzymes of carnitine O-palmitoyltransferase (CPT-I and CPT-II) present on both sides of the *mitochondrial* inner membrane react specifically with long-chain acyl-CoA compounds, while carnitine O-acetyltransferase I found in *microsomes, peroxisomes* and *mitochondria* accepts $C_2...C_{10}$ residues from acyl-CoA. There also exists a carnitine O-octanoyltransferase.

6.1.5 Fatty Acid Oxidation (Figs. 6.1-9 and 6.1-10)

The overall sequence of reactions for fatty acid oxidation is similar to a reversal of the fatty acid synthesis. During each reaction cycle, the acyl chain is shortened by removal of a two-carbon unit as acetyl-CoA (β-oxidation). However, there are some differences between the degradation and synthesis pathways, as shown in Table 6.1-1.

β-Oxidation in *animals* starts with the acyl-CoA dehydrogenase reaction. The removed hydrogens are taken up by FAD, which is reoxidized by the electron transferring flavoprotein, ETF (a FAD-linked matrix protein). ETF transfers the reducing equivalents via the ETF-ubiquinone dehydrogenase to the ubiquinone pool of the respiratory chain (16.1.2). The other reducing step of the fatty acid oxidation, the 3-hydroxyacyl-CoA dehydrogenase reaction yields NADH, which is reoxidized by the NADH dehydrogenase complex of the respiratory chain (16.1.2). The liberation of acetyl-CoA is performed by the acetyl-CoA C-acyltransferase ('general' 3-oxoacyl-CoA thiolase). Only the final cleavage of acetoacetyl-CoA is catalyzed by acetyl-CoA C-acetyltransferase (acetoacetyl thiolase). However in *liver*, a portion of acetoacetyl-CoA is used for the formation of cholesterol (7.1.1) and of ketone bodies (6.1.7).

There are multiple enzymes for each of the constituent steps of fatty acid degradation which vary in their chain-length specificity. In *eukarya*, the enzymes which turn over long chain substrates are bound to the inner *mitochondrial* membrane. The enoyl-CoA hydratase, 3-hydroxyacyl-CoA dehydrogenase and acetyl-CoA C-acyltransferase activities are combined into a trifunctional enzyme with the first 2 activities located on the α subunit and the third on the β subunit. The enzymes with specificity for shorter chain length are present in the *mitochondrial* matrix. They can be isolated as separate enzymes but may be organized as a multienzyme complex providing 'substrate channeling'. This question has not completely been solved. A model of the possible arrangement is shown in Fig. 6.1-10.

Odd-numbered fatty acids yield propionyl-CoA as the product of the final thiolysis step. Its conversion into succinyl-CoA is shown in Fig. 4.5-2.

Figure 6.1-8. Activation of Fatty Acids and Carnitine Shuttle
(Example: long chain fatty acids)

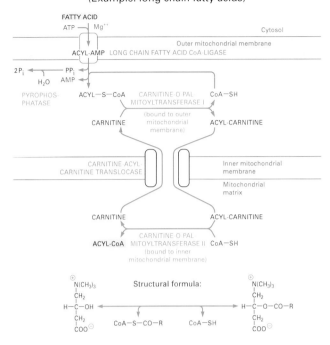

Figure 6.1-9. β-Oxidation of Fatty Acids. There are several enzymes catalyzing each step, which differ in their chain-length specificity

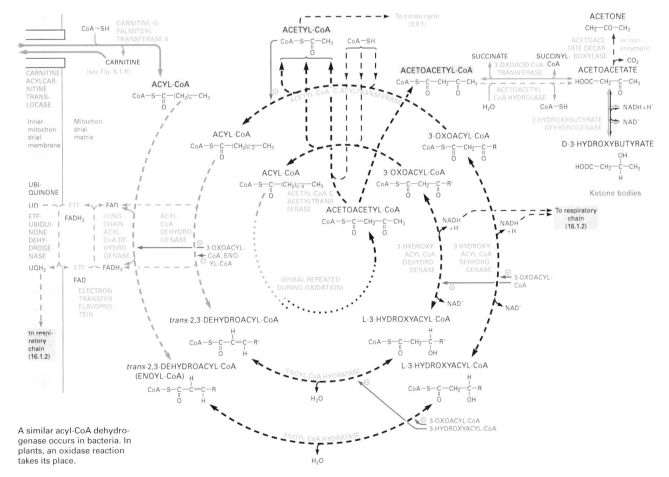

A similar acyl-CoA dehydrogenase occurs in bacteria. In plants, an oxidase reaction takes its place.

Figure 6.1-10. Possible Arrangement of the Enzymes for β-Oxidation *in Mammals* (modified after Eaton et al.).
Red arrows: reactions of long chain acyl derivatives. Orange arrows: reactions of medium and short chain acyl derivatives. Blue arrows: Transfer of hydrogen.

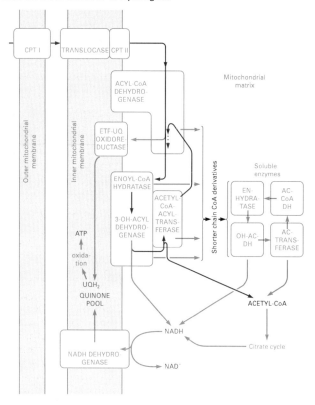

During the degradation of unsaturated fatty acids, an isomerase catalyzes the conversion of a *cis*-Δ^3 into a *trans*-Δ^2 double bond (compare Fig. 6.1-7), which is a regular member of the β-oxidation sequence. Further enzymes provide the reduction of additional double bonds in polyunsaturated fatty acids.

Medical aspects of fatty acid oxidation: Most common are deficiencies of carnitine O-palmitoyltransferase (CPT II) and of long-chain 3-hydroxyacyl-CoA and acyl-CoA dehydrogenases (ca 1 : 10 000 for homozygotes). They result in non-ketotic hypoglycemia, muscle breakdown, cardiomyopathy etc.

Regulation of fatty acid oxidation in *mammals*: The rate of fatty acid oxidation is primarily a function of the plasma concentration of unesterified fatty acids which is regulated by the action of hormones like insulin and glucagon (17.1.3) on the hormone-sensitive lipase (6.2.2).

In the *liver*, the rate of *mitochondrial* β-oxidation depends on the entry of acyl groups to the *mitochondria*. This process is regulated by modulation of the carnitine O-palmitoyltransferase I activity (Fig. 6.1-8), which is inhibited by malonyl-CoA. In addition, there is feedback inhibition of all of the fatty acid oxidation enzymes by the products.

If there is ample supply of glucose, acetyl-CoA and malonyl-CoA are formed and used for fatty acid synthesis. The counteracting fatty acid oxidation is switched off by malonyl-CoA. This inhibition ceases when the glucose level is low (see Fig. 3.1-7). Fatty acids are oxidized for the specific requirements of the organ and for the production of ketone bodies (6.1.7).

Initiation of fatty acid degradation by an oxidase reaction: In *animals*, there is also considerable fatty acid degradation in the *peroxisomes* of the *liver* and of other organs. In contrast to the *mitochondrial* system, the initial step of the degradation is catalyzed by the FAD-containing acyl-CoA oxidase and produces H_2O_2. This toxic compound is decomposed by catalase (4.5.8).

Acyl-CoA + O_2 = *trans*-2,3-dehydroacyl-CoA + H_2O_2
2 H_2O_2 = H_2O + O_2

The further reactions leading to acetyl-CoA correspond to the *mitochondrial* sequence. The peroxisomal degradation system shortens only long fatty acids and does not result in acetoacetate.

Peroxisomes do not possess a mechanism for the reoxidation of NADH which is produced by the 3-hydroxyacyl-CoA dehydrogenase reaction. They have to transfer the reduction equivalents into the cytosolic space.

Fatty acid oxidation in *plants* takes place in *glyoxisomes* analogously to animal *peroxisomes*. However, acetyl-CoA formed in this way can enter the glyoxylate cycle (3.9.1) and thus contributes to the formation of carbohydrates. This is not possible to *animals*.

The fatty acid oxidation in *E. coli* proceeds via the same steps as in animal *mitochondria*. On the other hand, however, *E. coli* can convert acetyl-CoA into oxaloacetate via the glyoxylate cycle in a way similar to *plants*.

6.1.6 Energy Yield of the Fatty Acid Oxidation

Oxidation of fatty acid provides a major source of energy, especially for the heart and skeletal muscle of *mammals*. The complete degradation of palmitoyl-CoA by β-oxidation proceeds as follows:

Palmitoyl-CoA + 7 FAD + 7 NAD^+ + 7 CoA + 7 H_2O =
= 8 Acetyl-CoA + 7 $FADH_2$ + 7 NADH + 7 H^+ $\Delta G'_0$ = −9797 kJ/mol.

Considering the ATP yield of the citrate cycle (3.8) and the respiratory chain (16.1; using the quotients 2.5 ATP/1 NADH; 1.5 ATP/1 FADH), a total of up to 108 ATP are formed. Since 2 energy-rich phosphate bonds are consumed in the activation of palmitate, the net yield is 106 mol ATP/mol palmitate. This corresponds to a total energy conservation of about 33 % under standard conditions.

6.1.7 Ketone Bodies

The acetoacetyl-CoA formed in fatty acid degradation is normally thiolyzed to acetyl-CoA (3.3.3). In *liver*, a major portion of acetoacetyl-CoA is converted into 3-hydroxy-3-methylglutaryl-CoA, a precursor of cholesterol and steroids (7.1.1). Some acetoacetyl-CoA is hydrolyzed, yielding the ketone bodies acetoacetate and its reduction product 2-hydroxybutyrate.

When in *mammals* the carbohydrate supply is limited (e.g. in starvation), the mobilization of depot lipids and as a consequence the fatty acid degradation increases (3.1.5). More ketone bodies are formed by the *liver*. As a substitute for glucose, these compounds deliver the life-sustaining energy supply especially to the *brain*. Also *muscle* cells can use ketone bodies. For utilization they are converted to the CoA-thioesters through a CoA transfer from succinyl-CoA.

Under extreme conditions, but also due to dysregulation in diabetes (17.1.3), the ketone body level in blood increases from normally < 3 mg/100 ml up to 90 mg/100 ml (ketosis). By decarboxylation of acetoacetate, acetone is produced, which cannot be reutilized. Excessive urinary excretion of ketone bodies takes place.

Literature:
Brownski, R.W., Danton, R.M. in Boyer, P.D., Krebs, E.G.: *The Enzymes*. Vol. XVIII. Academic Press (1987) 123–146.
Eaton, S. *et al.*: Biochem. J. 329 (1996) 345–357.
Joshi, A.K. *et al.*: Biochemistry 37 (1998) 2515–2523.
Neidhardt, F.C. et al. (Eds.): *Escherichia coli and Salmonella*. ASM Press (1996).
Vance, D.E., Vance, J.E. (Eds.): *Biochemistry of Lipids, Lipoproteins and Membranes. New Comprehensive Biochemistry* Vol. 31. Elsevier (1996).

6.2 Triacylglycerols (Triglycerides)

Triacylglycerols (triesters of fatty acids with glycerol) are highly concentrated stores of metabolic energy due to their nearly anhydrous storage form as well as to the high yield of free energy on oxidation (38 kJ/g). This is about 3 times the energy per gram as compared to glycogen, which is stored in hydrated form. They are also osmotically inert and without influence on the pH.

E.g., an average *human male* (70 kg) stores about 11 kg of triacylglycerols with an energy content of about $4 * 10^5$ kJ, while glycogen reserves (3.2) contribute less than 1 % of this value. The total muscle protein, which can only partially be turned over, has an total energy content of ca. 25 % of the triacylglycerol value.

Although in *mammals* most tissues are able to produce triacylglycerols, the major site of their accumulation is the *cytoplasm* of *adipose cells*. Both the synthesis of triacylglycerols as well as their hydrolysis to fatty acids serving as fuel molecules for other organs takes place there.

6.2.1 Biosynthesis of Triacylglycerols (Lipogenesis, Fig. 6.2-1)

Triacylglycerols are transported by chylomicrons and VLDL to the *cellular membrane* of *adipose cells* and to *muscle cells*. There they are hydrolyzed by lipoprotein lipase and resorbed as free fatty acids (18.2.2). Other cells mainly extract free fatty acids from blood, which are present there in albumin-bound form.

Inside of the cell, the fatty acids are converted into the acyl-CoA thioesters by fatty acid-CoA ligase (6.1.4). Glycerol-3-phosphate, the reaction partner for esterification, is provided by reduction of glycerone-P (dihydroxyacetone-P), an intermediate of glycolysis (Fig. 3.1-1).

Contrary to *adipose tissue, liver, kidney, intestinal mucosa* and the *lactating mammary gland* are additionally able to phosphorylate free glycerol, thus enabling the utilization of this compound (Fig. 3.1-1).

The acyl residues are linked to the *sn*-1 position of *sn*-glycerol 3-phosphate by *mitochondrial* or *microsomal* glycerol-3-phosphate O-acyltransferases to form acylglycerol 3-P (lysophosphatidate) and then to the *sn*-2 position by *microsomal* 1-acylglycerol 3-phosphate O-acyltransferase resulting in the formation of 1,2-diacylglycerol 3-P (L-phosphatidate). Usually, the *sn*-1 position is occupied by saturated acyl residues, while at *sn*-2 unsaturated residues predominate (probably due to the substrate specifity of the transferases).

Dephosphorylation catalyzed by phosphatidate phosphatase results in 1,2-diacylglycerols. These compounds are also the precursors of phosphatides (6.3). They may also arise from 1- or 2-monoacylglycerols, especially in the *intestinal mucosa*. The consecutive formation of triacylglycerols in various *mammalian* tissues catalyzed by diacylglycerol O-acyltransferase takes place in the *microsomal* fraction. They are then stored as triacylglycerol droplets or incorporated into secretory products, lipoproteins and milk, dependent on the tissue.

Regulation of triacylglycerol synthesis occurs in *mammals* at three steps. The glycerol-3P and diacylglycerol acyltransferases are regulated by hormonally controlled phosphorylation-dephosphorylation mechanisms. In addition, insulin increases the supply of glycerol 3-P.

During lipogenesis, adipose tissue secretes the peptide hormone leptin (*ob* gene product). It binds to a receptor in the *hypothalamus*. This diminishes the release of the neuropeptide Y (which expresses the feeling of hunger, releases insulin and glucocorticoids) and thus controls the uptake of food. In obesity of *humans*, apparently the receptor is defective.

6.2.2 Mobilization of Triacylglycerols (Lipolysis, Fig. 6.2-1)

The mobilization of triacylglycerols by hydrolysis to yield free fatty acids is catalyzed by lipases. In *mammals*, a hormone-sensitive lipase exists in *adipose tissue* which is interconverted by a phosphorylation-dephosphorylation cycle between an active (lipase A) and an inactive (lipase B) form. The phosphorylation is performed by protein kinase A, which is activated by catecholamines, glucagon and other hormones via cAMP (17.4.2). Insulin and growth hormone counteract this effect, possibly by activating the cAMP degrading phosphodiesterase (17.4.2). This enzyme is responsible for hydrolysis of the esters in position 1 and 3 of triacylglycerol. It also hydrolyzes cholesterol esters (7.1, 18.2.4). A second enzyme, 2-monoacylglycerol lipase, catalyzes the hydrolysis of the remaining ester to yield free fatty acids (which are released to supply other organs with fuel) and glycerol (which is shuttled back to the *liver*, phosphorylated and oxidized to glycerone-P = dihydroxyacetone-P for use in oxidation or gluconeogenesis).

The hormone-sensitive lipase also occurs in *brown adipose tissue*, which is rich in *mitochondria*. In these cells, the resulting free fatty acids are not released, but undergo β-oxidation (6.1.5). The fatty acids also bind to and activate an uncoupling protein in the *mitochondrial inner membrane*. This protein 'uncouples' the respiratory chain and the ATP synthesis, since it enables the return of protons to the mitochondrial matrix, bypassing the ATP synthase reaction. The energy of the electrochemical potential gradient is thus not converted into chemical energy, but into heat (thermogenesis, 16.1.4).

Enzymes, which are identical or closely related to the hormone sensitive lipase are present in steroid producing cells, e.g. in *adrenals* and *corpus luteum*. Other animal lipases are present in the *mouth* (lingual lipase formed in von Ebner's gland), in *milk* (bile salt-stimulated lipase) and *pancreas*. The pancreatic lipase needs a protein cofactor, colipase, for activity in the presence of bile salts.

Bacterial lipases are generally secreted into the medium.

Literature:
Boyer, P.D. (Ed.): *The Enzymes*. 3rd Ed. Vol. XVI. Academic Press (1983).
Rohner-Jeanrenaud, F., Jeanrenaud, B.: New Engl. J. Med. 334 (1996) 324–325.
Strahlfors, P. *et al.* in Boyer, P.D., Krebs, E.G. (Eds.): *The Enzymes*. 3rd Ed. Vol. XVIII. Academic Press (1987) 147–177.
Tartaglia, L.A. *et al.*: Cell 83 (1995) 1263–1271.

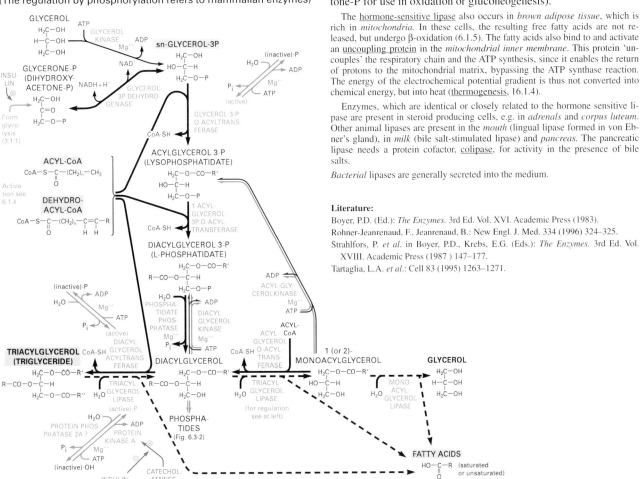

Figure 6.2-1. Formation and Degradation of Triacylglycerols
(The regulation by phosphorylation refers to mammalian enzymes)

6.3 Phospholipids

Phospholipids are important building blocks of cell membranes. They are derived from phosphate esters of glycerol (glycerophospholipids), glycerol ethers (plasmalogens) or sphingosine (sphingophospholipids), as shown in Fig. 6.3-1. Their basic structures and their chemistry are dealt with in 1.4.4 ... 1.4.6.

They show an amphipathic behavior, which is caused by the presence of

- a hydrophilic 'head group' (esterified with the phosphate residue) and
- a hydrophobic 'lipid tail' (long hydrocarbon chains of fatty acids, ethers or sphingosine)

Figure 6.3-1. Basic Structures of Phospholipids (Y = Head Group)

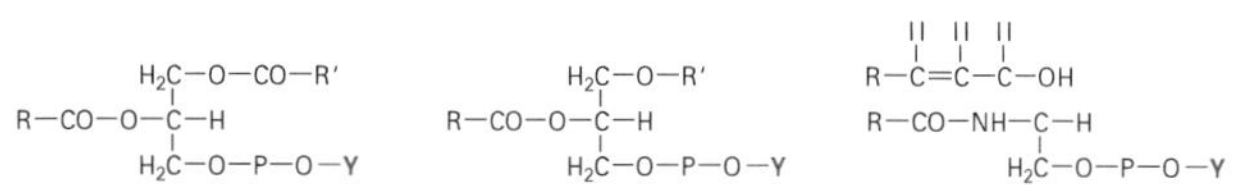

6.3.1 Occurrence of Phospholipids

Whereas *E. coli* possesses a very simple phospholipid composition, consisting primarily of phosphatidylethanolamine, phosphatidylglycerol and cardiolipin, the phospholipid composition of *eukaryotic* membranes is more complex and varies in different tissues (Table 6.3-1).

In addition to phospholipids, cholesterol (7.1.3) and many glycolipids (13.2) including glycoglycerolipids are present in *eukaryotic* membranes. The contribution of the various components to the membrane properties is discussed in 1.4.9.

6.3.2 Glycerophospholipids (Phosphoglycerides)

Biosynthesis (Fig. 6.3-2): For the phospholipid-producing condensation reaction, either the glyceryl moiety or the head group has to be activated. *Bacteria* use the first way, while *animals* and *plants* are able to use both pathways.

Bacteria: In *E. coli*, diacylglycerol 3-phosphate (L-phosphatidate, 6.2.1) is converted to a mixture of CDP-1,2-diacylglycerol and dCDP-1,2-diacylglycerol by a single enzyme, phosphatidate cytidyltransferase (CDP-diacylglycerol synthase). This compound then reacts with either serine or glycerol-3-P to form phosphatidylserine (PS) or 3-phosphatidyl-glycerol-1-P, respectively.

PS is only a minor membrane constituent of *E. coli*, since it is rapidly decarboxylated by PS decarboxylase to give phosphatidylethanolamine. PS decarboxylase (36 kDa) uses a pyruvoyl moiety as the prosthetic group. This way, the ethanolamine moiety is produced *de novo*. 3-Phosphatidylglycerol-1-P, on the other hand, gets dephosphorylated to 3-phosphatidylglycerol (PG). At least two different phosphatases exist which may catalyze this reaction. By condensation of two PG molecules and release of glycerol, cardiolipin is obtained.

Animals, plants and yeast: L-phosphatidate is hydrolyzed to 1,2-diacylglycerol (DG), catalyzed by two phosphatidate phosphatases (a *cytoplasmic* form bound to the *endoplasmatic reticulum*, and an enzyme bound to the plasma *membrane*). DG can react with CDP-ethanolamine (formed from ethanolamine by phosphorylation, followed by transfer of a CMP moiety from CTP) to yield phosphatidylethanolamine (PE). However, PE can also be formed by decarboxylation of phosphatidylserine, by an exchange reaction of ethanolamine with PS, or by re-acylation of lyso-PE.

Diacylglycerol can also react with CDP-choline to produce phosphatidylcholine (lecithin, PC) which, especially in *liver*, can also be formed from PE by methylation. The regulated step for PC (and likely) PE synthesis is the cytidylyl transferase reaction. The active enzyme is reversibly removed from the membrane and this way inactivated by the final product PC or by lack of the substrate DG.

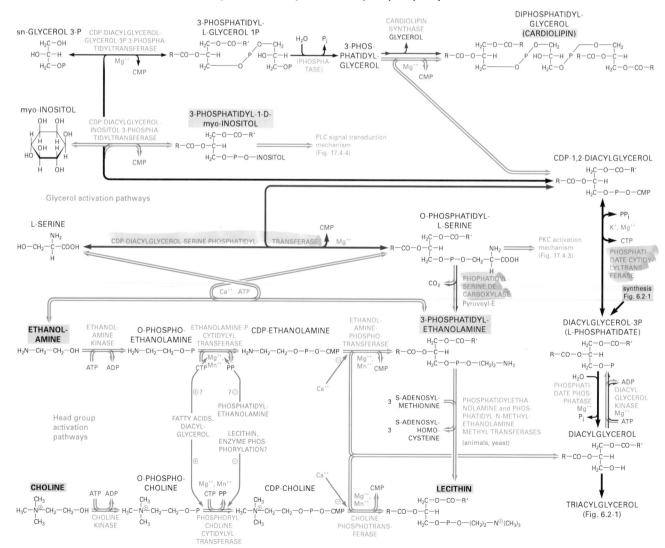

Figure 6.3-2. Biosynthesis of Glycerophospholipids

Table 6.3-1. Distribution of Phospholipids (PL) in Various Species (Harwood, 1980)

Phospholipid	'Head group' (Fig. 6.3-1) Y =	% of Total PL in *Animals*			% of Total PL in *Plants*		% of Total PL in *Bacteria* (ca.)
		Bovine Liver	Bovine Brain	Bovine Heart	Mitochondria Inner Membrane	Plasma Membrane	
Glycerophospholipids:							
Phosphatidate	H	2	1	2	0	0	0
Phosphatidylcholine (lecithin, PC)	choline	54	29	24	27	32	0
Phosphatidylethanolamine (PE)	ethanolamine	9	12	17	29	46	75
Phosphatidylinositol (PI)	inositol	8	6	4	0	19	0
Phosphatidylserine (PS)	serine	4	17	2	25	0	trace
3-Phosphatidylglycerol (PG)	glycerol	trace	trace	trace	0	0	15 ... 20
Cardiolipin (diphosphatidylglycerol)	phosphatidylglycerol	4	1	9	20	0	5 ... 10
Ether lipids:							
Choline plasmalogen	choline	2	trace	18			
Ethanolamine plasmalogen	ethanolamine	4	21	11			
Sphingophospholipids:							
Sphingomyelin	choline	6	13	12			
Sphingoethanolamine	ethanolamine						

In *eukarya*, phosphatidylserine (PS) is not formed through the pathway of glycerol activation (except in *yeast*), but by a base-exchange reaction from PE or PC. The catalyzing phosphatidylserine synthase requires Ca^{2+} and ATP and is located in the *microsomes*. Together with the decarboxylation step of PS this accounts for the net synthesis of ethanolamine or choline:

PC + serine = PS + choline   PE + serine = PS + ethanolamine
PS = PE + CO_2 PS = PE + CO_2
PE + 3 SAM = PC + 3 SAH
(SAM = S-adenosylmethionine,
SAH = S-adenosylhomocysteine)

Alternatively to the conversion into diacylglycerol (6.2.1), phosphatidate in *mammals* (as in *E. coli*) reacts with CTP to yield CDP-1,2-diacylglycerol. This activated compound then condenses with inositol to yield phosphatidylinositol or with glycerol-3-P, resulting in phosphatidylglycerol. The consecutive reactions leading to cardiolipin proceed differently to *bacteria* and employ the condensation with another molecule of CDP-diacylglycerol.

Remodelling of fatty acid substituents: Once a phospholipid is synthesized, the fatty acid substituents in the *sn-1* or *sn-2* positions can be remodelled via deacylation-reacylation reactions catalyzed by phospholipases A_1 and A_2, acyl-CoA: lysophospholipid 1-(or 2-)acyltransferases, and CoA dependent and independent transacylation systems. Their substrate specificity regarding acyl chain length and saturation is of crucial importance for the fatty acyl pattern of phospholipids.

Degradation of glycerophospholipids (Fig. 6.3-3): The hydrolysis of phospholipids is catalyzed by phospholipases (PL) which are classified according to their positional specificity (Fig. 6.3-3 and Table 6.3-2). Lysophospholipids, the products of phospholipase A_1 and A_2 reactions, are further hydrolyzed by lysophospholipases.

Phospholipases are abundant enzymes which have been found in *animals, bacteria,* and *plants* (PL D only). Their activity is restricted to the interface between the aqueous and lipid phases so that usual enzyme kinetics do not apply.

Metabolic role of glycerophospholipids and phospholipases: The role of phospholipids in *eukaryotic* membranes is not just a structural one. They take part in many metabolic processes.

Phosphatidylinositol is a key component in the PLC-PKC signal transduction mechanism (17.4.3). After two phosphorylations, it is cleaved upon arrival of hormonal signals. Both cleavage products act as second messengers: Inositolphosphates release calcium, diacylglycerol is an activator of protein kinase C. Phosphatidylserine is involved in this reaction.

Release of arachidonic acid from phosphatidylinositol by PLA_2 and other phospholipases, e.g., during an inflammatory process leads to the production of prostaglandins and leukotrienes (17.4.8).

The interaction of Ca^{++} with phospholipids is of importance for membrane fusion processes. Phosphatidylinositol functions also as a membrane anchor for proteins (13.3.4).

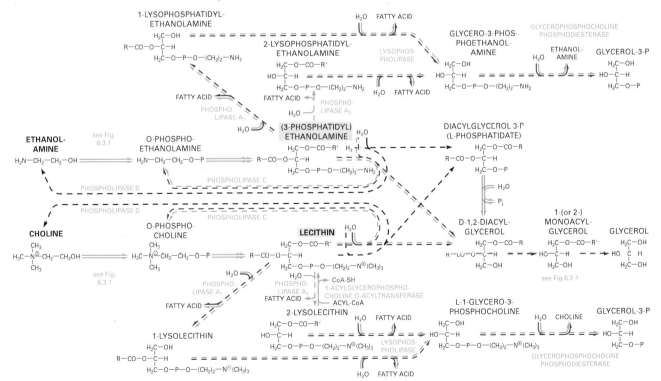

Figure 6.3-3. Degradation of Glycerophospholipids

With respect to foreign cells, phospholipases can be mediators of infectivity (secreted *bacterial* phospholipases) or act as toxins on membranes (e.g. in *snake venoms*).

6.3.3 Ether Lipids

Ether lipids of various configuration occur in membranes of higher and lower *animals*, of anaerobic *prokarya* (mostly *archaea*) and of some higher *plants* (as minor components). A high concentration of ether lipids is found in tumors. These compounds apparently decrease the permeability and lower the melting temperature of membranes. The platelet activating factor (PAF) is involved in many regulatory functions.

Biosynthesis (Fig. 6.3-4): The initial reaction is an unusual exchange reaction of a fatty alcohol with the acyl group of 1-acyldihydroxyacetone-P. Some details are still unknown. Then a NADPH-dependent reduction of the oxo group at C-2, acylation and dephosphorylation follow, resulting in alkylacylglycerol. Analogous to the reaction with diacylglycerol, CDP-ethanolamine and CDP-choline add the respective residues to yield plasmanylethanolamine and plasmanylcholine. A 1-2 double bond can be introduced into plasmanylethanolamine by action of the plasmanylethanolamine desaturase, resulting in plasmalogen.

The exchange of the long fatty acid (frequently arachidonate) esterified at the C-2 site of plasmanylcholine with an acetyl residue yields the platelet activating factor (PAF). This reaction takes place mainly in *leukocytes* and *endothelial cells* and is induced by thrombin, antigens, bradykinin, ATP etc. PAF is involved in the pathogenesis of asthma, hypertension, allergies and hypersensitivity, inflammation etc. and acts on specific receptors in a concentration of 10^{-10} mol/l (Table 6.3-3). The further mechanism in unknown. The compound is inactivated by removal of the acetyl or of the phosphocholine residue. Also a hydroxylation at the α-C atom of the long-chain alcohol, followed by hydrolytic removal of the formed aldehyde moiety takes place. This reaction is catalyzed by glyceryl-ether monooxygenase, a tetrahydrobiopterin dependent enzyme.

6.3.4 Sphingophospholipids

While a large number of glycosphingolipids exists in various organisms (13.2.1), the only major sphingophospholipid is sphingomyelin in *higher animals*. It is present in membranes, lung surfactant, lipoproteins (especially LDL, 18.2) and atherosclerotic plaques.

Biosynthesis (Fig. 6.3-4): The biosynthesis up to ceramide is common for both glycosphingolipids and sphingophospholipids and takes place in the *endoplasmic reticulum*. It starts with the condensation of palmitoyl-CoA and serine to give 3-dehydrosphinganine (3-ketosphinganine) in a pyridoxal-P dependent reaction (β-replacement, 9.4.2). The product is then reduced to sphinganine, acylated, and desaturated to yield ceramide. This compound is also the precursor of glycosphingolipids (13.2.1).

Table 6.3-2. Phospholipases

Enzyme	Source	Properties, Mol. Mass
Phospholipase A$_1$ (PL A$_1$)	*bacteria* *mammals*	mostly membrane-bound in lysosomes, various forms known
Phospholipase A$_2$ (PL A$_2$)	snake venoms (type I): *cobras* and *kraits*	secreted, ca. 13–15 kDa
	mammalian pancreas (type I)	secreted, ca. 13–15 kDa
	snake venoms (type II): *rattlesnakes* and *vipers*	secreted, ca. 13–15 kDa (see 17.4.8)
	mammalian tissues and synovial fluid (type II)	secreted, ca. 13–15 kDa
	Bee, lizard (type III)	secreted, 16–18 kDa
	mammalian tissues (type IV)	cytoplasmic; ca. 85 kDa; regulated by Ca^{++} concentration (see 17.4.8)
	mammals (heart muscle)	cytoplasmic; ca. 40 kDa; no requirement for Ca^{++}
	mammals (PAF acetyl hydrolase)	blood, ca. 60 kDa; no requirement for Ca^{++}
Lysophospholipase (LPL)	microorganisms, *bee* venom, *mammalian* tissues	variable in size and pH optima
Phospholipase C (PL C)	bacteria (e.g. *Bacillus cereus*), *mammals*	PC- and PI-specific enzymes known, many isoenzymes, role in signal transduction (17.4.3)
Phospholipase D (PL D)	plants, microorganisms, mammals	Functions as transphosphatidylase with various alcohol acceptors and in signal transduction

Table 6.3-3. Selected Cellular Functions Influenced by the Platelet Activating Factor (PAF). See 19.1.7, 20.4.

Bronchoconstriction ↑	Aggregation of neutrophils and platelets ↑
Pulmonary edema ↑	Degranulation of platelets, mast cells etc. ↑
Pulmonary blood pressure ↑	Chemotaxis of neutrophils ↑
But: systemic blood pressure ↓	Activity of
Heart rate ↑	– Protein kinase C ↑
Hypersensitivity response ↑	– G-protein receptor kinases ↑
Vascular permeability ↑	– Protein-tyrosine kinase ↑
	– PL-C ↑, leads to Ca^{++} uptake ↑
	Phosphoinositide turnover ↑
	Glycogenolysis ↑

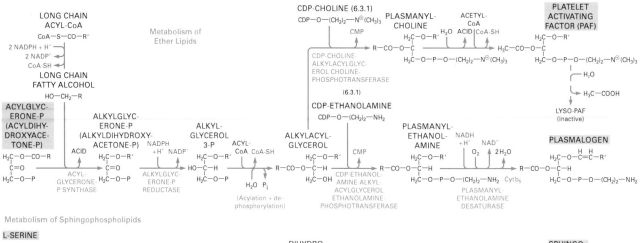

Figure 6.3-4. Metabolism of Ether Lipids and Sphingophospholipids

Sphingomyelin is synthesized in the *Golgi apparatus* of *liver* and in *plasma membranes of the nervous system* (oligodendrites and myelin membranes) by transfer of phosphorylcholine from phosphatidylcholine to ceramide, liberating diacylglycerol. Degradation begins with the removal of the head group (as choline or phosphocholine). This step is regulated by, e.g. tumor necrosis factor (TNF-R1), interferon-γ, nerve growth factor (NGF) etc. Depending on the location of this process (lysosomal or associated to the cellular membrane), the ceramide formed initiates signal cascades (17.5.3) leading to proliferation or to apoptosis. Many details are still unknown. Niemann-Pick patients have a deficiency of the lysosomal sphingomyelin phosphodiesterase. Further degradation steps involve deacylation to sphingosine (after which the whole group of compounds has been named) and its cleavage.

Literature:
Harwood, J.L., Sanchez, J.: Progr. in Lipid Res. 33 (1994) 1–202.
Kent, C.: Ann. Rev. Biochem. 64 (1995) 315–343.
Merrill, A.H. Jr., Jones, D.D.: Biochim. Biophys. Acta 1044 (1990) 1–12.
Snyder, J.: Biochim. Biophys. Acta 1254 (1995) 231–249.
Taguchi, H., Armarego, W.: Med. Res. Rev. 18 (1998) 43–89.
Testi, R.: Trends in Biochem. Sci. 21 (1996) 468–471.
Vance, D.E. & Vance, J.E. (Eds.) *Biochemistry of Lipids, Lipoproteins and Membranes*, New Comprehensive Biochemistry Vol. 31, Elsevier (1996).
Yamashita, A. et al.: J. Biochem. 122, 1–16 (1997).

6.3.5 Choline, Betaine, Sarcosine (Fig. 6.3-5)

Choline metabolism: In *plants* and to a limited extent in *animals*, phosphatidylcholine is synthesized from phosphatidylethanolamine by methylation. Its hydrolysis yields choline (6.3.2). Many microorganisms (e.g., *E. coli*) cannot synthesize choline.

Besides its presence in membrane phospholipids (6.3.2...4), choline plays an additional important role in the form of acetylcholine. This is an important neurotransmitter (17.2.3, Table 17.2-2). It is synthesized in the cells of the *cholinergic system* by choline acetyltransferase and is stored in *presynaptic vesicles*. Upon neuronal excitation, acetylcholine is released into the *synaptic cleft*, acts on *postsynaptic membrane* receptors and is inactivated within milliseconds by acetylcholine esterase, which is present in neighborhood of the receptors. Choline is thereafter taken up by the *presynaptic neurons* and recycled into acetylcholine.

Trimethylammonium compounds (e.g., physostigmin) inhibit the enzyme reversibly and thus enhance temporarily the excitatory acetylcholine effects, while most organophosphorus compounds (insecticides, e.g., E605, nerve gases, e.g., sarin, tabun) act irreversibly and cause cardiac arrest and respiratory paralysis. Acridinium derivatives (e.g., tacrine) are being tested for treatment of Alzheimer's disease.

Betaine metabolism: Betaine (= glycine betaine) is an atypical amino acid, since it contains a quaternary amino group. It is synthesized from choline. Betaine is nonproteinogenic, but is important as an osmoprotectant. Analogous derivatives of other amino acids are also called 'betaines'.

Biosynthesis: In *animals*, betaine is synthesized by a two-step oxidation of choline. This takes place in the *mitochondrial matrix* by the membrane-bound choline dehydrogenase (CDH, which is coupled to the respiratory chain, 16.1.2) and the soluble betaine aldehyde dehydrogenase (BADH). In the *chloroplast stroma* of spinach, the first oxidation step is catalyzed by the soluble ferredoxin dependent choline monooxygenase. In *microorganisms*, the first step is performed by CDH or via a soluble, H_2O_2 generating oxidase. Both enzymes are flavoproteins.

Osmoprotection: In *plants* and *microorganisms*, environmental changes such as salinity, cold or drought lead to osmotic stress. Many organisms can respond with the uptake or biosynthesis of osmoprotectants. These are substances with no adverse effects on the cell in concentrations of up to ca. 1 mol/l, that maintain the turgor of the cell in equilibrium with the environment (compatible compounds). Important osmoprotectants are glycine and proline betaines, others are proline (4.3), 3-dimethylsulfoniopropionate, non-reducing sugars (e.g. the disaccharide trehalose) and hexitols, e.g. sorbitol (3.5.5). In *animals*, osmoprotection of the kidney tubules against the increasing concentration of urine is of special importance.

Sarcosine metabolism: Betaine degradation leads to sarcosine and is apparently restricted to *animals* (mainly in *liver* and *kidney*), except for some *microorganisms*.

The first step involves a methyl transfer to homocysteine. This is the only way of methionine synthesis (4.5.4) in absence of methyltetrahydrofolate (9.6.2) and coenzyme B_{12} (9.5.2) and thus of major metabolic importance. The following two oxidative demethylation steps lead via sarcosine (N-methylglycine) to glycine (4.4.2) and are catalyzed in the *mitochondrial matrix* by dimethylglycine dehydrogenase and sarcosine dehydrogenase, which are connected to the respiratory chain. The methyl groups of betaine are oxidized and bound as methylene residues to tetrahydrofolate. Methylenetetrahydrofolate, in turn, may contribute to the conversion of glycine to serine by glycine hydroxymethyltransferase (4.4.2).

Some *bacteria* can obtain sarcosine by degradation of creatine (4.9.2) and oxidize it by sarcosine oxidase to glycine, formaldehyde and H_2O. Other, *anaerobic* bacteria reduce betaine, sarcosine and glycine via reductases containing selenocysteine (10.6.5) to trimethylamine, methylamine or ammonia, respectively. These reactions are coupled to energy conservation.

Sarcosine synthesis from glycine: Glycine N-methyltransferase (GNMT), an enzyme found in *animals* (*liver*, *kidney* and *exocrine cells*), catalyzes the S-adenosylmethione (SAM) dependent methylation (4.5.4) of glycine to sarcosine. This is the only example of an 'unimportant' methylation (no key metabolite synthesis or regulatory effect on the target takes place). The demethylation of SAM, however, enables the use of its sulfur for the synthesis of, e.g., acidic glycosphingolipids (sulfatides, 13.2.1) or glycosaminoglycan sulfates (13.1.5), and of the C-backbone for gluconeogenesis (3.1.3).

Literature:
Burg, M.B.: Am. J. Physiol. 268 (1995) F983–F996.
Lilius, G., Holmberg, N., Bülow, L.: Bio/Technology 14 (1996) 177–180.
Meyer, M. et al.: Eur. J. Biochem. 234 (1995) 184–191.
Miller, B. et al.: Biol. Chem. Hoppe-Seyler 377 (1996) 129–137.
Reuber, B.E. et al.: J. Biol. Chem. 272 (1997) 6766–6776.
Trossat, C. et al.: Plant Physiol. 113 (1997) 1457–1461.
Yeo, E.J., Wagner, C.: Proc. Natl. Acad. Sci. USA 91 (1994) 210–214.

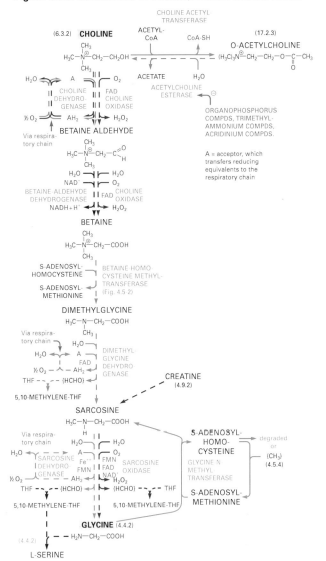

Figure 6.3-5. Metabolism of Choline, Betaine and Sarcosine

7 Steroids

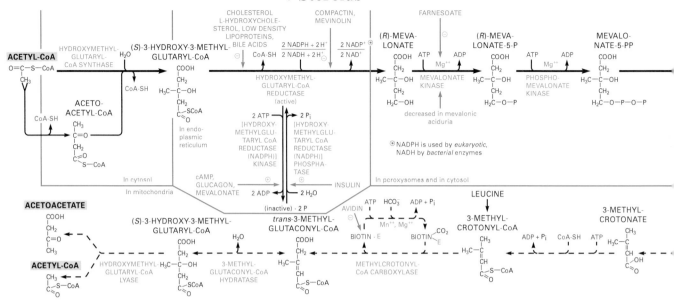

7.1 Cholesterol

Cholesterol is a compound of essential importance for *animals*. It is an integral member of cellular membranes and influences their fluidity. Furthermore, a wealth of biologically active compounds are derived from cholesterol or intermediates of its biosynthesis in various species:

```
                    Glyceraldehyde 3-P    hopanoids (pro-    steroid hormones (ver-
                    + pyruvate, 7.3       karya), 7.2        tebrates, insects), 7.4...8

Acetyl-CoA → early precursors → late precursors → cholesterol → bile acids
                                                                  (mammals),
                                                                  7.9
                    isoprenoids           steroids            steroids
                    (animals, plants,     (plants, fungi),    (plants), 7.2
                    bacteria), 7.3        7.2
```

Cholesterol occurs only rarely in *bacteria* and plants. However, they have functionally related structures (e.g. hopanoids, 7.2).

7.1.1 Biosynthesis (Figure 7.1-1)

The biosynthesis can be subdivided as follows:

$$C_2 (\times 3) \xrightarrow{①} C_6 (-CO_2) \xrightarrow{②} C_5 (\times 6) \xrightarrow{③} C_{30} \xrightarrow{④} C_{30} (-3 \times C_1) \xrightarrow{⑤} C_{27}$$

acetyl-CoA | hydroxymethylglutaryl-CoA | isopentenyl-PP | squalene | lanosterol | cholesterol

① Condensation of acetyl-CoA and acetoacetyl-CoA (formed from 2 acetyl-CoA) to hydroxymethylglutaryl-CoA (HMG-CoA). In *animals*, it occurs in the *cytosol* of *liver* (predominantly), *small intestine, adrenals* and *gonads*. The required acetyl-CoA is supplied from *mitochondria* by the citrate-pyruvate shuttle (Fig. 6.1-1). The reverse HMG-CoA lyase reaction is carried out by an *mitochondrial* enzyme and leads to ketone bodies (6.1.7).
② Reduction to mevalonate, catalyzed by HMG reductase, an enzyme embedded into the membrane of the *smooth endoplasmic reticulum* (but also present in *peroxisomes*). It is the key reaction for cholesterol biosynthesis and is therefore strictly regulated. The following reactions, pyrophosphorylation and decarboxylation to isopentenyl pyrophosphate (IPP) take place in *peroxysomes* (as major site) and in the *cytosol*. IPP, which is the origin of isoprenoids is, however, obtained in a different way (7.3).
③ Isomerization to an equilibrium with dimethylallyl pyrophosphate (Fig. 7.1-2). For the condensation to squalene, the activated compound dimethylallyl-PP acts as a primer and isopentenyl-PP is added via a 'head to tail'-nucleophilic substitution. Subsequently, the geranyl-PP formed becomes a primer for an analogous condensation yielding farnesyl-PP.

Figure 7.1-2. Condensation Reaction of Isopentenyl-PP and Dimethylallyl-PP to Geranyl-PP

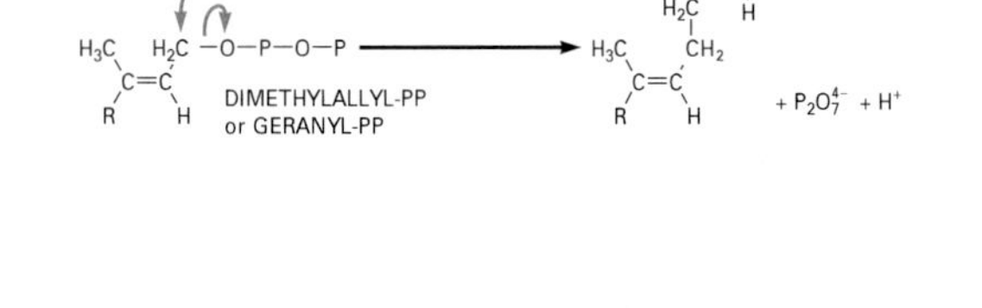

Squalene is synthesized by a 'head to head' condensation of 2 farnesyl-PP molecules. This, as well as the following reacions of cholesterol biosynthesis, take place in the *smooth endoplasmic reticulum*.
④ In *eukarya*, squalene is converted to (S-)squalene-2,3-epoxide by an oxygen and NADPH dependent reaction. This compound undergoes a cyclization to lanosterol (in all *animals* except *insects* and in *fungi*) or cycloartenol (in *plants*, 7.2) catalyzed by lanosterol or cycloartenol synthases, respectively (Fig.7.1-3).

Cyclization starts with an electrophilic attack by the enzyme at C-3, followed by electron shifts. The resulting C-20⁺ cation causes a backward rearrangement by H and CH₃ shifts. In case of *animals*, finally H at C-9 is eliminated, yielding lanosterol. In *plants*, nucleophilic interaction of the enzyme with C-9 causes hydrogen migration from C-9 to C-8 and finally elimination of H from C-19 with formation of a cyclopropane ring, resulting in cycloartenol. *Prokarya* cyclize squalene to hopanoids (7.2).

Figure 7.1-3. Cyclization of Squalene to Cycloartenol or Cholesterol

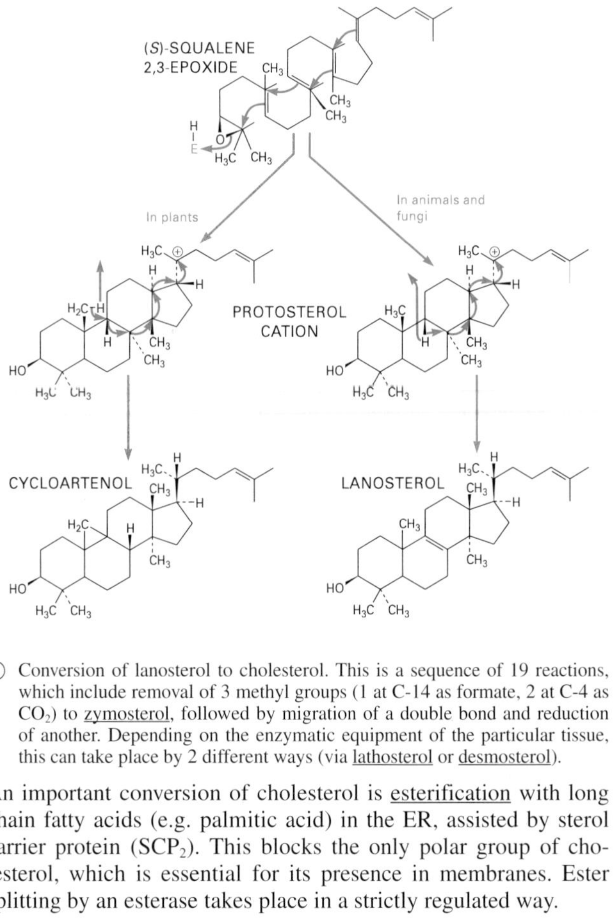

⑤ Conversion of lanosterol to cholesterol. This is a sequence of 19 reactions, which include removal of 3 methyl groups (1 at C-14 as formate, 2 at C-4 as CO₂) to zymosterol, followed by migration of a double bond and reduction of another. Depending on the enzymatic equipment of the particular tissue, this can take place by 2 different ways (via lathosterol or desmosterol).

An important conversion of cholesterol is esterification with long chain fatty acids (e.g. palmitic acid) in the ER, assisted by sterol carrier protein (SCP₂). This blocks the only polar group of cholesterol, which is essential for its presence in membranes. Ester splitting by an esterase takes place in a strictly regulated way.

Figure 7.1-1. Cholesterol Biosynthesis

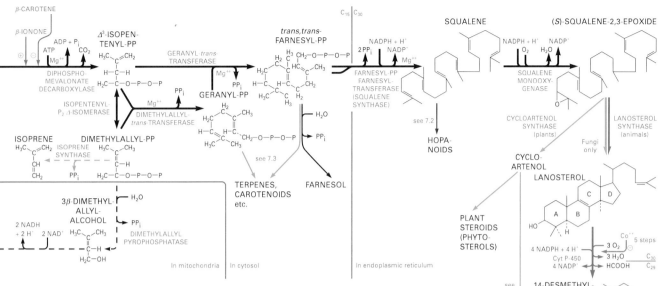

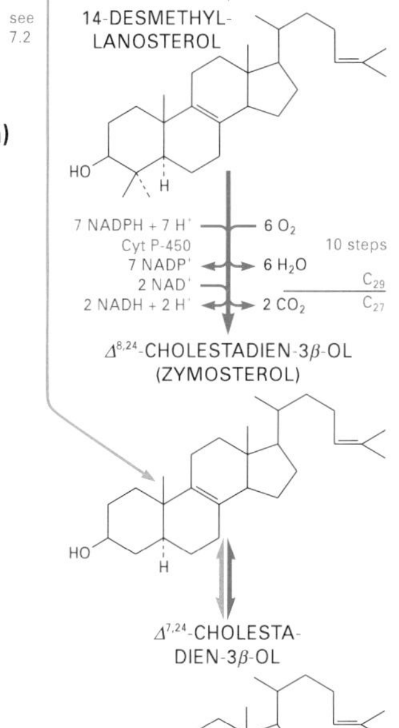

While cholesterol degradation in *mammals* leads to bile acids (7.9), *bacteria* may either reduce cholesterol to coprostanol (in the lower parts of the mammalian intestine) or oxidize it to Δ^4-cholesten-3-one.

7.1.2 Turnover of Cholesterol (Figure 7.1-4)

The daily demand of cholesterol in *humans* is 1...1.5 g. 50% or less is supplied by food intake. The intestinal resorption rate is only about 40%; it can be greatly reduced by administration of phytosterols (e.g. β-sitosterol). The endogenous cholesterol synthesis yields about 0.4...1.2 g/day. Cholesterol is degraded to bile acids (7.9) in the *liver* and excreted in the order of about 0.6 g/day. Small quantities are converted to steroid hormones and vitamin D_2. The flow of compounds in *humans* is shown in the graph.

Figure 7.1-4. Flow Sheet of Cholesterol in Humans

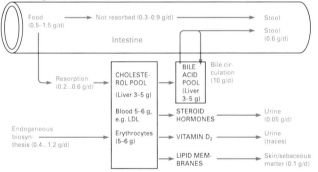

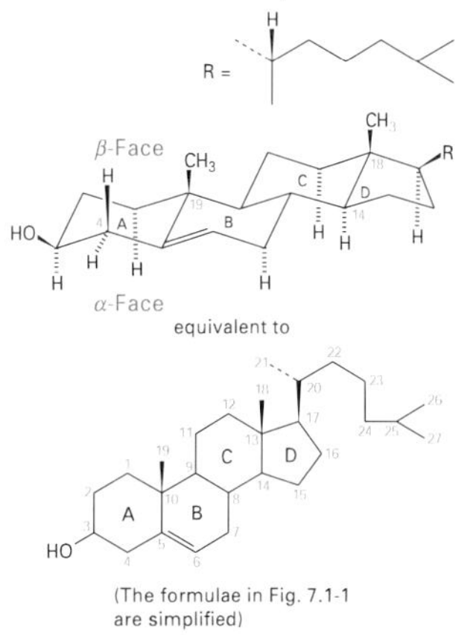

The intracellular 'cholesterol pool' (Table 7.1-1) consists of cholesterol esters and free cholesterol. Considerable quantities of esters besides free cholesterol are also present in blood lipoproteins (18.2), which effect the lipid transport. Mammalian plasma membranes contain 40...50% lipids, $1/2 ... 1/3$ of which is free cholesterol. In intracellular membranes (*mitochondria*, *ER*, *nucleus*), the cholesterol share is lower ($1/7 ... 1/10$). Cholesterol esters are practically absent from membranes. However, some precursors of cholesterol, such as lathosterol, 7-dehydrocholesterol and desmosterol can be found there.

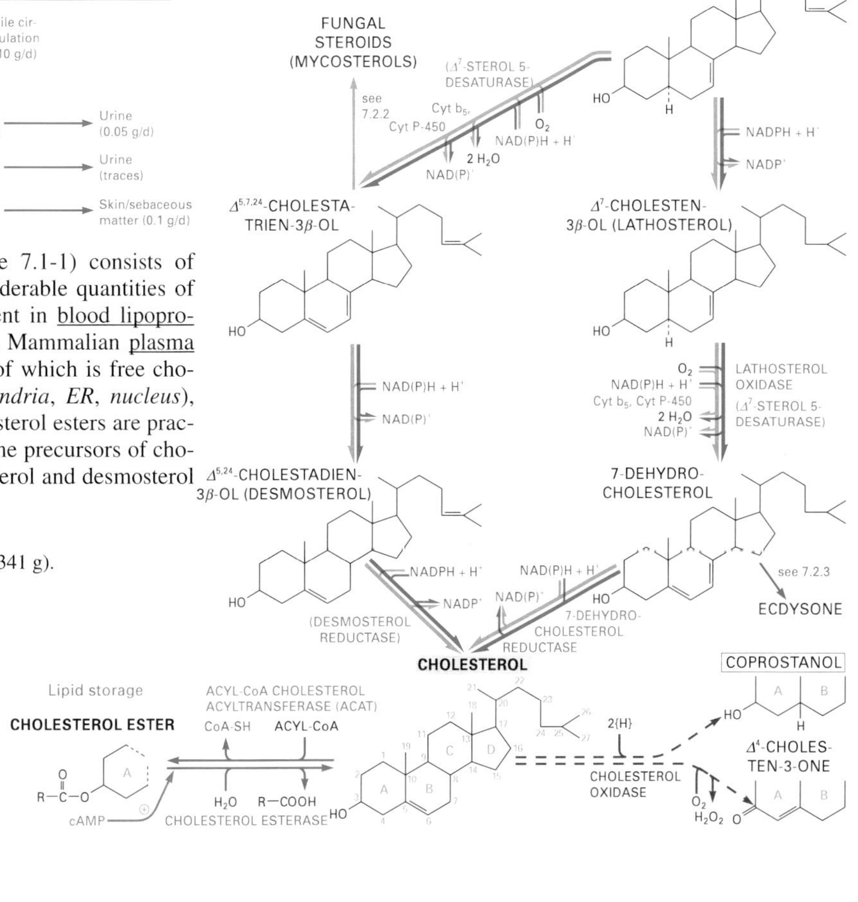

Table 7.1-1 Cholesterol Distribution (Rat, Weight 341 g).
(Jones, A., Glomset, J. in Danielsson *et al.*, p. 96)

	Free Cholesterol	Cholesterol Esters
Total	439 mg	60 mg
Plasma	0.4 %	8.7 %
Intestine	4.0 %	1.4 %
Liver	6.2 %	24.7 %
Skin	12.1 %	25.5 %
Muscle	16.9 %	2.5 %
Brain + nerves	21.2 %	0.2 %

Figure 7.1-5. Cholesterol Ring Structure (*all-trans* = chair representation)

7.1.3 Function of Cholesterol in Membranes

Unesterified cholesterol orients itself perpendicularly to the surface with the –OH group directed towards the aqueous phase. The planar, *all-trans* sterol structure (Figure 7.1-5) and the polarity promote its association with phospholipids and lipoproteins. Cholesterol lowers the membrane fluidity by restricting the movement of fatty acyl chains (above the transition temperature). Coated pits, which assemble LDL receptors (18.2.4), are cholesterol-poor membrane areas. Also other membrane functions are affected by variations in the cholesterol content.

7.1.4 Regulation of Cholesterol Synthesis (Figure 7.1-7)

The synthesis rate has to respond to the food intake and to the needed quantity of cholesterol, its metabolites and precursors. Thus it must be strictly regulated. Cholesterol synthesis exhibits a pronounced circadian rhythm with a maximum at midnight.

- Short term regulation (indicated by solid orange arrows) adapts itself within seconds to variations in supply or demand. This is primarily achieved by phosphorylation (inactivation) and dephosphorylation (activation) of hydroxymethylglutaryl-CoA(HMG-CoA) reductase, the rate-determining, initial enzyme of cholesterol biosynthesis (Figure 7.1-6). While insulin and thyroid hormones promote dephosphorylation, glucagon and cAMP enhance the (inactivating) protein kinase.

  The regulating protein kinase, in turn, is activated by (cAMP dependent) phosphorylation and inactivated by dephosphorylation, thus constituting a cascade mechanism (Fig. 7.1-6). The same protein kinase also regulates the activity of fatty acid synthesis by phosphorylating and inactivating the initial enzyme, acetyl-CoA carboxylase (6.1.2). Both pathways are coordinated this way.

Figure 7.1-6. Phosphorylation and Dephosphorylation of HMG-CoA Reductase

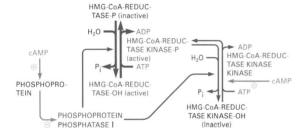

Furthermore, competitive feedback inhibition of several enzymes of cholesterol biosynthesis takes place. E.g., doubling of the physiological concentration of farnesyl-PP to 5 μmol/l inhibits mevalonate kinase more than 50 %. Still higher concentrations of farnesyl-PP (25 μmol/l) inhibit also its own conversion to squalene and favor its degradation (Fig. 7.1-7). Potent inhibitors of HMG-reductase are the fungal metabolites compactin and mevinolin.

- Long-term regulation (lasting hours, indicated by dashed orange arrows) occurs via changes in the transcription or translation rate or in the half-life time of mRNA or the enzymes.

  Thyroid hormones increase the biosynthesis rate for HMG-CoA reductase (7.1.1). On the other hand, oxidized metabolites of cholesterol (e.g. 25-hydroxycholesterol) diminish the transcription rate of this gene, of the genes for HMG synthase, geranyl transferase, farnesyl-PP, farnesyl transferase and LDL receptor.

  HMG reductase is regulated at additional levels: Glucocorticoids (being metabolites of cholesterol) and other compounds accelerate the mRNA breakdown. Farnesol, 25-hydroxycholesterol and LDL accelerate the breakdown of the enzyme protein. Thus, in *rat liver* the transcription rate of the HMG-CoA reductase gene can be modified 30 fold and the half-life time of the enzyme up to 35-fold, accounting well for the more than 450-fold change observed. However, no total blockade of the enzyme occurs, since sufficient intermediates have to be supplied for the synthesis of non-sterols (7.2, 7.3.). Similar regulation is exerted with HMG-CoA synthase and mevalonate kinase.

7.1.5 Cholesterol Homeostasis (Figure 7.1-7)

Besides controlling the cholesterol synthesis, regulation mechanisms have also to encompass all other cholesterol converting reactions.

- In addition to its effects on cholesterol synthesis, 25-hydroxycholesterol decreases the transcription rate of the LDL receptor gene (18.2.4), and thus coordinates cholesterol synthesis with its uptake into the liver.

  Involved in this regulation is a common *cis*-acting DNA sequence SRE-1 (sterol regulatory element 1) located in the 5′ flanking region of the genes for HMG synthase and reductase, geranyl transferase and the LDL receptor. In case of low cholesterol concentrations, the transcription factor SREBP-1 (sterol regulatory element binding protein, 68 kDa) is liberated by proteolysis, binds to SRE-1 and activates transcription.

- After LDL has entered the liver cells, the cholesterol esters are hydrolyzed by the lysosomal cholesterol esterase. Excess cholesterol formed this way activates the acyl-CoA cholesterol acyl transferase (ACAT) (18.2.4) of the *endoplasmic reticulum* and achieves reesterification this way. The esters are deposited as lipid droplets. Cholesterol can be mobilized again by the neutral cholesterol esterase, which (at least in some organs) appears to be regulated by cAMP dependent phosphorylation and dephosphorylation.

- Cholesterol regulates its own degradation by stimulating the transcription of cholesterol 7α-hydroxylase, the initial enzyme of bile acid production (7.9). Additionally, this enzyme is controlled by the bile acids returning to the liver after reabsorption. The 7α-hydroxylase appears to be regulated by phosphorylation (activation) and dephosphorylation (deactivation), the activating kinase being an cAMP dependent enzyme.

Disturbances in cholesterol metabolism cause various diseases. In familial hypercholesterolemia the uptake of cholesterol into the cells via LDL receptor is dimished (18.2.4). Therefore, in spite of high plasma cholesterol concentration, the (intracellularly regulated) cholesterol synthesis proceeds still at high speed.

Literature:

Botham, K.M.: Biochem. Soc. Transactions 20 (1992) 454–459.
Brown, M.S., Goldstein, J.L., Angew. Chem. 98 (1986) 579–599.
Danielsson, H., Sjövall, J. (Eds.): *Sterols and Bile Acids. New Comprehensive Biochemistry* Vol.12. Elsevier (1985).
Meigs, T.E. *et al.*: J. Biol. Chem. 271 (1996) 7916–7922.
Molowa, D.T., Cimis, G.M.: Biochem. J. 260 (1989) 731–736.
Rosser, D.S.E. *et al.*: J. Biol. Chem. 264 (1989) 12653–12656.
Vance, D.E., Vance, J. (Eds.): *Biochemitry of Lipids, Lipoproteins and Membranes. New Comprehensive Biochemistry* Vol. 31. Elsevier (1996).

Figure 7.1-7. Reactions Involved in Cholesterol Homeostasis

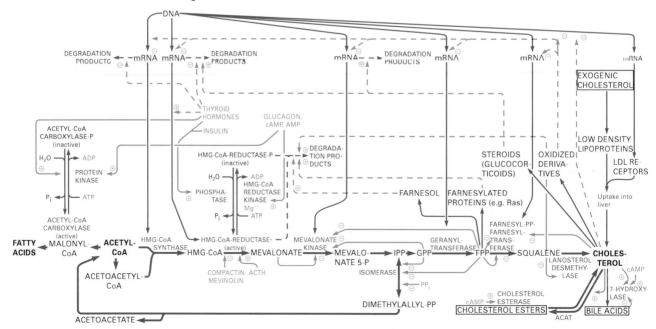

7.2 Hopanoids, Steroids of Plants and Insects

Late precursors of cholesterol biosynthesis, as well as cholesterol itself, can give rise to a large number of important compounds for *bacteria, plants, fungi* and *insects*. (For early precursors see 7.3.)

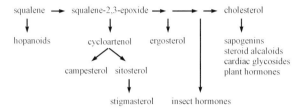

7.2.1 Hopanoids (Figure 7.2-1, left)

These pentacyclic compounds occur in all fossil sediments and petroleum deposits in huge quantities. They are widely distributed in *prokarya* as a membrane component and were found also in some *eukarya* (e.g. in the fungus *Tetrahymena* and in some *higher plants*). Apparently, this evolutionary very old group of compounds developed at a time, when the earth atmosphere still lacked oxygen. Whenever the epoxidation of squalene (leading to cholesterol or to cycloartenol) is impossible, squalene cyclizes in water directly to hopanoids (e.g. diploptene, diplopterol, tetrahymenol). Their side chains are then modified with ATP as energy source. Frequently, polyhydroxylated C_8-side chains are found, which are attached via ester or amide bonds to sugars, amino acids and nucleotides.

7.2.2 Phyto- and Mycosterols (Figure 7.2-1)

While in most *animals* cholesterol is the major steroid product, other sterols (substituted at C-24) are prevalent in *higher plants* and *fungi*. Well known sterols in *plants* are campestrol, sitosterol and stigmasterol and in *fungi* ergosterol. They act as membrane components. Other *plant* steroids (derived from cholesterol) are sapogenins (e.g. diosgenin), steroid alcaloids, cardiac glycosides and brassinosteroids (plant hormones).

Biosynthesis: In all *higher plants* and *algae*, 2,3 epoxysqualene is cyclized not to lanosterol, but to cycloartenol (Fig. 7.1-3). The 3-ring annelated to ring B is opened in later metabolic steps. Oxidative removal of the methyl groups at C-4 and C-14 proceeds analogously to the cholesterol biosynthesis. Different to this, side chain methylations (at C-24, sometimes at C-23 or C-25) take place and are frequently species specific. Cycloartenol can be converted into cholesterol, which is the source of other plant sterols.

Properties: β-Sitosterol decreases the intestinal resorption of cholesterol and is used for treatment of hypercholesterolemia. Ergosterol can be converted by illumination with ultraviolet light into ergocalciferol (vitamin D_2), an opening of ring B takes place. It has the same vitamin effects as cholecalciferol (vitamin D_3) (9.11.1). Sapogenins and steroid alcaloids are toxic and have hemolytic effects. Brassinosteroids cause cell elongation and cell division in *plants*. Cardiac glycosides [e.g. digoxin from *Digitalis* (*foxglove*), ouabain from *Strophantus*] are glycoside derivatives of the cardenolide group of plant steroids. They increase the contractility of the cardiac muscle (positive inotropic effect) by converting ATP more effectively into mechanical energy. Apparently the Na^+/K^+-ATPase (18.1.1) is inhibited, which causes intracellularly an increase of Na^+ and a decrease of K^+. The elevated Na^+ concentration, in turn, stimulates the Ca^{++}/Na^+ antiport mechanism and thus increases intracellular Ca^{++} (Table 18.1-3).

7.2.3 Ecdysone (Figure 7.2-1, right)

Insects are unable to synthesize cholesterol *de novo*. They satisfy their demand either by food intake of cholesterol or by conversion of phytosterols into it in unique reaction sequences. The critical step is the demethylation at C-24. Highly specific inhibitors of this step are effective insecticides, which do not disturb other organisms.

From 7-dehydrocholesterol, hydroxylations lead to α-ecdysone and further to 20-hydroxyecdysone, which are highly effective hormones for molting and pupating. They also control vital functions in *crustaceans, nematodes* and *molluscs* and are likely the phylogenetically oldest steroid hormones. They are found in some *plants*, too.

Literature:

Goodwin, T.W. in Danielsson, H., Sjövall, J. (Eds.) *Sterols and Bile Acids. New Comprehensive Biochemistry* Vol. 12. Elsevier (1985) 175–198.

Sahm, H. *et al*.: Adv. Microb. Physiology 35 (1993) 246–273.

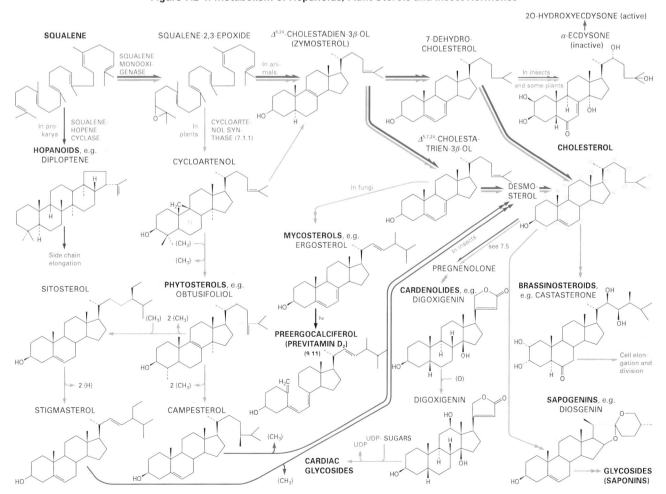

Figure 7.2-1. Metabolism of Hopanoids, Plant Sterols and Insect Hormones

7.3 Isoprenoids

Some early intermediates of the cholesterol biosynthesis are the origin of linear and cyclic polymers in *animals, plants and bacteria*:

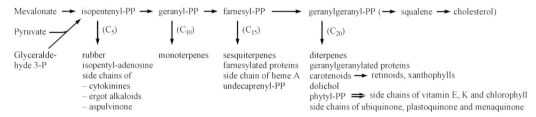

The common structural unit is isoprene ($CH_2=C(CH_3)–CH=CH_2$). In its activated form, isopentenyl-PP ($CH_2=C(CH_3)–CH_2–CH_2–PP$) is used by *animals* and *plants* for synthesis of linear and cyclic polymeres by prenyltransferases.

In *bacteria, green algae* and in chloroplasts of *higher plants*, isopentenyl-PP can be formed by a mevalonate-independent pathway, starting by condensation of (likely) hydroxyethyl-thiamine-PP (3.3.1) and glyceraldehyde 3-P and proceeding via the intermediate D-1-xylulose 5-P. The product is used for isoprenoid biosynthesis.

7.3.1 Terpenes (Table 7.3-1, Figure 7.3-2)

A very large group of organic *plant* compounds are terpenes, subdivided into monoterpenes (C_{10}), sesquiterpenes (C_{15}) and diterpenes (C_{20}).

The biosynthesis of terpenes generally takes place by intramolecular prenylation (analogous to Fig. 7.1.2, but within a single molecule), starting from geranyl-PP, farnesyl-PP or geranylgeranyl-PP, respectively. Consecutive rearrangements, introduction of hydroxy or carboxy groups etc. lead to an enormous number of compounds. Many of them are aromatic essences, some are of therapeutic importance (e.g. as antimalarial drugs).

7.3.2 *All-trans* Metabolites (Figure 7.3-2)

Carotenoids (C_{40}) are formed from 2 geranylgeranyl molecules.

The enzyme complex phytoene synthase in *higher plants, yeasts, algae* and *fungi* produces *cis*-phytoene (which is converted to the *trans* configuration in later steps), while most *bacteria* obtain *trans*-phytoene. Membrane associated phytoene desaturase oxidizes and cyclizises phytoene via several steps to *all-trans* α- or β-carotene. By stepwise oxidation, *plants* and *bacteria* convert them to xanthophylls, which act as light absorbing pigments in photosynthesis (16.2.1). They are also able to eliminate free radicals and act as membrane-protective agents (comparable to vitamin E in *humans*, 9.12). Catalyzed by specific dioxygenases, β-carotene is split symmetrically to *all-trans* retinal. This compound is essential for the visual process in *vertebrates* (17.4.6) as well as for photosynthesis in *halobacteria* (16.1.3). In cells, retinal is in equilibrium with *all-trans* retinol (vitamin A, 9.1).

Intercellular transport and distribution proceeds via specific retinol-protein complexes in blood (RBP, CRBP, CRBP II, IRBP), while storage and uptake from food takes place as retinol-fatty acid esters. On the other hand, small quantities of retinal are oxidized to retinoic acid. This compound is transported as a highly specific complex with cellular retinoic acid binding protein (CRABP). Retinoic acid shows hormone-like activity in cell development and cell division by acting on specific nuclear receptors (17.6). It is applied as a cancerostatic drug in promyelocytic leukemia. β-Carotenes are of importance for stimulation of immunological defense and prevention of arteriosclerosis.

Catabolism of all retinoids proceeds by conjugation with glucuronic acids and excretion in the urine.

7.3.3 *Poly-cis* Metabolites (Figure 7.3-2)

Dolichol-P ($C_{70}...C_{120}$) is an essential sugar carrier in *animals, plants* and *fungi* for formation of glycoproteins in the *endoplasmic reticulum* (13.3.1). Furthermore, dolichyl esters serve as a transport and storage form of fatty acids. *Cis*-geranylgeranyl-PP enters the pathway to dolichol via a highly affine polyprenyl-*cis*-transferase.

Bacteria use undecaprenyl-P (C_{55}) as carrier for glycopeptides in formation of cell wall murein (15.1) and as a sugar carrier for synthesis of other membrane components, like lipopolysaccharide or teichoic acids. Both compounds carry 2...3 double bonds in *trans*-configuration at their non-hydroxyl end.

Isopentenyl-PP is the origin of natural rubber (e.g. from *Hevea brasiliensis*), which consists of very long *all-cis* isoprenoid chains (ca. $C_{3500}...C_{25000}$).

7.3.4 Isoprenoid Side Chains (Figure 7.3-2)

Bacterial pigments and cytokines (*plant* hormones) contain isoprenoid side chains.

Phytyl-PP (C_{20}, formed from *all-trans*-geranylgeranyl-PP in *plants* and some *bacteria*) is a side chain in chlorophyll (5.4), α-tocopherol (vitamin E, 9.12) and phylloquinone (vitamin K_1). Longer isoprenoid side chains are present in menaquinone (vitamin K_2, $C_{20}...C_{30}$, 9.13), ubiquinone ($C_{30}...C_{50}$, 4.7.2) and plastoquinone (C_{45}, important in *plant* respiration).

Many proteins are attached to cellular membranes by isoprenoid anchors (via a thioether bond to mostly farnesyl, sometimes geranylgeranyl chains), e.g. Ras (17.5), G_γ proteins (17.4) and heme A (5.2.1). This prenylation determines the location and the function of the proteins. The enzymatic reaction is shown in Fig. 7.3-1:

Figure 7.3-1. Isoprenoid Anchors of Proteins
(C = cysteine, A = aliphatic amino acid)

Farnesylation takes place if X = serine, methionine or glutamine. Geranylgeranyl is attached if X = leucine or phenylalanine.

Proteins of the Rab group, which contain –C–C– or –C–X–C– sequences, are geranylgeranylated similarly. The enzyme is assisted by the Rep protein.

Literature:

Gershenzon, J., Croteau, R.B. in Moore, T.S. (Ed.): *Lipid Metabolism in Plants*. CRC Press (1993) 339–388.
Hemming, F.W. in Wiegandt, H.: *Glycolipids*. New Comprehensive Biochemistry Vol. 10. Elsevier (1985) 261–305.
Lichtenthaler, H.K. *et al.*: FEBS Letters 400 (1997) 271–274.
Packer, L. (Ed.): *Carotenoids. Part B. Methods in Enzymology* Vol 214. Academic Press (1993).
Packer, L. (Ed.): *Retinoids. Part A. Methods in Enzymology* Vol 189. Academic Press (1990).
Zhang, F.L., Casey, P.J.: Ann. Rev. Biochem. 65 (1996) 241–269.

Table 7.3-1. Terpenes

Class	Formed from	Structure	Examples	Occurrence	Comment
Monoterpenes (C_{10})	geranyl-PP, sometimes neryl-PP linalyl-PP	open chain monocyclic bicyclic tricyclic	geraniol limonene α-pinene, camphor teresantol	only in *higher plants*	Volatile, characteristic odor. Mostly not containing oxygen.
Sesquiterpenes (C_{15})	farnesyl-PP, sometimes nerolidyl-PP	open chain monocyclic bicyclic tricyclic	farnesol curcumene cadinene cedrene	all *tracheophyta, mosses, fungi, brown and red algae, insects*	In etheric oils, also plant hormones (e.g., phytoalexins, frequently oxidized); *insect* hormones (juvenile hormones)
Diterpenes (C_{20})	geranyl-geranyl-PP	open chain monocyclic bicyclic tricyclic tetracyclic	phytol kauric acid abietic acid	*plants, fungi*	In resins and balsams. Frequently highly oxidized. Gibberelic acid (formed via kaurene) is a growth hormone for *plants*.

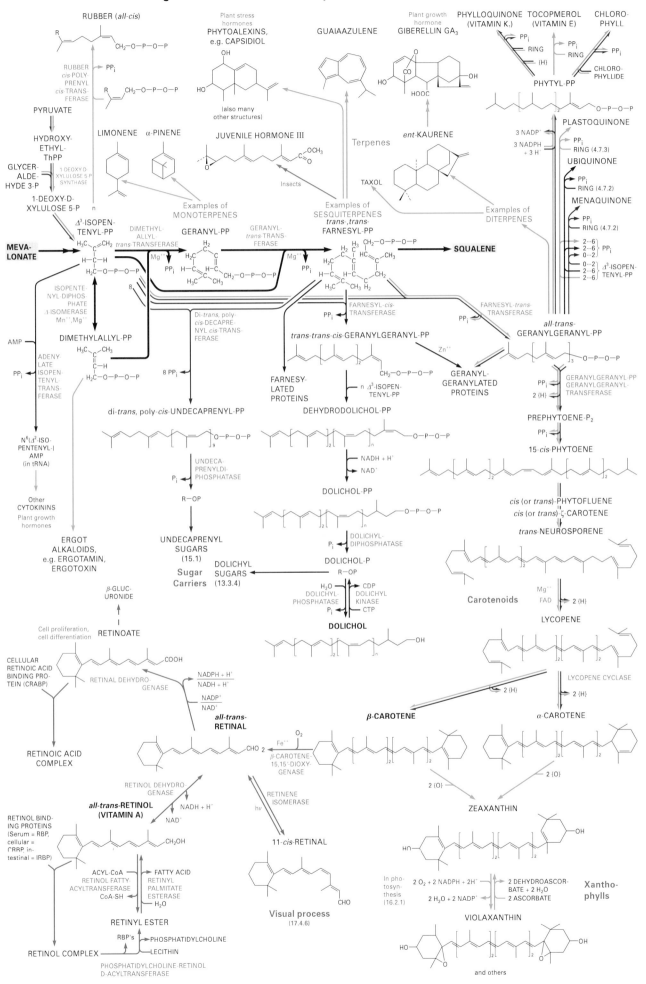

Figure 7.3-2. Metabolism of Terpenes, Carotenoids, Retinoids

7.4 Steroid Hormones

Cholesterol is the origin of a multitude of steroids, which are of high importance in metabolic processes. While the compounds in *plants* and *insects* are dealt with in 7.2 and 7.3, the following Section deals with *vertebrate* hormones.

7.4.1 Biosynthesis (Details on following pages)

The common synthesis pathway starts with the partial removal of the cholesterol side chain, yielding the gestagen progesterone. The further reactions depend on the enzyme equipment of the particular cells. Removal of the remaining side chain leads to androgens, a further aromatization of ring A yields estrogens. On the other hand, different hydroxylations of progesterone produce gluco- or mineralocorticoids.

The releasing and the trophic hormones are peptides and act on cell surface receptors, which transmit the signals intracellularly via various pathways (e.g. G-proteins and phosphorylation cascades, 17.4 and 17.5) to their site of action, e.g. gene expression in the *nucleus* (resulting in enzyme synthesis) or elevated NADPH production (which increases hydroxylation reactions).

Contrary to this 'second messenger' system, the steroid hormones produced by this cascade act directly. Since they are very lipophilic, they are able to diffuse through the membrane of their target cells. In the *nucleus*, they are bound to specific receptors with very high affinity ($K = 10^{-8}...10^{-10}$ mol/l). The hormone-receptor complexes attach themselves to special DNA sequences (hormone responsive elements) and enhance transcription (17.6). Since in different organs different genes are activated, the same hormone may control the synthesis of various proteins.

The secreted hormones exert a feedback action towards the *hypothalamus* and the *pituitary gland*. Therefore, the reaction cascade is activated only for short times. Furthermore, the steroid hormones show only short half-life times (10...90 min.). Both effects prevent prolonged, high concentrations of steroid hormones and allow the system to respond quickly to changing situations.

As an exception, aldosterone is regulated by the renin-angiotensin system, which is dependent on blood volume and plasma electrolyte concentration. (17.1.8)

Figure 7.4-1. Sites of Steroid Hormone Biosynthesis

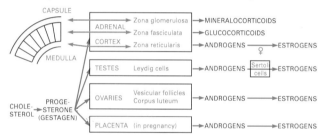

Reaction types: The major conversions are hydroxylations, reductions/isomerizations (usually combined), C-C bond splitting and aromatization (estrogen formation). Frequently, the enzymes do not show absolute specificity, so that several alternative reaction sequences can lead to the same end product.

Hydroxylations are mostly catalyzed by NADPH dependent mixed function steroid monooxygenases, which are associated with cytochrome P-450. They transfer one oxygen from O_2 to the steroid and the other one to hydrogen. They also play a major role in detoxification reactions.

$$R–H + O_2 + NADPH + H^+ \rightarrow R–OH + NADP^+ + H_2O$$

Figure 7.4-2. Mechanism of Cytochrome P-450 Dependent Hydroxylations

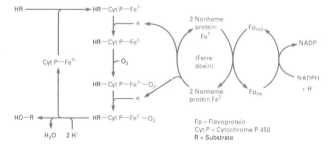

Other hydroxylases depend on different hydrogen donors, like NADH, $FADH_2$ or ascorbic acid.

7.4.2 Biological Activation and Regulation of Steroid Hormones

Activation of most steroid hormones in *mammals* is a three-step cascade starting at the hypothalamus (for a general discussion, see 17.1.5). An external stimulus causes the *hypothalamus* to secrete releasing hormones (activating liberins and inhibiting statins), which cause a specific activation of the *anterior pituitary lobe* (*adenohypophysis*). In turn, trophic hormones (tropins) are secreted, which are transported by the bloodstream to *peripheral tissues (adrenals, testis, ovary)* and cause there the secretion of the steroid hormones.

Figure 7.4-3. Regulation of the Synthesis and Secretion of Steroid Hormones (cf. Fig. 17.1-4)

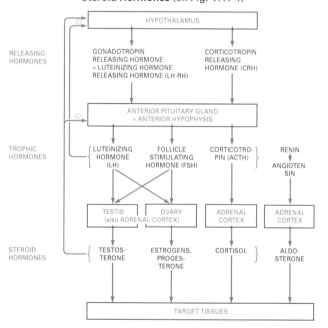

7.4.3 Transport of Steroid Hormones

For transportation in blood, the lipophilic steroid hormones are bound to plasma protein carriers (albumin, orsomucoid, transcortin, sex hormone binding globulin and α_1-fetoprotein). These complexes are physiologically inactive. At the surface of the target cell, the complexes release the hormone for diffusion into the cell.

7.4.4 Degradation of Steroids

Only *bacteria* are able to open the rings of the steroid skeleton.

In *animals*, deactivation and modification for excretion occur in *liver* and to a smaller degree in *kidneys* and *lung*. Major reactions, which increase the solubility in the aqueous phase are:
- reduction of the 3-oxo to the 3-hydroxy group (α or β) and hydrogenation of the double bond in ring A (the aromatic ring in estrogens is not modified)
- conjugation at the C-3 hydroxyl with sulfate (as phosphoadenylyl phosphosulfate) or glucuronide
- introduction of more hydroxy groups by cytochrome P-450 dependent enzymes
- oxidative degradation of the side chain in C_{21} steroids

Literature:

Cooke, B.A., King, R.J.B., van der Molen, H.J., *Hormones and their Actions*, part I and II. *New Comprehensive Biochemistry* Vol. 18 A and B. Elsevier, 1988.

Nebert, D.W., Gonzalez, F.J.: Ann Rev. Biochem. 56 (1987) 945–993.

7.5 Gestagen

7.5.1 Biosynthesis of Progesterone (Figure 7.5-2)

In *mammals*, the binding of follicle-stimulating hormone (FSH), luteinizing hormone (LH) or corticotropin (ACTH) to their receptors at the *adrenal cortex, ovary* or *testes* membranes starts a G-protein mediated reaction sequence (17.4). This leads to a transport of cholesterol from *cytoplasmatic inclusion droplets* into the *mitochondria*, assisted by sterol carrier protein SCP_2. The conversion of cholesterol to pregnenolone is effected by the trifunctional complex cholesterol monooxygenase, which is located at the *inner membrane of mitochondria*. The presumed reaction steps are 22-hydroxylation, 20-hydroxylation and side chain lyase reaction. The hydroxylations correspond to the monooxygenase scheme (7.4.1), using the $P450_{scc}$ variant and adrenodoxin as the Fe-S-protein.

Pregnenolone then passes through the mitochondrial membrane and is converted by a bifunctional enzyme complex to the gestagen hormone, progesterone (Figure 7.5-1). This reaction takes place at the *smooth endoplasmatic reticulum*. Pregnenolone and progesterone are also the origin of androgens (7.6), estrogens (7.7), gluco- and mineralocorticoids (7.8); pregnenolone also of cardiac glycosides in *plants*.

Figure 7.5-1. Conversion of Cholesterol to Progesterone

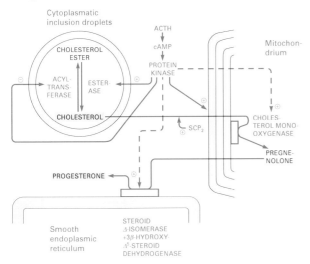

In *bacteria* and *plants*, progesterone biosynthesis proceeds analogously. However, the conversion of pregnenolone to progesterone is performed by 2 separate enzymes.

7.5.2 Gestagen Function, Transport and Degradation

In *female mammals*, progesterone plays an essential role as the only active gestagen. It is produced mostly in the *corpus luteum* of the *ovaries* during the second half of the cycle and in great quantities in the *placenta* during pregnancy. It binds tightly to receptor proteins in the *nucleus* and regulates the transcription to mRNA (11.4). Its physiological effects are

- during the second half of the female cycle:
  - preparation of the *uterus* for implantation of the fertilized ovum (modification of the endometrium)
  - general effects on the *CNS* (e.g. elevated body temperature)
- and during pregnancy:
  - preservation of the mucous coat of the uterus
  - prevention of further ovulations
  - formation of lactating alveoli in the *breasts* (synergistically with estrogen)

The estrogen-progesterone interaction is biphasic: High estrogen concentrations (e.g. during pregnancy) stimulate the formation of progesterone receptors, thus increasing progesterone effects. On the other hand, they also stimulate 17β-hydroxysteroid dehydrogenase, which converts estradiol into the much less active estrone. This decreases the progesterone receptor formation again.

In the *adrenal cortex* of both sexes and in *testis*, progesterone is only an intermediate in biosynthesis of sex hormones and of gluco- and mineralocorticoids (adrenals only).

Progesterone is transported in *serum* bound to albumin, transcortin (corticosteroid-binding globulin) or orsomucoid (acidic α_1 glycoprotein). Degradation (see Figure 7.8-2) involves reduction of the keto functions and the double bond and excretion as glucuronide.

The contraceptive RU 486 (Roussel-Uclaf) is an anti-progestin, which binds with high affinity to the progesterone receptors without producing the hormone effects.

Literature:
Cooke, B.A., King, R.J.B., van der Molen, H.J., *Hormones and their Actions*, part I and II. *New Comprehensive Biochemistry* Vol. 18 A and B. Elsevier, 1988.

Figure 7.5-2. Biosynthesis of Progesterone

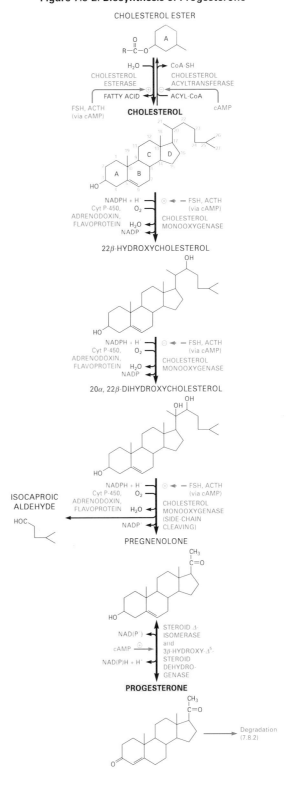

7.6 Androgens

Androgens are the determinating factors for male development and behavior in *vertebrates*.

7.6.1 Biosynthesis (Figure 7.6-1)

In male *mammals*, most of the androgens are produced in the *Leydig cells* of the *testis*. Additionally, androgens are synthesized in the *zona reticularis* of the *adrenal cortex* and in the *ovaries* of females. The respective enzymes are associated with the *smooth endoplasmic reticulum* of these cells. They are activated by luteinizing hormone (17.1.5), the concentration of which rises dramatically at puberty.

Either pregnenolone (Δ^5 pathway, e.g. in *humans* and *rabbits*) or progesterone (Δ^4 pathway, e.g. in *rats* and *mice*) is hydroxylated at the 17-position. The conversion between both pathways is catalyzed by a complex of 2 enzymes, it can take place at different levels of both pathways. Then the side chains are oxidatively removed, yielding 4-androstene-3,17-dione or dehydroepiandrosterone (DHA), respectively, which show weak androgen function. The splitting enzyme is closely linked with the 17-hydroxylase. Further redox reactions and isomerizations convert them into the important androgen, testosterone. 4-Androstene-3,17-dione can also be converted into adrenosterone, testosterone into 5α-dihydrotestosterone (5α-DHT), which shows even higher biological activity and is responsible for a series of specific androgen effects (see Table 7.6-1).

In *testes* and *ovaries*, most likely an additional pathway exists, which leads via 17α, 20α-dihydroxycholesterol directly to dehydroepiandrosterone. Up to 30% of the androgens are produced this way in these organs.
In *females*, 4-androstene-3,17-dione and testosterone are formed in the *theca interna cells* of the *ovaries* as intermediates for estrogen production (7.7.1).

7.6.2 Transport and Degradation

Since all steroid hormones are highly lipophilic, they have to be modified to a more hydrophilic form for transport purposes. DHA is transported in the circulation as (hormonally inactive) sulfate to its target organs, where it is desulfatized again. Another transport vehicle is the sex hormone binding globulin (SBG), which binds testosterone strongly and estradiol only slightly less.

The catabolism of androgens occurs in *kidneys* and *liver*. The compounds are hydrogenated at the Δ^4 double bond. The 3-oxo group is then reduced to the 3α hydroxy function in *humans* and to 3β in most other *vertebrates* and thereafter (mostly) glucuronized. The products are excreted.

7.6.3 Biological Function of Androgens (Table 7.6-1)

Androgens combine with nuclear receptors; the complexes act mainly through control of gene expression (17.6). This leads to growth and development of male sex characteristics and causes strong anabolic effects. Some androgen derivatives are even effective in nano- or picomolar concentrations.

Table 7.6-1. Effects of Androgens in Male Vertebrates

Species	Stage	Organ	Effects	Mainly involved
All	embryo	external genitals	sex differentation	testosterone, 5α-dihydrotestosterone
Most	embryo	liver, spleen	hemoglobin synthesis erythropoietin synthes.	5α-dihydrotestosterone 5α-dihydrotestosterone
Most	neonate	brain	sexual differentiation	5α-dihydrotestosterone
Most	neonate	liver	enzyme synthesis	androstenedione
All	puberty	testis	spermatogenesis	5α-dihydrotestosterone
Most	puberty	hair follicle	hair growth	5α-dihydrotestosterone
Most	puberty	vocal cords	thickening	5α-dihydrotestosterone
Most	puberty	muscle	growth, N-retention	testosterone
Human	puberty	sebaceous gl.	acne	5α-dihydrotestosterone
All	adulthood	brain	male libido	testosterone, estradiol
Birds	adulthood	brain	courtship display	
Deer	adulthood	antlers	growth	

7.6.4 Medical Aspects

Synthetic analogues of natural androgens are used as anabolic steroids for treatment of muscular dystrophy, psoriasis etc. Steroidal and non-steroidal antiandrogens block androgen receptors and are used, e.g., in tumor treatment.

Literature:
Cooke, B.A., King, R.J.B, van der Molen, H.J. (Eds.) *Hormones and their actions*. Part I. *New Comprehensive Biochemistry*. Elsevier (1988). Vol 18A, 3–38; 169–196.

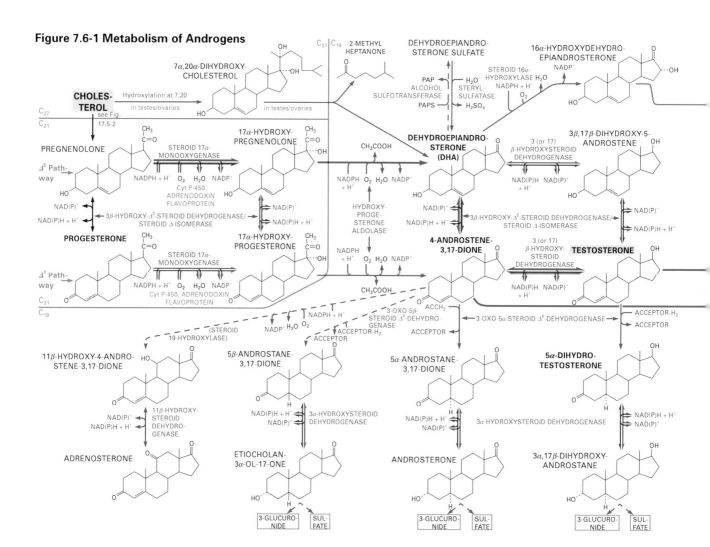

Figure 7.6-1 Metabolism of Androgens

7.7 Estrogens

Estrogens control the development of the reproduction system and the reproductive functions in *female vertebrates* (17.1.5).

7.7.1 Biosynthesis (Figure 7.7-1)

Estrone and estradiol are formed by oxidative removal of the C-19 methyl group from the androgens 4-androstene-3,17-dione and testosterone and subsequent aromatization of ring A in a concerted aromatase reaction. The estrogen synthesis is controlled by the luteinizing and the follicle stimulating hormones (17.1.5, 7.1.6).

The aromatase system contains cytochrome P 450 and NADPH-cytochrome P 450 reductase (7.4.1). The enzyme is associated with the *endoplasmic reticulum*. Two successive hydroxylations convert the C-19 methyl group via the hydroxy to the oxo level. The mechanism of the consecutive formate elimination has not been completely resolved yet.

The aromatization reaction is unique in *vertebrates*. It takes place mostly in the *granulosa cells* of the *ovaries*, but also in the *adrenal cortex* (of less importance in pre-menopausal women) and in the *Sertoli cells* of the *testis*. 'Peripheral estrogen synthesis' occurs in *adipose tissue* and *muscle* (in *females* prevalent after menopause) and in *liver* and *brain* (the most important estrogen source in *males*, ca. 60 %). In the *placenta* during pregnancy or in *breast cysts*, large quantities of estriol 3-sulfate are produced from dehydroepiandrosterone sulfate via initial 16α-hydroxylation, followed by aromatization. DHA originates both in the *fetal adrenal cortex* and in *maternal* circulation.

Estradiol and estrone can be interconverted by a redox reaction. Estriol is formed from estrone by 16-hydroxylation and 17-reduction reactions. 2-Hydroxylation of estrone and estradiol yields catechol estrogens, which are neuroendocrinologically active. This reaction can be performed by *placental* aromatase. Other steroids with estrogen function are found in some *animals*, e.g. equilin and equilenin in *horses*.

7.7.2 Transport and Degradation (Figure 7.7-1)

Estrogens are transported in blood by the sex-hormone binding globulin. Degradation takes place mostly in the *liver*.

Hydroxylation reactions of estrone and estradiol are performed by *microsomal* NADPH-dependent steroid hydroxylases. This takes place in different species at positions 6α, 6β, 7α, 11β, 14α, 15α, 16α, 16β and 18. Glucuronidization (prevalently) or sulfatation at the 3-position increases the solubility. Ca. 50 different metabolites are excreted in human urine. Some excretion also occurs via the *intestine*.

7.7.3 Biological Function of Estrogens (Table 7.7-1)

The control the *female* reproductive functions includes a multiplicity of estrogen effects on gene expression. Estrogens combine with high affinity in the *nucleus* with a specific estrogen receptor (65...70 kDa, 17.6). The complex effects an increase in the transcription rate (11.4.3) for proteins and peptides involved in growth control and DNA replication. This is in line with the estrogen dependency of some tumors. The estrogen activities decrease in the order estradiol > estriol > estrone. Generally, estradiol is active at concentrations, which are 100 to 1000 times lower than those required for hormone action of androgens.

Estrogens act anabolically on fat (causing reduced blood levels of cholesterol) and slightly anabolically on proteins. During menopause the estrogen level decreases and androgens from the *adrenal cortex* become more predominant.

Table 7.7-1. Effects of Estrogens in Female Vertebrates

Species	Stage	Organ	Effects
All	embryo	sex tract	development
Mammals	puberty	ovaries	estrus cycle
Mammals	puberty	endometrium of uterus	induction of cyclic proliferation
Mammals	puberty	epithelium of tubes and vagina	cyclic changes
Mammals	puberty	mammae	development
Human	puberty	hair, adipose tissue	adult distribution
Reptiles, amphibia, birds, fishes	adulthood	oviduct, liver	egg protein production

7.7.4 Medical Aspects

Estrogens are used for substitution therapy. Besides the natural hormones, ethinylestradiol and other estrogen derivatives, as well as stilbene derivatives can be applied (Natural stilbene derivatives occur in *plants*). Steroidal and nonsteroidal antiestrogens (e.g. tamoxiphen) compete with estrogens at the receptors and thus act as inhibitors (species-dependent). These compounds, as well as aromatase inhibitors are used for tumor treatment.

Literature:

Cooke, B.A., King, R.J.B, van der Molen, H.J. (Eds.): *Hormones and their Actions*. Part I. *New Comprehensive Biochemistry*. Elsevier (1988) Vol. 18A, 3–38; 197–215.
Raju, U., Bradlow, H.L., Levitz, M.: Ann. N.Y. Acad. Sci. 586 (1990) 83–87.
(Various authors): J. Steroid Biochem. Molec. Biol. 44 (1993) 321–691.

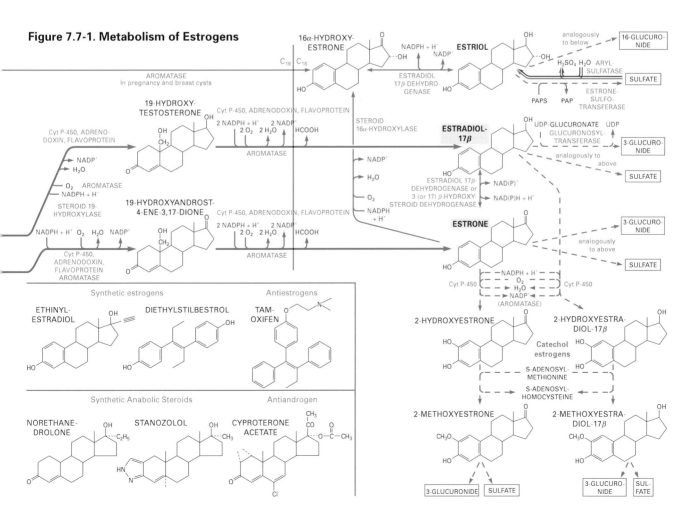

Figure 7.7-1. Metabolism of Estrogens

7.8 Corticosteroids

Corticoids are another class of hormones, which are produced by the *adrenal cortex* of *vertebrates*. They include the glucocorticoids, which are vital regulators of glucose metabolism and the mineralocorticoids, which control the mineral metabolism.

Table 7.8-1. Important Corticoids

Glucocorticoid Effects	Relative potency	Mineralocorticoid Effects	Relative potency
cortisol	1	aldosterone	1
cortisone	0.8	11-deoxycorticosterone	0.03
		cortisol	0.0003

7.8.1 Biosynthesis (Figure 7.8-2)

Formation of glucocorticoids in the *zona fascilata*: From pregnenolone, the 17α-hydroxysteroids 11-deoxycortisol, cortisol and cortisone are synthesized via hydroxylations at 17α, 21 and 11β. This occurs preferably in *humans* and *rabbits*. In *rats*, however, these compounds are formed from progesterone.

Formation of mineralocorticoids in the *zona glomerulosa*: Progesterone is the precursor of the 17-deoxycorticoids cortexone, corticosterone, 11-dehydrocorticosterone and aldosterone via hydroxylations at positions 21, 11β and 18. Oleyl-corticosterone is the active form of this hormone.

Although in both pathways the hydroxylations take place at the same positions 11β and 21, the enzymes are apparently different. Enzymatic defects lead to different diseases. In some species, a strict sequence of reactions does not exist due to only moderate specificity of the converting enzymes.

A striking phenomen is the repeated change between the compartments where the reactions take place (Figure 7.8-1). Cholesterol from the *cytosol* is converted to pregnenolone in the *mitochondria*, isomerized and further hydroxylated in the *smooth endoplasmic reticulum*, while final hydroxylations and dehydrogenase steps take place in the *mitochondria* again. Secretion into the *lymph* requires another passage through the *cytosol*. This separation of the synthesis sites likely allows fine tuning of the regulation.

Figure 7.8-1. Sites Involved in Corticosteroid Synthesis
(Numbers indicate positions of hydroxylation)

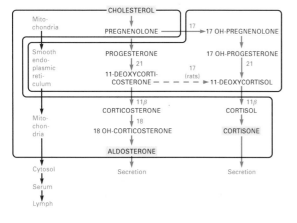

7.8.2 Transport and Degradation

In blood, cortisol is bound to transcortin (corticosteroid-binding globulin). The degradation of corticosteroids is generally initiated by hydration to tetrahydro derivatives (mostly 3α, 5β, but also 3α, 5α). Then usually follows glucuronidation and hydration to hexahydro derivatives, which are excreted in urine. The degradation of progesterone follows the same steps.

7.8.3 Biological Function

The steroid hormones pass through the *cellular membrane* and bind directly to *nuclear* receptor proteins, which induce DNA transcription (17.6).

Glucocorticoids are released from the adrenals in a diurnal rhythm: the blood level increases during the night and drops during the day. In stress situations, additional release takes place. Hormone actions proceed in almost all *mammalian* cell types:

- In *extrahepatic tissues (muscle, skin, adipose tissue, lymphocytes etc.)*, catabolic enzymes are induced. Glucose resorption and glycolysis decrease, while the blood glucose level, proteolysis and lipolysis (regulated via epinephrine) increase.
- In *liver*, due to anabolic effects, protein biosynthesis, gluconeogenesis and glycogen synthesis rise.
- In the *lymphatic system*, an immunosuppressive effect occurs via inhibition of the response of *T-lymphocytes* (19.1.4) and the activation of *macrophages*, as well as a decrease in the number of circulating *lymphocytes, eosinophils, monocytes* and *macrophages*.
- A strong suppressive effect on inflammation is exerted by high levels of (especially) cortisol and cortisone. This is due to:
  - biosynthesis of the glycoprotein lipocortin, which inhibits phospholipase A_2 and therefore decreases the liberation of arachidonic acid and its conversion into prostaglandins (17.4.8)
  - feedback inhibition of the *hypothalamus*, which suppresses the fever reaction to pyrogens
  - downregulation of NO synthase II (17.7.2)
  - inhibition of leukocyte immigration into the inflamed tissue (19.3)

Mineralocorticoids: Aldosterone promotes in the *proximal* and *distal kidney tubuli*

- reabsorption of Na^+ and Cl^-
- retention of water
- secretion of K^+, H^+ and NH_4^+ ions.

This is effected by increased expression of the Na^+/K^+channel, the Na^+channel and enzymes of the citrate cycle. Therefore, administration of aldosterone leads to a considerably decreased cytosolic K^+ level and to a tendency for edema formation. Similar effects are also shown by other corticoids, such as 11-deoxycorticosterone, cortisone, cortisol and corticosterone.

7.8.4 Medical Aspects

A number of serious diseases are caused by disturbances in the corticoid metabolism. Frequently, defective hydroxylases cause a shift in the corticoid hormone pattern (Table 7.8-2).

Glucocorticoids are used for substitution therapy of adrenal insufficiency. Synthetic glucocorticoid analogs with 2 double bonds in ring A (e.g. dexamethasone) do not show mineralocorticoidal side effects and are preferred. Artificial antiglucocorticoids are used for treatment of Cushing's syndrome and adrenocorticoidal carcinoma. Due to multiple side effects, only in very serious cases high doses of cortisol and cortisone are used as antiphlogistic drugs.

For treatment of essential hypertonia or of edema, antimineralocorticoids (e.g. spironolactone) are applied.

Literature:
Cooke, B.A., King, R.J.B., van der Molen, H.J., *Hormones and their Actions,* part I. *New Comprehensive Biochemistry* Vol. 18 A. Elsevier (1988).

Table 7.8-2. Diseases Caused by Defects in Corticoid Metabolism

Disease	Origin	Cause	Metabolic Effects	Clinical Picture
Cushing's syndrome	*hypothalamus / pituitary gland* (frequently *tumors*)	CRH or ACTH overproduced	cortisol ↑, ACTH ↑ (cannot be inhibited by cortisol)	steroid diabetes, muscular atrophy, bone destruction, decreased formation of antibodies
	adrenals, frequently *tumors*	cortisol overproduced	cortisol ↑ (no elevated ACTH)	same
Addison's disease	*adrenals*	chronic insufficiency of adrenal cortex	cortisol ↓ (in spite of elevated ACTH), Na^+ ↓, K^+ ↑	loss of water, myasthenia, hypoglycemia, blood pressure ↓, atrophy of genitals
Waterhouse-Friederichsen syndrome	*adrenals*	acute insufficiency of adrenal cortex	cortisol ↓↓	hemorrhagic necrosis due to meningococci-caused sepsis
Congenital adrenal hyperplasia	*adrenal cortex*	21-hydroxylase defective	plasma cortisol ↓, androgens ↑	androgenital syndrome, virilization, salt wasting syndrome
Conn syndrome (aldosteronism)	*adrenal cortex*, frequently *tumors*	aldosterone overproduced	aldosterone ↑, Na^+ ↑, K^+ ↓	elevated potassium excretion, alkalosis, edema
Secondary aldosteronism	*kidney blood supply*	arteriosclerosis, hypertonia, alkalosis	renin ↑, causing aldosterone ↑	same
Hypoaldosteronism	very rare	lack of 18-hydoxylase	Na^+ ↓, K^+ ↑	salt loss

Figure 7.8-2. Metabolism of Glucocorticoids and Mineralocorticoids

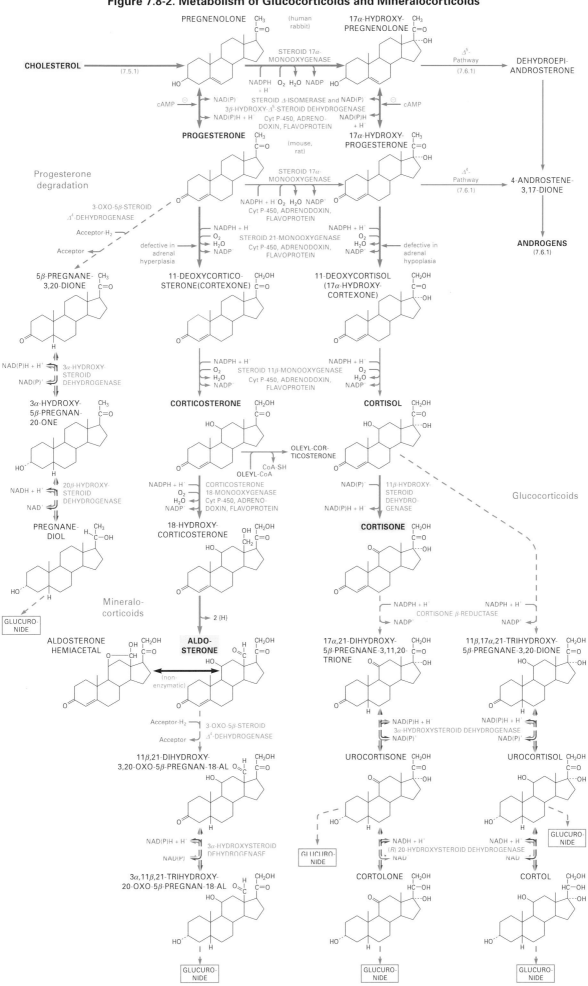

Figure 7.9.1 Metabolism of Bile Acids

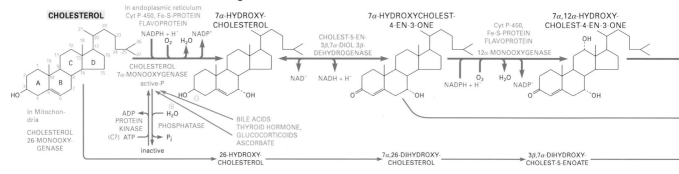

7.9 Bile Acids

In *vertebrates*, the major part of cholesterol is finally converted to bile acid conjugates in the *liver* and excreted via bile into the *small intestine*. The conjugates are bile acid amides with taurine or glycine or half-esters with sulfuric acid. Due to their low pK values (taurocholic acid ca. 2, glycocholic acid ca. 4), they dissociate completely at physiological pH values. Therefore they are good anionic detergents, which form mostly cylindric micelles. In the *intestine*, they act as emulsifiers on lipid food components (e.g. triglycerides, cholesterol and its esters). This enables their resorption by the *intestinal mucosa* (Fig. 18.2-2).

7.9.1 Occurence (Table 7.9-1)

Bile alcohol derivatives occur only in *fishes* and *amphibia* and have frequently 4 or more hydroxyl groups, one of which is esterified with sulfuric acid. C_{27}-bile acids are found in *amphibia* and *reptiles* (except *snakes*) as taurine conjugates. The largest group are the C_{24} bile acids, which are derivatives of 5β-cholanic acid (5α = *allo* derivatives are less frequent). Taurine conjugates are widely distributed, while glycine conjugates occur only in *mammals*. In *human* bile, the ratio of cholic acid / chenodeoxycholic acid / deoxycholic acid / lithocholic acid conjugates is typically 1 / 1 / 0.4 / traces.

7.9.2 Biosynthesis (Figure 7.9-1)

In the *endoplasmic reticulum* of *liver in rats* and likely in *humans*, cholic acid formation starts with hydroxylation at the 7-position of cholesterol, using the P-450 system (7.4.1). This is the rate-limiting step. Then the 3β-hydroxy-Δ^5 steroid is converted to the 3α-hydroxy-5β configuration. Small quantites of the 5α (*allo*) configuration are also formed, which constitute about 1...5 % of *mammalian* bile acids (Figure 7.9-2). After hydroxylation at C-12 and at C-26 in *mitochondria (in humans)*, oxidation to the carboxylic acid level takes place.

The removal of a part of the side chain possibly occurs in *peroxysomes* (or *mitochondria?*). A hydroxyl group is introduced at C-24; then in a reaction resembling the fatty acid degradation, propionyl-CoA and choloyl-CoA are formed. Propionyl-CoA enters a reaction sequence, which leads via methylmalonyl-CoA and succinyl-CoA to the citrate cycle (Fig. 4.5-2). Choloyl-CoA reacts with taurine or glycine to the respective conjugates. If the 12-hydroxylation does not take place, the reactions lead to chenodeoxycholyl- CoA and its conjugates.

Another pathway to chenodeoxycholate, which occurs completely in *liver mitochondria*, starts with 26-hydroxylation of cholesterol. Oxidation to the carboxyl function takes place early in the sequence, while the isomerization of the C-3 hydroxyl occurs later.

All bile acids, which are formed in the organism directly from cholesterol, are named primary bile acids (in *humans* mostly cholate, chenodeoxycholate and the corresponding *allo*-bile acids). Some *intestinal* bacteria are able to metabolize them into so-called secondary bile acids (in *humans* primarily deoxycholate, lithocholate and the corresponding *allo*-bile acids).

Figure 7.9-2. Configuration of Bile Acids

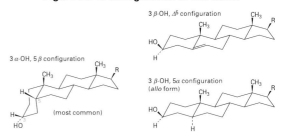

Bile acids are subject to enterohepatic circulation by secretion via the *gall bladder* into the *small intestine* (ca. 20...30 g/day in *humans*) and reabsorption in the *ileum* (Fig. 18.2-2). This way, about 90 % return via the *portal blood* to the *liver*. The rest is excreted in the feces.

7.9.3 Regulation of Biosynthesis (Figure 7.9-3)

The bile acid formation is regulated at the 7α-hydroxylase (= monooxygenase) step. This is coordinated with the regulation of the hydroxymethylglutaryl-CoA reductase (the key enzyme of cholesterol synthesis, 7.1.4). Both enzymes are closely associated with each other in the *endoplasmic reticulum*. High cholesterol concentrations stimulate the hydroxylase directly, while mevalonate, thyroid hormone, glucocorticoids and vitamin C increase its synthesis. The mRNA concentration and the enzyme activity follow a diurnal rhythm, which is in context with relatively short half-life times of mRNA and hydroxylase protein (a few hours).

On the other hand, high bile acid concentrations repress the synthesis of both 7α-hydroxylase and HMG reductase. A possible regulation of the 7α-hydroxylase by phosphorylation (activation) and dephosphorylation (deactivation) is under discussion.

Thyroid hormone also promotes the 26-hydroxylation (experiments in *rats*). Since a 26-OH group prevents consecutive hydroxylation at the 12-position, the ratio of chendeoxycholate / cholate rises.

7.9.4 Medical Aspects

A shift in the concentration ratio of cholesterol, bile acids and lecithin in the bile causes the formation of gallstones. Splitting of conjugates by bacterial infection of the *jejunum* leads to premature bile acid reabsorption and lack of conjugates in the *small intestine*. This causes a decrease of lipid resorption. If the intestinal pH value is lowered to such an extent, that no dissociation of bile acid conjugates takes place, similar deficits in lipid absorption occur (Zollinger-Ellison's syndrome). In Zellweger's syndrome, no peroxisomes are formed. Since no fatty acid oxidation is possible, 3α, 7α, 12α-trihydroxy-5β- cholestanoate accumulates.

Table 7.9-1. Occurence of Bile Acid and Bile Alcohol Conjugates (Examples)

Family	Trivial Name	Hydroxyl Groups at	Conjugated with	Species
Bile alcohols	5β-cyprinol	3α, 7α, 12α, 26	sulfuric acid	*fishes*
	scymnol	3α, 12α, 24, 26	sulfuric acid	*fishes*
	5α- or β-bufol	3α, 7α, 12α, 25, 26	sulfuric acid	*toads, lung fishes*
C_{27} Bile acids	3α, 7α, 12α, 22-tetrahydroxy-5β-cholestan-26-oic acid	3α, 7α, 12α,22	taurine	*turtles*
C_{24} Bile acids	cholic acid	3α, 7α, 12a	taurine, glycine	*snakes, birds, mammals*
	chenodeoxycholic acid	3α, 7α	taurine, glycine	*birds, mammals*
	ursodeoxycholic acid	3α, 7β	taurine	some *mammals*
	deoxycholic acid	3α, 12α	taurine, glycine	some *mammals, intestinal bacteria*
	lithocholic acid	3α	taurine, glycine	some *mammals, intestinal bacteria*
	α / β-muricholic acid	3α, 6β, 7α/ β		*rat, mouse*
	hyocholic acid	3α, 6α, 7α		*pigs*
	3α-hydroxy-7-oxo-cholanic acid	3α (+ 7-oxo)		some *birds* and *mammals*

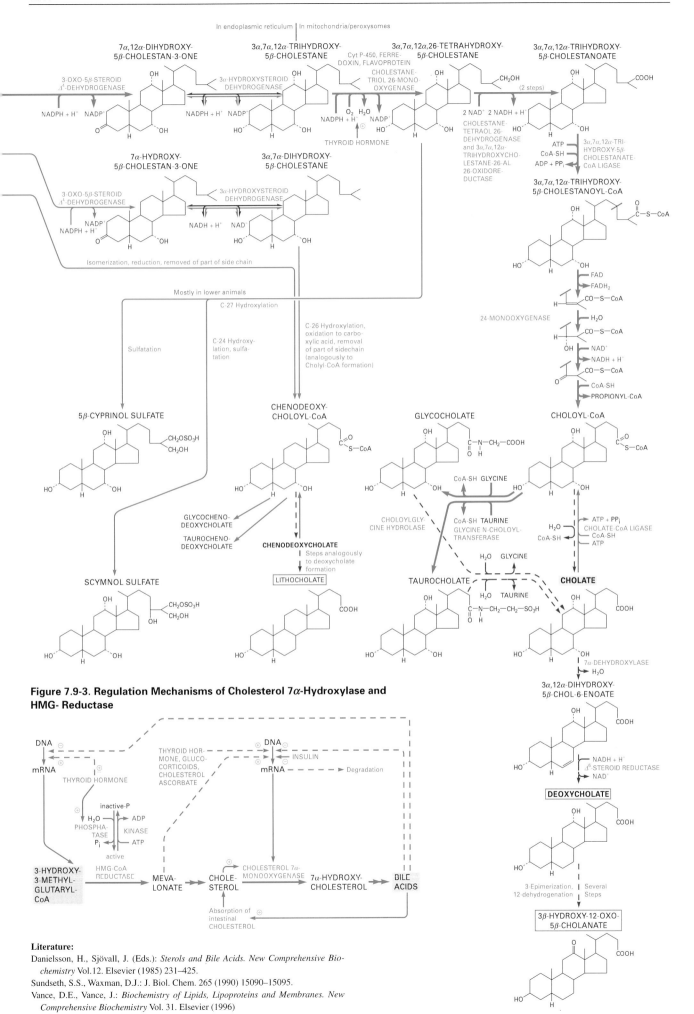

Figure 7.9-3. Regulation Mechanisms of Cholesterol 7α-Hydroxylase and HMG- Reductase

Literature:

Danielsson, H., Sjövall, J. (Eds.): *Sterols and Bile Acids. New Comprehensive Biochemistry* Vol.12. Elsevier (1985) 231–425.

Sundseth, S.S., Waxman, D.J.: *J. Biol. Chem.* 265 (1990) 15090–15095.

Vance, D.E., Vance, J.: *Biochemistry of Lipids, Lipoproteins and Membranes. New Comprehensive Biochemistry* Vol. 31. Elsevier (1996)

Figure 8.1-2. Synthesis of Purine Ribonucleotides

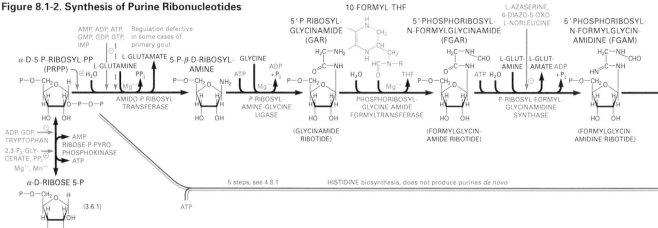

8 Nucleotides and Nucleosides

Nucleosides are composed of a purine or pyrimidine base linked to ribose or 2'-deoxyribose (ribo- and deoxyribonucleosides, respectively). Nucleotides are usually phosphorylated at the 5'-position of the pentose moiety. In a wider sense, structures with other N-containing ring systems are also named nucleotides, e.g., nicotinamide (9.9.1) or flavin derivatives (9.3.1). Nucleoside di- and tri- phosphates carry additional phosphate residues connected by pyrophosphate-type bonds. In cyclic nucleotides, the single phosphate forms a diester structure with two hydroxyl groups (mostly 3', 5') of the pentose. Dinucleotides are two nucleotide residues linked by a pyrophosphate bond, e.g., NAD$^+$ (9.9.1) or FAD (9.3.1). Fig. 8.1-1 shows the structures.

Nucleotides fulfill many important tasks:

- ATP is the universal transmitter of energy in biological systems (8.1.3).
- Nucleotides are involved in the regulation of metabolic processes. Examples are cAMP (17.4.2) and cGMP (17.7). Also the ratio of the ATP/ADP/AMP levels exerts a control function.
- Nucleotide derivatives represent the activated forms of many compounds during biosynthetic reactions (e.g., UDP-glucose, 3.2.2, CDP-choline, 6.3.2, phosphoadenylylsulfate, 4.5.6).
- Nucleotides are components of the coenzymes NAD$^+$ and NADP$^+$ (9.9), FAD and FMN (9.3), CoA (9.7).
- Ribonucleotides and deoxyribonucleotides are the building blocks of nucleic acids (10.2, 11.1).

The biosynthesis of the bases are multistep procedures. However, there are effective 'salvage pathways' permitting the recycling of the bases.

Figure 8.1-1. Structure of (Nucleo-)Bases, Nucleosides and Nucleotides

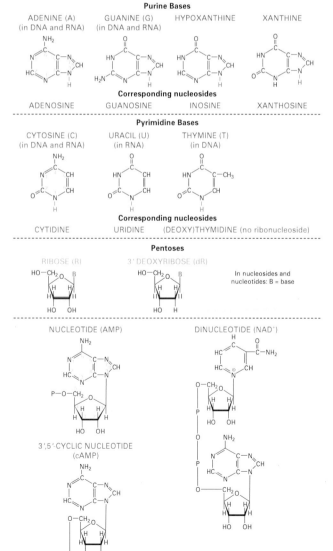

8.1 Purine Nucleotides and Nucleosides

8.1.1 Biosynthesis of Inosine-5'-Phosphate (IMP, Fig. 8.1-2)

The ribose component of purine ribonucleotides is derived from ribose 5-P, an intermediate of the pentose phosphate cycle (3.6.1). The atoms of the base moiety are contributed by many compounds (Fig. 8.1-3). They are added stepwise to the preformed ribose. There exist striking interrelationships with the pathway for histidine synthesis (4.8.1).

Figure 8.1-3. Origin of the Purine Ring Atoms

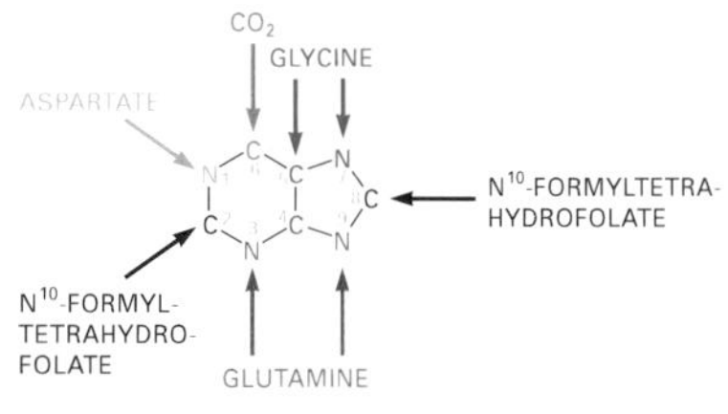

Ribose 5-P is activated by an unusual transfer of a pyrophosphate group, resulting in 5-phosphoribosyl-1-pyrophosphate (PRPP). Then an amino group from glutamine displaces the pyrophosphate group and yields 5-phosphoribosylamine. Hydrolysis of the released pyrophosphate pushes the reaction. This step also involves a conversion of the ribose from the α to the β anomeric form (1.2.1). The enzymes catalyzing the initial and the second reactions, ribose-P pyrophosphokinase and amido-P ribosyltransferase are regulated.

The first one, which takes also part in the histidine (4.8.1), tryptophan (4.7.1) and pyrimidine (8.2.1) synthesis, is feedback inhibited by ADP and GDP as well as by tryptophan. The second one is the commited (unambiguous) step for purine synthesis and is likewise feedback inhibited by purine nucleotides (Fig. 8.1-5). There exist two independent, but synergistic inhibitory sites, one of them binding AMP (preferred), ADP and ATP, the other binding GMP (preferred), GDP and GTP. Also, a feedforward activation by PRPP takes place at this step.

In the following reaction, which is energized by ATP hydrolysis via an acylphosphate intermediate, glycine is added. Then N^{10}-formyltetrahydrofolate (9.6.2) contributes a formyl group. In a glutamine and ATP dependent reaction, the oxo group at C-2 is replaced by an imino group. ATP energizes the consecutive closure and aromatization of the imidazole ring, resulting in 1(5′-phosphoribosyl)-5-aminoimidazole (AIR). An unusual carboxylation reaction, which does not use biotin, adds CO_2 and yields the carboxy derivative. This compound forms an amide bond with aspartate, followed by elimination of the aspartate carbon skeleton as fumarate. Thus, aspartate contributes only its amino group to the formation of 1(5′-phosphoribosyl)-5-amino-4-imidazolecarboxamide (AICAR, compare the aspartate cycle, 4.9.1). Another formylation by N^{10}-formyltetrahydrofolate contributes the last member of the C_6 ring. By dehydration, this ring is closed and inosine 5′-phosphate is obtained.

In *E. coli*, several of the genes of purine biosynthesis are combined in monocistronic operons. The P-ribosylaminoimidazolecarboxamide formyltransferase and the IMP cyclohydrolase functions are combined on a single peptide chain. In *vertebrates*, there exist even more multifunctional enzymes:

- The peptide chain encoded by the GART gene contains the P-ribosylamineglycine ligase, P-ribosylglycineamide formyltransferase and P-ribosylformylglycineamidine cycloligase functions,
- the chain encoded by the AIRC gene contains the P-ribosylaminoimidazole carboxylase and the P-ribosylaminoimidazole-succinocarboxamide synthase functions,
- the chain encoded by the IMPS gene contains the P-ribosylaminoimidazolecarboxamide formyltransferase and the IMP cyclohydrolase functions.

It has been found that at least in *E. coli*, the 'P-ribosylaminoimidazole carboxylase' consists of two enzyme entities. The 5′-phosphoribosyl-5-carboxyaminoimidazole synthase (NCAIR synthase) performs an ATP-assisted carboxylation of the amino group at C-5, while the NCAIR mutase (or isomerase) shifts it to the C-4 position.

AICAR is released during histidine biosynthesis in *plants* and *bacteria* (4.8.1). Since, however, its formation requires the ribose residue and part of the purine ring of ATP, this pathway does not lead to a net formation of purines. Rather, the steps from AICAR to purines serve only the reconstitution of ATP consumed before.

For the intermediates of the purine biosynthesis pathway, the terms of the IUB 'Enzyme Nomenclature' are used above. Since frequently other expressions occur in the literature, they are given in parentheses in Fig. 8.1-2.

8.1.2 Interconversions of Purine Ribonucleotides (Fig. 8.1-4, next page)

IMP does not accumulate in the cells. It is rapidly converted into adenosine and guanosine mononucleotides.

Adenosine 5′-monophosphate (AMP): Its synthesis involves an amino group transfer from L-aspartate in a two step pathway, which is energized by GTP hydrolysis. The cleavage of the intermediate adenylosuccinate liberates fumarate and yields AMP. This cleavage is performed by the same enzyme, which catalyzes an analogous reaction during IMP biosynthesis (8.1.1). Adenosine 5′-diphosphate (ADP) is obtained by a phosphate transfer from adenosine 5′-triphosphate (ATP), catalyzed by adenylate kinase (myokinase).

Besides being involved in the formation of AMP, the aspartate → fumarate interconversion has also the effect of increasing the amount of citrate cycle intermediates. This is of importance in *muscle*, where there is a low level of other anaplerotic enzymes (3.3.4). The fumarate formation from aspartate is enhanced by the AMP deaminase (adenylate deaminase) reaction, which converts the newly formed AMP back into IMP for another amination round (purine nucleotide cycle). Inherited myoadenylate deaminase insufficiency leads to cramps after muscle exercise.

Guanosine 5′-monophosphate (GMP): An initial dehydrogenation reaction yields xanthosine 5′-phosphate, which then receives an amino group from glutamine.

This reaction differs from transamination reactions (Fig. 4.2-3) in the absence of pyridoxal phosphate and the participation of ATP, the hydrolysis of which drives the reaction to completion. The use of glutamine as source of the amino group enables *animals* to avoid a high NH_4^+ level in blood. The *bacterial* enzyme also accepts NH_3 instead of glutamine.

The phosphate groups for the formation of guanosine 5′-diphosphate (GDP) and guanosine 5′-triphosphate (GTP) are supplied by ATP. While guanylate kinase is a highly specific enzyme, the diphosphate kinase is quite unspecific and accepts many ribo- and deoxyribo- di- and triphosphates.

Regulation of the purine nucleotide biosynthesis (Fig. 8.1-5): In *animals*, the first individual reactions leading to the formation of AMP and GMP are catalyzed by adenylosuccinate synthase and IMP dehydrogenase. These enzymes are feedback inhibited (competitively to IMP) by the end products of the pathway. On the other hand, GTP contributes to the formation of AMP and likewise ATP to the formation of GMP. By this reciprocal relationship, both individual pathways are coordinated, while the overall supply of the common precursor IMP is regulated by the combined action of adenosine and guanosine nucleotides. In *E. coli*, the purine repressor purR regulates the expression of all genes for purine and pyrimidine biosynthesis. Hypoxanthine and guanine serve as corepressors.

8 Nucleotides and Nucleosides

Figure 8.1-4. Interconversions and Degradation of Purine Ribonucleotides

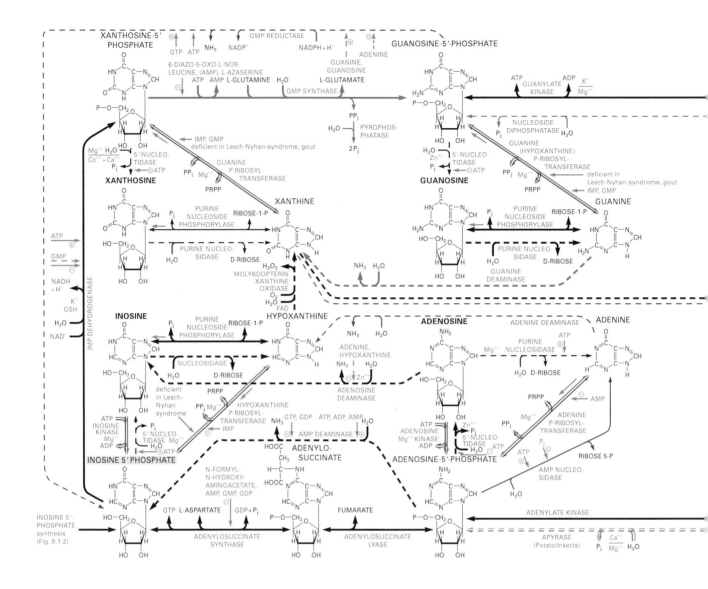

Salvage reactions: During degradation of RNA, besides mononucleotides (10.7.3), nucleosides and free bases are also obtained (for reactions, see Fig. 8.1-4). This formation of non-phosphorylated compounds is essential during digestion for resorption purposes, but also occurs during normal cellular metabolism. It is a prerequisite for the membrane passage, which proceeds in most cases through a common permease protein via facilitated diffusion (the nucleoside is phosphorylated after membrane passage and thus removed from the diffusion equilibrium, compare 15.3).

The purine bases can be reconverted into their respective mononucleotides by reaction with 5-P-ribosyl-PP (PRPP, 8.1.1, 'salvage pathway'). This is catalyzed by adenine P-ribosyl transferase and by hypoxanthine P-ribosyl transferase (HGPRT, which reacts with guanine and with hypoxanthine, the base moiety of IMP). For the phosphorylation of nucleosides, there are several kinases with different specificity.

8.1.3 ATP and Conservation of Energy

Chemical reactions proceed spontaneously only if there is a decrease in free energy ($\Delta G < 0$, see 1.5.1). Therefore, necessary endergonic reactions of living beings (e.g. biosynthesis, movements, signal transfer) have to be 'pulled' by coupling to exergonic reactions, which ususally employ 'high energy' compounds. The most versatile of them is adenosine triphosphate (ATP).

ATP is obtained

- by substrate level phosphorylation. This involves the transfer of phosphate from another energy rich bond to ADP, which can proceed aerobically as well as anaerobically. Examples are the reactions of pyruvate kinase (3.1.1) and bacterial fermentations (15.4).

- by phosphorylation of ADP via the H⁺-transporting ATP synthase (16.1.4), which is driven by a proton gradient. In many cases, this gradient is established by aerobic oxidation of substrates (16.1.2), but also anaerobic mechanisms occur (15.5).

ATP can drive other reactions by utilizing the energy stored in its pyrophosphate bonds ($\Delta G_0'$ for hydrolysis of ATP $\rightarrow$ ADP + P$_i$: -30.5 kJ/mol, for ATP $\rightarrow$ AMP + PP$_i$ -32.2 kJ/mol).

This high energy content is caused by the electrostatic repulsion of the negative charges on both sides of the bond and by the competition of the phosphoryl

Figure 8.1-5. Regulation of Purine Nucleotide Biosynthesis

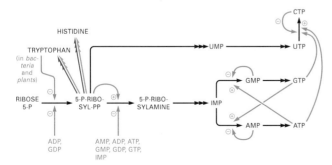

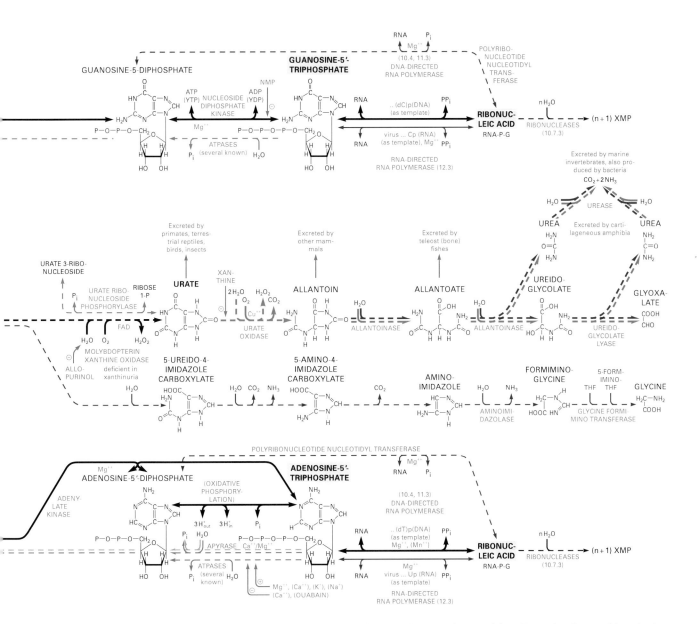

groups for the electrons at the bridging oxygen atom. The $\Delta G'_0$ refers to standard conditions (1.5.1) and varies with the actual concentrations of the reactants, the pH and the concentrations of divalent metal ions. Under physiological conditions, ΔG_{phys} is in the order of 50 kJ/mol (16.1.1).

The metabolic role of ATP consists mainly of the transfer of phosphate or of the adenylyl group to other compounds. This can effect an increase in reactivity. Example: fatty acid activation (6.1.4). The phosphorylation of proteins can have signalling purposes, e.g., in receptor-tyrosine kinase cascades (17.5.3). Hydrolysis of protein-bound ATP can effect changes in the protein configuration. Examples: muscle contraction, 17.4.5; ion transport by Na⁺/K⁺ exchanging ATPase, 18.1.1.

Figure 8.1-6. Reaction Mechanism of the Ribonucleoside-Diphosphate Reductase from *E. coli*

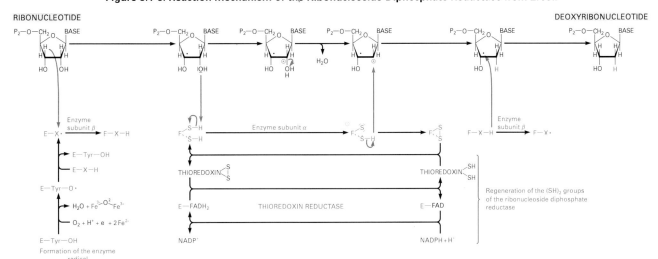

8.1.4 Ribonucleotide Reduction to Deoxyribonucleotides (Fig. 8.6-1)

The conversion of the purine diphosphoribonucleotides, as well as of pyrimidine diphosphoribonucleotides (NDP) into the respective diphospho-2'-deoxyribonucleotides (dNDP) is catalyzed by ribonucleoside-diphosphate reductase according to the general formula

$$\text{NDP} + \text{protein (SH)}_2 = \text{dNDP} + \text{protein} \genfrac{}{}{0pt}{}{\diagdown S}{\diagup S} + H_2O$$

After phosphorylation, the dNTP formed are used for incorporation into DNA. Control mechanisms acting on the reductase and the DNA polymerase keep the dNTP pools small. By additional regulation systems (Fig. 8.1-7), the optimum concentration ratios of all 4 deoxyribonucleotides have to be kept up, since this is a prerequisite for a high fidelity of DNA replication.

There are several classes of the reductase, which differ in the composition of the catalytic center.

Figure 8.1-7. Regulation of Ribonucleoside-Diphosphate Reductase *from E.coli* (Red: inhibition, green: activation)

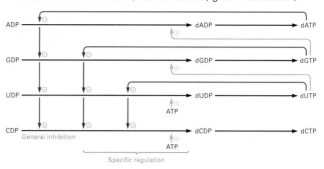

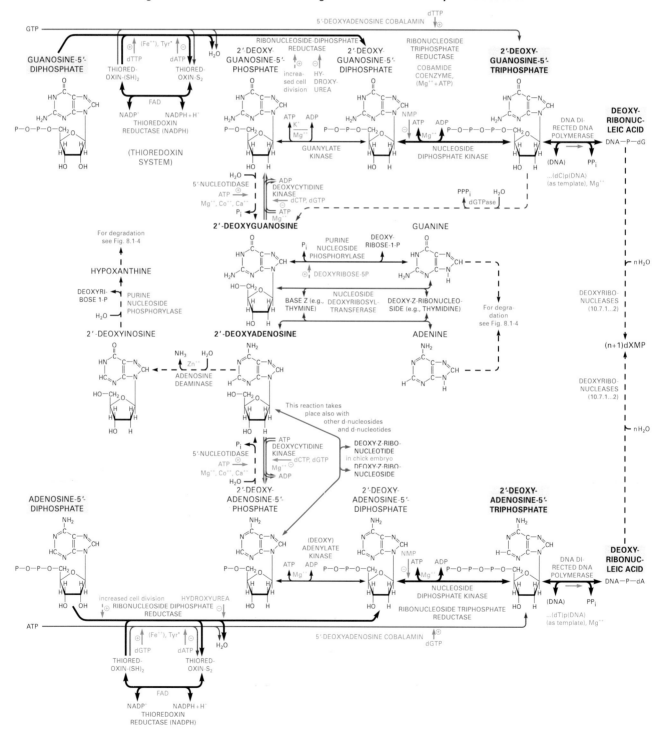

Figure 8.1-8. Interconversions and Degradation of Purine Deoxyribonucleotides

The *E.coli* ribonucleoside-diphosphate reductase (a class I enzyme) is composed of 2 dimers, α_2 (2 * 86 kDa) and β_2 (2 * 43 kDa). Each α subunit contains several -SH groups and 2 allosteric regulatory binding sites, while each β subunit contains a stable tyrosyl radical Y• with an unpaired electron, which is generated by interaction with an unusual iron-oxygen center Fe(III)-O^{2-}-Fe(III).

The reductase reaction takes place by a radical mechanism at the interface between the α and β subunits (Fig. 8.1-6). The enzyme's free radical abstracts (probably indirectly) the hydrogen from the 3′ position of the ribose, while a protonation from an enzyme -SH group leads to the removal of H_2O, leaving a cation radical at the 2′ position. This intermediate is reduced by a sulfhydryl group of the enzyme. Then hydrogen is returned to the 3′ position. This completes the formation of the deoxyribonucleotide. The sulfhydryl groups of the enzyme, which were oxidized during this process are reduced again by thioredoxin (12 kDa, contains 2 -SH groups in close proximity), which, in turn, is reduced by NADPH. (Thioredoxin is also involved in enzyme regulatory processes, e.g., 16.2.2.) In a number of *bacteria*, thioredoxin is replaced by glutaredoxin, which thereafter is reduced by glutathione (4.5.7).

The enzyme activity is allosterically regulated at two levels (Fig. 8.1-7): Binding of dATP (acting as a representative of deoxynucleotides) at the general regulatory site inhibits the overall activity. This is counteracted by ATP. This interplay responds to the needs of, e.g., DNA synthesis during cell division. On the other hand, binding of various deoxynucleotides to the specificity regulatory site serves the coordination of their production rates.

The expression of this enzyme in *eukarya* varies during the cell cycle. It increases greatly when cells leave the G_0 and enter the S-phase (Fig. 11.1-2). As a consequence, the dNTP pools increase likewise.

Class II enzymes are ribonucleoside-triphosphate reductases, which use a 5′-deoxycobalamin cofactor (9.5.2) for radical generation, while class III enzymes contain Fe-S centers and employ S-adenosylmethionine. These enzymes are active under anaerobic conditions.

8.1.5 Interconversions and Degradation of Purine Deoxyribonucleotides (Fig. 8.1-8)

Phosphorylation and degradation follow a similar pattern as the ribonucleotide analogues. In some cases, even the same enzymes react with ribo- and deoxyribo-substrates. There are also a number of base exchange reactions both at the nucleoside and mononucleotide level.

8.1.6 Catabolism of Bases (Figures 8.1-4 and 8.1-8)

Purine nucleosides and nucleotides to be degraded are at first cleaved, releasing the free bases (adenosine and AMP are previously deaminated to inosine and IMP, respectively). The bases hypoxanthine and xanthine are oxidized to urate by xanthine oxidase. This is the final product of purine degradation in a number of *animals*. In blood, urate is transported in protein-bound form. *Primates* excrete it in the urine.

Xanthine oxidase is a homodimeric enzyme (2 * 130 kDa), which contains molybdenum, FAD and 2 FeS clusters. In mammals, it is primarily present in the *liver* and in the *small intestinal mucosa*. The probable reaction mechanism is shown in Fig. 8.1-9. It starts by a nucleophilic attack of an enzyme residue on the C-8 position of xanthine and elimination of H as a hydride ion. This reduces the enzyme-bound Mo(VI) to Mo(IV). Water displaces the enzyme in the substrate complex and releases urate. Mo(IV) is oxidized by O_2. The H_2O_2 resulting from this reaction is decomposed by catalase.

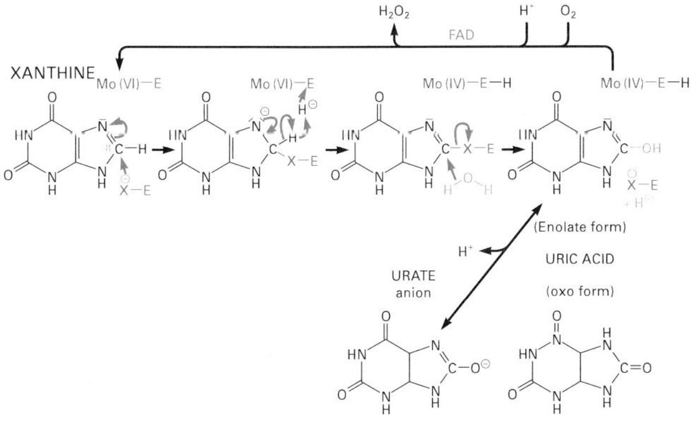

Urate is a very effective antioxidant. It destroys reactive oxygen species (HO•, $O_2^{•-}$, $^1O_2$, see 4.5.8) and reduces oxidized heme iron. In these respects, it is about as effective as ascorbate.

In *terrestrial reptiles, birds* and many *insects* (uricotelic animals), urate is also involved in the excretion of amino acid nitrogen, which ends up in this compound via purine intermediates. Due to its low solubility, it can be excreted as a paste of crystals and saves water this way. Non-primate *mammals* and other *animals* (except the uricotelic animals) degrade urate further and excrete allantoate, urea or NH_3 (Fig. 8.1-4).

8.1.7 Medical Aspects

A severe congenital deficiency of hypoxanthine P-ribosyl transferase (8.1.2, Lesch-Nyhan syndrome) leads to an accumulation of PRPP, since it is insufficiently consumed by the salvage pathway reactions. Since this compound feed-forward activates the purine biosynthesis (8.1.1), excessive quantities of purines and consecutively their degradation product urate are formed. This causes gout symptoms (see below) and (possibly in combination with additional reactions) mental retardation, aggressive behavior etc.

Health problems arise from the low solubility of urate. While uric acid is almost insoluble, urate (the prevalent form at physiological pH, pK = 5.4) is only little soluble. In blood, most of it is bound to proteins. Hyperuricemia (beyond the normal concentration of ca. 0.4 mmol/l) leads to deposition of sodium urate crystals in *joints, tendon sheaths, renal medulla* etc. (gout).

- Hereditary primary hyperuricemia is caused by
  – disturbed tubular secretion of urate (mostly) or by
  – elevated production of this compound (by the Lesch-Nyhan syndrome or by defective feedback inhibition of the amido-P ribosyltransferase etc.)
- Secondary hyperuricemia is caused, e.g., by elevated degradation of nucleic acids (leukemia, psoriasis).
- To some extent, meat-rich food may also contribute to the formation of gout.

One way of treatment is the inhibition of xanthine oxidase by allopurinol (an analog of hypoxanthine with interchanged C-8 and N-7). The more soluble substrates of xanthine oxidase pile up and are excreted.

A hereditary disease caused by defective adenosine deaminase is the severe combined immunodeficiency disease (SCID). Deoxyadenosine (and likewise adenosine, 8.1.6) accumulates and is phosphorylated, yielding dATP (up to 50 times the normal concentration), which is an allosteric inhibitor of ribonucleotide reductase (8.1.4). The inhibition of this enzyme leads to a lack of deoxynucleotides and to impaired DNA synthesis. This affects mostly the *lymphocyte* generation and causes severe malfunctions of B and T cells (19.1.2). This disease is lethal unless treated by administration of adenosine deaminase, which is protected against degradation by coupling to polyethylene glycol.

An imbalanced accumulation of dATP or dGTP occurs in some immune diseases. This leads according to the mutual regulation scheme of ribonucleoside-diphosphate reductase (Fig. 8.1-8) to a depletion of the other dNTPs, resulting in impaired DNA synthesis in S phase cells and insufficient DNA repair in resting lymphocytes, which results in cell death.

8.2 Pyrimidine Nucleotides and Nucleosides

8.2.1 Biosynthesis of Uridine 5′-Phosphate (UMP, Fig. 8.2-1, next page)

While during purine biosynthesis the components of the purine ring are assembled step by step at a preformed ribose moiety (8.1.1), during pyrimidine formation the base ring is formed first and ribose is added afterwards.

Aspartate and carbamoyl phosphate are the source of the ring atoms. In *eukarya*, the latter compound is also a member of the urea cycle (4.9.1), but both pathways are separated by compartmentization. While the carbamoyl-P formation and the consecutive reactions for pyrimidine synthesis takes place in the *cytosol*, the respective reaction of the urea cycle is a *mitochondrial* one. Also, the carbamoyl synthase II involved in pyrimidine biosynthesis has properties different from the urea cycle enzyme (4.9.1): it uses glutamine as a nitrogen source (instead of ammonia) and it is not allosterically activated by N-acetyl glutamate. In *bacteria*, it is the only carbamoyl synthase present.

The next reaction, which is catalyzed by aspartate carbamoyl transferase represents the committed step of pyrimidine biosynthesis and is the subject of multiple regulation mechanisms.

The allosteric *E.coli* enzyme is composed of 2 homotrimeric catalytic (c_3) and 3 homodimeric regulatory (r_2) subunits, which result in the structure c_6r_6 with threefold rotational symmetry. The reaction takes place at the interface between two c-chains. Substrate binding effects a 10° turn and an increase of the distance between the catalytic trimers. Also the r-chains turn. This corresponds to the transition from the T- to the more active R-state (2.5.2). Studies of the catalytic mechanism proved that this transition is in agreement with the symmetry model of allosteric interaction (all subunits change simultaneously from the T- to the R-state).

Figure 8.1-9. Reaction Mechanism of Xanthine Oxidase

8 Nucleotides and Nucleosides

Figure 8.2-1. Synthesis of Pyrimidine Ribonucleotides

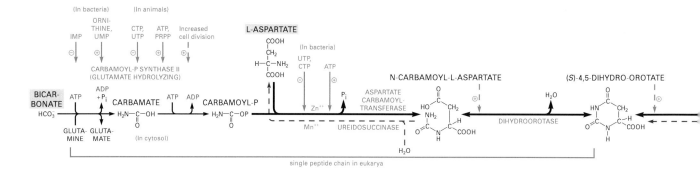

Figure 8.2-2. Interconversions and Degradation of Pyrimidine Nucleotides and Bases

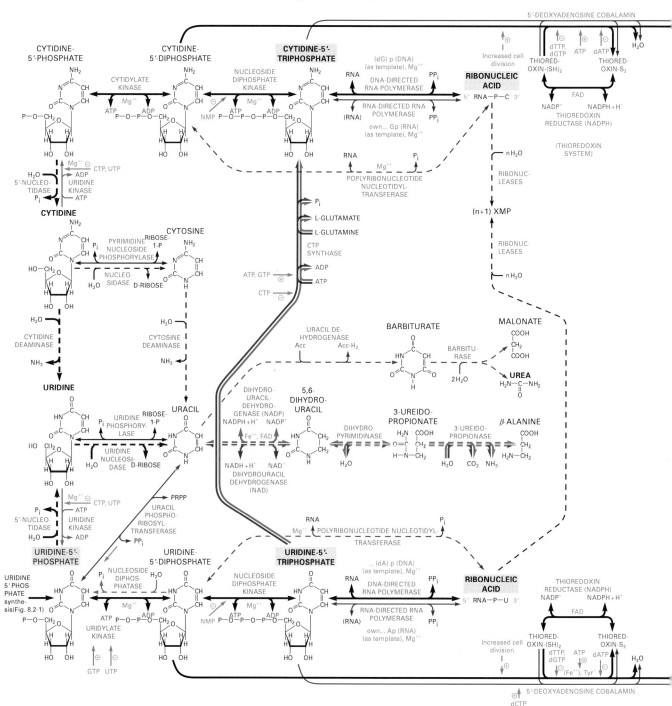

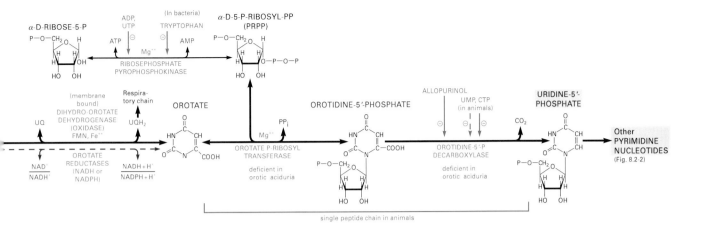
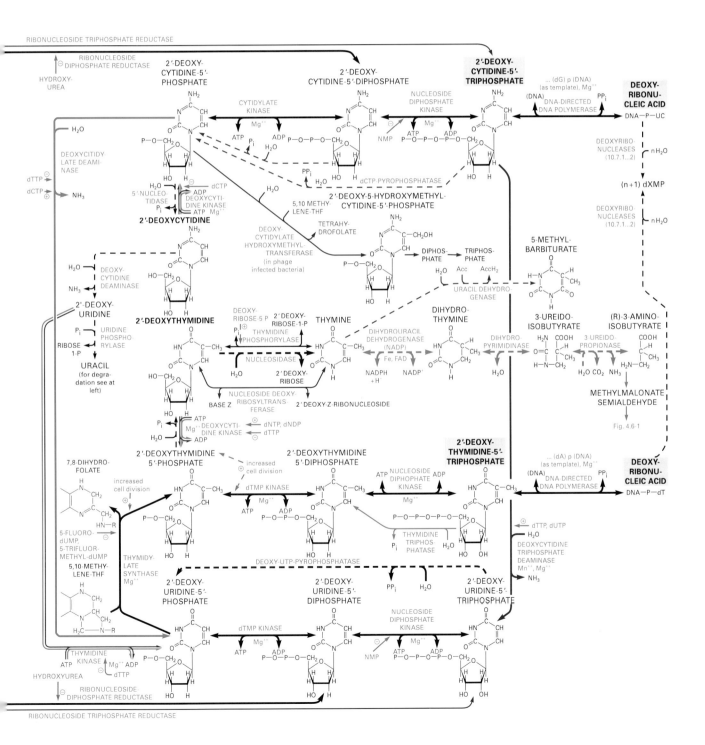

The *E.coli* aspartate carbamoyl transferase is feedback-inhibited by the pathway end product CTP. The previous enzyme carbamoyl-P synthase is also subject to feedback inhibition, however by UMP.

The next step is the ring closure by release of water, followed by the oxidation to orotate. In *eukarya*, the dihydroorotate oxidase (EC number 1.3.3.1, also called 'dehydrogenase') is located on the outside of the *mitochondrial* membrane and donates the reducing equivalents to the quinone pool (16.1.2). In *bacteria*, the enzyme is likewise membrane-bound and interacts with quinones. (There also exist *bacterial* NADH or NADPH dependent orotate reductases. The reactions catalyzed by these enzymes proceed mainly in the opposite direction.) The reaction with 5'-phosphoribosyl pyrophosphate (PRPP, compare 8.1.1) leads to the addition of the ribose moiety. Hydrolysis of the released PP_i drives this reaction to completion. Decarboxylation of orotidine 5-P is the last step in the biosynthesis of uridine 5-P.

While in *E.coli* the 6 enzymes of pyrimidine nucleotide biosynthesis are apparently individual entities, in *eukarya* the first three enzyme functions of this pathway are combined on a single peptide chain (CAD protein, 240 kDa). Likewise, the orotate P-ribosyltransferase and the orotidine-5'-P decarboxylase functions are located on a single chain. This enables a simple coordination of the synthesis and facilitates the substrate transfer from one enzyme function to the next one (analogous to, e.g., 6.1.1 and 8.1.1).

8.2.2 Interconversions of Pyrimidine Ribonucleotides (Fig. 8.2-2)

Pyrimidine mononucleotides are converted by phosphorylation reactions to di- and consecutively to triphosphates. Uridine 5'-triphosphate (UTP) is aminated in an ATP-energized reaction by glutamine (in *mammals*) or by NH_4^+ (in *E.coli*), yielding cytidine 5'-triphosphate (CTP, Fig. 8.2-3). The amination reaction proceeds analogously to the conversion of xanthosine 5'-phosphate (8.1.2), except for the release of P_i instead of PP_i.

A salvage reaction (of uracil) similar to purine metabolism (8.1.2) has been found in *bacteria*.

Figure 8.2-3. ATP-Dependent Replacement of an Oxo by an Amino Group in Nucleotides

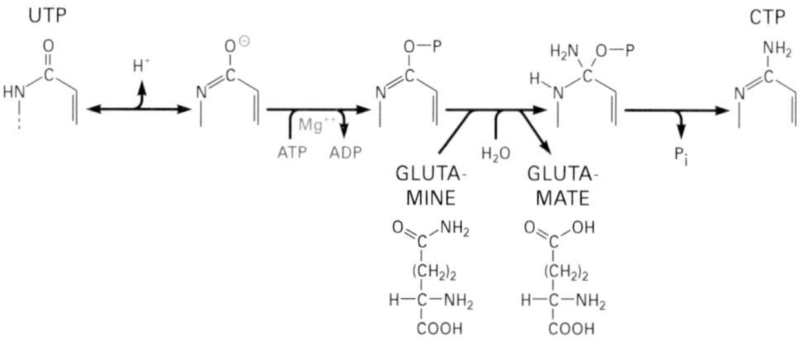

8.2.3 Ribonucleotide Reduction and Interconversions of Pyrimidine Deoxyribonucleotides (Fig. 8.2-2)

The same enzymes, which reduce purine ribonucleotides to deoxyribonucleotides, also act on pyrimidine nucleotides. These conversions are subject to mutual control by the other nucleotides (8.1.4).

Besides by reduction of UDP, deoxyuridine phosphates can also be formed via deoxycytidine phosphates. In *E.coli*, only ca. 25% of the dU and consecutively dT compounds (see below) are obtained by UDP reduction, while 75% are generated via dCDP intermediates. There are two major pathways for this conversion. In *bacteria*, they proceed via the steps dCDP → dCTP → dUTP (deamination) → dUMP (dephosphorylation to dUMP by deoxyuridine pyrophosphorylase). The deamination is feedback inhibited by dTTP and dUTP. In *eukarya* (and in *bacteriophage T2-infected E.coli*), the reaction sequence is dCDP → dCMP → dUMP (deamination). The deaminase is inhibited by dTTP and activated by dCTP. There exists also a conversion at the nucleoside level dCMP

→ dC → dU (deamination) → dUMP. The activities of the deaminase enzymes control the ratio dC/dT compounds available for DNA synthesis.

Deoxyuridine nucleotides do not occur in DNA. Their role is taken over by deoxythymidine nucleotides. Deoxythymidine 5'-monophosphate (dTMP) is obtained from deoxyuridine 5'-monophosphate by a methylation reaction, which is catalyzed by thymidylate synthase and uses N^5, N^{10}-methylene tetrahydrofolate (THF) as a methyl donor. The folate moiety is simultaneously oxidized to the dihydrofolate (DHF) level (9.6.1) and has to be reduced again by the NADPH dependent dihydrofolate reductase.

dUMP + N^5, N^{10}-methylene THF = dTMP + DHF
DHF + NADPH = THF + $NADP^+$ + H^+

Then dTMP is converted into deoxythymidine 5'-triphosphate (dTTP) by two phosphorylation steps. This compound is used for DNA synthesis. In order to prevent an accidental incorporation into DNA, the level of dUTP is kept low by the action of deoxyuridine pyrophosphorylase.

The activity of thymidylate synthase depends on the cell cycle and increases greatly when cells leave the G_0 and enter the S-phase (Fig. 11.1-2), similarly to ribonucleoside-diphosphate reductase (8.1.3).

Apparently the presence of thymine bases (instead of uracil bases) in DNA is a protective measure. U, which is accidentally present in the template strand pairs with A and effects its incorporation into the daughter strand during replication. Thus, if C is converted into U by deamination as a result of damage, replication would result in an U-A pair instead of a C-G pair, which by the next round would yield a A-T pair. This change of the codon would lead to grave consequences, when the genes are being transcribed and translated thereafter (10.1).

In *E.coli*, which is infected by *bacteriophages T2, T4* or *T6*, enzymes for hydroxymethylation of dCMP and its consecutive further phosphorylation are induced. The nucleotide triphosphate is introduced into DNA and pairs with G (Fig. 8.2-2).

8.2.4 Catabolism of Bases (Fig. 8.2-2)

Similarly to the purine analogues, the base residues of pyrimidine nucleotides and nucleosides have to be released for degradation. Cytosine compounds are converted into the respective uridine compounds by deamination. The general degradation pathway for uracil and thymine starts by reduction of the 5-6 double bond and consecutive hydrolytic ring opening. Deamination and decarboxylation lead to β-alanine and to 3-aminoisobutyrate, respectively.

An alternative pathway in *bacteria* begins with an oxidation reaction at C-6 to an oxo group, yielding barbiturate and to 5-methylbarbiturate, respectively. Then, a hydrolytic ring opening follows.

8.2.5 Medical Aspects

Since the only task of thymidylate synthase is the provision of deoxythymidine nucleotides for DNA synthesis, its inhibition efficiently blocks DNA replication and cell division ('thymineless death').

This inhibition is most effective, when a high division rate requires a continuous supply of deoxythymidine nucleotides. An effective inhibitor is 5-fluorouracil, where the methyl group of thymine is replaced by a fluorine atom. After application, it is metabolized into fluorodeoxyuridine 5'-monophosphate. During the thymidylate synthase reaction, a -SH group of the enzyme and methylene-THF form a covalent intermediary complex, which cannot dissolve in presence of the fluoro nucleotide and thus prevents further turnover. Another way to block the enzymatic reaction is the inhibition of dihydrofolate reductase, which reconstitutes the reaction partner tetrahydrofolate. This can be achieved by dihydrofolate analogues, e.g., aminopterine or methotrexate, which are strong competitive inhibitors of the *mammalian* enzyme ($K_I < 10^{-9}$ mol/l, see 1.5.4). These inhibitors are used as anticancer drugs, but act likewise upon other fast growing cells (e.g., *hair follicles*, *stem cells* of the *bone marrow*). They are therefore quite toxic to *humans*.

Other dihydrofolate reductase inhibitors (e.g., trimethoprim) affect the *bacterial* enzyme about 10^5 times more strongly than the *human* enzyme and can be used as antibacterial drugs.

Literature:
Jones, M.E.: Ann. Rev. of Biochem. 49 (1980) 253–279.
Neuhard J., Kell, R.A. in Neidhardt, F.C. (Ed.): *Escherichia coli and Salmonella.* ASM Press (1996) 580–599.
Reichard, P.: Ann. Rev. of Biochem. 57 (1988) 349–374.
Zalkin, H., Dixon, J.E.: Prog. Nucl. Acid Res. Mol. Biol. 42 (1992) 259–287
Zalkin, H., Nygaard, P. in Neidhardt, F.C. (Ed.): *Escherichia coli and Salmonella.* ASM Press (1996) 561–579.

9 Cofactors and Vitamins

Vitamins are compounds, which are required in small quantities for biochemical reactions to take place and which cannot be synthesized by *higher animals*. They have to be obtained by food intake. A number of them are precursors of cofactors for enzymatic reactions (vitamin B group), while others are involved in the visual process and regulation of transcription (vitamin A), redox reactions (vitamins C and E), bone formation (vitamin D), blood coagulation (vitamin K) etc.

The vitamins and cofactors belong to various chemical classes. A common way of classification refers to the solubility:

- Water-soluble vitamins: B_1, B_2, B_6, B_{12}, folate, pantothenate, biotin (= 'vitamin B complex'), C. They are poorly stored. Intake in excess leads to secretion in the urine.
- Fat-soluble vitamins: A, D, E, K. They can be stored. Intake in excess can lead to hypervitaminoses (especially with A and E).

Some of their biosyntheses and actions are dealt with in different chapters. This chapter gives a survey of their properties and provides cross-references to the other chapters.

9.1 Retinol (Vitamin A, Fig. 7.3-2)

Vitamin A is the common name for the alcohol retinol (axerophthol), the aldehyde retinal and retinoic acid, which are isoprenoids with lipid character (7.3). Several *cis-trans* isomers exist.

9.1.1 Biosynthesis and Interconversions

The biological source of the vitamin A group is the provitamin β-carotene, the biosynthesis of which takes place in *plants* and is shown in Fig. 7.3-2. *Animals* are able to cleave carotene into retinal by a dioxygenase reaction (already in the *intestine*), thus they can use it in place of the vitamin.

Vitamin A is resorbed together with fat and stored in the *liver* after reduction and esterification as retinyl palmitate. For transport to the various organs, the ester is hydrolyzed and retinol is released into the bloodstream, where it is bound to specific retinol binding proteins. In the organs, interconversions of retinol, retinal and retinoic acid proceed via NAD(P)$^+$-dependent reactions.

In *human* nutrition, carotene is mostly obtained from *plant* sources, while retinol is supplied by *animal* sources (e.g. fish liver oil). A daily intake of 1 mg retinol or 6 mg β-carotene (= 1000 retinol equivalents = 3333 IU) is required for adults. Avitaminoses are common in undernourished populations. They cause night blindness, disturbances in bone development and in differentiation of epithelial tissue, inhibition of growth and of the reproductive function.

9.1.2 Biochemical Function

Retinoic acid is a regulator of gene transcription. After passing the cellular membrane, it binds to specific receptors: *all-trans* retinoic acid to the RAR receptor, 9-*cis* retinoic acid to the RXR receptor (17.6). The heterodimeric retinoic acid-RAR-RXR complexes bind in the nucleus to DNA hormone response elements (consensus sequence AGGTCA, repeated).

Among the genes regulated in this way are the genes for retinol binding proteins (9.1.1), PEP-carboxykinase (3.3.4) and apolipoprotein A1 (18.2.1). Retinoic acid is involved in the control of embryo- and morphogenesis (even small doses of vitamin A during pregnancy are teratogenic!), growth, differentiation and fertility.

The light-induced interconversion of 11-*cis* and *all-trans* retinal forms the basis of the visual process in *animals* (17.4.6). *Halobacteria* use the light-induced interconversion of *all-trans* and 13-*cis* retinal for proton pumping (16.2.1).

Possibly, retinyl phosphate is involved in the biosynthesis of glycosaminoglycans (13.1.5) in epithelial cells by playing a similar role as dolichol phosphate (13.3.3). This way, it sustains indirectly the integrity of cellular and mitochondrial membranes, keeps up a normal epithelium of the skin and mucosa and prevents disturbances of growth.

Literature:
Friedrich, W.: *Vitamins.* De Gruyter (1988) 63–140.

9.2 Thiamin (Vitamin B1)

Vitamin B_1 (aneurin) is unphosphorylated thiamin. By pyrophosphorylation, it is converted into the coenzyme thiamin pyrophosphate (ThPP, codecarboxylase), which plays an essential role in oxidative decarboxylation and group transfer reactions.

9.2.1 Biosynthesis (Fig. 9.2-1)

Both components of thiamin, the pyrimidine and the thiazole moieties are synthesized separately and united thereafter. The pathways of biosynthesis differ among *pro-* and *eukarya* (*yeast, plants*). Details of the early steps are only partially known.

Apparently in *E. coli* and in *S. typhimurium*, 1(5′-phosphoribosyl-5-aminoimidazole (AIR, see purine synthesis, 8.1.1) is a precursor of the pyrimidine residue, while in *yeast* it is formed in another way. In *E. coli*, possible precursors of the thiazole moiety are pyruvate and D-glyceraldehyde (compare Fig. 7.3-2). They condense to 1-deoxy-D-xylulose (compare Fig. 7.3-2), which contributes the C-atoms 4′, 4, 5, 6, 7. Tyrosine is the origin of C-2′ and N-3′, while the sulfur may be provided by cysteine. The pyrophosphate ester of the pyrimidine moiety enters the coupling reaction. The conversion from the vitamin to the coenzyme is a 1-step pyrophosphorylation.

Rich sources of the vitamin are germinating *grain* and *yeast*. The recommended daily intake for *humans* is 1...1.5 mg. Avitaminosis leads to beriberi, which causes neurological disorders and affects the heart function.

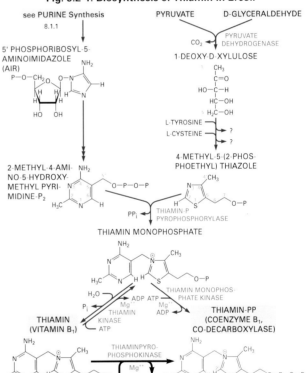

Fig. 9.2-1. Biosynthesis of Thiamin *in E. coli*

9.2.2 Biochemical Function (Figs. 3.3-4 and 9.2-2)

The essential function of the coenzyme is the transfer of activated (= ThPP bound) aldehyde groups. ThPP is the prosthetic group of pyruvate dehydrogenase (3.3.1) and pyruvate decarboxylase (3.3.6), 2-oxoglutarate dehydrogenase (3.8.1) and transketolase (3.6.1, 16.2.2); thus it is involved in the course of glycolysis and of the citrate, pentose-P and Calvin cycles.

The C-2′ atom has a very acidic character due to its proximity with N$^+$ and S. The loss of the proton results in a carbanion, which easily adds to carbonyl groups (Fig. 3.3-4). In the decarboxylase and dehydrogenase reactions mentioned above, this effects the decarboxylation of the 2-oxo acid and the formation of a resonance-stabilized ThPP-bound carbanion, which represents an activated aldehyde. This residue is either released (decarboxylase enzyme) or transferred to lipoamide with concomitant oxidation (dehydrogenase enzymes). In the transketolase reaction, the reaction partner is a keto sugar, which is cleaved. An aldehyde is released, while the ThPP-bound activated aldehyde is transferred to an incoming compound (Fig. 9.2-2).

Fig. 9.2-2. Reaction Mechanism of Transketolase
(T = pyrimidine residue, T' = C$_2$-PP residue)

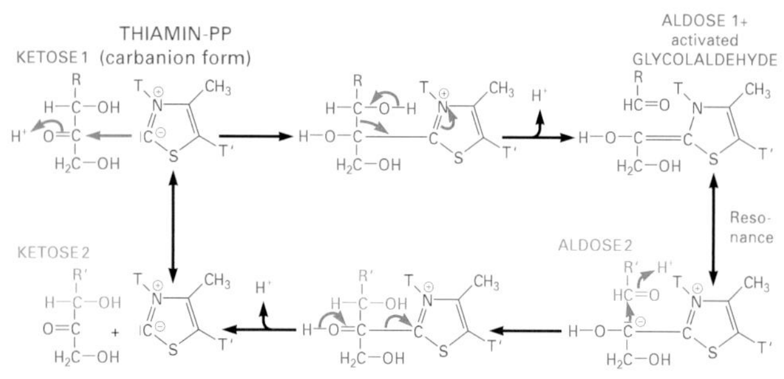

Literature:
Friedrich, W.: *Vitamins*. De Gruyter (1988) 339–401.
White R.L., Spenser, J.D. in Neidhardt, F.C. (Ed.).: *Escherichia coli and Salmonella*. 2nd Ed. ASM Press 1996, 680–686.

9.3 Riboflavin (Vitamin B$_2$), FMN and FAD

Besides the nicotinamide nucleotides NAD$^+$ and NADP$^+$ (9.9), the flavin nucleotides flavin-adenine dinucleotide (FAD) and flavin mononucleotide (FMN) are the other important group of hydrogen carriers. They take part in more than 100 redox reactions. These coenzymes are derived from the vitamin riboflavin by phosphorylation (FMN) and by a consecutive adenylylation (FAD).

9.3.1 Biosynthesis and Interconversions (Fig. 9.3-1)

The vitamin riboflavin is biosynthesized from GTP and ribulose 5-P in *plants*, *yeasts* and many *microorganisms*.

The pathway starts by the opening of the imidazole ring of GTP and removal of a pyrophosphate residue. Deamination, reduction and removal of the last remaining phosphate yields 5-amino-6-ribitylamino-2,4-pyrimidinedione. The reaction of this compound with 3,4-dihydroxy-2-butanone-4-P (originating from ribulose 5-P) results in the bicyclic compound 6,7-dimethyl-8-ribityllumazine. This compound undergoes an unusual dismutation by transfer of a 4-carbon unit, which leads to the tricyclic compound riboflavin and back to the pyrimidinedione. Phosphorylation of this vitamin finally results in flavinmononucleotide (FMN). A consecutive adenylylation yields flavin-adenine dinucleotide (FAD). The phosphorylation of riboflavin is strictly controlled in *animals*, e.g. by thyroid hormones and by aldosterone.

FAD and FMN are usually tightly bound to enzymes (flavoproteins). In some cases even a covalent bond exists (mostly to the 8α-methyl group, e.g. in succinate dehydrogenase). Thus, reduced flavin coenzymes have to be reoxidized 'on the spot' and cannot diffuse to another enzyme. Protein binding greatly diminishes the light sensitivity of the free flavin coenzymes.

The vitamin is present in many *vegetables* and in *meat*, less in *grain* products. During digestion in *animals*, the various flavoproteins from food are degraded and riboflavin is resorbed. The daily requirement of a *human adult* is about 1.4 ... 2 mg. Avitaminoses cause disturbances of growth and skin diseases. They are quite rare.

The major degradation and excretion product in *humans* is riboflavin. Other degradation steps (especially by *bacteria*) are hydroxylations of methyl residues and shortening of the ribityl chain.

9.3.2 Biochemical Function

The redox function of FMN and FAD is exerted by the tricyclic isoalloxazine ring. For full reduction, they take up two electrons and two protons (usually in the 1,5-positions, Fig. 9.3-1) and keep up electroneutrality. Contrary to NAD(P)$^+$, the flavin compounds can also take part in one-electron reactions by the formation of a semiquinone radical. Thus they can act as 'redox switches' between one- and two-electron mechanisms. The redox potential is more positive than the potential of NAD(P)$^+$ (Fig. 16.1-5) and is further modified by the protein binding of the coenzyme. This enables flavin coenzymes to perform other oxidation reactions than NAD(P)$^+$.

Frequently they react as intermediates in the transfer of redox equivalents between NAD(P)H and various acceptors, e.g., in the reactions catalyzed by:

- dihydrolipoamide dehydrogenase (3.3.2) or GSSG reductase (4.5.7) – transfer to or from -S-S- groups

Figure 9.3-1. Biosynthesis of Riboflavin, FMN and FAD

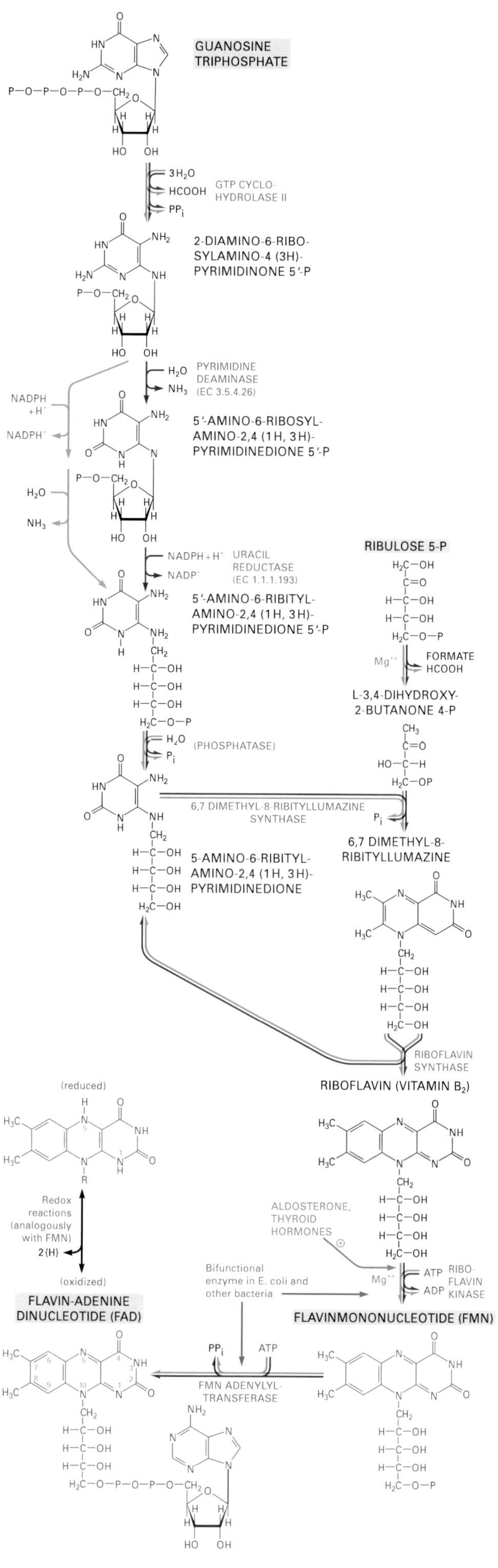

or in the transfer of redox equivalents between substrates and the respiratory chain (16.1), e.g., in the reactions catalyzed by:

- NADH dehydrogenase (Fig. 16.1-1) – transfer to FeS centers in complex I
- glycerol-3-P dehydrogenase as part of the glycerol 3-P dehydrogenase shuttle (Fig.16.1-2) – transfer to the ubiquinone (UQ) pool
- acyl-CoA dehydrogenases during fatty acid oxidation (6.1.5) or succinate dehydrogenase (3.8.1) – transfer to the UQ pool, formation of double bonds.

In reduced form a number of flavoproteins are able to react with molecular oxygen (for the reason see 4.5.8), e.g., in

- oxidation reactions with formation of H_2O_2 (2 e⁻ reactions):
  - oxidative deaminations, e.g., catalyzed by monoamine oxidase (4.7.4)
  - oxidation of aldehydes, e.g., catalyzed by by glucose oxidase (3.5.1)
  - oxidation of alcohols to yield aldehydes or ketones, e.g., catalyzed by bacterial alcohol oxidase.
- monooxygenase reactions (4 e⁻ reactions)
  - 'external' type: $X-H + FADH_2 + O_2 = X-OH + FAD + H_2O$ as in the kynurenine 3-monooxygenase reaction (Fig. 4.7.4), FAD is then again reduced by NADPH
  - 'internal' type: $X-CHOH-COO^- + O_2 = X-COO^- + H_2O + CO_2$ as in the lactate 2-monooxygenase reaction. The flavin acts as an intermediate carrier of redox equivalents.

The different oxidation states show characteristic spectra with the following λ_{max}: FAD 370 and 450 nm (yellow), semiquinone radical 370, 450 and 590 nm (red), $FADH_2$ 370 nm (pale yellow).

A related compound is factor F_{420} (5-deazaflavin, 15.5.2), which occurs in *methanogenic bacteria*.

Literature:
Bacher, A. et al. in Neidhardt, F.C. (Ed.).: *Escherichia coli and Salmonella*. 2nd Ed. ASM Press (1996) 657–664.
Friedrich, W.: *Vitamins*. DeGruyter (1988) 403–471.

9.4 Pyridoxine (Vitamin B_6)

Pyridoxine (pyridoxol), pyridoxal and pyridoxamine compose the vitamin B_6 group. Pyridoxal and pyridoxamine phosphates act as essential coenzymes in a large number of amino acid conversions, e.g. transaminase, decarboxylase and dehydratase reactions.

9.4.1 Biosynthesis and Interconversions (Fig. 9.4-1)

A number of *bacteria, yeast* and *plants* are producers of pyridoxine. Other *bacteria* and *animals* have to rely on external supply.

So far, the precursors for the pyridoxine atoms in *E. coli* have been identified, but enzymatic details are still unknown. The condensation product of pyruvate and D-glyceraldehyde, 1-deoxy-D-xylulose (compare 9.2 and Fig. 7.3-2), is the origin of the C-atoms 2', 2, 3, 4 and 4'. Erythrose 4-P is converted into 4-hydroxy-L-threonine, apparently in a reaction analogous to serine biosynthesis from 3-P-D-glycerate (Fig. 4.4-1). This compound is the precursor of the C-atoms 5', 5, 6 and N-1 of pyridoxine. The synthesis is feedback inhibited by pyridoxine. The dehydrogenation of unphosphorylated pyridoxine to pyridoxal is a *bacterial* reaction.

In *animals*, the vitamins pyridoxine, pyridoxal and pyridoxamine are phosphorylated after resorption. The kinase is subject to product inhibition. Then an oxidase reaction converts pyridoxine phosphate and pyridoxamine phosphate into pyridoxal phosphate. This oxidase is also product-inhibited.

The recommended daily intake for adult *humans* is ca. 2 mg. Deficiencies lead to neuronal disorders (neuritis), dermatitis and impaired amino acid metabolism (xanthurenate excretion, 4.7.3).

Degradation of the phosphorylated compounds begins with a phosphatase reaction. Further degradation of the vitamin proceeds by oxidation to the carboxylic acid 4-pyridoxate, which is biologically inactive.

9.4.2 Biochemical Function

Pyridoxal phosphate is the essential cofactor for reactions, which act on the C-2 (C_α) atom of amino acids and involve cleavage on any of its bonds. These bonds are labilized by the formation of a Schiff base between the coenzyme and the amino acid (Fig. 9.4-2).

① C_α–H:
  - transaminations (for mechanism see Fig. 4.2-3),
  - racemizations (e.g., alanine racemase, Fig. 15.1-1)

② C_α–COOH:
  - decarboxylation, e.g., glycine dehydrogenase (decarboxylating) (for mechanism see Fig. 4.4-3), histidine decarboxylase (Fig. 4.8-1)

③ C_α–R or C_β–C_γ:
  - β-eliminations, e.g., glycine hydroxymethyltransferase (Fig. 4.4-1), serine dehydratase (for mechanism see Fig. 4.4-2)
  - γ eliminations, e.g., cystathionine γ-lyase (Fig. 4.5-2)
  - β-replacement (e.g., cystathionine β-synthase, Fig. 4.5-2; the serine hydroxyl is replaced by the homocysteine residue)

Figure 9.4-1. Precursors (in *E. coli*) and Metabolism of Pyridoxine

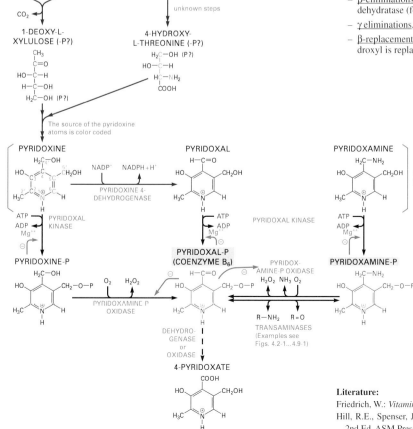

Figure 9.4-2. The Reactive Intermediate (Schiff Base) of Pyridoxal Catalyzed Reactions

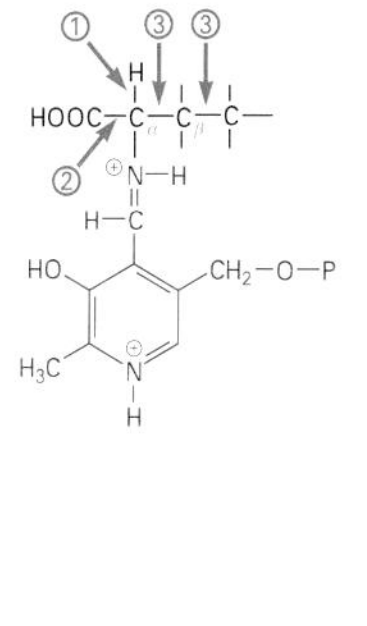

Literature:
Friedrich, W.: *Vitamins*. De Gruyter (1988) 543–618.
Hill, R.E., Spenser, J.D. in Neidhardt, F.C. (Ed.).: *Escherichia coli and Salmonella*. 2nd Ed. ASM Press 1996, 695–703.

Figure 9.5-1. Biosynthesis of Coenzyme B_{12} and Conversion of Vitamin B_{12} Into the Coenzyme

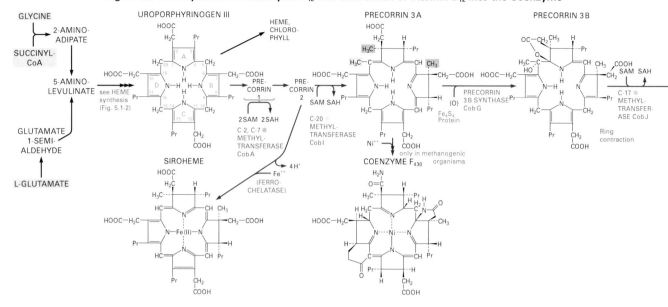

9.5 Cobalamin (Coenzyme B_{12}, Vitamin B_{12})

Cobalamin is one of the largest non-polymeric biological molecules. The central cobalt atom is coordinated with six ligands. Four of them are the pyrrols of the corrin ring, one is the unusual nucleotide dimethylbenzimidazole and the sixth can be hydroxyl (vitamin B_{12}), methyl (methylcobalamin) or 5′-deoxyadenosyl (deoxyadenosylcobalamin, coenzyme B_{12}). In the commercially available form the sixth ligand is cyanide.

9.5.1 Biosynthesis of the Coenzyme and Reduction of the Vitamin (Fig. 9.5-1)

Coenzyme B_{12} is biosynthesized exclusively by *microorganisms* in a multistep procedure. The first steps are identical to the pathways to heme and chlorophyll (5.1). There are several variants of the biosynthetic reactions. The pathway in *Pseudomonas denitrificans* and the names of its gene products are shown in Fig. 9.5-1.

After several methylations, ring narrowing from the porphyrin to the corrin ring takes place. After further methylations and amidations, Co^{++} is inserted, reduced to the Co^+ level and liganded with adenosine. Finally, α-ribazole is attached via a propionate-aminopropanol bridge and forms the 6th ligand of Co (which is in the Co^{+++} oxidation state in this coordination complex).

The needs of *mammals* (in *humans* < 10 µg/day) are met by the B_{12} production of intestinal *bacteria*.

A specific transporter protein (intrinsic factor, 50 kDa) is required for the resorption of vitamin B_{12} from the intestinal lumen. Insufficient production of this factor causes severe B_{12} deficiency (pernicious anemia). For the transport in blood, B_{12} is bound to different plasma globulins, called transcobalamins. B_{12} is taken up in the most oxidized form (hydroxycobalamin, Co^{3+}). After two reduction steps, the deoxyadenosyl form (coenzyme B_{12}) is restored (Fig. 9.5-1).

9.5.2 Biochemical Function

In *bacteria* many different reactions are dependent on the coenzyme B_{12}, whereas in *mammals* there are only two of them (see below).

The cobalamin-dependent reactions can be classified according to the way the Co-C bond is cleaved:

- **Homolytic cleavage**, employing deoxyadenosyl cobalamin:
  - *Intramolecular rearrangements*: The cleavage of the metalorganic bond ($Co^{3+} \rightarrow Co^{2+}$, Fig. 9.5-2) results in the formation of a free radical at the deoxyadenosyl group. This radical is used to exchange the positions of hydrogen and various substituents at neighboring atoms.

    Examples: methylmalonyl-CoA mutase in *mammals* (Fig. 4.5-2, exchanging H and a -CO-S-CoA group), propionate and glutamate fermentation by *bacteria* (Figs. 15.4-2 and 15.4-4).

  - *Reduction of ribonucleotides*: Some *bacterial* ribonucleoside triphosphate reductases (8.1.3) contain coenzyme B_{12} instead of the Fe(III)-O^2-Fe(III) center to generate a free radical.

- **Heterolytic cleavage**, employing methyl cobalamin:

  In some methyl transferases, methylcobalamin acts as the methyl group donor ($Co^{3+}-CH_3 \rightarrow Co^+ + CH_3$, Fig. 9.5-3).

  Examples are homocysteine methyltransferases in *mammals* and *bacteria* (Fig. 4.5-2). B_{12} is then remethylated by 5-methyltetrahydrofolate. An occasional oxidation of the cofactor from the intermediate Co^+ to the Co^{++} state requires a regeneration step, involving reduction by NADPH via flavo- or ferredoxin and methylation by 5-adenosylmethionine. Related reactions are methyl transfers in *archaeal* methanogenesis and *bacterial* acetogenesis (Figs. 15.5-2 and 15.5-3)

9.5.3 Siroheme and Coenzyme F_{430}

Siroheme is derived from precorrin 2 by introduction of an iron atom. It is the coenzyme of *bacterial* sulfite and nitrite reductases (15.5.1). Coenzyme F_{430} is a nickel-containing derivative of precorrin 3A. It acts as cofactor in the methyl-coenzyme M reductase reaction of methanogenesis (15.5.2) and functions similar to coenzyme B_{12} in methyl transfer reactions.

Literature:

Banerjee, R.: Chemistry and Biology 4 (1997) 175–186.
Blanche, F. *et al.*: Angew. Chem. Int. Ed. 34 (1995) 1001–1029.
Drennan, C.L. *et al.*: Science 266 (1994) 1669–1674.
Ludwig, M.L., Matthews, R.G.: Ann. Rev. of Biochem. 66 (1997) 269–313.
Stubbe, J.: Science 266 (1994) 1663–1664.

Figure 9.5-2. Mechanism of B_{12} Dependent Mutase Reactions Employing Deoxyadenosyl Cobalamin

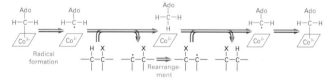

Figure 9.5-3. Reaction Mechanism of Homocysteine Methyltransferase Employing Methylcobalamin

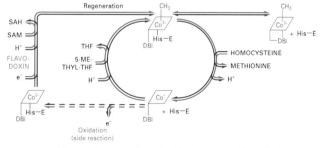

Abbreviations: Ado = adenosine, DBI = dibenzimidazole, SAM = S-adenosylmethionine, SAH = S-adenosylhomocysteine, THF = tetrahydrofolate, E = homocysteine methyltransferase, /Co/ = corrin ring

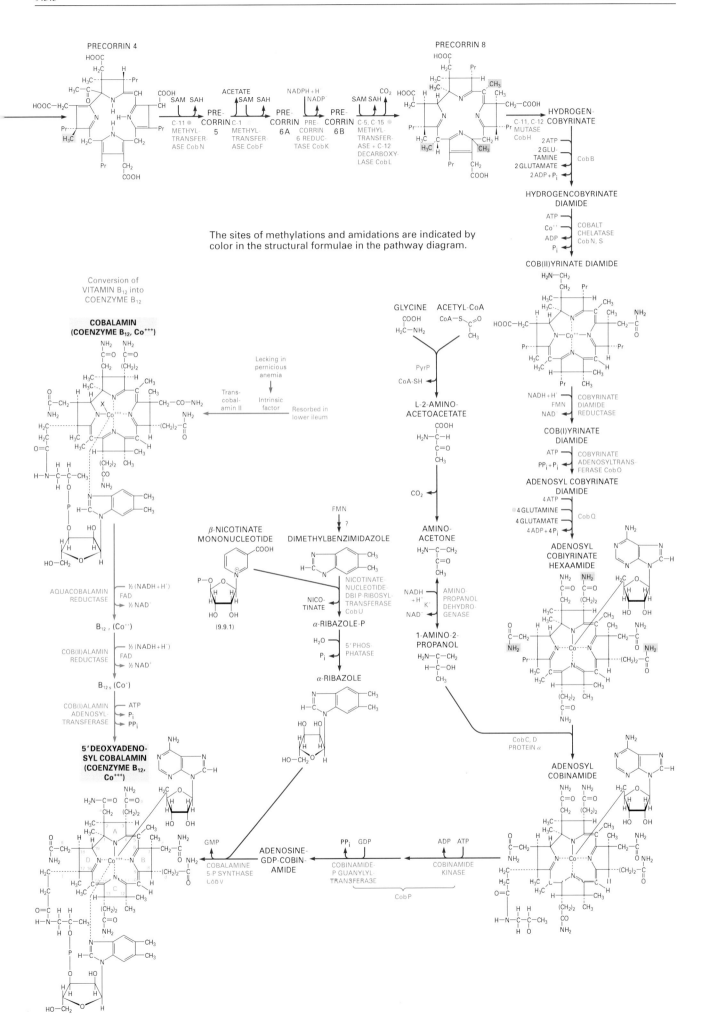

9.6 Folate and Pterines

A vitamin and several cofactors contain a pteridine ring system:
- Folate/tetrahydrofolate/tetrahydrofolylpolyglutamate (THF)
- Biopterin/tetrahydrobiopterin (THB)
- Molybdenum (and tungsten) cofactors (MoCo)
- Methanopterin/tetrahydromethanopterin (THMPT).

The synthesis of folate compounds is restricted to *plants* and *microorganisms*, but the product is required as a vitamin (vitamin B_c) by *animals*. Biopterin synthesis seems to be only performed by *animals*, whereas a molybdenum cofactor can apparently be synthesized by all living organisms. Methanopterin is formed exclusively by anaerobic *archaea* (*methanogens*).

9.6.1 Tetrahydrofolate/Folylpolyglutamate (Fig. 9.6-1)

Biosynthesis: The synthesis of folate starts with a two step conversion of GTP to dihydroneopterin-P_3 by GTP cyclohydrolase I (GTPCH I). C-8 of GTP is released as formate. The enzyme, in contrast to GTPCH II (riboflavin synthesis, 9.3), incorporates C atoms of ribose (1', 2') into the pteridine ring system. Dihydroneopterin-P_3 is also an intermediate of the pathways leading to biopterin (9.6.3) and to methanopterin (9.6.5).

The specific part of the tetrahydrofolate biosynthesis proceeds via a (so far unspecified) removal of all phosphates, side chain shortening and pyrophosphorylation to 2-amino-4-hydroxy-6-hydroxymethyl-7,8-dihydropteridine-P_2. The fusion of this intermediate with 4-aminobenzoate to 7,8-dihydropteroate is the target of the antibacterial action of sulfonamides. After the ATP dependent addition of glutamate, the resulting dihydrofolate is reduced to tetrahydrofolate. In *eukarya*, the synthesis is performed in *mitochondria*. Some steps are catalyzed by species specific multifunctional enzymes.

Polyglutamylation of reduced folate: Tetrahydrofolylpolyglutamate synthase catalyzes the sequential addition of usually 2...8 glutamate residues via γ-peptide linkages. Polyglutamylation increases or even enables the cofactor function of folates. It also inhibits the diffusion of folates through membranes and thus enables their accumulation and subcellular localization. For resorption of folates from the diet, these bonds must be hydrolyzed by pteroyl-poly-γ-glutamate hydrolase. The dietary requirement for *humans* is 0.4 mg/day.

Biochemical functions of folylpolyglutamate: Polyglutamylated tetrahydrofolate (THF) is the central cofactor of the one carbon (C_1) metabolism (9.6.2).

9.6.2 General Reactions of the C_1 Metabolism (Fig. 9.6-1)

C_1 metabolism is a key regulatory factor of catabolic and anabolic pathways by providing a controlled flux through the C_1 pool. It encompasses the transfer of C_1 compounds and their reduction or oxidation. CO_2 fixation and biotin dependent reactions (9.8.2) are not usually discussed in this context.

THF carries C_1 units at the oxidation levels of formate (as 5- or 10-formyl-THF, 5,10-methenyl- or 5-formimino-THF), formaldehyde (as 5,10-methylene-THF) and methanol (as 5-methyl-THF). The origin of the C_1 units entering at the various levels are
- 10-formyl-THF: formate
- 5-formimino-THF: degradation products of histidine (4.8.2) and xanthine (8.1.6)
- 5,10-methylene-THF (mainly in *mitochondria*): serine (4.4.1, main supplier of C_1 units), betaine degradation products (6.3.5), glycine (4.4.2).

C_1 Units are provided by
- 10-formyl-THF: for purine (8.1.1) and N-formylmethionyl-tRNA formation (in *bacteria, archaea, mitochondria* and *chloroplasts*, 10.6.2)
- 5,10-methylene-THF: for biosyntheses of thymidine 5'-P (8.2.3), coenzyme A (9.7.1) and serine (4.4.1, mainly in the *cytosol*)
- 5-methyl-THF: for methionine biosynthesis (4.5.4). Methionine in the form of S-adenosylmethionine serves as the donor of all energy driven methylation reactions.

The reversible interconversions of C_1 compounds take place by formate-THF ligase, methenyl-THF cyclohydrolase and methylene-THF dehydrogenase (NADPH, Fig. 9.6-1). In *animals* they are combined into a trifunctional enzyme. In *mitochondria*, the dehydrogenase activity is NAD-dependent and controls the metabolic flux from serine/glycine to formate (for protein synthesis and for cytoplasmic needs). Glycine hydroxymethyltransferase (4.4.2) and 5-formyl-THF cycloligase interconvert 5,10-methylene-THF and 5-formyl-THF. Irreversible reactions are catalyzed by 10-formyl-THF dehydrogenase, which releases THF and by methylene-THF reductase (NADPH), which leads to 5-methyl-THF. In the case of cobalamin deficiency, the 5-methyl-THF-homocysteine methyltransferase (4.5.4) is inoperative and a deleterious accumulation of 5-methyl-THF takes place.

Most regulation mechanisms are poorly understood. 5-Formyl-THF inhibits most enzymes of the C_1-THF interconversions and other C_1-THF dependent reactions. S-Adenosylmethionine inhibits the methylene-THF reductase.

9.6.3 Tetrahydrobiopterin (Fig. 9.6-1)

Biosynthesis: Tetrahydrobiopterin (THB) biosynthesis proceeds analogously to folate biosynthesis (9.6.1) up to the dihydroneopterin-P_3 level. The Zn-dependent 6-pyruvoyltetrahydropterin synthase catalyzes an internal redox transfer and eliminates the triphosphate residue. Two consecutive reduction steps are performed by the NADPH dependent sepiapterin reductase.

The rate limiting step of THB synthesis is the initial GTP cyclohydrolase I reaction. Feedback inhibition by THB is mediated by the GTP cyclohydrolase I feedback regulatory protein (GFRP, formerly p35). The *de novo* synthesis of the cyclohydrolase is suppressed by glucocorticoids and cell specifically stimulated by, e.g., interferon γ, interleukin-1β or other cytokines.

Biochemical functions of tetrahydrobiopterin: THB is the cofactor of the aromatic amino acid monooxygenases (phenylalanine, tyrosine and tryptophan monooxygenases, 4.7.3), of nitric oxide synthase (17.7.2) and of glyceryl-ether hydroxylase (e.g., PAF degradation, 6.3.3).

During these reactions, THB is oxidized to 4a-hydroxytetrahydrobiopterin (pterin-4a-carbinolamine). Regeneration to THB proceeds in two enzymatic steps via dehydration by pterin-4a-carbinolamine dehydratase and NADH dependent reduction by dihydropteroate reductase.

Since THB plays a role in the biosynthesis of serotonin (4.7.3) and catecholamines (4.7.4), defects in THB synthesis or regeneration may lead to, e.g., neurological diseases.

Additionally THB seems to be involved in other cofactor functions, e.g., in cytokine dependent proliferation of erythroid and T cells and in interactions with the IL-2 receptor.

Neopterin: This pteridine, together with dihydroneopterin, is secreted from *monocytes/macrophages* after stimulation of THB synthesis by interferon-γ.

Other pteridines: There is strong evidence that in *algae, fungi* and *plants*, a pteridine is one of the cofactors of the blue light receptor.

9.6.4 Molybdenum/Tungsten Cofactor (Fig. 9.6-1)

Biosynthesis: Molybdopterin (MPT) synthesis likewise seems to start from GTP. Ring opening and further steps (without release of formate) lead to the labile precursor Z, which is sulfurylated and rearranged by MPT synthase to MPT, forming the Mo binding dithiolene structure. Incorporation of Mo (delivered by the Mo uptake and transport system) into MPT yields the very unstable molybdenum cofactor (MoCo). The last step of the synthesis is the incorporation into the target enzyme, which effects the stabilization of the cofactor. In *bacteria*, MoCo is usually converted into a dinucleotide before incorporation into the enzyme takes place.

Biochemical function: All molybdoenzymes (except nitrogenase, 4.1) contain the organometallic molybdenum cofactor. *Mammalian* molybdoenzymes ere sulfite oxidase, xanthine oxidase (8.1.6) and aldehyde oxidase. Molybdopterin is degraded to urothione. Defects in molybdopterin biosynthesis in *humans* are very rare. They are lethal in early childhood.

In *plants* and *bacteria*, a broad range of reactions is catalyzed by molybdoenzymes. Three enzyme families were defined by functional criteria:

- DMSO reductase family in *bacteria*, e.g., DMSO reductase, nitrate reductase (dissimilatory, 15.5), formate dehydrogenases (one of them contains tungsten)
- Xanthine oxidase family, e.g., xanthine oxidase/dehydrogenase (8.1.6, additionally contains FAD), aldehyde oxidase (4.7.3)
- Sulfite oxidase family in *algae* and *higher plants*, e.g., sulfite oxidase/dehydrogenase (15.6), nitrate reductase (assimilatory)

Related are tungsten enzymes, which form the majority of the aldehyde ferredoxin oxidoreductase family, e.g., aldehyde ferredoxin oxidoreductase.

9.6.5 Methanopterin (Fig. 9.6-1)

Biosynthesis: The biosynthesis of tetrahydromethanopterin (THMPT) resembles that of tetrahydrofolate and apparently branches off at the dihydroneopterin-P_3 level. Then a S-adenosylmethionine dependent methylation at C-7 takes place. The source of the C-9 and C-9a atoms have not been conclusively proven. The side chain of methanopterin is formed by the condensation of 4-aminobenzoate (4.7.1) with ribose. The mechanism resembles the indole 3-glycerol-P synthase reaction (4.7.1). Ribosyl 5-P and 2-hydroxyglutarate (obtained by reduction of 2-oxoglutarate, Fig. 3.8-2) are added afterwards.

Biochemical function: The C_1 derivatives of tetrahydromethanopterin are analogous to those of tetrahydrofolate. They are involved in methane synthesis, as described in 15.5.2.

Literature:
Auerbach, G., Nar, H.: Biol. Chem. 378 (1997) 185–192.
DiMarco, A.A. et al.: Ann. Rev. Biochem. 59 (1990) 361–370
Kisker, C. et al.: Ann. Rev. Biochem. 66 (1997) 233–267.
Mendel, R.R.: Planta 203 (1997) 399–405
Neidhardt, F.C. et al. (Eds.): *Escherichia coli and Salmonella*. ASM Press (1996) 665–679.
Rébeillé, F. et al.: EMBO J. 16 (1997) 947–957.

Figure 9.6-1. Biosynthesis of Pterines

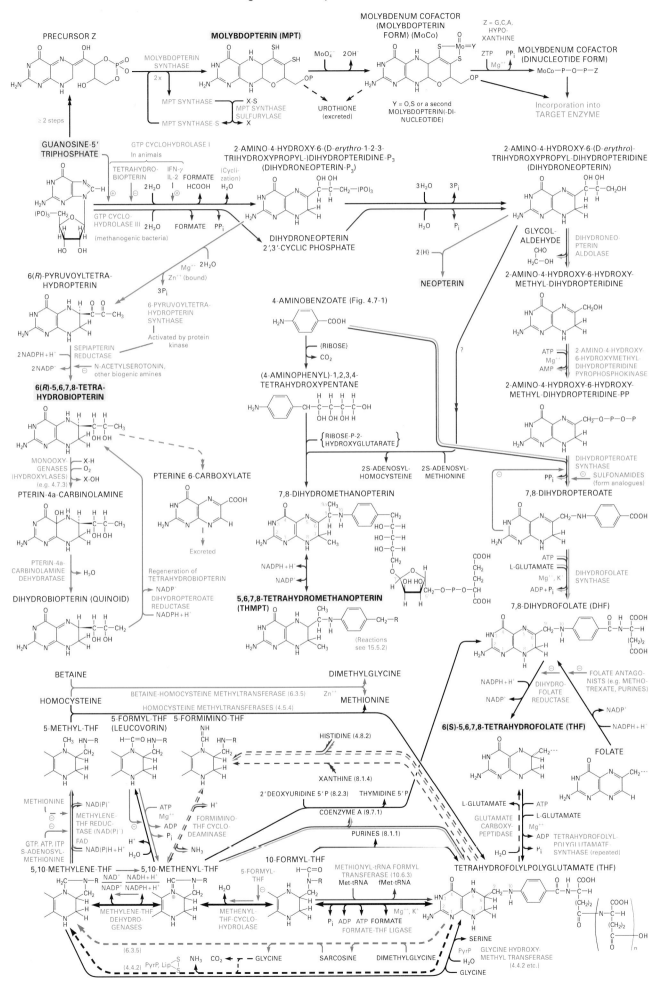

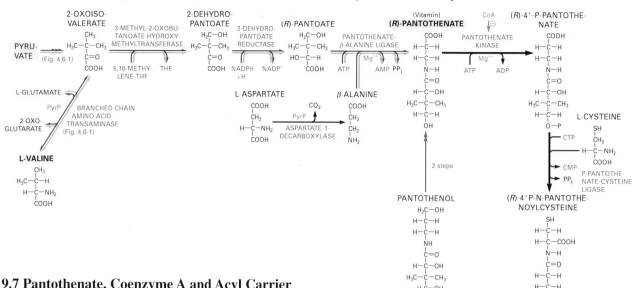

Figure 9.7-1. Biosynthesis of Pantothenate, Coenzyme A and Holo-Acyl Carrier Protein

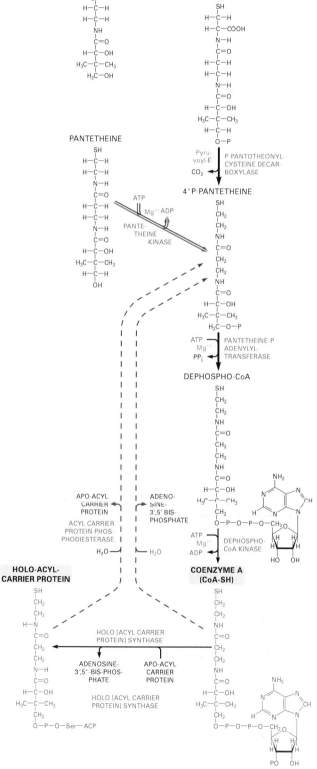

9.7 Pantothenate, Coenzyme A and Acyl Carrier Protein (ACP)

Panthothenate is a member of the B group of vitamins. It is an essential component of the important acyl group carriers coenzyme A (CoA or CoA-SH, 3.3.3) and acyl carrier protein (ACP), which are present in all cells and take part in more than 100 reactions. Among them are fatty acid synthesis (6.1.1) and degradation (6.1.5), pyruvate oxidation (3.3.1) and glyceride synthesis (6.2.1) etc.

9.7.1 Biosynthesis and Interconversions (Fig. 9.7-1)

Most research on the biosynthesis of the vitamin was done in *bacterial* systems.

2-Oxoisovalerate, the transamination product of L-valine, is methylated by 5,10-methylene-THF to yield 2-dehydropantoate, which is then reduced to (R)-pantoate. An ATP-dependent condensation with β-alanine (which is produced from aspartate by 1-decarboxylation, 4.2.4) leads to (R)-pantothenate. Since *intestinal E. coli* secrete considerable amounts of it, this becomes an important source of the vitamin to the *mammalian* host. Also, pantothenol can be oxidized to pantothenate. It is present in food, but is also used as a drug.

The synthesis of coenzyme A (CoA) from the vitamin pantothenate can be performed by all living beings and begins with its phosphorylation. This is the committed step (= first ambiguous step), which is regulated via allosteric inhibition by the end product CoA. Then follows the CTP-dependent condensation with L-cysteine and decarboxylation. Adenylylation and phosphorylation complete the synthesis of the coenzyme. In *mammals*, these two steps are carried out by a bifunctional enzyme, while in *bacteria* separate entities exist. The active holo-acyl carrier protein (ACP, 6.1.1) is obtained by transfer of the 4'-phosphopantetheine residue from CoA to a serine residue of apo-ACP.

Dietary requirement for *humans*: Since pantothenate is present in practically all biological material (name!) and is, in addition to food, supplied by *intestinal bacteria* (see above), avitaminoses are practically unknown. The daily requirement for *adults* is estimated to be in the order of 6...10 mg.

Degradation: CoA is cleaved by a phosphodiesterase, yielding 4'-phosphopantetheine and adenosine 3',5'-bisphosphate. Likewise, the cleavage of holo-ACP yields 4'-phosphopantetheine (in addition to apo-ACP).

9.7.2 Biochemical Function

The thioester bonds between acids and CoA or pantetheine are energy-rich bonds ($\Delta G'_0$ for hydrolysis = −31.5 kJ/mol, 3.3.3). Their formation corresponds to an activation of the acid for synthetic reactions, such as esterification, amidation (e.g. 3.7), anhydride formation (e.g. Fig. 15.4-4), C-C bond formation (e.g., 3.3.3) etc. Best known is acetyl-CoA, the 'activated acetic acid' (3.3.3).

R-COOH + ATP + CoA-SH = Acyl-CoA + AMP + PP$_i$

Acyl-CoA + X = Acyl-X + CoA-SH.

During fatty acid synthesis (6.1.1), the pantetheine residue of ACP acts as a 'movable arm' to move the bound acid from one catalytic center to the next. A similar task is performed by pantetheine 'arms' in non-ribosomal polypeptide synthesis (e.g., Fig. 15.8-1).

Literature:

Friedrich, W.: *Vitamins*. De Gruyter (1988) 809–835.

Jackowski, S. in Neidhardt, F.C. (Ed.).: *Escherichia coli and Salmonella*. 2nd Ed. ASM Press 1996, 687–694.

9.8 Biotin

Biotin is the essential cofactor for carboxylation reactions. It is covalently enzyme-bound by an amide bond to the ε-amino group of a lysine residue.

9.8.1 Biosynthesis and Interconversions (Fig. 9.8-1)

The biosynthesis is performed by *bacteria* (including *bacteria* in the intestines of animals), *yeasts* and *higher plants* in an identical way.

It starts from pimeloyl-CoA. The origin of this compound is not completely known. Possibly it results from the condensation of 3 molecules malonyl-CoA and the release of 2 molecules CO_2. The following condensation with L-alanine is pyridoxal-PP dependent and involves the loss of CO_2 and CoA-SH, similar to the δ-aminolevulinate synthesis (5.2.1). The product, 8-amino-7-oxopelargonate (KAPA) undergoes a transaminase reaction, yielding 7,8-diaminopelargonate. In this reaction, S-adenosylmethionine (4.5.4) acts as an amino group donor (instead of its more common role in transmethylation reactions). An ATP-dependent introduction of CO_2 closes the imidazolidone ring. Then a stereospecific introduction of S (originating from L-cysteine) and closure of the thiophane ring takes place. The result is (+)-biotin (one of 8 possible stereoisomers).

The binding of biotin to the respective enzymes requires an ATP-dependent activation, catalyzed by special ligases (Biotin-[enzyme] ligases). In *bacteria*, the ligases are allosterically regulated by several effectors. If all of the apoenzyme has been saturated, additional activated biotin is transferred to a repressor and by acting as corepressor, decreases the expression of the biosynthetic enzymes.

In *animal* food, protein-bound biotin is released by biotinidase and thereafter resorbed. Biotinylation of the carboxylase enzymes takes place mainly in the *liver*.

A supply of 100...200 µg/day is required for *human* adults. Avitaminoses are rare. However, genetic defects of all biotin-dependent enzymes have been observed, as well as of biotinidase, which causes insufficient biotin resorption. Biotin-binding proteins are present in whites and yolks of bird's eggs. These compounds form extremely tight complexes with biotin (notably avidin: $K = 10^{-15}$ mol/l). If native avidin is present in food, the avidin-biotin complex cannot be resorbed.

Degradation by microorganisms: In *Pseudomonads* (and probably in other microorganisms), β-oxidation shortens the side chain and removes the C atoms of the thiophane ring. The rest of the imidazolidone ring is released as urea. *Mammals* can only remove the side chain and form the sulfoxide compound. Most biotin is, however, excreted in unchanged form.

9.8.2 Biochemical Function (Fig. 9.8-2)

The essential function of biotin is the transfer of carboxyl groups. Enzyme functions involved in this process are biotin carboxylase (BC), which loads HCO_3^- on biotin, transcarboxylase (TC), which transfers CO_2 to an acceptor, biotin carrier protein (C), which provides an 'anchor' for the biotin side chain and biotin decarboxylase (BD). They can be present as a single multifunctional enzyme or as separate entitites. Various combinations of them carry out the following reaction types:

- Carboxylations (by BC, TC and C functions):

  The enzyme-catalyzed activation of bicarbonate leads via a carboxyphosphate intermediate to the carboxylation of biotin at the N-1 position. The side chain of biotin acts as a flexible arm, which enables the movement of the carboxylated biotin residue from the biotin carboxylase to the transcarboxylase reaction centers. There, CO_2 is transferred to an acceptor molecule (Fig. 6.1-2). Examples are pyruvate carboxylase (3.3.4), propionyl-CoA carboxylase (4.5.4) and methylcrotonyl-CoA carboxylase (4.6.2). Most important is acetyl-CoA carboxylase (6.1.1):

  HCO_3^- + ATP + biotin = ADP + P_i + biotin-CO_2

  Biotin-CO_2 + acetyl-CoA = biotin + malonyl-CoA

- Transcarboxylations (by 2 TC and 1 C functions):

  One carboxyl transferase function transfers CO_2 from a carboxyl donor to biotin, while the other moves it on to the carboxyl acceptor compound. No activating reaction is needed. An example is methylmalonyl-CoA carboxyltransferase (Fig. 15.4-2):

  Methylmalonyl-CoA + biotin = propionyl-CoA + biotin-CO_2

  Biotin-CO_2 + pyruvate = biotin + oxaloacetate

- Decarboxylations (by TC and BD functions):

  Biotin accepts CO_2 from a substrate, the decarboxylase function removes it. Examples are oxaloacetate decarboxylase or methylmalonyl-CoA decarboxylase (In *bacteria*, both enzymes energize sodium pumps in this way):

  Methylmalonyl-CoA + biotin = propionyl-CoA + biotin-CO_2

  Biotin-CO_2 + H_2O = biotin + H_2CO_3

Figure 9.8-1. Biosynthesis and Degradation of Biotin

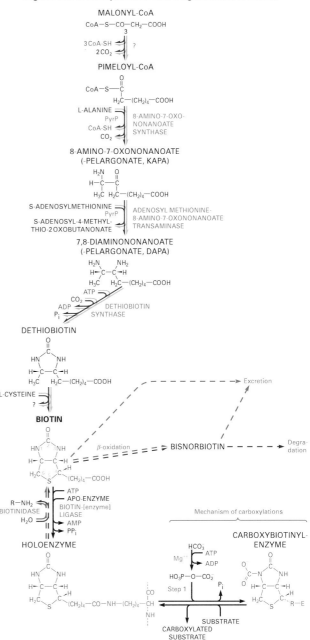

Figure 9.8-2. Reaction Types of Biotin

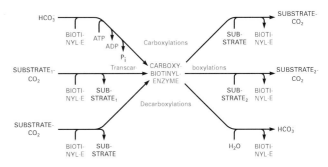

Literature:

Friedrich, W.: *Vitamins.* De Gruyter (1988) 752–805.

DeMoll, in Neidhardt, F.C. (Ed.).: *Escherichia coli and Salmonella.* 2nd Ed. ASM Press (1996) 704–709.

9.9 Nicotinate, NAD⁺ and NADP⁺

Nicotinate (niacin) and nicotinamide are precursors of the coenzymes nicotinamide-adenine dinucleotide (NAD$^+$) and nicotinamide-adenine dinucleotide phosphate (NADP$^+$), jointly known as nicotinamide (or pyridine) nucleotides. By interconversion with their reduced forms, NADH and NADPH, they are participating in several hundred redox reactions. These coenzymes therefore occupy a central role in metabolic processes of all living beings.

9.9.1 Biosynthesis and Degradation of NAD⁺ and NADP⁺ (Fig. 9.9-1)

Nicotinate is not a vitamin in a strict sense, since it can be synthesized in *animals* from tryptophan (which, however, is an essential amino acid).

An additional supply of nicotinate is only required when the supply of tryptophan is insufficient (e.g., by food containing mainly maize or sorghum → pellagra) or in order to satisfy peak demands. 18 mg/day are considered sufficient for *human* adults.

The *eukaryotic* catabolism of tryptophan yields 2-amino-3-carboxymuconate semialdehyde as an intermediate (Fig. 4.7-4). This compound cyclicizes nonenzymatically to quinolinate. In *E. coli*, aspartate oxidation produces iminoaspartate, whose reaction with dihydroxyacetone-P (glycerone-P) and a consecutive ring closure also results in quinolinate. The FAD-containing oxidase is feedback inhibited by NAD$^+$. Quinolinate synthase is O_2-sensitive. Decarboxylation and conversion to nicotinate mononucleotide are performed by a ribosyltransferase.

Adenylate transfer from ATP yields deamido-NAD$^+$, which is then converted into the amide, NAD$^+$. *Bacteria* use NH_3 for this reaction, while *eukarya* obtain the amino group from glutamine or use NH_3. An additional phosphorylation at a ribose moiety results in NADP$^+$. Also a reconversion to NAD$^+$ by a phosphatase reaction takes place.

Salvage Reactions: In aerobic *bacteria*, the half-life time of NAD$^+$ is only about 90 minutes. The valuable nicotinate moiety has to be recovered by pyridine nucleotide cycles. The primary step of NAD$^+$ degradation is mostly the cleavage by pyrophosphatase. Another NAD$^+$ degradation reaction is the direct removal of the nicotinamide residue. In both cases, consecutive steps lead to nicotinate, which reacts with phosphoribosyl pyrophosphate to yield nicotinate mononucleotide. Its reconversion to NAD$^+$ proceeds as described above.

Degradation: In *bacteria*, nicotinate is hydroxylated and then decarboxylated. Oxidative ring opening and further reactions lead to fumarate, which is a component of the citrate cycle (3.8.1).

9.9.2 Mechanism of the Redox Reactions, Stereospecificity

The reduction of the nicotinamide group proceeds by the uptake of 1 proton H$^+$ and 2 electrons e$^-$ (formally of a hydride ion, H$^-$). Another hydrogen of the substrate is released as a proton.

The positive charge on the pyridine nitrogen and the aromatic ring character is lost. This is accompanied by a spectral change (Fig. 9.9-2), which can be used for analytical work. The general reaction equation can be written as

NAD$^+$ (or NADP$^+$) + XH$_2$ = NADH (or NADPH) + X + H$^+$

This hydrogen transfer is a stereospecific one: the added H is located either above the plane of the nicotinamide ring (A side or pro-R side according to the Cahn-Ingold-Prelog 'RS'-system) or below (B side or pro-S side, Fig. 9.9-3). E.g., alcohol, malate and isocitrate dehydrogenases are 'A' enzymes, 2-oxoglutarate, glucose-6-phosphate and glutamate dehydrogenases are 'B' enzymes. Likewise, the removal of hydrogen from the substrate proceeds stereospecifically.

Figure 9.9-2. Absorption Spectra of NAD⁺/NADP⁺ and NADH/NADPH

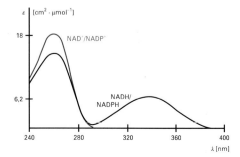

Figure 9.9-1. Biosynthesis of Nicotinamide, NAD⁺ and NADP⁺

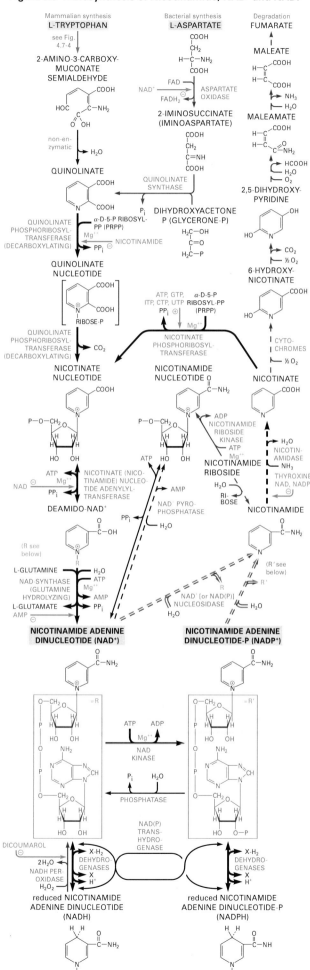

Figure 9.9-3. Stereospecifity of Dehydrogenase Reactions

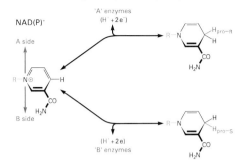

9.9.3 Biochemical Function of the Nicotinamide Coenzymes

Although the redox potentials of the $NAD^+/NADH$ and the $NADP^+/NADPH$ systems are almost identical ($E'_0 = -320$ and -324 mV), in living organisms a ratio of the oxidized/reduced dinucleotides of about 200...1000 with NAD and 0.01 with NADP is maintained.

- Via a membrane-bound transhydrogenase inside of *mitochondria* and *bacteria*, hydrogen exchange between internal NAD^+ and $NADP^+$ takes place (Fig. 16.1-1). The reaction is driven by the proton gradient across the membrane and is essential for maintaining the different redox states of both coenzymes:

    $NADH + NADP^+ + H^+_{outside} = NAD^+ + NADPH + H^+_{inside}$

- In other compartments, substrate concentrations cause the differences of the redox situation. The coenzyme specificity of most NAD^+ or $NADP^+$ employing enzymes prevents an equilibration of the redox states.

- Due to the prevalence of the reduced form, NADPH is well suited for reductive biosyntheses, e.g. of fatty acids and steroids. In *erythrocytes*, NADPH keeps up the glutathione (GSH) concentration, which is needed for the removal of membrane-damaging H_2O_2 (4.5.7). On the other hand, NADPH oxidation in activated *neutrophil granulocytes* and *macrophages* leads to aggressive superoxide radicals, which play a role in cellular defense (4.5.8).

- NAD^+ usually accepts hydrogen during oxidation of metabolites and delivers them to the cytochrome system for terminal oxidation (16.1) or transfers them to other substrates during fermentations (15.4). NAD^+ supplies the ribose moiety of α-ribazole during coenzyme B_{12} biosynthesis (9.5). It is also a substrate for the modification of proteins by ADP-ribosylation (e.g., 17.4.1).

A *bacterial* DNA ligase uses this reaction to energize the resealing of nicks in DNA repair (Fig. 10.3-1).

Figure 9.9-4. Ligase and ADP-Ribosylation Reactions Employing NAD^+

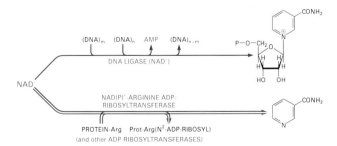

Literature:
Creighton, D.J., Murphy, N.S.R.K in Sigman, D.S., Boyer, P.D. (Eds.): *The Enzymes*. 3rd Ed., Vol. 19. Academic Press (1990) 323–421.
Dolphin, D.: *Pyridine Nucleotide Coenzymes: Chemical, Biochemical and Medical Aspects*. Wiley (1987).
Friedrich, W.: *Vitamins*. De Gruyter (1988) 475–542.
Penfound, D., Foster, J.W. in: Neidhardt, F.C. (Ed.): *Escherichia coli and Salmonella*. 2nd Ed. ASM Press (1996) 721–730.

9.10 Ascorbate (Vitamin C)

Vitamin C takes part in many reactions involving oxygen and frequently exerts a protective function.

9.10.1 Biosynthesis and Metabolism (Fig. 9.10-1, next page)

Ascorbate can be biosynthesized by *higher plants, algae, yeast* and most *animals*. This ability is lacking in some *mammals* (including *humans* and *guinea pigs*) and also in *insects*, *invertebrates* and most *fishes*. In those *mammals*, which require external ascorbate supply, the biosynthetic enzyme L-gulonolactone oxidase is absent.

The biosynthesis in *animals* and most *plants* proceeds from D-glucose via UDP-glucose, UDP-glucuronate, D-glucuronate, L-gulonate, L-gulonolactone and 2-dehydro-L-gulonolactone. Some *algae* proceed via the D-galacturonate pathway with essentially analogous steps.

After resorption of ascorbate from food, it is transported in blood as an albumin complex to the various organs. In some of them, uptake is an active process. Insulin promotes the uptake. The highest concentrations are found in the *adrenal* and *pituitary glands, eye, liver, pancreas, thymus* and *brain*. In *humans*, scurvy is prevented by an intake of 10 mg/day. Stress, smoking and pregnancy increase the demand. A daily allowance of ca. 70 mg is recommended, but this is under discussion. It is claimed that elevated doses of ascorbate strengthen the immune system, prevent easy tiring and bleeding and even decrease the risk of cancer. However, after reaching a pool size of 1500 mg by a daily intake of 50 mg, degradation and excretion of ascorbate increases greatly.

Degradation and excretion: In *primates*, part of the body ascorbate is excreted unchanged in urine. Other excretion products are oxalate and dioxogulonate (formed from dehydroascorbate by a nonenzymatic, irreversible reaction). Also, complete oxidation to CO_2 takes place (the preferred way of catabolism in *rodents*).

9.10.2 Biochemical Function (Fig. 9.10-2)

Ascorbate participates in many important redox reactions. Many of them involve oxygen. The E'_0 for the redox couple dehydroascorbate/ascorbate is $+58$ mV ($2e^-$ reaction). However, in most cases, a one-electron transfer takes place. This leads to the monodehydroascorbate radical, which then is reconverted to ascorbate either directly or via dehydroascorbate.

Important examples for ascorbate involvement are:

- Dioxygenases, e.g., proline and lysine dioxygenases (4.3, 4.5.2). If the substrates are not available, but the cofactor oxoglutarate is present, the enzyme-Fe^{++} is oxidized to Fe^{+++} and requires stoichiometric amounts of ascorbate to be reduced again. Under *in vivo*-conditions, smaller amounts suffice. A Fe-ascorbate-substrate complex as an intermediate of the enzymatic reaction has been discussed in the literature. Also other enzymes, such as 4-hydroxyphenylpyruvate dioxygenase (4.7.3) and homogentisate 1,2-dioxygenase (4.7.3) require ascorbate, although their involvement is less clear. Ascorbate also promotes the hydroxylation of xenobiotics.

- A number of monooxygenases, e.g., Cu^{++} containing dopamine β-monooxygenase (4.7.3). Stoichiometric amounts of ascorbate are required for reduction of the metal as part of the catalytic cycle.

- Reduction of heme proteins, which previously have been oxidized by H_2O_2 to the ferryl level (e.g., myoglobin).

- Direct reaction of ascorbate with activated oxygen species (4.5.8), mostly for detoxification and protection purposes. This way, the *eye* is protected against light-activated oxygen or *chloroplasts* against dangerous side reactions of the photosynthesis apparatus (generation of radicals or singlet oxygen). Non-enzymatic reactions are:

    – Removal of the superoxide radical

    $O_2^{\bullet-} + $ ascorbate $ + H^+ = H_2O_2 + $ monodehydroascorbate$^\bullet$

    – Removal of hydroxyl and peroxyl radicals

    $HO^\bullet + $ ascorbate $ = H_2O + $ monodehydroascorbate$^\bullet$

    $ROO^\bullet + $ ascorbate $ = ROOH + $ monodehydroascorbate$^\bullet$

    – Quenching of singlet oxygen $^1O_2$.

In *plants*, H_2O_2 is removed by the L-ascorbate peroxidase reaction (Fig. 4.5-6). Ascorbate also reacts with molecular oxygen, either nonenzymatically (especially in the presence of heavy metal ions) or catalyzed by ascorbate oxidase:

$O_2 + 4$ ascorbate $= 2 H_2O + 4$ monodehydroascorbate$^\bullet$

The monodehydroascorbate formed in these reactions is reconverted to ascorbate by NADH, by cytochrome b_5 (fatty acid desaturation system) or by ferredoxin (in *plants*, from photosystem I, 16.2.1). Alternatively, it disproportionates into ascorbate and dehydroascorbate. The latter is reduced to ascorbate by glutathione in a reaction catalyzed by dehydroascorbate reductase. The oxidized glutathione, in turn, is reduced again by NADPH.

Figure 9.10-1. Biosynthesis and Degradation of Ascorbate

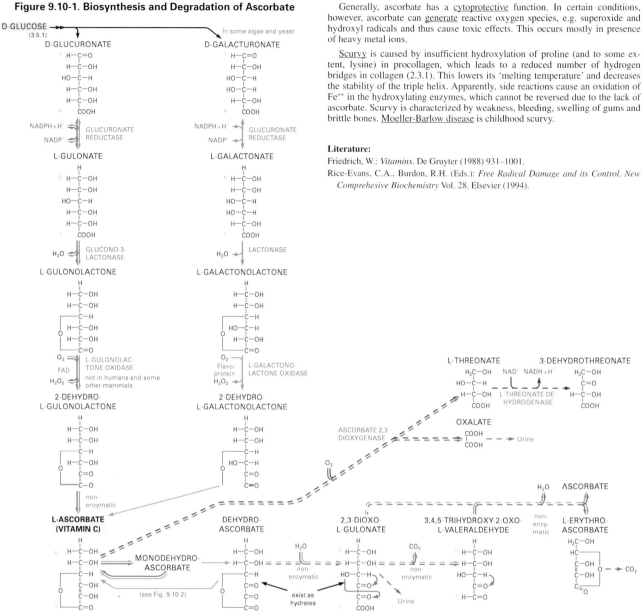

Generally, ascorbate has a cytoprotective function. In certain conditions, however, ascorbate can generate reactive oxygen species, e.g. superoxide and hydroxyl radicals and thus cause toxic effects. This occurs mostly in presence of heavy metal ions.

Scurvy is caused by insufficient hydroxylation of proline (and to some extent, lysine) in procollagen, which leads to a reduced number of hydrogen bridges in collagen (2.3.1). This lowers its 'melting temperature' and decreases the stability of the triple helix. Apparently, side reactions cause an oxidation of Fe^{++} in the hydroxylating enzymes, which cannot be reversed due to the lack of ascorbate. Scurvy is characterized by weakness, bleeding, swelling of gums and brittle bones. Moeller-Barlow disease is childhood scurvy.

Literature:
Friedrich, W.: *Vitamins*. De Gruyter (1988) 931–1001.
Rice-Evans, C.A., Burdon, R.H. (Eds.): *Free Radical Damage and its Control. New Comprehesive Biochemistry* Vol. 28. Elsevier (1994).

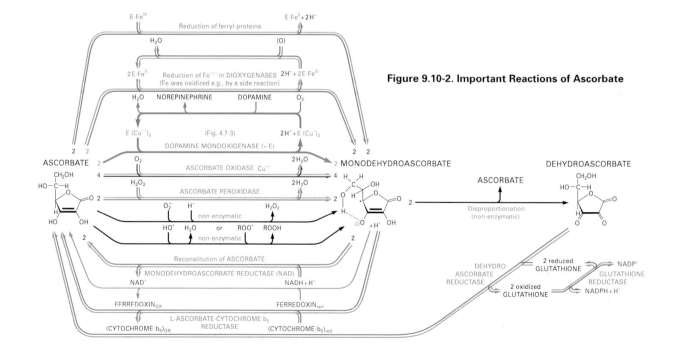

Figure 9.10-2. Important Reactions of Ascorbate

9.11 Calciferol (Vitamin D)

In *higher animals*, hydroxylation products of vitamin D (calciferol) play a central role in calcium metabolism.

Actually, the term 'vitamin D' encompasses a group of related compounds, but of different origin. They are not vitamins in the strictest sense, because one of them (D_3) can be formed in the *skin* by ultraviolet irradiation from a steroid (7-dehydrocholesterol, provitamin D), which is synthesized in the body.

9.11.1 Biosynthesis and Interconversions

The various D vitamins are formed from $\Delta^{5,7}$ steroids [which are intermediates in the biosynthesis of zoo- (7.1), phyto- and mycosteroids (7.2)] by light-induced opening of the B ring. They differ only in their side chains. The first two of the following are most important:

- Vitamin D_2 (ergocalciferol), obtained from ergosterol (7.2.2, present in *plants* and *fungi*)
- Vitamin D_3 (cholecalciferol), obtained from 7-dehydrocholesterol (7.1.1, present in *higher animals*)
- Vitamin D_4, obtained from 22-dihydroergosterol (present in *plants* and *fungi*)
- Vitamin D_5, obtained from 7-dehydrositosterol (present in *plants*).

Endogeneous formation: In a first reaction step, UV illumination causes an isomerization with the simultaneous opening of ring B of the sterol structure, yielding previtamin D. By another, temperature-dependent isomerization, vitamin D is obtained. This compound then enters the bloodstream and is bound to vitamin D binding protein (DBP). The optimum wavelength for the photoreaction is 295 nm. Irradiation at longer wavelengths leads to lumisterol, at shorter wavelengths to tachysterol, while too intense irradiation results in the formation of suprasterol I and II.

Supply by food: Although illumination can provide sufficient vitamin D, additional intake is recommended (for *human* adults 5 μg/day, for children 10 μg/day). After (bile acid assisted) resorption, dietary vitamin D is transported to the *liver* by *chylomicrons* or bound to DBP.

In the *liver*, hydroxylation at C-25 takes place through a cytochrome P-450 dependent enzyme (Fig. 7.4-2). The compound is then transferred to the *kidney*, where the biologically active 1α,25-dihydroxycalciferol (calcitriol, DHCC) is formed by another hydroxylation at C-1. An alternative hydroxylation leads to 24(R),25-dihydroxycalciferol. The hydroxylations also occur in the *placenta* and in *monocytes/macrophages*.

1α,25-dihydroxycalciferol is of great importance for the calcium homeostasis, therefore its formation is strictly regulated. Parathyroid hormone (parathormone, PTH, 17.1.7) stimulates the 1-hydroxylase and inhibits the 25-hydroxylase. High plasma levels of Ca^{++} repress the PTH synthesis and thus lower calcitriol formation. Calcitriol represses the synthesis of the 1-hydroxylase and enhances the synthesis of the 25-hydroxylase.

Degradation takes place by additional hydroxylations and/or oxidation of hydroxyl to carboxylic groups. These more polar compounds show only little biological activity and are excreted via stool or urine.

9.11.2 Biochemical Function (Fig. 9.11-1)

In *mammals*, 'vitamin D' (actually 1α,25-dihydroxycalciferol) regulates the calcium and phosphate metabolism (17.1.7).

It activates the (re)absorption of Ca^{++} and phosphate in the *intestine* and *kidney* and thus contributes to their deposition in *bones*, concomitant with PTH and calcitonin. On the other hand, it mobilizes Ca^{++} from *bone* for short time regulation of the blood Ca^{++} level.

Apparently this is effected by binding of the vitamin to vitamin D receptors (17.6). The complexes induce the transcription of genes for calcium binding proteins in *intestine, pancreas* and *bones* (CaBP, calbindin, osteocalcin etc.). These proteins, which contain γ-carboxyglutamate, are involved in calcium uptake and bone mineralization. This way, the vitamin acts in a hormone-like fashion. It is also involved in the differentiation of *macrophages* into *osteoclasts*.

To some extent, 24,25-dihydroxycalciferol is involved in these reactions.

Medical aspects: Diseases associated with abnormal levels of vitamin D are

- Rickets (diminished mineralization of the skeleton, mostly during childhood). This is caused by reduced resorption of Ca^{++} due to lack of vitamin D, sometimes also by deficient formation of calcitriol from calciferol.
- Osteomalacia (softening of bones) and osteoporosis due to impaired calcitriol formation. This is caused by kidney failure or by a reduced level of estrogens, which stimulate the hydroxylation step.
- D-hypervitaminosis. Demineralization of bones occurs by overdosing of vitamin D.

Literature:

Cancela, L. et al.: In Cooke, B.A., King, R.J.B, van der Molen, H.J. (Eds.): *Hormones and their actions*. Part I. *New Comprehensive Biochemistry*. Elsevier (1988). Vol. 18A, 269–289.

Darwish, H., DeLuca, H.F.: Crit. Rev. Eukar. Gene Expr. 3 (1993) 89–116.

Friedrich, W.: *Vitamins*. De Gruyter (1988) 143–216.

9.12 Tocopherol (Vitamin E)

The tocopherol group encompasses 8 compounds, which consist of a 8-chromanol ring and an isoprenoid side chain. The end product of biosynthesis is α-tocopherol (5,7,8-trimethyltocopherol), which also shows the highest bioactivity. Tocopherols protect lipids in *animals* and *plants* from oxidative damage.

The various tocopherols differ in the number and position of the methyl groups on the ring and of the double bonds in the side chain. All of them are very lipophilic.

The compounds are biosynthesized from homogentisate only in *plants*, mainly in *chloroplasts* of *leaves* (Fig. 4.7-3). They are mainly localized in the *cellular membranes* of *animals* and *plants* (e.g., of *mitochondria, erythrocytes* and *chloroplasts*).

The recommended intake for *human* adults is 8…12 mg α-tocopherol/day. It is higher when the food contains many polyunsaturated fatty acids. Avitaminoses in humans are rare, except in premature infants or during disturbances of lipid resorption. However, degeneration of various organs and infertility due to imbalanced food have been observed in animal breeding.

The biochemical function of tocopherols is mainly the protection of membrane lipids from peroxidation (in particular polyunsaturated fatty acids). Tocopherols are internal ethers of hydroquinones. They act as radical quenchers by transition to the tocopheroxy state (semiquinone, Fig. 9.12-1). This is an unreactive radical with

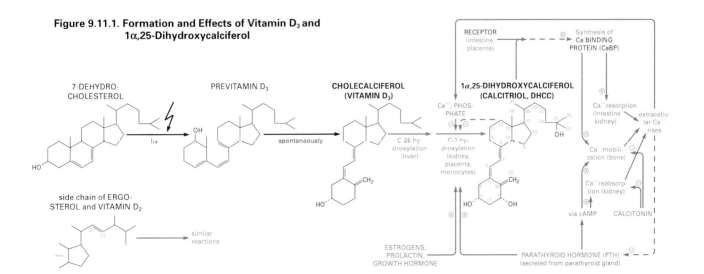

Figure 9.11.1. Formation and Effects of Vitamin D_3 and 1α,25-Dihydroxycalciferol

a half-life time of several hours, which interrupts chain reactions (4.5.8). Its formation is reversible. The reduction to tocopherol is most likely performed by ascorbate, which is oxidized in this way to monodehydroascorbate at the lipid-water interface (9.10.2). There is a synergism between both vitamins. Also β-carotene was proposed as a reductant. The further oxidation of the semiquinone to the quinone state involves a ring opening and is irreversible.

Literature:
Rice-Evans, C.A., Burdon, R.H. (Eds.): *Free Radical Damage and its Control. New Comprehensive Biochemistry* Vol 28. Elsevier (1994).
Friedrich, W.: *Vitamins*. De Gruyter (1988) 143–216.

Figure 9.12-1. Oxidation States of Tocopherol and its Function as a Radical Quencher

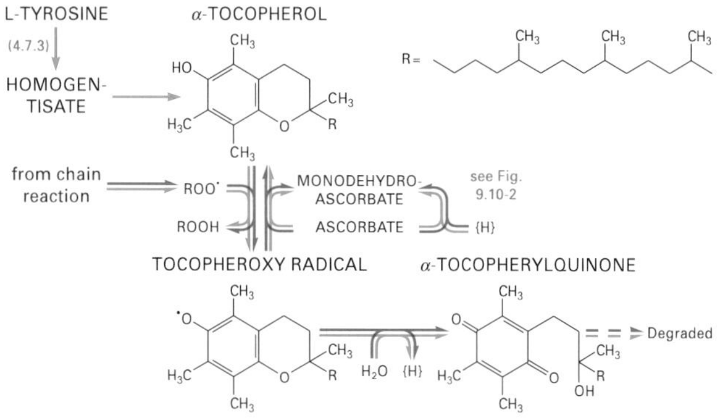

9.13 Phylloquinone and Menaquinone (Vitamin K)

Phylloquinone is a member of the photosystems in photosynthesizing *plants* (16.2.1). Menaquinone plays a role in anaerobic respiration of *bacteria* (16.1.3). In *animals*, both compounds (vitamins K_1 and K_2, respectively) act as cofactors for the γ-carboxylation of glutamate. They are monomethyl naphtoquinones with an isoprenoid side chain.

Phylloquinone and menaquinone differ only in the length and the number of double bonds in the side chain.

For the biosynthesis of menaquinone and phylloquinone see 4.7.2 and Fig. 4.7-1.

The recommended daily intake for human adults is 70...140 μg/day. It is usually met by food and by its production by intestinal bacteria. Degradation before excretion is usually limited to oxidative shortening of the side chain.

The biochemical function of these compounds in *plants* and *bacteria* are described in chapter 16. In *animals*, they are cofactors for the formation of γ-carboxyglutamate in proteins (Gla, Fig. 9.13-1) by a carboxylase activity of the *rough endoplasmic reticulum*. The proteins are secreted thereafter. This posttranslational modification is essential for Ca^{++} binding and for attachment of these proteins to membrane phospholipids via Ca^{++} bridges (e.g., during blood coagulation).

Vitamin K hydroquinone abstracts a proton from glutamate, which is carboxylated afterwards. Simultaneously, a K-quinone 2,3-epoxide is formed (details are still unknown). This epoxide is reduced by a dithiol-linked vitamin K epoxide reductase to the quinone and further on to the quinol hydroquinone level. The reductase step can be inhibited by coumarin derivatives, e.g. dicoumarol and warfarin, resulting in an anticoagulant action.

Gla-containing proteins are coagulation factors II, VII, IX, X, proteins C and S (20.3.1, synthesized in the *liver*), osteocalcin (regulates calcification of *bones*), ovicalcin (in *eggs*) and several others.

Literature:
Friedrich, W.: *Vitamins*. De Gruyter (1988) 143–216.
Suttie, J.W.: *Ann. Rev. of Biochem.* 54 (1985) 459–477.

9.14 Other Compounds

9.14.1 Lipoate as a Cofactor

Protein-bound lipoate is a cofactor of several dehydrogenase complexes, e.g. pyruvate dehydrogenase (3.3.1), 2-oxoglutarate dehydrogenase (3.8.1) and the glycine cleavage system (4.4.2).

The biosynthesis takes place by introduction of sulfur from cysteine into the hydrocarbon chain of octanoic acid (an intermediate of fatty acid biosynthesis, 6.1.1) at the positions C-6 and C-8. The reaction likely proceeds via a radical mechanism, but this has not been fully elucidated yet. Binding of ATP-activated lipoate to the ε-amino group of lysine in the respective enzymes proceeds in a way analogous to biotin (9.8). While most living beings can synthesize lipoate, it is required by some *bacteria* as a growth factor.

Biochemical function: The lipoate moiety accepts ligands at the oxo level and oxidizes them to the carboxylic function with concomitant formation of a thioester. Then, acting as a mobile arm (length 1.4 nm), it moves the acyl group to the next enzyme function (compare pantothenate, 9.7). After release of the ligand, lipoate disulfide is reconstituted (Fig. 9.14-1).

Figure 9.14-1. Structure and Biochemical Function of Lipoate (Examples: Pyruvate and 2-Oxoglutarate Dehydrogenases, Compare Fig. 3.3-4)

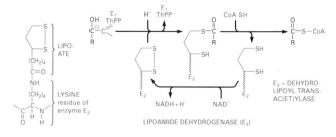

Literature:
Berg, A., de Kok, A.. *Biol. Chem.* 378 (1997) 617–634.

9.14.2 Essential Fatty Acids ('Vitamin F')

Mammals are unable to introduce double bonds into fatty acids beyond the Δ^9 position and have to obtain polyunsaturated fatty acids by food intake. Most important are linoleic and linolenic acids. For details see 6.1.1.

9.14.3 Essential Amino Acids

Mammals cannot synthesize 9 of the amino acids and are dependent on their supply in food. For details see 4.1.

Figure 9.13-1. Vitamin K Dependent Carboxylation Reactions

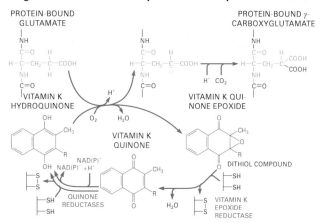

10 Nucleic Acid Metabolism and Protein Synthesis in Bacteria

10.1 Genetic Code and Information Transfer

This section deals with the general principles of protein synthesis in all kingdoms of biology (*archaea, bacteria, eukarya*).

10.1.1 From DNA to RNA

Deoxyribonucleic acid stores the genetic information encoded into a sequence of deoxyribonucleotides (A,C,G,T). Both strands of the double stranded DNA are linked by specific base pairing according to the Watson-Crick rules (A=T and G≡C, 2.6.1). Thus, each strand contains the complete genetic information. Specific base pairing is not only a safety measure to enable repair of damages (10.3 and 11.2), but also the tool to transmit the information during synthesis of more DNA (replication, 10.2 and 11.1) or of ribonucleic acid (transcription, 10.4 and 11.3).

During the transcription process, selected sequences of the DNA are copied into messenger RNA (mRNA). Other DNA sequences are transcribed into RNA with other functions. One of the two DNA strands acts as a template, the other (not involved in this procedure) carries the same information sequence as the synthesized RNA. Thus, the latter one is called the 'coding', 'sense' or (+) strand, while the template DNA strand has the sequence of a complementary antisense (−). In RNA, thymidine nucleotides are replaced by uridine nucleotides.

10.1.2 From Nucleic Acids to Proteins – The Genetic Code

The information for each polypeptide stored in a DNA sequence is named a gene (one gene – one polypeptide hypothesis). After transcription of this DNA sequence into mRNA, the information has to be translated into an amino acid sequence during protein synthesis. A group of 3 consecutive nucleotides (codon or triplet) represents one amino acid. The compilation of these relationships is the genetic code (Figure 10.1-3).

There are 20 'classical' amino acids, for which codons exist. The 21st amino acid, selenocysteine (Sec) uses the codon UGA, which otherwise has different functions (10.6.5). The triplet codons follow each other in 5′→3′ direction without interruption.

Therefore, when starting from a different nucleotide, one reads completely different codons for different amino acids. This requires that the starting point must be clearly recognizable. On the other hand, the same nucleotide sequence can code for more than one polypeptide, e.g. in some *viruses* (frameshift, 12.4.1). The other DNA strand can also be used as a template, it is read in the opposite direction (e.g. in some *viruses*, too).

Figure 10.1-1. Example of Different Reading Frames

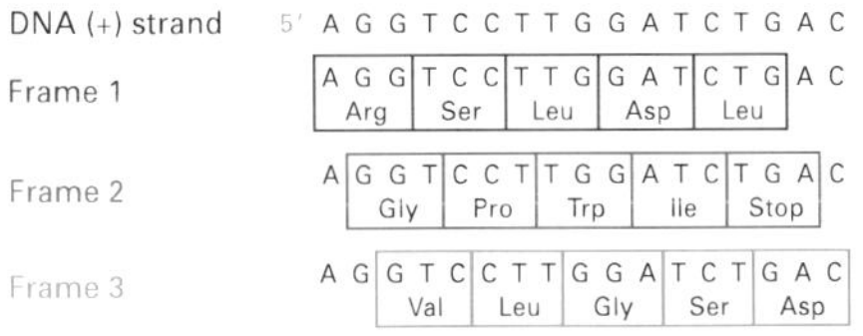

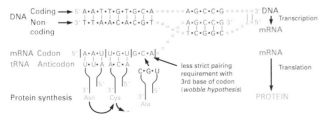

The translation of the nucleic acid information into the amino acid sequence of proteins needs a 'translator' or 'adaptor' system, which is competent in both 'languages'. For each amino acid there exist one or more tRNA molecules that can read the nucleic acid language via base pairing. Their specificity towards amino acids is effected by specific enzymes charging them with the cognate amino acid. Thus, the flow of information is as follows (only reverse transcriptase is an exception, 12.4):

Figure 10.1-2. Information Transfer During Protein Synthesis

Since there are $4^3 = 64$ possible nucleotide triplets, but only 21 amino acids (including Sec) plus some stop codons (see below), most amino acids are encoded by several nucleotide triplets (synonyms, at maximum 6 in the cases of Arg, Leu and Ser). Most variants occur in the 3rd position of the code ('wobble hypothesis'). The code is therefore called degenerate. The relationship between triplets and encoded amino acids is shown in Fig. 10.1-3.

Figure 10.1-3. The Genetic Code

The triplet sequence is read from the center outwards. The mRNA nucleotide terminology is shown. For DNA nucleotide sequences, replace U with T. The amino acids are given with their full names and with their three- and one-letter abbreviations. Basic amino acids are shown in blue, acidic amino acids in red, neutral amino acids in black and amino acids with uncharged, polar residues in orange. For details see 1.3.1.

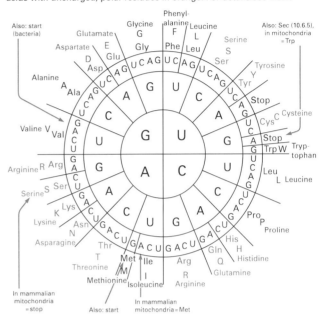

Although the code has been assumed to be universal among living beings, several exceptions in *mitochondria* have been found. Furthermore, in *mycoplasma*, UGA codes for Trp, while in some *ciliated protozoa*, the normal 'stop' codons UAG and UAA specify Gln and UGA specifies Cys instead.

Table 10.1-1 Codon Differences in *Mitochondria*

	UGA	AUA	CU(A,C,G,U)	AG(A,G)	CGG
General code	stop/Sec	Ile	Leu	Arg	Arg
Mitochondria					
Mammals	Trp	Met /start		stop	
Drosophila	Trp	Met/ start		Ser (AGA only)	
Protozoans	Trp				
Higher plants					Trp
S. cerevisiae	Trp	Met/ start	Thr		

Table 10.1-2 Codons with Special Functions

	Start codons	Stop codons
General code	AUG (codes also for Met)	UAG, UAA, UGA ('amber, ochre, opal')
Eukarya	AUG (codes also for Met), CUG (rare), ACG (rare), GUG (rare)	UAG, UAA, UGA
Bacteria	AUG, GUG, UUG (rare)	UAG, UAA, UGA
Mitochondria	AUA (codes also for Met), AUG	AGA, AGG (*mammals*)

10.1.3 Influence of Errors

Unrepaired changes in the DNA sequences (e.g. mutations) lead to variations in the expressed proteins, frequently with dramatic effects on the organisms and their progeny. Therefore, many repair systems exist to remove these damages (10.3, 11.2).

- Insertions or deletions of single nucleotides cause frameshifts (10.1.2). From this site on, the translated protein exhibits a different amino acid sequence. Also, stop codons might be either newly formed or removed.
- Modification of a single nucleotide influences only the particular triplet and thus may change a single amino acid. However, since many of the amino acids are encoded by several triplets (degenerate code, 10.1.2), such changes do not always take place.

Literature:

Khorana, H.G., in: *Nobel Lectures in Molecular Biology* 1933–1975, Elsevier (1977) 303–331.

Nirenberg, M., in: *Nobel Lectures in Molecular Biology* 1933–1975, Elsevier (1977) 335–360.

10.2 Bacterial DNA Replication

DNA, the carrier of genetic information, has to be replicated before cellular division takes place (except meiotic division in *higher eukarya*, 11.1). The semiconservative mechanism uses the parental DNA duplex strands as templates for replication; each parental strand forms a new duplex with a newly synthesized strand.

The mechanism of *bacterial* DNA replication described here applies to many bacterial chromosomes, episomes and plasmids. Unlike *eukaryotic* chromosomes (11.1.2), each of these circular, covalently closed, double-stranded molecules is replicated as a single unit (replicon). The same holds true for bacteriophage genomes of the circular duplex type (Table 12.1-1).

Replication of *bacterial chromosomes* starts from a single origin. Separation of both DNA strands yields two replication forks, which proceed bidirectionally around the molecule. Both strands are used as templates for synthesis of new DNA strands (Θ-mechanism, Fig. 10.2-1). In some *plasmids* and *bacteriophages*, replication is unidirectional: one of the circular DNA strands is opened and the other one is used as a template for replication, sometimes resulting in a series of consecutive strand copies (σ-mechanism, rolling circle mechanism). The newly synthesized strand then directs the synthesis of the complementary strand.

Figure 10.2-1. Mechanisms of *Bacterial* Replication

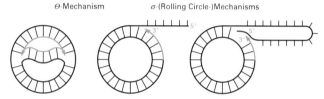

Table 10.2-1. Bacterial and Viral Replicons

Replicon	Size (base pairs)	Mechanism
Calothrix chromosome	$1.6 * 10^7$	Θ
Escherichia coli chromosome	$4.7 * 10^6$	Θ
Mycoplasma chromosome	$5.8 * 10^5$	Θ
Bacteriophage T4 (linear chromosome)	$1.7 * 10^5$	Θ
E. coli F-Episome	$1 * 10^5$	σ
Bacteriophage λ	$4.8 * 10^4$	σ
ColE1 plasmid	$6.6 * 10^3$	σ
Bacteriophage ΦX174	$5.4 * 10^3$	σ

Replication proceeds via a polymerization reaction in 5'→3' direction (Fig. 10.2-2):

$(dNMP)_n + dNTP \rightarrow (dNMP)_{n+1} + PP_i$; $\Delta G'_0 = +2$ kJ/mol,

which is driven by the subsequent hydrolysis of the liberated PP_i:

$PP_i + H_2O \rightarrow 2 P_i$; $\Delta G'_0 = -30$ kJ/mol.

The correct nucleotide (dATP, dCTP, dGTP, dTTP) is selected by its ability to form a Watson-Crick base pair (2.6.1) with the corresponding nucleotide at the template strand.

The reaction is a nucleophilic attack of the terminal 3' hydroxyl in the growing DNA chain on the α-phosphate of the incoming dNTP.

Figure 10.2-2. Reaction Mechanism of the DNA Replication

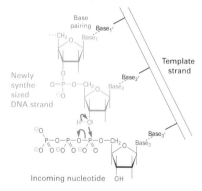

The replication speed in *bacteria* is about 500...1000 nucleotides/sec (vs. 30...50 nucleotides/sec in *eukarya*). The *Escherichia coli* chromosome is copied in about 40 minutes. Cells with doubling times < 40 minutes (e.g. *E. coli* 20 minutes) start another initiation round before the previous replication is completed. Thus, they are partially polyploid (having more than one DNA copy per cell).

10.2.1 Cell Cycle and Replication

Chromosomal DNA replication is linked to the cell cycle by a complex regulation system. Initiation, being the critical step, is tightly controlled. During the replication reaction, the *oriC* region (10.2.2) of the newly formed strand remains unmethylated and the duplex DNA binds to the cell envelope. Thus, premature DnaA binding is prevented (which would otherwise start another replication round). Only later in replication, the newly synthesized strand becomes also methylated. There are also some indications that the amount of DnaA protein influences initiation.

10.2.2 Initiation of Replication (Fig. 10.2-6, Table 10.2-2)

In *E. coli*, the single chromosomal origin of replication (*oriC*, length 245 bp, Fig. 10.2-3) contains a sequence of 4 identical binding sites (DnaA boxes, length 9 bp) for the DnaA initiator protein and an adjacent melting site (length 13 bp, repeated 3 times). The *oriC* region also contains no less than eleven 5'-GATC-3' boxes, which are target sequences for Dam methylase. Their methylation is involved in replication initiation (10.2.1).

Figure 10.2-3. Structure of *oriC* in *Escherichia coli*

For initiation, 10...30 molecules of DnaA bind to the DnaA boxes with simultaneous ATP hydrolysis (pre-initiation complex). The DNA is bent to a loop around the DnaA core. This is effected by binding of IHF protein (1 binding site), and/or FIS protein (4 binding sites) and by unspecific binding of HU protein. It is assisted by simultaneous transcription of genes in the *oriC* region. As a result, strand separation (melting) of the double helix occurs at the adjacent melting site, forming an open complex.

Hexameric helicase (DnaB protein) is complexed by accessory DnaC protein with simultaneous ATP hydrolysis and is loaded onto the single-stranded region of the DNA (prepriming complex). The rest of this region is covered by tetrameric single-stranded binding protein, SSB. Then, the primase enzyme (DnaG) joins to form the primosome (at some origins, this is assisted by Pri proteins).

The DNA directed DNA Polymerase III (Pol III) holoenzyme is assembled from 10 subunits in a defined order (Fig. 10.2-4). Assisted by the γ subunit complex, the dimeric β processivity factor (subunit β) forms a sliding ring around the DNA. Pol III* is finally loaded onto this ring, which clamps to the template strand. The addition of DNA gyrase completes the formation of the replisome.

Figure 10.2-4. Assembly of the DNA Polymerase Holoenzyme

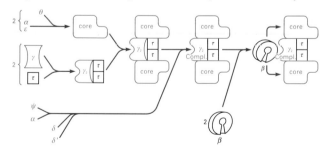

10.2.3 Elongation and Termination (Fig. 10.2-6, Table 10.2-2)

Gyrase (ATP-dependent topoisomerase type II, breaking and religating both strands) travels along the strands ahead of the helicase and introduces negative supercoils into the DNA (underwound state of the double helix). This assists the helicase (DnaB) in melting the parental helix by separating the strands. The energy is derived from simultaneous ATP hydrolysis. Reannealing of the strands is prevented by the SSB protein cover. DNA primase (DnaG), activated by helicase, then synthesizes RNA primers (length 10...12 nucleotides) as starting sequences for the action of Pol III. The γ-complex of Pol III* (see footnote to Tab. 10.2-2) recognizes the RNA primers and loads a sliding ring of dimeric β subunits onto the primer-template junction (ATP dependent). Then Pol III* moves to this β ring and starts DNA replication (Pol III can, however, carry out the primase reaction itself). Wrongly inserted nucleotides are removed by the 3′→5′ exonuclease function of Pol III.

Pol III (like all DNA polymerases) polymerizes dNTPs only in 5′→3′-direction, therefore only one template strand can be copied continuously this way (leading strand). The other template strand has a reverse orientation and can be replicated only discontinuously in 5′→3′-direction by ligating 1...2 kb long Okazaki fragments (lagging strand, Fig. 10.2-5). Thus multiple priming reactions are required.

A β-ring becomes attached to each new primer as described above. Upon completion of the previous Okazaki fragment, Pol III* moves to this β ring and synthesizes the next fragment until it reaches the primer of the previous one. Then it dissociates from DNA, leaving the β-ring behind and starts a new synthesis cycle. (The β-ring is later removed by Pol III, too).

Due to this mechanism, E. coli is able to perform synthesis of both strands with only 10...20 molecules Pol III* per cell. The coordination of the discontinuous lagging strand synthesis with the continuous leading strand synthesis is effected by the dimeric structure of Pol III (one enzyme for each strand) and by primase activation due to its association with helicase.

The RNA primer is then degraded by DNA directed DNA polymerase I (Pol I) and/or ribonuclease H (RNase H). In both cases, the degradation proceeds in the 5′→3′ direction. Pol I then fills the gap with deoxynucleotides. Finally, the DNA chains are joined by DNA ligase.

Termination: Replication terminates in a (non-essential) 0.5 kb termination region opposite of *oriC* at the chromosome. At 6 termination (Ter) sites, 6 Tus proteins are bound. Since the Ter sites are non-palindromic (they do not read identically in both directions), asymmetric complexes are formed, which block only one of the approaching replication forks by inhibiting strand separation by helicase (DnaB). This enables termination of replication at a definite site inspite of possible speed differences by both forks.

The parental strands are then unlinked by topoisomerases (decatenation, 11.1.4) and the daughter chromosomes are separated by binding of MukB and Par proteins (partition).

10.2.4 Fidelity of Replication

The overall error rate is only about 1 per $10^9...10^{10}$ nucleotides. This high replication accuracy is due to
- Twofold dNTP base selection mechanism by DNA Pol III at the binding and the catalytic step (error rate about 1 per 10^4 nucleotides)
- 3′→5′-exonuclease activities of Pol III and Pol I removing misincorporated nucleotides (200...1000 fold fidelity improvement)
- postreplicative mismatch repair (see 10.3)

Literature
Baker, T.A., Wickner, S.H.: Ann. Rev. of Genetics 26 (1992) 447–477.
Kelman, Z., O'Donnell, M.: Ann. Rev. of Biochem. 64 (1995) 171–200.
Johnson, K.A.: Ann. Rev. of Biochem. 62 (1993) 685–713.
Kornberg, A., Baker, T.A.: *DNA Replication*. Freeman & Co. (1992).
Marians, K.J.: Ann. Rev. of Biochem. 61 (1992) 673–719.

Figure 10.2-5. Synthesis of Okazaki Fragments (Lagging Strand, schematically)

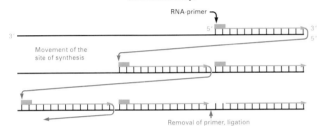

Table 10.2-2. Proteins Involved in *Bacterial* DNA Replication (*Escherichia coli*)

Type	Protein	No. subunits	Mol. mass (kDa)	Function
Initiator protein	DnaA	1	52	binds cooperatively to oriC; promotes double helix opening and DnaB loading
Integration host factor	IHF (heterodimeric)	2 (α, β)	11, 11	binds sequence-specifically to double stranded DNA, effects DNA bending
	FIS (homodimeric)	2	11 (*2)	binds sequence-specifically to double stranded DNA, effects DNA bending
Histone-like protein	HU	1	19	binds sequence-unspecifically to double stranded DNA, effects DNA bending
Helicase	DnaB	6	50 (*6)	unwinds parental strands; activates primase on single stranded DNA
Accessory protein	DnaC	6	29 (*6)	complexes DnaB; delivers DnaB to DNA
Primase	DnaG	1	60	DNA-dependent RNA-polymerase; synthesizes RNA primers
Single strand binding protein	SSB	4	19 (*4)	binds sequence-unspecifically, cooperatively to single stranded DNA, stimulates DNA-Pol III; facilitates DnaB loading
RNA-Polymerase	$\alpha_2\beta\beta' + \sigma$	4 (5)	40 (*2), 155, 160 + 70 (σ)	transcriptional activation of initiation; double helix destabilization
Type II topoisomerase	gyrase	2	97 (*2), 90 (*2)	ATP dependent topoisomerase type II, introduces negative supercoils into parental double helix ahead of the replication fork, removes positive supercoils. Inhibited by nalidixic acid and novobiocin.
DNA Polymerase III holoenzyme (core and τ act as dimers in replication) *	subunit α ⎫	1	130	catalyzes the main reaction of DNA replication. Long strand processivity.
	subunit ε ⎬ core	1	27.5	3′→5′-exonuclease activity provides proofreading of newly synthesized DNA.
	subunit θ ⎭	1	8.6	unknown, binds ε
	subunit τ	1	71	core dimerization. Gene *dnaX* encodes also subunit γ by translational frameshift
	γ-complex: γ (dimeric), δ, δ′, χ, ψ	4	47.5 (*2), 39, 37, 17, 15	loads β-subunit to a primed template ('matchmaker', ATP-dependent), recognizes RNA primers at the lagging strand.
	subunit β	2	40.6 (*2)	processivity factor, 'sliding ring', clamps Pol III to primed template, dissociates easily from rest of Pol III holoenzyme (→Pol III*).
DNA Polymerase I **		1	103	Trifunctional enzyme: removal of RNA primers from Okazaki fragments by its 5′→3′ exonuclease function, consecutive gap filling by polymerase action and proofreading by its 3′→5′ exonuclease activity. The enzyme also plays a major role in DNA repair (10.3). Protease splitting yields the 'Klenow fragment' (67 kDa, amino acid residues 324...928), which contains the polymerase and the 3′→5′ exonuclease function. It is widely used in research.
Ribonuclease	RNase H			removes RNA primer from Okazaki fragments
DNA Ligase			74	NAD dependent joining of Okazaki fragments
Terminator protein	Tus		36	inhibits helicase, replication fork arrest
Type II topoisomerase	topoisomerase IV	2	81, 67	decatenation/catenation, relaxation of supercoiled DNA
?	MukB (homodimeric)	2	177 (*2)	chromosome partitioning

* Terminology of DNA polymerase III: (subunits α + ε + θ) = core; [(core)$_2$ + (subunit τ)$_2$] = Pol III′; (Pol III′ + γ-complex) = Pol III*; (Pol III* + subunit β) = Pol III holoenzyme (HE).
** In *E. coli*, also DNA polymerase II has been found. This enzyme is likely to be involved in SOS DNA repair (10.3).

Figure 10.2-6. Bacterial DNA Replication
The newly synthesized DNA is drawn in green.

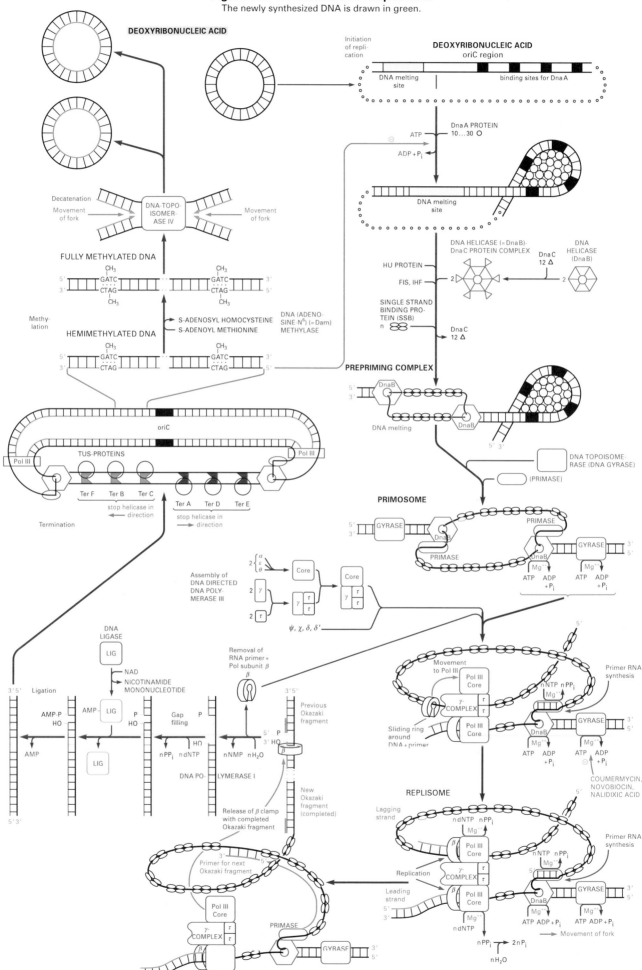

10.3 Bacterial DNA Repair

10.3.1 DNA Damage
The huge DNA molecule can suffer damage in many ways by exogeneous or endogeneous agents. In many cases the bases of DNA are affected. This may cause mutations or interfere with replication and transcription. The most important damages are:
- Alkylation: Alkylating compounds can modify nucleotides non-enzymatically. They can be of endogeneous (e.g., S-adenosyl-methionine) or exogeneous origin (e.g. N-methyl-N-nitrosurea). Products are, e.g., O^6-methylguanine (pairs with thymine!), 3-methyladenine or 2-methylcytosine.

O^6-Methylguanine   3-Methyladenine   2-Methylcytosine

The physiological enzymatic methylation of DNA strands, however, yields N^6-methyladenine, e.g. within the sequence 5' GATC 3' (in *bacteria*, 10.2) or 5-methylcytosine (in *eukarya*, 11.1), which are not subject to repair mechanisms.

- Pyrimidine dimerization: UV irradiation ($\lambda = 200...300$ nm) promotes covalent linking of pyrimidines, mostly thymine) via a cyclobutane ring. This distorts the DNA structure.

Cyclobutylthymidine dimer

- Spontaneous reactions: The most frequent spontaneous damage to DNA is hydrolytic loss of purine bases, leaving apurinic sites. Another hydrolytic reaction is the deamination of cytosine to uracil (U would be read as T in the next semiconservative replication!). Deaminations of adenine to hypoxanthine and guanine to xanthine occur more rarely. Also demethylation takes place (thymine to uracil).
- Oxidative damage: Radicals, especially of oxygen (caused e.g. by ionizing radiation) can induce unusual modifications of bases:

5-Hydroxycytosine   Thymine glycol   2,6-Diamino-4-hydroxy-5-formamido-pyrimidine

- Bulky adducts: Compounds which form adducts with bases (e.g. cis-platin•1,2-GpG, benzo[a]pyrene•guanine or psoralene•thymine) cause major distortions of the DNA helix.
- Double strand breaks: Free radicals induced by ionizing radiation or other agents can break both strands of a DNA molecule. This is highly hazardous to the cell, since without repair no replication of the DNA molecule can take place.
- Replication errors: In spite of proofreading by DNA polymerase (10.2), occasional errors take place during DNA replication and have to be eliminated by repair systems. Otherwise the fidelity would drop 100...1000 fold.
- In heavily damaged cells, the SOS-response (10.3.6) tolerates a high level of replication errors in order to enable replication at all.

DNA repair systems: DNA repair fulfills the vital task of maintaining the integrity of DNA structure and sequence. In *bacteria*, several systems for DNA repair exist, which can be redundant with respect to the lesion as well as to the implicated proteins. Generally, repair proceeds by one of the following strategies:
- direct reversal of damage
- excision of damage and resynthesis according to the information of the complementary strand (including repair of base pair mismatches)
- repair by transfer of sequences from a homologous strand (recombination)

The repair systems make use of the redundant information in the DNA duplex.

10.3.2 Direct Reversal of Damage (Figure 10.3-1)
Reversal of alkylation: Methylation of nucleotides is repaired by DNA repair methyltransferases (e.g. O^6-methylguanine-DNA-protein-cysteine S-methyltransferase). These proteins transfer the methyl group from DNA to a cysteine residue in their active site. Since this methyl group cannot be removed from the protein, the reaction leads to inactivation of the enzyme ('suicidal mechanism'). The inactivated enzyme, however, enhances transcription of new enzyme.

Photoreactivation: Cyclobutyl pyrimidine dimers are repaired in a photoreactivation reaction by deoxyribodipyrimidine photo-lyase, which catalyzes the cleavage of the cyclobutane ring in a light dependent reaction.

The enzyme contains 2 chromophores [FAD and in different species either 5,10-methenyltetrahydrofolate, λ_{max} ca. 380 nm (*E. coli, yeast*) or 8-hydroxy-5-deazariboflavin, λ_{max} ca. 440 nm], which absorb light and initiate dimer splitting by electron transfer to the pyrimidine dimer.

10.3.3 Excision Repair Systems
Base excision repair (Figure 10.3-1): Damage to single bases by oxidation, deamination, methylation or demethylation (dT→dU) or base-base mispairs are recognized by specific DNA glycosylases, which cleave the covalent sugar-base bond, leaving apurinic or apyrimidinic sites (AP sites). Examples are uracil, 3-methyladenine, hypoxanthine and formamidopyrimidine (= 8-hydroxyguanine) DNA glycosylases. AP sites may also have risen from spontaneous base losses. AP endonuclease cleaves the DNA backbone 5' of the AP site. Excision exonucleases (e.g. DNA deoxyribophosphodiesterase in *E. coli*) remove the damaged site. The gap is filled in and sealed by DNA polymerase (e.g. Pol I in *E. coli*) and DNA ligase.

The removal of the damaged site (and some additional nucleotides) may also be performed by the 5'→3' exonuclease activity of DNA polymerase I.

Alternatively to 'pure' DNA glycosylases, those with an associated DNA lyase activity carry out additional reactions. Since at the AP site the hemiacetal deoxyribose ring is at equilibrium with an open-chain aldehyde form, the latter can enter a β-elimination reaction leading to cleavage of the DNA chain 3' of the AP site. The biological importance of this pathway is uncertain.

Nucleotide excision repair (NER, Figure 10.3-1): Bulky adducts to DNA, as well as damages which cause minor distortions (e.g. pyrimidine dimers) can be removed by the ABC system, which responds to these distortions.

The *E. coli* enzyme activity 'ABC excinuclease' is carried out by the 3 proteins UvrA, UvrB and UvrC. Dimeric UvrA promotes complex formation between UvrB and the damaged DNA site and activates the ATPase/helicase activity of UvrB. ATP dependent unwinding and kinking of the DNA takes place. UvrC joins the UvrB/DNA complex. The 3' incision is apparently performed by UvrB, followed by the 5' incision by UvrC. The combined action of helicase II, UvrC and UvrB removes the damaged section (ca. 12 nt). DNA polymerase I (also II or III, albeit less effectively, needing accessory factors) fill the gap and ligase seals the nick.

The transcription-repair coupling factor (TRCF) connects NER with transcription. TRCF releases stalled RNA polymerase and the truncated RNA transcript from the damage site and recruits UvrA for consecutive repair as outlined before.

10.3.4 Mismatch repair
Very short patch repair (*E. coli*): This system corrects G•T mismatches, which occur by deamination of 5-methylcytosine in C(meC)(A/T)GG sequences of fully methylated DNA. The incision 5' of the mispaired T is catalyzed by Vsr protein, then DNA polymerase I removes the mispaired T and less than 10 more nucleotides by its 5'→3' exonuclease activity and fills the gap. The nick is then ligated. Mut S and Mut L proteins (see below) are also needed, possibly for damage recognition.

Long patch repair (Figure 10.3-2): Small insertion or deletion mispairs as well as base-base-mismatches are repaired by the methyl-directed pathway of *E. coli* (A related system is the Hex-dependent pathway of *Streptococcus pneumoniae*). This system takes advantage of the fact that for a short time after replication (10.2) the newly synthesized strand is not yet methylated by Dam-methylase, while the parental strand already carries methyl groups at adenine-N^6 in GATC (=Dam) sequences. This differentiates be-

10 Nucleic Acid Metabolism and Protein Synthesis in Bacteria

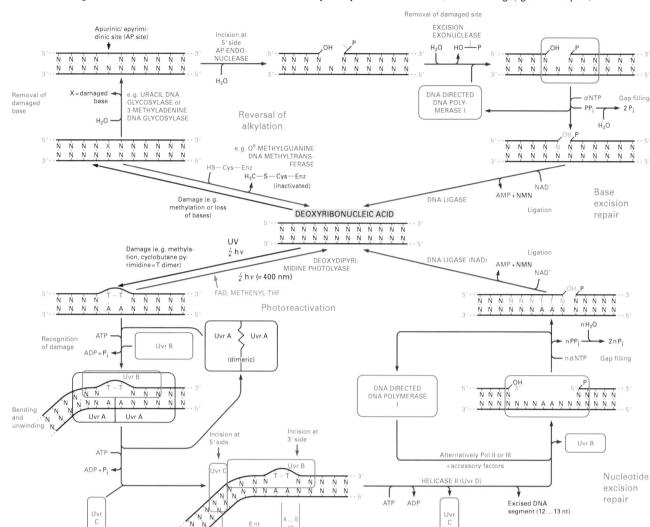

Figure 10.3-1. **Direct Reversal and Excision DNA Repair Systems in *Bacteria*** (red = damage, green = repair)

tween both DNA strands and insures that sequence information for repair is retrieved from the unmutated, parental strand.

According to the current model, MutS (95 kDa, oligomeric) recognizes the lesion, whereas MutH (25 kDa) binds to a hemimethylated GATC site up to 1000 bp away in either direction. Assisted by MutL (95 kDa), the proteins form a complex, which incises the non-methylated DNA strand 5′ from the GATC site. Subsequently, exonucleases degrade the faulty strand from here to a point beyond the site of the lesion. If the degradation occurs in 3′→5′ direction, DNA helicase II, SSB, exonuclease I, DNA polymerase III and DNA ligase take part in repair. Degradation in the other direction is performed by exonuclease VII or exonuclease RecJ.

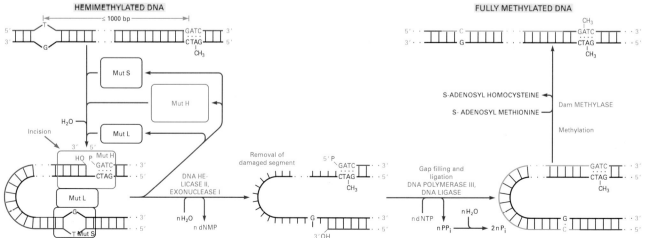

Figure 10.3-2. **Long patch mismatch repair** (Methyl Directed Pathway, *E. coli*)
(red = newly synthesized strand with mismatch, green = repaired segment)

10.3.5 Double-strand Repair and Recombination

A very difficult situation arises if double strand breaks occur. Similarly, if during replication a damaged, distorting template sequence is reached, the replication by DNA polymerase III is interrupted and reinitiated at some point beyond this site, leaving a gap in the daughter strand. In both cases the strand defects cannot be cured by excision repair, since no correct template sequence is available in the duplex DNA. However, by using the correct sequence from a homologous DNA to fill the gap, an intact DNA sequence is obtained, which can be used as a template for damage correction in the other strand (recombination or postreplication repair). This is possible because fast growing *bacterial* cells usually start another round of DNA replication before the previous one is completed and cell division has taken place (10.2), so that two or more copies of the same gene are available in one cell.

In case of double strand breaks, both free 5′ ends are degraded by exonuclease V (RecBCD) protein) until a 'chi-site' (sequence 5′GCTGGTGG3′, spaced at about 10 kb in *E. coli*) is reached, where RecBCD switches to a helicase activity. One single-stranded DNA 3′ end is helped by the nuclease function of RecA protein (38 kDa) to invade a homologous sequence (Figure 10.3-3). Expansion of the invaded stretch and gap filling lead to a double Holliday junction (crossover). Its subsequent resolution by RuvC protein brings about exchanges of the two homologous strands. Several variations of this reaction exist.

Similar mechanisms (see Figure 10.3-4) are used to fill the gap in the daughter strand opposite a defective site during replication. The defect can be repaired afterwards by the mechanisms described above. However, since mismatches are tolerated in the exchange reaction, recombinational repair is more error-prone than the other repair systems.

10.3.6 SOS Response (Damage Tolerance Mechanism, Figure 10.3-5)

Agents which cause intense damage to DNA induce a complex emergency repair system in *E. coli* and similarly in many other *bacteria*. The cells stop dividing and start a special DNA repair mechanism, sacrificing a high level of fidelity.

If the proceeding replication fork meets damaged DNA sequences, usually in- or post-replication repair mechanisms (as described before) are initiated. In case of frequent damages, however, DNA polymerase III action stops. Thus, single stranded DNA sequences result. RecA protein (see 10.3.5) binds to them and is converted into an activated form, which aids in autoproteolysis of the LexA repressor protein. Under physiological conditions, this repressor is bound to SOS boxes (consensus sequence CTGN$_{10}$CAG) at more than 20 genes and represses their expression considerably. Removal of this block leads to increased expression of (among others) UvrA, UvrB and UvrC (see above at NER), Lex A and RecA proteins as well as of helicase II, DNA polymerase III, UmuC and UmuD proteins. The latter two enable DNA polymerase III to resume replication and to proceed through sites of DNA damage despite missing or faulty information from the complementary strand in a way not known in detail yet. Although this results in a highly error-prone DNA replication, it enables replication at all.

Literature

Friedberg, E.C., Walker, G.C., Siede, W.: *DNA Repair and Mutagenesis*. ASM Press (1995).
Modrich, P., Lahue, R.: Ann. Rev. of Biochem. 65 (1996) 101–133.
Sancar, A.: Ann. Rev. of Biochem. 65 (1996) 43–81.
(various authors): Trends in Biochem. Sci. 20 (1995) 381–439.

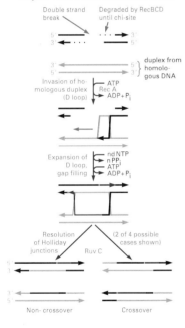

Figure 10.3-3. Recombination Repair of Double Strand Breaks

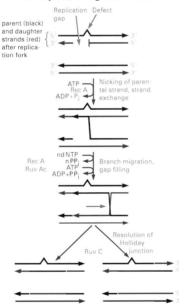

Figure 10.3-4. Recombination Repair of Gaps in Daughter Strands

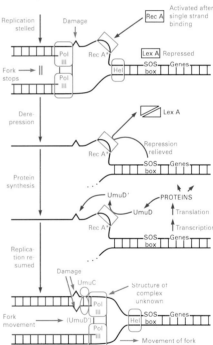

Figure 10.3-5. SOS Repair System

Table 10.3-1. DNA Repair in *Bacteria*

System	Damage Specifity	Implicated Repair Proteins *(E.coli)*	Remarks
Reversal of alkylation	O^6-methylguanine, O^4-methylthymidine	O^6-methylguanine DNA repair methyltransferase I (Ada, repairs also O^6-methylthymidine and methylphosphotriesters), O^6-methylguanine DNA repair methyltransferase II (Ogt)	"suicidal" mechanism (only one methyltransfer catalyzed by one protein molecule)
Photoreactivation	pyrimidine dimers	deoxyribodipyridine photo-lyase, carrying as chromophores methenyltetrahydrofolate + FAD	light dependent (λ = 300...500 nm)
Base excision repair	uracil, thymine glycols, hypoxanthine, 8-oxoguanine, 3-methyladenine	DNA glycosylases, apurinic/apyridinic endonuclease, excision exonuclease, DNA polymerase I, DNA ligase	length of repair tract $\leq$ 10 nucleotides
Nucleotide excision repair (Uvr)	bulky adducts, e.g cisplatin•GpG, pyrimidine dimers	UvrA, UvrB, UvrC, UvrD (Helicase II) proteins, DNA polymerase I, DNA ligase	length of repair tract 12–13 nucleotides
Very short patch mismatch repair	G•T mismatch (corrected to G•C) in 5′-CT(A/T)GG	Vsr, MutS, MutL proteins, DNA polymerase I	length of repair tract $\leq$ 10 nucleotides
Long patch mismatch repair (methyl-directed)	base-base-mismatch, small insertion/deletion mispairs	MutS, MutL, MutH proteins, DNA helicase II, single strand binding protein, exonuclease I (or exonuclease VII/exonuclease RecJ), DNA polymerase III holoenzyme, DNA ligase	length of repair tract ≈1000 nucleotides
Recombination repair	double strand breaks, distorting DNA sites	RecBCD, RecA, RuvC proteins	error-prone
SOS response	DNA polymerase stalled at lesion, DNA single strands remain	RecA, LexA, UmuC, UmuD proteins, DNA polymerase III	error-prone repair as part of the SOS regulon, lesion persists

10.4 Bacterial Transcription

Transcription is the copying of genetic information from DNA into single-stranded RNA, the base sequence of which is identical with that of one strand of the DNA duplex (except for U instead of T). This is effected by base selection according to the Watson-Crick pairing rules (2.6.1).

Figure 10.4-1. Principle of Transcription

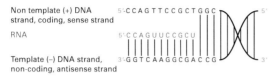

In *bacteria*, this leads to the synthesis of three classes of RNAs:
- messenger RNA (mRNA), coding for proteins
- combined ribosomal/transfer RNA (rRNA/tRNA) transcripts from which tRNAs and rRNAs are obtained by cleavage
- additional tRNA transcription units.

10.4.1 RNA Polymerase (Pol)

All classes of *bacterial* RNAs are synthesized by the same DNA-directed RNA-polymerase. 'Core' Pol consists of 4 subunits ($\alpha_2\beta\beta'$) and requires an additional σ- (sigma) factor for activity (Table 10.4-1). Unlike DNA polymerases (10.2.3, 11.1.3), no primer is needed for initiation.

Table 10.4-1. Components of the *Bacterial* RNA Polymerase Holoenzyme

Subunit	Copies in holoenzyme	Mol. mass kDa	Function
α	2	36.5	assembly of complex?
β	1	150	binds NTP, forms phosphodiester bonds
β'	1	155	binds to DNA, contains 2 Zn^{++}
σ	1	28...110 (mostly σ^{70})	recognizes the promoter and initiates synthesis (see Table 10.5-1)

10.4.2 Transcription (Figure 10.4-3)

The genetic unit of DNA involved in transcription is named operon and includes
- a promoter region, which initially binds Pol,
- an operator region, involved in regulation of transcription (optionally)
- the actual coding region (which frequently contains several contiguously arranged structural genes, 10.4.3), beginning at the start site (position +1)
- a termination sequence

The promoter region can be preceded by an upstream regulatory sequence, URS.

RNA-Pol starts transcription at the start site (initiation), proceeds through the coding region (elongation) and continues until a site for termination is reached. The regulation of *bacterial* transcription is discussed in detail in 10.5.

The mechanism of the actual bond forming reaction is analogous to that of DNA replication (Figure 10.2-2) except of involving ribonucleotides instead of deoxyribonucleotides.

Initiation: *Bacterial* promoters usually consist of 2 modules with the following consensus sequences (the subscripts indicate the frequency of occurrence):

–35 sequence (in highly effective promoters)	–10 sequence (Pribnow-box) (general)
$T_{0.69}T_{0.79}G_{0.61}A_{0.56}C_{0.54}A_{0.54}$	$T_{0.77}A_{0.76}T_{0.60}A_{0.61}A_{0.56}T_{0.82}$

The Pol holoenzyme binds weakly to the DNA and slides along this molecule until it recognizes the promoter region by its σ subunit. There it binds very tightly (K_M ca. 10^{-14} mol/l with effective promoters). Two turns of the DNA (from the –12 base pair to the +4 base pair) are unwound and both DNA strands are separated (initiation complex). The initial two nucleoside triphosphates are joined, the first being mostly ATP, less frequently GTP. Then the σ subunit dissociates from the complex.

Frequently, however, the σ subunit remains attached for several cycles, preventing Pol from leaving the promoter and entering the elongation mode (abortive initiation).

Besides the σ^{70} Pol subunit, which recognizes the above named promoters, there exist other proteins in *E. coli* with similar functions. E.g., σ^{32} recognizes heat shock promoters (10.5.1).

Elongation: RNA is synthesized at the template in the $5' \to 3'$ direction, while the transcription bubble moves along the DNA at a rate of 30...60 nucleotides/sec. 12 nucleotides of the newly formed RNA form a hybrid with the template DNA, before both strands become separated (elongation complex).

The movement of the transcription bubble is not uniform, since purified RNA Pol has been observed to pause *in vitro* for seconds or even minutes. Duration and frequency of these pauses can be decreased by binding of antitermination factors to RNAPol (e.g. NusA and G together with phage lambda N protein, 12.2.1). Transcription may also be resumed by action of GreA, which removes a short piece from the 3' end of the synthesized RNA.

Termination: There are two classes of termination sites:
ρ- (rho) independent ('intrinsic') DNA sites which do not require any additional factor. They consist of an palindromic, G•C-rich sequence followed downstream by an A (at the template strand)•T-rich region.

It is thought that the RNA-oligo U transcribed from the A rich sequence destabilizes the DNA/RNA hybrid. Simultaneously, the RNA forms a G•C stem-loop structure, which retracts RNA from the transcription bubble and causes the structure to disassemble.

No common DNA or RNA motifs were found for the much less abundant ρ-dependent termination sites, which require factor ρ for termination to occur.

The hexameric factor ρ (6 * 46 kDa, ATPase activity) presumably moves along nascent RNA in $5' \to 3'$ direction until it catches up with the RNA Pol stalled at a pausing site and releases the RNA from the enzyme.

10.4.3 Products of Transcription

RNA coding for proteins (mRNA): As mentioned before, *bacterial* genes are frequently organized in polycistronic operons (DNA sections coding for a series of proteins, which are most frequently involved in the same biochemical pathway). Each operon is transcribed as a whole, thus yielding a polycistronic mRNA chain. The consecutive translation of mRNA into proteins, however, occurs separately for each protein. Usually, no posttranslational processing of the mRNAs takes place in *bacteria*.

In a few cases, group II introns (11.3.3) have been found in *bacteria*. They have inherent ribozyme activity, which is involved in their splicing.

Untranslated RNAs (rRNAs and tRNAs, Figure 10.4-4): *E. coli* cells possess 7 separated operons containing the genes for these essential RNAs. The 30S primary transcripts are > 5500 nt long. They usually comprise one each of 16S rRNA, 23S rRNA and 5S rRNA sequences and also several (2...4) tRNA sequences, separated by spacers. The individual sequences are cut from the primary transcript already while transcription is still going on.

Ribonucleases (RNases) III, P, F cut at definite positions, producing pre-16S, 23S and 5S RNAs, as well as pre-tRNAs (Figure 10.4-2). Likely, stem-loops act as recognition sequences for RNase III. Both the 5' and the 3' ends of the pre-rRNAs are trimmed by action of the RNases M16, D, M23 and M5 to their final lengths. Then methylation takes place, yielding N^6, N^6-dimethyl adenine and other methylated bases.

The self assembly of the large and the small ribosome subunits from rRNAs and ribosomal proteins is a sequential process. Some proteins bind at distinct sites of the rRNAs, causing conformational changes and thus creating proper binding sites (scaffolds) for other ribosomal proteins.

Figure 10.4-2. Cleavage of the 30S Primary Transcript

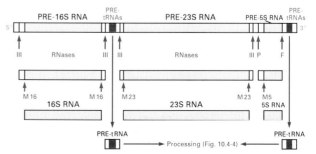

Figure 10.4-3. *Bacterial* Transcription

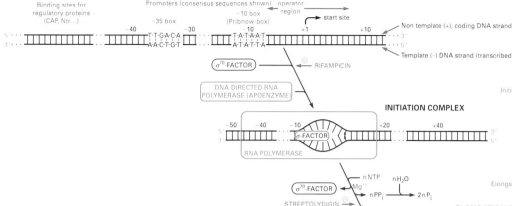

The pre-tRNA contains additional nucleotide sequences at both ends (Figure 10.4-4), which have to be removed.

The 19 nucleotide long 5' extension is trimmed by RNase P. This enzyme is noteworthy, since it contains a RNA component (125 kDa in *E. coli*, as compared to only 14 kDa protein content), which actually performs the cleavage. Thus, RNase P is a ribozyme (10.7.3). The 3' extension of tRNA is removed by RNases E or F and D. Finally, bases are modified, similar to *eukaryotic* tRNAs (11.3.8), but in smaller numbers.

Figure 10.4-5. Examples for Modification of tRNA

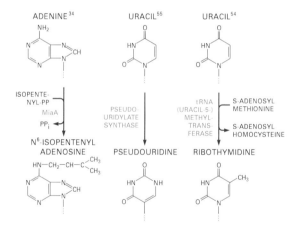

Additional tRNA genes occur in *bacterial* genomes, often in clusters. Their transcripts are processed the same way.

10.4.4 Accuracy of Transcription

The error rate of transcription is about 1 per 10^4 nucleotides, thus considerably higher than in DNA replication (10.2.4). This rate is a compromise between speed and accuracy and is apparently tolerable for the following reasons:
- The products of transcription are not transferred to the progeny
- Transcription of an encoding gene takes place repeatedly
- The genetic code contains synonyms for amino acids (10.1.2)
- Many amino acid substitutions do not affect protein activity

10.4.5 Inhibitors of Transcription

The antibiotics rifamycin B and rifampicin (semisynthetic) inactivate only *bacterial* Pol and block the initiation step, streptolydigin the elongation step. Actinomycin D binds to DNA and inhibits both *bacterial* and *eukaryotic* transcription.

Literature:
Björk, G.R., *et al.*: Ann. Rev. of Biochem. 56 (1987) 263–287.
Das, A.: Ann. Rev. of Biochem. 62 (1993) 893–930.
Eick, D., *et al.*: Trends in Genetics 10 (1994) 292–296.
Kornberg, A., Baker, T.A.: *DNA Replication*. Freeman & Co. (1992).
Neidhardt, F.C., *et al.* (ed.): *Escherichia coli and Salmonella typhimurium*, 2nd ed. (2 volumes) ASM Press (1996).

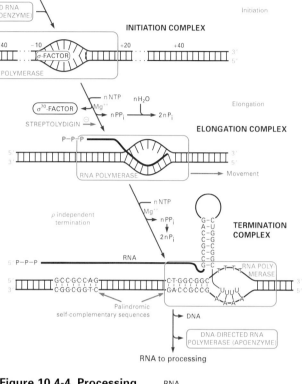

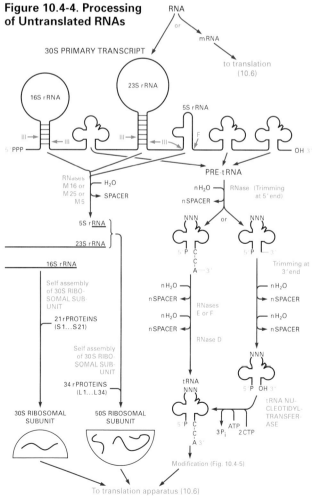

Figure 10.4-4. Processing of Untranslated RNAs

10.5 Regulation of Bacterial Transcription

In *bacteria*, translation of mRNA to protein already begins, before the transcription step is completed (contrary to *eukarya*, 11.5). Since both processes are very tightly coupled, most mechanisms for regulation of *bacterial* gene expression act at the transcription level: initiation, elongation and termination. Among them, initiation is most widely regulated.

10.5.1 Regulation at the Initiation Step

Regulation of promoter recognition: Core RNA-Pol (10.4.1) is not able to recognize promoters by itself. Binding of a σ-factor confers the ability of promoter recognition to the enzyme. The σ^{70} factor recognizes the consensus promoters for 'housekeeping' genes in *E. coli* (constitutively expressed genes, e.g. for ribosome components). In certain situations, cells produce alternative σ-factors, which recognize other promoter sequences. This leads to expression of a different pattern of proteins.

As an example, under heat shock conditions *E. coli* produces the σ^{32} factor. This factor assists Pol in initiating transcription of operons coding for other heat shock proteins, e.g. molecular chaperones (14.1.1).

In endospore forming *bacteria*, a whole cascade of σ-factors regulates the morphological and biochemical differentiation during sporulation. Different σ factors are also expressed during the infection of bacteria by phages (e.g., bacteriophage SP01).

Table 10.5-1. Examples of σ-Factors (*E. coli*)

Factor	Genes transcribed	Consensus Sequence of Promoters	
σ^{70}	most genes (–35/–10 box)	TTGACA	TATAAT
σ^{54}	nitrogen regulated genes (–24/–12 box)	CTGGCACN$_5$	TTGCA
σ^{32}	heat-shock regulated genes	CTTGAA	CCCATNTA
σ^{28} (σ^F)	flagellin and chemotaxis genes	TAAA	GCCGATAA

Regulation of promoter binding: Promoter binding by Pol can be influenced in two ways by regulatory proteins:
- Negative control: A repressor protein binds to or near a site overlapping the promoter area and prevents Pol binding
- Positive control: An activator protein binds near a promoter, from which otherwise transcription would not occur and assists in Pol binding or transcription initiation.

The DNA binding site for the repressor proteins is named operator region (Figure 10.5-1). The regulatory proteins detect and recognize it by sliding along the DNA strands. They show an extremely strong binding to this site (K_M typically 10^{-13} mol/l with high selectivity).

In more 'simple' systems, the activity of the regulatory protein is controlled by a compound of low molecular mass originating in or derived from the pathway that is regulated. Its binding to the regulatory protein either increases or decreases the affinity of binding of that protein to the operator.

Figure 10.5-1. Structure of the *Lac* Promoter/Operator Region (*E. coli*)

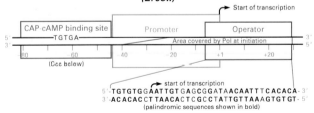

Examples for negative controls (Figure 10.5-2):
- Repression of the *lac* operon (coding for the enzymes of lactose metabolism): The tetrameric *lac* repressor binds to the *lac* operator in absence of lactose, preventing RNA-Pol from initiating transcription. If lactose becomes available, it is bound by the *lac* repressor which loses its DNA-binding capacity. Transcription of the *lac* operon starts. It is assumed that not lactose itself, but its intracellular transglycosylation product allolactose (Gal-1β-6-Glc) is the inducing agent. Additionally the expression of the *lac* operon is regulated by CAP protein (see below and 10.5.4).
- Repression of the *trp* operon (coding for enzymes of tryptophan biosynthesis): The mechanism is inverse. The free trp repressor has no affinity for DNA, leaving transcription unaffected. If tryptophan is present, it reacts (as a corepressor) with the repressor, which consecutively binds to the *trpO* operator region and stops initiation of transcription. Analogous reactions take place at the *trpP* and *aroH* operators, which regulate the expression of other genes of tryptophan biosynthesis. Additionally, tryptophan biosynthesis is regulated by attenuation (10.5.3).

Figure 10.5-2. Repression Mechanisms

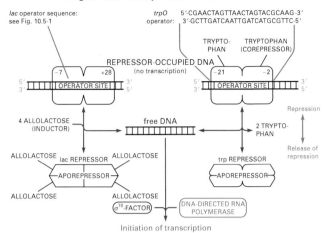

Example of a positive control (Figure 10.5-3):
- Activation of CAP (catabolite gene activator protein) controlled operons: If cyclic AMP, an intracellular signal molecule (17.4.2) is bound to CAP, this dimeric protein undergoes a conformation change, which allows it to bind to a site (TGTGA) 5' next to the (weak) promoter site. This increases transcription ca. 50-fold by facilitating the formation of a transcription complex by RNA-Pol.

In *bacteria*, during glucose starvation the cAMP level is increased. This causes production of enzymes for alternative catabolic pathways (10.5.4). The addition of the preferred nutrient glucose decreases the cAMP level again (catabolite repression, 15.3).

Figure 10.5-3. Activation by CAP Controlled Operons

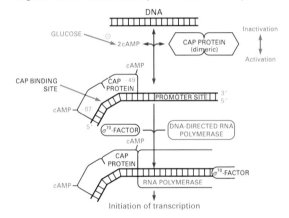

Stringent control: Starvation for an amino acid in *bacteria* causes a drastic decrease in transcription levels, caused by a signal from the translation step. In this stringent control mechanism, translating ribosomes whose A-sites are occupied by non-aminoacylated tRNAs bind the stringent factor enzyme, which then synthesizes the unusual nucleotide guanosine tetraphosphate (PP-5'-G-2'-PP). This compound, in turn, reduces transcription of rRNA and tRNA genes (and others) 10- to 20-fold by decreasing the affinity of Pol to the respective promoters via an unresolved mechanism. On the other hand, amino acid biosynthesis is stimulated.

Figure 10.5-4. Stringent Control

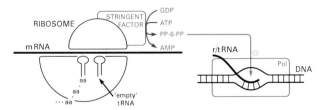

10.5.2 Regulation of Elongation

Only a few regulatory events during the elongation step are known so far. The overall RNA chain elongation rate, however, is proportional to the growth rate, ensuring that no excess RNA is produced when the translation system operates slowly under conditions of limited nutritional supply.

Figure 10.6-4. Bacterial Translation

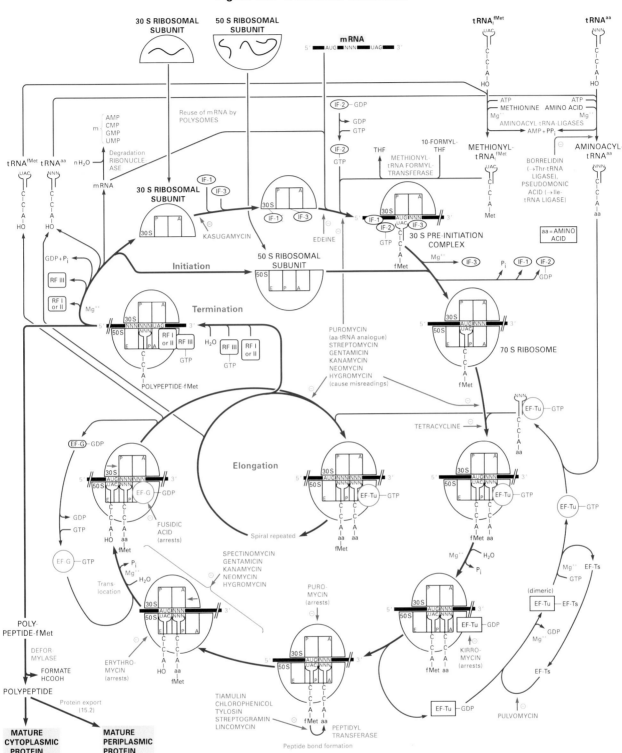

Ligases recognize the corresponding (= cognate) tRNA by various identity elements, which can be the anticodon, tertiary structural features, certain bases in the acceptor helix or specific modifications.

If a wrong tRNA is bound by a ligase, the aminoacylation is prevented or the already misacylated tRNA is immediately deacylated. E.g., tRNA^{Ile} is misacetylated with the structurally related valine instead of the cognate amino acid isoleucine only with a rate ratio of 1/50000. This proofreading is of high importance for translational fidelity. Only ligases for amino acids, which are easily discriminated (e.g. tyrosine) lack the proofreading ability.

10.6.3 Polypeptide Synthesis (Figure 10.6-4)

Bacterial translation starts soon after the Shine-Dalgarno sequence and the translation initiation codon (see below) have emerged from the RNA Pol during transcription.

Initiation: In *bacteria*, the first amino acid of a growing peptide chain is formyl-methionine (fMet). A specialized initiator tRNA (tRNA$_i^{fMet}$) is charged with methionine by the normal methionine-tRNA ligase. Then the methionine is formylated by 10-formyl-tetrahydrofolate in order to block the N terminus during peptide synthesis. The initiation complex is formed from this aminoacylated tRNA, mRNA, ribosome and initiation factors.

tRNA$_i^{fMet}$ is structurally different from tRNA^{Met}, therefore it cannot be bound by EF-Tu. In *E. coli*, 90% of all protein-encoding genes start with an AUG codon, 10% with GUG and 1% with UUG. These mRNA initiation codons are discriminated from 'regular' elongation codons (for Met, Val, Leu, respectively) by the preceding Shine-Dalgarno (SD) sequence GGAGG, which gets bound to the 3' end of the 16S rRNA. This sequence positions the initiation codon to the P-site of the ribosome. *Bacterial* mRNAs can have several translation initiation sites.

The mRNA and the complex fMet-tRNA$_i^{Met}$•IF-2•GTP bind to the 30S subunit. After positioning of the mRNA start codon and the tRNA anticodon in the peptidyl-(P) half site on the 30S subunit, the 50S subunit enters the complex. This promotes hydrolysis of GTP (bound to IF-2) and release of the initiation factors. This reaction is the rate limiting step in translation.

Elongation: All further aminoacyl-tRNAs are delivered to the ribosomal aminoacyl-(A) site by the elongation factor Tu (EF-Tu) in a ternary complex consisting of aa-tRNA•EF-Tu•GTP (exception: selenocysteine-tRNA^{Sec}, 10.6.4).

The ribosome induces the GTPase activity of EF-Tu. During this short time interval only strong binding (= correct) aa-tRNAs stay attached to the mRNA. This 'kinetic proofreading' increases transcription fidelity. After GTP hydrolysis, EF-Tu•GDP leaves the ribosome. 2 EF-Tu in complex with 2 EF-Ts promote the release of GDP. Subsequent rebinding of GTP induces a large conformational change in EF-Tu, which then again can bind aa-tRNA.

Peptide bond formation: Peptidyltransferase, an enzyme activity of the 23S rRNA, catalyzes the transfer of fMet (or of the peptidyl chain in the following cycles) from the tRNA at the P-site to the free amino group of the aminoacyl-tRNA at the A-site (Fig. 10.6-5).

The reaction is a nucleophilic displacement of the P site tRNA by the amino group of the aminoacyl-tRNA at the A site.

Fig. 10.6-5. Principle of Ribosomal Protein Synthesis

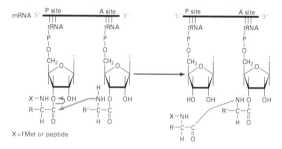

Translocation: The ribosomal movement to the next downstream codon of the mRNA is a two step process:
1) The acceptor ends of both tRNAs move to the next site at the 50S ribosome subunit (from A- to P-half sites and from P- to E-half sites, respectively), while their anticodons and the mRNA remain at their sites at the 30S subunit. Thus, the tRNAs are located in hybrid sites: A/P-site and P/E-site.
2) The tRNAs and the mRNA move one codon site relative to the 30S ribosome. The peptidyl-tRNA is now at the P/P site and the deacylated tRNA leaves the ribosome from the E-site. This step requires hydrolysis of GTP bound to EF-G; then EF-G dissociates from the complex. A new aa-tRNA can occupy the free A-site now, starting another elongation cycle.

It is assumed, that both steps involve conformation changes and movements of the ribosomal subunits relative to each other. Since EF-G•GTP has a similar tertiary structure as the aa-tRNA•EF-Tu•GTP complex, it may interact with the same ribosome site (molecular mimicry).

Termination: If a stop codon (UAA, UGA, UAG) arrives at the ribosomal A-site, protein synthesis terminates. Instead of another aa-tRNA, release factors RF-1 or RF-2 (depending on the codon) together with RF-3•GTP bind to the stop codon. After GTP hydrolysis, the polypeptide chain leaves the tRNA. Thereafter, EF-G and the ribosomal recycling factor (RRF) catalyze the liberation of mRNA and tRNA. Then the ribosomes separate into their subunits. The initiation factor IF-3 prevents their reassociation until it is released from the pre-initiation complex.

The formyl group (and in ca. 50% of all proteins also the N-terminal methionine) are enzymatically removed. Folding mechanisms, which produce the correct tertiary structure of the protein, are dealt with in 14.1.1.

Altogether, formation of a single peptide bond ($\Delta G_0' = 21$ kJ/mol) requires hydrolysis of 4 energy rich phosphate bonds ($\Delta G_0' = -124$ kJ/mol): 1 ATP → AMP for tRNA charging + 2 GTP → 2 GDP for delivering the aa-tRNA to the A-site and translocation.

Many inhibitors, most of them antibiotics, act at various steps of the protein synthesis (Figure 10.6-4).

10.6.4 Fidelity of Translation

The overall error rate of translation is ca. 1/10 000. Consequently, the probability for correct synthesis of a protein with 300 amino acids is 0.97.

Literature on Translation:
Agrawal, R.K., et al.: Science 271 (1996) 1000–1002.
Carter, C.W. jr.: Ann. Rev. of Biochem 62 (1993) 715–748.
Draper, D.E.: Ann. Rev. of Biochem 64 (1995) 593–620.
Moras, D.: Trends in Biochem. Sci. 17 (1992) 161–166.
Noller, H.: Ann. Rev. of Biochem 60 (1991) 191–227.
Weijland, A., Parmeggiani, A.: Trends in Biochem. Sci. 19 (1994) 188–192.

10.6.5 Selenocysteine (abbreviated Sec or U)

A number of *bacteria, archaea* and *eukarya* synthesize proteins which contain the amino acid selenocysteine. It is introduced into these proteins by an unusual decoding of mRNA. This process represents an extension of the genetic code. Selenocysteine is present in the catalytic center of some oxidoreductases and contributes to the catalytic mechanism by its high reactivity.

Table 10.6-3. Proteins Containing Selenocysteine (Selection)

Bacteria	Archaea	Eukarya
Seleno-P synthetase	Selenophosphate synthetase	Selenophosphate synthetase
Formate dehydrogenase	Formate dehydrogenase	Glutathione peroxidase
Hydrogenase	Hydrogenase	Thioredoxin reductase
Glycine reductase *	Heterodisulfide dehydrogenase	5'-Tetraiodothyronine deiodinase *
	Formyl-methanofurane dehydrogenase	Selenoproteins P and W

* no isoenzyme in which cysteine substitutes selenocysteine is known

In *bacteria*, selenocysteine is synthesized in a pyridoxal-P dependent reaction from serine bound to the special tRNA^{Sec} and selenophosphate (which contains the only selenium-phosphorus bond known in biochemistry) via an aminoacryl intermediate (Fig. 10.6-6).

During translation, selenocysteine is inserted into the polypeptide chain at an UGA codon, which in other context functions as a stop codon. In *bacteria*, a mRNA stem-loop structure adjacent to this codon determines UGA as selenocysteine codon. Instead of EF-Tu•GTP, a special elongation factor SelB•GTP (which additionally binds to the mRNA stem-loop) delivers selenocysteyl-tRNA^{Sec} to the ribosome (Fig. 10.6-5).

In *eukarya* and *archaea*, the corresponding special mRNA structure is located in the 3' untranslated region of the transcript.

Figure 10.6-6. Biosynthesis of Selenocysteine (Sec)

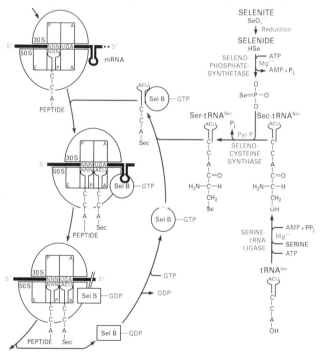

Literature on Selenocysteine:
Baron, C., Böck, A.: In Söll, D., RhayBandary, U. (eds.) *tRNA*. ASM Press (1995) 529–544.
Böck, A., Forchhammer, K., Heider, J., Baron, C.: Trends in Biochem. Sci. 16 (1991) 463–467.
Low, S.C., Bery, M.J.: Trends in Biochem. Sci. 21 (1996) 203–208.
Stadtman, T.C.: Ann. Rev. of Biochem. 65 (1996) 83–100.

10.7 Degradation of Nucleic Acids

Nucleases catalyze the cleavage of phosphodiester bonds in DNA (DNases) or in RNA (RNases). DNases play a decisive role during DNA synthesis, repair and recombination (10.2, 10.3, 11.2), for relief of obstructive superhelical tension as well as in apoptosis (11.6.6) and as a defense measure in restriction of foreign DNA (10.7.2). RNases are, e.g., of importance in RNA processing (10.4.3, 11.3.3) and also regulate the transcription by degradation of mRNA (11.5.5). Nucleases either remove terminal nucleotides (exonucleases) or act inside of the nucleic acid molecule (endonucleases).

10.7.1 Exodeoxyribonucleases (Exo-DNases, Table 10.7-1)

Exo-DNases are characterized by their cleavage direction, their preference for a single stranded or a double stranded substrate and by producing mono- or (more rarely) oligonucleotides. In a number of cases, they perform several functions (see below). The nucleases may dissociate after each catalytic event (non-processive or distributive action) or they may remain bound to the polymer until many reaction cycles are completed (processive or non-distributive action).

The reaction scheme of E. coli exonuclease III is schematically shown in Fig. 10.7-1. The enzyme is multifunctional: it acts also as an endonuclease specific for apurinic DNA sites (10.3.3) and as a 3′ phosphatase. Bacterial DNA polymerase I exerts a 3′ → 5′ exo-DNase activity for proofreading (similarly to eukaryotic DNA Pol δ and ε) and additionally a 5′ → 3′ exo-DNase/RNase function for DNA repair and for removal of Okazaki RNA primers (10.3.3; 10.2.3).

Figure 10.7-1. Reaction Scheme of E. coli Exonuclease III

10.7.2 Endodeoxyribonucleases (Endo-DNases, Table 10.7-2)

The endonucleases often show a strong preference for either single stranded or duplex DNA.

Endonucleases that function in repair of lesions identify the damaged DNA site and incise (nick) the DNA at one side of the lesion as a first step towards excision.

A second principal characteristic of these enzymes is the recognition of base sequences. A striking example is the cytosine-specific cleavage by T4 endonuclease IV. Pancreatic DNase, E. coli endonuclease I or spleen DNase produce oligonucleotide digests with characteristic sequence patterns at the 3′ and 5′ termini.

Restriction endonucleases occur in a variety of microorganisms. More than 2000 of them have been identified so far. They recognize sequences of 4...8 nucleotides in a DNA duplex with extraordinary accuracy and cleave both strands.

The organisms always produce a companion DNA methyltransferase, which recognizes the same sequence in endogeneous DNA and modifies it immediately after replication by methylation of A or C residues. This protects the organism's own DNA from degradation by the restriction enzyme. Thus, the restriction endonuclease and the cognate methyltransferase form a restriction-modification (R-M) system. Since it cleaves (restricts) infecting DNAs (e.g., viruses) and thus prevents them from parasitizing the cell, the R-M system is also called the 'immune system of the microbes'.

At least four different kinds of R-M systems exist:
- Type I enzymes carry methylase and nuclease activity on the same protein and require Mg^{++}, ATP and S-adenosylmethionine for cleavage. They cleave randomly and remote (> 400 bp) from the recognition sequence.
- Type II enzymes recognize mostly palindromic nucleic acid sequences and cleave within or near these sequences. (Palindromic sequences repeat each other at the other duplex strand in inversed order, resulting in twofold rotational symmetry. For examples see Fig. 10.7-2.) The enzymes require Mg^{++} for activity. Their homodimeric structure corresponds to the palindromic substrate. Many of them generate 'sticky end' ('cohesive end') duplex fragments with 5′ protruding termini (e.g. EcoRI) or with 3′ protruding tails (e.g. HhaI). Other enzymes cleave at the center of the recognition sequence and produce 'blunt end' fragments (e.g. HaeIII). The cognate methylase is a separate enzyme. The cleavage specificity is schematically shown in Fig. 10.7-2. These enzymes are indispensible tools for molecular cloning techniques and for DNA sequence analysis. They are named by the 3-letter abbreviation of the source organism.
- Type IIS enzymes recognize asymmetric sequences of 4...7 bp length. They cleave at a defined distance of up to 20 bp to one side of their recognition sequence.
- Type III enzymes have similar characteristics as Type I enzymes, but cleave at specific sites only a short distance (24...26 bp) away from the recognition sequence.

Figure 10.7-2. DNA Cleavage by Type II Restriction Endonucleases (◆ = twofold symmetry axis)

10.7.3 Ribonucleases (RNases, Tables 10.7-1 and 10.7-2)

Similarly to DNases, RNases differ by exo- and by endo-activity, preferences for termini (exo-enzymes) or in some cases for specific sequences (endo-enzymes). Some nucleases even cleave both DNA and RNA.

A number of RNase type reactions are not catalyzed by proteins, but rather by RNA sequences (ribozymes). E.g., the RNA component of RNase P catalyzes the processing of untranslated prokaryotic RNA (10.4.3, as well as some reactions in eukarya); the protein component has only assistant function. In a number of cases, eukaryotic group I or II introns are removed from the rRNA by its own action (self-splicing, e.g., in Tetrahymena, 11.3.3). It is speculated that RNA catalysis and self-replication preceded enzyme-protein catalysis during evolution.

Literature:

Lin, S.M., Roberts, R.J. (Ed.): Nucleases. CSH Monograph Series (1985).
Kornberg, A., Baker, T.A.: DNA Replication. 2nd Ed. Freeman (1992).
Roberts, R.J., Macelis, D.: Nucl. Acid Res. 24 (1996) 223–235.

Table 10.7-1. Examples for Different Types of Exonucleases

Specificity for	3′→5′ direction	Cleavage in 5′→3′ direction	either direction
Single stranded DNA	E. coli exonuclease I, mammalian DNase III	exo-DNase (phage SP_3 encoded)	E. coli exonuclease VII
Double stranded DNA	E. coli exonuclease III	exo-DNase (phage λ encoded), mammalian DNase IV	E. coli exonuclease V (ATP-dependent)
DNA or RNA	Venom exonuclease	Spleen exonuclease	
RNA	Exo-RNase II, RNaseQ, RNase BN	Yeast RNase	

Table 10.7-2. Examples for Different Types of Endonucleases

Specificity for	3′ P endproducts	Cleavage to 5′ P endproducts	other end products
Single stranded DNA	Aspergillus DNase K_1	DNase IV (phage-T_4-encoded), yeast DNase	Crossover junction endo-RNase (acts only on Holliday junctions)
Double stranded DNA	pancreatic DNase II	Type I, II and III restriction enzymes (10.7.2), pancreatic DNase I	
DNA or RNA	Micrococcal nuclease, spleen endonuclease	Aspergillus nuclease S_1, Mung bean nuclease, potato nuclease	
RNA-DNA duplex		Exo-RNase H (acts on RNA:DNA hybrids, degradation of Okazaki fragments)	
RNA	RNase T_2 = RNase II, pancreatic RNase Polynucleotide phosphorylase	RNase III and RNase P (processing of tRNA and rRNA precursors)	Bacillus subtilis RNase (yields 2′,3′-cyclic phosphates)

11 Nucleic Acid Metabolism, Protein Synthesis and Cell Cycle in Eukarya

11.1 Eukaryotic DNA Replication

DNA replication is a necessary prerequisite of cellular division (11.6), except the second meiotic division of germ cells.

Although in *eukarya* actually chromatin replicates (the complex of DNA + histones + other proteins, attached to the nuclear scaffold, 2.6.4), the following section concentrates on the replication of DNA. This proceeds in a series of integrated protein-protein and protein-DNA interactions and enzymatic reactions. Extreme accuracy of replication is required.

11.1.1 Cell Cycle and DNA Replication

DNA replication must be precisely linked to the cell cycle (summarized in Fig. 11.1-1). It depends on the passage through a special point in the G_1 phase, at which cells commit to another round of replication and cell division (see below). DNA is replicated exclusively in the S-phase. For the next round of replication, a passage through the M phase must have taken place.

Figure 11.1-1. Phases of the *Eukaryotic* Cell Cycle

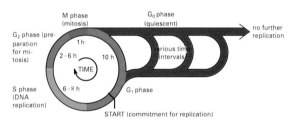

The mechanisms underlying these controls are only partially known. More details of the regulatory aspects are given in Section 11.6.
- The chromatin has to be in a replication competent state: A pre-replication complex of DNA and 'licensing factors' (11.1.2) must have been formed, which dissolves during the later phases of the cell cycle and thus prevents entering an untimely round of replication.
- Also, external signals (e.g. nutrient supply in *unicellular organisms* or hormonal messages in *multicellular organisms*) are required to initiate DNA replication and cellular division. In *yeast*, coordination of cell growth and DNA replication occurs at an unique point in the G_1 phase of the cell cycle, which is called START (Figure 11.1-2). This control step irreversibly commits the cell to another round of cell division, as most external cell cycle regulators have no effect on DNA replication beyond this point. Here, binding of cyclins (cell cycle regulating proteins, 11.6) to the cyclin dependent kinase Cdc 28 causes activation of transcription factors leading to expression of genes necessary for DNA replication as well as initiation of DNA replication. Similarly, cell cycles in *higher eukarya* depend on passage through the restriction or commitment point in G_1, which is the primary site of action for the major regulators of cell proliferation.

In many organisms active genes are replicated early in the S-phase, while inactive genes, telomeres, telomere-linked regions and heterochromatin are replicated later. The influence of various factors like histones, nucleosomes, the matrix and matrix associated enzymes (e.g. DNA topoisomerase II, acetyltransferase, deacetylase, poly-ADP-ribosyltransferase etc.) as well as the unfolding mechanism of the densely packed chromatin are only partially known and not presented here.

11.1.2 Initiation of Replication (Figure 11.1-2, Table 11.1-2)

Initiation of replication takes place at origins or initiation zones. These DNA regions are recognized by specific proteins or protein complexes which catalyze localized unwinding and load replication proteins onto the exposed single strands.

Unlike *bacterial* chromosomes, *eukaryotic* chromosomes contain multiple origins of DNA replication. On a typical *mammalian* chromosome these are spaced 50...100 kb apart. Not all replication origins are activated simultaneously. Some origins are used early, some late in S-phase and not all seem to be utilized in every cell or during every cell cycle.

Table 11.1-1. Replication of Genomes

Species	Size (bp, haploid)	Number of Chromosomes	Number of Replication Origins (total)	Duration of Genome Replication
Yeast	$1.4 * 10^7$	16	250...400	30 min
Human	$3 * 10^9$	$2 * 22 + 2$	20 000...50 000	8 h

In *yeast*, replication origins are named autonomously replicating sequences (ARS).

A stretch of at least 100...200 nucleotides is required for maximal function, arranged in domains A and B. Domain A contains an essential ARS consensus sequence, 5'-(A/T)TTTA(T/C)(A/G)TTT(A/T)-3'. Origins in *yeast* resemble promoters (11.2). Replication enhancers can function at distances up to 1 kb on either side of the start position and in both directions.

A complex of 6 essential proteins in *yeast* (origin recognition complex, ORC, homologues in other *eukarya*) binds in ATP-dependent manner to ARS and remains there throughout the cell cycle. The activating protein Cdc6 and a complex of Mcm proteins ('licensing factors') become available during mitosis (when the *nuclear membrane* dissolves) and are part of the pre-initiation complex at the end of M-phase, which is essential for initiation of replication. Thereafter, Cdc6 and the Mcm proteins leave the complex and are either degraded or sequestered for most of the cell cycle (possibly moving along with the replication fork). Complex formation and dissolving is controlled by the cyclin/cyclin dependent kinase complexes Clb1, 2, 3, 4-Cdc28 and Clb5, 6-Cdc28, respectively (11.6.2).

Identity and structure of replication origins in *higher eukarya* are probably similar. Initiation sites can be distributed over broad regions (initiation zones) of several kb length possibly due to more complex regulation requirements. The chromosomal structure seems to be a major influence.

Table 11.1-2. Essentials for Initiation of DNA Replication in *Yeast* (*Saccharomyces cerevisiae*)

Type	DNA Sequence/ Protein Factor	Mol. mass kDa	Function
Origin DNA element	ARS		Autonomously replicating sequence, core (A site) + 3' adjacent B sites
DNA element	Abf binding sequence		3' or 5' from core ARS, consensus sequence (A/G)TC(A/G)(T/C)NNNNNACG
DNA elements	MCB/SCB		Cell cycle boxes, located in the promoter of genes transcribed in late G_1 phase, bound by transcription factor Mbf and Sbf, respectively
Proteins:			
Origin recognition complex	ORC (subunits: Orc 1...6)	120, 68, 62, 56, 53, 50	Remains bound to ARS (core + 3' sequence) throughout the cell cycle
Activating proteins (in *nucleus* only in M + G_1 phases)	Cdc6 Cdc45	48	Activates ORC/ARS, required for binding of Mcm proteins and recruitment of Cdc45 Activates ORC/ARS, required for recruitment of Mcm-complex
Mcm proteins (in *nucleus* only in G_1 and S phases)	Mcm2 Mcm3	102 107	Activates ORC/ARS, possibly 'licensing factor' Binds 3' to ARS, activates ORC/ARS, opens duplex (helicase?). Analogy in *humans*: P1 = DNA Pol-primase accessory factor, targets Pol α?
also:	Mcm4 (Cdc54), Mcm5 (Cdc46), Mcm7 (Cdc47)		
Cyclins	G1 type: Cln1, 2, 3 B type: Clb5, 6 Clb1, 2, 3, 4	50, 44	G_1 phase cyclins, associate at START with kinase Cdc28, activate Mbf/Scf Late G_1 and S phase cyclins, associate with kinase Cdc28, inhibited by Sic1 G_2 and M-phase cyclins
Protein kinases	Cdc28 (cyclin-dependent, present throughout cell cycle)		a) Activation at START after association with G_1 cyclins Cln1, 2, 3, budding b) Association with B cyclins Clb5, 6: phosphorylates Cdc7, Swi6, Cdc10, RF-A Generally: causes DNA synthesis, spindle formation
	Cdc7	58	Expression depends on Mbf1 bound to MCB sequence, becomes activated by Dbf 4, phosphorylates Orc or other proteins in late G_1 phase
Activating protein	Dbf4	80	Targets Cdc 7 protein kinase to DNA
Inhibitor	Sic1 (p40)	32	Inhibits Cdc28–Clb5, 6 complex, degraded after phosphorylation by Cdc28/Cln
Ubiquitin-protein ligase	Cdc34	34	Ligates Sic1-P with Ub for degradation, thus activates Cdc 28/Clb5,6-complex
Transcription factors	Abf1	81	ARS binding factor, enhancer, required for origin function
	Mbf (subunits: Mbp1, Swi6) and Sbf	94, 91	Bind after phosphorylation to MCB and SCB DNA sequences, respectively, start transcription of ribonucleotide reductase, thymidylate synthase, Pol α-primase, Clb5, 6, Cdc6, 7, Cdc46, Dbf4, Orc6 etc.
Polymerase α/primase and repliction factor RF-A		see 11.3	

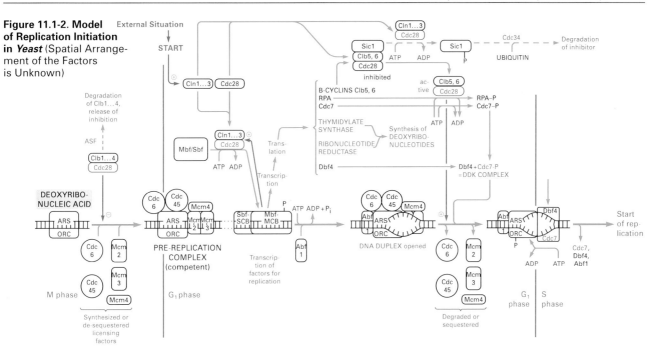

Figure 11.1-2. Model of Replication Initiation in Yeast (Spatial Arrangement of the Factors is Unknown)

11.1.3 DNA Polymerases (Pol, Table 11.1-3)

Eukarya have at least 5 different types of DNA polymerases. In yeast 3 DNA polymerases (α, δ, ϵ) appear to be essential for DNA replication. Their sequences are highly conserved (human and yeast DNA polymerase α show 31%, bovine and yeast DNA polymerase δ 44% homology).

Table 11.1-3. Eukaryotic DNA Polymerases

Polymerase*	Source	Number of subunits	Mol. mass kDa	Function
α (I)	yeast	4	180, 86, 58, 48 (primase)	Initiation of both strands. Subunit p86 is periodically phosphorylated in a cell-cycle dependent manner. Tightly associated primase activity. Starts Okazaki fragment synthesis. Synthesizes an RNA primer of 8...10 nt length, extends it with an initiator DNA of less than 50 nt. No exonuclease activity.
	human	5		
β (IV)	yeast	1	68 (?)	Involved in base excision repair (11.2.3). Essential in mammals (as shown in mice), not essential for viability in yeast.
	human	1	36...38	
γ (m)	mitochondria	1	144...150	Replication of mitochondrial DNA, similar enzyme in chloroplasts
δ (III)	yeast	2	125, 55	Synthesizes leading strand and completes Okazaki fragment synthesis. No primase activity. Requires PCNA (Table 11.1-4) for binding to DNA. Long strand processivity. Contains 3'→5' exonuclease for proofreading (by removing mononucleotides). Essential.
	human	2		
ϵ (II)	yeast	5	255, 80, 34, 30, 29	Similar to Pol δ. Long strand processivity even without presence of PCNA. 3'→5' exonuclease activity, degrades to 6...7 nt long oligonucleotides. Participates in DNA replication and repair. Essential, but mechanism unclear.
	human	5		

* A different nomenclature for yeast used by some authors is added in parentheses. Two other, non essential polymerases (PolV, 120...135 kDa and Pol Rev 3/Rev7, 173 kDa) have also been described in yeast.

Table 11.1-4. Other Important Proteins Involved in DNA Replication

Enzyme	Source	No. subunits	Mol. mass kDa	Function
DNA Helicases	eukarya	1	46...110	Associate with various DNA (and RNA) polymerases and catalyze DNA unwinding by disruption of hydrogen bonds between duplex DNA strands (ATP dependent). Action proceeds either in 5'→3' or in 3'→5' direction, yielding stretches of single stranded DNA for replication, repair (11.2), recombination, transcription and conjugation. More than 20 eukaryotic helicases have been characterized, their role and mechanism of action still remains to be elucidated in most cases.
DNA Topoisomerase I	eukarya	1	95...165	Removes positive superhelical tension by inducing a transient single strand nick into the DNA template ahead of the replication fork. Inhibited by camptothecin.
RPA (RFA)	yeast	3	69, 36, 13	Unspecific single strand and sequence-specific double-strand DNA binding protein. The 36 kDa subunit is periodically phosphorylated by a cyclin-dependent kinase during the cell cycle. The 69 kDa subunit contains single strand DNA binding and origin-unwinding activity.
	mammals	3	70, 34, 11	
RFC	yeast	4	120, 40, 37, 36	Essential accessory factor to DNA polymerases δ and ϵ. It binds to the primer terminus (ATP dependent) and recruits PCNA as a ring around DNA. The subunits are highly homologous among each other and between eukarya. Possibly analogous to the γ complex of E. coli DNA Pol III (10.2.2).
	mammals	3	140, 41, 37	
Proliferating cell nuclear antigen (PCNA)	eukarya	3 (homotrimeric)	26 (*3)	Essential accessory factor to DNA Pol δ. Binds possibly to DNA analogous to the β-subunit of E. coli DNA Pol III (10.2.2). DNA is thought to be threaded through a ring (sliding clamp) of PCNA, preventing DNA polymerase to dissociate from the DNA during synthesis. Participates in nucleotide excision repair of UV-damaged DNA (11.2.2).
Ribonuclease H				Degradation of RNA primers in Okazaki fragments.
Exonuclease MF1				Degradation of RNA primers in Okazaki fragments in 5'→3' direction. Cooperates with RNase H.
DNA ligase	yeast	1 (Cdc9)	ca. 85	Catalyzes the formation of phosphodiester bonds at single-strand breaks in double-stranded DNA. From the 3 mammalian DNA ligases, Ligase I is used in replication.
	mammals	1 (Ligase I)	102	
Telomerase	e.g., Tetrahymena	2	95, 80 (+55 RNA)	RNA dependent DNA polymerase, contains essential RNA. Extends 3' DNA strand ends by addition of telomeres.
DNA Topoisomerase II	eukarya	2 (homodimeric)	100...180 (*2)	Required for segregation of daughter DNA molecules after replication. Relaxes supercoils; acts via cutting and re-ligating of double strands. It is also important as a nuclear scaffolding protein.
Histones	eukarya	2* (2a, 2b, 3, 4)	14, 14, 15, 11	Basic chromosomal proteins, essential components of nucleosomes. Histone H1 binds to DNA between nucleosomes (2.6.4)

11.1.4 Replication Forks (Fig. 11.1-6, Tables 11.1-3 and 11.1-4)

The general aspects of DNA replication are dealt with in Chapter 10.2.

Eukaryotic replication forks move with a speed of about 30...50 nt/sec. (*in vitro*, on 'naked' DNA, about 500 nt/sec. are observed). The large number of replication origins shortens the replication time of *mammalian* chromosomes from more than 1000 hours (calculated for a single origin) to ca. 8 hours.

Although it is still generally assumed that the replication machinery travels along the DNA, there are also indications for a relatively immobile replication apparatus, where the DNA fork is being threaded through. (Analogous possibilities also exist for transcription complexes, 11.3.2).

After the initiation phase (which is presented in Section 11.6.2 for *yeast*), helicases unwind the DNA double strand and DNA topoisomerase I relieves the positive superhelical tension (Figure 11.1-3, one strand is nicked and released after passing past the other strand). Both reactions are energized by ATP hydrolysis. DNA directed DNA polymerase α/primase (Pol α/primase) attaches itself to the single strands, the still open portion of which is covered by RPA (= RFA). The primase synthesizes a short stretch of RNA complementary to the template DNA, which is extended as initiator DNA (ca. 34 nt) by Pol α. Then a polymerase switch takes place: RF-C displaces Pol α by binding to the 3' end of the initiator DNA. Then it recruits PCNA to form a trimeric ring around the DNA strand, which enables the DNA directed DNA polymerase δ (Pol δ) to join and to continue DNA replication. Pol δ is able to proofread the newly synthesized DNA strands.

Analogous to *bacteria* (10.2.3), replication of one template strand can proceed continuously in 5'→3' direction (leading strand as described above). Since the other template strand has a reverse orientation, its replication can be performed only discontinuously (lagging strand) by ligating short Okazaki fragments (about 100...200 nucleotides in *eukarya*), which are also synthesized in 5'→3' direction. Leading and lagging strand polymerases presumably act in concert resulting in the periodic formation of unreplicated looped lagging strand DNA (trombone model).

The Okazaki fragments are started likewise by RNA primers and initiator DNA. The Pol α/primase performing these reactions are consecutively replaced via a polymerase switch by Pol δ, which continues replication until the previous Okazaki fragment is reached. Removal of the RNA primers requires the combined action of the 5'→3' exonuclease MF1 and RNase H. The resulting gap is filled by a DNA polymerase and closed by DNA ligase I.

There is still some uncertainty about the mechanism of DNA replication. In particular, the exact spatial arrangement of the DNA replication system is unknown yet. Several models have been proposed, which differ in various aspects. Therefore, Figure 11.1-6 may not be exact in all details.

The completed DNA strands wind around histones and are further compacted, using scaffolding proteins (2.6.4). The relaxing of negative supercoil tensions, which occur during these reactions, is achieved by DNA topoisomerase II. Another function of this enzyme is the removal of 'knots' of the DNA strands (decatenation). The enzyme breaks both DNA strands and guides them through another DNA duplex.

Eukaryotic topoisomerase type IB (see above) acts preferentially on double stranded DNA. A tyrosine hydroxyl group of the enzyme temporarily forms a covalent bond with the 3' phosphate group of one of the DNA strands. After the free 5' end has turned in order to relax the twisting stress, releasing of the nick takes place. No supply of additional energy is required. (Contrary to this, in *bacterial* topoisomerase type IA the 5' phosphate group is temporarily bound by tyrosine.)

The dimeric topoisomerase type II breaks both strands of duplex DNA 4 bp apart. A covalent bond between 5' phosphate groups of the DNA and a pair of tyrosyl groups of the enzyme is formed. Then another unbroken DNA duplex passes through the gap. The reaction requires ATP in order to achieve the changes in the protein configuration, which allow the passage. [A similar mechanism is used by the *bacterial* gyrase (10.2.3), where a DNA duplex of about 140 bp is wrapped around the enzyme, crossing itself. Duplex strand passage through the gap produces negative supercoils.]

11.1.5 Telomeres (Figure 11.1-5)

In replicating the lagging strand of linear DNA a problem arises, when the 3' end of the template strand is reached. The last RNA primer cannot be replaced by DNA, since there is no further template for synthesizing another primer/Okazaki fragment. Therefore, with every round of replication shortening of DNA would occur.

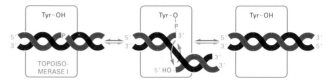

Figure 11.1-3. Mechanism of DNA Topoisomerases IB (3', Rotation Model)

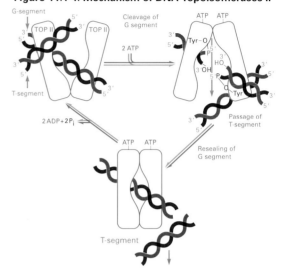

Figure 11.1-4. Mechanism of DNA Topoisomerases II

In order to avoid gradual loss of essential genes, linear DNA carries telomere extensions of up to 2000 short, tandemly repeated DNA sequences at its 3' end (species specific, 5 to 8 bases long, in *humans* TTAGGG).

In vitro, the G-rich sequences frequently fold back onto themselves by non-Watson-Crick G-G base pairing, forming hairpin loops and covering the strand ends. This way, telomeres might keep chromosomes from fusing end-to-end, thus preventing chromosome breakage and loss during cell division. Chromosomes lacking telomeres are not stably maintained.

The formation of these extensions is catalyzed late in S phase by telomerase. The enzyme contains an essential RNA component serving as template for the synthesis of the telomere repeats (159...1300 nt, in humans 450 nt). By synthesizing DNA from RNA templates, it resembles reverse transcriptase (12.4.1, except for using an internal template). The telomeres are bound by telomere binding proteins (TBP).

While *lower eukarya* like *yeast* and also the *germline cells* of *higher eukarya* contain telomerase, most *somatic cells* of *higher eukarya* do not express this enzyme (*Fetal cells*, *blood stem cells* and many *mouse cells* seem to be an exception). Therefore telomeres shorten with every round of DNA replication in *somatic cells*.

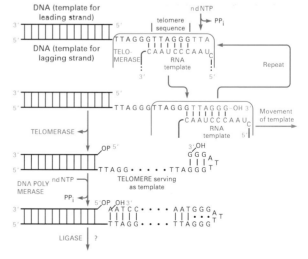

Figure 11.1-5. Formation of Telomeres

Figure 11.1-6. Eukaryotic DNA Replication (Details not completely known)

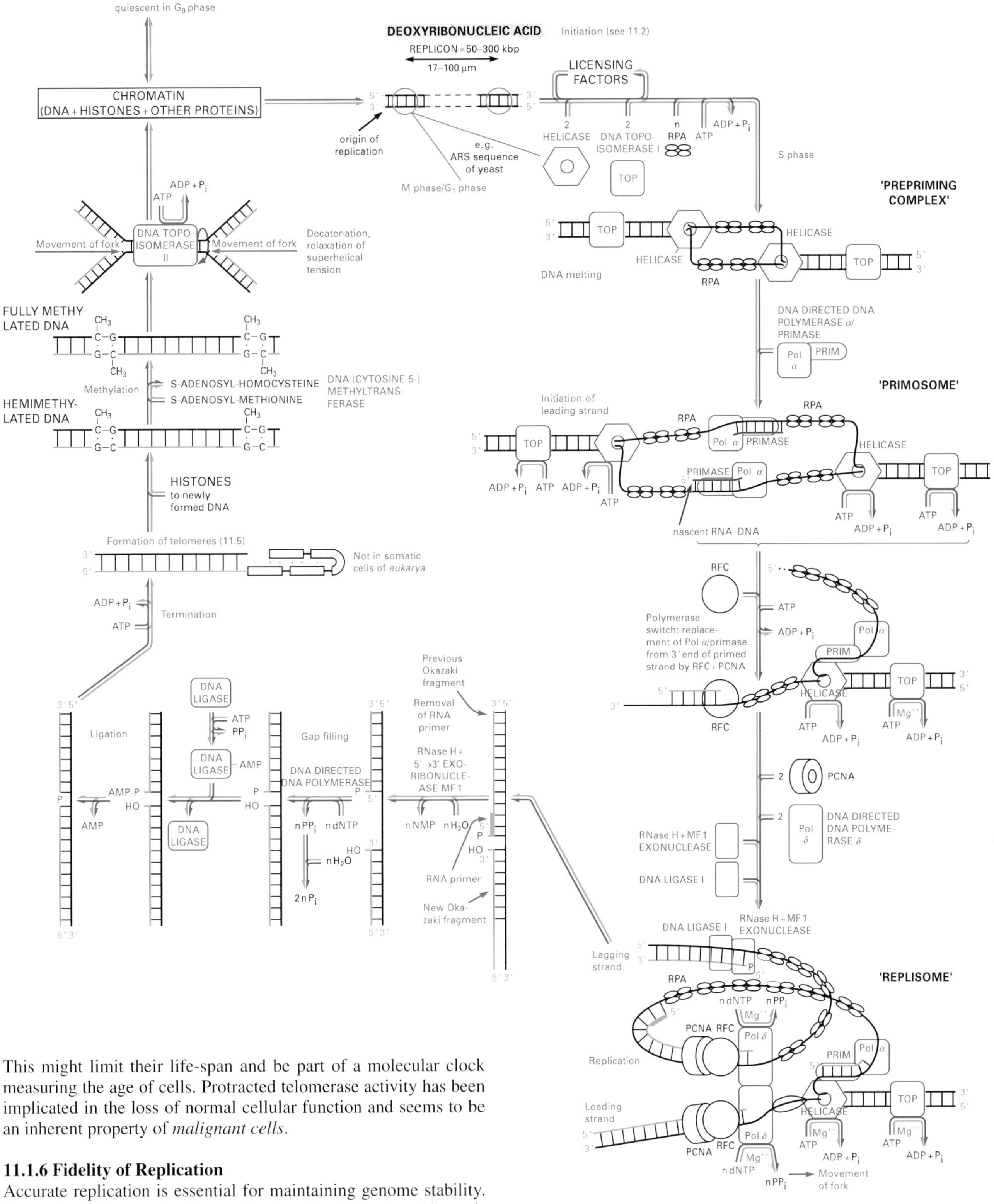

This might limit their life-span and be part of a molecular clock measuring the age of cells. Protracted telomerase activity has been implicated in the loss of normal cellular function and seems to be an inherent property of *malignant cells*.

11.1.6 Fidelity of Replication

Accurate replication is essential for maintaining genome stability. Base selectivity, proofreading and post-replicative mismatch repair (11.2-4) reduce the overall error rate to about 1 per 10^{10} nucleotides. This rate, as well as the relative contributions of the mechanisms for enhancing fidelity are similar to *bacteria* (10.2.4). The most frequent error is the deletion of a single nucleotide. Pol ε is the most accurate *eukaryotic* DNA polymerase, while pol α is the least accurate one (it lacks proofreading ability). In some hereditary forms of human cancer, impaired mismatch repair leads to several thousandfold higher error rates in DNA replication.

Literature:
Aparicio, O.M. *et al.*: Cell 91(1997) 59–69.
Campbell, J.L.: J. Biol. Chem. 268 (1993) 25261–25264.
Diller, J.D., Raghuraman, M.K.: Trends in Biochem. Sci. 19 (1994) 320–325.
Nasmyth, K.: Current Opin. in Cell. Biol. 5 (1993) 166–179.
Roberts, J.M.: Current Opin. in Cell. Biol. 5 (1993) 201–206.
Roca, J.: Trends in Biochem. Sci. 20 (1995) 156–160.
Sugino, A.: Trends in Biochem. Sci. 20 (1995) 319–323.
Stillman, B.: Cell 78 (1994) 725–728.
Toyn, J.H., *et al.*: Trends in Biochem. Sci. 20 (1995) 70–73.
Umar, A., Kunkel, T.A.: Eur. J. Biochem. 238 (1996) 297–307.
(various authors): Science 247 (1996) 1643–1677.
Waga, S., Stillman, B.: Nature 369 (1994) 207–212
Wang, T.A., Li, J.J.: Current Opin. in Cell. Biol. 7 (1995) 414–420.
Wang, T.S.F.: Ann. Rev. of Biochem. 60 (1991) 513–552.
Wang, J.C.: Ann. Rev. of Biochem. 65 (1996) 635–692.

11.2 Eukaryotic DNA Repair

11.2.1 DNA Damage and Principles of Repair

DNA is a highly vulnerable molecule due to its enormous size. It can be damaged by many external agents (e.g. environmental chemicals, radiation, UV irradiation). Also, spontaneous DNA damage arises from hydrolytic removal of bases, deamination of cytosine to uracil and (more rarely) adenine to hypoxanthine and guanine to xanthine, non-enzymatic DNA methylation by S-adenosyl methionine or damage by free oxygen radicals (4.5.8, 10.3.1). In *human cells* each of these types of spontaneous damage occurs at a daily rate in the order of 500...30000 per cell.

The ability to minimize the number of inheritable mutations caused by damage to DNA is essential for all organisms in order to maintain the genome integrity. Besides mutagenicity, DNA damage can also interfere with transcription and replication. Therefore, an intricate network of various systems for recognition and repair of DNA damage has evolved.

DNA repair in *eukarya* appears to follow the same basic reactions as in *bacteria* (10.3), but is more complex due to the larger genome and its more complicated structure. Similarities in DNA repair mechanism extend from *yeast* to *humans*. Some of the components can even be exchanged *in vivo*.

Repair is often associated with DNA replication and transcription when DNA is supposedly more accessible and repair is more essential than in a quiescent state. Transcriptionally active regions are repaired faster than 'silent' regions. Genes transcribed by RNA polymerases I or III are not subject to this specific type of DNA repair. *Eukaryotic* cells have checking mechanisms that can arrest cell cycle progress when damaged DNA is detected (11.6.2).

11.2.2 Direct Reversal of Damage (Fig. 11.2-1)

Reversal of alkylation: Homologues to the specialized alkyltransferases of *E. coli* for direct repair of alkylation damage (10.3.2) have also been found in *eukarya*. O^6-methylguanine-DNA-protein-cysteine S-methyltransferase (O^6-MGT I) is a 'suicide' enzyme that removes methyl groups from O^6-methylguanine to a cysteine-SH group of the enzyme at the expense of its own inactivation.

Photoreactivation: A deoxyribodipyrimidine photolyase (similar to *bacteria*, 10.3.2) has been found in *S. cerevisiae*, some other *fungi* and in *algae*, but so far not in *mammalian* cells. A similar enzyme in *animals* repairs the 6-4 photoproduct (another pyrimidine dimer).

11.2.3 Excision Repair Systems (Figure 11.2-1)

Excision repair relies on the redundant information in the DNA duplex to remove a damaged base or nucleotide and to replace it with the correct base by using the complementary strand as a template.

Base excision repair (BER): With this mechanism, damages caused by ionizing radiation, oxidizing and methylating agents (including products of the endogenous metabolism) and by loss of bases are repaired. It has only a limited substrate range since damaged bases have to be recognized specifically by the DNA glycosylases.

In base excision repair single altered bases are targeted. Each of the DNA-glycosylases (e.g. uracil DNA-glycosylase or 3-methyladenine DNA-glycosylase; the EC list uses 'glycosidase') recognizes specifically a particular type of modified bases and removes them by hydrolyzing the N-glycosidic bond between the base and the deoxyribose-P backbone of the DNA. The resulting 'apurinic' or 'apyrimidinic' site (AP site) is recognized by specialized AP-endonucleases which cut the DNA backbone 5' (upstream) to this location. A deoxyribophosphodiesterase (dRPase, an exonuclease) removes the base-free sugar phosphate. DNA polymerases (β, possibly also δ or ϵ) fill the gap with the correct nucleotide. Then the strand is closed by action of a DNA ligase. In *mammals*, this is DNA ligase III.

Nucleotide excision repair (NER): This mechanism is coupled to transcription with selective preference to the transcribed strand. Since NER responds to distortions in DNA structure and not to individually recognized lesions, it is an almost universal repair system which removes a wide spectrum of structurally unrelated damages.

Examples are most UV-induced photoproducts (e.g. cyclobutane pyrimidine dimers, 10.3.1), bulky chemical adducts (e.g. benzo[a]pyrene•guanine adducts caused by smoking, cis-platin•guanine adducts), methylated bases (e.g. O^6-methylguanine) and intrastrand as well as interstrand crosslinks. However, sometimes it also reacts (inefficiently) with mismatches with no strand preference, which can lead to fixation of mutations.

Although there are no sequence homologies to the *E. coli* UvrABC endonuclease system (10.3.3), the DNA strand in *eukarya* is cut similarly on both sides of the lesion by a specialized excision nuclease complex. *Eukaryotic* nucleases hydrolyze the 3rd to the 8th phosphodiester bond 3', followed by the 20th to 25th phosphodiester bond 5' of the lesion, removing a patch of 22 to 32 (in *humans* 27 to 29) nucleotides. The incision step alone requires many different proteins (≥ 10 in *yeast*, ≥ 16 in *humans*). A selection is shown in Figure 11.2-1 and Table 11.2-1. RFA temporarily covers the separated strands. The gap is then filled by DNA polymerase ϵ (or δ?) and the continuity of the phosphodiester backbone is restored by DNA ligase.

Some of the repair factors in *humans* are also part of the transcription machinery (11.3.2), e.g., TFIIH, TFIIJ. Coupling of repair to transcription is effected by CSA/ERCC8 and CSA/ERCC6 proteins, which temporarily replace stalled RNA polymerase by the NER system. However, many of the activities well known in *E. coli* and at least partially characterized in *S. cerevisiae* have not yet been found and/or characterized in *mammalian* cells.

Table 11.2-1. Protein Factors in *Eukaryotic* Nucleotide Excision Repair (Selection)

Gene product	Mol. mass kDa *human*	Function
Damage excision:		
XPA	31	recognition of damage (zinc finger)
RPA (RFA)	(see 11.1.3)	same, binds to single-stranded DNA
XPC/HHR23B	125, 58	stabilization of preincision complex
TFIIH	(see 11.3.2)	formation of preincision complex, transcription-repair coupling
XPB (ERCC3)	89	3'→5' helicase, TFIIH subunit
XPD (ERCC2)	87	5'→3' helicase, TFIIH subunit
XPG (ERCC5)	135	DNA incision, 3' side of lesion
XPF (ERCC4)/ ERCC1	112, 33 (1:1 complex)	DNA incision, 5' side of lesion
Gap filling:		
RFC	(see 11.1.3)	loads PCNA onto DNA (ATP-dependent)
PCNA, DNA polymerase ϵ or δ, DNA ligase – see 11.1.3		

11.2.4 Mismatch Repair

Base pairs in a double-stranded DNA helix not conforming to the Watson-Crick rules (A-T, G-C, 2.6.1) are called mismatches. Such incorrect base pairs arise for several reasons:

- occasional errors in proofreading mechanisms during DNA-replication
- formation of incorrect heteroduplexes between DNA-strands during recombination
- deamination of 5-methylcytosine to thymidine, causing a G-T mismatch.

Thus, this type of repair system is preferably connected to replication and recombination events. It follows the same principles as the excision repair systems mentioned before. The main problem is the distinction of the correct from the incorrect base, since both molecules are normally found in DNA. As in *E. coli*, mismatches in *mammalian* cells can be repaired by two different mechanisms depending on the sequence context:

- a short patch system with preference for G-T mismatches originating from deamination of (5-methyl-)C at CpG sites. It resembles base excision repair (11.2.3) and likewise shows restricted specificity. After thymine removal, 10 or less nucleotides (down to a single one in *mammals*) are excised. The resulting gap is closed by DNA polymerase β.

    Likely, the mismatch-specific thymine-DNA-glycosylase of *humans* is also able to remove dU generated at CpG sites by deamination. There are also hints for a mismatch repair mechanism in *mammals* which removes A (mispaired to G) by incisions at both sides.

- a long patch mechanism using MutS and MutL homologues similar to the *bacterial* methyl-directed pathway (10.3.4, recognition of the correct strand at methylated moieties after replication or of the defect strand at breaks). This system detects mismatches in many different contexts. Fairly large pieces of DNA (up to >1000 bases) are excised, this may occur in either direction. The gap filling is similar to nucleotide excision repair, possibly by DNA polymerase δ or ϵ.

11.2.5 Double-strand Repair and Recombination

One of the gravest lesions occurring in a cell is a DNA double-strand break, because it completely disrupts the integrity of the

Figure 11.2-1. Systems of *Eukaryotic* DNA Repair
The terminology of the human system is used

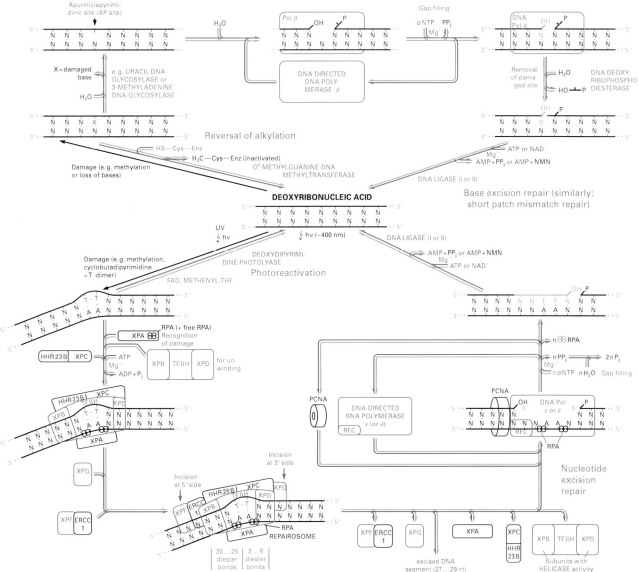

DNA molecule. This is the most frequent effect of ionizing radiation. Double strand ends of broken chromosomes easily cause DNA recombination and mutations. Double-strand breaks also occur during natural recombination.

Repair can be achieved by homologous recombination (cross-over between damaged DNA and correct 'sister' DNA). Several of the enzymes required for this repair type in simpler organisms are conserved in *eukarya*. E.g., in *yeast* RAD51 is a functional homologue to RecA of *E. coli*. Mainly, the RAD52 enzyme is involved in repair of double strand breaks. The repair mechanisms in *Saccharomyces cerevisiae* appear to be similar to *bacteria* (10.3.5). Generally, homologous recombination does not occur often in *higher eukarya*.

Although the genes involved in recombination are conserved, in *higher eukarya* sometimes 'illegitimate' events occur. These are reactions, which rely on very short stretches of homology or no homology at all between the recombining molecules.

Other enzyme activities: *Eukaryotic cells* contain a DNA-dependent protein kinase (DNA-PK, 465 kDa). The two Ku subunits of this enzyme (80 and 70 kDa) show ATP dependent helicase activity and bind to DNA ends, hairpins, nicks and gaps irrespective of the DNA sequence and thus likely recognize damage. Cellular targets of DNA-PK include many DNA binding regulatory proteins (e.g. c-Jun, c-Fos, c-Myc, SRF), TFIID, RNA polymerase II (CTD domain), DNA topoisomerases I and II etc. (11.1.4). It also autophosphorylates and deactivates itself. At present it is not well known how it contributes to the repair of double strand breaks. The enzyme is also involved in immunological processes [V(D)J recombination, 19.1-3]. In SCID cells (severe combined immunodeficiency 8.1.7) it appears to be defective.

Few other enzymes have been found so far that use damaged DNA as a signal for action. One is poly (ADP-ribose) synthase (PARS), which attaches itself to the damaged site and automodifies itself by formation and covalent binding of long poly (ADPribose) chains. Its function is unclear. Possibly it relaxes chromosome structures.

11.2.6 DNA Repair and Human Diseases
Defects in DNA repair cause severe diseases. It has been estimated that 80...90% of human cancers result from DNA damage.

Human DNA repair has mostly been studied in genetic diseases with impaired repair function. Examples are xeroderma pigmentosum (high sensitivity against UV radiation), Cockayne's syndrome and trichothiodystrophy, all of which are caused by defects in nucleotide excision repair. In ataxia telangiectasia the repair of double-strand breaks is defective, the patients are extremely sensitive to ionizing radiation. Damage in genes for long patch mismatch repair mechanisms (hMSH2 and hPMS1 on chromosome 2, hMLH1 on chromosome 3p21 or hPMS2 on chromosome 7) are associated with hereditary nonpolyposis colorectal cancer (HNPCC; Lynch syndrome II, Muir-Torre syndrome), which is characterized by instabilities of simple repeated sequences. This is one of the most common genetic diseases of humans (1 in 200 individuals).

Literature:
Carr, A.M., Hoekstra, M.F.: Trends in Cell Biology 5 (1995) 32–40.
Friedberg, E.C.: BioEssays 16 (1994) 645–649.
Friedberg, E.C., *et al.*: *DNA Repair and Mutagenesis*. ASM Press (1995).
Hanawalt, P.C., *et al.*: Current Biol. 4 (1994) 518–521.
Sancar, A.: Ann. Rev. of Biochem. 65 (1996) 43–81.
Umar, A., Kunkel, T.A.: Eur. J. Biochem. 238 (1996) 297–307.
(various authors): Trends in Biochem. Sci. 20 (1995) 381–439.
Weaver, D.T.: Trends in Genetics 11 (1995) 388–392.
Wood, R.D.: Ann. Rev. of Biochem. 65 (1996) 135–167.

11.3 Eukaryotic Transcription

Transcription is a process that transcribes genetic information from DNA into RNA. (For a drawing of the principle see Figure 10.4-1.) In *eukarya*, this takes place in the *nucleus*, in *mitochondria* and *chloroplasts*. Transcription is not self-contained. There is a close connection between transcription and other *nuclear* processes like DNA replication and DNA repair. The initial opening of the chromatin structure is not dealt with here.

11.3.1 RNA Polymerases (Pol, Table 11.3-1)

Transcription is performed by DNA-directed RNA polymerases. Unlike DNA polymerases, RNA polymerases do not need a primer to start the reaction. While *bacteria* contain only one RNA polymerase (10.4.1), there are three different RNA polymerases in *eukaryotic cells,* which transcribe different genes. All these polymerases are multisubunit complexes: two large polypeptides (ca. 200 and 140 kDa) are associated with about 12 smaller subunits, some of which are common to all three enzymes. Additionally, there are different Pol in *mitochondria* and *chloroplasts*.

Table 11.3-1. Eukaryotic Nuclear RNA Polymerases

Polymerase	Transcribes
Pol I	genes for ribosomal RNA (28 S, 18 S, 5.8 S rRNA)
Pol II	all protein-coding genes (yielding mRNA); genes for snRNAs U1, U2, U3, U4, U5
Pol III	genes for tRNA and 5 S ribosomal RNA, sn RNA U6

The basic components of the transcriptional machinery were well conserved in evolution. The order of the transcription complex assembly is basically the same for *yeast, Drosophila* and *humans*. The two large polymerase subunits are even homologous to the two largest subunits of the *E. coli* RNA polymerase (10.4.1).

11.3.2 mRNA Transcription by RNA Pol II (Figure 11.3-1)

mRNA synthesis is a complicated and time-consuming process which can take up to 40 minutes for a large gene (e.g. fibronectin). Since all proteins with their central role in cellular functions and structure are synthesized via mRNA, this process must be strictly regulated at different levels: at DNA binding sites (enhancers, silencers), via proteins (basic and specific transcription factors), by modification (capping, polyadenylation) and processing (splicing). The regulation is discussed in detail in 11.4.

Transcription factors: These proteins play essential roles in initiation and elongation. Only the so-called basic (= general) transcription factors (required for the transcription of virtually all genes) are listed in Table 11.3-2 and shown in Figure 11.3-1. Their size in the Figure does not always reflect the actual proportions. They provide the low-level, 'basic' rate of transcription. (For the regulating, gene-specific transcription factors see Table 11.4-2.)

Transcription factors often consist of a modular arrangement of distinct functional domains (for details, see 11.4.2):
- DNA-binding domains, e.g. zinc fingers or helix-turn-helix
- Transactivation domains, which mediate cooperative associations with other proteins

Initiation: In order to start transcription, a core promoter element at the DNA, e.g. the TATA box (11.4.1) must be recognized and bound by the TBP subunit of the basic transcription factor TFIID, introducing sharp kinks into the DNA. TFIIB and (in some cases) TFIIA stabilize this interaction, forming a pre-initiation complex. Frequently, these steps are controlled by activation or repression mechanisms (11.4).

In the case of promoters not containing the TATA-sequence, the initiator motif of DNA (Inr, which encompasses the start site) can mediate initiation. It is recognized by Inr-binding proteins (such as TFII-I to which TFIID associates) or by the subunits TAF_{II} 250 and TAF_{II} 150 of TFIID. This commonly occurs with 'housekeeping' genes (11.4.3).

RNA polymerase II (Pol II) together with transcription factors TFIIE, IIF and IIH is then recruited to form the initiation complex. TFIIB, assisted by TFIIF, acts as a bridge to Pol II. This is the rate-limiting step. At this stage the DNA strands start to become separated in an ATP requiring reaction forming an 'open initiation complex'.

The spatial arrangement of the components is only partially known yet; interacting factors are drawn in Figure 11.3.1 close to each other. It is still not clear if the complexes are (at least partially) preformed (holoenzyme assembly model) or if they assemble at the initiation site (ordered multistep model). Not all transcription factors are required for all types of promoters. The roles of histones and chromatin are not shown in this schematic drawing.

Elongation: The initiation complex becomes active when Pol II transcribes the first few bases close to the promoter, beginning at the transcription start site. The maximum length of the RNA-DNA hybrid is only 2...3 bp. RNA chain elongation involves a series of forward movements interspersed by pauses. The transition from initiation to elongation (promoter clearance by Pol II) is still poorly defined. For catalyzing elongation, Pol II gets highly phosphorylated at the carboxy terminal domain (CTD) of the largest subunit, which causes a conformation change and subsequent clearance from the TFIID complex. This reaction is catalyzed by TFIIH and assisted by TFIIE.

Table 11.3-2. Basic Transcription Factors for RNA Polymerase II

Name	Species	Subunits	Mol. mass (kDa)	Functions/Properties
TFIIA	*human*	3	37, 19, 13	Initiation, stabilizes TFIID association with promoter by increasing TBP affinity for TATA box, not essential for basic transcription *in vitro*.
	yeast	2	32, 14	
TFIIB	*human*	1	35	Required for binding of RNA Pol II to the initiation complex; functions in transcription start site selection. Target for steroid hormone receptors (17.6). Two domains, one interacts with TBP, the other with the small subunit of TFIIF.
	yeast (TFe)	1	38	
TFIID	*human*	TBP + 12 TAF_{II}	38 15...250	Central role in transcriptional activation. Binding of TBP (TATA-binding protein) with the TATA-box of DNA is thought to be the first step in the assembly of the pre-initiation complex. TFIID supports transcription also from TATA-less promoters. Individual TAF_{II} coactivators (TBP associated factors) are the specific target for many transcriptional activators and repressors which modulate TBP/TFIID binding to DNA (11.4.3). DNA topoisomerase I is associated with TFIID.
	yeast (TFd)	TBP + ≥ 8 TAF_{II}	27 18...250	
TFIIE	*human*	2	56, 34	Necessary for recruitment of TFIIH, stimulates TFIIH dependent kinase, ATPase and helicase activity. Involved in DNA strand separation. Not necessary for transcription at all promoters. Probably an $\alpha_2\beta_2$ tetramer.
	yeast (TFa)	2	66, 43	
TFIIF	*human*	2	70, 30	Essential for initiation of transcription and transcript elongation (suppresses pausing). Activated via phosphorylation by TAF_{II} 250. Stabilizes binding of Pol II and TFIIB to the TFIID-promoter complex. Significant homologies to *bacterial* σ factors.
	yeast (TFg)	3	105, 54, 30	
TFIIH	*human*	≥ 9	89...34	Shows helicase, DNA-dependent ATPase and kinase (= TFIIK) activity. Catalyzes phosphorylation of Pol II CTD, which plays a role in promoter clearance and possibly also in the elongation phase. Involved in DNA excision repair (11.2.3).
	yeast (TFb)	9	105...38	
TFII I	*human*	1	45	Similar to TFIIA, binds to DNA initiator motif (Inr), may help forming an alternative pre-initiation complex; interacts with regulatory proteins.
	yeast	–	–	
TFIIS = S II	*human*	1	38	Transcription elongation factor, facilitates passage of Pol II through pause sites by removing a sequence from the 3' end of the nascent mRNA transcript. Highly conserved.
S III / elongin	*human*	3	110, 18, 15	Important for elongation (suppresses pausing) and termination. Activated by phosphorylation. Partial homology to *bacterial* ρ protein.

Figure 11.3-1. Transcription of mRNA (Ordered Multistep Model)

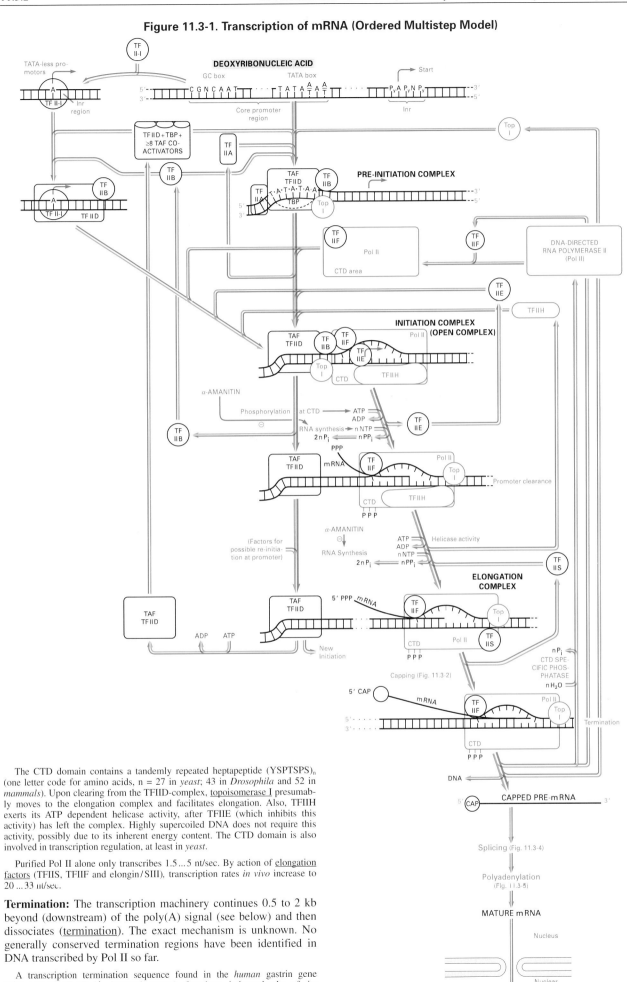

The CTD domain contains a tandemly repeated heptapeptide $(YSPTSPS)_n$ (one letter code for amino acids, n = 27 in *yeast*; 43 in *Drosophila* and 52 in *mammals*). Upon clearing from the TFIID-complex, topoisomerase I presumably moves to the elongation complex and facilitates elongation. Also, TFIIH exerts its ATP dependent helicase activity, after TFIIE (which inhibits this activity) has left the complex. Highly supercoiled DNA does not require this activity, possibly due to its inherent energy content. The CTD domain is also involved in transcription regulation, at least in *yeast*.

Purified Pol II alone only transcribes 1.5...5 nt/sec. By action of elongation factors (TFIIS, TFIIF and elongin/SIII), transcription rates *in vivo* increase to 20...33 nt/sec.

Termination: The transcription machinery continues 0.5 to 2 kb beyond (downstream) of the poly(A) signal (see below) and then dissociates (termination). The exact mechanism is unknown. No generally conserved termination regions have been identified in DNA transcribed by Pol II so far.

A transcription termination sequence found in the *human* gastrin gene $(5'-T_9A_2T_5AT_4AT_4AT_5-3'$, inverted repeat) functions independently of its distance from the promoter but is strongly orientation dependent. Termination takes place immediately upstream of this sequence.

11.3.3 Processing of mRNA

During the course of transcription, mRNA is also processed by capping, splicing and polyadenylation. These are coupled reactions that influence each other. Capping and polyadenylation are important for the efficacy of translation, but not always absolutely required. Only spliced, 'mature' mRNA is transported to the *cytoplasm* where it gets translated into protein by ribosomes (11.5).

Capping (Figure 11.3-2): Nascent mRNA's and snRNAs (11.3.4) are modified at their 5' ends by addition of a 'cap', mostly 7-methylguanosine, attached via a 5'-5' triphosphate bridge. This cap is important for mRNA stability and may be essential for recognition and removal of the first intron.

Figure 11.3-2. Capping of mRNA

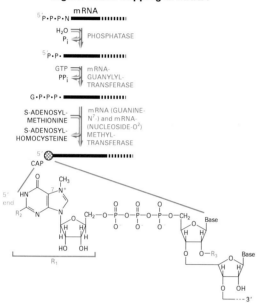

Cap structure	R_1	R_2	R_3	Present in
7mG cap	see Figure	NH_2	CH_3	mRNA
m_3G cap	see Figure	$N(CH_3)_2$	CH_3	U 1, 2, 4, 5
γ-Monomethyl-P cap	CH_3	—	H	U 6

Splicing (Figure 11.3-4): Splicing is an integral and essential step of gene expression in *eukarya* (and in a few cases also in *archaea* and *bacteria*). In this procedure, the non-translated portions (introns) are removed from the *nuclear* pre-mRNA and the remaining translated exons are joined with each other.

Split genes have been identified in all types of *eukaryotic* cells. There can be more than 50 exons in a single gene, some of them as short as 10 nucleotides. The number of introns in pre-mRNA varies from none to dozens (average ca. 8), their length from ca. 70 up to 200 000 nucleotides (average in *vertebrates* ca. 137). In *lower eukarya*, introns are short and often flanked by large exons. In *higher eukarya*, large introns separate normally short exons. This may lead to different mechanisms for splicing ('intron definition' in *lower* vs. 'exon definition' in *higher eukarya*). Almost all protein-coding genes in *vertebrates* have introns (notable exception: histones). The primary transcription unit is typically four to ten times larger than the final mRNA which has an average length of 1 to 2 kb (Figure 11.3-3).

Accuracy of splicing is crucial for cell function. Many *human* diseases are caused by mutations that interfere with RNA splicing. For example, approximately 25% of the human globin gene mutations in thalassemia occur in sequences responsible for correct splicing. On the other hand, alternative splicing can effect genetic variability (e.g., formation of different immunoglobulins, 19.1.3 or MHCs, 19.1.5) and produce proteins with different functions (e.g. sexual dimorphism in *Drosophila*).

Figure 11.3-3. Effect of mRNA Splicing (Ovalbumin Gene)

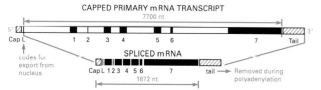

Splicing can occur co- and post-transcriptionally and takes place within the spliceosome, which is a large RNA-protein complex (60S). The spliceosome is a dynamic structure, which is assembled stepwise on the pre-mRNA at each individual intron.

It consists of small nuclear ribonucleoproteins, snRNPs (U1, U2, U4, U5, U6; see below, containing U-rich RNA of 57...217 nucleotides and about 80% protein), several minor snRNPs and spliceosome associated proteins, SAP, which perform important auxiliary functions (not shown in Figure 11.3-4). The RNAs are essential for the catalytic activity.

Highly conserved consensus sequences exactly indicate the exon-intron boundaries. The consensus sequences in *higher eukarya* show the following structure (invariant nucleotides are printed in bold, the branching nucleotide is underlined):

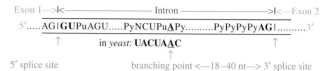

The first step in splicing is the recognition of these sequences at the 5' splice site and at the branch-point by complementary sequences of U1 and U2 snRNPs. The intron excision proceeds in two consecutive transesterification reactions. A nucleophilic attack by the 2'-OH group of the branch point-A on the phosphate at the 5' splice site causes exon-intron separation and ATP-dependent lariat formation by the intron. After U1 has left the complex, U4, U5 and U6 bind. Thereafter, the newly formed 3'-OH at the 5' splice site exerts a nucleophilic attack on the phosphate at the 3' splice site. This causes elimination of the intron and ligation of the two exons. Then the snRNPs dissociate from the mRNA. The lariat is later linearized by an RNA debranching enzyme and degraded.

Figure 11.3-4. Splicing Mechanism for mRNA

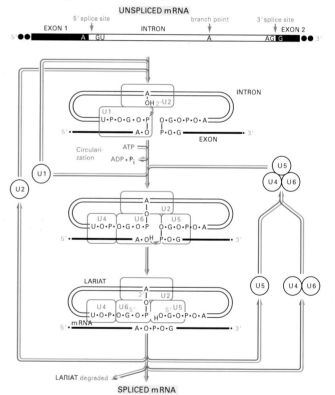

Besides this mechanism (group III introns) other splicing mechanisms exist. They proceed with an external guanosine nucleotide instead of the branching-point A (group I, rRNA) or with a ligation procedure after endonuclease splitting of the pre-mRNA (group IV, tRNA). Splicing of group II introns (e.g. *mitochondria* and *chloroplast* mRNA) resembles the group III mechanism, but without participation of snRNPs. Frequently, the group I or II procedures are performed by RNA activity only (self-splicing, e.g. rRNA in *Tetrahymena*). Exons of different RNA strands can also be combined by splicing mechanisms (*trans*-splicing, e.g. mRNA in *Trypanosoma*).

Polyadenylation (Figure 11.3-5, Table 11.3-3): Posttranscriptional addition of poly(A) to the 3' end of mRNA is an important biological process conserved from *bacteria* to *humans*. In *eukarya*, polyadenylation is essential for the stability of many mRNAs and also influences the efficiency of translation. Most histone-mRNAs and all sn-RNAs (except U6), however, are not polyadenylated.

The start of polyadenylation takes place ca. 10 to 30 nucleotides downstream from the polyadenylation signal (consensus sequence: AAUAAA). Also necessary are downstream elements, which consist of poorly defined G/U-rich sequences. Sometimes an U-rich region upstream of the polyadenylation signal acts as an enhancer of polyadenylation. The array of protein factors involved in polyadenylation has been called a 'poly(A)osome'. The total number of factors involved is still unknown.

Polyadenylation is initiated by cleavage of the mRNA which has been transcribed considerably beyond the adenylation start site. Involved are TFIIS (see above) and CPSF (cleavage and polyadenylation specificity factor), which binds to the polyadenylation signal and stimulates polyA-polymerase (PAP). When the growing poly(A) tail has reached a length of 10 to 12 nucleotides, nuclear poly(A) binding protein II (PAB II) also binds. This strongly stimulates polyadenylation and enables poly(A)-polymerase to synthesize a stretch of $(A)_{200...250}$ (in *mammals*) in a single processing event. Poly(A) tails in *yeast* have a length of ca. $(A)_{70...90}$, *bacterial* (*E. coli*) tails of about $(A)_{15...40}$. Later on, the poly(A) tail gets shortened in the *cytoplasm*.

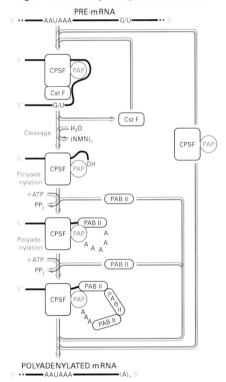

Figure 11.3-5. Polyadenylation of mRNA

Table 11.3-3. Some Factors Involved in *Mammalian* Polyadenylation

Name	Subunits	Mol. mass (kDa)	Function
Poly(A) polymerase (PAP)	1	82 (*yeast* 64)	involved in cleavage and polyadenylylation
cleavage/polyadenylation specificity factor (CPSF)	4	160, 100, 73, 30	binds specifically to the AAUAAA signal, stimulates poly(A) polymerase
Cleavage stimulating factor (CstF)	3	77, 64, 50	binds to U or G/U rich upstream elements
poly(A) binding protein II (PAB-II)	1	49	binds to the growing poly (A) tail, stimulates polyadenylation to $(A)_{200...500}$

The mature mRNA moves to the *nuclear* surface (probably by diffusion) and is exported through nuclear pores. This is an energy requiring, probably carrier-mediated process as is the export of ribosomal subunits and tRNAs.

11.3.4 snRNA Transcription

snRNAs (small nuclear RNAs, e.g. U1...6) are constitutively expressed ($10^5...10^6$ copies per *human* cell). Initiation of transcription starts at a proximal sequence element (PSE, at position ca. –50). After transcription by Pol II, the sRNAs U1, U2, U4 and U5 are transported to the *cytosol*, where they acquire a modified 2,2,7-trimethylguanosine cap structure (m_3G cap) at their 5′ termini (similar to mRNA, Figure 11.3-2). Addition of proteins, leading to snRNPs (small nuclear ribonucleoparticles) also takes place there. The hypermethylated cap is essential for transport back to the *nucleus*. U6 is transcribed by Pol III, obtains a γ-monomethyl triphosphate cap and remains in the *nucleus*.

11.3.5 rRNA Transcription by RNA Pol I (Figure 11.3-6)

Eukaryotic ribosomes (80S) consist of two subunits (60 and 40S), which contain rRNAs (large subunit: 28S, 5.8S and 5S rRNA; small subunit: 18S rRNA) and ribosomal proteins (large subunit ca. 45, small subunit ca. 33).

Ribosomal RNAs are transcribed by RNA polymerase I (Only 5S rRNA is transcribed by RNA polymerase III). Many transcription units for rRNAs are arranged in tandem arrays. They are localized in a structure called the nucleolar organizer region (NOR). There, a special structure (nucleolus) forms where most of the ribosome biogenesis takes place: Transcription of rRNA genes, processing of the transcripts to mature rRNAs and assembly with proteins to form both ribosome subunits.

Within the nucleus of *higher eukarya*, several nucleoli are present, in *yeast* only one. Nucleoli contain, in addition to ribosomal proteins, a large number of non-ribosomal proteins involved in ribosome biogenesis and maintenance of nucleolar structure. During mitosis, nucleoli disassemble in the prophase and reassemble during telophase. DNA topoisomerase I, RNA polymerase I and UBF remain associated with the NOR, but no transcription takes place during that interval.

The promoter for rRNA synthesis consists of at least 2 elements, a GC rich upstream control element (UCE, 11.4.1) and a core region at the transcription start site. Both elements are recognized by the upstream binding factor (UBF, Table 11.3-3), which causes sharp bends to the DNA. Transcription factor IB (TFIB, in *humans* SLI, consisting of TBP and 3 TAF_I) binds to the core region followed by TFIC recruitment, leading to the pre-initiation complex. Pol I then joins, forming the initiation complex. Elongation proceeds upon binding of DNA topoisomerase I.

Eukaryotic transcription units for ribosomal genes carry short DNA sequences at their 3′ ends which contain a SalI site ('Sal-Box') and cause termination of transcription. TTF-I protein binds to this region as a monomer and causes DNA bending. Terminator sites for RNA polymerase I function only in one orientation.

Table 11.3-4. Transcription Factors for RNA Polymerase I

Name	Species	Subunits	Mol. mass (kDa)	Functions/Properties
TFIB	*human* (SL1)	TBP + 3 TAF_I	38 95, 64, 53	Initiation, promoter selection, recruitment (together with UBF) of RNA polymerase I. Equivalent to TFIID.
TFIC	?	?	?	Role in formation of pre-initiation complex
UBF (upstream binding factor)	*human*, highly conserved	dimer of differently spliced gene products	97, 94	Required for the formation of stable initiation complexes by RNA polymerase I and TFIB, specific for large rRNA genes. Causes DNA bending.
TTF1	*human*	1	130	Termination factor, abundant

11.3.6 Processing of rRNA (Figure 11.3-6)

After termination, the 80S RNP precursor contains a 5′ leader sequence and 18, 5.8 and 28S rRNAs separated by spacer RNAs. It is processed by cleavage of the 5′ leader, splicing and nucleolytic degradation of the spacer RNA. As in mRNA, splicing is directed by small nuclear ribonucleoproteins (snRNPs). This leads to 20S (containing 18S rRNA) and 32S intermediates (containing 5.8 and 28S rRNAs) which are further processed to yield mature 28, 18 and 5.8S rRNAs.

Post-transcriptional modification of the nucleotides results in methylation of about 100 nucleotides per ribosome at the 2′ OH of ribose and isomerization of more than 100 uridine residues per ribosome to pseudo-uridine (Ψ; see tRNA-modification, 11.3.8).

The rRNAs are complexed with ribosomal proteins in a self-organizing mode, forming both the large and small ribosomal subunits which are then separately transported to the *cytoplasm*.

11.3.7 tRNA Transcription by RNA Pol III (Figure 11.3-7)

As in *bacteria*, there are multiple tRNA genes in *eukarya* (e.g. about 1300 in a human cell). Mature tRNAs are mostly 75 to 80 nucleotides long. They are transcribed by RNA polymerase III.

tRNA gene promoters consist of 2 separated 10 bp elements (Box A and B) located downstream of the transcription start site. TFIIIC binds to box B, then box A orients TFIIIC towards the start site. TFIIIC causes correct positioning of TFIIIB (preinitiation complex), which then recruits Pol III. This DNA/

Figure 11.3.6. Transcription of rRNA

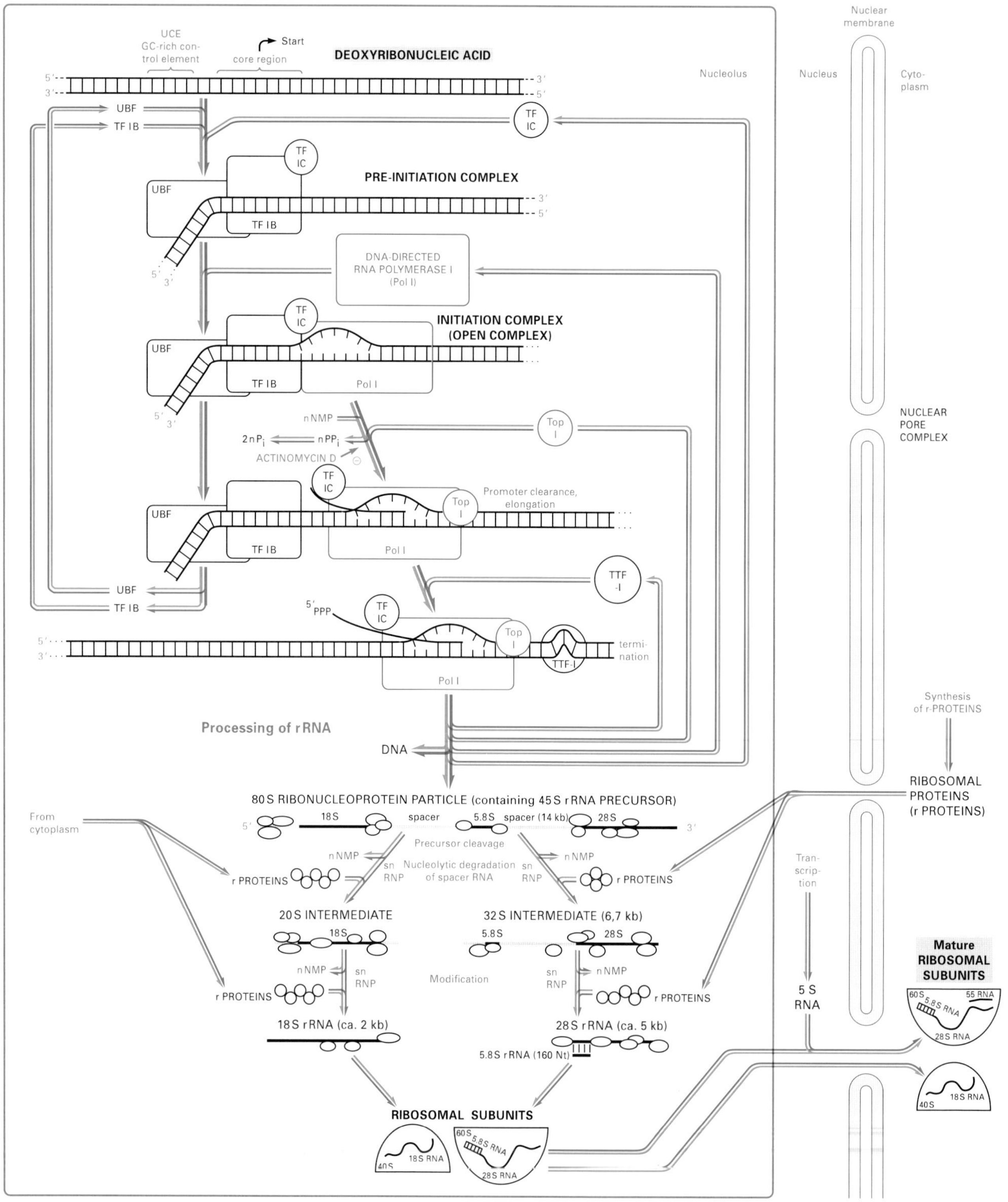

TFIIIB/Pol III initiation complex is very stable and may pass through many rounds of tRNA transcription. Transcription and elongation start immediately after assembly of the initiation complex. The transcription of snRNA U6 (11.3.4) proceeds similarly.

11.3.8 Modification/Processing of tRNAs (Figure 11.3-8)
Pre-tRNAs are processed by cleavage of a 5′ leader sequence and by splicing to remove an intron close to the anticodon loop. Upon maturation the UU sequence at the 3′ end is replaced by CCA.

Eukaryotic tRNAs contain a large variety of modified nucleotides for fine tuning of activity, fidelity and stability, which are formed post-transcriptionally. The 2′ hydroxyl groups of about 1% of all riboses are methylated. There are between 7 and 15 unusual bases per molecule, e.g. methylated or dimethylated A, U, C or G residues and pseudo-uridine (Ψ).

Mature tRNAs are then transported to the *cytoplasm*. The three-dimensional structure of *eukaryotic* tRNAs is similar to *bacterial* tRNAs (Figure 10.6-1).

11.3.9 5S rRNA Transcription by RNA Pol III (Figure 11.3-9)
5S rRNA is a short (120 nt) molecule, which is highly conserved in sequence and structure (5 stem loops). It is transcribed from a group of tandemly arranged genes outside of the *nucleolus*. The procedure is similar to tRNA transcription, starting with binding of TFIIIA, followed by TFIIIB and TFIIIC and Pol III recruitment. After elongation and termination, the primary transcript undergoes only minor processing, e.g. removal of 10...50 nucleotides from the 3′ end. Surplus 5S RNA is degraded in the *nucleus*.

Figure 11.3-7. Transcription of tRNA

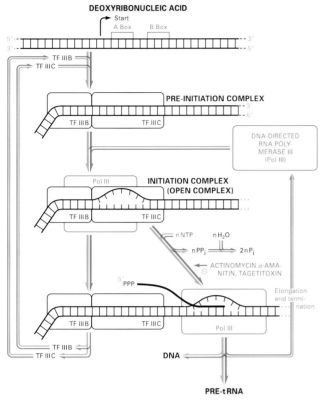

Figure 11.3-9. Transcription of 5S rRNA

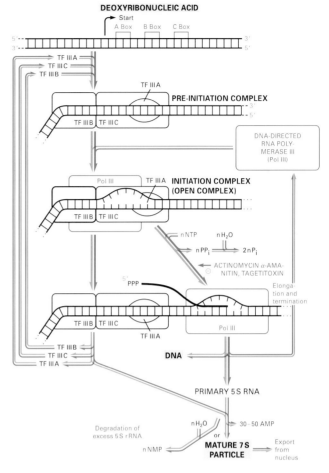

Table 11.3-5. Transcription Factors for RNA Polymerase III

Name	Species	Subunits	Mol. mass (kDa)	Functions / Properties
TFIIIA	human	1	42	Assembly factor for positioning of TFIIIB; binds to an internal control region of 5S rRNA genes. Role in export of 5S rRNA.
	yeast	1	10	
TFIIIB	human	TBP +TAF$_{III}$	–	Required for expression of all Pol III transcribed genes. Role in initiation, binds upstream of transcription start site. Bends DNA upon binding. Equivalent to TFIID.
	yeast	TBP + 2 TAF$_{III}$	90, 67	
TFIIIC	human	6	230...55	Binds to 2 intragenic promoter elements of tRNA genes. Causes positioning of TFIIIB; not always required for transcription.
	yeast	6	145...55	
PBP	human	1	90	Proximal element binding protein

Figure 11.3-8. Modification of tRNA (Example: tRNA^{Tyr} from yeast)
The sites to be modified are shown in red

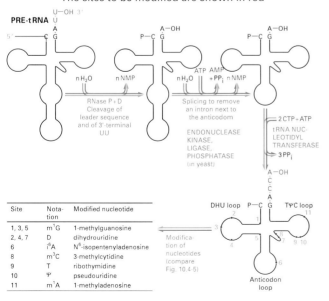

11.3.10 Inhibitors of Transcription

Actinomycin D binds tightly and specifically to double-stranded DNA and stops transcription in general. α-Amanitin inhibits the translocation step in the elongation process. Some RNA polymerases, however (e.g. RNA polymerase II from *Aspergillus nidulans*) are resistant to α-amanitin. RNA polymerase I is markedly less sensitive to α-amanitin. This property can be utilized to distinguish between different RNA polymerases. Tagetitoxin (from *Pseudomonas*) inhibits Pol II and III transcription at concentrations that leave Pol I unaffected.

Literature:
Aso, T., *et al.*: FASEB J. 9 (1995) 1419–1428.
Berget, S.M.: J. Biol. Chem. 270 (1995) 2411–2414.
Dahmus, M.E.: Biochim. Biophys. Acta 1261 (1995) 171–182.
Geiduschek, E.P., *et al.*: Curr. Opin. in Cell Biol. 7 (1995) 334–351.
Koleske, A.J., Young, R.A.: Trends in Biochem. Science 20 (1995) 113–116.
Krämer, A.: Ann. Rev. of Biochem. 65 (1996) 367–409.
Maldonado, E., Reinberg, D.: Curr. Opin. in Cell Biol. 7 (1995) 352–361.
Mélère, T., Xue, Z.: Curr. Opin. in Cell Biol. 7 (1995) 319–324.
Morrissey, J.P., Tollervey, D.: Trends in Biochem. Sci. 20 (1995) 78–82.
Newman, A.: Curr. Opin. in Cell Biol. 6 (1994) 360–367.
Pugh, B.F.: Curr. Opin. in Cell Biol. 8 (1996) 303–311.
Scheer, U., Weisenberger, D.: Curr. Opin. in Cell Biol. 6 (1994) 354–359.
Sharp, P.A.: Cell 77 (1994) 805–815.
(various authors): Trends in Biochem. Science 21 (1996) 320–364.
Wahle, E.: Biochim. Biophys. Acta 1261 (1995) 183–194.
Zavel, L., Reinberg, D.: Ann. Rev. of Biochem. 64 (1995) 533–561.

11.4 Regulation of Eukaryotic Transcription

Eukaryotic transcription (11.3) is regulated by an interplay between specific DNA sequence elements (promoters, enhancers, silencers) and special proteins (transcription factors), which recognize these DNA regions.

- Core promoter DNA elements are always required for accurate and efficient initiation of transcription. They are recognized by the basic transcription factors (Table 11.3-2); the complex provides the basic transcription rate.
- DNA response elements (enhancers, silencers, Fig.11.4-1 and Table 11.4-2) increase or decrease the basic transcription rate of a given promoter. Specific transcription factors can bind to these 'cis-acting' DNA elements, modulating the rate of transcription by interacting either directly or via co-activators with the general transcription apparatus.

A differentiation is to be made between *trans*-acting protein factors (transcription factors etc.), which can interact with every site of a genome containing their recognition sequence, and *cis*-acting DNA regions (enhancers, silencers), which only react with their corresponding promoter.

Figure 11.4-1. Regulating DNA Sequence Elements (Any Order is Possible)

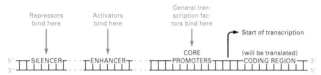

In a number of cases it has been shown that in transcriptionally active genes the sequences at the start site and up to some distance upstream of it (DNase I hypersensitive sites) are free of nucleosomes (2.6.4), allowing free access of the protein factors to the promoters.

11.4.1 Structure of Core Promoter DNA Elements

These DNA regions determine the starting point and the basic initiation frequency of transcription. In *eukarya*, different core promoters exist for each of the RNA polymerases (11.3.1).

RNA polymerase II core promoters (Fig. 11.4-2): Polymerase II core promoters consist of either a so-called TATA box (consensus sequence: $T_{0.82}A_{0.97}T_{0.93}A_{0.85}A_{0.63}/T_{0.37}A_{0.88}A_{0.50}/T_{0.37}$; the indices indicate the frequency of the nucleotide) or an initiator element (Inr, consensus sequence PyPyANTAPyPy) or both. The TATA box is usually located at about –30 nucleotides (upstream of the transcription start site) whereas the Inr element includes the start site.

Human snRNAs, which are transcribed by polymerases II and III, use a TBP (see 11.3) – TAF (TBP associated factors) complex called small nuclear RNA activating protein complex (SNAPc). This complex binds specifically to the proximal sequence element at the DNA (PSE, position ca. –50), which is a core promoter element common to snRNA genes. In case of transcription by Pol II, the TATA-box is missing.

Figure 11.4-2. Examples of RNA polymerase II core promoters

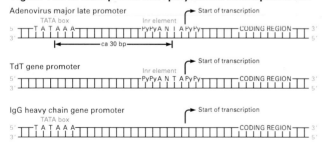

RNA polymerase I core promoters (Fig. 11.4-3): RNA polymerase I promoters (and the vast majority of RNA polymerase III promoters) lack TATA boxes. The core promoter (in *mammals* and *amphibians*) consists of two critically spaced sequences, the upstream control element (UCE, location at –200...–100 nucleotides from the start site) and the core region (location at –50...+20 nucleotides, therefore enclosing the start site). Only little transcriptional control is exerted on pre-rRNA synthesis.

The upstream binding factor (UBF protein) binds to both sequences and greatly enhances recruitment of the 'TATA binding protein complex' (TBP, a part of transcription factor TFIB, 11.3.5), which even in the absence of TATA supports transcription *in vitro*. If, however, the core region is lacking, no transcription takes place; without the upstream element, the transcription rate drops to 5…10 %.

Figure 11.4-3. Example of RNA polymerase I core promoters

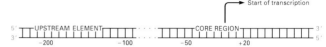

RNA polymerase III core promoters (Fig. 11.4-4): Several types of polymerase III promoters exist. They contain downstream elements within the genes [A and B boxes (which are each 10 bp long and bind transcription factor TF IIIC) and also I and C (internal control region) boxes]. Many polymerase III promoters (e.g. t-RNA and 5S rRNA promoters) have no TATA boxes.

Figure 11.4-4. Examples of RNA polymerase III core promoters

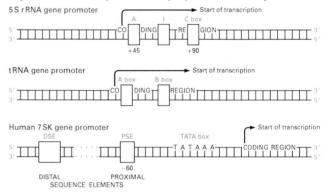

11.4.2 Structure of Specific Transcription Factors

The basal activity of promoters can be modulated by additional, specific transcription factors (11.4.3). They contain these functional domains:

- a DNA binding domain that recognizes specific recognition or response sequences within its target promoters. The structures (zinc fingers, helix-turn-helix etc.) are frequently highly conserved between species. These domains exist also in many basic transcription factors (11.3.2, 11.3.6, 11.3.8).
- an activation domain of variable composition (acidic, glutamine-rich, proline-rich, serine/threonine-rich etc.) that is required for transcription stimulation. This activation domain makes either direct protein-protein contact with components of the basic transcription machinery or acts via co-activators.
- in some cases (e.g. nuclear receptors) a hormone binding domain.

A number of specific transcription factors react as homo- or heterodimers. Their components are connected by dimerization domains.

Table 11.4-1. Examples of Domains in Transcription Factors

Functional Domain	Structure, Action	Occurs in Transcription Factor (Examples)
DNA binding Domains:		
Zinc finger	1 Zn is complexed by Cys_2-His_2 or by Cys_2-Cys_2 2 Zn are complexed by Cys_6	TFIIIA, Sp1 steroid receptors Gal 4 (*yeast*)
Helix-turn-helix	Pair of tilted α-helices, binds via hydrogen bonds, salt bridges etc.	homeodomain proteins
Dimerization Domains:		
Leucine zipper	Dimerizes proteins with α helices by hydrophobic attraction of several Leu on one side of the α helices.	Fos, Jun, Myc, CREB = ATF-2, CREM
Basic helix-loop-helix	A basic region followed by 2 helices, connected by a loop.	Max, Myc, MyoD

11.4.3 Modulation of the Transcription Rate

The modulating specific transcription factors bind to DNA response elements (enhancers, repressors, silencers). These response elements can be located within the genes or upstream or downstream up to several 1000 bp away in either orientation. This often requires the bending of the DNA to enable contact between the specific and the basic transcription factors or coactivators (Fig. 11.4-5). Since frequently more than one response element belongs to a given promoter, several transcription factors can interact by enhancing or attenuating each other's effects.

Figure 11.4-5. Contact of Upstream Elements With Core Promoters (Schematically)

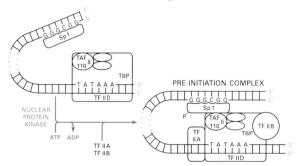

The specific transcription factors have to be activated by binding of ligands or by phosphorylation in order to bind to their target DNA sites. Some examples for this mechanism are shown in Fig. 11.4-6. For steroid receptors, see Section 17.6.

The (more frequent) enhancement or the repression of gene expression proceeds by influencing the initiation (in most cases), promoter clearance and/or elongation steps (11.3).

E.g., enhancers of RNA polymerase II often act on the TAF_{II} coactivator subunits of transcription factor TFIID (11.3.2) to stabilize preinitiation complexes or to assist in recruiting Pol II into preinitiation complexes. Regulation of gene expression can also be exerted by controlling the rate of the transcription factor synthesis or transcription factor access to the nucleus.

These mechanisms effect transcription of genes
- either constitutively in all cell types ('housekeeping genes')
- or only in certain organs or cell types
- or on demand under certain conditions

In *eukarya*, the expression ratio of 'turned on' and 'turned off' genes can be up to $10^9/1$ (In *bacteria*, the rate is about $10^3/1$).

In addition to regulation at the transcriptional level, gene expression can also be modulated by the degradation rate of mRNA (11.5.5), by different splicing (e.g., 11.3.3), by the translocation rate to the *cytosol* or by modification of the translation rate (11.5).

Literature:
Beato, M.: Cell 56 (1989) 335–344.
Bohmann, D.: Cancer Cells 2 (1990) 337–344.
Cowell, I.G.: Trends in Biochem. Sci. 19 (1994) 38–42.
Guarante, L.: Trends in Biochem. Sci. 20 (1995) 517–521.
Pahl, H.L., Baeuerle, P.A.: Current Opin. in Cell Biol. 8 (1996) 340–347.
Teisman, R.: Trends in Biochem. Sci. 17 (1992) 423–426.
Vandromme, M., *et al.*: Trends in Biochem. Sci. 21 (1996) 59–64.

Table 11.4-2. Examples of Eukaryotic Enhancers and Repressor Elements

Response Element	Consensus Sequence (Palindromes in bold letters)	Binds Specific Transcription Factors	Modulates Expression of Genes Coding for (Examples)
GC box	GGGCGG	stimulatory or specificity protein 1 (Sp1)	enhancer for housekeeping genes transcribed by Pol II and Pol III, regulated by Rb protein (11.6.4).
cAMP response element (CRE)	**TGACGTCA** (similar: in TRE element)	CRE binding protein (CREB = ATF-2). Similar: CREM. Requires association with CBP.	somatostatin, c-Fos, PEPCK, VIP, PTH, tyrosine hydroxylase, fibronectin. Absent in some other genes responsive to cAMP, e.g. growth hormone, prolactin.
Serum response element (SRE)	**GTCCATATTAGGAC**	Serum response factor (SRF, assoc.with TCF, Elk-1, Sap-1)	c-Fos
Phorbol ester (TPA**, AP-1) response element (TRE)	**TGACTCA** (also part of VDRE)	AP-1 (Jun•Fos)	collagenase, stromelysin, c-Myc, c-Sis, Pro-1
Vitamin D receptor response element (VDRE)	a**GGTGACTCACCt** *	vitamin D receptor (17.6)	osteocalcin, osteonectin, calbindin, calreticulin, alkaline phosphatase
Glucocorticoid receptor response element (GRE)	**TGTTCT** (palindromic half site)	glucocorticoid receptor (17.6)	growth hormone, bone sialoprotein, *chicken* lysozyme
Estrogen receptor response element (ERE)	a**GGTCA**NNN**TGACCt** *	estrogen receptor (17.6)	estrogen responsive genes, e.g. osteocalcin, *chicken* ovalbumin
CCAAT box	g**GCCAAT**ct *	CCAAT-binding transcription factor (CP1), CTF	α-globin and many other products
CACCC box / Rb control element (RBE)	gc**CACCC** *	Rb protein (tumor suppressor, 11.6.4) and others	globin
Myc control site	**CACGTG**	Myc/Max, Mad/Max, Mxi/Max, Max/Max dimers	regulators of cell growth and differentiation
κB sequence motif	**GGGA**NN**PyTCC** (**GGGAAATTCC**)	nuclear factor kB (after release from IkB complex)	immunoglobulins, IL-2, IL-2 receptor, IFN-β, GM-CSF, TNF-β (in *B cells*, certain *T cells* and *monocytes*)
Octamer DNA bind. motif	ATGCAAAT	Oct-1, 2, 3, 4 (OTF-1, 2, 3, 4)	snRNA (by Oct-1), β-globin (by Oct-2, in lymphoid cells)
E-Box	ccgaa**CACATGTG**cccgc *	basic helix-loop-helix transcription factors	scleraxis (bone development, e.g.in *rats*)

* Less conserved bases are printed with lower case letters
** TPA (12-O-tetradecanoyl-phorbol-13-acetate) is a potent tumor promoter, potentiating the effect of a subcarcinogenic dose of an initiating carcinogen

Figure 11.4-6. Activation and Binding of Regulatory Proteins to Modulating DNA Elements (Examples, Schematically)

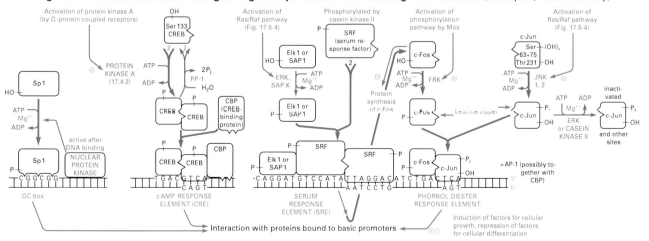

11.5 Eukaryotic Protein Synthesis

Translation of the genetic code (i.e. the biosynthesis of proteins from amino acids as encoded in the RNA message) proceeds on ribosomes and uses charged tRNs as activated forms of the amino acids. In *eukarya*, it takes place in the *cytosol* or at the membranes of the *endoplasmic reticulum* (13.3) or of the *nucleus*.

The steps of *eukaryotic* protein synthesis resemble the *bacterial* ones (10.6), but are more complicated. On the other hand, the protein synthesis machinery in *mitochondria* and *chloroplasts* is simpler than in *bacteria*.

11.5.1 Components of the Translation System

Transfer RNAs: The general structure of eukaryotic tRNAs is analogous to *bacterial* tRNAs (Figure 10.6-1). The tRNAs are charged at their 3' ends (sequence -CCA) with their cognate amino acids by specific aminoacyl-tRNA ligases (also named 'synthetases'). This reaction requires ATP and takes place in two steps: Activation of the amino acid followed by charging of the tRNA (see 10.6.1 and Fig.10.6-3).

Contrary to *bacteria*, in many higher *eukarya* a number of ligases associate to a multienzyme particle or even fuse to a single polypeptide (GluPro-ligase).

Messenger RNAs: While *bacterial* mRNA is usually translated without modification, *eukaryotic* mRNAs are extensively processed before leaving the *nucleus*, resulting in a complicated structure (Fig.11.5-1). They contain
- a methylated 5' cap (for details, see Fig. 11.3-2)
- a 5' untranslated leader sequence (5'-UTR, usually < 100 bases), involved in regulation of transcription initiation and mRNA degradation
- the coding sequence, to be translated into proteins
- a 3' untranslated sequence (3'-UTR, length up to 1000 bases), involved in mRNA localization, initiation (or repression) of translation and mRNA degradation (by controlling the decapping rate, 11.5.5)
- the poly (A) tail, essential for the stability of mRNA (see Fig.11.3-5).

Ribosomes: *Eukaryotic* ribosomes (80S or 4300 kDa) consist of two subunits of 40S (1400 kDa) and 60S (2900 kDa). They contain RNA and many different proteins (Table 11.5-1). Although their structure is more complex than that of *bacterial* ribosomes, the basic features are similar. This includes the general shape (Fig. 10.6-2) and the presence of the aminoacyl- (A), peptidyl- (P) and exit (E) interaction sites with tRNA and mRNA (10.6.1).

Table 11.5-1. Composition of *Eukaryotic* Ribosomes (*Rat liver*)

	Small subunit (40S)	Large subunit (60S)
RNA	18S, 1874 nt, 700 kDa	28S, 4718nt + 5.8S, 160 nt + 5S, 120 nt; total 1820 kDa
Protein	33 polypeptides, 700 kDa	49 polypeptides, 1000 kDa

Translation factors (Table 11.5-2): In *eukaryotic* translation, a large number of protein factors are involved. E.g., there are at least 10 initiation factors (eIF) in *eukarya* as compared to 3 in *bacteria*. While *bacteria* regulate protein synthesis almost completely at the transcriptional level, *eukarya* also exert control during translation involving such factors.

11.5.2 Polypeptide Synthesis (Figure 11.5-2)

Initiation: *Eukaryotic* protein synthesis starts with an initiator tRNA$_1^{Met}$ which is always charged with methionine (not formylmethionine as in *bacteria*) and is structurally different from the regular tRNA^{Met}. It is, however, charged by the normal ligase.

After completing a round of polypeptide synthesis, the ribosomes dissociate into the subunits. Reassociation is prevented by binding of initiation factors eIF-1A and eIF-3 to the 40S and eIF-6 to the 60S subunit. Then a 43S initiation complex forms from the 40S subunit, eIF-1A, eIF-2•GTP, eIF3 and Met-tRNA$_1^{Met}$. This complex localizes the capped 5' end of the mRNA and migrates along the mRNA to the first AUG codon (scanning model). Met-tRNA$_1^{Met}$ and the start codon of mRNA are placed at the P-half site of the 40S ribosomal subunit.

Initiation is facilitated if AUG occurs in the context GCCGCC(A/G)CC AUGG. In less than 10% of all cases, initiation at non-AUG codons (CUG, ACG, GUG) or at downstream AUG codons takes place, possibly caused by secondary structural features of the mRNA.

Eukaryotic mRNAs apparently lack a sequence like the *bacterial* Shine-Dalgarno-Sequence (10.6.3), which is specifically recognized by the ribosome. Consequently, they are mostly monocistronic (coding for only 1 gene).

Figure 11.5-1. Structure of Eukaryotic mRNA

Table 11.5-2. *Eukaryotic* Translation Factors

Factor	Subunits	Mol. Mass (kDa)	Prokaryotic homologue	Function
eIF-1A	1	17		Binds to the 40 S ribosomal subunit, prevents ribosome reassociation in the absence of an initiation complex.
eIF-2	3 (α, β, γ)	52, 38, 35		Binds initiator tRNA (Met-tRNA$_1$) in GTP-dependent manner. Very important control reaction of translation. Its activity is shut off by phosphorylation of Ser 51 in the α subunit, which is catalyzed, e.g., by PKR (double-stranded RNA dependent kinase) or HRC (heme controlled repressor).
eIF-2B	5	82, 67, 58, 39, 26		*Mammalian* guanyl nucleotide exchange factor (= GEF), regenerates GTP at eIF-2.
eIF-3	8...10	500...750	IF-3	Cooperates with eIF-1A in prevention of ribosome reassociation.
eIF-4A	1 (2 isoforms)	45		Subunit of eIF-4F. Assembles with eIF-4E and eIF-4G at the mRNA cap, melts secondary intermolecular structures which otherwise prevent binding of the 43S pre-initiation complex. Single-stranded RNA dependent ATPase and helicase.
eIF-4B	2 (homodimer)	69		Stimulates the RNA-dependent ATPase and helicase activity of eIF-4A
eIF-4C	1	17		Binds after ribosome separation to the 40S subunits, facilitates consecutive steps
eIF-4D	1	16		Appears to function late in initiation
eIF-4E	1	24		Subunit of eIF-4F. Cap binding protein (CBP). Cooperates with eIF-4A for mRNA melting. One of the least abundant eIF's
eIF-4F	3	100...150 (eIF-G), 45 (eIF-4A), 24 (eIF-4E)		The functions of subunits 4A, 4E and 4G are described in the respective lines. *Polio-, rhino-* and *hoof-and mouth-disease viruses* split the eIF-G subunit, causing inactivation of translation in the host and stimulation of *viral* mRNA translation.
eIF-4G	1	100...150		Other names: p420 and eIF-4γ. Cooperates with eIF-4A for mRNA melting.
eIF-5	1	49		Possibly effects release of eIF factors by triggering hydrolysis of eIF-2 bound GTP after the ribosome joins the AUG triplet.
eIF-6	1	25		Binds to the 60S ribosomal subunit, apparently prevents ribosome reassociation in the absence of an initiation complex (similar to eIF-1A, -3).
eEF-1α	1	50...60	EF-Tu	Catalyzes GTP-dependent binding of aa-tRNA to ribosomes. Very abundant protein.
eEF-1βγ	2 (β, γ)	48, 35	EF-Ts	Facilitates nucleotide exchange at eEF-1α
eEF-2	1	95	EF-G	Responsible for GTP dependent translocation step during elongation
eEF-3	1	125		Ribosome-dependent nucleotidase (ATP and GTP). Found so far only in *fungi*
eRF-1	2 (homodimer)	49...55	RF-1, RF2	Release factors for peptide chain with a structure analogous to tRNAs. eRF-3 binds GTP. The homology to *bacterial* release factors is only with respect to function.
eRF-3	1	80	RF-3	

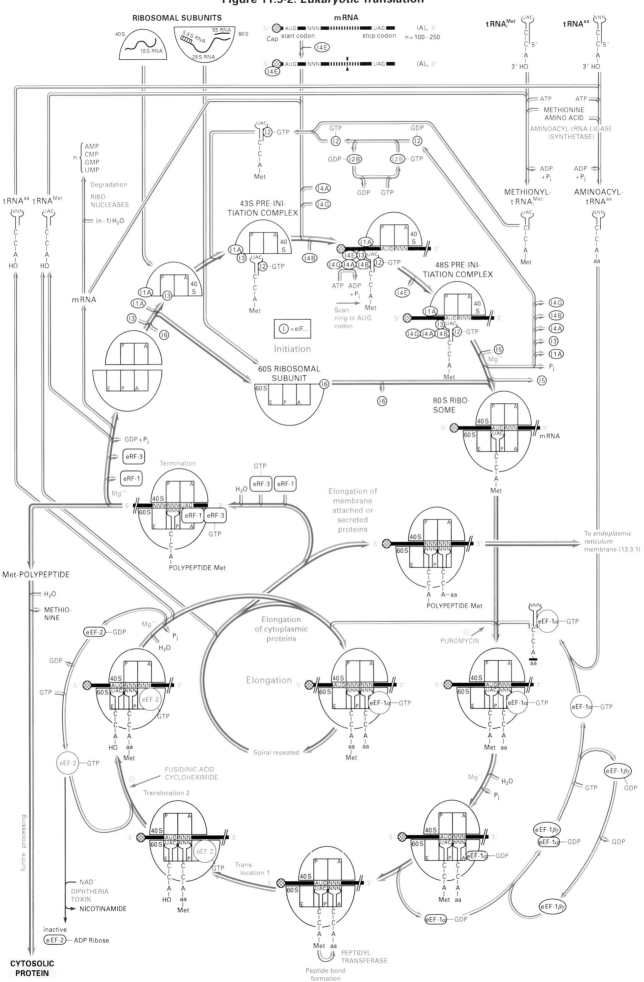

Figure 11.5-2. *Eukaryotic* Translation

eIF5 joins the 43S initiation complex and triggers hydrolysis of eIF-2 bound GTP, which causes eIF-2•GDP and the other initiation factors to leave. Then the 40S ribosomal subunit combines with the 60S subunit to form the 80S initiation complex. eIF2-GTP is regenerated from eIF2 by the action of eIF-2B (GEF = guanine nucleotide exchange factor). The eIF2 cycle is a control point in protein synthesis: Phosphorylation of the α subunit prevents regeneration of the GTP form.

Elongation: The steps of peptide bond formation and movement to the next codon proceed analogous to the mechanism in *bacteria* (10.6.3). The anticodon sequences of all further charged tRNAs recognize (by base pairing) complementary codon sequences in the mRNA (10.1). This takes place at the aminoacyl (A) site of the ribosome. tRNA also interacts with the ribosome itself.

Due to the flexibility in base pairing at the third position of a codon (wobble hypothesis, see Fig.10.1-2) only ca. 35 tRNAs are needed to translate all 61 codons specifying amino acids.

The rRNA of the large ribosomal subunit exerts peptidyltransferase activity, which catalyzes the transfer of Met (or of the peptidyl chain in the following cycles) from the tRNA at the P-site to the N end of the aminoacyl residue attached to the tRNA at the A-site (see Fig. 11.5-2). Then the amino acid acceptor ends of both tRNAs move at the 60S ribosomal subunit from A- to P-half sites and from P- to E-half sites, respectively, while their anticodons and the mRNA remain at their sites at the 40S subunit. Thus, the tRNAs are located at hybrid sites: A/P-site and P/E-site. In a second step, both tRNAs and the mRNA move one codon site relative to the 40S ribosome. The peptidyl-tRNA occupies now the P site in both ribosomal subunits, while the deacylated tRNA leaves the ribosome from the E-site. A new charged tRNA can now bind to the free A-site, thus starting another elongation cycle.

The movements of the ribosome involve configuration changes. The energy comes from GTP hydrolysis which occurs after binding of eEF-2•GTP to the ribosome. After the translocation step, eEF-2•GDP is released. It reacts with free GTP to regenerate the complex for another reaction cycle. Cycloheximide stops the translocation step by inhibition of GTP binding, diphtheria toxin inactivates eEF-2 by ADP-ribosylation (compare 17.4.1).

Many glycosylated proteins (e.g. membrane-bound, secreted or lysosomal ones) are synthesized in the *endoplasmic reticulum* (13.3). The N-terminal parts of these proteins, which are still formed by free ribosomes in the *cytoplasm*, have the function of a signal sequence. This sequence is recognized by a signal recognition particle (SRP, 13.3.1), which targets the whole translation complex to the *endoplasmic reticulum*. The growing peptide chain is threaded into the lumen of this compartment, where posttranslational modifications take place (13.3.1, also see below).

Termination: Like *bacteria*, *eukarya* use the stop codons UAA, UAG and UGA to signal the end of the coding region of a gene (10.6.3). There are two *eukaryotic* release factors, eRF-1 (codon specific) and eRF-3 (codon unspecific, binds GTP).

Apparently the eRFs recognize the stop codon by mimicking a tRNA anticodon structure. This prevents further transfer of the polypeptide chain, which is then hydrolytically removed from the tRNA. Puromycin causes premature chain termination by acting as an analogue of aminoacyl-tRNA.

A high amount of free energy is spent on achieving the peptide bond formation and the reaction specificity (for details, see 10.6.3).

11.5.3 Posttranslational Protein Processing

The mechanism of protein folding is dealt with in Section 14.1.

Many hydrolytic processes take place:
- In most cases, the N-terminal methionine group of proteins is enzymatically split off. Frequently, additional N- or C-terminal sequences are also removed by specific peptidases. Examples are targeting (signal) sequences after the protein has reached its destination (13.3.2). In about half of all proteins, the N-terminus becomes then acetylated.
- Another proteolytic process is the conversion from inactive precursors into active proteins (e.g., 17.1.9).

Group transfer reactions include:
- Glycosylation in the *endoplasmic reticulum* and in the *Golgi apparatus* (13.3 and 13.4), which is important for modifying the protein properties and in cellular recognition processes.
- Phosphorylation reactions, which frequently regulate enzyme activities, calcium binding ability etc. (17.4, 17.5).
- Membrane attachment. This is effected by binding of myristic or palmitic acid to glycosylphosphatidyl residues or to isoprenyl residues (13.3.4 and 13.3.5).

A number of non-standard amino acids is formed by posttranslational modification. Examples are carboxylation of glutamate (4.2.2), methylation of lysine (4.5.2) and hydroxylation of proline (4.3) and lysine (4.5.2).

11.5.4 Translational Regulation

In contrast to *bacteria*, regulation of translation by repressor proteins is a common feature in *eukarya*.

E.g., by acting as an ion sensor, iron response element binding protein (IRE-BP) regulates the production of the iron sequestering compound ferritin. When iron concentration is low, IRE-BP binds to ferritin mRNA; this inhibits ferritin translation. Iron excess causes its dissociation from the mRNA and resumption of translation. Similar mechanisms exist for stabilization of mRNAs.

Phosphorylation of initiation factors is a common way to regulate translation, e.g. for coordination of hemoglobin synthesis.

In the absence of heme, *reticulocytes* generate the heme controlled repressor from a precursor (Fig. 11.5-3). This repressor phosphorylates the α subunit of eIF-2, which then binds tightly to eIF-2B, thus stopping the regeneration cycle of eIF-2•GTP (Fig. 11.5-2) and thus the initiation of globin transcription. Heme inhibits the repressor formation and thus allows globin synthesis. Already phosphorylated eIF-2B can be reactivated by a phosphatase.

Figure 11.5-3. Control of Translation in *Reticulocytes* by Heme

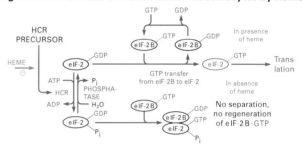

Translation can re-start at sites downstream from the first AUG. This re-initiation process is a control mechanism often encountered in translation of important cellular genes like oncogenes, growth factor genes etc. Another method of regulation is the competition of different mRNAs for mRNA specific initiation factors. Also, temporary masking of mRNA by association with proteins occurs, especially in unfertilized eggs of several species.

11.5.5 mRNA Degradation

mRNA stability is an important control factor in *eukaryotic* protein synthesis. The half-life of *eukaryotic* mRNA can vary from a few minutes to months, while it is generally very short in *bacteria*. Structures which influence the degradation rate can occur in all parts of the molecule (cap, 5' and 3'-UTR, coding region).

For degradation of a number of *yeast* mRNAs, the 3' poly(A) tail, which is covered by poly(A) binding protein 1 (Pab1), gets shortened in the *cytoplasm* by the nuclease PAN. When a length of $A_{10...15}$ is reached, Pab1 cannot be bound any more. This enables the Dcp1 enzyme to decap the 5' end. The free terminus is now open for attack by 5'→3' exonucleases, e.g. Xrn1 (Fig. 11.5-4). In other *eukarya*, the procedure is probably similar.

Also, endonucleolytic cleavage can cause exonucleolytic degradation, beginning at the unprotected end. Histone mRNAs are not polyadenylated but possess a stem-loop structure at their 3' ends. The initial decay event removes nucleotides at the 3' end resulting in disruption of the stem-loop.

Figure 11.5-4. mRNA Degradation in *Yeast*
Pab1 = Poly(A) binding protein, PAN = Pab1 dependent poly(A) nuclease

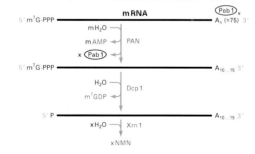

Literature:
Caponigro, G., Parker, R.: Microbiol. Reviews 60 (1996) 233–249.
Carter, C.W., Ann. Rev. of Biochem. 62 (1993) 715–748.
Decker, C.J., Parker, R.: Curr. Opin. in Cell Biol. 7 (1995) 386–392.
Nakamura, Y., et al.: Cell 87 (1996) 147–150.
Pain, V.M.: Eur. J. Biochem. 236 (1996) 747–771.
Ross, J.: Microbiol. Reviews 59 (1995) 423–450.
Ross, J.: Trends in Genetics 12 (1996) 171–175.

11.6 Cell Cycle in Eukarya

When cells divide, the correct transmission of genetic information to the next generation has to be insured. In contrast to *prokarya*, where periods of DNA replication and chromosome segregation overlap, *eukaryotic cells* duplicate and segregate their chromosomes in a strictly sequential order.

The cell cycle of *somatic cells* of *eukarya* (Figure 11.6-1) is divided into
- M phase (mitosis), the phase of chromosome and cell separation (Fig. 11.6-4)
- interphase, the interval between succeeding rounds of mitosis, consisting of
  - G_1 phase, the gap between mitosis and the beginning of DNA replication
  - S phase, the period of DNA synthesis
  - G_2 phase, the gap between S phase and mitosis.

While the mechanism of *eukaryotic* DNA replication is described in 11.1, this section deals with regulatory effects during the cell cycle, especially in the G_1 and M phases. However, many aspects are still only partially known.

11.6.1 Cyclins and Cyclin-Dependent Kinases

The *eukaryotic* cell cycle is regulated by activity waves of different cyclin-dependent kinases (CDKs). CDKs are protein serine/threonine kinases of 30...36 kDa sharing more than 50% identity with each other. A prerequisite for activity is the complex formation between their catalytic subunit and highly unstable proteins, which are therefore called cyclins (Fig. 11.6-1). Binding also contributes to substrate specificity of the kinases.

Cyclin generation and degradation provide the decisive control mechanism for stepwise and unidirectional progress through the various cell cycle phases.

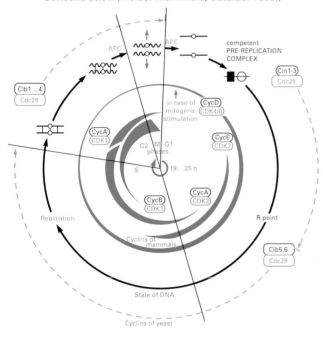

Figure 11.6-1. *Eukaryotic* Cell Cycle: State of DNA and Cyclin Concentration (inside: *Mammals*, outside: *Yeast*)

Regulation of Cyclin-CDK Activity (Fig. 11.6-2): Cyclins share a conserved sequence of 100 amino acids called the cyclin box, which interacts with the CDK subunit. This association causes a conformational change within the CDK, which leads to partial activation of CDKs. A subsequent phosphorylation of a key threonine residue (T160 in *mammalian* CDK2) by a CDK activating kinase (CAK, Table 11.6-1) is required for full activation and stabilization of the cyclin-CDK complex. The active complex then catalyzes the phosphorylation of a number of substrates leading to cell cycle progression.

Reversible inactivation of this complex takes place by additional phosphorylation of a tyrosine residue (Y15 in CDK2) in response to a cell cycle checkpoint (11.6.6). Dephosphorylation is mediated by the Cdc25 phosphatase. Another inactivation of the cyclin/CDK complex is achieved by association with members of the Kip/Cip family of inhibitory proteins (p21, p27, p57, Table 11.6-1) in response to DNA damage (p21) or environmental changes (p27, Fig. 11.6-4). This is reversed by degradation of this inhibitory protein.

Degradation of the cyclin and dephosphorylation of the CDK protein results in irreversible inactivation (mostly at the end of a cell cycle transition).

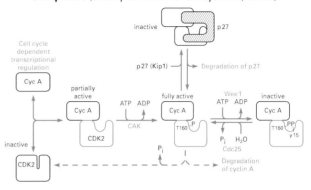

Figure 11.6-2. Activation and Deactivation of Cyclin-CDK Complexes (Example: *Mammalian* Cyclin A/CDK2)

11.6.2 Regulation of G_1 to S Phase Transition in *Yeast* (Fig. 11.6-3, see also 11.1.2 and Fig. 11.1-2)

The early steps of DNA replication are best known in *yeast* (*Saccharomyces cerevisiae*). Functional analogies to *mammals* are shown in Table 11.6-2.

Cell division depends on the environmental state (e.g. food supply). If sufficient quantities are available and the cell has grown beyond a critical size, a START signal is generated (equivalent to the restriction point in animal cells), resulting in the activation of the G_1 cyclin Cln1,2,3-cyclin-dependent kinase Cdc28 complex in the late G_1 phase. The activated kinase performs three functions:

- It phosphorylates and activates the transcription factors Sbf and later Mbf which promote transcription of enzymes necessary for DNA replication.
- It also phosphorylates Sic1 (p40), an inhibitor of S-phase CDK activities. This targets Sic1 for degradation by Cdc34 (ubiquitin-protein ligase), which conjugates it with activated ubiquitin (Ub) and thus initiates its degradation (11.6.7). This way, the previously inhibited S-phase (B-type) cyclin complex Clb5,6–Cdc28 is activated. It phosphorylates and thus activates factors, which then enable the entrance into the S-phase (Fig. 11.1-2).
- The G_1-phase cyclins Cln1, 2, 3 are marked analogously for degradation (Fig. 11.6-2). This terminates the G_1 phase.

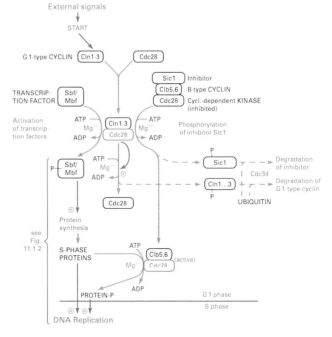

Figure 11.6-3. Regulation of DNA Replication by Cyclins in *Yeast*

11.6.3 Control of the Pre-Replication Complex Assembly in *Yeast* (Fig. 11.1-2)

In the S phase, chromosomes are duplicated resulting in two identical chromatides. The replication process starts at autonomously replicating DNA sequences (ARS, 11.1.2) by assembling a pre-replication complex (pre-RC). This complex is compiled at the end of M phase from the proteins Orc1..6 (which remain there throughout the cell cycle), the activating enzyme Cdc6 and several Mcm proteins (see Table 11.1-2). The presence of these compounds ('licensing factors') renders chromatin competent to replicate upon the arrival of the START signal. At the beginning of the S phase the complex dissolves, effected by the Clb5,6–Cdc28 activity (11.6.2). A new complex cannot be formed, before the Clb5,6–Cdc28 and later the Clb1...4–Cdc28 activities cease (at the end of M phase). Thus replication starts only once during a cycle.

11.6.4 Regulation of G_1 to S Phase Transition in *Mammals* (Fig. 11.6-4) – The Role of the Rb Protein

In multicellular organisms, cells respond to special extracellular signals (+: mitogens together with growth factors, –: anti-proliferative cytokines) and
- either pass a restriction point R in G_1 phase committing them to cell division (11.1.1) or
- exit from the cell cycle and enter a resting phase (G_0).

Upon mitogenic stimulation by hormones (17.5.1), D type cyclins are induced during the first part of the G_1 phase and assemble with their catalytic counterparts, the cyclin dependent kinases CDK4 and CDK 6. For further activation, the cyclin D-CDK complexes must be additionally phosphorylated on a single threonine residue by a CDK activating kinase (CAK, 11.6.1).

Upon withdrawal of mitogens, cyclin D synthesis is downregulated (Cyclin D has a half-life time of only 15 minutes).

One important substrate of the active cyclin D-CDK complex is the retinoblastoma tumor suppressor protein (Rb, 110 kDa). Unphosphorylated Rb binds to a heterodimeric transcription factor, E2F-DP1, and blocks its activity. Consecutive hyperphosphorylation of Rb by cyclin D-CDK 4,6 and cyclin E-CDK2 leads to dissociation of E2F-DP1 from Rb. E2F-DP1 is then able to transactivate the expression of genes, whose products are required for entry into S phase (Fig. 11.6-4, Table 11.6-1) and the consecutive DNA replication (11.1).

In addition, Rb generally represses the transcription of rRNA, tRNA and snRNA genes by DNA-directed RNA polymerases I and III and thus inhibits cell growth. It is assumed, that Rb blocks the action of the transcription factors UBF (in case of Pol I, 11.3.5) and TFIIIB (in case of Pol III, 11.3.7 and 11.3.9).

Elimination of the Rb block corresponds to the passage through the restriction point R (11.1.1) and marks the shift from mitogen-dependent (G_1, cyclin D driven) to mitogen-independent (S, G_2, M, cyclin E, A, B driven) cell cycle phases. Rb remains hyperphosphorylated until reentry in G_1 phase.

The proteins p107 and p130 are closely related to Rb both in sequence and function.

The expression of cyclin E peaks at the $G_1 \rightarrow$ S transition. This expression is stimulated by free E2F, forming a positive feedback loop. Following entry into S phase, cyclin A-CDK2 phosphorylates the DP-1 component of the E2F-DP1 heterodimer. This inhibits its DNA binding ability and thus stops further expres-

Table 11.6-1. *Mammalian* Proteins Involved in Cell Cycle Regulation

Proteins	Expression/Activation and Inhibition	Partners/Targets
Cyclins and cyclin-dependent kinases		
Cyclin A	Periodic expression, peak at G_2/M transition	CDK2 (CDK = cyclin-dependent kinase) in S-phase; CDK1 in G_2/M transition
Cyclin B (1, 2)	Periodic expression, peak at G_2/M transition	CDK1 (analogous to Cdc2 of *S. pombe*). It is involved in nuclear envelope breakdown, chromosome condensation, alignment of chromosomes on the metaphase plate.
Cyclin C	?	CDK8
Cyclin D (1, 2, 3)	Mitogen-responsive expression in G_1 phase; cyclin D1 and CDK4 are overexpressed in many tumors	CDK4,6: The cyclin-CDK complex phosphorylates Rb protein
Cyclin E	Periodic expression, peak at G_1/S transition, stimulated by E2F, mitogen-independent	CDK2
Cyclin F	Periodic expression, peak at G_2/M transition	?
Cyclin G	Does not vary during cell cycle, transcription is activated by p53	?
Cyclin H	Protein levels do not vary during cell cycle	CDK7 (MO15): cyclin H-CDK7 is also termed 'CDK-activating kinase' (CAK). CAK phosphorylates CDK2, CDK4, CDK6
Regulatory proteins		
E2F-DP1 (heterodimeric, several isoforms: E2F 1, 2, 3, 4, 5; DP1, 2)	Phosphorylation of DP1 by cyclin A-CDK2 in S phase, inhibits E2F binding to DNA and thus activation of further transcription	Associated with Rb. Dissolution of this complex by Rb phosphorylation in late G_1 phase enables free E2F to activate transcription of genes encoding e.g. dihydrofolate reductase, thymidine kinase, thymidylate synthase, DNA-Pol-α, CDK1, E2F-1, b-Myb, c-Myc, cyclins E and A etc. – E2F-1, 2, 3 bind to Rb, E2F-4, 5 bind to p107 and p130.
Rb (retinoblastoma tumor suppressor protein)	Phosphorylated by CDKs 4,6,2 in late G_1 phase. Remains hyperphosphorylated until reentry into G_1 phase. Connects cell cycle with transcriptional machinery. Mutations leading to loss of function as tumor suppressor protein are involved in tumorigenesis. Closely related molecules: p107, p130	Binds to and negatively regulates E2F-DP1, also binds to D-type cyclins, CDKs, Elf-1 #, MyoD, PU1, ATF-2 *, c-Abl °, Mdm2 @, UBF §
p53	Expression induced by, e.g., ionizing radiation	Tumor suppressor protein, activates expression of p21 (Cip1).
Cip/Kip family: p21 (Cip1)	Induced by tumor suppressor protein p53, e.g. after radiation. Potent and universal inhibitor of CDK activity, capable of inducing cell cycle arrest	Inhibits activated cyclin-CDK4, 6; CDK2 and CDK1 complexes
p27 (Kip1, kinase inhibitory protein 1)	Low level expression during cell cycle, increases upon mitogen deprivation. This induces cell quiescence ($\rightarrow G_0$ phase). Also induced by TGFβ, cAMP or contact inhibition.	Inhibits activated G_1 cyclin-CDK complexes by formation of a heterotrimeric complex.
p57 (Kip2)		Inhibits activated cyclin-CDK4, 6 and CDK2 complexes
Ink4 family: p16 (Ink4a), p15 (Ink4b), p18 (Ink4c), p19 (Ink4d)	Frequent mutations of the the tumor suppressor protein p16 play a role in *human* tumorigenesis	Inhibit cyclin D-CDK4, 6 by competition with cyclin D
APC (anaphase promoting complex)	Activated in anaphase via phosphorylation after activation of M-phase cyclins A,B-CDK1, active until accumulation of G1-CDKs	Essential for sister chromatid separation. Promotes ubiquitination of proteins containing a destruction box (e.g. M-phase cyclins, in *yeast*: Pds1, Cut2).

Elf-1 = Lymphoid- specific transcription factor, association with Rb blocks G_0 exit; * ATF-2 = Activating transcription factor-2 (Table 17.5-3); ° c-Abl = Transcripton factor, promotes the phosphorylation of the CTD domain of RNA Pol II (Table 17.5-3); @ Mdm2 = Mouse double minute-2; § UBF = Upstream binding factor, enhances Pol I transcription (11.4.1).

sion of cyclin E and S phase genes. Simultaneously, cyclin E is phosphorylated by its own counterpart CDK2, which renders it a target for ubiquitin-dependent proteolysis (11.6.6) and thus terminates its activity.

Modulation of the Rb control system: Several growth inhibitory factors prevent phosphorylation of Rb and thus block entrance into the S phase.
- TGFβ (transforming growth factor β, 19.1.6) induces expression of p15 (Ink4b) and p16 (Ink4a) proteins, which compete with D type cyclins for binding to CDK 4,6 and thus inhibit the kinase activity. The ability of members of the Ink family to inhibit the cell cycle depends on functional Rb.
- TGFβ, cAMP (17.4.2) and cell contact (contact inhibition) elevate the level of p27 (Kip1), which inhibits cyclin-CDK complexes by formation of a heterotrimeric complex (Fig. 11.6-2). Accordingly, terminally differentiated and quiescent cells have high p27 levels.
- The inhibition of Rb phosphorylation by p21 after DNA damage is described in Section 11.6.6.

Pathological aspects: Uncontrolled cell proliferation is a hallmark of cancer.
- Some oncoproteins encoded by tumor viruses either sequester unphosphorylated Rb or effect its phosphorylation and thus eliminate the inhibitory function of Rb on the cell cycle.
- Tumor cells frequently escape cell cycle control by modification of genes, which encode cell cycle regulating factors. Examples are:
  – functional loss of tumor suppressor proteins, e.g. Rb or Ink4a (p16)
  – upregulation of cell cycle promoting factors, e.g. CDK4/6 or cyclin D. Within a given tumor cell, only one of these four proteins is modified. This underlines their critical function in the regulation of the Rb dependent G$_1$ checkpoint.

Figure 11.6-4. Regulation of DNA Replication in *Mammals*

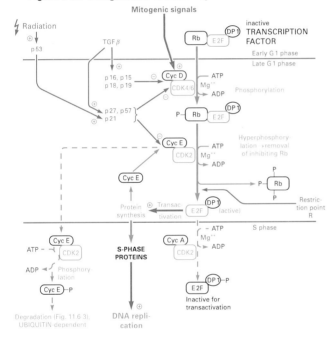

Table 11.6-2. Functional Analogies/Homologies Between Cell Cycle Proteins

	Mammals	Yeast (S. cerevisiae)
Cyclin dependent kinases	CDK1/Cdc2, CDK2, CDK3, CDK4, CDK6, CDK7	Cdc28 (cell division cycle) = p34 (Cdc2 in *S. pombe*) Kin28
Cyclins	G1 type cyclins: cyclin D, E S phase cyclins: cyclin E, A Mitotic cyclins: cyclin A, B	G1 type cyclins: Cln1, 2, 3 B type cyclins: Clb5, 6 B type cyclins: Clb1, 2, 3, 4
Ubiquitin-protein ligase	Human Cdc34 APC	Cdc34 APC
Anaphase inhibitor	–	Pds1 (Cut2 in *S. pombe*)
DNA damage checkpoint (11.6.7)	p53 p21 ATM (kinase, ataxia telangiectasia mutated) ATR (FRP), Function?	– – TEL1 (regulates telomere length) Mec1 (ESR1, SAD3)
Spindle assembly checkpoint	hsMad2 (*human* homologue of Mad2)	Mad2 (mitotic arrest deficient)

11.6.5 Regulatory Mechanisms During M Phase (Mitosis)

A short survey of the mitotic cell division is given in Fig. 11.6-5. For further morphological details, as well as for the meiotic division of *germ cells*, refer to textbooks of biology.

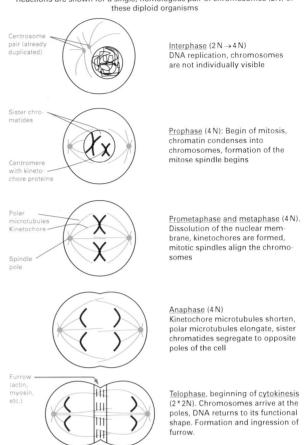

Figure 11.6-5. Summary of Mitotic Cell Division in *Animals*
Reactions are shown for a single, homologous pair of chromosomes (2N) of these diploid organisms

DNA replication (11.1) yields pairs of sister chromatids that are linked to each other at the centromere DNA section (ca. 110 bp in *yeast*, much longer in *mammals*). Protein complexes (kinetochores) assemble on either side of the centromeres during metaphase.

Simultaneously, M phase cyclin-CDKs are activated by dephosphorylation (Fig. 11.6-2). They promote the outgrowth of the bipolar mitotic spindle structure from both poles (which have formed and moved apart during the prophase). The spindle consists mainly of microtubules (see below and 18.1.3). A number of them attach to the sister kinetochores after the nuclear membrane has dissolved. They align the sister chromatides in the equatorial plane (metaphase plate) of the cell.

Attachment is the product of a chance encounter of the growing spindle extending in all directions from each pole with a correctly assembled kinetochore. In improper attachments, no tension is exerted on the chromosome, which results in instability of the microtubules until the proper attachment is reached. Other (so-called 'polar') microtubules originating at both poles just contact each other.

Besides promoting spindle assembly, M phase cyclin-CDKs also activate the anaphase-promoting complex (APC or cyclosome, 20S in *yeast*, 8…9 subunits) by phosphorylation. APC, in turn, enables the M phase cyclins A and B to ligate with ubiquitin (similarly to Cdc34, 11.6.2) and marks them for destruction (11.6.7). Likewise, it initiates destruction of anaphase inhibitory proteins (cohesive proteins in the kinetochore, Pds1, Cut1) and enables chromatid separation. The kinetochore microtubules then shorten while the polar microtubules elongate. This separates the sister chromatids and extends the cell before the cleavage process takes place. This mechanism ensures that each daughter cell obtains an identical set of chromosomes during mitosis. APC remains active until G$_1$ cyclin-CDKs have accumulated, which in turn, inactivate it.

In *animals*, the microtubule spindle also directs the accumulation and the alignment of contractile proteins (filaments of actin, myosin II and other proteins) around the metaphase plate underneath the plasma membrane by an unknown signalling process. The actomyosin contraction mechanism (partially resembling muscular contraction, 17.4.5) pulls the cellular membrane inwards (furrow ingression). As the furrow gets narrower, the filaments are gradually disassembled. Cell separation (cytokinesis) is achieved by the resolution of one membrane into two distinct membranes. The mechanism is unknown.

In *plants* and *fission yeast*, the division plane is filled with new membrane and cell wall components, starting at the center. They are transported there in vesicles from the Golgi apparatus (13.4) along microtubules. The accuracy of cellular division in *yeast* is about 1 error/10^5 divisions.

Movement of mitotic spindles (Fig. 11.6-6): The mitotic spindles are highly dynamic structures. They are largely composed of microtubules (polymerized heterodimeric α/β tubulin subunits, forming a hollow tube, 18.1.3) and display structural and kinetic polarity: Tubulin units are added and lost rapidly at the plus ends and slowly at the minus ends (which are attached to the spindle poles or centrosomes) in a GTP-dependent mode (Fig. 18.1-2). This dynamic instability is involved in the generation of pulling and pushing forces. Likely, a number of molecular motors (kinesin like proteins, dynein) contribute to these procedures, although many details are still unknown. The microtubules of mitotic spindles assemble at the time of cell division and disassemble upon completion of cytokinesis.

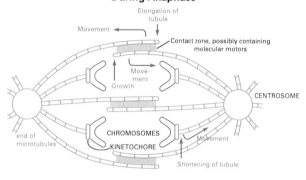

Figure 11.6-6. Elongation and Shortening of Microtubules During Anaphase

11.6.6 Cell Cycle Checkpoints

For the precise timing and order of events during the cell cycle, control mechanisms (checkpoints) have to ensure that an earlier event is completed before starting a later event.

DNA damage checkpoint (Figure 11.6-7): DNA damage (11.2, caused by irradiation or chemical mutagens) or interferences with the DNA metabolism (e.g. lack of nucleotides) generate a signal that

- arrests cells either in G_1 or G_2 phase and decreases the rate of DNA synthesis in S phase (dependent on the phase in which the cell was damaged) and
- induces the transcription of some repair genes prior to continuation of the cell cycle, e.g. DNA ligase or O-methylguanine DNA methyltransferase (11.2.2). Most other repair genes are constitutively expressed.

In *mammals*, the arrest is caused by expression of the tumor suppressor protein p53. Possibly the ATM kinase (Table 11.6-2) acts as a damage sensor. It phosphorylates and thereby activates p53. p53, in turn, modulates the expression of several target genes. The synthesized proteins cause the following effects:

- p21 (Cip1) stalls DNA replication in two ways: It inhibits the cyclin dependent kinases CDK4, 6 and CDK2 and thereby prevents Rb phosphorylation and the passage through the R point (11.6.4). It also forms a complex with PCNA, a processivity factor of DNA polymerase δ or ε for DNA replication and repair (11.1.4, 11.2.3). This prevents further DNA synthesis, likely without inhibiting DNA repair. p21 also reduces the proliferation of senescent cells. Additionally, p53 and p21 play a role at the mitotic spindle checkpoint (see below).
- GADD45 (growth arrest and DNA damage) is induced in cells with DNA damage by p53 dependent and independent pathways. It binds to PCNA and possibly depletes it in replication complexes. Thus, it becomes a potent inhibitor of cell proliferation. It also enhances nucleotide excision repair (11.2.3).
- Bax homodimers start a cellular suicide program (apoptosis) in case of excessive DNA damage in order to prevent gross chromosomal aberrations in the daughter DNA. Bax activates the cysteine proteases caspases (formerly named ICE family), which promote apoptosis. The activation of a cascade of proteases and nucleases results in the condensation of nucleus and cytoplasm, fragmentation of DNA and phagocytosis of the membrane-bound apoptotic bodies. Apoptosis also plays a role in maturation and differentiation processes, e.g. with *lymphocytes* (19.1.4).
- Bcl-2 antagonizes Bax effects. Consequently its expression is downregulated by p53 when Bax is induced. Bcl-2 shows homology with Bax and is able to form heterodimers with it, which inhibit the apoptosis mechanism. Bcl-2 is important for cell survival.

p53 can also induce apoptosis independent of transcriptional activation.

Pathological aspects: The central role of p53 in protecting a cell from DNA damage is shown by the fact that mutations of p53 occur in more than 50% of all human tumors. Lost or defective p53 predisposes cells to gene amplifications and aberrant chromosomal separation during mitosis. Thus, p53 is dubbed the 'guardian of the genome'.

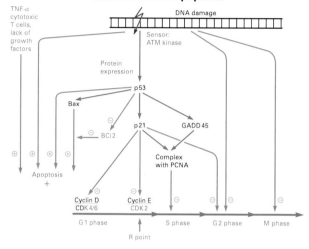

Figure 11.6-7. The DNA Damage Checkpoint in *Mammals* and Initiation of Apoptosis

Spindle assembly checkpoint: This checkpoint stops the separation of chromatides (i.e. progression into anaphase, 11.6.5) as long as it receives signals that sufficient attachment of the bipolar spindle microtubules to the kinetochores and correct alignment of chromatides on the metaphase plate has not yet been achieved.

Signal transduction at this checkpoint has been best characterized in *S. cerevisiae*. It involves the activation of the kinetochore protein Mad2 (mitotic arrest deficient) by binding with another protein. The complex generates an inhibitory signal, which propagates through the cell and stalls premature separation of the chromosomes. It also prevents activation of the APC complex, which is essential to promote anaphase (11.6.5). After correct attachment of the spindle this signal ceases. In *mammals*, p53 has been shown to participate in the mitotic spindle checkpoint, while p21 seems to be responsible for coupling S phase to mitosis.

11.6.7 Protein Degradation

The ubiquitin system provides controlled protein degradation in all *eukarya* (Section 11.6.4, Fig. 11.6-4). The enzymes Cdc34, APC (see above) or others ligate activated ubiquitin with the ε-amino group of lysine in the 'condemned' protein. Mostly polyubiquitin chains are formed. Protein degradation then occurs within a large complex (28S proteasome).

The Ub degradation apparatus recognizes appropriate proteins by their N-terminal amino acid (fast degradation with R, K, D, L or P- one letter abbreviations) or by internal PEST sequences (stretches of amino acids rich in P, E, S, T). In some proteins (mitotic cyclins, Pds1 etc.) a 'destruction box' (D box) RAALGNISN close to the N-terminus is present. Recognition of some substrates (Sic1, Cln2) takes place by a Cdc53–Cdc4–Skp1 complex, which is associated with Cdc34.

Literature:
Bartek, J., et al.: Curr. Opin. in Cell Biol. 8 (1996) 805–814.
Barton, N.R., et al.: Proc. Natl. Acad. Sci. USA 93 (1996) 1735–1742.
Enoch, T., Norbury, C.: Trends in Biochem. Sci. 20 (1995) 426–430.
Martin-Castellanos, C., et al.: Trends in Cell Biol. 7 (1997) 95–98.
Nicklas, B.R.: Science 275 (1997) 632–637.
Nicolescu, A.B. III et al.: Mol. and Cell. Biol. 18 (1988) 629–643.
Pines, J.: Seminars in Cell Biology 6 (1994) 399–408.
Rudner, A.D., Murray, A.W.: Curr. Opin. in Cell Biol. 8 (1996) 773–780.
Sanchez, I., Dynlacht, B.D.: Curr. Opin. in Cell Biol. 8 (1996) 318–324.
Sherr, C.J., Roberts, J.M.: Genes & Development (1995) 1149–1163.
Taya, Y.: Trends in Biochem. Sci. 22 (1997) 14–17.
(various authors): Science 274 (1996) 1643–1677.
(various authors): Science 279 (1998) 447–448, 509–533.
Wang, T.A., Li, J.J.: Curr. Opin. in Cell Biol. 7 (1995) 414–420.
White, R.J.: Trends in Biochem. Sci. 22 (1997) 76–80.

12 Viruses

12.1 General Characteristics of Viruses

Viruses are supramolecular structures of nucleic acids, proteins and sometimes lipids or carbohydrates. With a minimum of genetic information (4... >100 genes) they redirect the metabolism of infected host cells to perform their own reproduction. Due to this dependency, they can infect only a limited range of hosts. Besides the genome, the virus usually contains some enzymatic functions, supplementing the host enzyme activities. The free, infectious virus particle is named virion.

Still smaller are viroids, which consist of a single stranded, covalently closed RNA of 250...400 nt with multiple base pairing, but without a protein content. They reproduce in *plants* by using host proteins (perhaps DNA-directed RNA polymerase II, 11.3.2) and have pathogenic effects. An example is potato spindle tuber viroid.

12.1.1 Genomic Characteristics

Viruses can be classified according to their genome and their mode of replication as DNA-, RNA- and retroviruses. The genomic nucleic acids can be single- (ss) or double-stranded (ds). Single-stranded viruses are subdivided into (+)viruses (their nucleic acid is analogous to mRNA and to the coding = nontranscribed strand of DNA) and (−)viruses (only RNA viruses, their nucleic acid equals a complementary strand to mRNA). Their way of reproduction is schematically shown in Figure 12.1-1.

12.1.2 Structure

Usually, the genomic nucleic acid is surrounded by a protein coat (capsid). This consists of multiple copies of one or a few proteins and is arranged in the shape of a cylinder or of a regular icosahedron. In many cases, some enzyme proteins are associated with the nucleic acid. More complex viruses have additionally a lipid membrane (envelope, which originates from the cellular membrane of the host with interspersed viral proteins) or specialized tails etc. Some of the most important families of viruses are listed in Table 12.1-1. Due to the large variety of viruses it is only possible to discuss one representative of each group on the following pages.

Since the host organisms have developed a variety of antiviral defense systems (e.g. restriction endonucleases in bacteria, complex immune system in vertebrates), viruses responded by evolutionary strategies to escape these control mechanisms either by fast and massive replication or by persistent and latent infections.

Literature:
Cann, A.J.: *Principles of Modern Virology.* Academic Press (1993).
Levine, A.J.: *Viruses.* Scientific American Library (1992).
Riesner, D., Gross. H.J.: Ann. Rev. of Biochem. 54 (1985) 531–564.
Voyles, B.A.: *The Biology of Viruses.* Mosby (1993).

Figure 12.1-1. Flow of Information in Virus Reproduction (red arrows = virus reactions, blue arrows = host reactions)

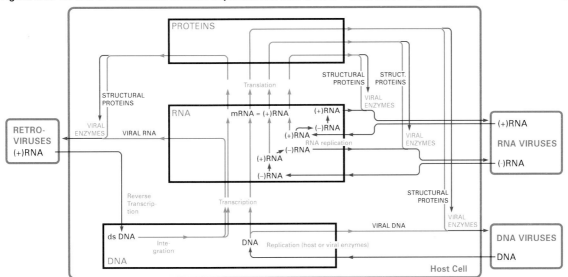

Table 12.1-1. Important Viruses and Their Families

Genome, Strands	Family	Species (Examples)	Genome Size (ca. kb), Structure	Size (nm)	Enveloped	Morphology	Host	Disease
ds DNA	papovaviridae	papilloma virus	9; closed circle, supercoiled	55	−	icosahedral	*mammals*	warts, assoc.w. cervix cancer
	adenoviridae	adenovirus	36; linear	70...90	−	icosahedral	*mamm., birds*	respiratory inf.
	herpesviridae	herpes simplex virus	200; linear	200	+	icosahedral	*vertebrates*	genital herpes, chicken pox
	poxviridae	vaccinia virus	200; linear, sealed ends	300	+	complex	*vertebrates*	cowpox
	bacteriophages	T4	167	95...65 (head); 17*115 (tail)	−	icosahedral (head), helical (tail), 6 tail fibers		*E. coli*
	bacteriophages	λ	48; linear, cohesive ends	55 (head); 15*135 (tail)	−	icosahedral (head), helical (tail)	*E. coli*	
ds DNA- RNA*	hepadnaviridae	hepatitis B virus	3; circular, single strand ends	42	+	spherical	*human, duck, woodchuck*	hepatitis B
(+)ss DNA	parvoviridae	adeno-assoc. virus	5; linear	18...26	−	icosahedral	*vertebrates*	
	bacteriophages	ΦX174	5.4; circular	25	−	icosahedral	*E. coli*	
ds RNA	reoviridae	reovirus	25; 10 lin.segm.	60...80	−	spherical	*mammals*	
(+)ss RNA	picornaviridae	poliovirus	7.4; linear	30	−	icosahedral	*primates*	poliomyelitis
	bacteriophages	Qβ	4.2	25	−	polyhedral	*E. coli*	
	togaviridae	Semliki forest v.	12	60...70	+	icosahedral	*vertebrates*	
	(tobacco mosaic group)	tobacco mosaic virus	6.4; helical	300*18	−	tubular rod	*plants*	
(−)ss RNA	orthomyxoviridae	influenza virus	16; 8 lin.segm.	80...120	+	helix in spher. envelope	*mammals*	influenza A
	paramyxoviridae	mumps virus	15; linear	200	+	pleomorphic	*human*	mumps
	rhabdoviridae	rabies virus	12...15	80...180	+	helix in spher. envelope	*mammals*	rabies
	arenaviridae	lassa fever virus	10...14; 2 segm.	50...300	+	pleomorph	*humans*	Lassa fever
(+)ss RNA → DNA	retroviridae	HIV	9.2; linear diploid	100	+	conical core in spherical envelope	*humans*	AIDS

* Replication via a RNA intermediate, which undergoes reverse transcription.

12.2 DNA Viruses

The replication of the DNA virus genome can occur in many ways, depending on the DNA structure. If the DNA has a circular shape, the replication methods resemble those of bacterial genomes, either by the Θ- or by the σ (rolling circle) mechanism (Figure 10.2-1, e.g., *SV 40* and *ΦX* 174, respectively). Linear DNA raises the question of the priming step for autonomous replication (independent of the host DNA). Some of these viruses have proteins attached to the DNA termini, which interact with the DNA polymerase (e.g. adenovirus). Other double-stranded viruses have their ends covalently sealed by a loop structure (e.g., poxvirus), while still others have 'cohesive' (complementary) ends, which are sealed into a loop after entering the host by a host enzyme (e.g., bacteriophage λ). The viral DNA can also be incorporated into the host genome. In multicellular organisms, this can cause cell transformation and tumorigenesis by interference with regulatory mechanisms (DNA tumor viruses, 17.5).

12.2.1 Bacteriophage λ

The genome of this complex virus is a linear double-stranded DNA (48 502 bp) with complementary single-stranded ends (length 12 nt.). The genome of the phage λ contains approximately 50 genes and numerous control sequences. Most genes with similar functions are organized in groups which are transcribed as single operons. The genome is embedded in an icosahedral head consisting of 420 copies of glycoproteins gpE and gpD each and 6...15 copies of gp's B, C, FII and W. The movable tail (32 rings consisting of a total of 192 gpV) ends into a single protein fiber of gpJ and contains also gp's G, H, L, M, U and Z (Figure 12.2-1).

Figure 12.2-1. Structure of *Bacteriophage* λ

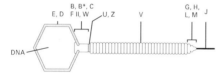

Infection of the bacterial cell (Figure 12.2-3): Adsorption to the *E. coli* host cell takes place via specific interaction of the phage tail fiber and a maltose group of the bacterial outer membrane. Then the phage DNA is injected through the tail into the host cell. The linear DNA is transformed into a cyclic form, whereupon the host DNA ligase closes the phage genome ring covalently and the DNA gyrase produces supercoiled phage DNA. Now the phage enters one of two different states:

- the lysogenic cycle (prophage stage, integration into the host DNA, passive replication together with the host genome) or
- the lytic cycle (virulent, reproduction of the virus, destruction of the host cell).

More than 90% of all known phages can shift between both states, depending on the nutritional state of the host and the degree of infection (temperent phages). The thoroughly investigated λ phage is an important model for understanding transcription control.

Figure 12.2-2. Genes Transcribed in the Lysogenic and Lytic Phases of *Bacteriophage* λ

The outer ring displays the gene map of phage λ with the genes coding for enzymes (blue), structural proteins (black) and regulator proteins (red). The sites of promoters are indicated in green.

The second ring (green) shows the genes transcribed in the establishment of the lysogenic phase, while the 3 innermost rings (blue) indicate the genes transcribed in the early, delayed early and late phases of the lytic cycle.

During packaging of the DNA into the virion head, the DNA is cut at the *cos* site by action of Nu I.

Lysogenic phase (Figure 12.2-3, right side): In the lysogenic state the phage DNA is integrated into the *host* chromosome by recombination at the homologous attachment sequences attP (in the *phage*)/attB (in the *bacterium*). This is effected by cooperation between the phage integrase and the bacterial integration host factor (IHF).

A complex interplay between λ-repressor and Cro-protein (Cro: control of repressor and other things) underlies the selection of lysogenic or lysis pathways. Each protein blocks the synthesis of the other one, when bound to the operator site of the other's gene (10.5.1).

The lysogenic state is induced by high concentrations of the 'early' gene product cII which stimulates transcription from the promoters p_I (for integrase) and p_{RE} (for repressor establishment). The λ repressor obtained this way prevents the transcription of the genes which initiate the lytic cycle (Figure 12.2-2). The lysogenic life cycle is preferred when the host is growing well and is then very stable (only 1 induction in 10^5 cell divisions). Damage to the DNA activates the repair RecA protein (10.3.5), which also splits the λ repressor. Consecutively, integrase and excisionase are synthesized and cause excision of the phage DNA, which is then ready to enter the lytic cycle.

Lytic infection cycle (Figure 12.2-3, lower left side): The lytic cycle can be divided into different transcription phases:

- In the early transcription phase the host RNA polymerase starts synthesizing mRNA in both directions (on opposite strands) from promoter p_L and from the promoters p_R and p'_R, respectively. The mRNA codes for control proteins (including N and Cro).
- The delayed early transcription starts when a large amount of the N protein (transcription anti-terminator) has accumulated during the early phase. This overrides termination sequences and causes expression of auxiliary proteins, which stimulate production of phage DNA in concatamer form (= several genomes in sequence). The Cro protein prevents overexpression of early genes.
- In the late transcription phase a large amount of head and tail proteins is synthesized, together with gene products which catalyze the lysis of the host cell (gpR: endolysin, which cleaves a peptide bond in the peptidoglycan of the host cell wall and gpS, which induces membrane pores). Important gene sequences, which are transcribed in the different phases, are shown in Figure 12.2-2.

Assembly of virus particles: Head and tail (Figure 12.2-1) are assembled independently. For correct assembly the individual steps in both reaction pathways have to occur in a certain, regulated order.

Three phage proteins (gpB, gpC and gpNu3) and two host proteins (GroEL, GroES, 14.1.1) are involved in the assembly of the initiation complex for the pro-head. Proteolytic processing (enzymes unknown) and other maturation processes lead to a stabilized head containing the binding site for the tail. DNA is 'pumped' into the head by an ATP-consuming process. This results in a tightly packaged DNA inside the head. The concatamer DNA is cut at the *cos*-sequences. The tail is assembled in three steps and finally connected with the head, thereby generating a new infectious phage.

Literature:
Adhya, S.L. *et al.*: Progr. Nucl. Acid Res. Mol. Biol. 26 (1981) 103–118.
Das, A.: Ann. Rev. of Biochem. 62 (1993) 908–918
Herskowitz, I., Hagen, D.: Ann. Rev. of Genetics 14 (1980) 399–445.
Murialdo, H.: Ann. Rev. of Biochem. 60 (1991) 125–153.

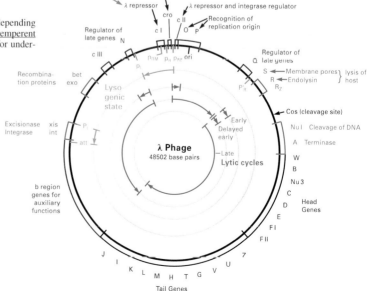

Figure 12.2-3. Life Cycle of Bacteriophage λ (red arrows: virus reactions, blue arrows: host reactions)

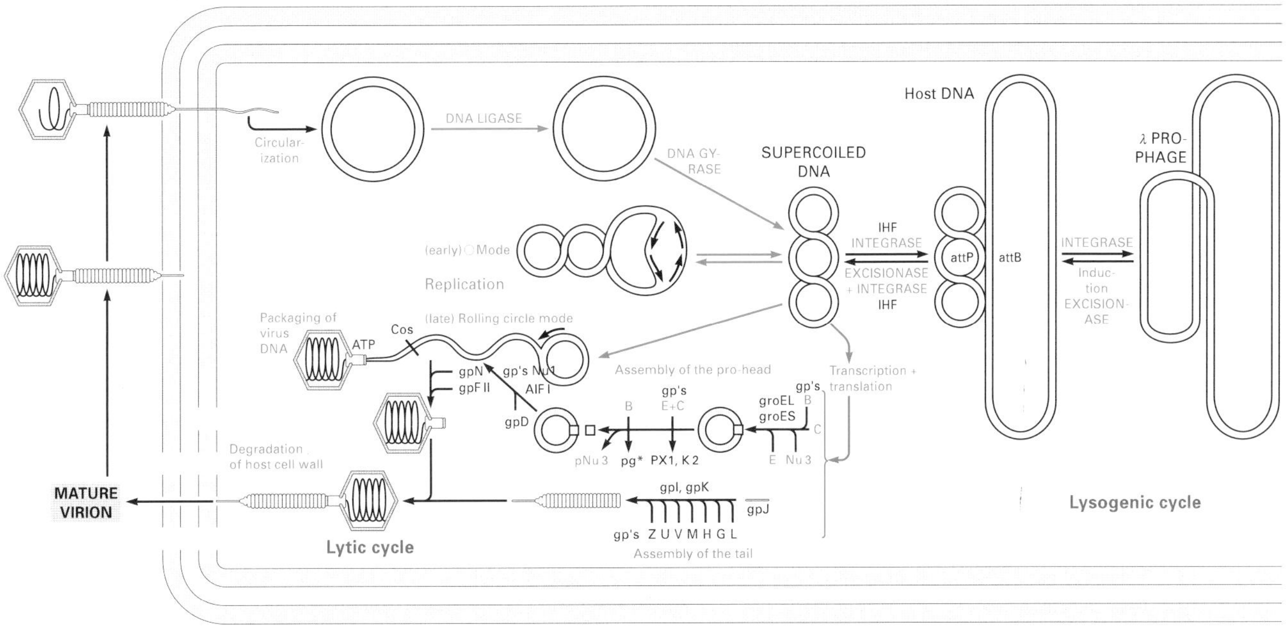

12.3 RNA Viruses

RNA viruses encompass a large variety of RNA types and protein arrangements. Apart from retroviruses (12.4), the major differentiation is between (+) and (–)RNA viruses. In both cases, the viral RNA is used as a template for its replication. A second round of replication then produces viral RNA in multiple copies. The RNA replication is error-prone, similar to the error rate of DNA transcription into mRNA (about $1/10^4$ nucleotides).

With (+)viruses, the original viral RNA can also serve as mRNA for formation of the proteins necessary for virus replication by the host cell. (–)Viruses, however, have to carry along the necessary enzymes in the virion (at least the RNA-directed RNA polymerase), since most host cells are unable to produce (+)RNA (= mRNA) from the (–)viral RNA (Fig. 12.1-1).

12.3.1 Tobacco Mosaic Virus

A very simply constructed virus is the tobacco mosaic virus (TMV). It was the first known example of self-assembly of an active biological structure. The 2130 identical coat protein subunits (of 158 amino acids each) are closely packed in a helical array around a single-stranded RNA containing 6390 nucleotides. This way, the RNA is completely protected from RNase attack.

Virus formation (Figure 12.3-1): The protein subunits arrange themselves first at adequate pH and ionic strength into 'lockwasher' protohelices of about 39 subunits, which form a little more than two turns. One of them attaches itself to a preformed hairpin loop of the RNA (containing a G-rich sequence). The RNA (at the 3'-side of the loop) forms a helix at the inner hole of the protein arrangement. 3 Bases each attach to one protein subunit and surround one of their helices. Then more of the double rings are added and the process repeats itself.

Cell-to-cell spread of viral genomes (Figure 12.3-2): After infecting one cell, TMV like other plant viruses establishes a systemic infection by moving through plasmodesmata, the cytoplasmic interconnections between cells, which resemble nuclear pores (14.3-2). These openings normally allow only passage of molecules below approx. 1 kDa. The virus, however, provides its own transport vehicle by transiently expressing a 30 kDa movement protein. This protein binds cooperatively to single-stranded nucleic acids, stretching them into an elongated shape. It also increases the effective plasmodesmal channel size, possibly by interacting with actin filaments located in the channel opening. These functions form the basis of a model for viral movement as shown.

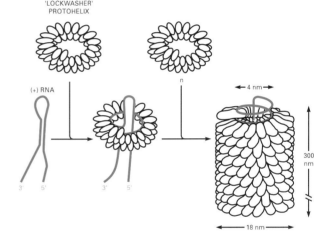

Figure 12.3-1. Self-Assembly of the Tobacco Mosaic Virus

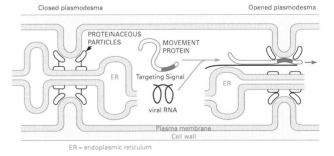

Figure 12.3-2. Systemic Infection by TMV Passage Through Plasmodesmata

ER = endoplasmic reticulum

Literature:
Butler, P.J.G. et al.: J. Gen.Virol. 65 (1984) 253–279.
Ding, B.: Trends in Cell Biol. 7 (1997) 5–9.
Raghavendra, K. et al.: Biochemistry 27 (1988) 7583–7588.
Waigman, E. et al.: Proc. Natl. Acad. Sci. (US) 91 (1994) 1433–1437.
Waigman, E., Zambryski, P.: Curr. Biol. 4 (1994) 713–716.
Wu, X., Xu, Z., Shaw, J.G.: Virology 200 (1994) 256–262.

12.4 Retroviruses

Retroviruses have the unique property of transcribing their RNA genome into DNA. As an example, the intensively investigated human immunodeficiency virus (HIV-1) is described here. This virus is the causative agent of the acquired immunodeficiency syndrome (AIDS). A related virus, HIV-2, apparently causes a different form of this disease. Some other retroviruses contain additional genes (oncogenes), which modify expression of gene products or encode modified gene products and can transform normal cells into tumor cells (oncorna viruses, tumor viruses, e.g., Rous sarcoma virus, 17.5.1 and Table 17.5-5).

12.4.1 Human Immunodeficiency Virus (HIV)
Genome structure of HIV-1 and protein processing (Figure 12.4-1): The 9.7 kb, dimeric (+)RNA genome of HIV-1 contains gag-pol-env genes coding for the essential enzymes and structural protein precursors (common to all retroviruses) and additionally, a number of regulatory genes. This RNA sequence is flanked by 3' and 5' enhancer and promoter sequences as well as untranslated (U3, U5) and repetitive sequences (R), which are of importance in the reverse transcription step. All of the possible three reading frames are used in an overlapping way to derive a maximum of information out of the quite small genome.

During transcription, differentially spliced mRNA's are synthesized to be translated into the regulatory protein factors and the gag, gag-pol and env polyprotein precursors. This is shown in the scheme below.

HIV Replication cycle (Figure 12.4-2): The initial event in the replication cycle of HIV is the binding of the viral envelope glycoprotein gp120 to the CD4 receptor at the surface of host cells (*CD4-positive target cells, lymphocytes, macrophages, dendritic cells*). This is followed by a second contact of gp120/gp41 with other receptors (e.g., the chemokine receptors fusin = CXCR4; CCR5 etc.). After membrane fusion of the virus envelope and the host cell, the virus becomes partially uncoated.

The initiation complex for reverse transcription [consisting of the *viral* enzyme reverse transcriptase (RT), the nucleocapsid proteins and the *viral* primer t-RNA^{Lys3}] is being activated by the *viral* protease. Reverse transcription of the (+)RNA (Figure 12.4-3) results in double-stranded DNA which enters the *nucleus* and is integrated at random sites into the host DNA by the *viral* integrase enzyme. In this 'proviral state' HIV can stay dormant for long periods unless the provirus is being activated by cellular gene activators such as IL-2, T-cell mitogens, cytokines or TNFα.

Transcription is initiated by host transcription factors, e.g., NF κB (11.4.3) and yields at first highly spliced mRNA, which codes for regulatory proteins (e.g., Rev, Tat). mRNA produced later is unspliced and codes for polyproteins, using different reading frames. These mRNA's as well as the viral RNA dimer are exported from the *nucleus* to the *cytoplasm*.

After translation, the viral proteins (in the form of the polyprotein precursors p55, p160) become esterified with membrane-attached myristic acid and are located at the membrane assembly site. The viral glycoprotein precursor gp160 is synthesized by ribosomes at the *rough endoplasmatic reticulum*. Cleavage to the external gp120 and the transmembrane gp41 glycoproteins is accomplished by a *host* protease. After virus assembly (packaging of the RNA dimer) and budding from the *host cell membrane*, cleavage of the precursor polypeptides by the HIV protease p11 results in mature and infectious virions.

Mechanism of reverse transcription: Retroviral reverse transcriptases (RT) are multifunctional enzymes exhibiting three enzymatic activities: RNA-dependent DNA polymerase, ribonuclease H and DNA-dependent DNA polymerase. These activities are necessary for copying the (+)RNA genome into (–)strand DNA, for removal of the RNA template and synthesis of the second (+)strand of DNA using the (–)DNA as a template. RT does not have proof-reading functions, therefore its actions are error-prone (*in vitro* ca. $1/10^4$). This might cause the high mutation rate of retroviruses. The current model for the process of reverse transcription is shown in Figure 12.4-3.

Literature
Bukrinsky, M.I. *et al.*: Nature 365 (1993) 666–669.
Darlix, J.L.: J. Mol. Biol. 254 (1995) 523–537.
Katz, R.A., Skalka, A.M.: Ann. Rev. Biochem. 63 (1994) 133–173.
Litvak, S. *et al.*: Trends in Biochem Sci. 19 (1994) 114–118
Marsh, M. *et al.*: Trends in Cell Biology 7 (1997) 1–4.

Figure 12.4-1. Gene Products and Protein Processing of HIV-1 (Red arrows: virus reactions, blue arrows: host reactions)

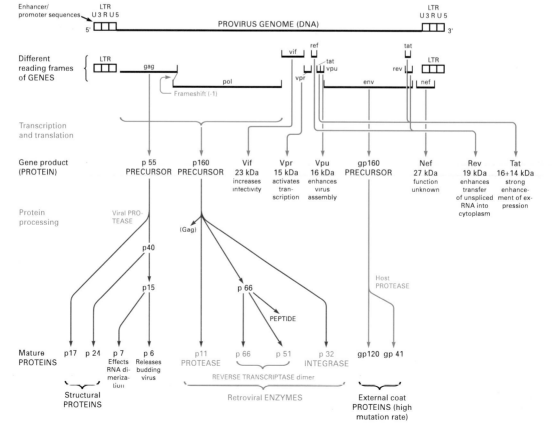

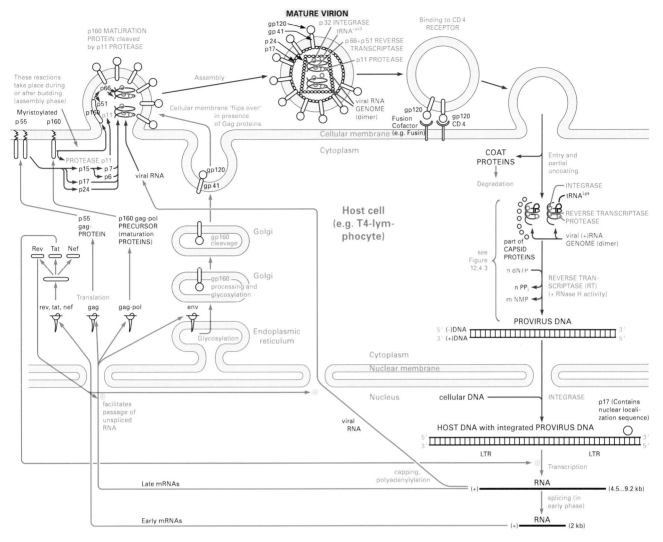

Figure 12.4-2. Mechanism of HIV Replication (red arrows: virus reactions, blue arrows: host reactions)

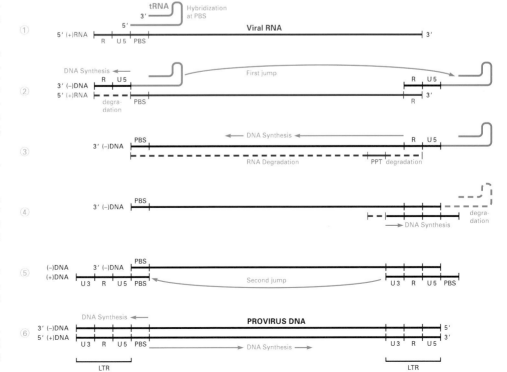

Figure 12.4-3. Formation of Provirus DNA From HIV-1 RNA (red: RNA, blue: tRNA^{Lys3}, black: DNA)

① Viral t-RNA^{Lys3} primer is annealed to the primer binding site (PBS, length 18 nucleotides).

② (–)Strand DNA synthesis is started, the corresponding RNA strand gets afterwards degraded. The R region of the newly formed DNA binds to the R region at the 3' end of the same or the second RNA strand (first jump).

③ The (–)DNA strand gets extended, most RNA is degraded.

④ The remaining RNA (PPT = polypurine tract) acts as primer for (+)strand DNA synthesis and is degraded afterwards, as well as the tRNA primer.

⑤ The (+)DNA strand binds to the other end of the (–)DNA strand at the PBS region (second jump).

⑥ Both linear strands are completed, the provirus formed contains long terminal repeats (LTR) at both ends.

13 Glycosylated Proteins and Lipids

13.1 Glycosylated Proteins and Peptides

Glycosylated derivatives of proteins and peptides (as well as of lipids) play a major role in celllular metabolism and recognition.

Table 13.1-1. Classification of Glycosylated Proteins

	Glycoproteins	Proteoglycans	Peptidoglycans
Occurrence	animals, plants and fungi	animals	bacteria
Saccharide chain	Short oligosaccharide chains of different units (frequently branched, frequently terminating with sialic acid, n = 2...20) are covalently linked to protein (N- and O-glycosylation, 13.3 and 13.4).	Long sequences of many repeating disaccharide units containing amino sugars and frequently uronic acids (average of n = 80, called glycosaminoglycans or muco-polysaccharides) are bound to the protein through 'core' carbohydrate structures analogous to glycoproteins	Long unbranched chains of repeating disaccharide units consisting of N-acetyl-glucosamine and acetylmuramic acid
Peptide chain	Many variations	Core protein, up to $3 * 10^5$ Da, forming together with the attached sacccharide chains a 'bottle brush' structure; additional link proteins. These structures are bound non-covalenty to hyaluronic acid.	Short chains of amino acids interconnecting the saccharide chains.
Total mol. mass	Varying, mostly $< 3 * 10^5$ Da	$2 * 10^3 ... > 10^7$ Da	Very large
Examples	Almost all exported proteins (e.g., secreted immunoglobulins and blood group substances, transport proteins, proteohormones, collagen, mucin); cell membrane components (e.g., MHC compounds, receptors, bound immunoglobulins, fibronectin), many cytoplasmic enzymes and other proteins.	Constituents of the extracellular matrix, e.g., chondroitin sulfate, dermatan sulfate, heparan sulfate, heparin, keratan sulfate, hyaluronic acid (as backbone of many proteoglycans).	Bacterial cell walls

The structures of the saccharide chains are generally determined by enzyme action and not by genetic matrices. The glycan chains are bound to asparagine and sometimes to the ε-group of lysine (N-glycosylation) or to serine or threonine, occasionally also to 5-hydroxylysine, hydroxyproline or tyrosine (O-glycosylation). In eukarya, the N-glycosylation of glycoproteins and proteoglycans begins in the endoplasmic reticulum (13.3) and is continued in the Golgi apparatus (13.4), while the O-glycosylation usually starts in the Golgi. The transport between the various cellular compartments and to the cell surface proceeds via vesicles (14.2). In some cases, specific carbohydrate sequences 'target' the glycoprotein to its destination (e.g. terminal mannose 6-phosphate guides to lysosomes (14.2.1). In its absence due to a genetic defect, the glycoprotein is secreted: I-cell disease).

13.1.1 Glycoproteins (for Structures, see Figure 13.4-2)

The carbohydrate content varies within a wide range (from almost 0 to 85%). Both linear and branched carbohydrate chains occur. Frequently, the glycosidic groups are essential for protein function, e.g., in cell-cell interactions (cellular attachment, immunological reactions, formation of neuronal interconnections). Many soluble plasma glycoproteins carry terminal sialic acid (N-acetylneuraminic acid, 3.7.1). Degradation starts after removal of the neuraminic acid by neuroaminidase. The protein is recognized by a asialoglycoprotein receptor of the liver, internalized and degraded.

Mucins are very large (ca. 10^7 Da) glycoproteins. They carry a huge number of mostly O-linked glycosidic chains; the saccharide residues are often sulfated. They protect epithelial layers (e.g., of the stomach). In plant cell walls, hydroxyproline-rich, rod-shaped glycoproteins carry mostly side chains of arabinose tetrasaccharide units (3.6.3).

13.1.2 Proteoglycans

These compounds consist of a 'core' protein with many attached carbohydrate side chains (Figure 13.1-1). Their length increases from the N to the C terminus of the protein. In longer chains, there is usually closest to the protein a short sequence of sugars, amino sugars and N-acetylneuraminic acid, followed by many repeating disaccharide units containing amino sugars (glycosaminoglycans, mucopolysaccharides, 13.1.5 and Fig. 13.1-2). The carbohydrate content can amount up to 95%. The covalent binding of the glycosidic side chains to the 'core' protein takes place in the Golgi apparatus of fibroblasts, chondroblasts (cartilage), osteoblasts (bone) and similar cells.

These structures are secreted into the extracellular space, where they frequently get non-covalently linked to hyaluronic acid and become part of the extracellular matrix.

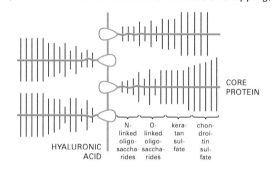

Figure 13.1-1. Structure Model of Cartilage Proteoglycan
(The areas of the various side chains are overlapping)

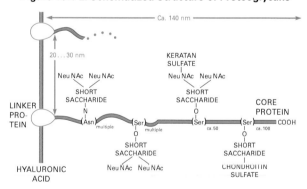

Figure 13.1-2. Schematized Structure of Proteoglycans

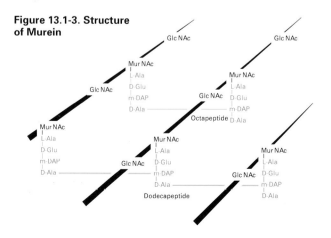

Figure 13.1-3. Structure of Murein

13.1.3 Peptidoglycans (Figure 13.1-3)

Murein is a most important component of practically all *bacterial cell walls*. The glycosidic chains of repeating N-acetylglucosamine (GlcNAc) and N-acetyl muramic acid (MurNA) disaccharides are interconnected by short peptide chains. For synthesis see Figure 15.1-1.

13.1.4 Glycoprotein Degradation Diseases and Mucopolysaccharidoses

The saccharide chains of glycoproteins and proteoglycans are degraded in *lysosomes*. Lack of enzymes due to genetic reasons is the cause of various lysosomal storage diseases. They are usually characterized by accumulation of the undegraded compounds in tissues and urine, skeletal deformations etc.

Since the specificities of the enzymes usually refer only to the structure of the particular bond, a single enzyme defect can cause various diseases (e.g. Morquio syndrome and G_{M1} gangliosidosis). Typical glycoprotein degradation diseases are fucosidosis and aspartylglucosaminuria (failure to cleave the N-acetylglucosamine-asparagine bond). Mucopolysaccharidoses are mentioned in 13.1.5.

13.1.5 Repeating Units of Glycosaminoglycans as Components of Proteoglycans
(The attack site of degradative enzymes is shown by an arrow)

CHONDROITIN 4-SULFATE
[D-glucuronate ($\beta 1 \rightarrow 3$) D-GalNAc (4-sulfate) ($\beta 1 \rightarrow 4$)]$_n$

The sulfate group may also be at 6-position. These units are attached to a central sequence of Ser-O-Xyl-Gal-Gal-. $5*10^3 \ldots 5*10^4$ Da. The biosynthesis proceeds via the consecutive action of the enzymes with EC-numbers 2.4.2.26, 2.4.1.133...135, (2.4.1.174 and 2.4.1.135?)$_n$, 2.8.2.5 or 2.8.2.17. Large quantities are present in *cartilage, aorta, connective tissue, bone, skin*.

Lack of β-glucuronidase for degradation = mucopolysaccharidosis VII.

DERMATAN SULFATE
Same structure as chondroitin sulfate + about 20% of structure
[L-iduronate ($\alpha 1 \rightarrow 3$) D-GalNAc (4-sulfate) ($\beta 1 \rightarrow 4$)]$_n$

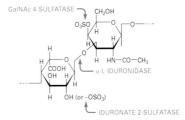

$1.5*10^4 \ldots 4*10^4$ Da, also contains some D-Gal, D-glucuronate, D-xylose. To some extent, sulfate is attached to the 2-position of L-iduronate. In *skin, heart, blood vessels*.

Lack of α-L-iduronidase for degradation = Hurler's syndrome (mucopolysaccharidosis I). Lack of iduronate 2-sulfatase = Hunter's syndrome (mucopolysaccharidosis II). Lack of N-GalNAc 4-sulfatase = Maroteaux-Lamy's syndrome (mucopolysaccharidosis VI). Dermatan sulfate also inhibits blood coagulation (20.3).

HEPARIN
[D-glucuronate (2-sulfate) ($\alpha 1 \rightarrow 4$) N-sulfo-D-glucosamine (6-sulfate) ($\alpha 1 \rightarrow 4$)]$_n$ (mostly)

$6*10^3 \ldots 2.5*10^4$ Da, contains also some D-Gal and D-xylose. Large quantities are found in *mast cells, lung* and *liver*. Heparin is an inhibitor of blood coagulation (20.3) by promoting complex formation between active proteases (IIa, IXa, Xa) and antithrombin III.

HEPARAN SULFATE
Similar structure to heparin, but with more N-acetyl groups and less O- or N- attached sulfate groups

$5*10^3 \ldots 1.2*10^4$ Da. Ubiquitously present at *cell surfaces* and *laminae*.

Lack of sulfamidase for degradation = Sanfilippo's syndrome A. Lack of α-N-acetylhexosamidase = Sanfilippo's syndrome B (mucopolysaccharidosis III). Like heparin, it inhibits blood coagulation (20.3).

KERATAN SULFATE
[D-Gal ($\beta 1 \rightarrow 4$) D-GlcNAc(6-sulfate) ($\beta 1 \rightarrow 3$)]$_n$

The units are attached to a central sequence of
Ser-O-GalNAc — Gal-NeuNAc-... / Gal-NeuNAc...

$5*10^3 \ldots 2*10^4$ Da, also contains D-mannose and L-fucose. Large quantities occur in *cartilage, cornea, vertebrate disks*.

Lack of β-galactosidase for degradation = Morquio type B syndrome (mucopolysaccharidosis IV). Deficiency in fucosidase = fucosidosis.

HYALURONIC ACID
[D-glucuronate ($\beta 1 \rightarrow 3$) GlcNAc ($\beta 1 \rightarrow 4$)]$_n$

$4*10^3 \ldots 8*10^6$ Da. Large quantities in *connective tissues, skin, cartilage, synovial fluid*. Hyaluronic acid forms the backbone of many proteoglycans. E.g., in the proteoglycan of *cartilage*, core proteins (which bear side chains of keratan sulfate and chondroitin sulfate) are non-covalently attached to hyaluronic acid with their N-terminus. The binding site is stabilized with linker proteins (Fig. 13.1-2).

Degradation takes place by hyaluronidases (a mixture of the enzymes with EC-numbers 3.2.1.35 and 36), which occur in many *animal tissues, snake* and *insect toxins* and in *bacteria*.

Literature:
Montreuil, J., Vliegenthart, J.F.G., Schachter, H.: Glycoproteins. New Comprehensive Biochemistry Vol. 29a. Elsevier (1995).
Ruoslahti, E.: Ann. Rev. of Cell Biol. 4 (1988) 229–255.

13.2 Glycolipids

Similar to many glycosylated proteins (13.1), glycolipids are members of the glycocalyx layer covering the cellular membrane. The lipid part forms the anchor in the membrane (Figure 13.2-1), to which the (frequently antigenically reactive) glycosidic chain is bound. Three different types are distinguished (Table 13.2-1):

Table 13.2-1. Types of Glycolipids

	Glycosphingo-lipids	Glycoglycerolipids	Glycosylphospho-polyprenols
Membrane anchor (lipid)	ceramide or other sphingo-sine derivatives	mono- and diacylglyc-erol, glycerol ethers and phosphatidic acid	isoprenoids, e.g. doli-chol-P, ficaprenol-P, undecaprenol-P
Lipid-sugar bond	O-glycosidic	O-glycosidic or phosphate ester	phosphate ester
Function	membrane con-stituent	membrane constituent	precursor of glycosy-lated proteins
Site of synthesis	*Golgi appara-tus, bacterial membranes*	*Golgi apparatus, chloroplasts, bac-terial membranes*	*endoplasmic retic-ulum, bacterial membranes*

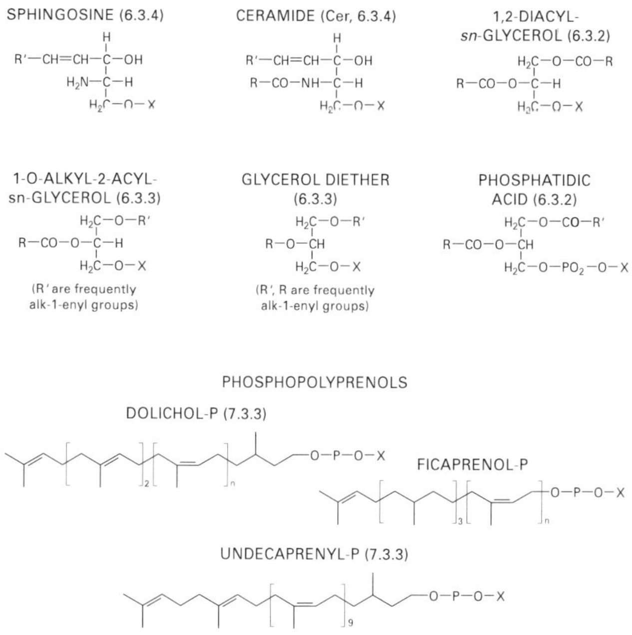

13.2.1 Glycosphingolipids

Glycosphingolipids occur in *vertebrates* and *lower animals*, as minor components in *plants* and (only rarely) in *bacteria*.

Many diverse structures exist, they are species and organ specific. Some of them vary with age. They are classified by the type of the 3 sugars next to the lipid moiety ('root structure', Table 13.2-2). Frequently they carry acidic groups (sulfate, uronic acids, sialic acid etc.). In membranes they tend to aggregate.

Figure 13.2-1. Structure of Common Membrane Anchors
(The carbohydrate chains are bound at -X)

Table 13.2-2. Series of Glycosphingolipids (for symbols, see Fig. 13.2-2)

Series	Abbr.	Structure	Present in	Examples	Derivatives
Globo	Gb	R—⬠—⬠—△—Cer	*mammalian extraneural tissue* (major glycolipid)	R= ■— : main component of *human* erythrocyte membrane (= globoside), R= ■—■— : Forssman antigen (in some *mammals*, also in *bacteria*)	gangliosides
Isoglobo	iGb	R—⬠—⬠—△—Cer			
Lacto	Lc	R—□—⬠—△—Cer	*erythrocytes (non-human)*	R = ◇— : precursor of ABO blood groups and Le antigens (not in *humans*)	gangliosides, sulfate derivatives
Neolacto	nLc	R—□—⬠—△—Cer	*erythrocytes (human, animals), spleen*	R = ◇— : precursor of ABO blood groups and Le^x antigens (*humans* and *others*). R = ◇—◇— : P antigen	gangliosides, sulfate derivatives
Ganglio	Gg	R—■—⬠—△—Cer	mostly in *brain gangliosides*, also *extraneural*		gangliosides, sulfate derivatives
Gala	Ga	R—⬠—⬠—Cer	*animals, fungi*	in *mammals* only R = H- or ■—■—	esters (C_{16}, C_{18})
Arthro	Ar	R—⬠—●—△—Cer	*arthropods*	R = □— or ■—◇—	
Mollu	Ml	R—●—●—△—Cer	*mollusks, plants*	R = □— or -xylose-	
(Glucosylceramide)		△—Cer	*animals, plants, fungi*	intermediates of synthesis for globo, lacto, ganglio series	esters
(Lactosylceramide)		⬠—△—Cer	*animals, plants, fungi*		esters

Some fucose derivatives have blood group specificity (13.5).

Gangliosides are glycosphingolipids, which also contain sialic acid (N-acetylneuraminic acid or its 9-O acetyl derivative in *humans*, additional derivatives in *other mammals, fish, birds* etc.). They are found primarily in the *central nervous system*, but also in other organs of *vertebrates* at the *outer side* of the *plasma membrane* (Table 12.2-3).

They mediate interactions between cells and with the *extracellular matrix*, cell-cell communication, immunological functions, *leukocyte* differentiation, oncogenic reactions etc. During embryogenesis and thereafter, the ganglioside pattern changes. Gangliosides (and other glycosphingolipids) influence cell differentiation and the development of the *neuronal* network.

Table 13.2-3. Gangliosides in *Mammals*

Series	Occurrence e.g. in	Examples
Ganglio	*brain grey matter*	sialyl$_{1...4}$ gangliotetraosyl ceramide
	brain white matter	sialyl$_1$ gangliotetraosyl ceramide
	erythrocytes (rat)	sialyl$_2$ gangliotetraosyl ceramide
Neolacto	*erythrocytes (man, dog, pig)*	sialyl$_{1...2}$ lactosyl ceramide
	peripheral nerves	sialyl$_{1...2}$ lactosyl ceramide
		sialyl$_{1...2}$ lactotetraose ceramide

Gangliosides of other series occur in *lower animals*.

Biosynthesis: Ceramide is synthesized in the *endoplasmic reticulum* (6.3.4). The primary glycosylation steps, which take place mainly in the *Golgi apparatus*, are described in 13.4. They are analogous to the biosynthesis of glycosylated proteins by ligand transfer from XDP-sugars, XDP-aminosugars or CMP-sialic acid to ceramide.

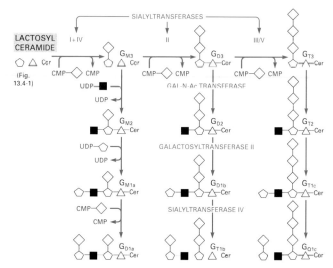

Figure 13.2-2. Synthesis of Higher Gangliosides
(Examples only, many variants exist. For basic structures, see Fig.13.4-1)

Symbols

△	Glc	glucose	■	Gal Nac	N-acetylgalactosamine
□	GlcNAc	N-acetylglucosamine	●	Man	mannose
⬠	Gal	galactose	◇	NeuNAc	N-acetylneuraminic acid

Higher gangliosides (Fig.13.2-2) are formed by addition of more sialyl and glycoside groups to the basic structures shown in Fig.13.4-1. The nomenclature system of of Svennerholm is used.

Degradation: The *lysosomes* are the site of glycolipid degradation. Exoglycosidases sequentially remove single sugar moieties from the end. Special proteins facilitate interaction of the enzymes with their lipid substrate.

Disturbances of degradation are most striking with gangliosides (Figure 13.2-3). Deficiencies in the degradation process lead to various inherited diseases, which are manifested by severe neurodegenerative defects. They are characterized by accumulation of those glycolipids, which are substrates of the defective enzyme.

Tay-Sachs and Sandhoff diseases are two forms where a deficiency in the enzyme β-hexosaminidase leads to the accumulation of ganglioside G_{M2}. In Tay-Sachs, the α subunit of this enzyme and in Sandhoff the β subunit is affected. Some other diseases are caused by a defect in activator proteins for β-hexosaminidase. In Gaucher's disease, the terminal glucose residue cannot be removed due to a defect in β-glucosidase. Fabry's disease originates in a defect in α-galactosidase A, while deficiency of β-galactosidase leads to G_{M1} gangliosidosis. General decrease of gangliosides is observed in Creutzfeld-Jacobs encephalopathy. Reduced activities of biosynthetic enzymes are presumably the cause of incomplete glycolipid chains in some transformed cells.

13.2.2 Glycoglycerolipids (Table 13.2-4)
Glycoglycerolipids occur in all kingdoms of living beings. Their structure is species specific.

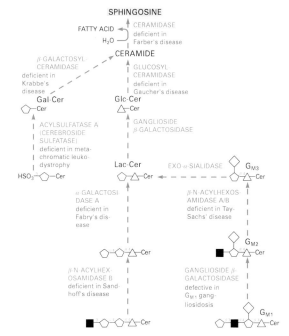

Figure 13.2-3. Degradation of Glycosphingolipids

Table 13.2-4. Examples of Glycoglycerolipids

Primary Lipophilic Residue (Formulae see Fig.13.2-1)	Sugars or Derivatives	Ligands of Sugar	Occurrence
Neutral glycoglycerolipids			
Diacyl- or monoacylglycerol	(-Glc, -Gal)$_1$		*plants* (esp. in *chloroplasts*), *animal nerves*, few *bacteria*
	(-Glc, -Gal)$_{2\ or\ higher}$		*bacteria* (also with -Man), *plants* (esp. in *chloroplasts*), *animal nerves*, brain, saliva
Alkylacylglycerol	-Gal, -(Glc)$_n$		*mammalian spermatozoa and brain* (-Gal), *saliva* (-Glc$_n$)
	(-Gal, -Glc)$_n$		*bacteria*
Glycerol diethers (with phytyl or other residues)	(-Man, -Glc)$_{1-24}$		*archaea* (also tetraethers with 2 glycerol residues opposing each other → monolayer membrane in some thermophilic species)
Acidic glycoglycerolipids			
Diacyl- or monoacylglycerol	glucuronic or galacturonic acids, sialic acid		some *bacteria*
	-Gal, -Glc-Gal	-HSO$_3$	*animal brain*
	-Gal, -Glc	-HSO$_3$	*archaea*, some *bacteria*
	(-Glc, -Gal)$_n$, also acylated	-(glycerol-P)$_n$	*bacteria*: lipoteichoic acids (immunogenic!)
Alkylacylglycerol	-Glc	-HSO$_3$	*human gastric mucus*
	-Gal	-HSO$_3$	*mammalian spermatozoa* (seminolipid)
Glycerol diethers (with phytyl or other residues)	(-Gal, -Man, -Glc)$_n$	-HSO$_3$	*archaea*
Phosphatidic acid = (Acyl)$_2$ glycerol-P	-inositol-GlcN-(Man)$_3$	-P-ethanolamine-protein	*protozoa, higher animals* (membrane anchor of proteins)

Contrary to glycosphingolipids, they play a major role in *bacteria* (e.g., polymeric structures are members of the *bacterial* cell wall, 15.1). In *plants*, they are specifically enriched in the thylakoid membrane of *chloroplasts*, while in *mammals*, they are major constituents of the myelin sheath of *neurons* and occur in *mucous layers* and in *germ cells*.

Biosynthesis and biodegradation: The lipophilic part is synthesized in *mammals* and in *plants* at the membrane of the *endoplasmic reticulum*. See 6.2.1 for glycerol esters and 6.3.3 for glycerol ethers.

With glycolipid-anchored proteins *in animals*, the further steps take place in the *endoplasmic reticulum* (13.3). However, the transfer of sugars and sulfate groups in *animals* usually occurs in the *Golgi apparatus* (13.4), while in *plants*, this is performed preferably in the *chloroplast envelope* (The completed compounds are later transferred into the *thylakoids*).

In *bacteria*, biosynthesis and sugar transfer are both located in the membrane. The reactions follow the same principles as with glycosphingolipids (13.2.1), using activated XDP-sugars. For *bacterial* macromolecules, see 15.1.

Biodegradation in *mammals* takes place by a stepwise removal of the terminal residues by specific *lysosomal* enzymes. Lack of sulfatase causes metachromatic leukodystrophia.

13.2.3 Glycosylphosphopolyprenols
These compounds function as carriers of glycosyl groups. The synthesis of phosphorylated proteins in the *endoplasmic reticulum* of *eukarya* uses dolichol-P. Undecaprenyl-P functions as a carrier in the formation of *bacterial* cell wall murein as well as in biosynthesis of the O-antigen part of lipopolysaccharide (*Gram negative bacteria*) and of teichoic acids (*Gram positive bacteria*).

Both polyprenols are membrane anchors for a growing glycosidic chain. They have *tri-trans or di-trans, poly-cis* configuration. Their biosynthesis is described in 7.3, while their function as carriers of glycosidic chains is dealt with in 13.3 and 15.1.

Literature:
Bell, R.M., *et al.* (Eds.): *Advances in Lipid Research* Vols. 25 and 26: *Sphingolipids.* Academic Press (1993).
Gieselmann, V.: Biochim. biophys. Acta 1270 (1995) 103–136.
Merill, A.H, Sweeley, C.C., in Vance, D.E., Vance, J. (Eds.): *Biochemistry of Lipids, Lipoproteins and Membranes.* New Comprehensive Biochemistry Vol. 31. Elsevier (1996) 309–339.

13.3 Protein Processing in the Endoplasmic Reticulum

The *endoplasmic reticulum* (ER) is an important organelle of *eukarya*. With respect to proteins, its major functions are folding, glycosylation, preparation for targeting and membrane attachment.

In *eukarya*, glycosylation is very frequent among membrane associated and secreted proteins. The main glycosylation machinery is located in the *endoplasmic reticulum* and the *Golgi apparatus* (13.4). The protein backbone is synthesized by *ribosomes* (10.5.2) attached to the *ER membrane* and threaded into the *ER lumen*. In case of N-glycosylation, the first steps of sugar attachment take place here (cotranslational glycosylation). This is continued in the *Golgi* (13.4).

Only with some secretory proteins, the protein chain is synthesized *cytoplasmatically* and then translocated through the ER-membrane in a way similar to *mitochondrial* import (14.4), followed by glycosylation (mostly in *yeast*, e.g., pre-pro-α-factor).

One type of O-glycosylation, the addition of single N-acetylglucosamine residues to serine, takes place in the *cytosol*. This modification occurs with *cytosolic* and *nuclear* proteins and is as abundant and dynamic as protein phosphorylation (2.5.2, 17.5).

13.3.1 Protein Synthesis and Import into the Endoplasmic Reticulum (Figure 13.3-1, upper part)

Most *eukaryotic* proteins to be glycosylated carry a signal sequence of varying structure near the N-terminus, containing several positively charged amino acids, followed by 7...20 hydrophobic amino acids and another charged sequence. After translation of this sequence at ribosomes within the *cytosolic space*, the signal recognition particle (SRP) binds to the translation complex. It arrests the polypeptide chain elongation by fixation of the signal sequence to its subunit p54. The whole complex of ribosome, mRNA, nascent polypeptide chain and SRP moves to the ER and 'docks' at the *ER-membrane* by interaction of the SRP with the SRP receptor. GDP bound to the SRP-receptor (and also to the SRP) is then replaced by GTP. This leads to separation of SRP both from the signal sequence and the ribosome. Thereafter, hydrolysis of GTP removes SRP also from the SRP receptor and enables it to start a new reaction cycle.

The ribosome with the nascent protein chain is transferred to the Sec61 complex, a heterotrimer which is the core component of the translocation system (translocon). The ribosome receptor is part of the Sec61 protein. The signal sequence is bound to the transmembrane TRAM protein. Most likely, the Sec61 complex forms an aqueous membrane pore, through which the polypeptide chain is pushed. The protein synthesis is then resumed.

13.3.2 Location of ER proteins

- When secretory proteins pass through the pore, the signal sequence is cleaved off by a signal peptidase and the whole protein is located in the *ER lumen* to be exported later.
- The proteins can also be transferred to phosphatidylinositol membrane anchors.
- Transmembrane proteins contain additional stop transfer effector (STE) sequences, which mark the preceding hydrophobic areas to become transmembrane sequences. The STE sequences transiently stop the translocation and mediate the transmembrane integration of the protein (while the protein synthesis continues). In case of multiple-spanning transmembrane proteins, this procedure is repeated.
  - Frequently, the signal sequence is located at the N-terminus of the nascent protein. When it is cleaved off by a signal peptidase, the newly formed N-terminus is oriented towards the *ER lumen*, while the C-terminus remains at the *cytosolic side* of the membrane (type I membrane proteins).
  - If an uncleavable signal sequence (signal anchor sequence) is present in the protein, it remains anchored in the membrane, while the nascent protein is pushed into the lumen. This yields a protein, in which the N-terminus is located in the *cytosol* and the C-terminus in the *ER lumen* (type II membrane proteins).

Mechanism for retaining proteins in the ER: While correctly folded secretory and membrane proteins (14.1.2) leave the ER in *transport vesicles* (14.2), special mechanisms exist for proteins destined to remain in the *ER*:
- Most membrane-bound ER proteins are retained by interaction with cellular structures on the cytosolic surface, probably due to a consensus motif (2 Lys, 3 and 4 or 5 positions away from the C terminus, KKXX motif).
- Soluble ER proteins, like protein disulfide isomerase or binding proteins (BiP) have special amino acid sequences at the C-terminus, like KDEL, KEEL or HDEL (in *yeast*). These sequences most likely interact with a receptor in the *cis-Golgi network* ('salvage compartment'), which returns them to the ER.

13.3.3 Synthesis of Dolichol-bound Oligosaccharides and N-Glycosylation (Figure 13.3-1, lower part)

The stepwise construction of glycosyl chains attached to membrane-bound dolichyl-P (7.3.3) up to Dol-PP-GlcNAc$_2$-Man$_5$ takes place on the *cytosolic side* of the *ER-membrane*. The 'flipping

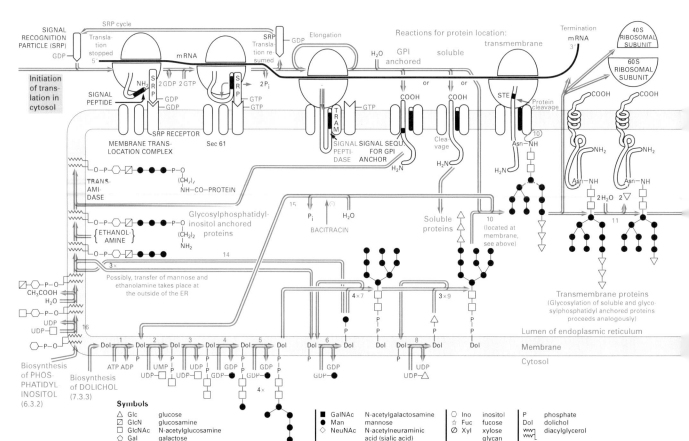

over' of this structure to the *luminal* side has been postulated, but not been shown so far. Dol-P-mannose carries mannose units through the *ER-membrane*, which are used to extend the glycosyl chains both of the dolichol and the glycosylphosphatidylinositol anchors (13.3.4). Generally, nucleoside diphosphate glycosides cannot pass through the *ER membrane*, although they are able to do this with the *Golgi membrane* by an antiport mechanism (13.4.1). The preformed oligosaccharides are transferred from dolichol to an asparagine residue of the protein chain. This amino acid has to be a member of the consensus sequence Asn-X-Ser/Thr (X may be any amino acid except proline). The oligosaccharide transferase complex is associated with the protein translocation machinery.

The structure of the glycoside chains plays an important role in the folding mechanism of glycoproteins (14.1), which also takes place in the ER.

13.3.4 Formation of Lipid-anchored Proteins in the ER (Figure 13.3-1, left side)

Synthesis of glycosylphosphatidylinositol (GPI) in *eukarya*: The synthesis starts at the *outside* of the *ER* by adding N-acetylglucosamine residues from UDP-GlcNAc to phosphatidylinositol (6.3.2), followed by deacylation. The question of whether the following transfer of 3 mannoses take place at the *outside* or the *inside* of the *ER-membrane*, is still under discussion. Phosphoethanolamine is finally added, possibly from phosphatidylethanolamine (6.3.2). This structure may be modified by the addition of more mannose residues or galactosyl side chains to the core carbohydrates or of fatty acids to inositol etc.

Protein transfer to the GPI anchor: Proteins to be transferred own a signal peptide for GPI addition near their C terminus with three distinct regions: a signal-cleavage/GPI attachment site (including Ala, Asn, Asp, Cys, Gly and Ser, but without strict sequence requirement), a spacer region of about 7...14 amino acids and a stretch of ca. 20 hydrophobic amino acids. While the proteins are still attached to the membrane with their C terminus after synthesis, the appropriate peptide bond is probably subjected to a nucleophilic attack by the NH_2 group of ethanolamine-GPI, resulting in transfer of the protein to the anchor.

GPI acts as an anchor for many proteins involved in cell adhesion, membrane signalling events, T-cell activation (e.g. LFA-3; 18.1.4), catalysis (e.g. *human erythrocyte* acetylcholinesterase, 6.3.5 etc.

13.3.5 Acylation of Proteins

In about half of all proteins, acetyl- or myristoyl-CoA reacts with the N-terminus of the nascent polypeptide in an enzyme-catalyzed mode:

CH_3-CO-SCoA + H_2N (Ser/Ala/Met)-protein → CoA-SH + acetyl-protein

$C_{13}H_{27}$-CO-SCoA + H_2N(Gly)-protein → CoA-SH + myristoyl-protein

Myristoylated proteins bind to specific membrane receptors and become membrane-attached this way. Palmitoylation (mostly of a cysteine close to the N-terminus), however, takes place posttranslationally at locations outside of the ER. This reversible reaction may play a role in regulation (e.g. of receptors and of G-proteins, 17.4.1).

Table 13.3-1. Components of Protein Processing in the ER

Unit	Components	Function	Found in
Signal recognition particle (SRP)	associated peptides of 72, 68, **54**, 19, 14, 9 kDa, 7S RNA (ca.300 nt)	54 kDa subunit binds signal sequence of the nascent protein	*mammals*, homologous to other *eukarya* and *bacteria*
Membrane translocation complex (translocon)	SRP receptor (SR), dimeric (α, β/72, 30 kDa)	SRP binds to a subunit, GTPase	same
	Sec61 complex, trimeric (α, β, γ/54, 14, 8 kDa)	Integral membrane complex forming the translocation channel, binds ribosomes	same
	Translocating chain associating membrane protein (TRAM/36 kDa glycoprotein)	Spans membrane, transport of some proteins, orientation of signal sequence?	*mammals*
'Signal sequence receptor complex'	subunits α, β, γ, δ	Spans membrane at translocation site, receptor function?	*mammals, birds, fish*
	Sec62, Sec63	Transport of proteins, interaction with Bip*	*yeast*
Signal peptidase complex	peptides of 25, 23, 22, 21, 18, 12 kDa	Spans membrane, protease function, removes signal peptide within the lumen	*mammals, yeast*
Oligosaccharyltransferase complex	48 kDa protein, ribophorins I (65 kDa) and II (63 kDa)	Spans membrane, transfers within the lumen oligosaccharides to nascent protein chain	*mammals, yeast*

* Bip = binding protein (Hsp70), ATPase function, present in the lumen of the *ER*

Figure 13.3-1. Protein Processing in the Endoplasmic Reticulum

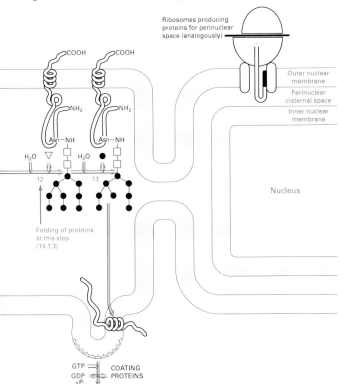

Table 13.3-2. Enzymes Involved in Glycosylation Reactions in the ER (EC numbers in parentheses)

1. Dolichol kinase (2.7.1.108)
2. UDP-N-acetylglucosamine-dolichyl-P N-acetylglucosamine-P-transferase (inhibited by tunicamycin, 2.7.8.15)
3. N-acetylglucosaminyldiphosphodolichol N-acetylglucosaminyltransferase (2.4.1.141)
4. Chitobiosyldiphosphodolichol α-mannosyltransferase (2.4.1.142)
5. Glycolipid mannosyltransferases (several)
6. Dolichyl-P β-D-mannosyltransferase (inhibited by amphomycin, 2.4.1.83)
7. Dolichyl-P-mannose-glycolipid α-mannosyltransferase
8. Dolichyl-P β-glucosyltransferase (2.4.1.117)
9. Dolichyl-P-glucose-glycolipid α-glucosyltransferase (not characterized yet)
10. Dolichyl-diphosphooligosaccharide-protein glycotransferase (2.4.1.119)
11. Mannosyl-oligosaccharide glucosidase (glucosidase I; inhibited by castanospermine, 3.2.1.106)
12. Glucosidase II (no EC number yet; inhibited by 1-deoxynojirimycin)
13. Mannosyl-oligosaccharide 1,2-α-mannosidase (3.2.1.113)
14. Dolichyl-P-mannose – glycolipid-α-mannosyltransferase
15. Dolichyl diphosphatase (inhibited by bacitracin, 3.6.1.43)
16. Phosphatidylinositol N-acetylglucosaminyltransferase (2.4.1.198)

Literature:

Casey, P.J.: Science 268 (1995) 221–224.
Chen, W., *et al.*: Proc. Natl. Acad. Sci. USA 92 (1995) 6229–6233.
Doering, D.L., *et al.*: J. Biol. Chem. 265 (1990) 611–614.
Hartl, F.-U., Hlodan, R., Langer, T.: Trends in Biochem. Sci. 19 (1994) 20–25.
Hirschberg, C.B., Snider, M.D., Ann. Rev. of Biochem. 56 (1987) 63–87
Low, M.G.: Biochim. biophys. Acta 988 (1989) 427–454.
Lütcke, H.: Eur. J. Biochem. 228 (1995) 531–550.
Rapoport, T.A.: Science 258 (1992) 931–936.
Reithmeier, R.A.F., in Vance, D.E., Vance, J. (Eds.): *Biochemistry of Lipids, Lipoproteins and Membranes. New Comprehensive Biochemistry* Vol. 31. Elsevier (1996) 425–471.
Takeda, J., Kinoshita, T.: Trends in Biochem.Sci. 20 (1995) 367–371.

13.4 Glycosylation Reactions in the Golgi Apparatus

In *eukarya*, the *Golgi apparatus* continues the protein and lipid glycosylation (which was initiated in the *ER*) and effects the sorting of proteins to guide them to their destination.

The *Golgi apparatus* is arranged asymmetrically. Its parts have the following functions:
- *Cis-Golgi network:* recognition and recycling of proteins, which falsely escaped from the *endoplasmic reticulum* (*ER*, 13.3)
- *Cis-, medial-* and *trans-Golgi compartments:* glycosylation reactions
- *Trans-Golgi network:* numerous sorting events. Proteolytic cleavage as a final processing step starts here with some proteins, and is continued in the *vesicles* (e.g., some hormones and neuropeptides, 17.1.5).

All compartments are characterized by different sets of enzymes (glycosidases and glycosyltransferases, Tables 13.4-1 and 13.4-2). The proteins are transported between the compartments by *vesicles* (14.2).

The carbohydrate chains shown in Figures 13.4-1 and 13.4-2 are a selection of basic structures. They occur in glycoproteins and proteoglycans (13.1) as well as in glycolipids (13.2). In glycoproteins, they are frequently modified further, often depending on the individual (e.g. blood group types, 13.5), the organ or the metabolic situation (e.g. in activated *lymphocytes*, O-glycosylation core I is transformed into core II, while in resting *lymphocytes*, core I is retained). In proteoglycans, the structures depicted form the linkage region between the protein and the extended, outer glycan chains (see Figures 13.1-1 and 13.1-2).

13.4.1 Synthesis of Glycoproteins (Fig. 13.4-2, Table 13.4-2)

While protein synthesis uses product specific templates (mRNA) and a constant set of enzymes to obtain a definite product, the final product of protein glycosylation is determined by cooperation of different, specific enzymes.

All sugar moieties in the *Golgi* are added stepwise directly from XDP-sugars or CMP-NeuNAc. Specific carriers exist in the *Golgi membrane* for all required XDP-sugars, which transport them from the *cytosol* into the *Golgi lumen*. Contrary to the *ER*, no oligosaccharide precursors are synthesized at anchors like dolichylphosphate.

N-glycosylated proteins imported from the *ER* have a 'core' glycostructure (13.3.3), which is initially trimmed in the *cis-Golgi* compartment by action of glycosidases. The amount of this shortening determines the further fate of the compound. During the following journey through the *medial* and *trans* compartments the various complex, hybrid and high mannose structures are formed (Pathways A...D in Fig.13.4-2).

This process effects the microheterogeneity of carbohydrate structures, since only part of all possible reactions occur at an individual protein. This is (at least partially) caused by the particular protein structure. In addition, the degree of glycosylation depends on the transit time in the compartments, which is influenced by the translation rate and the secretion efficiency of the cell. The precise function of the individual structure is not always known.

O-glycosylated proteins: In *mammals*, the pathway for their synthesis starts in the *Golgi apparatus* (Pathway E in Fig.13.4-2). Here, the first sugar (usually N-acetylgalactosamine ■) is added to serine or threonine, in collagen to hydroxylysine. In *yeast* and *green plants*, however, the first sugar (mannose ●) already becomes attached in the *ER* and the chain is continued in the *Golgi* by adding up to 3 more mannose residues (Pathway F in Fig. 13.4-2).

13.4.2 Synthesis of Proteoglycans

The short, inner branches of proteoglycans (see Fig.13.1-2) have the structure of N- or O-glycoproteins and are synthesized the same way as described above. The synthesis of the repeating units of the long, outer glycosaminoglycan branches of proteoglycans also takes place in the *Golgi apparatus*, but the exact localization of the reactions is not known so far. This is also valid for the sulfatation reactions, which are not shown in Fig.13.4-2.

13.4.3 Synthesis of Glycolipids (Fig.13.4-1, Table 13.4-1)

Glycolipids contain a lipid anchor (e.g. ceramide or acylglycerol, Fig. 13.2-1) in the membrane with an attached glycoside chain. The addition of the first sugar residue is supposed to take place at the *cytosolic side* of the *Golgi membrane*. The monoglycosylated ceramide is then transferred to the *luminal side*, where all the following reactions occur. Up to G_{M3} the glycosylations take place in the *cis compartment* of the *Golgi*, thereafter in the *medial* and *trans compartments*. Finally the glycolipids are transferred to the *cellular surface*, where they arrange themselves in clusters. Only the initial reactions, forming the 'root structures' are shown in Fig.13.4-1, for larger structures see Fig.13.3-2.

Literature:
Goldberg, Kornfeld, S.: J. Biol. Chem. 258 (1983) 3160–3167.
Hirschberg, C.B., Snider, M.D.: Ann. Rev. of Biochem. 56 (1987) 63–87.
Montreuil, J., Vliegenthart, J.F.G., Schachter, H.: *Glycoproteins. New Comprehensive Biochemistry* Vol. 29a. Elsevier (1995).
Vance, D.E., Vance, J. (Eds.): *Biochemistry of Lipids, Lipoproteins and Membranes. New Comprehensive Biochemistry* Vol.31. Elsevier (1996).

Figure 13.4-1. Synthesis of Glycosphingolipids in the *Golgi Apparatus*

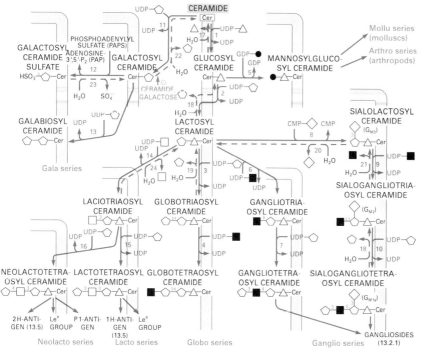

Table 13.4-1. Well Characterized Enzymes in Glycosphingolipid Synthesis
(Common names and EC- numbers in parentheses)

1. Ceramide glucosyltransferase (2.4.1.80)
3. Lactosylceramide β-1,3 (or 4) galactosyl-transferases (2.4.1.179)
8. Lactosylceramide α-2,3-sialyltransferase (2.4.99.9)
9. N-acetylneuraminylgalactosylglucosylceramide N-acetyl-galactosaminyltransferase (2.4.1.92 or 165)
10. Ganglioside galactosyltransferase (2.4.1.62)
11. Ceramide galactosyltransferase
12. Galactosylceramide sulfotransferase (2.8.2.11)
14. Lactosylceramide β-1,3-N-acetyl-β-D-glucosaminyltransferase (2.4.1.206)
15. Glucosaminylgalactosylglucosylceramide β-galactosyltransferase (2.4.1.86)
17. Glucosylceramidase (3.2.1.45)
18. Ganglioside β-galactosidase
19. α-Galactosidase (3.2.1.22)
20. Exo-α-sialidase (neuraminidase) (3.2.1.18)
21. β-N-Acetylhexosaminidase A (3.2.1.52)
22. Galactosylceramidase (lysosomal) (3.2.1.46)
23. Cerebroside sulfatase (arylsulfatase A) (3.1.6.8)

Figure 13.4-2. Synthesis of Glycoproteins in the *Golgi Apparatus*

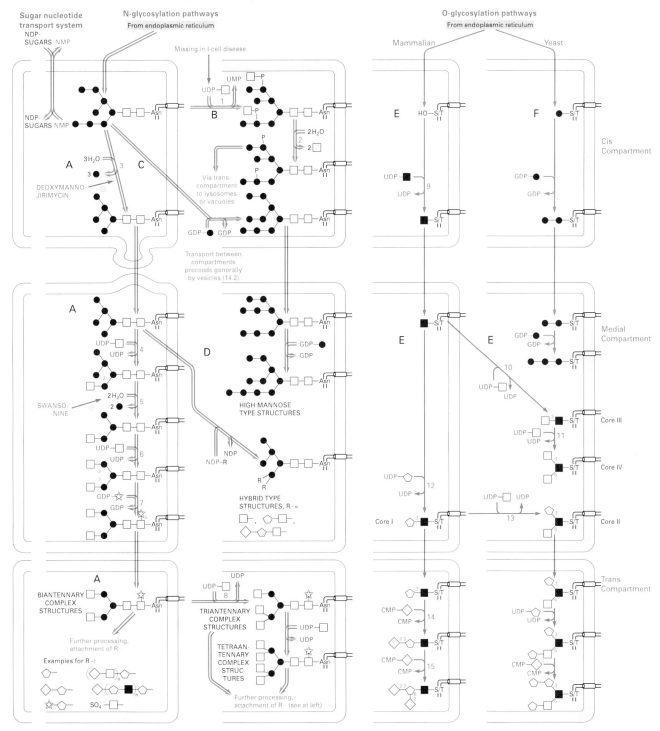

Symbols

- △ Glc glucose
- □ GlcNAc N-acetylglucosamine
- ◇ Gal galactose
- ■ GalNAc N-acetylgalactosamine
- ● Man mannose
- ◇ NeuNAc N-acetylneuram. acid
- ☆ Fuc fucose
- ⊨⊏ transmembrane domains

Pathways of Protein Glycosylation (The letters correspond to the pathways in Figure 13.4-2):

A) N-glycosylation pathway leading to complex structures. Additional glycoside units (besides Man and GalNAc) are added, leading to final structures shown (R-). They are also present in proteoglycans.
B) pathway for glycoproteins destined for lysosomes (general) or vacuoles (yeast). Mannose units become phosphorylated and are able to react with the mannose 5-P receptor at the Golgi membrane.
C) pathway to high-mannose type carbohydrate structures.
D) pathway to hybrid structures.
E) O-glycosylation pathway in *mammalian cells*. These structures are also present in proteoglycans.
F) O-glycosylation pathway in *yeast*.

Table 13.4-2. Enzymes in Glycoprotein Synthesis
(Common names and EC-numbers in parentheses)

1. UDP-N-Acetylglucosamine-lysosomal enzyme N-acetylglucosaminephospho-transferase (2.7.8.17)
2. N-acetylglucosamine-1-phosphodiester α-N-acetylglucosaminidase (phosphodiester glycosidase, 3.1.4.45)
3. Mannosyl-oligosaccharide 1,2-α-mannosidase (mannosidase I, 3.2.1.113)
4. Mannosyl-glycoprotein β-1,2-N-acetylglucosaminyltransferase (N-acetylglucosaminyltransferase I, 2.4.1.101)
5. Mannosyl-oligosaccharide 1,3-1,6-α-mannosidase (mannosidase II, 3.2.1.114)
6. α-1,6-Mannosyl-glycoprotein β-1,2-N-acetylglucosaminyltransferasae (N-acetylglucosamine transferase II, 2.4.1.143)
7. Glycoprotein 6-α-L-fucosyltransferase (2.4.1.68)
8. β-1,4-Mannosyl-glycoprotein β-1,4-N-acetylglucosaminyltransferase (N-acetylglucosamine transferase III, 2.4.1.144)
9. Polypeptide N-acetylgalactosaminyl transferase (2.4.1.41)
10. Acetylgalactosaminyl-O-glycosyl-glycoprotein β-1,3-N-acetylglucosaminyltransferase (2.4.1.147)
11. Acetylgalactosaminyl-O-glycosyl-glycoprotein β-1,6-N-acetylglucosaminyltransferase (2.4.1.148)
12. Glycoprotein-N-acetylgalactosamine 3-β-galactosyltransferase (2.4.1.122)
13. β-1,3-Galactosyl-O-glycosyl-glycoprotein β-1,6-N-acetylglucosaminyltransferase (2.4.1.102)
14. β-Galactoside α-2,3-sialyltransferase (2.4.99.3)
15. α-N-acetylgalactosaminide α-2,6-sialyltransferase (2.4.99.3)

13.5 Terminal Carbohydrate Structures of Glycoconjugates

The surface of *eukaryotic cells* carries many glycoproteins and glycolipids with characteristic arrangements of carbohydrate chains. Such arrangements can also be found on secreted proteins. These carbohydrate structures can be antigenic, the respective determinants are usually located in the terminal glycosidic groups. Some of them play a role in cell-cell recognition and interaction (e.g. sialylated Le^x groups as part of ligands for selectins, 18.3). Antigenic properties of cell surfaces can, however, also originate in protein structures.

13.5.1 Blood Groups (Figure 13.5-1)

A number of antibodies in *human/animal serum* are directed against antigens at the *erythrocyte* surface of other members of the same species (but also against similar antigens in other species, e.g. *animals, microorganisms, viruses*). The antigen-antibody reaction causes blood agglutination and cell lysis. This happens when foreign antigens are introduced into the circulation (e.g. by incompatible blood transfusion). The presence or absence of this reaction in individuals of the same species allows classification of blood groups.

The human blood group antigens are located in the *layer* of the extracellular matrix (2.2.4) not only at the surface of *erythrocytes*, but also of most other cells of the body. In about 80% of the population, they occur also in soluble form in the *body fluids*. The antigens at the surface of erythrocytes have been classified in (so far) 19 blood group systems. Only the ABO, Hh, Ii, P and Sid systems are based on carbohydrate structures, while others (e.g. the MN and the rhesus antigens) originate in the expression of antigenic protein epitopes.

The MN antigens are variants of the protein sequence of glycophorin (an erythrocyte-attached glycoprotein), while the carbohydrate structure is identical. The protein chains of the rhesus antigens are encoded by 3 closely coupled gene pairs; details of their structure are not known yet. Antibodies against them are normally absent in blood, they are raised only after sensibilisation.

The cell-bound antigens of the ABO and other systems are glycoproteins and glycosphingolipids, whereas the soluble ones are glycoproteins. The terminal antigenic determinants are carried by short chains (e.g. R = Cer-■■, see 13.4.3) or by long sequences of (frequently repeated) glycosidic groups, which are bound via a carbohydrate core to a protein (e.g. anion transporter or glucose transporter of *erythrocytes*). The Lewis antigens are glycosphingolipids, which are not expressed at erythrocytes, but get acquired from plasma lipoproteins (HDL and LDL).

The pedigree of *human* blood group antigens is given in Figure 13.5-1. Starting from a basic structure (R-■■), fucose is either bound to the Gal ○ moiety, leading to the ABO blood group substances and after a second fucosylation to the Lewis systems Le^b and Le^y, or to the GlcNAc □ moiety, leading to the Lewis systems Le^a and Le^x. At present 7 types of fucosyl transferases with varying specificity have been isolated, therefore the pathways of biosynthesis are not identical in different cell types. Lewis structures are frequently masked by terminal sialylation. Other additions to the basic structure lead to the I/i, P etc. blood groups.

Literature:
Feizi, T.: Nature (London) 314 (1985) 53–57.
Hakomori, S.: Adv. Cancer Res. 52 (1989) 257–332.
Monteuil, J., Vliegenthart, J.F.G., Schachter, H.: *Glycoproteins. New Comprehensive Biochemistry* Vol. 29a. Elsevier (1995).
Muramatsu, T.: Glycobiology 3 (1993) 291–296.

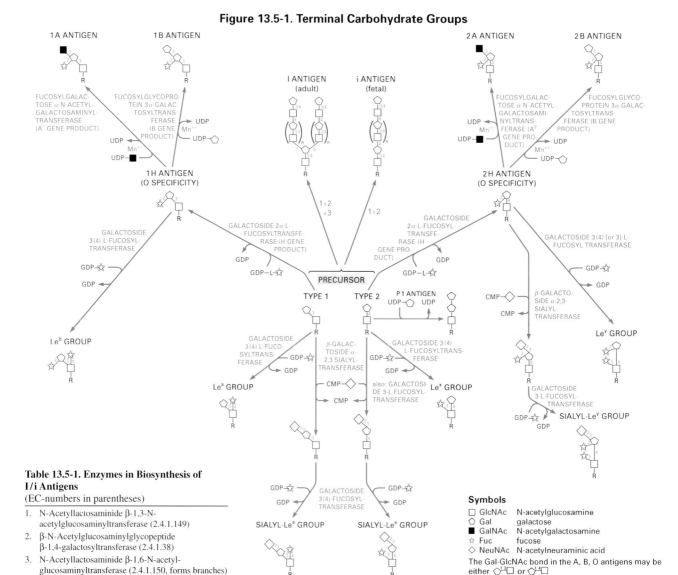

Figure 13.5-1. Terminal Carbohydrate Groups

Table 13.5-1. Enzymes in Biosynthesis of I/i Antigens
(EC-numbers in parentheses)
1. N-Acetyllactosaminide β-1,3-N-acetylglucosaminyltransferase (2.4.1.149)
2. β-N-Acetylglucosaminylglycopeptide β-1,4-galactosyltransferase (2.4.1.38)
3. N-Acetyllactosaminide β-1,6-N-acetyl-glucosaminyltransferase (2.4.1.150, forms branches)

Symbols
□ GlcNAc N-acetylglucosamine
○ Gal galactose
■ GalNAc N-acetylgalactosamine
☆ Fuc fucose
◇ NeuNAc N-acetylneuraminic acid
The Gal-GlcNAc bond in the A, B, O antigens may be either ○^{1,3}□ or ○^{1,4}□

14 Protein Folding, Transport and Degradation

14.1 Folding of Proteins

The tertiary structure of proteins (2.3.1) is determined by their amino acid sequence. Nevertheless, most proteins do not fold spontaneously into their final three-dimensional structure.

Polypeptides emerge from ribosomes (10.6.3, 11.5.2, 13.3.1) in an unfolded state. The formation of secondary structures (α helices and β sheets, 2.3.1) is a spontaneous process, which leads to so-called 'molten globules'. These folding intermediates still contain unstructured regions and are devoid of a regular tertiary structure. While hydrophobic amino acids are usually buried in the core of correctly folded soluble proteins, they are still exposed to the solvent in these globules. This leads to inter- and intramolecular aggregation, which inhibits further folding.

Specialized proteins (chaperones) in *bacteria* and in *eukarya* are involved in folding proteins into their correct structure.

14.1.1 Protein Folding in Bacteria (Table 14.1-1, Fig. 14.1-1)

In *bacteria*, one group of molecular chaperones prevents undesired aggregation by binding and stabilizing unfolded polypeptide regions (DnaK/J, GrpE). The other group provides an environment for the folding intermediate, that allows a controlled folding process (GroEL/ES complex).

Many chaperones belong to the group of heat-shock proteins (Hsp), since they also respond to cellular stress situations. The chaperones not only prevent aggregation of polypeptides, they also can resolve aggregated proteins and protein complexes, which occur e.g. by thermal denaturation. DnaKJ/GrpE also repress excessive gene transcription of the heat shock proteins by binding the σ^{32} factor (10.4.1).

The nascent polypeptide evolving from the *ribosome* is bound by DnaJ, which targets DnaK•ATP to the protein. ATP hydrolysis stabilizes the complex of the polypeptide chain–DnaK•ADP–DnaJ. This prevents the polypeptide from aggregation with other unfolded polypeptides. The polypeptide is released as a molten globule after GrpE has resolved the complex by ADP/ATP exchange. If the polypeptide still contains unfolded regions or has adopted an unfavorable structure, the cycle is repeated.

For many proteins, the consecutive folding of the molten globule into its final tertiary structure takes place inside of the GroEL/ES complex. The folding process requires a sequence of binding steps and final release of the polypeptide. It is dependent on ATP.

14.1.2 Protein Folding in the Eukaryotic Cytosol (Table 14.1-1)

In the *cytosol* of *eukarya*, the process proceeds similarly to *bacteria*, employing the respective chaperone homologues (which are highly conserved in all organisms). The evolving protein chain immediately binds Hsp40 and Hsp70 and then enters the Hsp60-type TRIC structure (GroEL/ES homologue).

Similar mechanisms take place in folding of proteins imported into *mitochondria* and *chloroplasts* (14.4, 14.5).

14.1.3 Protein Folding in the Eukaryotic Endoplasmic Reticulum (ER, Table 14.1-2, Figure 14.1-3)

The ER differs from the *cytosol* by providing an oxidizing environment (resulting in protein-S-S-bonds) and by glycosylation reactions on numerous proteins. Specific enzymes acting on the glycan chains contribute to the folding process of glycoproteins.

Glucosidase I removes 2 of the 3 glucose residues of the glycoprotein precursor (Fig.13.3-1). The chaperone-like transmembrane protein calnexin (p88, IP90) then binds specifically to the remaining glucose residue. Other chaperones (BiP/Grp78 and Grp94) cooperate in the folding mechanism. Upon completion of the folding process, the single remaining glucose is removed by glucosidase II, and the proteins are transported to the *Golgi apparatus*. However, incorrectly folded proteins undergo a reglucosylation cycle, catalyzed by a specific UDP-glucose: glycoprotein glucosyltransferase, which retains them in the ER until correct folding is achieved (quality control). Protein disulfide isomerase (PDI) contributes to correct folding by thiol oxidation and reshuffling of disulfide bonds (Fig.14.1-2). Similar enzymes (DsbA, DsbE) exist in the *periplasmic space* of Gram-negative bacteria.

Table 14.1-2. Chaperones Involved in Protein Folding in the *ER*

Chaperones	Family	Function
Calnexin (p88: *mice*, IP 90: *humans*)	(Hsp90)	Transmembrane phosphoprotein, binds glycoproteins, binds Ca^{++}
BiP/ Grp78	Hsp70	Soluble, ATPase, binds and retains proteins in the ER (DnaK homol.), involved in posttranslat. import
Grp94?	Hsp90	Endoplasmin, soluble, ATPase, binds proteins, dimeric?

Literature:
Bergeron, J.J.M. *et al.*: Trends in Biochem. Sci. 19 (1994) 124–128.
Friedman, R.B. *et al.*: Trends in Biochem. Sci. 19 (1994) 331–333.
Frydman, J. *et al.*: Nature 370 (1994) 111–117.
Hartl, F.U. *et al.*: Trends in Biochem. Sci. 19 (1994) 20–25.
Mayhew, M., Hartl, F.U.: Science 271 (1996) 161–162.

Table 14.1-1. Molecular Chaperones and Enzymes Involved in Protein Folding in *E. coli* and Homologues in *Eukarya*

E. coli Components	Properties and Function	Homologue in *Eukarya*
DnaJ (41 kDa)	Binds to polypeptide, mediates DnaK binding, induces ATPase action of DnaK, resolves protein aggregates	Hsp40. In *mitochondria* Mdj1 (14.3.2)
DnaK (ca. 65 kDa)	Stabilizes nascent, unfolded or partially folded polypeptides, binds and hydrolyzes ATP, resolves protein aggregates, controls heat-shock response	Hsp70 (Hsc70, BiP). In *mitochondria* mtHsp70 (14.3.2)
GrpE (20 kDa)	ADP/ATP exchange factor, effects release of the polypeptide from the DnaK/J complex, resolves protein aggregates.	– (In *mitochondria* mtGrpE, 14.3.2)
GroEL (14 · 57 kDa)	2 barrel-shaped stacked rings formed by 14 subunits with a hydrophobic inner surface, catalyzes folding of cytoplasmic proteins, ATPase	TCP-1 ring complex = TRIC (ca 10-fold hetero-oligomer). In *mitochondria* mtHsp60
GroES (7 · 10 kDa)	Ring of 7 subunits capping one end of the GroEL 'barrel', releases proteins from GroEL	– (In *mitochondria* Hsp10)
Dsb proteins (A…E)	Reformation of -S-S- bonds	Protein disulfide isomerase (in *ER*)
Rot A, Sur A, FkpA (*periplasmic*), trigger factor (*cytoplasmic*)	Isomerization of proline. In *eukarya*, also an essential subunit of prolyl-4-hydroxylase, being involved in collagen synthesis (3 families)	Peptidyl-proline *cis-trans* isomerase (in *ER*)
Thioredoxin system (e.g., 8.1.4)	Reversible reduction of disulfide bonds	Thioredoxin system (e.g., 8.1.4 and 16.2.2)

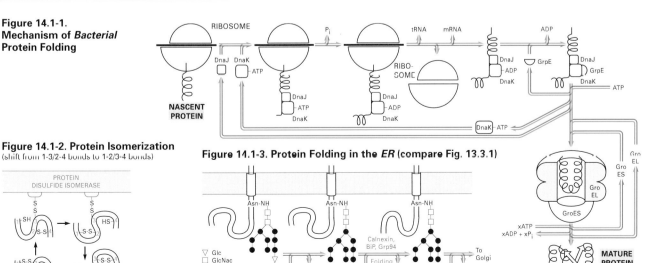

Figure 14.1-1. Mechanism of *Bacterial* Protein Folding

Figure 14.1-2. Protein Isomerization (shift from 1-3/2-4 bonds to 1-2/3-4 bonds)

Figure 14.1-3. Protein Folding in the *ER* (compare Fig. 13.3.1)

14.2 Vesicular Transport and Secretion of Proteins

14.2.1 Pathways of Transport

Proteins to be secreted from *eukaryotic* cells or to become members of the cellular membranes are transported by vesicles to their destination. This is a conserved mechanism in all *eukarya*. For secretion, the proteins follow either the constitutive pathway for permanent secretion or the regulated pathway. In this case, they are stored intracellularly in secretory vesicles, which fuse with the membranes and discharge their content only after receiving specific signals (e.g. in presynaptic vesicles of the *nervous system*, 17.2.3).

Most secreted proteins are biosynthesized in the *endoplasmic reticulum* (ER, 13.3) and further modified in the *Golgi apparatus* (13.4). Secretion ist thought to be in general a bulk-flow process from *ER* via *Golgi* to the *cell surface*. All deviations from this pathway are signal encoded.

As an example, the mannose groups of glycoproteins destined for lysosomes are phosphorylated (Fig. 13.4-2). They become attached to a mannose-6-P-receptor, which causes the protein to enter the lysosomes. Thereafter, the receptor is recycled to the *Golgi*.

Vesicular transport operates also in the other direction from the *plasma membrane* back to the *Golgi* (endocytosis) and even farther on to the ER (18.1.2). An example is described in 18.2-4.

Figure 14.2-1. Pathways of Vesicular Transport

14.2.2 Transport Vesicles (Table 14.2-1)

Different coated vesicles take part in the various steps. In the *ER* to *Golgi* and *intra-Golgi* steps COP proteins are involved. In *yeast* 2 different coats (COP I and COP II) have been described. Most *yeast* proteins are interchangeable with *mammalian* proteins in the various *in-vitro* transport assays. Figure 14.2-2 shows the main features of *mammalian* intra-Golgi transport. This mechanism ist typical for vesicular traffic. The specificity of the fusion between vesicle and target membrane is determined by the properties of v- and t-SNARE receptors (Proteins with analogous function in other transport steps are noted below in Table 14.2-1). ATP hydrolysis at NSF supplies the energy for the fusion step.

The endocytotic pathway, as well as the pathway from the *trans-Golgi network* to *endosomes* proceeds via clathrin-coated vesicles (18.1.2).

Table 14.2-1. Proteins With Functions in the Intra-Golgi Transport System

	Mammalian		Homology in *Yeast*
COP	Coat proteins β, β', γ α, δ, ε, ζ	+	Sec26, 27, 21 (COP I) Ret1, Ret2, –, Ret3
–	–		Sec23, 24, 13 (COP II)
Arf1	ADP ribosylating factor (GTP/GDP binding)	++	Sar1 (part of COP II)
GAP	GTPase activating proteins		
GEF	Guanine nucleotide exchange factor		
Rab	GTP/GDP binding		Ypt1, Sec4
Sec1	?		Sly1, Sec1
NSF	N-ethylmaleinimide sensitive protein		Sec18
SNAP	NSF attachment protein (α, β, γ)		Sec16 (homol. to SNAP α)
SNARE	SNAP receptor (v-SNARE = vescile-SNARE t-SNARE = target SNARE)	* **	Sec22, Bos1, Bet1 Sec9, Sed5, Sso1

+ similar: COP II in transport from *ER* to *Golgi*
++ similar: Sar1p in transport from *ER* to *Golgi*
* similar: VAMP (synaptobrevin) in *presynaptic vesicles*
** similar: syntaxin in *presynaptic membranes*, stabilized by SNAP-25

In *neurons,* synaptic vesicles containing neurotransmitters (17.2.3) are stored near the *presynaptic membrane* after their synthesis and active transport to this site (18.1.3). They are attached to the *cytoskeleton* by synapsin I and to the *presynaptic membrane* by interaction between the membrane receptors syntaxin (t-SNARE analogue) and SNAP-25 and the vesicle receptor synaptobrevin. However, the vesicle membrane component synaptotagmin (p65) prevents formation of the fusion particle, possibly by blocking SNAP access to syntaxin. Increase of intracellular Ca^{++} due to nerve stimulation releases the attachment to the cytoskeleton by phosphorylation of synapsin I. It also removes the synaptotagmin block, permits the formation of the fusion complex and concomitant release of the vesicle contents into the synaptic cleft.

Literature:
Bernett, M.K., Scheller, R.H., Ann. Rev. Biochem. 63 (1994) 63–100.
Fischer von Mollard, G. *et al.*: Trends in Biochem Sci. 19 (1994) 164–168.
Nuoffer, C., Balch. W.E.: Ann. Rev. Biochem. 63 (1994) 949–990.
Pryer, N.K. *et al.*: Ann. Rev. Biochem. 61 (1192) 471–516
Rothman, J.E.: Nature 372 (1994) 55–63.
Söllner, T.: FEBS Letters 369 (1995) 80–83.
Trasdale, R.D., Jackson, M.R.: Ann. Rev. Cell Dev. Biol 12 (1996) 27–54.

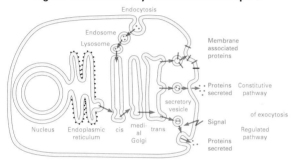

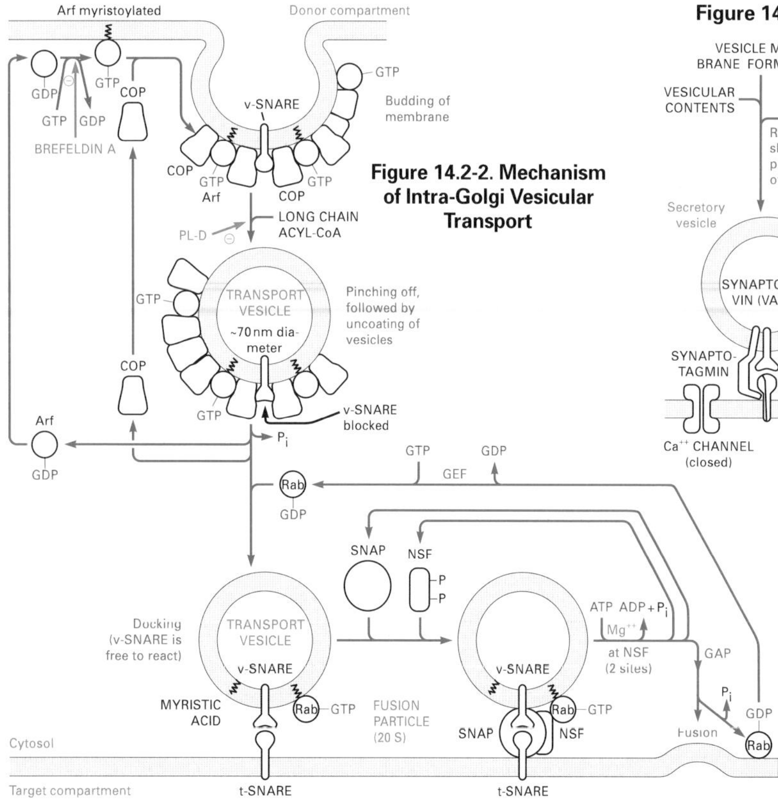

Figure 14.2-2. Mechanism of Intra-Golgi Vesicular Transport

Figure 14.2-3. Action of Synaptotagmin in Nerve Terminals

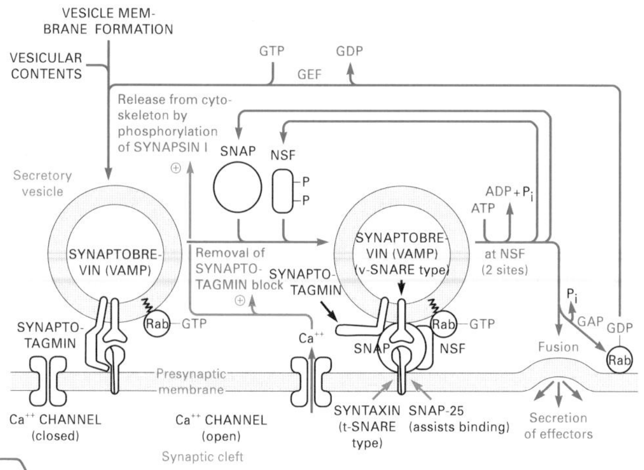

14.3 Protein Transport Into the Nucleus

While in *bacteria* only protein export to the *extracellular* and into the *periplasmic space* takes place (15.2), *eukaryotic cells* require besides processes for protein export (14.2) also mechanisms for transporting proteins into intracellular organelles [*nucleus, mitochondria* (14.4), *chloroplasts* (14.5)].

Import of proteins into the *nucleus* is a signal- and energy-dependent, saturable process, which is highly conserved in *eukaryotic* species. Examples for transported components are DNA/RNA polymerases, transcription factors, histones, viral proteins. In contrast to transport of proteins across other membranes (e.g. *endoplasmic reticulum, mitochondria, chloroplasts*, but also *bacteria*), it comprises several unique features:

- the carriers are soluble and are shuttling forth and back
- folded proteins can be transported
- the transport occurs via a nuclear pore complex (NPC) of more than 100 proteins, large enough to allow passive diffusion of proteins <60 kDa. The total mass is ca. $1.25 * 10^8$ Da with eightfold rotational symmetry; the outer diameter is ca. 125 nm; the channel diameter 9 nm. The channel expands up to 26 nm for active protein transport. It disassembles during mitosis.

The import machinery consists of *cytosolic* components [importin α and β, Ran (a GTP binding protein), GTPase interacting factor and optionally a Hsp70-homologue] and subunits of the *nuclear pore complex* (special O-glycosylated nucleoporins, NUP). A selection is listed in Table 14.3-1.

There exists also an important mechanism for export of RNA, RNP (11.3.3...8) and proteins. Many details are still unknown, however, there are striking similarities to the import process.

14.3.1 Targeting Mechanism (Figure 14.3-1, upper part)

Proteins to be transported ('cargo') carry a signal for selective transport (nuclear localization sequence, NLS), which is a continous or bipartite sequence within the protein of 5...15 mostly basic amino acids with no obvious homologies. The NLS sequence of the cargo protein forms a complex with importin α. Promoted by importin β, this complex binds to special nucleoporins (e.g. NUP358) on the cytosolic side of the NPC.

If the nuclear localization sequence is masked by regulatory components, e.g., Hsp90 for the steroid hormone receptor (17.6), IkB for the transcription factor NF-kB (Table 11.4-2), its attachment to importin α and the protein import is prevented.

14.3.2 Transport Mechanism (Figure 14.3-1, lower part)

This step requires Ran and the hydrolysis of GTP bound to it. The cargo/importin α/β-complex moves from the periphery of the nuclear pore complex to its central region and from there to its nucleoplasmic side. This stepwise transport into the *nucleoplasm* (= *nuclear matrix*) possibly occurs by multiple dissociation-association cycles of the cargo complex with various nucleoporins lining the pore complex and is controlled by GTPase cycles involving Ran and Ran-interacting proteins (RanGAP1, RanBP1, NTF-2, Rcc1, cf. 17.5.3). In the *nucleoplasm*, the cargo/importin α/Ran-GDP complex becomes disassembled, the imported protein is released and the assisting import components are recycled into the *cytoplasm*.

The molecular mechanism of the transport steps (including interaction with the central channel, channel gating, translocation, disassembly and recycling) is not known in all details. Hsp70-like proteins do not seem to have the same prominent role as in other systems.

Literature:
Floer, M. *et al.*: J. Biol. Chem 272 (1997) 19538–19546.
Garcis-Bustos, J. *et al.*: Biochim. Biophys. Acta 1071 (1991) 82–101.
Melchior, F., Gerace, L.: Curr. Opin. in Cell Biology 7 (1995) 310–318.
Nigg, E.A.: Nature 386 (1997) 779–787.
Panté, N., Aebi, U.: Current Opin. in Cell Biology 8 (1996) 397–406.
Rout, M.P., Wente, S.M.: Trends in Cell Biol. 4 (1994) 357–365

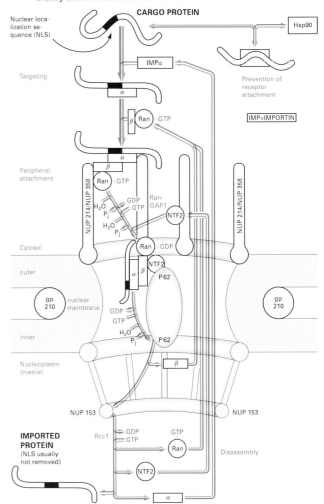

Figure 14.3-1. Mechanism of Protein Import Into the *Nucleus*
The spatial arrangement and the sequence of some steps have not been exactly demonstrated. The nomenclature for *vertebrates* is used.

Table 14.3-1. Components for Import Into the Nucleus

Localization	Selection of Components (*Species*)	Characteristics, Function
Cytosol	Importin α [karyopherin 60 (*frog*); p54/p56 (*bovine, rat*); Kap60 (*yeast*); NPI-1/Rch1 (*human*)]	NLS receptor, binds to NLS of the cargo and to importin β. Enters *nucleus* together with cargo, Ran, NTF2.
	Importin β [karyopherin 90 (97 kDa, *frog*); p97 (*bovine*); Kap95 (*yeast*)]	Adaptor, promotes docking of cargo-importin α complex to binding site at the NPC (NUP 358, NUP 214, NUP 153), Remains at the nuclear envelope.
	Ran [TC4 (25 kDa, *frog, mammals*); Gsp1 (*yeast*)]	'Small' G-protein (Table 17.5-2). Binds as Ran•GTP to Nup358, hydrolyzes GTP. Mediates the committed step from docking NPC to translocation. Very abundant.
	RanGAP1; Rna1p (*yeast*)	Binds to NUP's. Enhances GTPase activity of Ran on bound GTP.
	RanBP1	Interacts with Ran, supports recycling of the Ran•GTP/importin β system by RanGAP1
	NTF2 (15 kDa, *human*); p10/B-2 (10 kDa, *frog*)	Transfer of the transport complex from docking site to gated channel (p62)
	Hsc70 (Hsp 70-like protein, *human*)	Chaperone, facilitates cargo binding to importin α (not a general import factor)
Nuclear pore complex		
cytoplasmic side	NUP358, NUP214 (*rat*); NUP159 (*yeast*)	Special O-glycosylated nucleoporins, docking sites for cargo-NLS-receptor-adaptor complex, involved in stepwise transport of the cargo through the NPC; proposed binding sites for Ran and importin β.
nucleoplasmic side	NUP98 (*rat*); NUP153 (*rat*)	
unknown	p265 (*rat*); NSP1(*yeast*)	
both	p62 (*rat, frog, mouse*); NUP1/NUP2 (*yeast*)	
Nucleoplasm	Rcc1	GDP/GTP exchange factor for Ran

14.4 Protein Transport Into Mitochondria

More than 90% of the *mitochondrial* proteins are encoded on the nuclear DNA and have to be imported posttranslationally into the organelle after synthesis on *cytosolic* ribosomes. Examples are subunits of the respiratory chain (e.g. ATPase subunit $F_1\beta$, 16.1.4), translocators (e.g. ADP/ATP-carrier, 16.1.4) and matrix enzymes (e.g. alcohol dehydrogenase III, ornithine carbamoyltransferase, 4.9.1).

14.4.1 Targeting Mechanism (Figure 14.4-1, upper part)

The typical imported protein is a precursor containing a N-terminal targeting presequence (15...70 amino acid residues), which is removed enzymatically after import. Some presequences, however, are not cleavable or the targeting signal is located internally.

Presequences do not share sequence homologies, but show generally a high content of basic and hydroxylated amino acids and a lack of acidic residues. Targeting is achieved by specific binding of the presequence to receptors at the *outer membrane*. Additional sequences can target the protein to the final subcompartments (outer membrane, intermembrane space, inner membrane).

14.4.2 Transport Mechanism (Figure 14.4-1, lower part)

The transport of the various proteins takes place by a single type of import machinery located in the *outer* (TOM-complex) and *inner* (TIM-complex) *membrane* and in the *mitochondrial matrix* (chaperones). The precursor protein can be transported simultaneously across both membranes at so-called translocation contact sites, where both membranes are in close vicinity. The process is thought to take place through an aqueous protein conducting channel formed by the TOM and/or TIM-complex (Table 14.4.1).

The precursor maintains a loosely folded, import competent conformation by binding to chaperone molecules (e.g. *cytosolic* Hsp70), or to a presequence binding factor (e.g. MSF). Then the presequence becomes attached to receptors of the TOM-complex. After transport across the *outer mitochondrial membrane*, another temporary binding of the presequence takes place at the inner (*trans*-)side of the outer membrane.

Many precursors bind first to the Tom20/22 receptor subcomplex for recognition, followed by transfer to the translocation channel. For MSF-bound precursors (e.g. carrier proteins or alcohol dehydrogenase III) the primary recognition is achieved by binding to Tom70/Tom37 (not shown in Fig. 14.4-1).

The energy for transport of the positively charged targeting sequence through the TIM-complex of the *inner membrane* is supplied by the electric membrane potential ($\Delta\Psi$, 1.5.3 and 17.2.1).

Within the *matrix*, the protein is at first attached to Tim44. mtHsp70•ATP binds to the protein. According to the 'Brownian ratchet' model, mtHsp70 traps the protein and makes inward Brownian movements of the protein irreversible. In addition or alternatively, the ATP hydrolysis (assisted by Mdj1) could cause a conformational change of Hsp70, which pulls the protein inward ('translocation motor model', analogous to the 'power stroke' of myosin, 17.4.5). mtGrpE (Mge1) removes ADP from Hsp70 and causes its dissociation from the protein. Then the cycle begins anew.

During the import step, the targeting sequence is removed by a specific peptidase. Folding and assembly of matrix proteins into complexes is thought to be facilitated by molecular chaperones (e.g. mt-Hsp70 and Hsp60/10), energized by ATP-hydrolysis.

Proteins destined to other subcompartments (*inner membrane* and *intermembrane space*) are sorted either during transport or are delivered by a second transport event after they have entered the *matrix*. This sorting step is energized by the pH-gradient across the inner membrane and by matrix ATP (see also *chloroplasts*, 14.5 and *bacteria*, 15.2).

Table 14.4-1. Components for Import into *Mitochondria*

Location	Consensus Nomenclat.	Component N. crassa	Component yeast	Function, Characteristics (compare Table 14.1-1 for chaperones)
Cytosol		n.i.*	ctHsp70	import competence, ATPase, DnaK homologue
		n.i.	Vdj1	import competence, DnaJ homologue
	MSF *(rat)*	n.i.	(homol. to MSF)	mitochondrial import stimulating factor, dimeric, targets precursor, dissolves aggregates, ATPase
Outer membrane	Tom70	Mom72	Mas70	receptor, recognizes MSF-bound precursors
	Tom71	n.i.	Tom71	associated with Tom70 and Tom37
	Tom37	n.i.	Mas37	associated with Tom70 and Tom71
	Tom22	Mom22	Mas22	receptor, essential for recognition of precursors
	Tom20	Mom19	Mas20	receptor, associates with Tom22
	Tom40	Mom38	Isp42	translocation channel
	Tom7	Mom7	Mom7	translocation channel regulator
	Tom6	Mom8b	Isp6	associated with Tom40
	Tom5	Mom8a	Mom8	associated with Tom40
Inner membrane	Tim23	n.i.	Mim23	form a complex, achieve transport of precursor across inner membrane, exact location?
	Tim17	n.i.	Mim17	
	Tim33		Mim33	
	Tim14		Mim14	
	Tim11			
	Tim44	n.i	Mim44	peripheral prot., transfer of precursor to mtHsp70
Matrix		mtHsp70		translocation and sorting, ATPase, DnaK hom.
		n.i.	Mdj1	recycling of Hsp70, DnaJ homologue
		n.i	mtGrpE/Mge1	recycling of Hsp70, GrpE homologue
		mtHsp60/10		folding and assembly, GroEL/ES homologue
		MPP	Mas1/2	enzymatic removal of targeting presequence

* not identified

Literature:

Lill, R., Neupert, W.: Trends in Cell Biology 6 (1996) 56–61.
Horst, M. *et al.*: Biochim. Biophys. Acta 1318 (1997) 71–78.
Pfanner, N. *et al.*: Trends in Biochem. Sci. 21 (1996) 51–52.
Ryan, K.R., Jensen, R.E.: Cell 83 (1995) 517–519.
Schatz, G.: J. Biol. Chem. 271 (1996) 31763–31766.
Wickner, W.T.: Science 266 (1994) 1197–1198.

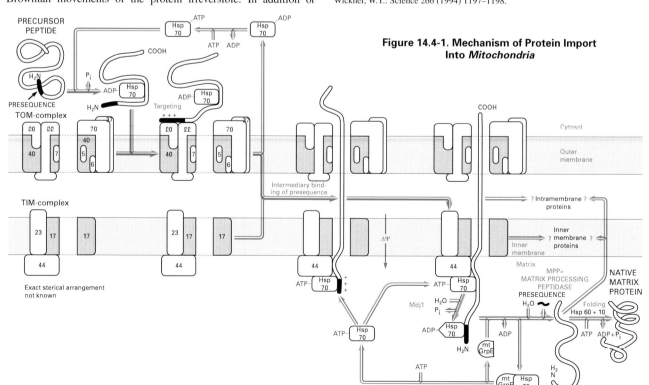

Figure 14.4-1. Mechanism of Protein Import Into *Mitochondria*

14.5 Protein Transport Into Chloroplasts

About 80% of the *chloroplast* proteins are encoded in nuclear DNA and have to be imported posttranslationally into the organelle after being synthesized on cytosolic ribosomes. Examples are subunits of the photosynthetic machinery (e.g. subunits 17, 23 and 33 of photosystem II and plastocyanin) and stromal enzymes.

14.5.1 Targeting Mechanism (Figure 14.5-1, upper part)

The typical imported protein is a precursor containing a N-terminal extension (transit peptide, targeting presequence; 35...>100 amino acids), which is removed after import. Some transit peptides, however, are not cleavable. Transit peptides do not share sequence homologies, but generally show a high content of aromatic and hydroxylated amino acids and a lack of acidic residues.

Targeting to the organelle is mediated by binding of the transit peptide to specific receptors in the *outer envelope* of the *chloroplast* (Table 14.5-1). Additionally, the precursor contains sequences effecting further targeting and sorting to the final subcompartment (*outer envelope, intermembrane space, inner envelope, stroma, thylakoid membrane* and *thylakoid lumen*).

14.5.2 Transport Mechanism (Figure 14.5-1, lower part)

The transport of the various proteins into the *stroma* takes place via a single type of import machinery, composed of two individual protein complexes located in the *outer* and *inner envelope*. The precursor protein can be transported simultaneously across both envelopes at translocation contact sites, where the envelopes are in close vicinity to each other. Transport presumably takes place through an aqueous protein conducting channel. Energy is provided by hydrolysis of GTP and ATP. GTP binding proteins and Hsp70-like chaperones seem to play a central role in the recognition of the precursor transit sequence (e.g. proof-reading) and in translocation events.

The preprotein maintains a loosely folded, import competent conformation by interactions with *cytosolic* (Hsc70, SRF) and/or envelope associated (Com70) components. Specific binding of the transit peptide to *outer envelope* receptors (members of the OEP-complex) is followed by transfer to the protein-conducting channel (possibly formed by OEP75), requiring GTP hydrolysis.

When the transported preprotein emerges at the *trans* side of the *outer envelope*, it binds to the molecular chaperone Hsp70 IAP. Thereafter presumably transfer of the transit peptide to a different receptor takes place, which causes opening of the membrane channel in the *inner envelope* (possibly involving IAP100 and IAP36). It is not certain, whether a molecular mechanism similar to the mitochondrial Hsp70-Tim44 system (14.4.2) also exists in the *stroma* of *chloroplasts*.

During and subsequently to import, *stromal* proteins are processed (removal of the transit peptide by a stromal peptidase). Folding and assembly into complexes is facilitated by molecular chaperones (Hsp70 and Cpn60/10 = Rubisco binding protein) in an ATP-dependent process, similar to folding after protein synthesis (14.1).

For non-stromal proteins further sorting is again energy- and signal dependent (e.g. lumen targeting domain, LTD). For the import of proteins into the thylakoid lumen, apparently 3 ways exist:

- Similar to the *bacterial* SecA dependent pathway (15.2), requiring ATP and ΔpH
- Similar to the *bacterial* (15.2) and *eukaryotic* (13.3.1) SRP dependent pathways, requiring GTP, but acting post-translationally
- Driven only by ΔpH

Sorting into other compartments (e.g. *intermembrane space* or *inner envelope*) is still less well characterized.

Differences of protein import into *chloroplasts* as compared to *mitochondria* are:

- Almost no homology of envelope components involved in translocation
- Involvement of more than two different Hsp70-like proteins
- ATP-dependent binding of precursors to the import machinery
- Involvement of at least two GTP-binding proteins
- Requirement of NTP-hydrolysis in the intermembrane space
- No need for a membrane potential ($\Delta\psi$) across the inner envelope.

Table 14.5-1. Components for Import into *Chloroplasts*

Location	Consensus nomenclat.	Other names, Mol. Mass in kDa	Function/Characteristics (compare Table 14.1-1 for chaperones)
Cytosol		SRF	signal recognition factor
		Hsc70	import competence, Hsp homologue
Outer envelope	Toc86	OEP86/IAP86	precursor recognition/GTP-binding
	Toc34	OEP34/IAP34	precursor recognition/GTP-binding
	Toc75	OEP75/IAP74	channel/precursor binding, ATP-dependent
	Toc36	OEP36/IAP36	interaction with transit peptide
		Com70 (72 kDa)	chaperone, Hsp70-homologue/peripheral protein at *cis* side, ATPase
		Hsp70-IAP(75kDa)	chaperone, Hsp70-homologue/integral membrane protein at *trans* side, ATPase
Inner envelope	Tic110	IAP100/Cim97	precursor recognition; channel?
		IAP36	unknown
		IAP25	interaction with transit peptide, location?
		IAP21	interaction with transit peptide, location?
		Cim44+Com44	unknown, location?
Stroma		Hsc70 (75 kDa)	transport?, folding; Hsp70-homologue
		Cpn60/10	chaperone, folding, assembly
		SPP (140 kDa)	stromal processing peptidase (Zn protease), cleavage of transit sequence
		CPSecA	transport into *thylakoids*, SecA homol. (*bact.*)
		CPSecY	transport into *thylakoids*, SecY homol. (*bact.*)
		CP54	transport into *thylakoids*, SRP homologue(*ER*)

Literature:
Cline, K., Henry, R.: Ann. Rev. Cell Dev. Biol. 12 (1996) 1–26.
De Boer, A.D., Weisbeek, P.J.: Biochim. Biophys. Acta 1071 (1991) 221–253.
Gray, J.C., Row, P.E: Trends in Cell Biology 5 (1995) 243–247.
Kouranov, A., Schnell, D.J.: J. Biol. Chem. 271 (1996) 31009–31012.
Schnell, D.J.: et. al.: Trends in Cell Biology 7 (1997) 303–304.

Figure 14.5-1 Mechanism of Protein Import Into *Chloroplasts*

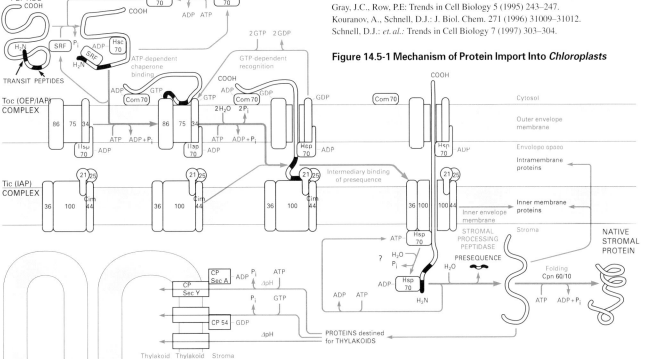

14.6 Protein Degradation

Peptidases is the general term for enzymes, which catalyze the cleavage of peptide bonds (as defined by the IUB). The terms 'proteases' or 'proteinases' are equivalent to 'endopeptidases', they refer only to internal cleavage of longer peptide chains (proteins), although in practice frequently 'proteases' and 'peptidases' are used exchangibly. These enzymes serve a large number of functions in living organisms:

- Digestion of dietary proteins
- Intracellular protein turnover including termination of enzyme action, e.g. cyclins and other cell cycle factors (11.6)
- Maturation of newly synthesized proteins; e.g. removal of signal or targeting sequences (13.3.2, 14.3...5)
- Processing of precursor proteins and peptides, e.g. of organelle proteins or hormones (preproteins, e.g. 17.1.3, 17.1.5) or
- Activation of zymogens (proproteins, inactive precursors of proteases, 17.1.9)
- Participation in biological processes requiring cleavage of peptide bonds, e.g., coagulation (20.3), fibrinolysis (20.5), control of blood pressure (Fig. 17.1-9), complement activation (19.2), inflammation (19.1.7), cell division (11.6.5), tissue disintegration and remodelling, apoptosis (11.6.6), sporulation, fertilization, tumor cell invasion etc.

Due to the wealth of their essential functions, peptidases are ubiquitous in all living cells and organisms.

14.6.1 Classification of Peptidases
By Type of Catalyzed Reaction (Table 14.6-1): There are two groups of peptidases, comprising the exopeptidases (cleaving near the N- or C-terminus of the peptide) and the endopeptidases.

By Cleavage Site Specificity (Table 14.6-2): Often, the cleavage specificity of peptidases is directed only towards the amino acid residue(s) immediately neighbouring the cleavage site in the substrate protein. In other cases, the specificity depends on longer amino acid sequences ('restriction endoproteinases').

By Catalytic Mechanism (Table 14.6-3, Figs. 14.6-1...3): Most frequently, peptidases are classified according to their catalytic mechanism or the amino acids in the active center, which are involved in catalysis. There are four major categories and a few with other reaction types (e.g., threonine proteases, 14.6.7).

All mechanisms involve the attack of an activated oxygen (or S^- in case of cysteine peptidases) on the carbonyl-C of the substrate (which becomes the center of a tetraedric intermediate) and the consecutive cleavage of the C-N bond with concomitant proton binding to the substrate-N.

14.6.2 Reaction Mechanism of Serine Peptidases (Fig. 14.6-1)
As an example for serine peptidases, the mechanism of chymotrypsin (an endo peptidase) is described. The reaction proceeds in two steps:

- Acylation step (upper line): The reaction starts with a nucleophilic attack of the serine 195-hydroxyl-O on the carbonyl-C of the substrate, while the proton is transferred to His 57. The resulting histidine cation is stabilized by the neighboring carboxyl group of Asp 102. Another transfer of this proton to the amide-N of the substrate causes the cleavage of the peptide bond. The amino component of the substrate is then released.
- Deacylation step (lower line): This reaction is a reversal of the acylation step, in which water plays the role of the amino component of the substrate, resulting in release of the carboxy component of the substrate.

The enzyme cleaves besides peptide also ester bonds by a similar mechanism. Its formation from an inactive precursor after secretion is described in 17.1.9. Regulatory serine peptidases contain an additional peptide chain, which interacts with cofactors essential for regulation.

14.6.3 Reaction Mechanism of Cysteine Peptidases
The mechanism of cysteine peptidases (e.g. papain) is similar to serine peptidases, with cysteine-S^- and protonated histidine replacing the serine-OH and the nonprotonated histidine.

14.6.4 Reaction Mechanism of Aspartate Peptidases (Fig. 14.6-2)
As a representative for aspartate peptidases, the catalytic mechanism of pepsin (an endopeptidase) is described. It depends on the γ-carboxyl groups of two aspartate residues in the active center (Asp 32 and Asp 215), one of them has to be ionized. This environment activates a water molecule, the oxygen of which performs a nucleophilic attack on the carbonyl-C of the substrate, while the proton released from the water combines first with the ionized carboxyl of aspartate and then with the amide-N of the substrate, resulting in peptide bond cleavage. The other, initially unionized aspartate carboxyl group acts first as proton donor and thereafter as proton acceptor.

Table 14.6-1. Groups of Peptidases

Group	EC-Subclass	Action (Symbols see below)	
		N-terminus	C-terminus
Exopeptidases			
Aminopeptidases	3.4.11	●-l-○-○-○-○.....	
Dipeptidyl-peptidases	3.4.14	●-●-l-○-○-○.....	
Tripeptidyl-peptidases	3.4.14	●-●-●-l-○-○-○.....	
Carboxypeptidases	3.4.16		.....○-○-○-○-○-l-●
Peptidyl-dipeptidases	3.4.15		.....○-○-○-○-l-●-●
Dipeptidases	3.4.13	●-l-●	
Omega peptidases	3.4.19	☆-●-l-○-○-○-○..... or	.....○-○-○-○-○-l-●-☆
Endopeptidases (Proteinases)			
	3.4.21...3.4.24, 3.4.99	.....○-○-○-○-l-○-○-○-○.....	

Symbols: ○: internal, ●: N- or C-terminal amino acid residue(s),
☆: residue bound in linkage other than peptide bond

Table 14.6-2. Preferred Cleavage Specificity of Peptidases

Peptidases (Examples)	Cleavage Specificity [X:Any Amino Acid]
Trypsin (17.1.9)	...Lys-l-X... or ...Arg-l-X...
Chymotrypsin (17.1.9)	...hydrophobic-l-X..., e.g.
	...Phe-l-X... or ...Tyr-l-X...
Thrombin (20.3)	...Arg-l-X...
Endoproteinase Glu-C from *Staphylococcus aureus* V8	...Glu-l-X... or ...Asp-l-X...
Asp-N-metallo-endoproteinase from *Pseudomonas fragi*	...X-l-Asp... or ...X-l-cysteic acid...
Coagulation Factor Xa (20.3)	...Ile-Glu-Gly-Arg-l-X...
Enteropeptidase (enterokinase)	...Asp-Asp-Asp-Asp-Lys-l-X...
Kexin (*yeast* KEX2 protease)	...(Lys or Arg)-Arg-l-X...

Table 14.6-3. Catalytic Mechanism of Peptidases

Class	Examples
Serine peptidases (Fig. 14.6-1)	trypsin, chymotrypsin, elastase, subtilisin, coagulation factors, t-PA, yeast carboxypeptidase Y, hepatitis C virus NS-3 protease
Cysteine peptidases (Fig. 14.6-1)	papain, bromelain, cathepsins B, H and L, calpain, hepatitis A virus 3C protease
Aspartate peptidases (Fig. 14.6-2)	pepsin, chymosin, renin, cathepsins D and E, HIV-1 protease
Metallopeptidases (Fig. 14.6-3)	*bacterial* and *mammalian* collagenases, thermolysin, carboxypeptidase A

14.6.5 Reaction Mechanism of Metallopeptidases (Fig. 14.6-3)
Carboxypeptidase A, another digestive enzyme, is an example of Zn^{++} dependent enzymes. The activation of water to enable the nucleophilic attack of its oxygen on the carbonyl-C of the substrate is effected by its environment: it is located between the complexed Zn^{++} ion and the ionized carboxyl group of Glu 270. The Zn^{++} ion also stabilizes the negatively charged tetraedric intermediate, assisted by Arg 127 (not shown). The ionized Glu-carboxyl acts as a proton acceptor and later as a proton donor analogously to one of the Asp residues in aspartate proteases.

An additional feature of carboxypeptidase A is the formation of a 'hydrophobic pocket' by movement of Tyr 218, which temporarily seals the reaction center, promoting the activity and contributing to the specificity of the enzyme.

14.6.6 Peptidase Inhibitors
Many biological and synthetic inhibitors exist. They are either active-site specific, which cause irreversible modifications, or special proteins, which act as pseudosubstrates (most natural inhibitors are of this type).

Unspecific inhibitors are, e.g., α-macroglobulin (acting by steric hindrance) or peptide aldehydes (non-cleavable peptide bond analogues). Class specific for serine peptidases are, e.g., organophosphates and coumarin derivatives (irreversible inhibition) and α_1 proteinase inhibitor (α_1 antitrypsin, 20.3.4). Pepstatin (a *bacterial* peptide with an unusual amino acid) inhibits the aspartate peptidase class and metal chalators inhibit metallopeptidases. Aprotinin (Trasylol®), soybean trypsin inhibitor and serpins form tight, but reversible complexes with individual peptidases.

14.6.7 Protein Degradation by the Ubiquitin (Ub) System
This system provides controlled protein degradation in all *eukarya* and exerts important regulatory functions in cell division (11.6.5), DNA replication (11.1.2), transcription, receptor function (17.1.2), cell differentiation, apoptosis (11.6.6) etc. At first, the protein has to be marked for degradation by attachment of ubiquitin (Ub, 8.5 kDa), which is the most conserved *eukaryotic* protein.

Thioester-activated ubiquitin is bound to the ε-amino group of a lysine residue in the protein to be degraded, forming an isopeptide (Fig. 14.6-4). The enzymes involved are, as an example taken from the cell cycle mechanism (11.6.5), Cdc34 and its homologues (E_2, possibly also E_3 function) together with Cdc53 (E_3 function) or APC (anaphase-promoting complex, E_3 function. Polyubiquitin chains are frequently formed by reaction of the Ub C-terminal glycine with an internal lysine of another Ub molecule.

Figure 14.6-1. Reaction Mechanism of Chymotrypsin (Example of a Serine Peptidase)

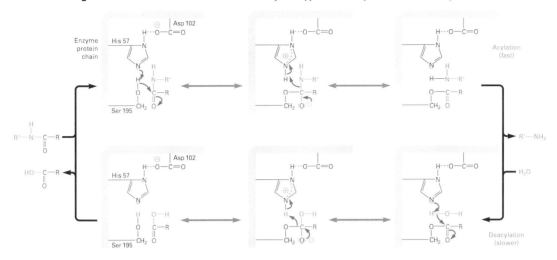

Figure 14.6-2. Reaction Mechanism of Pepsin (Example of an Aspartate Peptidase)

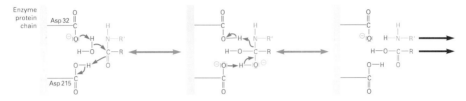

Figure 14.6-3. Reaction Mechanism of Carboxypeptidase A (Example of a Metallopeptidase)

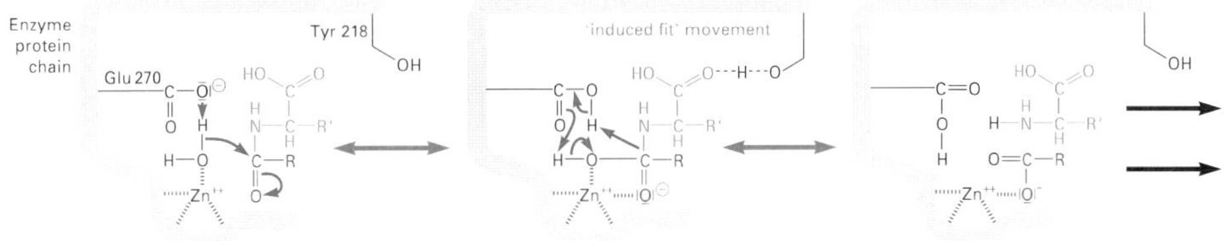

The protein recognition by the Ub system (and hence the degradation rate) depends on the N-terminal amino acid or on internal sequences of the protein (for details see 11.6.7). In some cases it was shown that a complex of several proteins associates with activated ubiquitin and takes part in the recognition of proteins to be degraded. A number of proteins require previous phosphorylation for their Ub dependent degradation.

Degradation of the protein marked by ubiquitination occurs within a large multicatalytic proteinase complex (28S proteasome, a threonine peptidase). The Ub is spared from destruction and reused.

Literature:

Barrett, A.J.: *Methods in Enzymology* Vols. 244 and 248. Academic Press (1994 and 1996)
Barrett, A.J., McDonald, J.K.: *Mammalian Proteases: A Glossary and Bibliography*, Vols. 1 and 2. Academic Press 1980 and 1986.
Barrett, A.J., Salvesen, G.: *Proteinase Inhibitors*. Elsevier (1986).
Beynon, R.J., Bond, J.S.: *Proteolytic Enzymes: A Practical Approach*. IRL Press (1989).
Christianson, D.W.: Adv. Prot. Chem. 42 (1991) 281–355.
Ciechanover, A.: Biol. Chem. Hoppe-Seyler 375 (1994) 565–581.
Davies, D.R.: Ann. Rev. of Biophys. Biophys. Chem. 19 (1989) 189–215.
Gerlt, J.A., Gassman, P.G.: Biochemistry 32 (1993) 11943–11952.
Jackson, P.K.: Current Biology 6 (1996) 1209–1212.
King, R.W. *et al.*: Science 274 (1996) 1652–1658.
Philips, M.A., Fletterick, R.J.: Curr. Opin. Struct. Biol. 2 (1992) 713–720.

Figure 14.6-4. Regulated Proteolysis by the Ubiquitin System

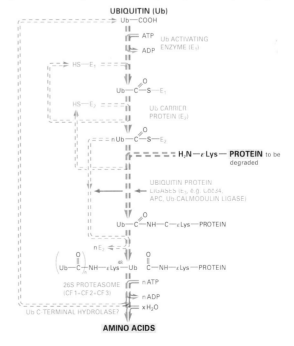

15 Special Bacterial Metabolism

15.1 Bacterial Envelope

The versatile cell envelope allows *bacteria* to adapt to many different environments. It fulfills a number of tasks:
- mechanical and chemical protection
- osmotic barrier
- adhesion to surfaces, e.g., other cells
- detection and selective uptake of substrates
- exchange of genetic material (conjugation)

The envelope consists of the cytoplasmic (inner) membrane, the peptidoglycan (=murein) and (in case of *Gram negative bacteria*) the outer membrane. Structural differences of the envelope, which can be assayed by Gram staining are important for bacterial taxonomy.

Gram negative bacteria (**Figure 15.1-1, left**): The inner membrane is surrounded by one giant molecule of peptidoglycan (13.1.3). This sac-like structure is extremely rigid and shapes the cell (e.g. rods, cocci, spirilloids, filaments). It is covered by the outer membrane, which differs chemically from all other biological membranes, because the phospholipids in the outer leaflet are replaced by the unique lipopolysaccharide (LPS).

Its constituents are:
- Lipid A: an unusual glycolipid, containing, e.g., 3-hydroxymyristic acid (C_{14})
- Core: a saccharide chain containing oxo-desoxy-octonic acid, heptose, hexoses and phosphoethanolamine
- O-specific side chain (O-antigen): a polysaccharide chain

This highly polar saccharide surface renders *Gram-negative bacteria* resistant to detergents, e.g., bile salts. The LPS is antigenic, pyrogenic and cytotoxic. The O-antigen units are assembled at undecaprenol membrane anchors (7.3) and are transferred to the LPS core. Since the outer membrane is a barrier even for many substrates, it is penetrated by porins, which allow diffusion of low molecular weight substances (15.3).

Gram-positive bacteria (**Figure 15.1-1, right**): They possess a multi-layer peptidoglycan. The site of synthesis is at the cytoplasmic membrane. The murein wall grows from the innermost to the peripheral layer. An outer membrane is absent, but a number of proteins and accessory polymers like polyanionic teichoic acids, lipoteichoic acids or teichuronic acids are found at the surface of *Gram-positive bacteria*. The murein wall is no barrier for small molecules such as metabolites.

Murein synthesis (Figure 15.1-1, lower part): Murein (peptidoglycan) consists of linear polysaccharide strands of alternating N-acetyl-glucosamine □ and N-acetyl-muraminic acid ⊗ (MurNAc) units. The strands of about 30 units are cross-linked by the short peptides attached to the lactyl group of MurNAc. In *E. coli* about one half of the peptide chains are cross-linked to octapeptides, 5% form dodecapeptides, whereas the rest is not crosslinked. The glycan strands run around the cell axis, the peptide crosslinks parallel to the axis (Figure 15.1-2). The units of the chains are synthesized at undecaprenyl membrane anchors (7.3) and transferred to the growing chain.

Figure 15.1-2. Growth of the Murein Layer

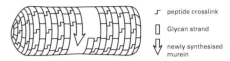

During cell wall synthesis, new material is inserted around the equator of the cell. A number of transglycosylases and transpeptidases, which are penicillin-binding proteins (PBPs), are involved in chain elongation, cross-linking and also invagination of the wall (septum) in the division zone of a dividing cell.

Serotyping, Diagnosis: Due to antigenic surface components, some bacterial species, especially pathogens, can be classified into different immunological groups. For example, the Gram positive *Streptococci* can be differentiated according to the polysaccharides attached to murein (Lancefield groups A, B, C, F, G), teichoic acids (D) and lipoteichoic acids (N). Differences in the sugar composition of the O-specific side chains allow the identification of several thousand serotypes of the Gram negative *Salmonellae*.

Literature:
Cooper, S.: Microbiol. Reviews 55 (1991) 649–674.
Neidhardt, F.C., Ingraham, J.L., Schaechter, M.: *Physiology of the Bacterial Cell.* Sinauer Assoc. (1990).

Figure 15.1-1. Bacterial Cell Envelope

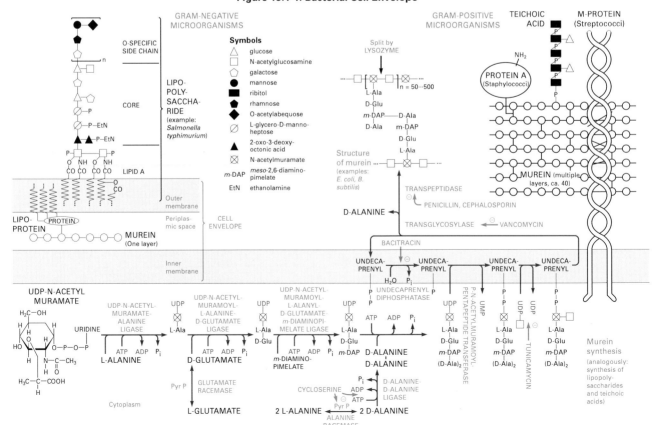

15.2 Bacterial Protein Export

Extracellular, periplasmic and outer membrane proteins are translocated across the *cytoplasmic membrane* either co- or posttranslationally. These proteins are synthesized as preproteins, which contain a N-terminal signal peptide (composed of a positively charged N-terminus, followed by a hydrophobic core and a polar region). This peptide sequence is recognized by the translocation apparatus. The different preproteins are delivered to the translocase complex (SecYEG) either through the SecA or the signal recognition particle (SRP) dependent pathway.

SecA dependent pathway (Figure 15.2-1, upper part): Posttranslational translocation of proteins is dependent on the chaperone SecA (Table 15.2-1). This peripheral subunit of the translocase complex binds the signal peptide of a preprotein after its release from the ribosome. SecA initiates the translocation: Upon binding of ATP, SecA inserts a domain into the *cytoplasmic membrane* and pushes a loop of the preprotein across the membrane. Translocation of the remaining protein through the translocase channel is dependent on the proton motive force (PMF, 1.5.3). As soon as the C-terminus of the signal peptide reaches the *periplasmic surface* of the membrane, it is cut off by the signal peptidase.

In *E. coli*, posttranslational translocation of preproteins also requires the *cytosolic* chaperone SecB. It stabilizes the unfolded state of the preprotein and delivers it to SecA. Most *bacteria* lack a SecB homologue, however.

Signal recognition particle (SRP)-dependent pathway (Figure 15.2-1, lower part): Cotranslational protein export and integration of membrane proteins into the cytoplasmic membrane is initiated by the *bacterial* signal recognition particle (SRP). This SRP is smaller than the *eukaryotic* SRP, which mediates protein translocation into the *endoplasmic reticulum* (13.3, compare Table 15.2-1 with Table 13.3-1). The SRP, composed of Ffh and 4.5 S RNA, arrests the synthesis of the nascent preprotein at the ribosome by binding to its signal peptide. The complex then binds to the SRP-receptor, which in *bacteria* consists of only one peripheral membrane protein, FtsY. After GTP-hydrolysis, the ribosome-preprotein complex is transferred to the translocase SecYEG, independent of SecA.

Archaea contain genes homologous to the *bacterial* genes encoding signal peptidase, SRP, SRP receptor and SecY, but not genes for SecA and SecB.

Some *bacterial* proteins are secreted by special mechanisms different from the Sec-system (e.g. *Yersinia* outer proteins, YOP; pullulanase, PUL).

Table 15.2-1. Factors Involved in Protein Export of *Escherichia coli*

Component	Factor	Mol. Mass (kDa)	Function
Export chaperone	SecB	18	Binds and stabilizes unfolded preprotein
Export ATPase	SecA	102	Peripheral membrane protein; associated with SecY chaperone, binds signal peptide of preprotein, directs it to translocase; initiates translocation.
Translocase	SecY	48	Trimeric integral membrane complex, forming the translocation channel (homologous to *eukaryotic* Sec61complex)
	SecE	14	
	SecG	15	
	SecD	67	Integral membrane proteins, facing the *periplasm*. Regulate deinsertion of SecA
	SecF	39	
Signal peptidase	LepB	36	Membrane bound protease facing the *periplasm*, cleaves signal peptide
Signal recognition particle (SRP)	P48 = Ffh	48	Chaperone, binds signal peptide of nascent protein; GTPase; activates GTPase of FtsY (homologous to *eukaryotic* SRP 54)
	RNA	4.5S	
SRP receptor	FtsY		GTPase; interacts with SRP and activates SRP GTPase; homologous to *eukaryotic* SRα

Literature:
Economou, A. *et al.*: Cell 83 (1995) 1171–1188.
Lütcke, H.: Eur. J. Biochem. 228 (1995) 531–550.
Randall, L.L., Hardy, S.J.S.: Trends in Biochem. Sci. 20 (1995) 65–69.
Wolin, S.L.: Cell 77 (1994) 787–790.
Wickner, W.T.: Science 266 (1994) 1197–1198.

Figure 15.2-1. Pathways of Bacterial Protein Export

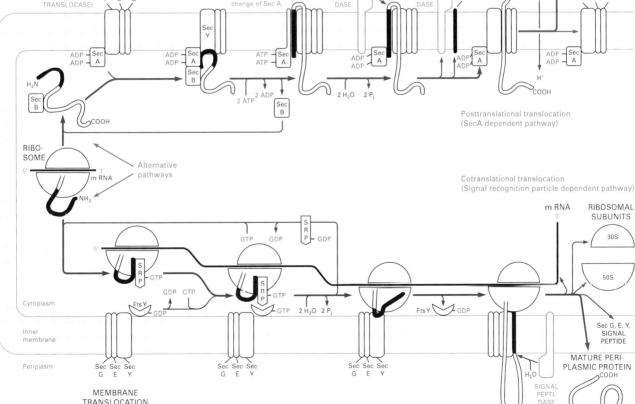

15.3 Bacterial Transport Systems

Bacterial membranes are a permeation barrier to nutrients, except for small molecules like O_2, H_2O, CO_2 or NH_3. *Bacteria* have developed various mechanisms to transport nutrients into the cell.

Hydrophilic molecules smaller than 600 Da cross the *outer membrane* of *Gram negative bacteria* via protein channels (porins). This process proceeds via passive diffusion and thus requires the nutrient concentration in the *periplasm* to be kept lower than in the *extracellular space*. Since in most natural environments microorganisms live under substrate limitation, this requirement can only be fulfilled by removal of these compounds from the periplasm by active (= energy dependent) transport across the *inner membrane*.

Transport in *Gram positive bacteria* involves only the *inner membrane*, because murein is no barrier for nutrients (except for polymers like starch and cellulose. They must be degraded extracellularly).

15.3.1 Types of Active Transport (Figure 15.3-1)

The primary transport of protons across the membrane, e.g. by respiration processes (16.1) establishes a proton motive force (PMF [mV]), which is composed of the chemical pH gradient (ΔpH; $c\,[H^+_{outside}] > c\,[H^+_{inside}]$) and the electrical membrane potential ($\Delta\Psi$ [mV]; $\Psi_{outside}$ +, Ψ_{inside} –; see 1.5.3):

$$\begin{aligned}\text{PMF [mV]} &= (-2.303 * R * T/F) * \Delta pH + \Delta\Psi \text{ [mV]} \quad [15.3\text{-}1]\\ &= -59 \quad\quad * \Delta pH + \Delta\Psi \quad \text{(at 25 °C)}.\end{aligned}$$

Secondary active transport systems use the energy of this disequilibrium to transfer nutrients into the cytoplasm or to secrete metabolic end products. *Bacteria* use various mechanisms:

- Symport: The uptake of nutrients is coupled to a component with a favorable electrical ($\Delta\Psi$) or chemical (e.g., ΔpH) gradient, such as lactose (in) / H^+ (in); lactate$^-$ (in) / H^+ (in).
- Antiport: Contrary to symport, uptake is coupled to extrusion of a second component, or vice versa, e.g., Na^+ (out) / H^+ (in).
- Uniport: Only the membrane potential $\Delta\Psi$ is used to transport cations into the cell or anions out of the cell, e.g., lysine$^+$ (in).

A number of permeases are not coupled to the proton gradient, but to a secondary Na^+ gradient. *Bacteria* living in alkaline environments cannot establish a H^+ gradient across the cytoplasmic membrane, therefore they exclusively use a Na^+ gradient.

Phosphotransferase system (PTS): Substrates that enter the cells by this system are phosphorylated, yielding membrane impermeable derivatives in the *cytoplasm* (e.g. glucose → glucose 6-P). As compared to other uptake mechanisms, this system is energy conserving, because phosphorylation is also required for further metabolization of the nutrients. Therefore it is the predominant uptake system in *anaerobes*.

As energy source, phosphoenolpyruvate is used. In *Gram-negative bacteria*, phosphate transfer is mediated by a sequence of enzymes: EI, histidine protein Hpr, EIIA, EIIB and the transmembrane protein EIIC. Often, EIIA, EIIB and EIIC are fused to a single polypeptide.

In *E. coli*, EIIA of the glucose system regulates other uptake systems: If glucose transport is high, EIIA$_{Glc}$ is dephosphorylated. This form inhibits allosterically other uptake systems, like glycerol kinase and lactose permease. If glucose transport is low (glucose shortage), EIIA$_{Glc}$ is phosphorylated and activates adenyl cyclase (17.4.2), thus more cAMP is produced. The catabolite repressor protein (CRP) binds cAMP. This enables the transcription of genes, which encode enzymes involved in uptake or metabolization of alternative nutrients (10.5.1).

Formerly the terminology was different (the regulatory protein of the glucose system was named III).

Binding proteins dependent transport system (BPDS): This system belongs to the large class of ABC (ATP binding cassette) transporters, which are also present in *eukarya*. In the *periplasm* of *Gram negative bacteria*, the substrate is bound by a binding protein with high affinity (K_m = 0.1...1 µmol/l) and high specificity, which transfers the substrate to the membrane permease. ATP hydrolysis effects a conformational change of the permease, which allows the substrate to enter the *cytoplasm* (Example: maltose, bound in the *periplasm* to Mal E). *Gram positive bacteria* lack an outer membrane and a periplasm, therefore the binding protein is anchored in the inner membrane.

Facilitated diffusion: The uptake of a nutrient can also be achieved by diffusion, if it is rapidly metabolized. For example, glycerol diffuses into the cytoplasm via a facilitator (channel protein) and is immediately phosphorylated. As a consequence, the concentration of glycerol in the *periplasm* is kept lower than that in the *extracellular space*. In contrast to *eukarya* (18.1.1), facilitated diffusion is rarely found in *bacteria*.

Literature:
Booth, I.B. in Anthony, C. (Ed.): *Bacterial Energy Transduction.* Academic Press (1988) 378–426.
Meadow, N.D. *et al.*: Ann. Rev. of Biochem. 59 (1990) 497–542.
Saier, M.H. *et al.*: Trends in Biochem. Sci. 20 (1995) 267–271.

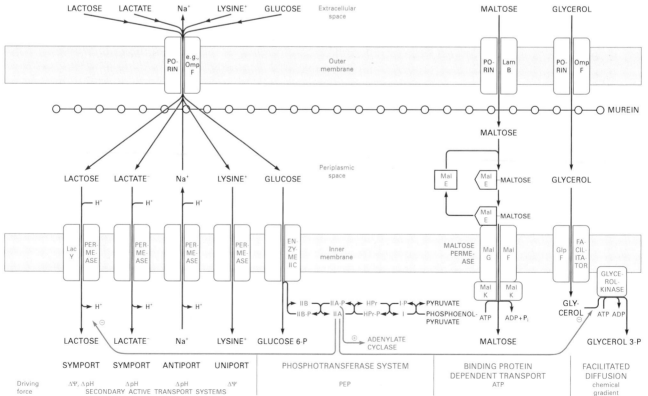

Figure 15.3-1. Examples for Bacterial Transport Systems

15.4 Bacterial Fermentations

Aerobic and anaerobic respiration implies the oxidation of substrates and the transfer of electrons via the respiratory chain to external acceptors, which are hereby reduced. The free energy $\Delta G_0'$ obtainable decreases in the following order: O_2 / nitrate / fumarate / sulfate (Fig. 16.1-5). In the absence of these acceptors, fermentative *bacteria* gain free energy by reduction of internally generated acceptors, e.g. pyruvate ($\rightarrow$ lactate), acetyl-CoA ($\rightarrow$ ethanol), H^+ ($\rightarrow H_2$). NADH (which is generated during the first catabolic steps of the substrates) transfers the reducing equivalents to these acceptors, is reoxidized and thus able to enter another redox cycle (Fig. 15.4-1). The end products of metabolism are usually excreted. Frequently, no exact stoichiometries for the reactions can be given, since alternative pathways are selected during different growth conditions. Many fermentations have economic importance.

During fermentation, ATP can only be generated by substrate-level phosphorylation. This results in quite a low ATP yield and therefore requires a high substrate throughput. E.g., in homolactate fermentation, 2 mol ATP are obtained from 1 mol glucose, while aerobic respiration theoretically results in 33 mol ATP and about 29 mol ATP under practical conditions (see 3.8.3 and 16.1.1).

Figure 15.4-1. General Principle of Fermentation

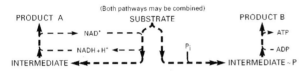

Only some specialized fermentations, e.g. diacetyl fermentation by *Leuconostoc spec.*, have been found to involve ion transport mechanisms. These generate a membrane potential, which can be used for ATP synthesis by membrane bound ATP synthase (16.1.4).

Fermentation of sugars (Figures 15.4-2 and 15.4-3, Table 15.4-1): Breakdown of glucose is initiated by reactions of the glycolytic pathway (3.1) or, less frequently, by the Entner-Doudoroff pathway (3.6.4). An important intermediate is pyruvate. In some cases, e.g., in mixed acid fermentation of *Enterobacteria*, two (or more) phases of fermentation can be observed. In the first phase, mainly organic acids are produced and excreted, leading to acidification of the medium. When a critical pH is reached, additional enzyme systems leading to pathways with neutral end products are expressed (e.g., the formate : hydrogen lyase complex in *E. coli* or acetolactate synthase in *Enterobacter aerogenes*).

It has to be emphasized that not all branches of the pathways can be found in the organisms mentioned in Table 15.4-1. Among mixed acid fermenters, *E. coli* (and other species) do not produce acetoin or butanediol. Various species of *Clostridia* will show different patterns of organic acid production.

Other fermentations: Fermentations of non-sugar carbon sources and especially of nitrogenous compounds involve highly specialized and complicated pathways.

With amino acids, the critical step of amino group removal by *anaerobic microorganisms* is performed in many different ways, such as reductive cleavage of the C-N bond, C-C rearrangement, α,β-elimination of ammonia, shift of -NH_2 from α- to β-position, conversion to oxo acids and consecutive reduction. See Figure 15.4-4 for examples.

Literature:
Buckel, W. in Hanska, E., Thauer, R (Ed.): *The Molecular Basis of Bacterial Metabolism.* 41st Colloquium Mosbach. Springer (1990) 21–30.
Gottschalk, G.: *Bacterial Metabolism.* Second edition. Springer (1986).
Konings, W.N., Lolkema, J.S., Poolman, B.: Arch. Microbiol. 164 (1995) 235–242.
Schlegel, G.: *General Microbiology.* 7th ed. Cambridge University Press (1993)

Table 15.4-1. Fermentation of Sugars (Examples)

The stoichiometries given in both tables refer only to the energy metabolism. Due to anabolic reactions the actual yield of end products is lower.

Type		Gross Reaction, ATP yield	Pyruvate Converting Enzyme	Organisms (Examples)
Glycolytic Fermentations and Lactate Converting Fermentations				
Alcohol fermentation	#	glucose $\rightarrow$ 2 ethanol + 2 CO_2 (+ 2 ATP)	pyruvate decarboxylase	*Sarcina ventriculi, Saccharomyces cerevisiae*
Homolactate fermentation	#	glucose $\rightarrow$ 2 lactate (+ 2 ATP)	lactate dehydrogenase (L or D)	*Lactobacillus lactis, Lactococcus lactis, Enterococcus faecalis*
Butanol / butyric acid fermentation	#	glucose $\rightarrow$ butyrate, acetate, H_2, CO_2 (+ $\geq$ 2 ATP) or 2 glucose $\rightarrow$ butanol + acetone + 4 H_2 + 5 CO_2 (+ 4 ATP) *	pyruvate: ferredoxin oxidoreductase	*Clostridium spec.*
Mixed acid fermentation	#	glucose $\rightarrow$ formate, ethanol, acetate, lactate, fumarate, CO_2, H_2 **	pyruvate formate-lyase (formate C-acyltransferase)	*Escherichia spec., Salmonella spec.*
Mixed acid fermentation	#	same + acetoin, 2,3-butanediol, no fumarate **	pyruvate formate-lyase	*Enterobacter spec., Erwinia spec.*
Homoacetate fermentation	#	glucose $\rightarrow$ 3 acetate (+ 4 ATP)	pyruvate: ferredoxin oxidoreductase	*Clostridium thermoaceticum, Acetobacterium woodii*
Propionate fermentation: succinate pathway	#	3 lactate $\rightarrow$ 2 propionate + acetate + CO_2 (+ 3 ATP) ***	lactate dehydrogenase (cytochrome)	*Propionibacterium spec.*
Propionate fermentation: acrylate pathway	#	3 lactate $\rightarrow$ 2 propionate + acetate + CO_2 (+ 1 ATP) via different intermediates	pyruvate: ferredoxin oxidoreductase	*Clostridium propionicum, Peptostreptococcus elsdenii*
Non-glycolytic Fermentations				
Alcohol fermentation via Entner-Doudoroff pathway	#	glucose $\rightarrow$ 2 ethanol + 2 CO_2 (+ 1 ATP)	pyruvate decarboxylase	*Zymomonas spec.*
Heterolactate fermentation	#	glucose $\rightarrow$ lactate + ethanol + CO_2 (+ 1 ATP) 3 fructose $\rightarrow$ lactate + acetate + 2 mannitol + CO_2 (+ 2ATP)	lactate dehydrogenase lactate dehydrogenase	*Leuconostoc mesenteroides Lactobacillus brevis*
Diacetyl fermentation	#	3 citrate $\rightarrow$ 3 acetate + diacetyl + lactate	pyruvate decarboxylase, pyruvate dehydrogenase, lactate dehydrogenase	*Lactococcus spec., Leuconostoc spec.*

Table 15.4-2. Fermentation of Nitrogenous Compounds (Examples)

Compound Fermented		Gross Reaction	Remarks	Organisms (Examples)
Fermentation of Amino Acids				
Alanine fermentation		3 alanine $\rightarrow$ 2 propionate + acetate + CO_2 + 3 NH_3	via acrylate pathway	*Clostridium propionicum*
Glycine fermentation		4 glycine $\rightarrow$ 3 acetate + 4 NH_3 + 2 CO_2		*Peptostreptococcus micros*
Glutamate fermentation	#	5 glutamate $\rightarrow$ 6 acetate + 2 butyrate + 5 CO_2 + H_2 + 5 NH_3	via mesaconate, citramalate	*Clostridium tetanomorphum*
Glutamate fermentation		2 glutamate $\rightarrow$ butyrate + 3 acetate + 2 NH_3	via 2-hydroxyglutarate, glutaconyl-CoA	*Acidaminococcus fermentans*
Cofermentation of pairs of amino acids	#	e.g., alanine + 2 glycine $\rightarrow$ 3 acetate + CO_2 + 3 NH_3	'Stickland reaction'	*Clostridium botulinum, C. sticklandii*
Fermentation of Heterocyclic Compounds				
Fermentation of urate		urate $\rightarrow$ acetate + 4 CO_2 + 4 NH_3	all heterocyclic compounds are degraded via xanthine	*Clostridium purinolyticum, C. acidiurici*

\#   Reaction sequence shown in Figures 15.4-2 ... 15.4-4.
*   The end products vary, depending on the fermentation conditions. Instead of acetone, also 2-propanol can be formed.
**   The ratio of the end products depends mostly on the pH of the medium.
***   Acetyl-CoA is formed in variable amounts by pyruvate decarboxylase, it is hydrolyzed to acetate.

Figure 15.4-2. Glycolytic Fermentations and Lactate Converting Fermentations

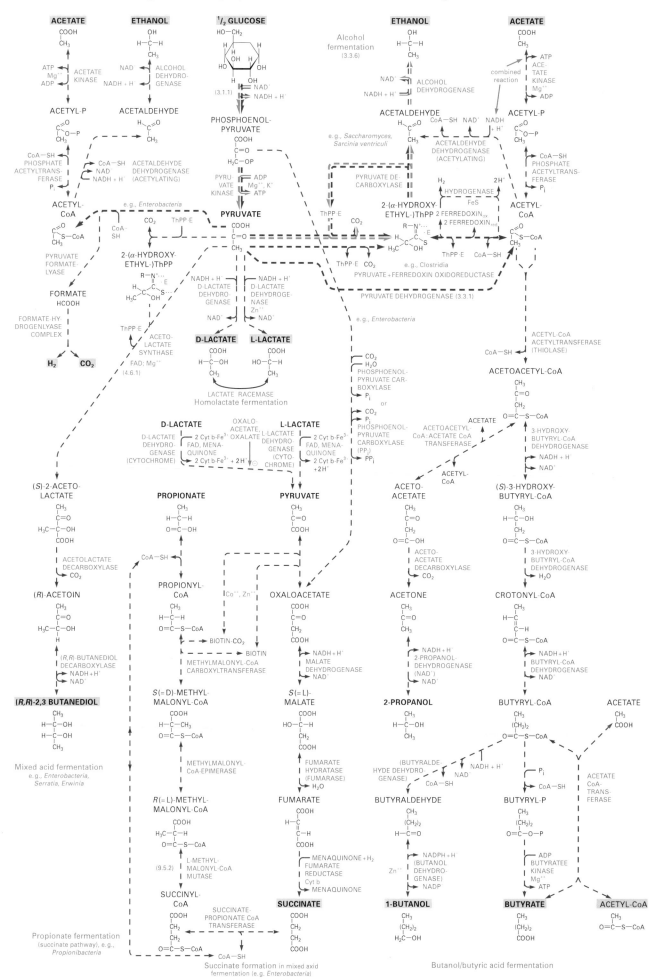

Figure 15.4-3. Non-Glycolytic Carbohydrate Fermentations

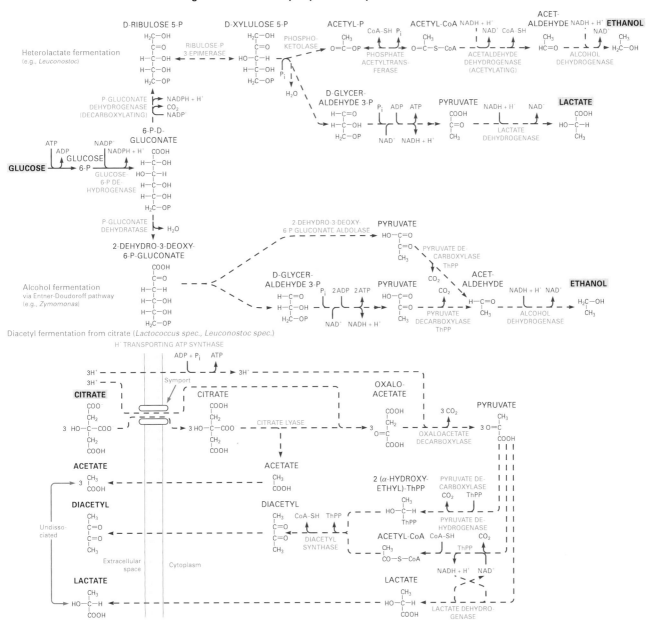

Figure 15.4-4. Fermentations of Nitrogenous Compounds (For Other Bacterial Degradations of Amino Acids, see Chapter 4).

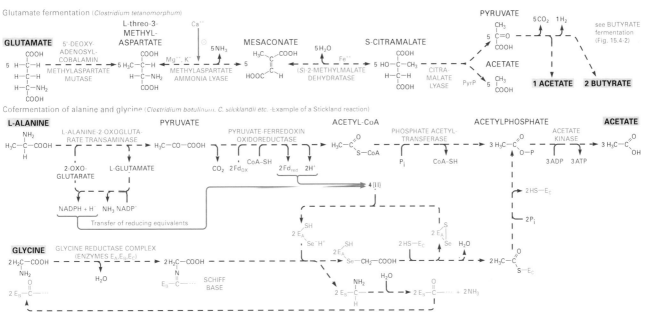

15.5 Anaerobic Respiration

Anaerobic respiration consists of the utilization of compounds other than O_2 as terminal electron acceptors in the respiratory electron transport chain. This mode of energy generation occurs in *microorganisms* living in oxygen-free habitats (sediments, soil, water, anaerobic cavities of animal bodies). Most of the electron acceptors are inorganic compounds in oxidized states such as CO_2, sulfate, nitrate or Fe^{3+}. Fumarate is an important organic electron acceptor. In most cases, the energy source (electron donor) is a reduced organic compound (carbohydrate, fatty acid, other organic acid). Molecular hydrogen (H_2) is a widely used inorganic donor (anaerobic lithotrophy).

15.5.1 Redox Reactions and Electron Transport (Figure 15.5-1)

Electrons from the oxidation of the energy source (electron donor reactions) are fed into membrane-bound electron transport chains and transferred to the acceptor substrate by specialized terminal reductases, which take the role of the oxidase in aerobic respiration. The system of *Wolinella succinogenes* is shown in Figure 16.1-1.

Energy yield and conservation: The theoretical energy yield depends on the difference of the redox potentials of the electron donor and the electron acceptor (Table 15.5-1). Since almost all electron acceptors have redox potentials lower than that of O_2/H_2O, anaerobic electron transport is less energy efficient than aerobic respiration (16.1). As in aerobic respiration, energy conservation takes place via chemiosmotic coupling: The electron flow causes proton translocation out of the cells. The electrochemical gradient built up this way drives the H^+-dependent ATP synthase (16.1.3).

Table 15.5-1. Redox Potentials

	Redox Couple	E'_0 [mV]
Typical electron donors	$2 H^+/H_2$	−410
	pyruvate/lactate	−197
	fumarate/succinate	+33
Typical electron acceptors	CO_2/CH_4	−244
	HSO_3^-/HS^-	−110
	$APS/HSO_3^- + AMP$	−60
	fumarate/succinate	+33
	NO_2^-/NH_3	+340
	NO_3^-/NO_2^-	+430
	NO_3^-/N_2	+740
	N_2O/N_2	+1355
For comparison: aerobic	$½ O_2/H_2O$	+816

In anaerobic respiration, the most important electron acceptor conversions are:
- CO_2 reduction to methane by *methanogenic archaea* (15.5.2). *Homoacetogenic bacteria* reduce CO_2 to acetate (carbonate respiration, 15.5.3).
- Sulfate and sulfur respiration by a variety of unrelated, strictly *anaerobic microorganisms*. Both acceptors are reduced to hydrogen sulfide, H_2S (sulfidogenesis). Sulfate needs to be activated before reduction; adenosine phosphosulfate (APS) is the true electron acceptor. Sulfite reductases contain siroheme, an iron-tetrahydroporphyrin (9.5.3). *Sulfidogenic bacteria* often use H_2 as energy source.
- Fumarate respiration is a wide-spread process, both of *obligate* and *facultative anaerobes*.
- Reduction of nitrogenous oxides by many *facultative anaerobes*. Two pathways are known:
  - Denitrification: $NO_3^- \to NO_2^- \to NO \to N_2O \to N_2$. All steps are exergonic and can be used for energy conservation.
  - Nitrate ammonification: $NO_3^- \to NO_2^- \to NH_4^+$ The nitrate reduction is coupled to energy conservation. While the mechanism of nitrite reduction in, e.g., *E. coli* and in *Salmonellae* is also coupled to energy conservation, this does not take place in other organisms.
- Reduction of metal ions to lower oxidation states (e.g. $Fe^{3+} \to Fe^{2+}$ in *Thiobacillus ferrooxidans*; sulfur oxidation supplies the reduction equivalents).
- Reduction of metal oxides to elemental metal (U, Te, Se). These are recently discovered important reactions.
- Other electron acceptors are glycine, dimethylsulfoxide (DMSO) and trimethylamine-N-oxide (TMAO).

Regulation: *Facultative anaerobes* (e.g., *E. coli*) can switch from oxygen to the use of alternative electron acceptors. Since the energy yield depends on the available redox potential difference, these organisms utilize electron acceptors in the order $O_2 > NO_3^- >$ fumarate (see 15.4 'Fermentations'). The synthesis of these alternative

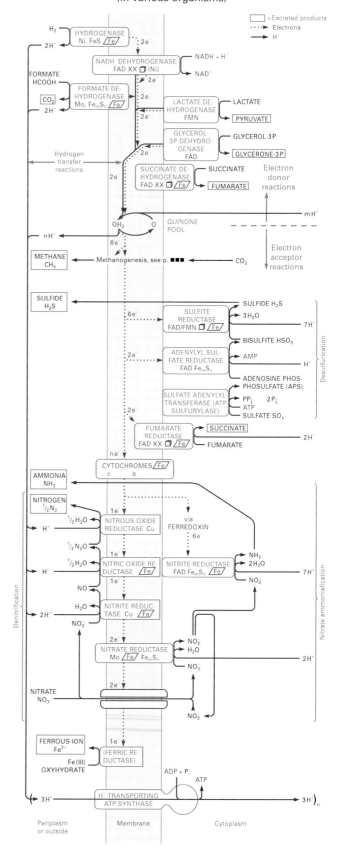

Figure 15.5-1. Systems of Anaerobic Respiration
(in various organisms)

electron transport systems is transcriptionally regulated. In general, the energetically more favorable electron acceptor represses expression of the genes necessary for growth with less favorable ones.

Literature

Anthony, C.: *Bacterial Energy Transduction*. Academic Press (1988).
Moodie, A.D., Ingledew, W.J.: Adv. Microb. Physiol. 31 (1990) 225–269.

15.5.2 Methanogenesis (Figure 15.5-2)

Methanogenesis is the reduction of CO_2 or small organic molecules (e.g. methanol, acetate) to methane (CH_4). It is only found in *strictly anaerobic archaea* (*methanogens*) and is therefore a type of anaerobic respiration. Very often H_2 is utilized as the reductant (electron donor). These reactions employ coenzymes unique to *methanogens*.

Reduction of CO_2 (oxidation state +4) to CH_4 (–4) occurs in several steps. The carbon atom remains covalently bound to special one-carbon carriers. At first, CO_2 is reduced to the formyl group (+2) bound to methanofuran. The formyl group is transferred to tetrahydromethanopterin (H_4-MPT, THMPT, 9.6.5). Water is subtracted to yield the methenyl group, which is then reduced to the methylene level (0). The 5′-deazaflavin coenzyme F_{420} (9.3.2) serves as electron carrier from the hydrogenase. Further reduction (involving F_{420}) yields methyl-H_4-MPT (–2) from which the methyl group is transferred to coenzyme M (2-mercaptoethanesulfonic acid). The resulting methyl-CoM is reduced to methane; the nickel porphinoid coenzyme F_{430} (9.5.3) serves as cofactor. The actual reductant is 7-mercaptoheptanoylthreonine phosphate (HS-HTP, coenzyme B) which is oxidized and forms a heterodisulfide with coenzyme M. The reduced forms of CoM and HTP are regenerated by reduction of the disulfide via disulfide reductase, using H_2 as primary reductant.

Energy Yield and Conservation: Since $\Delta E'_0$ (CO_2/CH_4) is quite negative (–244 mV), a strong reductant is needed, e.g. hydrogen [$\Delta E'_0 (2H^+/H_2)$: –410 mV].

$$CO_2 + 4\ H_2 = CH_4 + 2\ H_2O \qquad \Delta G'_0 = -128\ kJ/mol.$$

In methanogenic habitats, however, H_2 concentration is very low and ΔG_{eff} is much less negative. Thus, the decrease in free energy allows only the generation of < 2 moles ATP per mol CH_4 produced.

The energy is conserved via chemiosmotic coupling: the methyl-CoM reductase is linked to a membrane-bound H^+-pump which sets up an electrochemical potential gradient (1.5.3). Na^+-electrochemical potentials are also involved in energy conservation. Substrate level phosphorylation (8.1.3) is absent in methanogens.

Literature
DiMarco, A.A. et al.: Ann. Rev. Biochem. 59 (1990) 355–394.
Ferry, J.G. (Ed.): *Methanogenesis*. Chapman & Hall (1993).

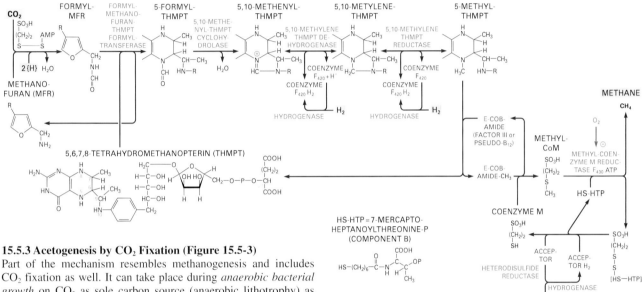

Figure 15.5-2. **Methanogenesis** (e.g. in *Methanobacterium thermoautotrophicum*)

15.5.3 Acetogenesis by CO_2 Fixation (Figure 15.5-3)

Part of the mechanism resembles methanogenesis and includes CO_2 fixation as well. It can take place during *anaerobic bacterial growth* on CO_2 as sole carbon source (anaerobic lithotrophy) as well as part of homocetate fermentation (e.g., *Clostridium thermoaceticum* uses CO_2 and H_2 formed by the pyruvate: ferredoxin oxidoreductase and hydrogenase reactions (Fig. 15.4-2) for acetate synthesis).

Some of the enzymes involved are different in various organisms, e.g. formate dehydrogenases or methylene-tetrahydrofolate reductases, which can use ferredoxin, cytochromes b or NADH as electron donors. The carbon monoxide dehydrogenase is a tetrameric, multifunctional enzyme which catalyzes the addition of a carboxyl group from CO_2 or from CO to the CH_3 moiety. It contains Ni, Zn and Fe-S centers.

Additionally, various other fermentations yield acetate by different pathways (15.4).

Literature
Drake, H.A.: Acetogenensis. Chapman & Hall (1994)

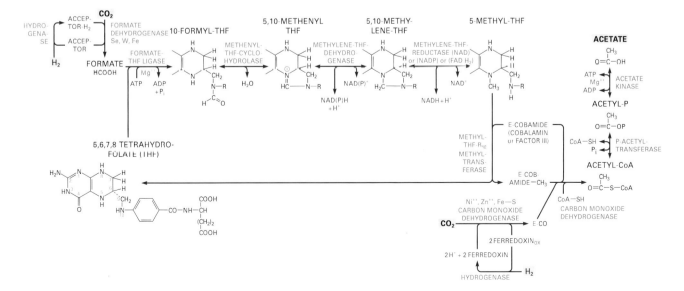

Figure 15.5-3. **Acetate formation** (e.g. in *Clostridium thermoaceticum*)

15.6 Chemolithotrophy

Chemolithotrophic organisms (*bacteria* and *archaea*) derive energy from the oxidation of reduced inorganic compounds, e.g., sulfide, hydrogen, ammonia or Fe^{++} (rather than from organic molecules like hexoses. This is called chemoorganotrophy).

15.6.1 Redox Reactions and Electron Transport (Figure 16.6-1)

Electrons are released during the oxidation of the donor substrate and are fed into a membrane-associated respiratory chain (16.1). This occurs at the redox level corresponding to the redox potential of the substrate / product couple (Figure 15.6-2, Table 15.6-1, columns 1...3). Finally, the electrons are transferred to the terminal electron acceptor, which is reduced this way. In most cases, this acceptor is oxygen. Some species, however, use nitrate or sulfate instead.

Example for the calculation of the theoretical energy yield (Definitions see 1.5.2): Oxidation of nitrite to nitrate; the electrons are taken up by oxygen (2-electron reaction):

$E'_0 [NO_3^-/NO_2^-] = +430$ mV $E'_0 [O_2/H_2O] = +816$ mV

$\Delta G'_0 = n * F * \Delta E'_0$ [1.5 – 6 b]
$\Delta G'_0 = 2 * 0{,}0965 [kJ * mV^{-1} * mol^{-1}] * (430 – 816) [mV]$
$\Delta G'_0 = -75 [kJ * mol^{-1}]$ (siehe Tabelle 15.6.1).

The first step of ammonia oxidation (to hydroxylamine) has a high positive redox potential and requires electrons as reductant (Table 15.6-1, Figure 15.6-3). Thus it has to be 'primed' by the NADH dehydrogenase reaction, while later on the electron supply is met by the consecutive hydroxylamine oxidoreductase reaction.

Sulfite oxidation may either proceed directly to sulfate or with simultaneous adenylylation yielding adenosine phosphosulfate as an intermediate (APS reductase pathway, named after the inverse reaction, Figure 15.6-4). APS is then phosphorylytically cleaved to sulfate and ADP, which disproportionates to ATP and AMP. This way, half of the original AMP is converted to ATP, causing additional substrate level phosphorylation.

Energy conservation and reductive power: Energy conservation takes place via chemiosmotic coupling (16.1). At various reaction steps external excess H^+ is generated. Additionally, protons are translocated out of the cell, e.g., by the terminal oxidase (16.1.2). This builds up an electrochemical gradient, which drives the H^+-dependent ATP synthase (16.1.3).

In many cases, the redox potential of the substrate couple is more positive than that of $NAD^+/NADH$ (–340 mV). Therefore, direct reduction of NAD^+ (or $NADP^+$) in order to supply reducing equivalents for the anabolic reactions is not possible. Rather, reversed electron pumping (to a more negative redox potential) is necessary in order to produce NADH (Figure 15.6-5). This reaction is also driven by the electrochemical gradient, i.e. protons flow back into the cell and thus partially reverse the H^+ outflow by the oxidation reactions.

$$n\, H^+_{out} + NAD^+ + 2\, e^- = (n – 1)\, H^+_{in} + NADH$$

Most chemolithotrophic bacteria derive the carbon for biosynthesis from CO_2 (autotrophy). Fixation of CO_2 takes place via the ribulose bisphosphate pathway (see 'Calvin cycle', 16.2.2):

Ribulose 1,5-P_2 + CO_2 + H_2O = 2 3-P-D-glycerate

Literature:
Anthony, C.: *Bacterial Energy Transduction.* Academic Press (1988).

Table 15.6-1 Redox Reactions in Chemolithotrophic Bacteria

Reductant	Redox Couple*	E'_0 [mV]	Reactions	$\Delta G'_0$ [kJ*mol^{-1}]**	Enzymes; Cofactors	Organisms
Carbon monoxide	CO_2/CO O_2 (0.21 atm)/H_2O	–540 +816	$CO + H_2O \to CO_2 + 2\,H^+ + 2\,e^-$ $\frac{1}{2} O_2 + 2\,e^- + 2\,H^+ \to H_2O$ $\Sigma = CO + \frac{1}{2} O_2 \to CO_2$	 –261	carbon monoxide dehydrogenase	'Carboxidobacteria', e.g. *Pseudomonas*
Hydrogen	$2\,H^+/H_2$	–410	$H_2 \to 2\,H^+ + 2\,e^-$ $\frac{1}{2} O_2 + 2\,e^- + 2\,H^+ \to H_2O$ $\Sigma = H_2 + \frac{1}{2} O_2 \to H_2O$	 –237	hydrogenase, [NiFe] or [Fe]	'Knallgas' bacteria
Sulfide	S_0/HS^-	–260	$HS^- \to S_0 + H^+ + 2\,e^-$ $\frac{1}{2} O_2 + 2\,e^- + 2\,H^+ \to H_2O$ $\Sigma = HS^- + \frac{1}{2} O_2 + H^+ \to S_0 + H_2O$	 –207	'sulfide oxidase'	*Thiobacillus; Beggiatoa; Wolinella succinogenes* (uses NO_3^- as el.acc.)
	HSO_3^-/HS^-	–110	$HS^- + 3\,H_2O \to HSO_3^- + 6\,H^+ + 6\,e^-$ $\frac{1}{2} O_2 + 2\,e^- + 2\,H^+ \to H_2O$ $\Sigma = HS^- + 1\frac{1}{2} O_2 \to HSO_3^-$	 –536	sulfite reductase; siroheme, FeS	*Thiobacillus; Sulfolobus*
Sulfur	HSO_3^-/S_0	–45	$S_n + O_2 + H_2O \to S_{n-1} + HSO_3^- + H^+$	–332	'sulfur dioxygenase'; FeS	*Thiobacillus*
Sulfite	SO_4^{2-}/HSO_3^-	–520	$HSO_3^- + H_2O \to SO_4^{2-} + 3\,H^+ + 2\,e^-$ $\frac{1}{2} O_2 + 2\,e^- + 2H^+ \to H_2O$ $\Sigma = HSO_3^- + \frac{1}{2} O_2 \to SO_4^{2-} + H^+$	 –258	sulfite: cytochrome c oxidoreductase; heme-Fe, Mo^{++}	*Thiobacillus* (*T. denitrificans* uses NO_3^- as electron accept. anaerobically)
	APS/HSO_3^-	–60	a) $HSO_3^- + AMP \to APS + 2\,e^-$ b) $APS + P_i \to SO_4^{2-} + ADP + H^+$ c) $ADP \to \frac{1}{2} AMP + \frac{1}{2} ATP$ $\frac{1}{2} O_2 + 2\,e^- + 2\,H^+ \to H_2O$ $\Sigma = HSO_3^- + \frac{1}{2} O_2 + P_i + \frac{1}{2} AMP + H^+ \to$ $SO_4^{2-} + \frac{1}{2} ATP + H_2O$	~–227	a) adenosine phosphosulfate reductase; FAD, FeS, b) sulfate adenylyl transferase; Mg^{++}, c) adenylate kinase, Mg^{++}	*Thiobacillus*
Ammonia	NH_2OH/NH_3 NO_2^-/NH_2OH $\Sigma = NO_2^-/NH_3$	+900 +60 +340	a) $NH_3 + O_2 + 2\,H^+ + 2\,e^- \to NH_2OH + H_2O$ b) $NH_2OH + H_2O \to NO_2^- + 5\,H^+ + 4\,e^-$ $\frac{1}{2} O_2 + 2\,e^- + 2\,H^+ \to H_2O$ $\Sigma = NH_3 + 1\frac{1}{2} O_2 \to NO_2^- + H_2O + H^+$	 –276	a) NH_3 monooxygenase; Cu, b) hydroxylamine: cytochrome c_{554} oxidoreductase, heme-Fe, Mo^{++}	nitrosofying bacteria, e.g. *Nitrosomonas*
Nitrite	NO_3^-/NO_2^-	+430	$NO_2^- + H_2O \to NO_3^- + 2\,H^+ + 2\,e^-$ $\frac{1}{2} O_2 + 2\,e^- + 2\,H^+ \to H_2O$ $\Sigma = NO_2^- + \frac{1}{2} O_2 \to NO_3^-$	 –75	nitrite: cytochrome c_{550} oxidoreductase; heme-Fe, Mo^{++}	nitrifying bacteria: e.g., *Nitrobacter*
Fe^{2+} at low pH (ca. 2.0)	Fe^{3+}/Fe^{2+} O_2 (0.21 atm)/H_2O	+770 +1100	$Fe^{2+} \to Fe^{3+}$ $\frac{1}{4} O_2 + e^- + H^+ \to \frac{1}{2} H_2O$ $\Sigma = Fe^{2+} + \frac{1}{4} O_2 + H^+ \to Fe^{3+} + \frac{1}{2} H_2O$	 ~–32	rusticyanin; Cu	*Thiobacillus ferrooxidans, Sulfolobus*
Fe^{2+} at ca. neutral pH	$Fe(OH)_3$ (sat.)/Fe^{2+} (10 µmol/l)	– 150	$Fe^{2+} + \frac{1}{4} O_2 + 2\frac{1}{2} H_2O \to Fe(OH)_3 + 2\,H^+$	?	?	*Gallionella*
Mn^{2+} at ca. neutral pH	MnO_2/Mn^{2+}	?	unknown	?	?	marine *Pseudomonas*

* These reactions proceed from reduced to oxidized state
** Calculated from $\Delta E'_0$ or from free energies of formation per mole of substrate oxidized

Figure 15.6-1. General Chemolithotrophic Reactions (in various organisms)

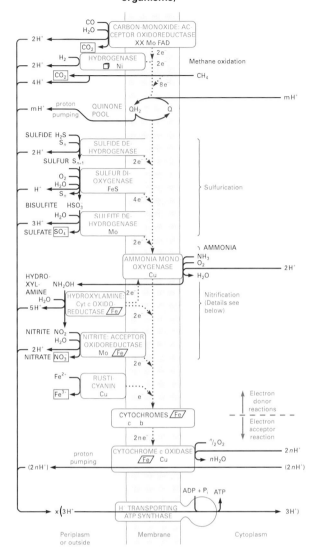

Figure 15.6-3. Details of Electron Flow and Energy Conservation During Oxidation of Ammonia in *Nitrosomonas*

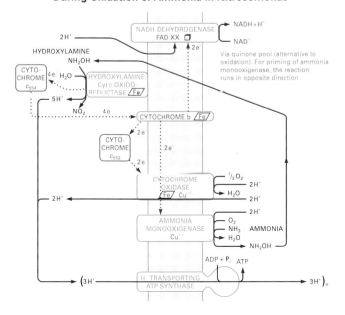

Figure 15.6-2. Redox Potentials E'_0 [mV, 25 °C, pH = 7.0] of Substrate Couples

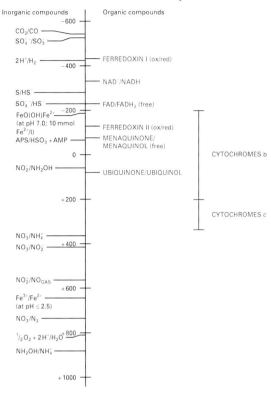

Figure 15.6-4. Electron Flow and Substrate Level Phosphorylation During Sulfite Oxidation by the 'APS Reductase Pathway' in *Thiobacillus*

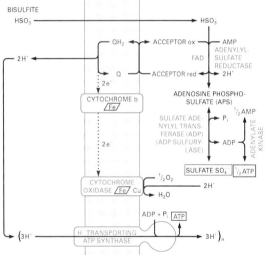

Figure 15.6-5. Reversed Electron Flow Driven by the Electrochemical Gradient µ

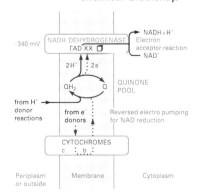

15.7 Alkane and Methane Oxidation, Quinoenzymes

Alkane oxidation (Figure 15.7-1): The capacity for utilization of aliphatic hydrocarbons is quite widespread among *bacteria* and *fungi*. n-Alkanes with chain lengths of $C_{10}...C_{18}$ are among the most frequently occurring substrates, they are rapidly oxidized. Since these substances are insoluble in water, uptake and enzymatic attack are difficult. Many *bacteria* synthesize specialized lipids (e.g. trehalo- and rhamnolipids) as constituents of their cell walls, in which these hydrocarbons 'dissolve' and in which they are transported to the *membrane*, where they are oxidized at the *cytoplasmic side*. The fatty acids formed this way enter the usual degradation pathway via β-oxidation (6.1.5), supplying acetyl-CoA as source of cell constituents as well as NADH for reduction reactions.

Methane oxidation (Figure 15.7-2): This reaction takes place in *obligate methylotrophs*, i.e. *Gram negative bacteria*, which solely grow at the expense of compounds containing no C-C bond.

Facultative methylotrophs, on the contrary, are organisms, which are able to use a variety of carbon sources including C_1-compounds (e.g. methanol). They never use methane as source of C.

The oxidation to CO_2 supplies reducing equivalents and (during growth on C_1-compounds) cell carbon by CO_2 fixation via the serine-isocitrate lyase pathway (Figure 15.7-3, compare 3.3.4 and 4.4.2) or the ribulose monophosphate pathway (Figure 15.7-4).

Quinoenzymes: The redox carrier pyrrolo-quinoline quinone (PQQ) is the coenzyme of methanol dehydrogenase in *methylotrophs* and of alkan-1-ol dehydrogenase and aldehyde dehydrogenase (PQQ) in alkane oxidizing *bacteria*. It performs similar functions in *bacterial* glucose dehydrogenase (EC 11.99.17). The biosynthesis of PQQ is initiated by condensation of dopaquinone (4.7.3) with glutamate.

A number of previously assumed PQQ functions are most likely performed by other quinone derivatives of amino acids. Examples are:
- Tryptophantryptophylquinone (TTQ) in *bacterial* methylamine oxidase (formed by posttranscriptional condensation of two tryptophan molecules)
- Trihydroxyphenylalanine quinone (TPQ, topa quinone) in *mammalian* serum amine oxidase, *pea* and *bacterial* amine oxidases (originating from tyrosine).

Literature
Klinman, J.P., Mu, D., Ann. Rev. of Biochem. 63 (1994) 299–344.
Klinman, J.P.: J. Biol. Chem. 271 (1996) 27189–27193.

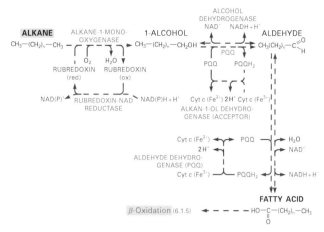

Figure 15.7-1. Alkane Oxidation

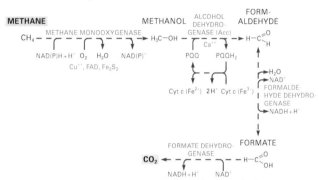

Figure 15.7-2. Methane Oxidation

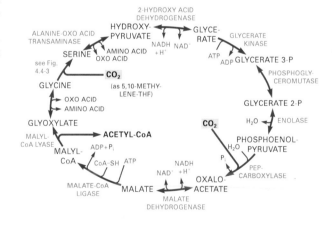

Figure 15.7-3. Serine- Isocitrate Lyase Pathway

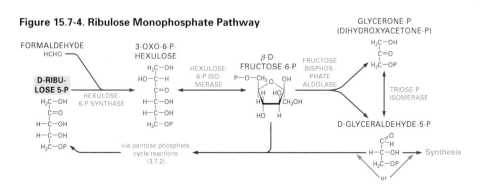

Figure 15.7-4. Ribulose Monophosphate Pathway

15.8 Antibiotics

Antibiotics are substances with low molecular mass (below a few kDa) which inhibit the growth of microorganisms (*bacteria, fungi, viruses*) at low concentrations. They belong to various chemical classes. Some important antibiotics are discussed in detail below.

Table 15.8-1. Classification of Antibiotics (AB) (acc.to Berdy, 1985)

Chemical Class	Subgroups (Selection)	Antibiotic (Example)	Producing Organisms	Site of Antibiotic Effect
1 Carbohydrate derived AB	aminoglycoside AB	streptomycin	bacteria, e. g. *Streptomyces griseus*	bacterial protein biosynthesis, inhibits the function of the 30 S ribosomal subunit by causing translational errors
2 Makrocyclic lactone	macrolide AB	erythromycin	bacteria, e. g. *Saccharopolyspora erythrea*	bacterial protein biosynthesis, inhibits the function of the 50 S ribosomal subunit
3 Quinones and related AB	linearly condensed polycyclic compounds	tetracycline	bacteria, e. g. *Streptomyces aureofaciens*	bacterial protein biosynthesis*, inhibits the function of the 30 S ribosomal subunit
4 Amino acid and peptide AB	β-lactam AB	penicillin	fungi, e. g. *Penicillium chrysogenum, Emericella nidulans*	biosynthesis of the bacterial cell wall
		cephalosporin	bacteria, e. g. *Streptomyces clavuligerus*; fungi, e. g. *Acremonium chrysogenum* (syn. *Cephalosporium acremonium*)	
	amino acid derivatives	chloramphenicol	bacteria, e. g. *Streptomyces venezuelae*	bacterial protein biosynthesis, inhibits the function of the 50 S ribosomal subunit
5 N-containing heterocyclic AB	nucleoside AB	puromycin	bacteria, e. g. *Streptomyces alboniger*	eukaryotic and bacterial protein biosynthesis
		nikkomycin	bacteria, e. g. *Streptomyces tendae*	fungicide and insecticide, inhibits the biosynthesis of chitinous cell walls (chitin synthase)
6 O-containing heterocyclic AB	polyether AB	monensin	bacteria, e. g. *Streptomyces cinnamonensis*	cytoplasmic membrane, acts as ionophore
7 Alicyclic AB	steroid AB	fusidic acid	fungi, e. g. *Fusidium coccineum*	bacterial and eukaryotic protein biosynthesis
8 Aromatic AB	benzofurane derivatives	griseofulvin	fungi, e. g. *Penicillium griseofulvum*	fungicide, effective against fungi with chitinous cell wall
9 Aliphatic AB	polyene AB	fumagillin	fungi, e. g. *Aspergillus fumigatus*	inhibition of the eukaryotic DNA synthesis
10 AB with unusual structures				

* Deactivates also eukaryotic ribosomes, but is incorporated quantitatively only into *bacteria*.

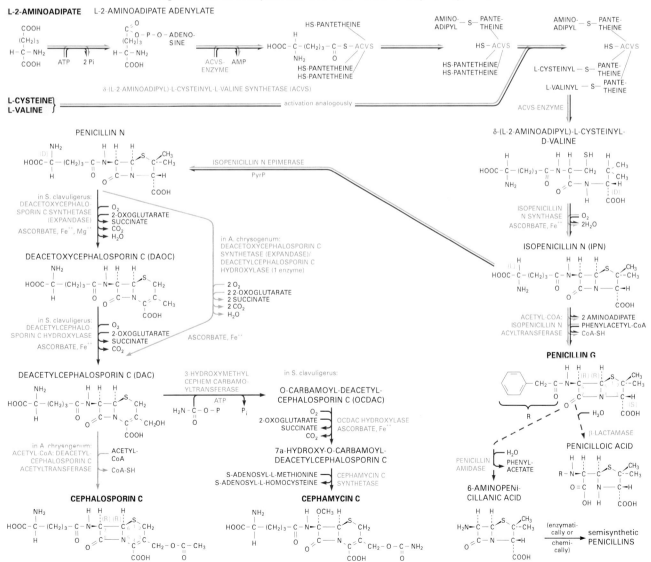

Figure 15.8-1. Biosynthesis of Penicillins and Cephalosporins

15.8.1 Penicillin and Cephalosporin (Figure 15.8-1)

Both antibiotics are derivatives of tripeptides, which are converted to typical ring structures: penicillins show a five-membered thiazolidine ring, cephalosporins a six-membered dihydro-thiazine ring, which are fused to a four-membered β-lactam ring. Because of the ring substituents, penicillins which are secreted are hydrophobic, whereas cephalosporins are hydrophilic. Their activity spectrum is determined mostly by the structure of the acylamino side chain.

Biosynthesis: The biosynthesis is an example of a non-ribosomal peptide synthesis by a multienzyme complex. As far as is known, all naturally occurring penicillins and cephalosporins are synthesized from the same three amino acids, L-2-aminoadipic acid, L-cysteine and L-valine. They are condensed to the tripeptide δ-(L-2-aminoadipyl)-L-cysteinyl-D-valine (ACV). These reactions are catalyzed by the single enzyme δ-(L-2-aminoadipyl)-L-cysteinyl-D-valine synthetase. The bicyclic isopenicillin N (IPN) is then formed from the linear ACV tripeptide. IPN is the branch point of penicillin and cephalosporin biosynthesis.

In the penicillin biosynthesis (only in *fungi*), the hydrophilic L-2-aminoadipic acid side chain of isopenicillin N is exchanged for a hydrophobic acyl group, e. g., phenoxyacetyl in penicillin V or phenylacetyl in penicillin G.

The biosynthesis of cephalosporins has been reported in *bacteria* (cephamycin C) and *fungi* (cephalosporin C). The L-2-aminoadipic acid side chain of IPN is epimerized to the D-enantiomer to give penicillin N. Then ring expansion to deacetoxycephalosporin C (DAOC) occurs. Oxidation of the methyl group at C-3 yields deacetylcephalosporin C. In *A. chrysogenum*, this hydroxy group is then acetylated to cephalosporin C, while in *bacteria* different groups are attached. E.g., in *S. clavuligerus*, several additional steps lead to the end product cephamycin C (7-methoxycephalosporin).

For the synthesis of semisynthetic penicillins, biosynthetic penicillins (e.g., penicillin G) are cleaved *in vitro* either enzymatically by penicillin amidase or chemically. The resulting 6-aminopenicillanic acid is chemically modified by addition of different side chains leading to a variety of different penicillins.

Resistances: Resistance against β-lactam antibiotics is mainly due to hydrolytic cleavage of the β-lactam ring by β-lactamases or removal of the side chain by acylhydrolases. These enzymes are secreted by resistant microorganisms into the *periplasmic space*. In addition, a decrease in the penetrability of the external cellular membrane of *Gram negative bacteria* or modifications of the target sites (penicillin binding proteins in *Gram positive* and *negative bacteria*) can cause resistance.

15.8.2 Streptomycin (Figure 15.8-2)

Antibiotics of the aminoglycoside antibiotic group frequently contain cyclitols in addition to amino sugars. They are broad spectrum antibiotics, effective against both *Gram negative* and *Gram positive bacteria*. Streptomycin is a water soluble molecule exhibiting a basic pH.

Biosynthesis: Streptomycin is synthesized from the precursors dTDP-dihydrostreptose, NDP-N-methyl-L-glucosamine and streptidine-6-P, which are formed by independent biosynthetic pathways. All of the precursor molecules are derived from D-glucose.

For formation of dTDP-dihydrostreptose, glucose is converted to the dTDP derivative. The hydroxy group at C-4 is oxidized to the oxo function, followed by reduction of the hydroxy group at C-6. Then, epimerization at C-3 and C-5 and finally a ring narrowing to a five-membered ring takes place. The biosynthesis of N-methyl-L-glucosamine requires the transformation of a D-sugar to its L-enantiomer. Some steps have not yet been completely resolved. The pathway also needs activation of the sugar as a nucleoside diphosphate derivative (compare 3.2.2). It is unclear whether this occurs with UTP or CTP. Streptidine 6-P is synthesized via *myo*-inositol (reaction see 3.5.4) which is then oxidized at C-1 and transaminated to give inos-1-amine. After phosphorylation, the compound is transamidinated by arginine. The same procedure is repeated at the C-3 position. For the final steps of streptomycin biosynthesis, streptidine-6-P and dTDP-streptose most likely react first to an intermediate named pseudo-disaccharide-6-P. Addition of N-methyl-L-glucosamine yields dihydrostreptomycin-6-P. This is the last soluble intermediate which can be found within streptomycin producing cells. Oxidation to streptomycin-6-P takes place at the *cytoplasmic membrane*. NAD$^+$ serves as an electron acceptor. The phosphorylated product is still inactive; dephosphorylation to the active antibiotic is achieved by an extracellular alkaline phosphatase.

Resistances: Resistance of *bacteria* against streptomycin can be due to decreased penetration into the cells, a mutated target site or to bacterial modification of the streptomycin molecule. So far, phosphorylation at the C-6 bound hydroxyl group of the streptidine moiety or at the C-3″ bound hydroxyl group of the N-methylglucosamine moiety has been described, as well as a mannosylation at the C-4″ bound hydroxyl of the N-methylglucosamine residue.

Figure 15.8-2. Biosynthesis of Streptomycin

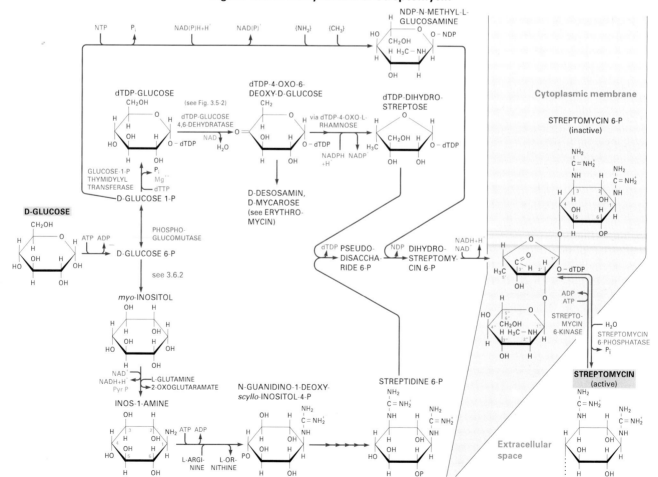

15.8.3 Erythromycin (Figure 15.8-3)

Erythromycin is a macrolide antibiotic consisting of a 14-membered ring. Two sugar substituents (D-desosamine and L-cladinose) are essential for its biological activity.

Biosynthesis: The biosynthesis of macrolides proceeds via the polyketide pathway. To 1 molecule of propionyl-CoA, 6 units of 2-methylmalonyl-CoA are consecutively added. During each condensation step, 1 C-atom is eliminated via CO_2. Apart from the lack of reductive steps, these reactions resemble those of the fatty acid synthesis (6.1.1). From the instable intermediates, upon ring closure the first stable intermediate 6-deoxy-erythronolide B is formed. After oxidation at C-6, glycosylation with both (dTDP activated) L-mycarose and D-desosamine yields erythromycin D. By methylation of the bound L-mycarose moiety to L-cladinose and oxidation at C-12, erythromycin A is formed. These two last steps may also proceed in reverse order. Erythromycin E results from oxidation of both the methyl group at C-2 of the macrolide ring and the C-1 of the L-cladinose ligand.

Resistances: Resistances of *bacteria* are mostly due to modification of the target site of the antibiotic (e. g. MLS-resistance by N^6-dimethylation of the base 2058 of 23 S rRNA).

Figure 15.8-3. Biosynthesis of Erythromycin

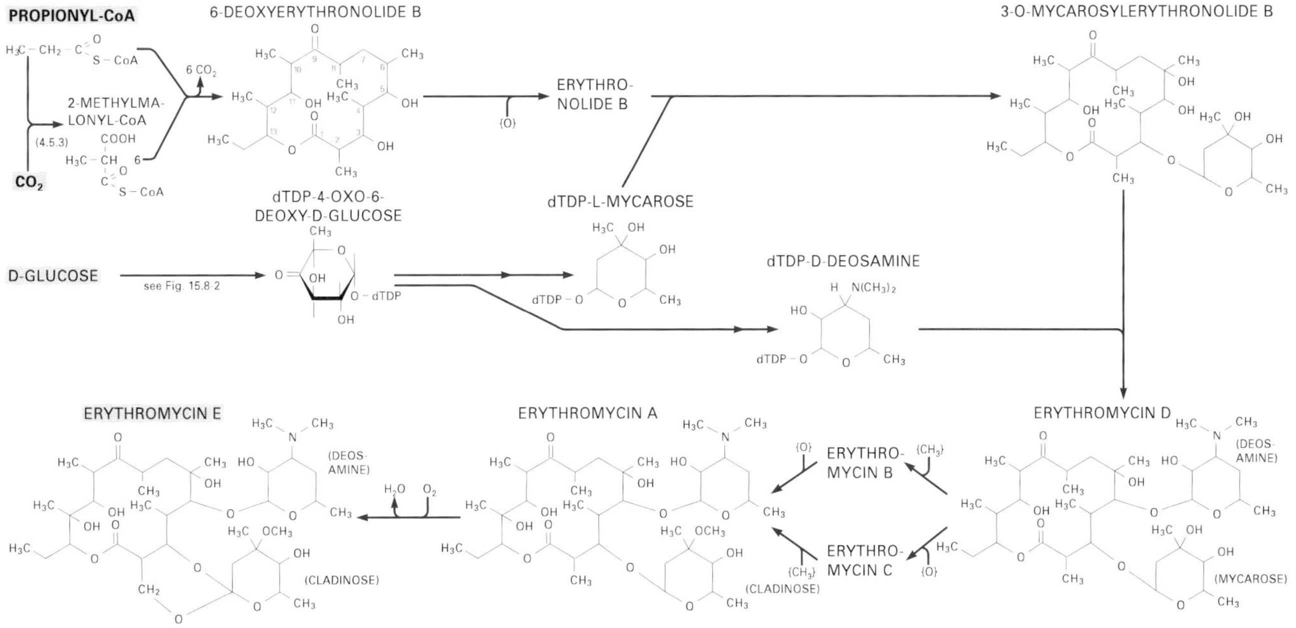

15.8.4 Tetracycline (Figure 15.8-4)

Biosynthesis: As described for the biosynthesis of erythromycin, tetracyclines are also synthesized via the polyketide pathway.

Malonamoyl-CoA is formed from asparagine via L-oxosuccinamate (4.2.4). Enzyme-bound malonamoyl-CoA is condensed with malonyl-CoA (formed from acetate, 5.1). Consecutively, 7 more malonyl-CoA units are added, with the elimination of 1 C-atom as CO_2 in each step. The first intermediate is 6-methylpretetramide, which has been demonstrated by the use of block mutants. After hydroxylation and oxidation reactions, 4-dedimethylamino-4-oxo-anhydrotetracycline is a branch point of the biosynthetic pathway. Halogenation by chloride haloperoxidase leads via several following steps to chlorotetracycline. For formation of tetracyclines an amino group is introduced, followed by the addition of 2 methyl groups and subsequent hydroxylation and reduction reactions. Another branch of the pathway at the stage of 5α (11α)-dehydro-tetracycline leads to formation of oxytetracycline. Some steps of the biosynthetic pathway have not been clarified yet.

Resistances: Tetracycline enters bacteria by an active transport mechanism through the *cytoplasmic membrane*. The major mechanisms of resistance are decreased uptake and active transport out of the cells.

Literature:

Aharonowitz, Y. *et al.*: Bio/technology 11 (1993) 807–810.
Berdy, J. In: Verrall, M. S. (ed.) *Discovery and Isolation of Microbial Products*. Ellis Horwood Publishers, (1985) 9–31.
Brakhage, A. A.: Microbiology and Molecular Biology Reviews. In press (1998).
Kleinkauf, H., von Döhren, H.: Eur. J. Biochem 236 (1996) 335–351.
Piepersberg, W. In: Vining, L., Stuttard, C. (eds.) *Biochemistry and Genetics of Antibiotic Biosynthesis*. Butterworth, Heinemann, Stoneham (1994) 71–104.

Figure 15.8-4. Biosynthesis of Tetracycline

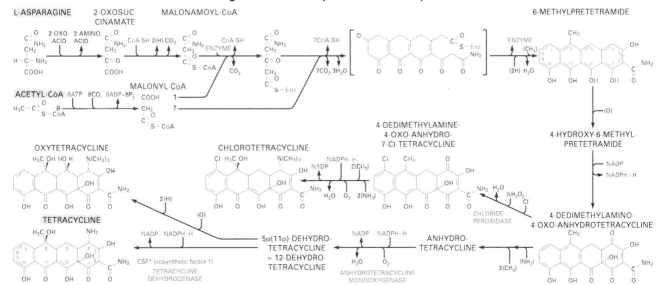

16 Oxidative Phosphorylation and Photosynthesis

16.1 Oxidative Phosphorylation

The oxidative phosphorylation systems have the task of supplying ATP. They are present in all kingdoms of life. The oxidation of compounds establishes a proton gradient across the membrane, which provides the energy for ATP synthesis (chemiosmotic coupling). The systems encompass
- the respiratory chain (which pumps H$^+$ out) and
- the H$^+$ transporting ATP synthase (which couples H$^+$ uptake to ATP synthesis).
- In *mitochondria*, the system is enlarged by carriers for the import of phosphate and ADP and for the export of ATP.

Whereas in *mitochondria* the electron acceptor is primarily oxygen, in *anaerobic living bacteria* (e.g., *Wolinella succinogenes*) organic and inorganic electron acceptors occur. Consequently, the construction of the electron transport system is uniform in *mitochondria*, but highly variable in *bacteria*, depending on the nature and energy difference of the reductant and oxidant.

16.1.1 Energy Balance and Reaction Yield

Energy Turnover of the Reactions: The change of free energy for any combination of reductant and oxidant can be calculated from the difference ΔE of their redox potentials E according to Eq. [1.5-6 a].

If the standard conditions for the reaction are used, the oxidation potential ΔG_{ox} is obtained:

$$\Delta G_{ox} = n * F * \Delta E'_0, \qquad \Delta G_{ox} [kJ * mol^{-1}] = n * 0.0965 * \Delta E'_0 [mV] \quad [16.1-1]$$

In *mitochondria*, according to the formula

$$NADH + H^+ + \tfrac{1}{2} O_2 = NAD^+ + H_2O,$$

an oxidation potential of $G_{ox} = -219$ [kJ/mol] results from $\Delta E'_0 = -1135$ mV (pH = 7.0 with n = 2, equal concentrations of NAD$^+$ and NADH and atmospheric O$_2$ pressure at 37 °C). For other conditions, $\Delta E'_0$ in the above formula has to be modified according to Eq. 1.5-7:

$$\Delta G_{ox} = n * F * \Delta E'_0 + R * T * 2.303 * \log \frac{[NAD^+]}{[NADH] * [\tfrac{1}{2} O_2]}. \quad [16.1-2]$$

This oxidation potential has to be matched with the requirement of free energy for the phosphorylation of ADP according to the formula

$$ADP + P_i + H^+ = ATP + H_2O,$$

By convention, the phosphorylation potential ΔG_{ATP} is defined in the opposite direction, i.e. for hydrolysis of ATP and depends on the concentration ratio of the reactants according to Eq. [1.5-2] as

$$\Delta G_{ATP} = \Delta G'_0 + R * T * 2.303 * \log \frac{[ADP] * [P_i]}{[ATP]}, \quad [16.1-3]$$

where $\Delta G'_0 = -30.5$ kJ/mol. It varies between $\Delta G_{ATP} = -46$ to -67 kJ/mol from *bacteria* to *mitochondria*. $\Delta G_{ATP} > -59$ kJ/mol are obtained by the export of ATP from *mitochondria* to the *cytosol* of *eukaryotic* cells.

ATP yield: The number of the H$^+$ generated in the respiratory chain by NADH oxidation divided by those consumed in the ATP synthesis gives the maximum number of ATP molecules obtainable by oxidation of 1 molecule NADH. In the *mitochondrial* respiratory chain the stoichiometry is n$^{H+}/_{NADH}$ = 10 H$^+$/NADH and for the ATP synthase n$^{H+}/_{ATP}$ = 3 H$^+$/ATP. In addition, 1 H$^+$/ATP is consumed for the export of ATP from mitochondria (see below). Thus, for *cytosolic* ATP the ratio is n$^{H+}/_{ATP}$ = 4 H$^+$/ATP and the oxidative phosphorylation ratio P/O = 2.5 ATP/NADH ('P/O quotient').

The maximum free energy ΔG_{max} available from NADH oxidation or stored in the synthesized ATP is calculated by multiplication of the number of H$^+$ generated by oxidation of NADH n$^{H+}/_{NADH}$ = 10 or the number of H$^+$ consumed by the ATP synthase n$^{H+}/_{ATP}$ = 3, respectively, with the electrochemical proton gradient $\Delta\mu_{H^+}$ (1.5.3) across the *inner membrane*. With the value of $\Delta\mu_{H^+} = -23$ kJ/mol (measured in the direction of import) one obtains

$$(\Delta G_{ox})_{max} = n^{H+}/_{NADH} * \Delta\mu_{H^+} = -230 \text{ kJ/mol}. \quad [16.1-4]$$

$$(\Delta G_{ATP})_{max} = n^{H+}/_{ATP} * \Delta\mu_{H^+} = -69 \text{ kJ/mol}. \quad [16.1-5]$$

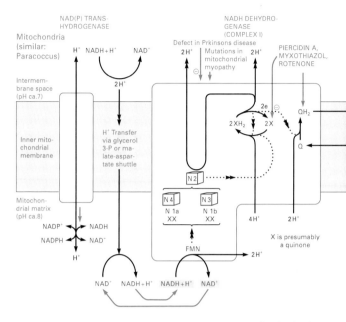

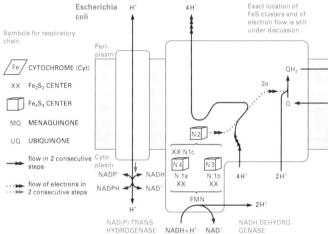

16.1.2 Electron Transport System in Mitochondria (Fig.16.1-1)

The respiratory chain consists of 3 multienzyme complexes and succinate dehydrogenase (Table 16.1-1), which (except succinate dehydrogenase) are transmembrane proteins of the *inner mitochondrial membrane*. They contain as active redox components flavin mononucleotides, iron sulfur proteins, cytochromes (5.2.4) and Cu centers.

Table 16.1-1. Complexes of the Respiratory Chain in *Mammalian Mitochondria*

Complex	Mol. Mass (kDa)	Sub-units	Active Groups	Inhibitors
I: NADH dehydro-genase (ubiquinone)	907	43	FMN, 2 Fe$_2$S$_2$ and 3 (4?) Fe$_4$S$_4$ centers, ubiquinone	rotenone, piercidine
II: Succinate dehydro-genase (ubiquinone)	130	5	FAD, 2 Fe$_2$S$_2$ and 1 Fe$_4$S$_4$ centers, Cyt b$_{558}$	malonate
III: Ubiquinol-cyto-chrome-c reductase (Cyt bc$_1$ complex)	248	11	Cyt b$_{566}$, Cyt b$_{560}$, Cyt c$_1$, Fe$_2$S$_2$ center (Rieske), ubiquinone	antimycin, myxothiazol etc.
IV: Cytochrome-c oxi-dase (Cyt aa$_3$ complex)	210	13	Cyt a, Cyt a$_3$, 2 Cu$_A$, Cu$_B$	CN$^-$, SH$^-$, N$_3^-$, CO

Complex I performs the oxidation of NADH. It consists of a peripheral part extending into the *mitochondrial matrix*, which contains FMN and all FeS clusters and an extremely lipophilic part, which is embedded in the *membrane*. Apparently the FeS centers transmit the electrons from the NADH oxidation to a putative qui-

Figure 16.1-1. Respiratory Chain and Oxidative Phosphorylation in *Mitochondria* (Top) and *Bacteria* (Below)
Model of the mitochondrial complex I mechanism according to Brandt, Biochim. Biophys. Acta. 1318 (1997) 92.

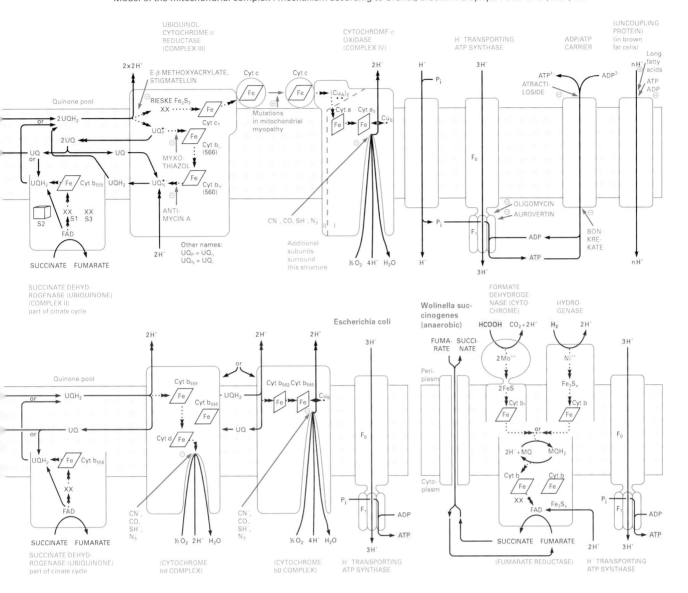

none (X), which is involved in pumping of 4 H$^+$ to the *cytosol* as well as in reduction of 1 ubiquinone (UQ, 4.7.2) to ubihydroquinone (ubiquinol) (UQH$_2$) per 1 NADH oxidized. The UQH$_2$ then enters the so-called quinone pool within the *membrane*, which transports by diffusion hydrogen from complexes I and II (see below) to complex III.

The exact structure of complex I, as well as the mechanism of H$^+$ and electron transfer has not been resolved. A model is shown in Fig. 6.1-1. Alternatively, double presence of the reaction centers has been suggested.

Representative for several other FAD containing dehydrogenases, complex II (succinate dehydrogenase) oxidizes organic metabolites directly. The reducing equivalents liberated by these reactions are taken up by ubiquinone which also enters the quinone pool.

In the 'UQ cycle', ubiquinone from the pool interacts with the ubiquinol-cytochrome c reductase complex III at two sites, which are located close to the two cytochrome b heme centers:
- UQH$_2$ is oxidized by the Fe$_2$S$_2$-'Rieske' protein to the low potential semiquinone UQ$_P^{•-}$ (1 electron transfer), releasing 2 H$^+$ to the *cytosol*. UQ$_P^{•-}$ reduces the low potential Cyt b$_L$ (Cyt b$_{566}$, $\Delta E_0' = -100$ mV), which then transfers electrons to the high potential Cyt b$_H$ (Cyt b$_{560}$, $\Delta E_0' = -50$ mV).

  The Rieske protein releases the obtained electrons to cytochrome c, which acts as a shuttle and transmits them to complex IV.
- Cyt b$_H$ reduces UQ to high potential semiquinone UQ$_N^{•-}$, which disproportionates into UQ and UQH$_2$ under uptake of 2 H$^+$ from

the *matrix side*. The UQH$_2$ enters the quinone pool and is turned over as described above, resulting in release of 2 more H$^+$ into the *cytosol*.

The cytochrome c oxidase complex IV uses the received electrons for formation of water from ½ O$_2$ and 2 H$^+$, which were taken up at the *matrix* side. O$_2$ reduction occurs at the bimetallic Cyt a$_3$/Cu$_B$ center. Further, 2 H$^+$ are pumped from the *matrix* into the *cytosol* by complex IV. This results in a total transfer of 10 H$^+$ into the *cytosol* and formation of 1 H$_2$O per 1 oxidized NADH + H$^+$.

Actually, complex IV collects 4 electrons in order to reduce O$_2$ according to O$_2$ + 8 H$^+_{inside}$ + 4e$^-$ = 2 H$_2$O + 4 H$^+_{outside}$. The electrons are at first distributed between Cu$_A$ (a binuclear center), Cyt a, Cyt a$_3$ and Cu$_B$. The 2-electron formulation shown in Fig. 16.1-1 is for keeping up a consistent stoichiometry along the respiratory chain (based on oxidation of 1 NADH + H$^+$).

The atomic structure of the complex IV has been resolved for the complex from *bovine heart mitochondria* (13 subunits) and from *Paracoccus denitrificans* (4 subunits). Subunits I and II carry the electron transfer centers.

Inhibitors specific for these complexes are given in the table. The inhibitors of complex I appear to act on the ubiqinone sites. Complex III is inhibited by several antibiotics, some of which have also herbicidal activity. These inhibitors bind selectively to either one of the two ubiquinone centers in the complex III.

Supply of hydrogen to the respiratory chain: Hydrogen is provided either from substrates by intramitochondrial dehydrogenases or by extramitochondrial NADH, which has to deliver its hydrogen into the matrix by hydrogen shuttles across the *inner mitochondrial membrane*. Most important are the pathways mediated by glycerol-3-phosphate and by aspartate-glutamate-malate (Figure 16.1-2).

Figure 16.1-2. Carrier Systems for Transport of Extramitochondrial Hydrogen to the Respiratory Chain

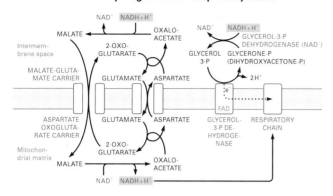

Hydrogen transfer from intramitochondrial NADH to $NADP^+$ is catalyzed by a transhydrogenase: $NADH + NADP^+ = NAD^+ + NADPH$. Although $\Delta E'_0$ of both compounds are almost identical, the actual ΔE of $NADPH/NADP^+$ is more negative, since the reduced compound prevails (9.9.3). The additional energy requirement for the transfer reaction is met by influx of 1 proton (Fig.16.1-1).

16.1.3 Bacterial Electron Transport Systems (Figure 16.1.1)

In *bacteria*, the electron transport is highly diversified. The enzymes for oxidation and reduction of substrates are expressed according to the environmental conditions, using systems described in 10.5. They exchange electrons either with the quinone pool or directly with cytochromes, depending on their redox potentials.

The electron transport complexes contain fewer subunits than in *mitochondria* (Table 16.1-2). The *E. coli* NADH-dehydrogenase has 14 subunits. UQH_2 is oxidized directly (without a Cyt bc_1 complex) alternatively

- by the cytochrome bo complex, which (similarly to the cytochrome aa_3 complex of *mitochondria*) contains a Cu-heme center or
- by the cytochrome bd complex which contains heme proteins (Cyt b_{558} and b_{595}) and a chlorine-Fe protein (Cyt d, Fig. 5.1-2).

At high oxygen concentrations, the Cyt bo complex is operative, while at lower pressures the Cyt bd complex is activated.

Table 16.1-2. Complexes of the Respiratory Chain in *Escherichia coli*

Complex	Subunits	Active Groups	Inhibitors
NADH dehydrogenase (ubiquinone)	14	FMN, 2 (3?) Fe_2S_2 and 3 Fe_4S_4 centers (550 kDa)	rotenone, piercidine
Succinate dehydrogenase (ubiquinone)	?	FAD, 1 FeS center, Cyt b_{556}	
Cytochrome bd complex	2	Cyt b_{558}, Cyt b_{595}, Cyt d	
Cytochrome bo complex	3	Cyt b_{562}, Cyt b_{555} = Cyt o, Cu_B	CN^-, SH^-, N_3^-, CO

Obligate anaerobic bacteria lack an oxidase-type complex for obvious reasons. Many different combinations of redox reactions have been found. A survey is given in section 15.5 'Anaerobic respiration'.

As an example, *Wolinella succinogenes* uses fumarate as terminal electron acceptor (Fig.16.1.1). The dehydrogenases for H_2 (hydrogenase), for formate and other substrates reduce menaquinone (MQ, 4.7.2), which is reoxidized by fumarate reductase.

The formate dehydrogenase of *W. succinogenes* contains 2 Mo and the hydrogenase Ni in the active centers. Because of the low redox potential of the system, menaquinone (with a 110 mV lower redox potential, 4.7.2) is used as carrier instead of ubiquinone.

In some *bacteria* living in alkaline or Na^+-rich environments, NADH oxidation can be used to pump Na^+ instead of H^+. Other *bacteria* employ decarboxylation reactions to establish a Na^+ gradient. This gradient is used to energize ATP synthesis by a Na^+-transporting ATP synthase (instead by the H^+ transporting enzyme, below) or by secondary transport processes (15.3.1).

16.1.4 H^+-Transporting ATP Synthase (Figure 16.1-3)

The enzyme is a multiprotein complex consisting of a hydrophilic domain (F_1) protruding into the matrix space and a membrane bound part (F_0). Each of them is composed of several subunits (Table 16.1-3).

The membrane bound section F_0 is not yet well defined. In *mitochondria*, it probably contains 10 different subunits and in *E. coli* 3 different subunits, of which the c subunit occurs in 9 to 12 copies per F_0.

The structure and the subunits of F_1 have been resolved at atomic resolution. The γ subunit forms a tilted quasi axis within the ring of 3 αβ subunits. The α, β, γ subunits of the *mitochondrial* F_1 part are highly homologous to the *chloroplast* and to the *bacterial* enzymes.

Table 16.1-3. Complexes of the ATP Synthase

Complex	Subunits	Mol.mass (kDa)
Membrane bound part F_0	*Mitoch.*: a, b, c, d, e, f, g, F_6, A6L, OSCP	?
	E.coli: a, b_2, $c_{9...12}$	30, 17, 8
Matrix part F_1	*Mitoch./E. coli*: $α_3$, $β_3$, γ, δ, ε	55, 50, 31, 19, 14

The ATP synthesis is driven by the transport of 3 H^+/ATP through F_0 to F_1, resulting in a 3 step rotation of the asymmetric γ subunit within F_1. This causes sequential conformation changes in the 3 β subunits which contain the catalytic centers (Fig. 16.1-4).

During the reaction cycle, they pass through different conformations:
- One β subunit binds ADP + P_i while entering the 'loose' (L) conformation,
- the next one synthesizes ATP from ADP + P_i in the 'tight' (T) stage
- the third releases ATP, while being in the 'open' (O) stage. This is the energy consuming step.

The final step in the oxidative phosphorylation system of *mitochondria* is the export of ATP against the import of ADP through the inner mitochondrial membrane by the nearby located ATP/ADP carrier. The electrical exchange of ATP^{4-} against ADP^{3-} is driven by the membrane potential $\Delta\Psi$ consuming one H^+ per ATP, which is reintroduced into the matrix by the parallel phosphate carrier (Fig.16.1-1).

Heat generation in *brown adipose tissue mitochondria* takes place when protons pumped out by the mitochondrial chain are recycled into the matrix by the uncoupling protein. It requires free long-chain fatty acids as activators and is inhibited by ATP and ADP (6.2.1).

Figure 16.1-3. Structure of the ATP Synthase *(E. coli)*

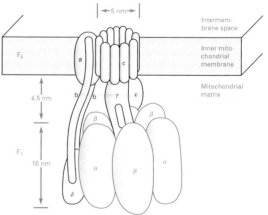

Figure 16.1-4. Rotational Conformation Changes of the ATP Synthase (as seen from the matrix side)

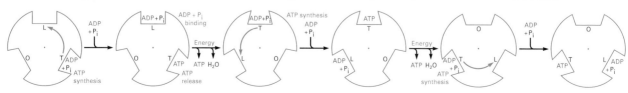

16.1.5 Redox Potentials in the Respiratory Chain (Fig. 16.1-5)

The graph 16.1-5 gives a survey on the standard potentials E'_0 of electron carriers and substrates under standard conditions (concentrations 1 mol/l, gases at 101.3 kPa (1atm) pressure, pH = 7.0, temperature = 25 °C = 298 K). For other concentrations, the redox potentials can be calculated according to Eq. 1.5-8:

$$E = E'_0 + \frac{R*T}{n*F} * 2.303 * \log \frac{[A_{ox}]}{[A_{red}]}$$

Quinone/hydroquinone potentials refer to the free compounds. In bound state, the potentials may differ up to 200 mV from these figures.

At the left of the graph, the data for both *bacterial* systems shown in Fig.16.1.1 are listed.

<u>Anaerobic</u> bacteria, e.g. *Wolinella succinogenes,* can use only the moderate differences in redox potentials (and thus a change in free energy $\Delta G'_0$) between electron donors like formate or hydrogen and electron acceptors, such as fumarate or nitrate (15.5). This is in contrast to <u>aerobically</u> living organisms, which make use of the much larger difference between the redox potentials of $NAD^+/NADH$ and $½ O_2 + 2 H^+/H_2O$. Additionally, succinate acts as electron donor in the succinate dehydrogenase reaction, which is part of the citrate cycle (3.9).

The redox potentials of *mitochondrial* components are shown at the right of the graph. They are differentiated according to the complexes of the respiratory chain. The potentials increase from complexes I...IV, corresponding the approach to the terminal oxidase.

Literature:
Abrahams, J.P. et al.: J.E.: Nature 370 (1994) 621–628.
Albracht, S.P.J. et al.: Biochim. biophys. Acta 1318 (1997) 92–106.
Boyer, P.D.: Ann. Rev. of Biochem. 66 (1997) 717–749.
Brandt, U.: Biochim. biophys. Acta 1318 (1997) 79–91.
Iwata, S. et al.: Nature 376 (1995) 660–669.
Kröger, A. et al.: Arch. Mikrobiol. 158 (1992) 311–314.
Leif, H. et al.: Friedrich, T.: Eur. J. Biochem. 230 (1995) 538–548.
Trumpower, B.L., Gennis, R.B.: Ann. Rev. of Biochem. 63 (1994) 675–716.
Tsukihara, T. et al.: Science 269 (1995) 1069–1074.

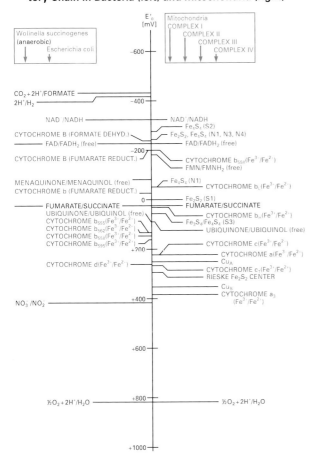

Figure 16.1-5. Redox Potentials of Components of the Respiratory Chain in *Bacteria* (left) and *Mitochondria* (right)

16.2 Photosynthesis

Photosynthesis is the light-energized formation of organic compounds (initially carbohydrates) from H_2O and CO_2 by *green plants* and by specialized *bacteria*. This is the original source for the organic material composing living organisms.

$$6 CO_2 + 6 H_2O = 6 O_2 + C_6H_{12}O_6 \qquad \Delta G'_0 = 2840 \text{ kJ/mol}$$

With an energy input of ca. 10^{18} kJ/year, about 10^{11} t C/year are converted this way into biomass on the earth. 3.5...2.0 billion years ago, early forms of life changed by this reaction the original no-O_2, high-CO_2 atmosphere into the present high-O_2, low-CO_2 atmosphere.

Photosynthesis is composed of two parts: In the <u>light reaction</u> (16.2.1), electrons are energized by light of definite wavelengths for transport of protons (enabling, e.g., ATP formation) or for supplying reducing compounds (NADPH or NADH), while in the <u>dark reaction</u> (16.6.2), the synthetic conversions take place.

16.2.1 Light Reaction (Figure 16.2-1)

Pathways of electron flow (compare 16.1.1): In *chloroplasts* of *green plants* (and similarly in *Cyanobacteria*), the energy of the electrons can be used in 2 ways (Fig. 16.2-2):

- <u>Cyclic electron flow</u>: Protons are moved from the *stroma* to the *thylakoid space* (2.2.3) forming a <u>proton gradient</u> which allows the operation of the <u>H^+-transporting ATP synthase</u>:

$$ADP + P_i + 3 H^+_{thyl.\ space} = ATP + 3 H^+_{stroma}.$$

The synthase structure and the reaction mechanism corresponds closely to the *mitochondrial* enzyme (16.1.3).

When the proton gradient becomes too small, the *chloroplast* enzyme is inhibited in order to avoid ATP degradation by the reverse reaction

- <u>Noncyclic electron flow</u>: The electrons are used for reduction of NADP:

$$NADP^+ + H^+ + 2 e^- = NADPH \qquad \Delta G'_0 = 61.7 \text{ kJ/mol}, \Delta E'_0 = -320 \text{ mV}$$

Electrons consumed in the latter reaction are replaced by the cleavage of water:

$$H_2O = ½ O_2 + 2 H^+ + 2 e^- \qquad \Delta G'_0 = 157 \text{ kJ/mol}; \Delta E'_0 = -815 \text{ mV}$$

resulting in an overall reaction of:

$$H_2O + NADP^+ = NADPH + ½ O_2 + H^+$$
$$\Delta G'_0 = 219 \text{ kJ/mol}; \Delta E'_0 = -1135 \text{ mV}$$

Since the photon energy amounts to

$$E = h*v*N = \frac{h*c*N}{\lambda} = \frac{119625}{\lambda} \text{ [kJ/Einstein]},$$

where 1 Einstein = 1 mol = N = $6.022 * 10^{23}$ photons, h = Planck's constant ($6.626 * 10^{-34}$ J/sec), c = speed of light ($2.998 * 10^8$ m/sec), λ = wavelength of the light (m),

light of the wavelength 700 nm has an energy content of 171 kJ/Einstein. This is light of the longest wavelength (lowest energy), which still can be used by *plants* for photosynthesis. Actually, in order to reduce 1 molecule of $NADP^+$ according to the above formula, 4 photons are required. About $1/3$ or less of their energy is used for this purpose and additionally $1/8$ or less for proton movement into the *thylakoid space*, which enables ATP synthesis. These reactions take place in 2 different photosystems.

Photosynthetic *purple bacteria* (e.g. *Rhodobacter*) have also two mechanisms for electron flow (Fig.16.2-3):

- The <u>cyclic electron</u> flow for proton translocation from *cytoplasm* to *periplasm* exists in a form different to green plants.
- Since the bacteria have only one photosystem and are unable to split water to supply electrons for NAD^+ reduction, they must obtain them from, e.g., H_2S or $H_2S_2O_3$ (reverse = <u>noncyclic electron flow</u>). Thus, they are not fully <u>photoautotrophic</u> as *green plants* are.

Halobacteria have only a light-driven proton transport system, which operates by a different mechanism (Figure 16.2-7).

Light absorption step: The <u>light harvesting complexes</u> (LHC) or <u>antenna complexes</u> surround the photosynthetic reaction centers in a concentric ring-like fashion (Table 16.2-1). They contain besides <u>chlorophyll</u> (5.4) also <u>carotene</u> and <u>xanthophylls</u> (7.3.2). *Cyanobacteria* use <u>phycocyanobilin</u> and <u>phycoerythrobilin</u> instead (5.3.2). Their absorption spectra are given in Fig. 16.2-4.

An absorbed light quantum excites an electron in one of the LHC molecules, which transfers its energy ('<u>exciton</u>') by resonance interaction via other LHC molecules quickly ($\approx 10^{-13}$ sec,

16 Oxidative Phosphorylation and Photosynthesis

Figure 16.2-1. Photosynthetic Systems in *Green Plants* and *Cyanobacteria* (Top); in *Purple Bacteria* (Below)

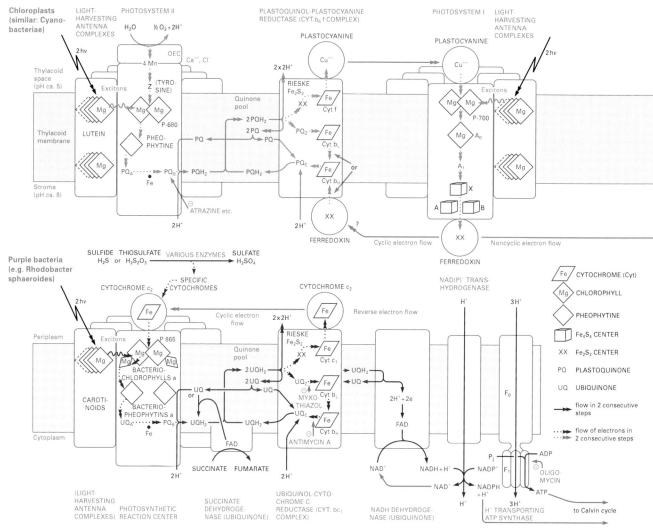

Fig. 16.2-2. Electron Flow in *Green Plants* and in *Cyanobacteria*

Cyclic electron flow   Noncyclic electron flow

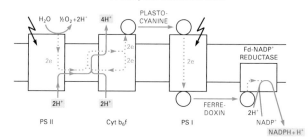

Fig. 16.2-3. Electron Flow in *Purple Bacteria*

Cyclic electron flow   Noncyclic electron flow

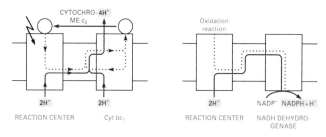

>90% efficiency) to the reaction center. In photosytem II of *plants* and *Cyanobacteria* or in the only reaction center of *purple bacteria*, it excites a pigment special pair, which in turn donates an electron extremely fast to a primary acceptor (pheophytine, 5.4 or bacteriochlorphyll, 5.4), causing the reaction to become irreversible. The electron finally reaches phylloquinone B (PQ_B, 4.7.2) or ubiquinone B (UQ_B, 4.7.2), respectively, where 2 electrons and 2 protons (from the *cytoplasm*) accumulate, forming a hydroquinone (quinol) (Fig. 16.2-5).

Table 16.2-1. Composition of the Light Harvesting Complexes (LHC)

Purple bacteriae	Plants	Cyanobacteria (Blue Algae)
(1 reaction center)	(2 photosystems)	
32 bacteriochlorophylls, 16 carotenes	Ph. Sys. I: ca. 200 Chl., a>b Ph. Sys. II: ca. 250 Chl., a>b 50 carotenes each	phycocyanobilin phycoerythrobilin

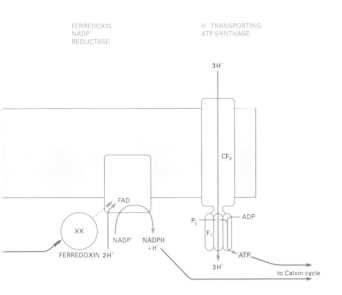

lakoid space or to the *periplasm*, respectively. These complexes resemble the *mitochondrial* ubiquinol-cytochrome c reductase (complex III). Analogously, a 'Q cycle' (16.1.2) operates for transfer of additional protons to the *thylakoid space* or to the *periplasm*, respectively. Its quantitative contribution is under discussion.

The corresponding electrons are finally transferred to photosystem I (in *plants* via plastocyanine) or returned to the reaction center (in *purple bacteria:* cyclic electron flow via cytochrome c_2).

NAD^+ or $NADP^+$ reduction: *Plants* and *cyanobacteria* energize electrons additionally in photosystem I. These are then conferred via ferredoxin either to the $NADP^+$ reductase (noncyclic electron flow), or alternatively back to the cytochrome b_6f-complex for additional proton transfer (cyclic electron flow). This allows a fine adaptation to the requirements of the cell, since NADPH reduction equivalents or ATP energy can be supplied in variable ratios. The graph of the reduction potential of the steps passed through resembles a 'Z' (Fig. 16.2-6, for details of the redox potentials see 16.1.5).

As described above, *purple bacteria* cannot follow this mechanism. They have to obtain reducing power from the environment to be able to reduce $NADP^+$ (noncyclic = reverse electron flow, since the electrons have to flow 'uphill' the redox potential).

Fig. 16.2-4. Absorption Spectra of Light Absorbing Compounds
(Line colors: green-*plants*, blue-*Cyanobacteria*).

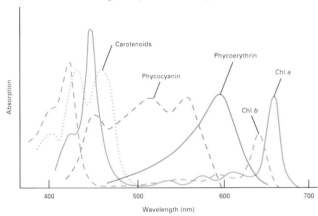

Regeneration of the reaction center: In *plants and cyanobacteria*, the 'special pair' $P680^+$ replaces the lost electron by abstraction of another electron from a Mn_4-protein complex (oxygen evolving complex, OEC) via a tyrosine residue, Z. After 4 repeats, OEC^{4+} reacts with water and is reduced again.

$OEC^{4+} + 2 H_2O = OEC^0 + 4 H^+ + O_2$

In *purple bacteria*, in the case of 'cylic electron flow', the lost electron of the special pair ($P865^+$) is returned from the cytochrome bc_1 complex via diffusing cytochrome c_2. No extra reducing power for other purposes becomes available this way.

Fig. 16.2-5. Time Course of Electron Transfer in *Purple Bacteria*

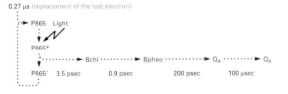

In case of 'Noncyclic electron flow' in these bacteria, an oxidation reaction (of H_2S, S, $H_2S_2O_3$, succinate etc.) takes place:

$H_2S = S_{solid} + 2 H^+ + 2 e^-$ (in *periplasm*)

or succinate = fumarate + $2 H^+ + 2 e^-$ (in *cytoplasm*).

The liberated electrons enter the reaction center via a bound cytochrome complex (e.g., in *Rhodopseudomonas viridis*) or via soluble cytochrome c_2 (e.g., in *Rh. sphaeroides*) and reduce the special pair.

Cytochrome b_6f and bc_1 complexes: The hydroquinone (quinol) formed in the primary photosynthetic reaction transfers its hydrogen via the 'quinone pool' to the cytochrome complexes b_6f (in *plants*) or bc_1 (in *bacteria*), where protons are released to the *thy-*

Fig. 16.2-6. Standard Redox Potentials in Photosynthesis (*Purple Bacteria/Plants* and *Cyanobacteria*).
In vivo, the actual potentials can differ due to protein binding, variant concentration ratios etc.

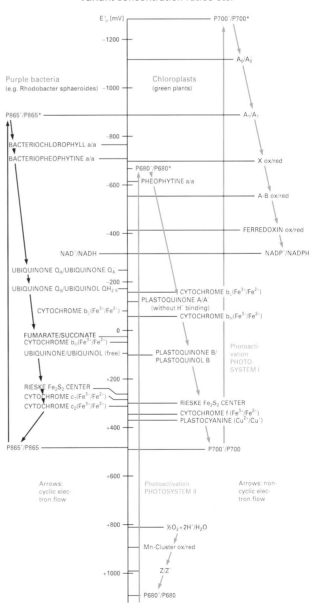

Halobacteria (Fig. 16.2-7): The photosystem of these *archaea* is unrelated to photosynthesis in higher plants. It uses bacteriorhodopsin, a small retinal protein (26 kDa) with 7 transmembrane passes, which pumps one proton per absorbed photon through the membrane.

It is mediated by light- induced *trans-cis* isomerization of the retinyliden chromophore and involves the following steps:

- Isomerization of retinal from the *all-trans* to the 13 *cis*-configuration [BR_{568} to J state (0.5 psec) and on to K and L states]
- Transfer of a proton from the protonated Schiff base (SBH) to the carboxylate of Asp85 (L to M_I states), followed by its release to the *extracellular medium*
- Modification of chromophore/protein structure. This changes the accessibility from the extracellular side to accessibility from the cytoplasmatic side (M_I to M_{II} states)
- Transfer of a proton from Asp 96 to the Schiff base (M to N state, several msec)
- Thermal *cis-trans* reisomerization and restoration of the inital state (N to O to BR_{568} states, several msec)

Isomerization of retinal (11-*cis* ↔ *all-trans*) plays also a role in the visual process of *vertebrates* (17.4.6).

Figure 16.2-7. Photosynthesis and Reaction Mechanism in *Halobacteria*

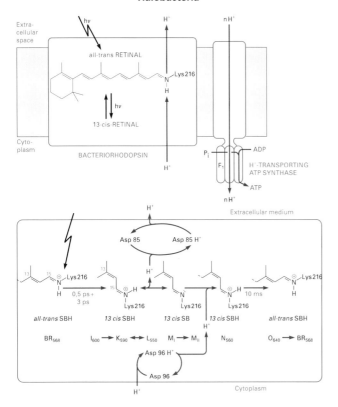

Literature:
Andersson, B.; Franzén, L.G. in Ernster, L. (Ed.): *Molecular Mechanisms in Bioenergetics.* Elsevier (1992) 103–120.
Arlt, T. *et al.*: Proc. Natl. Acad. Sci. (US) 90 (1993) 11757–11761.
Deisenhofer, J., Michel, H.: EMBO J. 8 (1989) 2149–2169 (Nobel Lecture).
Goldbeck, J.H.: Ann. Rev. Plant Physiol. Plant Mol. Biol. 43 (1992) 293–324.
Haupts, U. *et al.*: Biochemistry 36 (1997) 2–7.
Jansson, S.: Biochim. biophys. Acta 1184 (1994) 1–19.
(various authors): Israel J. of Chem. 35 (1995) No. 3–4.

16.2.2 Dark Reactions

As described above, the light reactions provide both the energy carrier ATP and the reductant NADPH. For the consecutive synthesis of biological material (initially carbohydrates), also CO_2 and water is required.

Calvin Cycle (Fig. 16.2-8): CO_2 fixation takes place in a cyclic process within the *stroma* by carboxylation of ribulose 1,5-diphosphate and concomitant cleavage into two 3-phosphoglycerate molecules. This is followed by phosphorylation and reduction reactions. Then an aldol condensation and a series of transfer reactions takes place, using mostly reactions closely related to the pentose phosphate cycle (3.6.1). As a result, the carboxylation of 6 C_5 molecules yields 1 C_6 molecule (glucose-P or fructose-P) and the reconstitution of the original 6 C_5 molecules:

$$6 C_5 + 6 CO_2 \rightarrow 6 C_5 + 1 C_6$$

according to the overall reaction of the Calvin cycle

$$6 CO_2 + 12 H_2O + 18 ATP^{4-} + 12 NADPH =$$
$$C_6H_{12}O_6 + 18 ADP^{3-} + 18 P_i^{2-} + 12 NADP^+ + 6 H^+.$$

The produced hexose is converted in *chloroplasts* into starch (3.2.2) or in the *cytosol* into sucrose (3.4.1).

The enzyme ribulose bisphosphate carboxylase/oxygenase (Rubisco) catalyzes the key reaction of the Calvin cycle

Ribulose bisphosphate + CO_2 + H_2O = 2 3-phospho-D-glycerate
$$\Delta G'_0 = -35.1 \text{ kJ/mol}.$$

The enzyme is apparently the most abundant enzyme in the biosphere. It consists of 8 large and 8 small subunits (51...58 and 12...18 kDa). It has a low catalytic efficiency (k_{cat} = 3.3 sec^{-1} per large subunit). Although the carboxylase reaction is usually preferred, it performs also an oxygenase side reaction (Fig. 16.2-9, see also 'photorespiration', below).

Figure 16.2-9. Carboxylase (Top) and Oxygenase (Below) Reaction Mechanisms of Rubisco

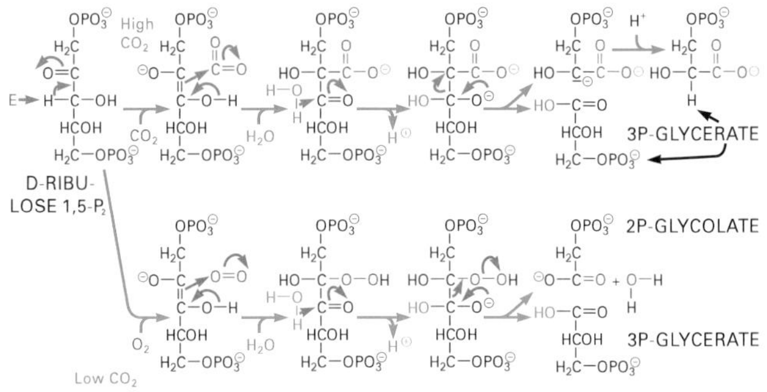

Regulation of the Calvin cycle: The cycle has to operate only if sufficient NADPH and ATP from the light reaction are available in order to prevent useless degradation reactions. This is performed by light-induced activation of rubisco, fructose bisphosphatase (FBPase) and sedoheptulose bisphosphatase (SBPase).

- The pH in the *stroma* increases during the light reaction (16.2.1), since protons are pumped out. It approaches the pH optimum of rubisco, FBPase and SBPase.
- Reduced ferredoxin, the reaction product of photosystem I, reduces thioredoxin, which in turn activates FBPase and SBPase by reduction of enzyme -SS- bridges (Fig.16.2-5). Simultaneously, phosphofructokinase (3.1.2) is deactivated by this reduction and thus decreases the competing glycolysis reaction (3.1.1).
- Mg^{++}, which flows into the *stroma* during illumination, activates rubisco, FBPase and SBPase.
- NADPH, which is produced by the light reaction, activates FBPase and SBPase.

During dark, these reactions are switched off. The energy supply of photosynthesizing cells is provided the same way as in non-photosynthesizing cells by glycolysis (3.1.1), pentose phosphate cycle (3.6.1) and oxidative phosphorylation (16.1).

Photorespiration and C_4 cycle: The rubisco side reaction with O_2 yields at first 3-phosphoglycerate and 2-phosphoglycolate, which later on is partially oxidized, resulting in CO_2 liberation (photorespiration, Fig. 16.2-8, see also 3.9.2). This counteracts photosynthesis and requires additional energy input for recycling. The rate of this reaction increases relatively to the rate with CO_2 at higher temperatures and at low CO_2 concentration at the site of synthesis (e.g., at hot, bright days) and limits the growth rate of plants.

A number of plants (C_4 plants, mostly tropical ones) developed a mechanism for increasing the CO_2 concentration in the fluid phase of *chloroplasts* from ca. 5 µmol/l to ca. 70 µmol/l (Fig.16.2-10). So-called mesophyll cells surround the bundle-sheath cells, which contain the Calvin cycle enzymes. The *mesophyll cells*, which lack rubisco, perform a CO_2 fixation by the highly exergonic (and thus practically irreversible) reaction (3.3.4):

Phosphoenolpyruvate + HCO_3^- → oxaloacetate + P_i

and transfer this bound CO_2 through a number of further reactions to the *chloroplasts* of *bundle-sheath* cells, where it is released to be used in the Calvin cycle. Several reaction types exist (Fig. 16.2-10). These reactions require 5 energy-rich P bonds/ CO_2 (instead of 3 in the Calvin cycle alone). Therefore, this mechanism is of advantage only in hot, sunny climates.

Literature:
Furbank, R.T., Taylor, W.C.: The Plant Cell 7 (1995) 797–807.
Gutteridge, S., Gatenby, A.: The Plant Cell 7 (1995) 809–819.
Heldt, H.W., Flügge, U.I. in Tobin, A.K.: *Plant Organelles.* Cambridge University Press (1992).
Heldt, H.W.: *Plant Brochemistry and Molecular Biology.* Oxford University Press (1998).
Portis, A.R.: Ann. Rev. Plant Physiol. Plant Mol. Biol. 43 (1992) 415–437.

Figure 16.2-8. CO₂ Fixation by the Calvin Cycle and its Regulation

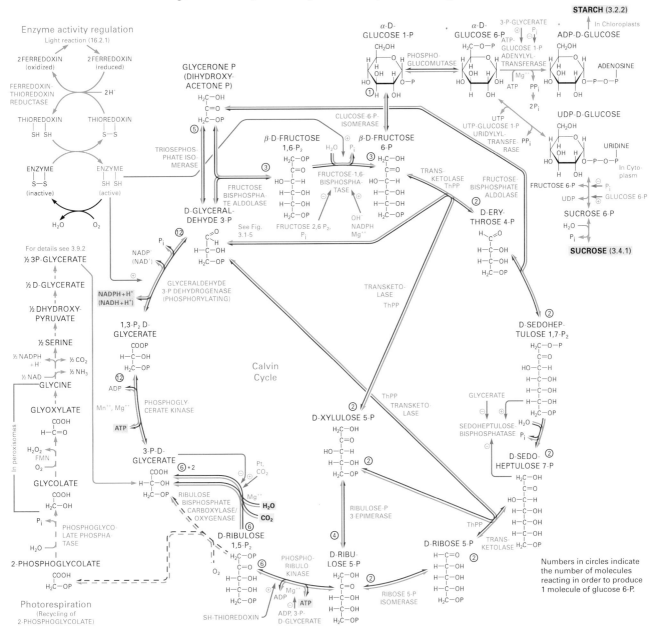

Figure 16.2-10. CO₂ Pumping by the C₄ Cycle (NADP⁺-Malate Enzyme Type, e.g. in *Maize* and *Sugar Cane*)

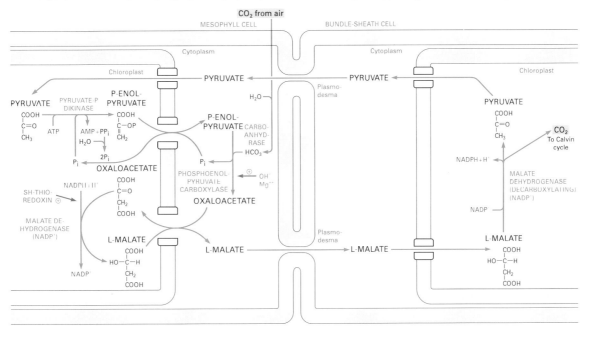

17 Cellular Communication

Multicellular organisms require signal transduction systems to coordinate the cellular activities. This can take place the humoral way (hormones in *plants* and *animals*) or the neuronal way (only in *animals*). This chapter deals with the situation in *vertebrates*.

Both communication systems cannot be exactly separated; there are various connections between them. In the following, hormones are dealt with first (17.1), which act mainly extracellularly both on non-neural cells and within the nerve system. Thereafter, the neuronal signal transmission (17.2) and finally the intracellular signal processing (17.3...17.7) are described.

17.1 Intercellular Signal Transmission by Hormones

17.1.1 General Characteristics of Hormones (Table 17.1-1)

Hormones are chemical signal transmitters, which are transferred humorally to their target cells. They are synthesized in
- hormone glands (glandular hormones, this Section)
- various cells [non-glandular hormones, such as tissue hormones (17.1.9, 17.4.8, 20.2.2) and cytokines, 18.1)]
- nerve cells (neurohormones, 17.2).

They act on the targets in different ways:
- autocrine: acting on the same cell
- juxtacrine: membrane bound hormones need direct contact to the target cell
- paracrine: acting on neighboring cells
- endocrine: acting on distant cells

Hormones belong to various chemical classes. Peptide hormones are frequently synthesized as prohormones, which are later cleaved to yield the final hormone. Many hormones are stored in vesicles (secretory granula) and released upon arrival of appropriate signals, e.g., changes of the intracellular Ca^{++} concentration (17.4.4). The hydrophobous steroid and thyroid hormones are bound to carrier proteins for their transport in blood.

The hormone concentrations are usually very low ($10^{-7}...10^{-12}$ mol/l). Frequently, there exists a hormone cascade, where an endocrine hormone causes the generation of another one in a different tissue and so on. A functional grouping of hormones is given in Table 17.1-1.

17.1.2 General Characteristics of Receptors

After reaching the target cells, the hormones (being agonists) bind specifically and usually with high affinity to receptors, which thereafter initiate specific intracellular actions. These are described in the following sections. There are 3 types of receptors:
- transmembrane proteins, where binding of the hormone to the extracellular portion causes a conformation change, which effects a transmembrane signal transduction resulting in an intracellular response via second messengers (17.4) or via phosphorylation cascades (17.5).
- ligand-gated ion channels. After binding of the agonist, these transmembrane proteins either open or close integral ion channels. They play a central role in nerve conduction (17.2). The agonist may be an extracellular hormone or an intracellular agonist of the second messenger type, e.g. in regulation of the intracellular Ca^{++} concentration (17.4.4), sensing of light (17.4.6), odors and tastes (17.4.7).
- intracellular proteins. The hormones, e.g., steroids, thyroid hormones and retinoic acid pass through the cellular membrane and meet their receptors intracellularly in the *cytosol* or in the *nucleus* (17.6). The receptor then modulates the transcription of genes.

In all cases, a deactivating mechanism is required for termination of the signal effect. This can be
- inactivation of the receptor by, e.g., specific phosphorylation
- desensitization of the receptor
- internalization of the receptor-ligand complex (similar to the lipoprotein-receptor complex, Fig.18.2-3, but with degradation of receptor and ligand).

Kinetics of hormone binding: Similarly to enzymes (1.5.4), the hormone H and the receptor R can reach an equilibrium with the hormone-receptor complex HR. This can be written as a concentration equation:

$$[H] + [R] \underset{k_{-1}}{\overset{k_1}{\rightleftharpoons}} [HR] \quad \text{with the dissociation constant} \quad K_D = \frac{[H] * [R]}{[HR]} = \frac{k_{-1}}{k_1}$$

Using $[R_t]$ for the total receptor concentration $[R] + [HR]$, a derivation analogously to the Michaelis-Menten equation [1.5-19] yields:

$$[HR] = \frac{[H] * [R_t]}{K_D + [H]} \quad \text{or} \quad \frac{[HR]}{[R_t]} = Y = \frac{[H]}{K_D + [H]}.$$

where Y is the share of the receptor, which has bound the hormone.

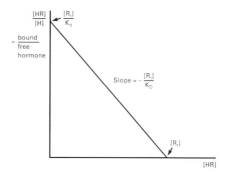

Figure 17.1-1. Scatchard Plot of a Hormone-Receptor Interaction

Table 17.1-1. Hormones Grouped According to Function

	Group	Examples	Steroids, Lipids	Amino Ac.deriv	Polypeptides	Section	Acting via (Receptor = R)	Section
I	Cytokines (non-glandular growth and differentiation hormones)	PDGF, EGF, FGF, IGF-1+2, TGF-α GM-CSF, G-CSF, EPO, interferons, interleukins, TNF			* *	17.1.5 19.1.6	R Tyr kinases Tyr kinase assoc. R	17.5.3 17.5.4
II	Glandular growth and differentiation hormones	growth hormone (GH) thyroid hormone sex hormones, glucocorticoids	*	*	*	17.1.5 17.1.5 7.4 ... 8, 17.1.5	Tyr kinase assoc. R thyroid hormone R steroid hormone R	17.5.4 17.6 17.6
III	'Fast' hormones	insulin glucagon catecholamines[2]		*	* *	3.2.3, 17.1.3 3.2.4, 17.1.3 4.7.4, 17.1.4	R Tyr kinases[1] G-protein R/cAMP G-protein R/cAMP	17.5.3 17.4.2 17.4.2
IV	Gastrointestinal hormones (digestion and resorption)	secretin gastrin			* *	17.1.9 17.1.9	G-protein R/cAMP G-protein R/PLCβ	17.4.2 17.4.3
V	Ca^{++} and phosphate metabolism regulating hormones	parathormone, thyreocalcitonin vitamins D	*		*	17.1.7 9.11, 17.1.7	G-protein R/cAMP steroid hormone R	17.4.2 17.6
VI	H_2O and electrolyte metabolism regulating hormones	vasopressin atrial natriuretic factor angiotensin mineralocorticoids	*		* * *	17.1.8 17.1.8 17.1.8 7.8.3, 17.1.8	G-protein R/cAMP guanylate cyclase G-protein R/PLCβ steroid hormone R	17.4.2 17.7 17.4.3 17.6
VII	Releasing and inhibit. horm.	CRH, TRH; ACTH, TSH, FSH, LH etc.			*	17.1.5 ... 8	G-prot.R/PLC, cAMP	17.4.2
VIII	Neurotransmitters	acetylcholine, dopamine, glutamate etc. endorphins, enkephalins		*	*	17.2.3 17.2.3	various R opiate R	17.2 17.2.3
IX	Others							

[1] Act also on cAMP phosphodiesterase antagonistically to glucagon (17.4.2)   [2] Also neurotransmitter (17.2.3)

This can be written as

$$\frac{[HR]}{[H]} = -\frac{1}{K_D}([HR] - [R_t]), \quad \text{where } \frac{[HR]}{[H]} \text{ is the ratio of bound/free hormone.}$$

A plot of this ratio [HR] / [H] versus the concentration of bound hormone [HR] yields theoretically a straight line with the slope $-(1/K_D)$, the abscissa intersection [R_t] and the ordinate intersection [R_t]/K_D (Scatchard plot, Fig. 17.1-1). This allows, e.g., the calculation of the dissociation constant and the receptor concentration. However, frequently the plot is not linear, e.g., if a population of receptors with different dissociation constants is present.

17.1.3 Insulin and Glucagon

Insulin is the most important anabolic hormone and acts primarily on *muscle, adipose tissue* and *liver*. Its binding to the insulin receptor induces the synthesis of glycolytic and represses the synthesis of gluconeogenetic enzymes (Table 3.1-1, 17.5.3). The fast metabolic effects of insulin (e.g., direct control of the blood glucose level, carbohydrate and fatty acid metabolism, 3.1.5) proceed by influencing the cAMP level, since insulin inhibits the adenylate cyclase and activates the 3′,5′-cyclic nucleotide phosphodiesterase. Details are described in 3.2.4, 6.1.2 and 17.4.2.

Insulin is synthesized in the *β cells* of *Langerhans islets* in the *pancreas*. Its expression is regulated by the glucose concentration. Insulin is stored in granula of the β cells. It is released, when the cellular glucose 6-P concentration rises (sensor: ATP/ADP ratio, transmitted via the K_{ATP} channel and voltage-gated Ca^{++} channels, Table 18.1-1). The insulin secretion is enhanced by the gastric inhibitory peptide and decreased by epinephrine and somatostatin.

The primary transcription product (prepro-insulin, 104...110 amino acids, species dependent) is converted in the *endoplasmic reticulum* into proinsulin (removal of the signal peptide, formation of 3 S-S bonds) and in the *Golgi apparatus* (removal of the intermediary C peptide) into mature insulin.

Figure 17.1-2. Structure of Insulin and Its Precursors

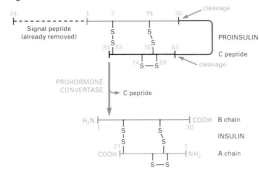

In diabetes type I (juvenile diabetes), the production of insulin is disturbed due to autoimmune diseases or virus infections, while in diabetes type II either the insulin release mechanism does not respond properly to elevated blood glucose levels or there are defects in the insulin receptors. Diabetes is characterized by elevated blood sugar and by lipolysis, formation of ketone bodies (6.1.7), loss of electrolytes and acidosis.

Glucagon, being a catabolic hormone, is the metabolic opponent of insulin. It is synthesized in the α cells of Langerhans islets in the pancreas. It is released when the glucose concentration drops. Upon the binding to its receptor in the liver, it stimulates adenylate cyclase via G-proteins (17.4.2). The cAMP formed, in turn, activates a protein kinase, which increases the activity of phosphorylase and decreases the activity of glycogen synthase (3.2.4).

Analogous to insulin, the synthesis proceeds via the inactive precursor prepro-glucagon and several intermediates. In the *intestine*, prepro-glucagon is cleaved differently, yielding the glucagon-like peptides GLP-1 and GLP-2, which promote insulin secretion after food intake.

17.1.4 Epinephrine and Norepinephrine (Catecholamines)

These compounds are synthesized from tyrosine (4.7.4) in *postganglionic nerve terminals* and in the *adrenal medulla* (Fig. 17.1-3). The catecholamines are stored in specific granula as a complex with ATP and Mg^{++} and are released after arrival of neuronal signals, which are transmitted by acetylcholine. This takes place in physical and psychic stress situations.

The catecholamines act on several adrenergic receptor types, producing multiple effects (Table 17.1-2). The mechanism of action is described in detail in Sections 17.4.2 and 17.4.3.

The expression of the biosynthetic enzymes tyrosine 3-monooxygenase (hydroxylase) and dopamine β-hydroxylase is induced by neuronal signals, the expression of phenylethanolamine N-methyltransferase by glucocorticoids. Since catecholamines, in turn, promote the synthesis of regulators of glucocorticoid formation (CRH, ACTH), this results in an upregulating circuit (Figure 17.1-3).

Figure 17.1-3. Regulation of Catecholamine Biosynthesis

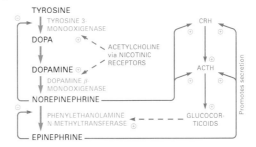

Table 17.1-2. Metabolic Effects of Catecholamines

Effect	Adrenergic Receptors	Acting via
Glycogenolysis, gluconeogenesis	α_1; β_1	PLCβ↑; adenylate cyclase ↑
Lipolysis (*adipose tissue*)	β_2, β_3	adenylate
Increased contractility of the *heart muscle*	β_1	adenylate cyclase ↑
Vasodilatation in the *skeletal muscle*	β_2	adenylate cyclase ↑ *
Vasoconstriction in the *intestinal area*	α_1	PLCβ↑ → Ca^{++}↑
Decreased insulin secretion	α_2	adenylate cyclase ↓

* cAMP activates PKA, this promotes the uptake of Ca^{++} into the *sarcoplasmic reticulum* (17.4.4) and decreases the *cytosolic* Ca^{++} concentration, which controls contraction. Additionally, myosin light chain kinase is phosphorylated and deactivated (17.4.5).

17.1.5 Hypothalamus-Anterior Pituitary Hormone System (Fig. 17.1-4)

This complex system is arranged as a cascade: one hormone puts another into action and so on (by enhancing the synthesis and/or the release). Each step occurs in a different organ. The system regulates many important functions (growth, overall metabolic rate, reproduction, pain control etc.).

External and internal stimuli are processed in the *hypothalamus* and lead to the secretion of various releasing and inhibitory hormones (liberins and statins), which are transported by the bloodstream to the nearby *anterior pituitary lobe* (also called *adenohypophysis*). There, they bind to specific receptors, which regulate the formation and release of trophic hormones (tropins). These trophic hormones, in turn, act on receptors at *peripheral tissues* and stimulate them to synthesize and release hormones, which thereafter perform their respective functions in their target tissues. The hormones also control the synthesis or release of their own regulators in a feedback fashion (see Fig. 17.1-4).

The hypothalamus (and most of the pituitary) hormones are synthesized as long preprohormones and posttranscriptionally processed to the prohormone and further on to the actual hormone (similarly as described above for insulin). In some cases, several hormones are cleaved from a common predecessor. All hypothalamus hormones are amidated at the C terminus.

The melanocyte stimulating hormones (α- and β-MSH) are synthesized in the *intermediate part* of the *pituitary gland*. The *posterior pituitary lobe* (*neurohypophysis*) performs the final processing of oxytocin and vasopressin, which are synthesized in the *hypothalamus* (17.1.8). Some releasing hormones are additionally produced in other organs besides the *hypothalamus*.

Hypothalamo-pituitary-adrenal axis: 7...10 times a day, but increased during stress situations, the corticotropin releasing hormone (CRH) is secreted from the *hypothalamus* and acts on the *pituitary gland*. Here the proopiomelanocortin (POMC) is synthesized and processed into corticotropin (ACTH, stored in secretory granula), endorphins and α-, β- and γ-melanocyte stimulating hormone (MSH). The CRH and the ACTH secretions occur spasmodically and follow the same rhythm.

Cortisol: In the *adrenal cortex*, ACTH activates (via cAMP/protein kinase A, 17.4.2) the hydrolysis of cholesterol esters (7.1.5) and the *de novo* synthesis of cholesterol (7.1.1). It also induces the hydroxylases involved in the conversion of cholesterol to cortisol (7.8.1) and effects the cortisol secretion. This steroid is bound to transcortin during transport in blood.

Almost all cells of the organism are targets for cortisol action, where the hormone binds to the internal cortisol receptor (17.6). This induces transcription of genes coding for many catabolic enzymes in *extrahepatic tissues* (as an antagonist to insulin) and for gluconeogenetic enzymes in *liver*. It is also an immunosuppressor and a suppressor of inflammation (7.8.3).

The maximum of the wavelike diurnal cortisol secretion is in the morning, the minimum in the evening, with additional spikes corresponding to the CRH/ACTH secretion.

Endorphins and related neuropeptides: Several endorphine isoforms and Met-enkephalin are cleaved from POMC in the *pituitary gland*, but also in the *stomach* and *intestine, placenta, lung* etc. Dynorphin and neoendorphins are cleaved from other precursors. They act on opioid receptors (Table 17.2.2), cause analgetic effects and are involved in body temperature regulation.

Other cleavage products of POMC are melanotropins (MSH, melanocyte stimulating hormones), which promote skin darkening in *amphibia* and *fishes* by activating tyrosinase and thus increasing the melanin synthesis (4.7.3).

Hypothalamo-pituitary-thyroid axis: The TSH releasing hormone (TRH) of the *hypothalamus* and other organs stimulates the release of thyrotropin (thyroid stimulating hormone, TSH) from the *pituitary gland*. This heterodimer (the α subunit is identical with the subunits of LH and FSH) binds to the TSH receptor on the *thyroid gland*, which is of the heterotrimeric G protein associated type (17.4.1). The IP_3 mechanism (17.4.3) activates the iodination of thyreoglobulin at multiple tyrosine residues in the *thyroid follicles* and the *lysosomal* degradation of the protein, which results in the liberation of thyroid hormones triiodothyronine (T_3) and thyroxine (T_4, for metabolism see 4.7.5). The release of these compounds is controlled by cAMP (17.4.2).

During their transport in blood, most of the thyroid hormones are bound to the thyroxine binding globulin (TBG) or other proteins. Only the free compounds are hormonally active. Inside of the target cells, T_3 combines with its receptor (17.6, T_4 has to be deiodated first) and induces the expression of many enzymes and regulatory proteins.

In *higher vertebrates*, the agonist binding leads to enhanced glycolysis, gluconeogenesis and liponeogenesis. The activity of Na^+/K^+-ATPase (18.1) and of lysosomal enzymes is increased. The expression of growth hormones (STH, EGF, IGF, NGF etc.) promotes growth and cellular differentiation. The increase in adrenergic β receptors enhances catecholamine actions (e.g., yielding positive inotropic and chronotropic *heart* effects, increased thermogenesis).

Abnormal levels of thyroid hormones cause common diseases. Hypothyroidism can originate from mutations at every level of the hormone cascade (causing se-

Figure 17.1-4. Hypothalamus-Anterior Pituitary Controlled Hormones

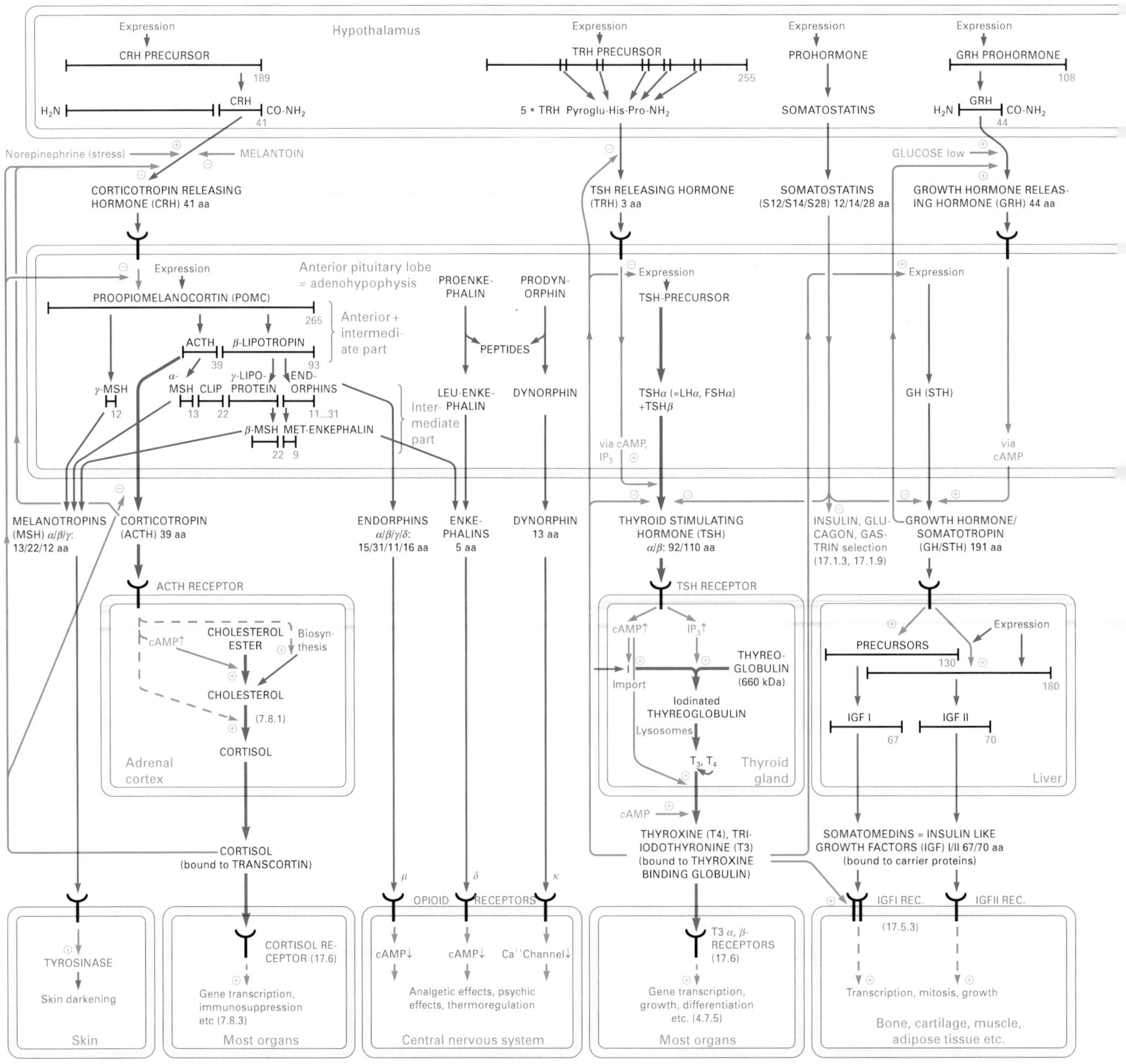

rious physical and mental retardation of infants, cretinism) or by formation of autoantibodies against thyroid enzymes. Lack of iodine supply leads via feedback regulation to an increased secretion of TSH and consecutively to hypertrophy of the thyroid gland (goiter). The most common cause of hyperthyroidism is the presence of autoantibodies, which permanently activate the TSH receptor (Basedow's disease).

Hypothalamo-pituitary-liver/bone axis: The glucose concentration in *plasma* is sensed by *neural* glucoreceptors. A decrease leads to secretion of the growth hormone releasing hormone (GRH) by the *hypothalamus*. This effects the secretion of somatotropin (STH, also named growth hormone, GH) from the *pituitary gland* (the greater portion during the night). The STH-expression is promoted by triiodothyronine (see above).

The release is antagonized by somatostatin (growth hormone release inhibiting hormone), which originates in many organs (including the *hypothalamus*) and also inhibits various other reactions (gastrointestinal secretions, release of TSH, insulin, glucagon etc.).

In the *liver*, STH stimulates the synthesis and release of somatomedins (insulin-like growth factors, IGF I and II), which bind to specific receptors and promote growth of *bone* and *cartilage*, likely also of *muscle* and *adipose tissue*.

The IGF I receptor resembles the insulin receptor, while the IGF II receptor is a tyrosine kinase associated receptor. Both initiate reaction sequences leading to increased transcription (17.5.3).

Inactivating mutations can occur at every level of the cascade and cause dwarfism. The eosinophilic adenoma (a pituitary tumor) permanently activates the adenylate cyclase and thus the STH release, resulting in gigantism (in children) and in acromegalia (in adults).

Hypothalamo-pituitary-testis axis: The *hypothalamus* regulates both male and female sex hormone production by the formation and intermittent secretion of the gonadotropin releasing hormone (GnRH, also called luteinizing hormone releasing hormone, LHRH) about 12...18 times daily. This causes the release of the gonadotropins [follicle stimulating hormone (FSH) and luteinizing hormone (LH)] from the *anterior pituitary lobe*.

In the *Leydig cells* of *male testes*, LH stimulates the biosynthesis of testosterone and 5α-dihydrotestosterone (7.6.1). These hormones bind to intracellular receptors of *Sertoli cells* and promote the mitosis and meiosis steps of spermatogenesis, which are also supported by FSH. They also cause many other physical and psychic effects (Table 7.6-1).

An additional control factor is prolactin, a product of the *anterior pituitary lobe*, which potentiates the LH action on the *Sertoli cells* and supports the testosterone effects. Its secretion is promoted by endorphins and inhibited by GAP (another part of the LHRH prohormone).

Early in *embryogenesis*, the expression of the SRY gene on the Y chromosome results in synthesis of the testis determining factor, which acts as a switch for sexual differentiation. It activates the transcription of the Müllerian inhibi-

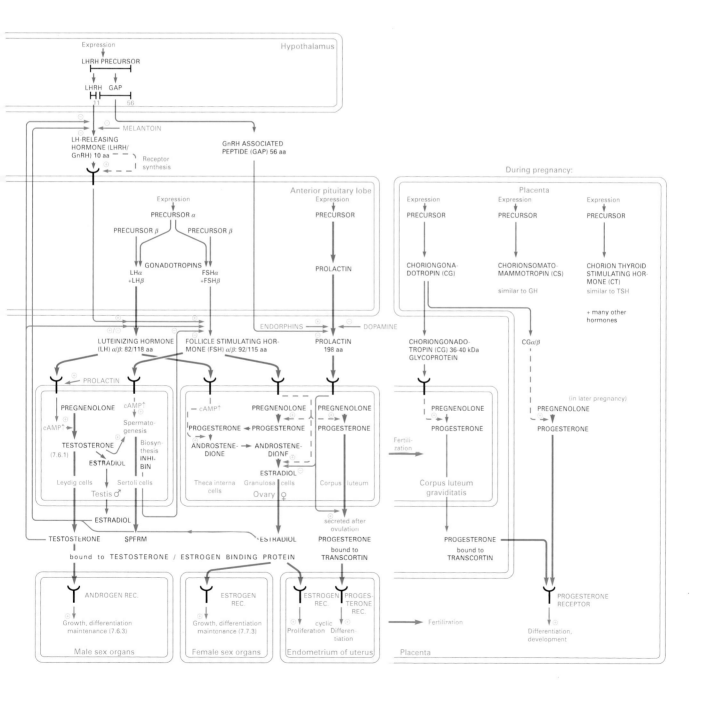

tory substance (MIS), which suppresses the development of female genitalia. Then testosterone and 5α-dihydrotestosterone induce the differentiation of the Wolffian ducts into the male reproductive organs. Later on, these hormones direct the development of many other male sex characteristics (Table 7.6-1).

Hypothalamo-pituitary-ovary/uterus axis: The control of female sex hormone production depends also on the the pulsed release of GnRH, LH and FSH (see above). Estrogens (estradiol, estrone) are synthesized via androgen intermediates. This takes place in the *granulosa* and in the *theca interna* cells of the *follicle*, which are under control of FSH and LH, respectively (Fig. 17.1-5). Formulas and details of the metabolism are shown in Figures 7.6-1 and 7.7-1.

Fig. 17.1-5. Synthesis of Estrogens in the Ovary

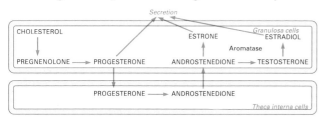

Puberty marks the beginning of the menstrual cycle with a characteristic pattern of the plasma hormone fluctuations (Fig. 17.1-6), which are additionally modulated by the diurnal pulsating secretions. This periodicity is regulated by a complicated interplay of forward activations and feedback inhibitions within the hormone cascade.

Before ovulation, the *granulosa cells* of a *follicle* express at first receptors for FSH and thereafter intracellular receptors for estradiol, progesterone, testosterone and cortisol (17.6). The binding of these hormones results in enzyme induction (e.g. of aromatase) and in cell proliferation. Other enzymes of estrogen biosynthesis are induced by LH binding to its receptor on the *theca interna cells*. The estrogen production increases and exerts a feedback inhibition on the FSH secretion, which blocks the maturation of additional follicles. The sharp peak of LH concentration at the midpoint of the cycle is associated with the formation of hydrolytic enzymes, which results in ovulation. Thereafter *granulosa* and *theca interna* cells form the *corpus luteum*, which is sustained by LH and prolactin hormones. The *corpus luteum* secretes mainly progesterone.

The estrogens also cause a proliferation of the *endometrium* and other responses of the *uterus* in the first phase of the menstrual cycle. After ovulation, progesterone modifies the endometrium.

Without fertilization, the decrease of hormone production at the transition from the proliferative to the secretory state causes luteolysis and sloughing off of the uterine lining (menstruation).

Generally, estrogens are responsible for formation and upkeep of female sex characteristics (Table 7.7-1).

Figure 17.1-6. Hormone Concentration in Plasma During the Menstrual Cycle

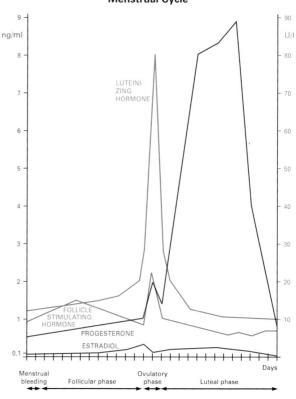

Figure 17.1-7. Regulation of the Ca++ Concentration in the Extracellular Space
The antagonistically acting calcitonin reactions are shown in red.

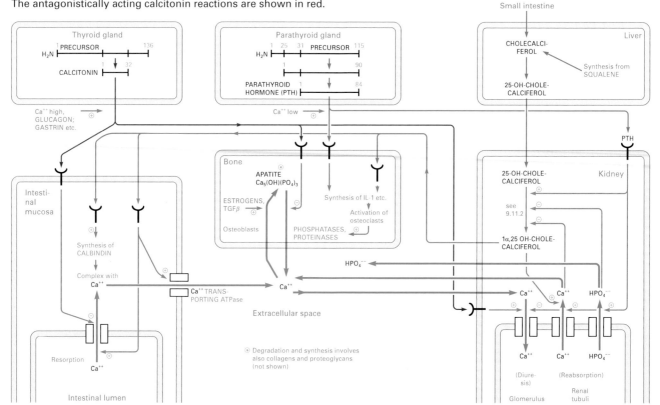

During the embryonic state, estrogens control the development of the female gonads. FSH and LH plasma levels increase slowly until puberty, when they commence the characteristic concentration changes of the menstrual cycle. After menopause, the decreasing estrogen production no longer exerts a feedback inhibition, FSH and LH levels rise without periodicity.

17.1.6 Placental Hormones

After fertilization, the *corpus luteum* increases the progesterone output, but later on this hormone is provided by the *placenta* (7.5.2).

Estrogens prepare the *uterus epithelium* by induction of growth factors (e.g., EGF, TGF-α etc.), cytokines and adhesion molecules (lectin-like compounds, integrins etc., see Table 19.3-1). The *early embryo* (*blastocyte* state) also produces various cytokines, adhesion molecules etc. Their interaction with receptors results in implantation of the fertilized ovum.

The developing *placenta* produces a series of hormones, which are analogous to *hypothalamus* or *pituitary* hormones:
Choriongonadotropin (CG) ⟷ LH
Chorionsomato(mammo)tropin (CS) ⟷ STH
Chorionthyrotropin (CT) ⟷ TSH,
Additionally it supplies LH-RH, TRH, somatostatin etc.

In the earliest phase of pregnancy, the *blastocyte* provides CG (human form: hCG). It prevents the regression of the *corpus luteum* and its progesterone production, until the *placenta* can take over. The hCG determination in plasma or urine can be used as a pregnancy test.

17.1.7 Hormones Regulating the Extracellular Ca^{++}, Mg^{++} and Phosphate Concentrations (Fig. 17.1-7)

Calcium is involved in many important biochemical functions, e.g., muscle contraction (17.4.5), nerve conduction (17.2) and blood coagulation (20). The intracellular mechanisms are described in the respective sections. The extracellular calcium concentration likewise has to be strictly controlled; the more, since only a small portion of the total body calcium is in the liquid phase (99% is fixed in bones, teeth etc, but can be mobilized) and since a large concentration gradient across membranes has to be kept up (17.4.4).

The Ca^{++} flow in an adult human is shown in Fig. 17.1-7. The Ca^{++} metabolism is closely coupled with the phosphate metabolism. Solubility of the salts plays a major role.

Figure 17.1-8. Flow Sheet of Calcium in an Adult Human (70 kg)

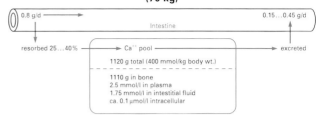

The resorption, the mobilization from the *skeleton* and the excretion of Ca^{++} are regulated by the parathyroid hormone (PTH) and 1α, 25-dihydroxycholecalciferol (increasing the concentration in the *extracellular space*) and by calcitonin (decreasing the concentration). The regulatory mechanisms and the biosynthesis of PTH and calcitonin are shown in Fig. 17.1-7, while the biosynthesis of cholecalciferol and other compounds of the vitamin D group is dealt with in 9.11.1.

Magnesium is a cofactor of many reactions involving phosphate transfer, e.g., kinases (as Mg-ATP complex), phosphatases, nucleic acid synthesis etc. Little is known about the intestinal resorption mechanism and the concentration control in body fluids. During transfer through ion channels it acts as an antagonist to Ca^{++}.

17.1.8 Hormones Regulating the Na^+ Concentration and the Water Balance (Fig. 17.1-9)

While sodium is the major cation in the *extracellular space*, potassium is prevalent in the *intracellular space* (Table 17.1-3). These dysequilibria are produced by the ubiquitous Na^+/K^+-exchanging ATPase (Fig. 18.1-1). They are essential for cellular function (membrane potential, 17.2.1; Na^+ driven symport and antiport channels, Table 18.1-3). Na^+ also plays an important role in the osmoregulation of the cells (see below).

Table 17.1-3. Concentration of Na^+ and K^+ Ions

		Na^+	K^+
Plasma	mmol/l	135...145	3.5...5.5 (normal 4.0)
Interstitial fluid	mmol/l	144	4.0
Intracellular	mmol/l	10	ca. 150 (tissue dependent)

Figure 17.1-9. Regulation of the Na^+ Concentration and the Extracellular Water Volume
The antagonistically acting ANF reactions are shown in red.

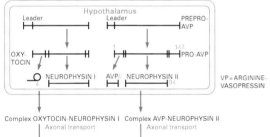

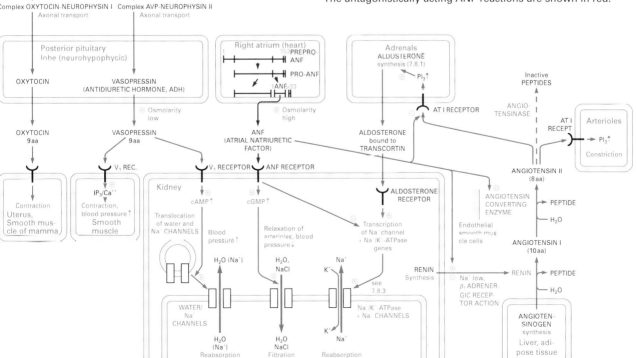

Na⁺ turnover: The *intestinal* Na⁺ resorption proceeds via a co-transport mechanism with glucose (Table 18.1-3). The excretion takes mainly place in the urine and is the balance between *glomerular* filtration and *tubular* reabsorption. The reabsorption is an active transport process employing both the Na⁺/K⁺-ATPase and Na⁺ channels. The expression of both proteins is hormone controlled.

Water turnover: Water resorption from the *intestine* follows the osmotic gradient established by the Na⁺ resorption. The major excretion occurs via the *kidney*, second to it are *lung* and *skin*. Renal reabsorption is facilitated by hormone-controlled translocation of water channels from internal vesicles to the membrane of *epithelial cells* of the *collecting tubules*.

The regulation of the extracellular volume and the interrelated control of blood pressure is performed by hormones: vasopressin and the renin-angiotensin system increase the blood pressure, the atrial natriuretic factor (ANF) antagonizes it. They respond either to the osmotic pressure or to the blood pressure as signals for their release. For biosynthesis and mechanism of action see Figure 17.1-9. Aldosterone (7.8.1) promotes the Na⁺ and water retention.

Oxytocin (Ocytocin, 9 aa), chemically closely related to vasopressin, contracts the uterus during labor and later on the breast smooth muscle (milk secretion).

17.1.9 Hormones of the Gastrointestinal Tract

The secretion of digestive fluids, the secretion and reabsorption of water and electrolytes and the intestinal movements are regulated by hormones. Some of them are neuronally controlled.

Gastric hormones (Fig.17.1-10): Neuronal impulses (acetylcholine, or from the *parasympathetic neurons*: gastrin releasing peptide), as well as signals from the *stomach* (high pH value, stretching) promote the secretion of gastrin, which causes release of HCl into the stomach. The same neuronal impulses inhibit the secretion of somatostatin, which blocks the HCl release.

Gastrin, cholinergic impulses or high acidity of the gastric milieu cause secretion of the digestive prohormone pepsinogen and of the epithelium-protective compound mucin. Pepsinogen cleaves itself autocatalytically at low pH into the active pepsin.

Pancreatic hormones (Fig.17.1-11): The secretion of pancreatic enzymes and proenzymes is also under multiple control. *Neuronal* impulses are transmitted by acetylcholine. Hormonal signals are initiated by contents of the *duodenum* (peptides, amino acids, fatty acids), which cause release of cholecystokinin (CCK, also called pancreozymin, PZ) into the *intestine*. Other hormones are secretin and the related vasoactive intestinal peptide (VIP). They stimulate both the secretion of pancreatic (pro-)enzymes and of water. The endocrine hormones of the *pancreas* are described above (17.1.3).

Literature:
Barnard, E.A.: Trends in Biochem. Sci. 17 (1996) 305–309.
Baxter, J.D. *et al.*: Rec. Progr. Horm. Res. 47 (1991) 211–258.
Brent, G.A.: New Engl. J. Med. 331 (1994) 847–853.
Brown, E.: Physiol. Rev. 71 (1991) 371–411.
Cooke, B.A., King, R.J.B., van der Molen, H.J. (Eds.): *Hormones and Their Activation. New Comprehensive Biochemistry* Vol. 18A, 18B. Elsevier 1988.
Efrat, S. *et al.*: Trends in Biochem. Sci. 17 (1994) 535–538.
Werner, M.H. *et al.*: Trends in Biochem. Sci. 21 (1996) 302–308

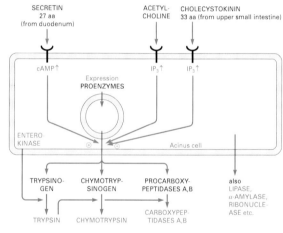

Figure 17.1-11. Regulation of Pancreatic Enzyme Secretion

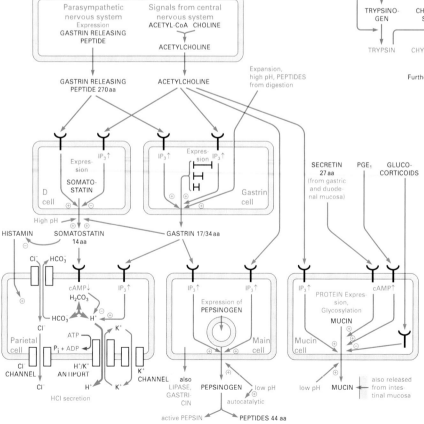

Figure 17.1-10. Regulation of Gastric Secretion

17.2 Nerve Conduction and Synaptic Transmission

Nerve cells (*neurons*) receive, process and transmit signals. While transport (= conduction) and processing within a single cell proceed electrically, the connection with other cells for transmission of signals at *synapses* may be electrical or chemical.

17.2.1 Membrane Potential

Neuronal membranes contain channels which allow passage of ions at different rates. Due to this selectivity, the ion concentrations on both sides of the cellular membrane are different, causing an electric potential difference across the membrane.

The equilibrium potential $\Delta\Psi$ of a cell is calculated according to the Goldman equation (compare 1.5.3). This form is valid for singly charged ions. P = permeability coefficient, ca = cations, an = anions:

$$\Delta\Psi \, [mV] = \frac{1000 * R * T}{F} * 2.303 * \log \frac{\Sigma \, [P_{ca} * c_{ca\text{-outside}}] + \Sigma \, [P_{an} * c_{an\text{-inside}}]}{\Sigma \, [P_{ca} * c_{ca\text{-inside}}] + \Sigma \, [P_{an} * c_{an\text{-outside}}]}$$

The equilibrium potential of *mammalian* cells is ca. -83 mV. It is composed mainly of the individual equilibrium potentials of $K^+ = -91$ mV and $Na^+ = +55$ mV. The Na^+/K^+-ATPase (18.1.1), however, moves Na^+ out of the cell and K^+ into it, causing a resting potential of an unstimulated cell of -65 mV. (The values for these parameters differ somewhat in the literature.) Opening of an ion channel allows the particular ion to approach its equilibrium state and therefore changes the resting potential towards the equilibrium potential of this particular ion (action potential).

Due to the capacitor effect of the membrane, only ca. $^1/_{2000}$% of the Na^+ ions present must diffuse to depolarize a cell body from -66 to $+45$ mV.

17.2.2 Conduction of the Action Potential Along the Axon (Figure 17.2-1)

An endogenous or exogenous stimulant shifts the membrane potential away from its resting state ①. If this change is beyond a threshold value, it is recognized by voltage sensitive structures in neighboring channel proteins and leads to a conformational change, causing the opening of closed Na^+ and K^+ channels ②. At first, the predominant influx of Na^+ shifts the membrane potential to positive values ③ (depolarization). After the closing of the Na^+-channel, the K^+ outflux through the still open K^+ channel reverses the potential, even beyond the resting potential ④ (hyperpolarization). As soon as the K^+-channel closes, the potential returns to the value of the resting potential ⑤. This way, the action potential travels along the *axon* at constant speed to the synapse (ca. 0.5 m/sec in unmyelinated cells, up to 100 m/sec in myelinated cells, also depending on their diameter).

The general structures of the voltage-gated Na^+, Ca^{++} and K^+ channels are similar (Figures 17.2-2 and 17.2-3). The actual channel consists of 4 circularly arranged domains (K^+ channel: 4 separate protein chains) with 6 transmembrane segments each ($S_1...S_6$). The P domains (hairpin loops between S_5 and S_6, with different amino acid sequences for each channel type) are part of the channel lining and likely effect the ion specificity of the channel. The segments S_4 with a regular arrangement of positively charged amino acids may act as voltage sensors, causing the conformation changes. Additional subunits regulate or locate the channel protein(s). Closing of the K^+ channel (K_A type) is apparently effected by the contact of the N terminus of the protein chain with receptor structures at the intracellular mouth of the channel. The Na^+ channel may be closed when the short intracellular loop between S_6 (domain III) and S_1 (domain IV) covers the channel mouth like a lid.

17.2.3 Transmitter Gated Signalling at the Synapse (Table 17.2-1)

Presynaptic reactions: Presynaptic vesicles are synthesized in the *ER/Golgi system* (13.3 and 13.4) and are transported by axonal transport mechanisms (17.2.6) to their location near the *presynaptic membrane*, where they stay attached to the cytoskeleton. During maturation, they pick up the respective neurotransmitters by specific, ATP-dependent uptake mechanisms.

When the action potential reaches the nerve terminal, it triggers the process of synaptic transmission. Voltage-sensitive Ca^{++}-chan-

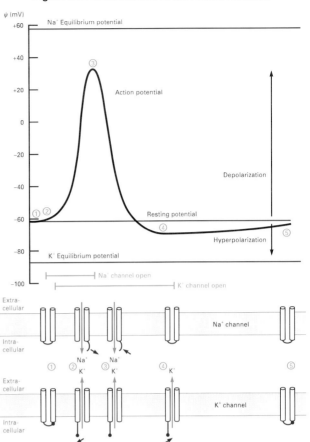

Figure 17.2-1. Generation of the Action Potential

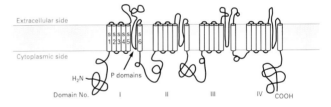

Figure 17.2-2. Protein Segments of Voltage Gated Na^+ and Ca^{++} Ion Channels (α subunit, additionally there are several other subunits). In the K^+ Channel there are 4 separate α protein chains.

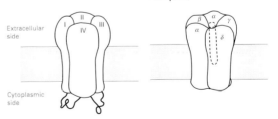

Figure 17.2-3. Structure of Voltage Gated Ion Channels (Na^+, Ca^{++}, possibly also K^+)

Figure 17.2-4. Structure of Directly Transmitter Gated Ion Channels (Type I, nicotinic acetylcholine receptor)

Figure 17.2-5. Protein Segments of Directly Transmitter Gated Ion Channels at the Postsynaptic Membrane (Nicotinic acetylcholine receptor)

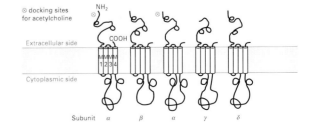

nels open and Ca^{++}-influx into the active zone of the presynaptic membrane takes place (Fig 17.2-7). The increased Ca^{++} concentration causes phosphorylation of Ca^{++} sensors (e.g., synapsin) by protein kinases, which in turn effects vesicle separation from the cytoskeleton, fusion of the vesicle with the presynaptic membrane and neurotransmitter release into the *synaptic cleft* (for mechanism see 13.5). The neurotransmitters diffuse to specific *postsynaptic* receptors (17.2-5).

Postsynaptic reactions: Binding of the neurotransmitters to the receptors (Table 17.2-3) changes the excitability of the *postsynaptic membrane*. This leads to one of the following effects:

- Direct transmitter gating (receptors type I, ionotropic receptors): Ion conducting channels are opened.

  The ion channels (Figures 17.2.4 and 17.2.5) consist of 5 separated subunits with 4 transmembrane segments each (e.g., nicotonic acetylcholine receptor: subunit structure $\alpha_2\beta\gamma\delta$, both α subunits bind acetylcholine). The M_2 segments of the subunits are thought to line the pore. The amino acids contained therein (acidic or basic) determine the selectivity of the channel. However, new work on the glutamate receptor finds only 3 transmembrane segments and a loop between the first and second segment.

- Indirect transmitter gating (receptors type II, metabotropic receptors): Heterotrimeric G proteins as second messenger systems are activated. Details are described in 17.4.

  The receptors (Figure 17.2-6) usually contain 7 transmembrane helices (e.g., muscarinic acetylcholine receptor, β-adrenergic receptor). Their activity can be modulated by phosphorylation.

Depending on the type of transmitters and receptors, these signals can be either excitatory (at the *muscular* endplate and in the *central nervous system, CNS*, causing depolarization) or inhibitory (only in the *CNS*, causing hyperpolarization).

The transmitter compounds finally return either by high- or low-affinity transport processes to the presynaptic vesicles (e.g. catecholamines) or are enzymatically inactivated (e.g. acetylcholine, 6.3.5); their metabolites are retrieved for resynthesis.

As an example of direct transmitter gating, synaptic excitation at the *nerve-muscle synapse* is mediated by acetylcholine, which is released from presynaptic vesicles out of *synaptic buttons* at the *muscular endplate* (Fig. 17.2-8). It binds to the *postsynaptic* nicotinic acetylcholine receptor, which causes the opening of Na^+ and K^+ channels and thus changes the end-plate potential (EPP). This is converted into an action potential (17.2.2) by voltage gated Na^+-channels located in the membranes of the muscle cell. The action potential propagates to the Ca^{++} stores. Release of Ca^{++} via voltage gated channels leads to muscle contraction (17.4.5).

Integration of signals: If signals from a number of afferent nerve fibers arrive at a neuron of the CNS (usually causing differences of less than 1 mV each) they are correspondingly added or subtracted, also taking into account time and space effects (neuronal integration). Only if the accumulated input causes membrane depolarization in excess of the threshold value of −55 mV (a change of +10 mV from the resting potential), an action potential is initiated. This is in contrast to the *nerve-muscular synapse*, where simultaneous action of 300…400 release sites of the single nerve and many postsynaptic receptors cause changes of ca. +70 mV, thus always surpassing the threshold value (except in pathological conditions).

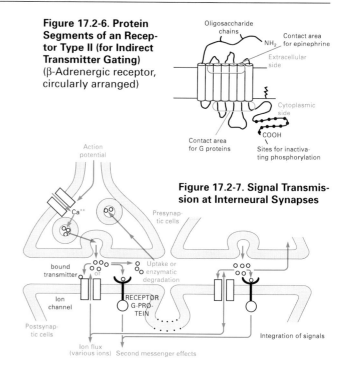

Figure 17.2-6. Protein Segments of an Receptor Type II (for Indirect Transmitter Gating) (β-Adrenergic receptor, circularly arranged)

Figure 17.2-7. Signal Transmission at Interneural Synapses

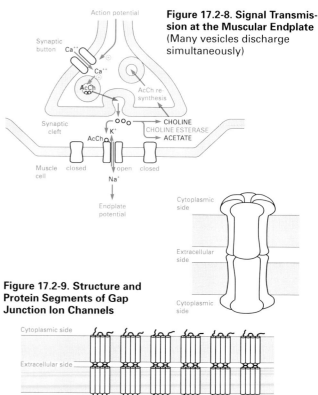

Figure 17.2-8. Signal Transmission at the Muscular Endplate (Many vesicles discharge simultaneously)

Figure 17.2-9. Structure and Protein Segments of Gap Junction Ion Channels

Table 17.2-1. Types of Synaptic Transmission (For a List of Ion Channels, see Table 18.1-1)

Type	Characteristics	Mechanism	Location
Directly gated transmission	Chemical synapses: cleft 30–50 nm, presynaptic active zones, postsynaptic receptors, unidirectional transmission. Chemical neurotransmitters: acetylcholine, GABA, glycine, glutamate and aspartate.	The released neurotransmitters act directly on ionophoric postsynaptic receptors. Synaptic delay 1-5 msec. In case of excitatory effects, Na^+ and K^+ channels are opened. The depolarization, in turn, opens neighbored voltage gated channels, producing an endplate-potential in *muscles* and excitatory effects in the *CNS*. In case of inhibitory effects, voltage-gated Cl^- channels are opened, causing hyperpolarization.	*Nerve-muscle synapse* (nicotinic acetylcholine receptor). Aminergic, cholinergic and amino acid dependent *neurons* in the *CNS*.
Indirectly gated transmission	Chemical synapses: cleft 30–50 nm, presynaptic active zones, postsynaptic receptors. Chemical neurotransmitters: norepinephrine, dopamine, serotonin, acetylcholine, neuropeptides.	The released neurotransmitters bind to specific postsynaptic receptors. This activates ion channels indirectly via heterotrimeric G-proteins, engaging second messengers. Synaptic delay usually >10 msec.	α- and β-adrenergic receptors, serotoninergic, dopaminergic, peptidergic and muscarinic acetylcholine receptors in the *CNS* (e.g., in the *cerebral cortex*).
Voltage gated transmission	Electrical synapses: cleft 3.5 nm, gap-junction channels. Modulation by phosphorylation.	The action potential causes depolarization of the presynaptic cell. The ionic current (Na^+ and Ca^{++} influx, K^+ efflux) causes postsynaptic electrical changes. No synaptic delay, rapid bidirectional transmission.	Intracellular conduction of the action potential in *neurons* requiring rapid and synchronous synaptic transmission (*neurons, heart*, also in *invertebrates*).

17.2.4 Voltage Gated Signalling at the Synapse (Fig. 17.2-9)

Voltage gated transmission is performed in the *nervous system* and the *heart* and especially in lower organisms (*invertebrates*) for fast and synchronous dissemination of the action potential.

Gap-junction channels appear as a pair of pores directly connecting the pre- and postsynaptic membranes. Both channels consist of connexins (6 separated subunits with 4 transmembrane segments each); the corresponding ones in both membranes are in close connection.

The characteristics of neuronal transmission and of neurotransmitters are compiled in Tables 17.2-1, 17.2-2 and 17.2-3.

Table 17.2-2. Selection of Neurotransmitters and Receptor Subtypes
(Some neurotransmitter compounds are also acting as hormones in the *bloodstream*. For G-protein dependent reaction mechanisms, see 17.3. Prostanoid and leukotriene receptors are dealt with in 17.4.8)

Name	Receptors (Receptor Type)	*Location* / Effects of the Neurotransmitters
Acetylcholine (6.3.5)	nicotinic receptor (I)	Binding to ionophoric receptors permeable to Na^+ and K^+ → → *nerve-muscle synapse* ('muscle type receptor'): activation of endplate potential → *CNS* ('neuronal type receptor'): excitatory and inhibitory actions
	muscarinic M_1 (neuronal), M_3 (glandular), M_5 receptors (II)	Binding to receptor linked with $G_{q\alpha}$-protein → activation of phospholipase C → formation of inositol-P_3 and diacylglycerol → → *parasympathetic system*: mediation of pre- and postganglionic synaptic transmission → *sympathetic system*: mediation of preganglionic synaptic transmission → *CNS*: predominantly excitatory transmitter, e.g., activation and maintenance of cognitive function in *brain*
	muscarinic M_2 (cardiac), M_4 receptors (II)	Binding to receptor linked with $G_{i2\alpha}$-protein[2] → inhibition of adenylate cyclase → decrease of cAMP and opening of Ca^{++} channels → *heart*: negative inotropic and chronotropic effects
Biogenic amines Norepinephrine, epinephrine (4.7.4)	adrenergic β_1, β_2, β_3 receptors (II)	Binding to β receptors linked with $G_{s\alpha}$-protein → activation of adenylate cyclase → increase of cAMP → → *target organs of the sympathetic system*: mediation of postganglionic action potential, metabolic effects (e.g., positive inotropic and chronotropic effects on the heart, relaxation of smooth muscles, vasodilatation) → *CNS*: excitatory actions, in *brain*: autonomic control of cardiovascular function, activation of cortical "arousal" behavior
	adrenergic α_1 receptor (II)	Binding to α receptors linked with $G_{q\alpha}$-protein → activation of phospholipase C → formation of inositol-P_3 → release of Ca^{++} from the *endoplasmic reticulum* → → *target organs of the sympathetic system*: metabolic effects
	adrenergic α_2 receptor (II)	Binding to receptor linked with $G_{o\alpha}$-protein[2] → inhibition of adenylate cyclase → decrease of cAMP → → *target organs of the sympathetic system*: mediation of postganglionic action potential antagonistic to β receptors, metabolic effects (e.g., vasoconstriction) → *CNS*: central regulation of blood pressure
Dopamine (4.7.3)	D_1, D_5 receptors (II)	*CNS*: binding to receptors linked with $G_{s\alpha}$-protein → activation of adenylate cyclase → increase of cAMP → excitatory cortical actions
	D_{2a} receptor (II)	*CNS*: binding to receptor linked with $G_{i\alpha}$-protein[2] → inhibition of adenylate cyclase → decrease of cAMP → inhibitory autoregulation of neuron firing rate and dopamine release
	D_{2b} receptor (II)	*CNS*: binding to receptor linked with $G_{q\alpha}$-protein → activation of phospholipase C → formation of inositol-P_3 and diacylglycerol → excitatory cortical actions
	D_3, D_4 receptors (II)	*CNS*: binding to receptor linked with $G_{q\alpha}$-protein → activation of phospholipase C → formation of inositol-P_3 and diacylglycerol → excitatory mesolimbic actions (weakly expressed)
Serotonin (5-hydroxytryptamine, 5-HT, 4.7.3)	5-HT_1 receptor (II)	*CNS*: binding to receptor linked with $G_{i\alpha}$-protein[2] → inhibition of adenylate cyclase → decrease of cAMP → inhibitory actions, e.g., sleep induction
	5-HT_2 receptor (II)	*CNS*: binding to receptor linked with $G_{q\alpha}$-protein → activation of phospholipase C → formation of inositol-P_3 and diacylglycerol → excitatory actions involved in emotional behavior
	5-HT_3 receptor (I)	*CNS*: binding to ionophoric receptor located on channel permeable to Na^+ and K^+ → different excitatory actions in emotional behavior and memory processing
Amino acid transmitters γ-Aminobutyrate (GABA, 4.2.2, major inhibitory transmitter in the CNS)	$GABA_A$ receptor (I)	*CNS*: binding to ionophoric receptor (similar to nicotinic acetylcholine receptor) → influx of Cl^- → hyperpolarization → inhibitory actions (e.g., of sensory transmission in the *spinal cord*), serotonin release, indirect dopamine release (Benzodiazepine and barbiturates increase Cl^- influx).
	$GABA_B$ receptor (II)	*CNS*: binding to receptor linked with $G_{i\alpha}$-protein[2] → inhibition of adenylate cyclase → decrease of cAMP → inhibitory actions
Glycine (4.4.2)	glycine receptor (I)	*CNS*: binding to ionophoric receptor (similar to GABA receptor) → influx of Cl^- → hyperpolarization → inhibitory action (e.g., of sensory transmission in the *spinal cord*)
Glutamate (4.2.2), major excitatory transmitter in the CNS) and aspartate (4.2.4)	NMDA (N-methyl-D-aspartate) receptor (I)	*CNS*: a) binding to ionophoric receptor located on channel permeable to Ca^{++}, K^+ and Na^+ → excitatory actions (e.g., contribution to long term potentiation for memory processing); b) activation of NO synthetase → activation of guanylate cyclase → increase of cGMP (17.7.2)
	kainate receptor[1] (I)	*CNS*: binding to ionophoric receptor located on channel permeable to K^+ and Na^+ → excitatory actions (e.g., activation of motor neurons)
	AMPA receptor (quisqualate receptor)[1,3] (I)	*CNS*: binding to ionophoric receptor located on channel permeable to K^+ and Na^+ → excitatory actions (fast depolarizing of excitatory synapses)
	$mGlu_{2,3}$ receptor (II)	*CNS*: binding to receptor linked with G-protein → activation of phospholipase C → formation of inositol P_3 and diacylglycerol → opening of channel permeable to Na^+, K^+ and Ca^{++} → excitatory actions
Neuropeptides (Some examples of this very heterogeneous group)		
Enkephalins, dynorphine and β-endorphin	δ, μ, κ opioid receptors (II)	*CNS*: binding to receptor linked with $G_{i\alpha}$-protein → inhibition of adenylate cyclase → decrease of cAMP → inhibitory action, e.g., pain inhibition, inhibition of glutamate effects and of transmitters (κ receptor: closing of Ca^{++} channels)

[1] named after agonists for the glutamate receptor used in research
[2] effects also opening of K^+ channels, closing of Ca^{++} channels
[3] α-amino-3-hydroxy-5-methylisooxazole-4-propionic acid

17.2.5 Postsynaptic Receptors

The receptors are divided in many subtypes. Agonists and antagonists perform transmitter-specific actions with different affinities to the receptors. This allows discrimination of the receptors in pharmacological and toxicological studies and development of specific pharmaceuticals. A selection of them, including synthetic compounds of high specificity is listed below.

Table 17.2-3. Agonists and Antagonists for Postsynaptic Receptors

Receptor (Selection)	Agonists (physiological agonists in bold type)	Antagonists
Acetylcholine receptors: nicotinic receptor	**acetylcholine**, nicotine	curare, α-bungarotoxin (muscle type rec.), κ-bungarotoxin (neuronal type rec.)
Muscarinic M_1 (neuronal), M_3 (glandular), M_5 receptors	**acetylcholine**, muscarine, carbachol	atropine, pirenzepine
Muscarinic M_2 (cardiac) and M_4 receptors	**acetylcholine**, muscarine, carbachol	atropine, methoctramine
Adrenergic β_1 receptor	**norepinephrine ≈ epinephrine**; isoproterenol	betaxolol, atenolol
Adrenergic β_2 receptor	**norepinephrine ≥ epinephrine**; isoproterenol	butaxamine
Adrenergic α_1 receptor	**norepinephrine, epinephrine**	
Adrenergic α_2 receptor	**norepinephrine, epinephrine**	prazosin
Dopamine receptors: D_1, D_5 receptors (II)	**dopamine**	SCH23390
D_2 receptors (II)	**dopamine**, quinpirole	(–) sulpiride, haloperidol
D_3, D_4 receptors (II)	**dopamine**	
Serotonin receptors: 5-HT_1 receptor (II)	5-carboxamidotryptamine > LSD > **serotonin**	spiperone, spiroxatrine
5-HT_2 receptor (II)	DOI, LSD, **serotonin**	ketanserin
5-HT_3 receptor (I)	**serotonin**	ondansetron
GABA receptors: $GABA_A$ receptor	**γ-aminobutyrate (GABA)**, muscimol; benzodiazepine and barbiturates effect potentiation	biculluline, picrotoxin
$GABA_B$ receptor	**γ-aminobutyrate (GABA)**, L-baclofen	2-hydroxy-S-saclofen
Glycine receptor	**glycine, β-alanine, taurine**	strychnine
Glutamate receptors: NMDA (N-methyl-D-aspartate) receptor	NMDA, **glutamate** (coagonist **glycine**)	2-amino-5-phosphono-pentanoic acid
Kainate receptor	kainate, **glutamate**	
AMPA receptor (quisqualate receptor)	AMPA, quisqualate, **glutamate**	NBQX; 2,3-benzodiazepines
$mGlu_{2,3}$ receptor (quisqualate B receptor)	1-amino-cyclopentane-1,3 dicarboxylic acid (APCD), quisqualate, **glutamate**	2-amino-3-phosphonopropionic acid
Neuropeptide receptors: δ-receptor	**Leu-enkephalin, Met-enkephalin**	naltrindole
κ-receptor	**dynorphin A**	nor-binaltorphimine
μ-receptor	**β-endorphin**, morphine, L-polamidone	CTOP

17.2.6 Axonal Transport (Table 17.2-4)

Maintenance of neuronal function requires continuous supply of metabolites from the cell body to the nerve terminal and vice versa. Axonal transport occurs along tracks formed by microtubules and by the *smooth endoplasmic reticulum* (18.1.3). Fast axonal transport is rapidly established by the microtubuli associated ATPases dynein (MAP1C) and kinesin, which act as motor molecules. Alkaloids, which disrupt microtubuli (e.g. vinblastin and colchicine) block this process immediately. In contrast to this, slow axonal transport involves physical translocation of the cytoskeleton. It provides mainly cytoskeletal elements for the nerve terminal.

Literature:

Barnard, E.A.: Trends in Pharmacol. Science 17 (1996) 305–309.
Bajjalieh, S.M., Scheller, R.M.: J. Biol. Chem. 270 (1995) 1971–1974.
Bennett, M.K., Scheller, R.H.: Ann. Rev. of Biochem. 63 (1994) 63–100.
Catteral, W.A.: Ann. Rev. of Biochem. 64 (1995) 493–531.
Ghosh, A.; Greenberg, M.E.: Science 268 (1995) 239–274.
Hucho, F. (Ed.): *Neurotransmitter Receptors. New Comprehensive Biochemistry* Vol. 24. Elsevier (1993).
Kandel, E.R. in: *Principles of Neural Science*, 3rd Edition (Eds: Kandel, Schwartz and Jessel), Elsevier (1991) 194–213; Koester, J.: *Ibid.*, 104–120.
Kennedy, M.B.: Ann. Rev. of Biochem. 63 (1994) 571–600.
Strange, P.G.: *Brain Chemistry and Brain Disorders*, 1st Ed. Oxford University Press (1992).
(various authors): Pharmacol. Reviews 45 (1994) 121–224.
Watson, S., Girdlestone, D.: *Receptors and Ion Channels Supplement*. Trends in Pharmacol. Science 17 (1996)

Table 17.2-4. Components and Characteristics of Axonal Transport

Type		Mechanism	Speed [mm/day]	Transported Components
Fast axonal transport of membrane-associated organelles	anterograde	saltatory transport along linear arrays of microtubules forming stationary tracks; motor molecules: ATPases, e.g. kinesin	200...400	synaptic vesicles and related metabolites, membrane associated proteins and lipids, neurotransmitters (i.e. aminergic transmitters), enzymes
	mitochondrial	mitochondrial components transported along central microtubuli clusters	50...100	mitochondria and mitochondria-associated proteins and lipids
	retrograde	components packed in large membrane-bound organelles. Motor molecules: dynein, MAP1C	100...200	lysosomes, growth factors (e.g. NGF)
Slow axonal transport of cytoskeleton- and cytoplasmic components	type A	slow axoplasmic flow. Components move in polymerized form with regulatory and cross-linking proteins	2...6	actin, chlathrin, spectrin, calmodulin, glycolytic enzymes
	type B	same	0.1...1	neurofilaments, microtubules

17.3 Principles of Intracellular Communication

This section deals with the intracellular communication of higher multicellular organisms, mostly *vertebrates*.

The chemical information arriving at the cell surface by hormones or cytokines (17.1) or by discharge of neurotransmitter compounds (17.2) is recognized by specific transmembrane receptor proteins. Binding of these agonists causes a structural change in the *intracellular* section of the receptors, which starts chemical reactions leading to the activation or release of second messengers. These transmit the information to the final site of action, either directly or via a sequence (cascade) of further reactions. In this case, the additional steps allow modification or integration of various signals.

As an exception, steroid, thyroid and retinoic acid receptors receive their agonists intracellularly and act directly on the final target (see below).

Receptor antagonists bind to the same site as agonists, but do not initiate the reaction sequence as agonists do.

There are several major chains of events:
- Directly transmitter gated ion channels are activated by acetylcholine, serotonin, amino acids etc. They change the intracellular ion concentration (Receptors type I, already described in 17.2.3).
- Receptors, which are coupled to heterotrimeric G-proteins (Type II, see 17.2.3 and 17.4.1) bind hormones or indirectly gating neurotransmitter compounds. They can even be activated by light.
  This leads to
  - activation of adenylate cyclase (17.4.2), including olfactory and gustatory sensing (17.4.7)
  - activation of phospholipase C (17.4.3)
    * causing activation of proteinkinase C (17.4.3)
    * causing release of inositol phosphates (17.4.4), resulting in, e.g., muscle contraction (17.4.5)
  - activation of cGMP phosphodiesterase (visual process, 17.4.6)
  - activation of phospholipase A_2 (17.4.8) etc.
- Receptors, which activate tyrosine kinase cascades, bind growth hormones, cytokines etc. They regulate transcription (17.5).
- *Nuclear* or *cytoplasmatic* receptors bind steroid or thyroid hormones or retinoic acid. They act directly as transcription factors (17.6).
- Guanylate cyclase can also be a hormone receptor. It activates various processes (17.7).

The reaction chains are interconnected in multiple ways. The general aspects of these pathways are described in Fig. 17.3-1.

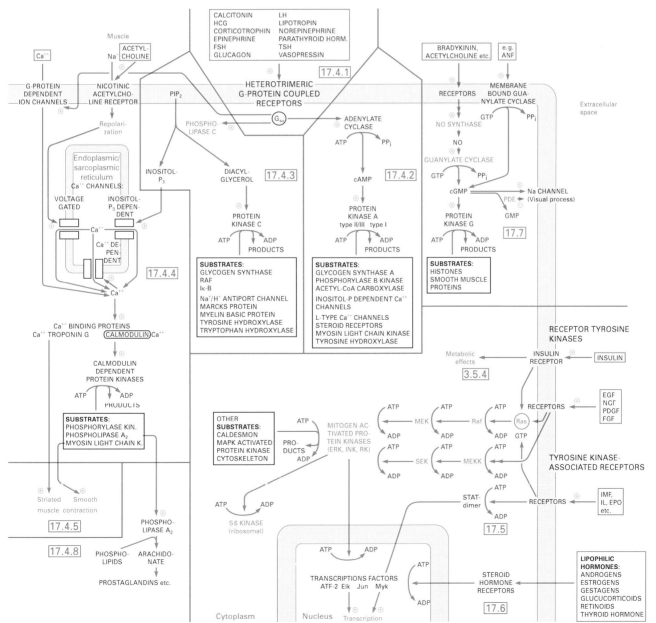

Figure 17.3-1. Receptors and Intracellular Transmission of Messages

17.4 Receptors Coupled to Heterotrimeric G-Proteins

An essential role in signal transduction through the cellular membrane *of animals* is played by receptors, which are coupled to heterotrimeric G-proteins (Table 17.4-1. They were found also in *plants*, precursors exist already in *yeast*). A common structural feature of this group of many hundred receptors are seven transmembrane passes of the glycoprotein chain ('serpentine receptors', Figure 17.2-6). Otherwise, the whole system shows a high diversity of all members.

The agonists (ligands, which specifically activate the receptor) can be small biogenic amines (e.g. histamine, 4.8.2 and epinephrine, 4.7.4) or peptides (e.g. bradykinin, 19.2.2) as well as large glycoproteins (e.g. luteinizing hormone, 17.1.5). Even light (in the visual process) can cause receptor activation (17.4.6).

The activated receptors, in turn, activate G-proteins, which as 'second messengers' *intracellularly* continue the signal transmission chain. Several different receptors may activate the same G-protein type, thus integrating their effects.

The various heterotrimeric G-proteins consist of one each of α, β and γ subunits (40...46, 35...38 and 6...9 kDa, respectively). G_α carries the GTP-binding site and is reversibly membrane anchored by myristoylation, frequently also by palmitoylation (13.3.5). G_γ is prenylated (7.3.4). The G_β and G_γ subunits are tightly, but noncovalently associated. In *mammals* there exist at least 17 genes for G_α subunits, 5 genes for β and 9 genes for γ subunits. Still more G-proteins exist due to differential splicing and post-translational modification. Although only a moderate number of preferred combinations of G proteins and receptors exist, high flexibility of signal transmission is provided.

The heterotrimeric G-proteins are a subgroup of the GTP- binding proteins (for other members of this group, see Table 17.5-2 and Fig.11.5-2).

17.4.1 Mechanism of Heterotrimeric G-Protein Action (Figure 17.4-1)

After forming a complex with the agonist, the receptor undergoes a conformation change. This allows the binding of the heterotrimeric G-protein $G_\alpha \bullet GDP$-G_β-G_γ to its intracellular domain. The contact effects structural changes in the $G_{\beta\gamma}$ subunit. This 'open state' permits the escape of GDP from the G_α subunit, temporarily causing high affinity association of the $G_{\alpha\beta\gamma}$ complex to the receptor. The consecutive binding of GTP by G_α reverses the situation: It causes the separation of G_α and $G_{\beta\gamma}$ subunits and the dissociation of both from the receptor. The receptor is then able to activate more G-proteins (signal amplification). The G subunits act as activators or inhibitors on cellular components:

- **G_α subunits (Table 17.4-1):** $G_{s\alpha} \bullet GTP$ activates the cAMP pathway via adenylate cyclase (17.4.2). On the other hand, $G_{i\alpha}$ inhibits the adenylate cyclase activity. $G_{q\alpha} \bullet GTP$ activates the protein kinase C (17.4.3) and the inositol (17.4.4) pathways via phospholipase Cβ. Still other effects are caused by other G_α subunits (17.4.6...7).
- **$G_{\beta\gamma}$ subunits (Table 17.4-2):** Their effects include activation of adenylate cyclase II and IV, phospholipase C subtypes β1, β2 and β3, phospholipase A_2 (17.4.8), K^+ and Ca^{++} channels and of Raf and Arf. These latter reactions form a link to the tyrosine kinase pathways (17.5.3). The mechanism of correlation with the G_α effects is only partially known.

After autocatalytic hydrolysis of $G_\alpha \bullet GTP$ to inactive $G_\alpha \bullet GDP$ ($t_{1/2}$ ca. 20 seconds to several minutes), the G_α subunit reassociates with $G_{\beta\gamma}$ to the inactive heterotrimeric protein, which then may enter another signal transduction cycle.

The G-protein coupled receptor activity can be terminated by feedback inactivation (desensitization) caused by the $G_{\beta\gamma}$ complex (Fig. 17.4-1, bottom left). $G_{\beta\gamma}$ binds the β-adrenergic receptor kinase (βARK) and guides it to the cellular membrane, where the kinase phosphorylates only ligand-carrying receptors at serine or threonine groups. This, in turn, decreases the receptor's affinity for G-proteins or allows the binding of inhibitory factors (e.g. arrestin), which cause inactivation of the receptor. The phosphorylated receptor undergoes endocytosis (18.1.2) into vesicles which are lacking the second messenger systems. When the ligand concentration decreases, the phosphate is removed by a phosphatase and the receptor returns to the cellular membrane.

Pathogenic reactions (Figure 17.4-1): Cholera toxin or heat labile enterotoxin of *E. coli* specifically transfers the ADP-ribose residue of NAD^+ via the ADP ribosylating factor (Arf) to the activating subunit $G_{s\alpha} \bullet GTP$. This prevents its dephosphorylation to $G_{s\alpha} \bullet GDP$ and its successive reassociation with Gβγ. Thus, $G_{s\alpha}$ permanently remains able to activate adenylate cylase. The intracellular cAMP level in *epithelial cells* of the *intestine* increases about 100 fold.

Pertussis toxin transfers the ADP-ribose residue of NAD^+ to the (normally inhibiting) subunit $G_{i\alpha} \bullet GDP$ (in the trimeric complex). This modified $G_{i\alpha}$ subunit is no longer able to inhibit adenylate cyclase. This results in an elevation of the intracellular cAMP level.

Literature:
Hepler, J.R., Gilman, A.G.: Trends in Biochem. Sci. 17 (1992) 383–387.
Hamm, H.E., Gilchrist, A.: Curr. Opin. in Cell Biology 8 (1996) 189–196.
Kissilev, O., *et al.*: Proc. Nat. Acad. Sci USA 92 (1995) 9102–9106.
Neer, E.: Cell 80 (1995) 249–257.
Premont, R.T., Inglese, J., Lefkowitz, R.J.: FASEB J. 9 (1995) 175–182.
Ray, K., *et al.*: J. Biol. Chem. 270 (1995) 21765–21771.
Reus-Domiano, S., Hamm, H.E.: FASEB J. 9 (1995) 1059–1066.
Strader, C., *et al.*: Ann. Rev. of Biochem. 63 (1994) 101–132.

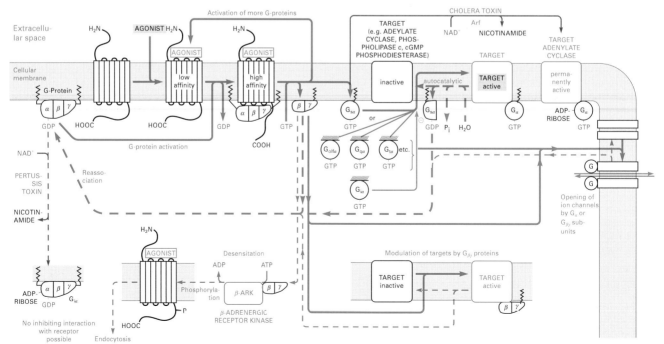

Figure 17.4-1. Activation Mechanism by G_α-Proteins

Table 17.4-1. *Mammalian* G_a Proteins (For indirectly gating nerve transmission systems, compare Table 17.2-2)

G_α Protein	Receptors (R, Examples)	Agonists (Examples)	Effects	Location
G_s Family (stimulatory G_a-proteins, 44...46 kDa, react with cholera toxin, palmitoylated at Cys)				
$G_{s\alpha}$ (Large or small splicing variants)	β-adrenergic R, ACTH R, FSH R, LH R, PTH R, TSH R.	norepinephrine, epinephrine ACTH, FSH, LH, PTH	adenylate cyclase type I and III ↑, II and IV (synergistically with $G_{\beta\gamma}$) ↑, L-type Ca^{++} channel ↑, Na^+ channel ↓	ubiquitous
$G_{olf\ \alpha}$	olfactory R.	odorants (aromatic essences)	adenylate cyclase type III ↑	*olfactory epithelium (nose)*
G_i Family [inhibitory G_a-proteins, 40...41 kDa, react with pertussis toxin (except $G_{z\alpha}$), myristoylated at N-terminal Gly, frequently also palmitoylated at Cys]				
$G_{i1\alpha}$	somatostatin R	somatostatin	adenylate cyclase ↓	ubiquitous
$G_{i2\alpha}$	acetylcholine R. (muscarinic type M_2, M_4)	acetylcholine	Ca^{++} channel ↓, K^+ channel ↑	ubiquitous
$G_{i3\alpha}$	prostaglandin EP_1 R	prostaglandin E	phospholipase A_2 ↑	nearly ubiquitous
$G_{oαA}$ } (splice variants)	Met-enkephalin R	Met-enkephalin	Ca^{++} channel ↓, K^+ channel ↑	*brain, nerves*
$G_{oαB}$	$α_2$ adrenergic R	norepinephrine, epinephrine	adenylate cyclase ↓, N- type Ca^{++} channel ↓	*CNS, sympathetic nervous system*
$G_{z\alpha}$	unknown	?	adenylate cyclase ↓, Ca^{++} channel ↓	*brain, adrenals, platelets*
$G_{t1\alpha}$ (transducin)	rhodopsin	light	cGMP phosphodiesterase γ ↑	*rods of retina*
$G_{t2\alpha}$ (transducin)	cone opsin	light	cGMP phosphodiesterase γ ↑	*cones of retina*
$G_{g\alpha}$ (gusducin)	gustatory R	food	adenylate cyclase ↓	*taste buds*
G_q Family (42...43 kDa, palmitoylated at Cys, no reaction with pertussis toxin)				
$G_{q\alpha}$ (formerly $G_{p\alpha}$)	$α_1$ adrenergic R acetylcholine R (muscarinic type M_1, M_3, M_5)	norepinephrine, epinephrine acetylcholine	phospholipase Cβ ↑ phospholipase Cβ ↑	*muscle, sympathetic nerves CNS, sympathetic and parasympathetic nerve system*
$G_{11\alpha,\ 14\alpha,\ 15\alpha,\ 16\alpha}$	serotonin R, IL-8 R, C5a R	serotonin ($G_{6\alpha}$), IL-8, C5a	phospholipase Cβ ↑	nearly ubiquitous
G_{12} Family (44 kDa, palmitoylated at Cys, no reaction with pertussis toxin)				
$G_{12\alpha,\ 13\alpha}$	?	?	phospholipase Cβ1...3 ↑	ubiquitous ($G_{13\alpha}$ in platelets)

Table 17.4-2. *Mammalian* $G_{\beta\gamma}$ Proteins (Tightly Associated Complexes)

G_β Proteins (35...38 kDa)	G_γ Proteins (6...9 kDa)	Receptors (R, Examples)	Effects	Remarks	Location
$β_1$	$γ_1$ *	rhodopsin	cGMP phosphodiesterase γ ↑	Action independent of activation by G_α	*rods of retina*
$β_1$	$γ_2$ #	β adrenergic R	adenylate cyclase type II/IV ↑	Action independent of activation by G_α	*brain, adrenals*
$β_1$	$γ_2$ #	muscarinic acetylcholine R: M_2	K^+_{IR} channel ↑ (Table 18.1-1)		*heart*
$β_1$	$γ_3$ #	somatostatin R	Ca^{++} channel (N type) ↓		*brain*
$β_2$	$γ_2$ #		adenylate cyclase type II ↑	Inhibited by G_α	*brain, adrenals*
$β_2$	$γ_3$ #		adenylate cyclase type II ↑		*brain, testes, heart*
$β_3$	not specif.		inhibits G_α activation		*brain, testes, heart*
not specif.	not specif.	$α_1$ adrenergic R	phospholipases Cβ2,3 ↑	Action independent of activation by G_α	*muscle, sympathetic nervous system*
not specif.	not specif.	$α_{2A}$ adrenergic R	tyrosine protein kinase ↑, Ras dependent (mechanism?)		*CNS, sympathetic nervous system*
not specif.	not specif.	β adrenergic R	$G_{s\alpha}$ activated adenylate cyclase type I ↓	Inhibition only by surplus of $G_{\beta\gamma}$	*ubiquitous*
not specif.	not specif.	(locate receptor-specific kinases of β-ARK fam.)	phosphorylates muscarinic and β-adrenergic receptors	Inactivation of receptors	*ubiquitous*

<u>Membrane anchored:</u> * by farnesyl at C-terminal serine, # by geranylgeranyl at C-terminal leucine
<u>Also known</u>: $β_4$, $β_5$, $γ_4$ (#, pairs with $β_1$ and $β_2$, widespread occurrence), $γ_5$ (#, pairs with $β_1$ and $β_2$, widespread occurrence), $γ_7$ (#, pairs with $β_1$ and $β_2$, widespread occurrence), $γ_8$ (#, in olfactory epithelium), $γ_{10}$ (#, pairs with $β_1$ and $β_2$), $γ_{11}$ (*, similar to $γ_1$, pairs with $β_1$, widespread occurrence)

17.4.2 cAMP Metabolism, Activation of Adenylate Cyclase and Protein Kinase A (Figure 17.4-2)

cAMP (3′,5′-cyclic AMP, adenosine 3′,5′-monophosphate) is a messenger molecule in all kingdoms of life. In *bacteria*, it signals the nutritional status of the cell (10.5.1). In *fungi*, it regulates, e.g., glycolysis, while in *higher plants*, only few regulatory functions have been demonstrated so far (e.g., Ca^{++} influx, enzyme induction during seed germination). In *vertebrates*, its major role as 'second messenger' in signal transfer is initiated by the G-protein system.

Adenylate cyclase: After leaving the receptor (17.4.1), $G_{s\alpha}$•GTP binds to adenylate cyclase and activates it to perform the reaction

$$ATP = cAMP + PP_i \qquad ΔG_0' = + 6.7\ kJ/mol.$$

On the other hand, $G_{i\alpha}$ inhibits the adenylate cyclase (17.4.1). For other inhibitors and activators see Figure 17.4-2.

The slow hydrolysis of bound GTP by G_α terminates the reaction (17.4.1). cAMP itself is hydrolyzed by 3′,5′-cyclic nucleotide phosphodiesterase (PDE) to 5′-AMP. This enzyme is activated by insulin or by calmodulin + Ca^{++} and inhibited by methylxanthines, e.g. theophylline or caffeine.

There are 8 known isoforms of *mammalian* adenylate cyclase with different properties. All forms contain 12 transmembrane domains and are activated by $G_{s\alpha}$ and by forskolin (a diterpene). Adenylate cyclase type I is also activated by calmodulin/Ca^{++} (17.4.4) but is strongly inhibited by $G_{\beta\gamma}$. Types II and IV are stimulated by $G_{\beta\gamma}$ in the presence of $G_{s\alpha}$. Types V and VI are inhibited by $G_{i\alpha}$ and by Ca^{++} (Tables 17.4-1 and 17.4-2).

Protein kinase A (PKA): cAMP activates protein kinase A, which, in turn, phosphorylates and modifies this way the activity of key enzymes in many essential metabolic pathways, both in the *cytoplasm* and in the *nucleus* (Table 17.4-3). The regulatory effects are terminated when specific phosphatases (e.g. phosphoprotein phosphatase-1, PP-1 in glycogen metabolism, 3.2.4) dephosphorylate these enzymes. Some of the phosphatases are regulated themselves (e.g., PP-1 by insulin stimulated protein kinase).

Inactive protein kinase A from *muscle* consists of 2 regulatory and 2 catalytic subunits (C_2R_2). Binding of cAMP dissociates the complex by an allosteric mechanism. The free subunits R_2 are enzymatically active. Important substrates of protein kinase A are shown in Table 17.4-3. The substrate proteins are phosphorylated at Ser or Thr residues in a consensus sequence Arg-Arg-short amino acid-(Ser/Thr)-hydrophobic amino acid.

Figure 17.4-2. Metabolism and Effects of cAMP and Protein Kinase A in *Vertebrates*

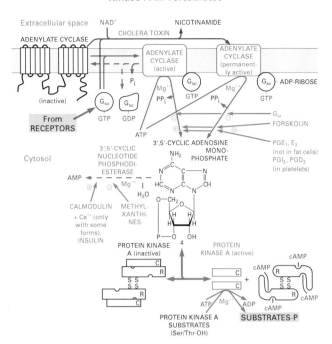

Table 17.4-3. Substrates of Protein Kinase A (Examples)

Substrates	Effect of Phosphorylation	Section
Glycogen synthase a	synthase a → b, activity ↓	3.2.4
Phosphorylase kinase	activity ↑ (if Ca^{++} present)	3.2.4
Phosphatase inhibitor protein 1	activity ↑	3.2.4
Phosphofructo-2-kinase / fructose 2,6-bisphosphatase	kinase ↓ / bisphosphatase ↑	3.1.2
Acetyl-CoA carboxylase	activity ↓	6.1.1
Triacylglycerol lipase	activity ↑	6.2.2
Cholesterol esterase	activity ↑	7.1.5
Inositol-P dependent Ca^{++} channels	close	17.4.4
Phospholamban	Ca^{++} transporting ATPase ↑	17.4.4
L-Type Ca^{++} channels (*muscles*)	open	17.2.4
Steroid receptors (*nucleus*)	transcription ↑	11.4.3
CREB (*nucleus*)	transcription ↑	11.4.3
Myosin light chain kinase	smooth muscle relaxation	17.4.5

Literature:
Fischer, E.H.: Angew. Chem. 105 (1993) 1181–1188 (Nobel Prize Lecture)
Walsh, D.A., van Patten, S.M.: FASEB J. 8 (1994) 1227–1236.

17.4.3 Activation of Phospholipase C and Protein Kinase C (Figure 17.4-3)

Phospholipases C (PLC): This family of enzymes is an important control center for intracellular signal processing. Most of these enzymes are located at the *inner surface* of the *cellular membrane* and become activated by binding of subunits of heterotrimeric G-proteins (17.4.1) in presence of Ca^{++}. The activated G-protein subunit $G_{q\alpha} \bullet GTP$ binds to the phospholipase C isoenzymes of the PLCβ group and activates them to hydrolyze membrane bound 1-phosphatidyl-1D-*myo*-inositol (4,5)-P_2 (PIP_2), yielding two second messengers:

- Inositol (1,4,5)-P_3 (IP_3) opens Ca^{++}-channels at the *sarcoplasmic* and *endoplasmic reticulum*, thus achieving an elevation of the *cytosolic* Ca^{++}- concentration (17.4.4).
- 1,2-Diacylglycerol (DAG, mostly 1-stearoyl-2 arachidonyl-*sn*-glycerol) remains in the plasma membrane and, in combination with phospholipid and Ca^{++}, activates membrane-bound protein kinase C (PKC). This enzyme is a connecting link to the signal transduction pathways mediated by tyrosine kinases (17.5).

Diacylglycerol can also arise from other sources, e.g. from phosphatidyl-choline by phospholipase C or by phospholipase D + phosphatidate phosphatase (6.3.2).

Besides $G_{q\alpha} \bullet GTP$, the G-protein subunits $G_{\beta\gamma}$ are able to activate the isoenzymes PLC-β2 and -β3 (Table 17.4-2).

Other phospholipases C: The activation of the isoenzymes PLC-γ1 and PLC-γ2 is independent of G-proteins and takes place by receptor tyrosine kinases (e.g., PDGF and NGF receptors, 17.5.3), which bind these *cytosolic* enzymes (at their SH2 groups) and phosphorylate several of their tyrosine residues. The enzyme thus becomes membrane-attached close to its substrate PIP_2 (Fig. 17.4-3, bottom left). In *muscle*, the degradation of PIP_2 dissolves PIP_2-profilin complexes this way. The liberated profilin associates with actin and disassembles actin filaments (17.4.5).

There also exist PLC-δ isoenzymes with a yet unknown activation mechanism and role. Their structures resemble PLC-β.

Protein kinases C are a large family of serine/threonine kinases (so far 12 known in *mammals*, 68...84 kDa), which phosphorylate a wide range of substrates, including numerous receptor tyrosine kinases and transcription factors (Table 17.4-4). The 'conventional' PKCα, β, γ are activated by diacylglycerol (supplied by PLC), membrane phosphatidylserine and Ca^{++}. This involves a transfer from the *cytosol* to the *membrane*. The tumor promoter phorbol is an efficient activator in place of diacylglycerol.

Other isoenzymes of PKC: The 'novel' PKCδ, ε, η, μ, σ only need diacylglycerol and phosphatidylserine for activation. It is assumed, that these PKCs are stimulated

Figure 17.4-3. Activation of Phospholipase C and Protein Kinase C

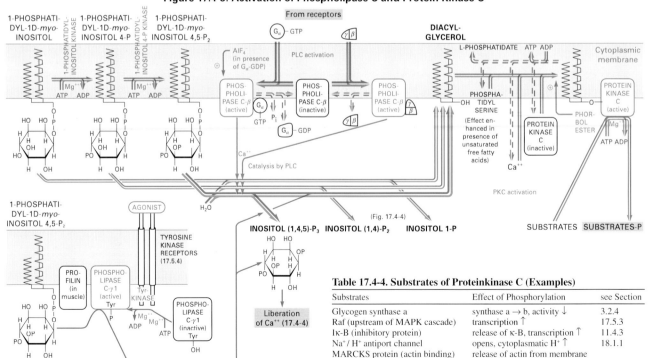

Table 17.4-4. Substrates of Proteinkinase C (Examples)

Substrates	Effect of Phosphorylation	see Section
Glycogen synthase a	synthase a → b, activity ↓	3.2.4
Raf (upstream of MAPK cascade)	transcription ↑	17.5.3
Iκ-B (inhibitory protein)	release of κ-B, transcription ↑	11.4.3
Na^+ / H^+ antiport channel	opens, cytoplasmatic H^+ ↑	18.1.1
MARCKS protein (actin binding)	release of actin from membrane	
Tyrosine hydroxylase (*brain*)	activity ↑	4.7.3
Tryptophan hydroxylase (*brain*)	activity ↑	4.7.3

by DAG originating from phospholipase D action and thus are not involved in the G-protein regulatory system. PKCε plays a role in the contraction of smooth muscles (17.4.5). During activation, both the 'conventional' and the 'novel' PKC isoenzymes are bound to the *cytoplasmic* side of membranes. Another group, the 'atypical' PKCλ and ζ isoenzymes remain soluble and do not require diacylglycerol and Ca^{++} for maximum activity. They are possibly involved in the MAPK cascade (17.5.3).

Literature:
Jaken, S.: Current Opinion in Cell Biol. 8 (1996) 168–173.
Lee, S.B., Rhee, S.G.: Current Opinion in Cell Biol. 8 (1996) 183–189.

17.4.4 Metabolic Role of Inositol Phosphates and Ca^{++}

Calcium ions play a decisive role in the regulation of many cellular processes (Table 17.4-5). Inositol phosphates have a transmitter function in one of the Ca^{++} release mechanisms.

Ca^{++} concentration in cells: The concentration in the *cytosol* of all *eukaryotic* cells is very low (ca.10^{-7} mol/l), while *extracellularly* and in the *smooth endoplasmic reticulum* (*ER*, in *muscles: sarcoplasmic reticulum, SR*) it is above 10^{-3} mol/l. This gradient is kept up by a Ca^{++}/Na^+ antiport mechanism in the *cellular membrane* and also by Ca^{++} transporting ATPases in the *cellular membrane* ① and in the *ER/SR membrane* ② of *nerve* and *muscle cells* (Fig. 17.4-4). In *plants*, Ca^{++} is accumulated in *vacuoles*.

In *slow skeletal* and in *cardiac muscles*, the pumping into the SR is positively regulated by phospholamban, which has been previously phosphorylated by PK-A or by calmodulin dependent protein kinase. Inside of the SR, Ca^{++} is sequestered by calsequestrin (55 kDa, highly acidic, > 40 Ca^{++} binding sites / molecule) and calreticulin.

Release of Ca^{++} from intracellular stores in *animals* (Fig. 17.4-4, left): In the *ER/SR membrane* of various organs there exist two types of Ca^{++} channels, which open after receiving appropriate signals and release Ca^{++} into the *cytosol*.

- Inositol (1,4,5)-P_3 receptor Ca^{++} channels (IP$_3$R, ③, e.g. in *Xenopus oocytes*): They open after binding of IP$_3$, which is generated by PLC action as described above (17.4.3).

- Ryanodine receptor (RYR) Ca^{++} channels open
  – after sensing a slight elevation of the *cytosolic* Ca^{++} concentration caused by Ca^{++} entry through voltage-gated dihydropyridine receptor Ca^{++} channels (DHPR) at the *cellular membrane*, (④a, e.g., in *cardiac muscle*) or
  – after sensing a conformation change in the DHPR Ca^{++} channels effected by depolarization of the cellular membrane (④b, e.g., in *skeletal muscle*).
  Ryanodine and dihydropyridine are compounds used in research to characterize the channel types.

- Both types of channels are present e.g. in *heart atrium, vascular smooth muscle, pituitary gland* and *neurons*.

With both channel types, the opening is promoted by the rising cytosolic Ca^{++} concentration up to a level of about 300 nmol/l, while higher concentrations are inhibitory and close the channels. This feedback mechanism causes the formation of Ca^{++} waves, which spread through the *cytosol,* into organelles (*nucleus, mitochondria*) and even into neighboring cells. Frequently, this occurs in a series of 'spikes'.

Both channel types are structurally and functionally closely related. They consist of 4 protein chains, each with 8 transmembrane domains near the C terminus (forming the Ca^{++} channel) and a long N-terminal sequence located in the *cytosol* (important for signal sensing, IP$_3$ binding site).

The emptying of the Ca^{++} stores initiates a Ca^{++} influx through the *cellular membrane*. This is effected either by a signal transfer from the receptors on the *ER/SR membrane* to Ca^{++} channels on the *cellular membrane* (⑤, 'capacitative' Ca^{++} influx) or through channels regulated by the phosphorylation product of IP$_3$, inositol (1,3,4,5)-P_4 (IP$_4$) and perhaps also by IP$_3$ itself ⑥. The Ca^{++}-transporting ATPase ② immediately transfers the Ca^{++} from the *cytosol* into the ER/SR and reconstitutes the low *cytosolic* Ca^{++} concentration.

Role of calmodulin: Calmodulin (CaM, 17 kDa) is an *eukaryotic*, highly conserved acidic protein with 4 Ca^{++}-binding sites of high affinity (10^{-6} mol/l). The Ca^{++}/CaM complex activates a large number of enzymes (Table 17.4-5, mostly CaM dependent Ser/Thr kinases). CaM can be described as a 'decoder' of the Ca^{++} information, which plays a central role in cellular regulation.

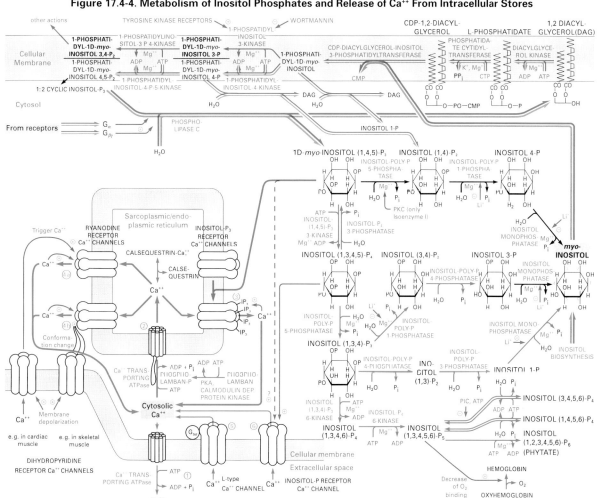

Figure 17.4-4. Metabolism of Inositol Phosphates and Release of Ca^{++} From Intracellular Stores

The 4 Ca^{++} binding sites of CaM are located in pairs in helix-loop-helix motifs ('EF hand') at both ends of the molecule. Ca^{++} binding causes a conformational change in CaM. This exposes 2 hydrophobic patches which are suitable to bind to other proteins in a regulating way, e.g. by removing autoinactivating protein segments from the active center of kinases. The Ca^{++} signal is terminated when the Ca^{++} concentration in the *cytosol* drops by export as described above. The activated, multifunctional CaM kinase II, however, autophosphorylates itself, thus extenting its activity until dephosphorylation ('molecular memory', Table 17.4-5).

Table 17.4-5. Targets Regulated by Binding of Calmodulin-Ca^{++}

Target (Examples)	Effect	Section
Phosphorylase kinase	glycogen degradation ↑	3.2.4
CaM kinase II, phosphorylates:		
Acetyl-CoA carboxylase	activity ↓	6.1.1
ATP citrate lyase		3.8.1
Glycogen synthase	activity ↓	3.2.4
Phospholipase A$_2$	activity ↑	17.4.8
N type Ca^{++} channels (*neurons*)	open, cytosolic Ca^{++} ↑	18.1.1
Inositol P$_3$ kinase	IP$_4$ + cytosolic Ca^{++} ↑	17.4.4
Ca^{++} pumping ATPases *	cytosolic Ca^{++} ↓	17.4.4
Myosin light chain kinase	smooth muscle contract.	17.4.5
CaM kinase IV, phosphorylates: CREB	transcription ↑	11.4.3
Protein phosphatases (e.g., calcineurin)	transcription ↑	17.5.4
Adenylate cyclase I (*brain*)	activated, cAMP ↑	17.4.2
cAMP phosphodiesterase	activated, cAMP ↑	17.4.2
Soluble NO synthase Type III	guanylate cyclase ↑	17.7

* Plasma membrane enzyme is directly activated, ER/SR enzymes via CaM kinase

Metabolism of inositol phosphates (Figure 17.4-4, right): The second messenger IP$_3$ loses its regulatory function within seconds by removal of the 5-phosphate group. Further dephosphorylation leads to *myo*-inositol. Different inositol phosphates are formed by phosphorylation and dephosphorylation steps. The pathways described here have been demonstrated mostly in *animals*. The metabolism in *plants* is less known.

The action of inositol polyphosphate 1-phosphatase on inositol (1,3,4)-P$_3$ or (more weakly) on inositol (1,4)-P$_2$ as well as the action of inositolmonophosphate phosphatase are inhibited by Li$^+$ ions. Possibly this is the base for treatment of manic-depressive conditions by Li$^+$.

Higher inositol phosphates: Several phosphorylation steps, starting from inositol (1,3,4)-P$_3$ lead to inositol tetra-, penta- and hexaphosphates and compounds with additional pyrophosphate groups. Inositol (1,3,4,5,6)-P$_5$ and inositol-P$_6$ form the bulk of inositol in *mammals*, but are synthesized slowly. Inositol (1,3,4,5,6)-P$_5$ decreases the O$_2$ affinity of hemoglobin in *birds* similar to 2,3-bisphosphoglycerate (5.2.3) in other animals. The details of the hexaphosphate (phytate) biosynthesis in *plants* are not fully known. The large quantities of phytate in *seeds* are a storage pool.

Reconstitution of phosphatidylinositol phosphates: Diacylglycerol is phosphorylated to L-phosphatidate and then converted to CDP-diacylglycerol. This activated compound reacts with *myo*-inositol to 1-phosphatidylinositol (6.3.2). Phosphorylation by a 4- and a 5-kinase to 1-phosphatidyl-1D-*myo*-inositol (4,5)-P$_2$ closes the inositol reaction circle (Fig. 17.4-4).

A phosphatidylinositol 3-kinase/protein kinase with a catalytic (p110) and a regulatory (p85) subunit is involved in mitogenic signalling via protein tyrosine kinase receptors (17.5.3) and receptors associated with tyrosine kinases (17.5.4), in cell motility/adherence, in vesicle trafficking from the *Golgi apparatus* to lysosomes and membranes (14.2) and in protein secretion. Some details have not been resolved yet.

Literature:
Berridge, M.J.: Nature 361 (1993) 315–325.
Carpenter, C., Cantley, L.C.: Curr. Opinion in Cell Biol. 8 (1996) 153–158.
James, P., *et al.*: Trends in Biochem. Sci. 20 (1995) 38–42.
Majerus, P.W.: Ann. Rev. Biochem. 61 (1992) 225–250.
Menniti, F.S., *et al.*: Trends in Biochem.Sci. 18 (1993) 53–56.
Miyazaki, S.: Curr. Opinion in Cell Biol. 7 (1995) 190–196.
(various authors): Adv. in Sec. Mess. Phosphoprot. Res. 30 (1995) 1–174.

17.4.5 Muscle Contraction

Although muscular contraction depends generally on an interaction between myosin and actin, controlled by the *sarcoplasmic* Ca^{++} concentration, the mechanisms differ in the various muscle types (striated muscle, cardiac muscle, smooth muscle).

Structure of striated (voluntary) muscles (Figure 17.4-5, Table 17.4-6): Striated muscles of *vertebrates* and *arthopods* contain bundles of muscle fibers. Each fiber is a long, multinucleated cell which can extend through the whole length of the muscle. It is formed by fusion of precursor cells (syncytium). Within these cells there exists an arrangement of parallel myofibril bundles. These are composed of partially overlapping thick and thin filaments containing myosin and actin, respectively, as major components. They cause the striated (striped) appearance in the microscope. The movement of these filaments relative to each other effects the muscular contraction.

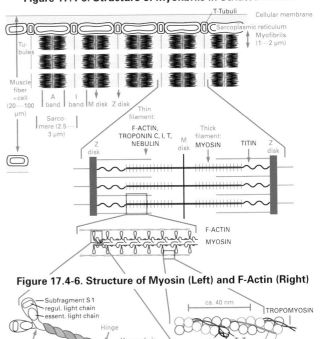

Figure 17.4-5. Structure of Myofibrils in Striated Muscle

Figure 17.4-6. Structure of Myosin (Left) and F-Actin (Right)

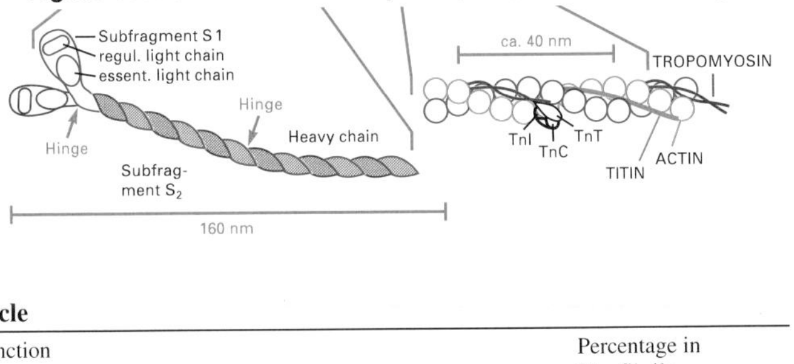

Table 17.4-6. Proteins of the Contractile System in *Vertebrate* Striated Muscle

Unit	Component	Subunits	Mol.Mass (kDa)	Structure, Function	Percentage in Myofibrils
Thick filament	myosin	heavy chain	2 * 230	Globular N-terminal head with long α-helical tail, which associates with a second chain into a coiled coil. Many coils aggregate into thick filaments. ATPase energizes muscular contraction.	44...50
		essential light chain	2 * 15...22	Associates with the globular head of heavy chain	
		regulatory light chain	2 * 15...22	Associates with the globular head of heavy chain	
	titin	heterodimer	1200/1400	Connects thick filament with Z-disk (spring function?)	9
Thin filament	F-actin	polymer of G-actin	n * 42	Polymerization with ATPase action. The polymer is helically arranged. Each monomeric unit binds 1 head of the myosin heavy chain near the cleft between the domains.	22
	tropomyosin	homodimer	2 * 33	Coiled coil of α-helices, wrapped around F-actin helix	5
	troponin	TnC	18	TnC: Ca^{++} binding, homologous to calmodulin. TnI: actin binding.	5
		TnI	23	TnT: tropomyosin-binding.	
		TnT	31	Together with tropomyosin they regulate the muscular contraction.	
	nebulin	3	3 * ca. 250	Actin-binding, coils around thin fiilament	3
M disk	C protein		150	Assists assembly of thick filaments	1
	M-protein	–	100	Assists assembly of thick filaments	1
Z disk	α-actinin	homodimer	2 * 95	Interior of Z-disk, symmetrically binds the ends of F-actin	1
	desmin		50	Peripheral zone of Z-disk	
	vimetin		52	Peripheral zone of Z-disk	

Contraction of striated muscles (Figure 17.4-7): The contraction starts upon arrival of a neuronal signal, which is transmitted in a Ca^{++} dependent way (see below). The contraction mechanism resembles a 'rowboat' movement with the following steps:

① ATP binding to the myosin head S1 forces open a cleft in its globular structure, which results in separation of actin and myosin.
② The consecutive change in myosin configuration effects hydrolysis of the myosin-bound ATP. The myosin head is activated ('cocked') and turns more towards the Z disk.
③ The myosin head binds weakly to a new actin site closer to the Z disk. Release of P$_i$ strengthens the actin-myosin binding by closing the actin cleft.
④ The 'power stroke' pulls the thick filament ca. 6 nm closer to the Z disk.
⑤ Release of ADP enables another round of ATP binding and repetition of the cycle.
⑥ ADP is rephosphorylated by phosphocreatine, which represents a large pool of energy-rich phosphates (4.9.2).

While isolated myosin has only low ATPase activity (turnover number 0.05 sec^{-1} due to slow release of the reaction products), the rate increases to 10 sec^{-1} in presence of actin. The proteins form ion pairs between Lys residues of myosin and Asp and Glu residues of actin during the respective phases of the contraction cycle. Also hydrophobic areas in both molecules contact each other. The myosin light chains stabilize the α helix in the heavy chain head.

Figure 17.4-7. Mechanism of Muscle Contraction

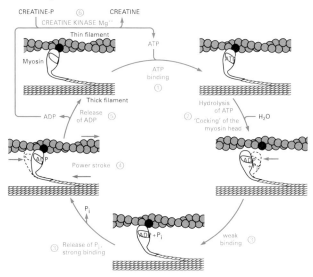

Regulation of striated muscle contraction (Figure 17.4-8): Ca^{++} plays a decisive role in muscular contraction. After the arrival of neuronal impulses at the neuromuscular endplate (17.2.3) the muscle cell depolarizes. This action potential propagates along the T-tubuli (membrane invaginations) to contact sites of the membrane with the *sarcoplasmic reticulum* (SR). The voltage-sensitive DHP-receptors of the cellular *membrane* transmit the signal to the ryanodine receptor Ca^{++} channels of the SR membrane (Fig. 17.4-4). Ca^{++} is released from this store and increases the *cytosolic* Ca^{++} concentration greatly.

The contact of actin and myosin can only take place above a threshold concentration of Ca^{++}. In this case, Ca^{++} binds to troponin C. This, in turn, causes, an allosteric interaction with tropomyosin, which moves deeper into the groove of the actin helix and uncovers the contact sites for myosin. After termination of the neuronal stimulus, Ca^{++} is pumped out of the *cytosol* again and the initial state is resumed.

Cardiac (heart) muscle: This muscle has a similar structure and contraction mechanism as voluntary *striated muscles*, although the major signals for contraction originate in the heart itself.

The ryanodine receptor Ca^{++} channels of the *SR membrane* are Ca^{++} gated, however. They respond to the small influx of Ca^{++} into the *cytosol*, which occurs after stimulation of the DHPR Ca^{++} channels by membrane depolarization (Fig. 17.4-4). The heart depends strictly on aerobic ATP synthesis via the respiratory chain (16.1.4), while *striated muscle* can provide extra ATP temporarily by glycolysis in addition to the aerobic metabolism (3.1.5).

Figure 17.4-8. A Model for Triggering of Striated Muscle Contraction by Ca^{++}

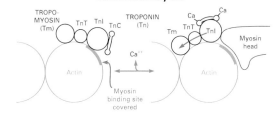

Structure of smooth muscles: The muscle cells are spindle shaped and mononuclear. The filaments do not form myofibrils and are differently (e.g., diagonally) arranged. Since they contain no troponin, other mechanisms regulate the contraction.

Smooth muscle myosin has only $1/10$ of the ATPase activity of its striated muscle counterpart. It is suited for strong, but slow contractions.

Contraction of smooth muscles: Smooth muscles are controlled by the autonomic (involuntary) nervous system. Only one of the two principles of its contraction is Ca^{++} regulated.

• Calcium-dependent contraction (Figure 17.4-9): Neuronal or hormonal stimulation of smooth muscle cells leads to an increase in *cytosolic* Ca^{++} by inflow from the *extracellular space* via voltage- or receptor-operated Ca^{++} channels (17.2.3) or from the *sarcoplasmic reticulum* via inositol-P$_3$-receptor Ca^{++}-channels (17.4.4). The Ca^{++} ions bind to calmodulin (17.4.4), which, in turn, becomes attached to myosin light chain kinase (MLCK) and activates it to phosphorylate the myosin light chain LC$_{20}$. This causes at first the aggregation of the myosin molecules and then their association with actin. The contraction cycle proceeds similarly as above.

The ability of MLCK to interact with calmodulin is abolished by phosphorylation of the enzyme. This reaction is catalyzed by protein kinase A and thus depends on the epinephrine-cAMP pathway (17.4.2). Therefore, epinephrine can relieve asthmatic brochoconstriction. A similar phosphorylation takes place by calmodulin dependent protein kinase II (not shown in Fig. 17.4-9) effecting desensitization at prolonged high Ca^{++} concentrations.

When the signals for elevation of the *cytosolic* Ca^{++} cease, Ca^{++} is pumped again out of the *cytosol* (17.4.4). The decreasing concentration causes dissociation of the Ca^{++}/calmodulin/myosin light chain kinase complex and inactivation of the kinase. The myosin light chains are dephosphorylated by myosin light chain phosphatases (e.g. PP1$_M$). Muscle relaxation takes place.

Figure 17.4-9. Regulation of Smooth Muscle Contraction

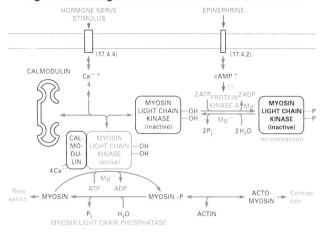

• Calcium-independent contraction: The mechanism has not yet been fully elucidated. A key role is played by Ca^{++} independent protein kinases C, especially PKCε (17.4.3). After activation by diacylglycerol (as second messenger) they either directly phosphorylate calponin or indirectly activate caldesmon via the Ras/Raf/MEK/MAPK phosphorylation cascade (17.5.3). Since both proteins are associated with thin filaments, they may play a role in smooth muscle contraction.

Nonmuscle cells: The contraction of actomyosin fibers involves myosin light chain phosphorylation and proceeds similarly as described above. Examples are the contractile ring during cell division (11.6.5), adhesion belts in *myoepithelial* and *myofibroblast cells* and cellular *microfilaments*.

Literature:
Allen, B.G., Walsh, M.G.: Trends in Biochem. Sci. 19 (1994) 362–368.
Hibbert, M.G., et al.: Ann. Rev. Biophys. Biophys.Chem. 15 (1986) 119–161.
Rayment, I., et al.: Science 261 (1993) 50–65.
(various authors): Curr. Topics in Cell. Reg. 31 (1990) 1–271.

17.4.6 Visual Process

The eyes of *vertebrates* contain photopigments in stacked membranes (discs) inside of the photoreceptor nerve cells. In the *retina* of *higher vertebrates*, both rod cells (for vision at low light levels) and cone cells (responsible for color view) are present.

Rods (Figure 17.4-10 and 17.4-11): The light sensitive rhodopsin is composed of the 7 transmembrane helix protein opsin and of 11-*cis*-retinal (7.3.2) which has an absorption spectrum corresponding to the solar spectrum with a large absorption coefficient ($\varepsilon_{500\,nm} = 4 * 10^4$ cm^{-1} * mol^{-1}, Figure 17.4-12). Light isomerizes this compound within picoseconds to *all-trans*-retinal, resulting in a series of conformation changes of the protein and leading to the photoexcited metarhodopsin II (R*).

Figure 17.4-10. Structure of Retinal Rods

Figure 17.4-11. Visual Cycle in Retinal Rods
(500 nm etc: wavelength of absorption maxima)

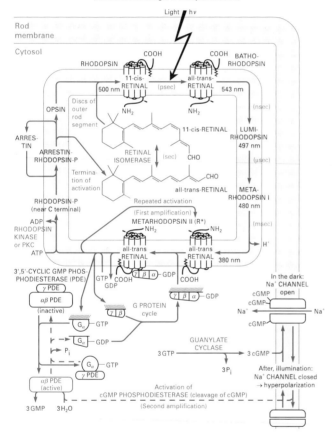

Metarhodopsin II, acting like other serpentine receptors (17.4), activates the heterotrimeric G-protein transducin ($G_{t1\alpha} \bullet G_{\beta\gamma}$). The liberated $G_{t1\alpha} \bullet$GTP binds to the inhibitory γ subunit of cGMP phosphodiesterase (PDE) and removes it. The PDE is now able to hydrolyze cGMP.

In the dark, binding of cGMP keeps the Na$^+$ channels of the *rod cells* open (Table 18.1-1) and the cells in a depolarized state. This effects the opening of voltage-gated Ca^{++} channels and a high intracellular Ca^{++} level. The hydrolysis of cGMP following illumination closes the channels and stops the Na$^+$ (and indirectly also the Ca^{++}) inflow. This leads to a hyperpolarization of these cells (see Figure 17.2-1), which changes the transmission of glutamate-mediated neuronal signals.

Reconstitution of cGMP: The GTP hydrolysis terminates the PDE activation by $G_{t1\alpha} \bullet$GTP. Guanylate cyclase (17.7.1) elevates the cGMP level again.

Reconstitution of rhodopsin: The photoexcited metarhodopsin II gets phosphorylated by rhodopsin kinase or PKC (17.4.3) and then binds to arrestin, which stops its ability to activate transducin. The Schiff base structure connecting opsin and *all-trans*-retinal is deprotonated within a few seconds, resulting in the separation of both moieties. *All-trans*-retinal is reconverted in a dark reaction to 11-*cis*-retinal and then binds again to the protein. Illumination can now initiate another reaction cycle.

Cones: The chromophore compound is likewise 11-*cis*-retinal. However, its protein partners shift the absorption maximum. The photoreceptors of *cones* absorb in the blue, green and red portion of the spectrum (Figure 17.4-12). The reaction mechanisms following illumination are apparently similar to the one in *rods*.

Figure 17.4-12. Absorption Spectra of Rhodopsin (in *Rods*) and of the Color Receptors (in *Cones*)

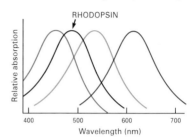

Rhodopsin is also used in the visual process of *mollusks* and *arthropods*. *Higher plants, mosses* and *algae* use the phytochrome system (with a tetrapyrrole chromophore) as light sensor for regulating metabolic processes. Phototaxis of *bacteria* and *algae* depends on retinal related or flavin sensors.

Interconversion of 13-*cis*- and *all-trans*-retinal in bacteriarhodopsin takes place during proton pumping of *halobacteria* (*archaea*, Figure 16.2-2), while *higher plants* and some *bacteria* use pyrrole derivatives for photosynthesis (16.2.1).

Literature:
Khorana, H.G.: J. Biol. Chem. 267 (1992) 1–4.
Nathans, J.: Ann. Rev. Neurosci. 10 (1987) 163–194.
Stryer, L.: Ann. Rev. Neurosci. 9 (1986) 87–119.
Wald, G.: Nature 219 (1968) 800–807 (Nobel Lecture).

17.4.7 Olfactory and Gustatory Processes

Olfactory Process: *Vertebrates* sense odors by the *olfactory epithelium* of the nose. Cilia, which project from each sensory nerve cell into the nasal cavity, carry a multitude of different G-protein coupled olfactory receptors.

Contact of an odorant with the receptor starts one of the following mechanisms:

- Adenylate cyclase is activated via $G_{olf\alpha}$ (Table 17.4-1). The cAMP produced causes opening of Na$^+$ ion channels and influx of Na$^+$ ions. This depolarizes the nerve cell and increases the frequency of action potentials.

- Phospholipase C is activated. The inositol-P$_3$ (or inositol-P$_4$) formed opens Ca^{++} channels in the *cellular membrane* (17.4.4). The Ca^{++} influx hyperpolarizes the nerve cell and decreases the frequency of action potentials.

It is yet unclear, how both mechanisms interact with each other and how different olfactory receptors respond to a particular odorant. Both mechanisms are effective only for short time intervals (ca. 50 msec). The receptor is desensitized by phosphorylation (17.4.1). The odorants are inactivated by degradation (mostly by P-450 catalyzed oxidation). The transmission of the action potentials to the brain is described in 17.2.2.

The mechanism for the hypersensitive recognition of specific odors by *insects* is still not completely known.

Literature on Olfaction:
Breer, H. (ed.): *Biology of Olfaction*. Academic Press 1994.
Farbman, A.I.: *Cell Biology of Olfaction*. Cambridge University Press 1992.
Liu, M., *et al.*: Science 266 (1994) 1348–1354.

Gustatory Process: The gustatory perception of *higher vertebrates* differentiates only sour, salty, bitter and sweet tastes. The receptors are located in the membrane of nerve cells present in the *taste buds* of the *tongue*.

It is assumed, that salty tastes increase Na$^+$ flux through respective channels, sour tastes effect a blockade of Na$^+$ or K$^+$ channels by H$^+$ ions, while bitter tastes cause a Ca^{++} release via the G-protein/PLC/IP$_3$ pathway (17.4.3) and sweet tastes increase the cAMP level via adenylate cyclase (17.4.2). In this process, apparently a heterotrimeric G-protein ($G_{g\alpha} \bullet G_{\beta\gamma}$, gustducin) is involved, which is closely related to transducin of the visual process (17.4.6). However, many details are still unknown.

Literature on the Gustatory Process:
McLaughlin, S.K. *et al.*: Nature 357 (1992) 563–569.

17.4.8 Arachidonate Metabolism and Eicosanoids

In all *mammalian* cells except *erythrocytes*, arachidonic acid (*all-cis*-5,8,11,14-eicosatetraenoic acid, 1.4.1) is converted to the eicosanoide group of compounds, which exert important hormone-like effects (Tables 17.4-7 and 17.4-8).

Although these compounds are effective in very low concentrations, due to their short lifetime they can act only on the originating cell (autocrine) or on neighboring cells (paracrine). Thus they are local mediators.

Release of arachidonic acid (Fig. 17.4-13): Arachidonic acid is present in membrane phosphatidylinositol (17.4.3), -ethanolamine and -choline (6.3.2), esterified to the 2-position of glycerol. It is released by hormone controlled enzymes:

- The cytoplasmic phospholipase A_2 (cPLA$_2$, 85 kDa) is activated by a rise in the intracellular Ca^{++} level (17.4.4) and by MAPK-catalyzed phosphorylation (17.5.3), initiated by, e.g., bradykinin (20.2.2) or angiotensin II (17.1.8).
- The non-pancreatic, secretory phospholipases A_2 (sPLA$_2$, 13…15 kDa) are secreted following a hormone stimulus (e.g. by TNF, IL-1, IL-6). They are only activated by the higher level of extracellular Ca^{++} as compared to intracellular Ca^{++}.
- Phospholipases C and D (6.3.2) can also release arachidonate.

Figure 17.4-13. Liberation and Metabolism of Arachidonic acid
(PL = phospholipid, LPL = lysophospholipid, MAPK = mitogen activated protein kinase)

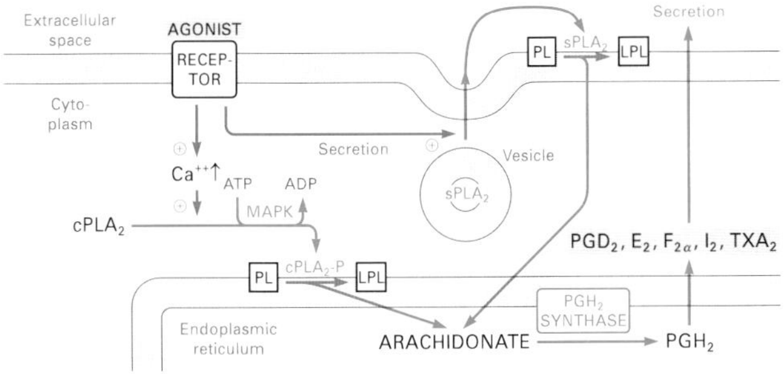

Biosynthesis of prostanoids by the cyclooxygenase pathway (Figure 17.4-15, lower right): Prostaglandin H_2 synthase (PGH$_2$ synthase, 72 kDa) produces the cyclic compound PGH$_2$. This dimeric, heme-containing enzyme is firmly anchored to the *luminal* side of the *endoplasmic reticulum* membrane (and to the *outer membrane* of the *nuclear envelope*). It possesses a cyclooxygenase ($\rightarrow$ PGG$_2$, Figure 17.4-14 left) and a peroxidase (PGG$_2 \rightarrow$ PGH$_2$) activity.

Figure 17.4-14. Conversion of Arachidonic Acid by the Cyclooxygenase Reaction (Left) and the Lipoxygenase Reaction (Right)

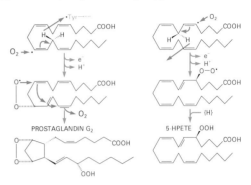

The cyclooxygenase reaction starts with abstraction of the 13-*pro*-S hydrogen, followed by attack of oxygen, rearrangements resulting in ring closure and finally reaction of a second oxygen with the resulting radical. There exist 2 isoenzymes, only one of them is inducible. The function of the enzyme is inhibited by steroidal and nonsteroidal antiinflammatory drugs. Acetylsalicylic acid (aspirin) acetylates a serine residue in the active center, while indometacin, phenylbutazone, acetaminophen etc. compete with the substrate.

The PGH$_2$ formed is the origin of the series-2 prostaglandins (with 2 double bonds in the side chains), which are synthesized on the *cytoplasmic side* of the ER membrane. In cells of the *vascular*

Table 17.4-7. Series-2 Prostanoids: Prostaglandins, Prostacyclins and Thromboxanes[1]

Name[1]		Receptor	Receptor Occurrence	Transduction System	Major Metabolic Effects in *humans*
Prostaglandin G$_2$	PGG$_2$				(unstable)
Prostaglandin H$_2$	PGH$_2$				(unstable)
Prostaglandin D$_2$	PGD$_2$	DP	platelets, vascular smooth muscle, nerves	$G_{s\alpha} \rightarrow$ cAMP$\uparrow$	Inhibition of platelet aggregation, relaxation of smooth muscle, renal vasodilatation, increased water reabsorption in the small intestine
Prostaglandin E$_2$	PGE$_2$	EP$_1$	rodent smooth muscle, kidney, lung	$Ca^{++}\uparrow$ (not via IP$_3$?)	Inhibition of platelet aggregation
		EP$_2$	smooth muscle, thymus, lung, spleen, heart, uterus	$G_{s\alpha} \rightarrow$ cAMP$\uparrow$	Potentiation of platelet aggregation, relaxation of smooth muscle, vasodilatation, secretion by epithelial cells, inhibition of mediator release by inflammatory cells, activation of sensory nerves
		EP$_3$ (splice variants)	ubiquitous: smooth muscle, adipocytes, kidney, uterus, thymus, spleen, lung, brain	a) $G_{i\alpha} \rightarrow$ cAMP$\downarrow$ b) $G_{q\alpha} \rightarrow$ IP$_3\uparrow$	Inhibition of autonomic neurotransmitter release, of lipolysis in adipocytes, of water reabsorption in renal medulla, suppression of gastric acid secretion contraction of smooth muscle, potentiation of platelet aggregation
		EP$_4$	veins	$G_{s\alpha} \rightarrow$ cAMP$\uparrow$?	?
Prostaglandin F$_{2\alpha}$	PGF$_{2\alpha}$	FP	corpus luteum, smooth muscle, kidney, lung etc.	$G_{q\alpha} \rightarrow$ PI$_3\uparrow$	Increase of uterus muscle tone, contraction of bronchial smooth muscle, inhibition of water reabsorption in the small intestine
Prostacyclin I$_2$	PGI$_2$	IP	arterial endothelium, platelets, nerves	$G_{s\alpha} \rightarrow$ cAMP$\uparrow$	Vasodilatation, inhibition of platelet aggregation, increase of water reabsorption in the small intestine
Thromboxane A$_2$	TXA$_2$	TP	vascular tissue, platelets, bronchial muscle, thymus	$G_{q\alpha} \rightarrow$ PI$_3\uparrow$	Vasoconstriction, induction of platelet aggregation and release reactions, strong contraction of bronchial smooth muscle (asthma)

[1] The series and the index numbers indicate the number of double bonds in the molecule. The series-1 prostanoids are synthesized analogously from 8,11,14-eicosatrienoic acid, the series-3 prostanoids from 5,8,11,14,17-eicosapentaenoic acid.

Table 17.4-8. Series-4 Leukotrienes and Precursors[1]

Name[1]		Receptor	Receptor Occurrence	Major Metabolic Effects in *humans*
5-hydroperoxyeicosatetraenoic acid	5-HPETE		leukocytes, mast cells, lung, spleen, brain, heart	Regulation of ion channels, histamine release, insulin and renin release
Leukotriene A$_4$	LTA$_4$		neutrophils, macrophages, monocytes, mast cells	(unstable)
Leukotriene B$_4$	LTB$_4$	LB$_4$	wide distribution	Contraction of vascular and intestinal smooth muscles, chemotaxis and adhesion of granulocytes, immunomodulation, liberation of lysosomal enzymes, initiation of inflammatory reactions, cAMP $\uparrow$
(Peptido-) Leukotriene C$_4$	LTC$_4$	LD$_4$? LC$_4$?	granulocytes, platelets, monocytes, mast cells, lung	SRS-A[2], bronchoconstriction (immediate hypersensitivity, asthma), mucus secretion, inflammatory reactions, increase of vascular permeability
(Peptido-) Leukotriene D$_4$	LTD$_4$	LD$_4$	granulocytes, platelets, monocytes, mast cells, lung	same
(Peptido-) Leukotriene E$_4$	LTE$_4$	LD$_4$?	granulocytes, lung	same

[1] The series and the index numbers indicate the number of double bonds in the molecule.
[2] Member of the slow reacting substances of anaphylaxis (SRS-A), acting at ca. 10^{-10} mol/l

endothelium, PGH$_2$ is converted by prostacyclin synthase to prostacyclin I$_2$ (PGI$_2$) and in *platelets* and *lung* by thromboxane synthase to thromboxane A$_2$ (TXA$_2$), which exert opposite effects on the circulation (Table 17.4-7). Other prostanoids are formed by isomerization or reduction reactions. The prostanoids leave the cell via a carrier-mediated transport mechanism.

The action of prostanoids takes place via specific receptors at the cell surface, which are coupled to G-proteins (17.4.1) and have the usual seven transmembrane helix-structure (Fig. 17.2-6).

The receptor mediated reactions are complex, since they may be different in various parts of the same organ or under different physiological conditions. There is also a considerable variation of effects in different species.

Biosynthesis of leukotrienes (LT) by the lipoxygenase pathway (Figure 17.4-15, left): The bifunctional 5-lipoxygenase converts arachidonic acid to 5(*S*)-hydroperoxy-6,8,11,14-eicosatetraenoic acid (5-HPETE, Figure 17.4-14 right) by its dioxygenase function and further on to a highly unstable epoxy product, leukotriene A$_4$ (LTA$_4$) by its LTA$_4$ synthase activity.

The LTA$_4$ hydrolase reaction results in the dihydroxy compound LTB$_4$. LTA$_4$ glutathione transferase (LTC$_4$ synthase) converts LTA$_4$ into LTC$_4$. After removal of the terminal glycine, LTD$_4$ is obtained.

5-Lipoxygenase is present primarily in the *cytoplasm* of *neutrophils*. It contains a non-heme, non FeS iron and is activated by Ca^{++}. The reaction is not inhibited by the above named nonsteroidal antiinflammatory drugs. The '5-lipoxygenase-activating protein' (FLAP, 18 kDa) possibly assists the activity.

Alternatively to LTA$_4$ formation, the intermediate 5-HPETE can be converted by a two-electron reduction reaction into 5-hydroxy-6,8,11,14-eicosatetraenoic acid (5-HETE). The enzyme has not been characterized yet.

There also exist 8-lipoxygenases (in *skin*), 12-lipoxygenases (mostly in *platelets*) and 15-lipoxygenases (in *granulocytes* and *epithelial cells* etc). The products of these enzymes have only modest biological effects.

Epoxygenase P-450 pathway (Figure 17.4-15, upper right): This term comprises all the various reactions which introduce a single oxygen into arachidonic acid metabolites by the various cytochrome P-450 monooxygenases (for mechanism compare 7.4.1).

5,6-, 8,9-, 11,12- or 14,15-epoxidation yield various *cis*-epoxy-eicosatrienoic acids (EPETREs) besides monohydroxy-eicosatetraenoic acids (HETEs, which are also produts of the lipoxygenase pathway, see above). This takes place in *liver, kidney, lung, eye* etc. These epoxy compounds are quickly converted by hydration to *vic*-diols (DIHETRE). Biochemical reactions of the epoxides include peptide hormone release, modulation of Ca^{++} efflux, vasodilatation, inhibition of Na$^+$/K$^+$ ATPase etc.

P-450 dependent hydroxylation reactions of many eicosanoids are also part of their degradation pathways (e.g. in *liver*, Figure 17.4-15, lower left).

Literature:
Coleman, R.A., *et al.*: Pharmacol. Reviews 46 (1994) 206–229.
Ford-Hutchinson, A.U., *et al.*: Ann. Rev. Biochem. 63 (1994) 383–417.
Smith, W.L., Fitzpatrick, F.A. in Vance, D.E., Vance, J. (Eds.): *Biochemistry of Lipids, Lipoproteins and Membranes.* New Comprehensive Biochemistry Vol. 31. Elsevier (1996) 283–339.
Waite, M.: *ibid.*, 211–236.

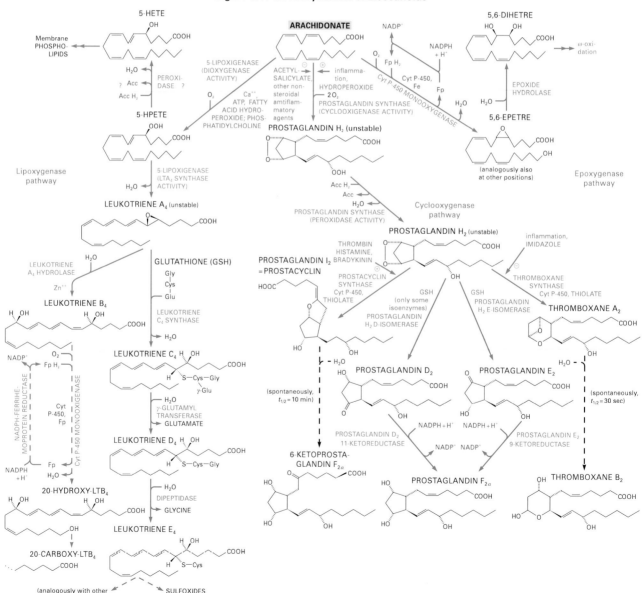

Figure 17.4-15. Biosynthesis of Eicosanoids

17.5 Receptors Acting Through Tyrosine Kinases

17.5.1 Regulatory Factors for Cell Growth and Function

Cell growth and differentiation in *multicellular animals* (*metazoans*) is regulated by a large number of extracellular polypeptide hormones (e.g. growth hormones, cytokines, 19.1.6), but also by steroids (dealt with in 17.6). Among the polypeptides (terminology of the *human* system) are:

- competence factors for transition from the resting phase G_0 to the G_1 phase or for bypassing the resting phase altogether (e.g., EGF, TGF-α, FGF, PDGF and IL-2).
- progression factors for passage through the restriction (commitment) point in the G_1 phase (11.6.2) and initiation of DNA replication and cellular division (e.g., IGF-1 or insulin in high concentrations).
- antagonists to these actions, e.g., transforming growth factor-β (TGF-β), interferons or tumor necrosis factor α (TNFα).
- Compounds, which additionally cause different effects [e.g., insulin, regulating glucose levels (3.2.4) or cytokines, controlling chemotaxis (Fig.19.1-16)].

The regulation of the cell cycle is described in 11.6. The discussion below deals with the mechanism of signal transmission from these agonists to the *nucleus*, where they primarily influence gene transcription (11.4).

The protein factors bind as ligands to receptors, which start a signal cascade via protein phosphorylation by tyrosine kinases, finally leading to the nuclear effects. This is, however, not a simple input-output sequence. Rather, there are interactions of different members of the signal cascade (17.5.2), causing amplification and modulation of the external signal. A complex interplay of phosphorylations and dephosphorylations is involved in regulation.

The downstream signal transmission mechanism of T and B cell receptors is similar to that of other receptors dealt with here. For this reason they are included in this section, although peptides presented by other cells and not hormones/cytokines activate these receptors (19.1.4, 19.1.3).

Oncogenes: Mutations in the genes of the growth hormones, the receptors or the downstream members of the signal cascade lead to malfunction of the encoded proteins, to loss of control and possibly to tumorgenesis. The pathologically modified, tumor-generating genes are therefore called oncogenes. They can be of cellular (c-oncogenes) or viral (v-oncogenes) origin. The corresponding physiological genes are named proto-oncogenes. Genes for tumorgenesis suppressing proteins are referred to as antioncogenes.

17.5.2 Components of the Signal Cascades

The essential participants of the signal cascades (in *vertebrates*, also homologues in other *eukarya* including *Drosophila* and *yeast*) are:

- membrane associated receptors, which contain an extracellular ligand binding domain, a transmembrane domain and either an intracellular tyrosine kinase domain or a binding site for tyrosine kinase (Tables 17.5-1 or 17.5-3, respectively)
- protein-tyrosine kinases (Table 17.5-4)
- protein-serine/threonine kinases (Table 17.5-5)
- protein phosphatases, counteracting the kinases (Table 17.5-5)
- small G-proteins (Table 17.5-2, different from trimeric G-proteins, 17.4)
- GDP/GTP exchange and guanosine nucleotide releasing factors (Table 17.5-2)
- transcription factors (Table 17.5-6)

Each of these protein groups is composed of several families with many members. The multitude of these factors permit a variety of cellular responses to extracellular signals. However, general principles of action exist.

There also exist a number of transmembrane protein-tyrosine phosphatases, which structurally resemble the receptor tyrosine kinases, except for the different enzymatic activity. Their role in signalling is not well understood (except CD45, 17.5.4).

17.5.3 Receptor Tyrosine Kinases (RTK)

Receptor structure (Table 17.5-1, Fig. 17.5-1): The receptors are usually monomers. After ligand binding, the receptors of *vertebrates* dimerize and undergo a conformation change, which enables their tyrosine kinase activity to autophosphorylate the intracellular receptor domains in several tyrosine residues. The different phosphorylated tyrosine residues are specific binding sites for the Src homology domains 2 (SH2) of various signalling proteins, which become activated this way. Such proteins are Grb and GAP (see below), PLC-γ1 (connecting to the PIP_2 pathway, 17.4.3), phosphatidylinositol 3-kinase (17.4.4), the protein-tyrosine kinase Src (Table 17.5-5), protein-tyrosine phosphatases (e.g., Shp = SH-PTP1, Syp), the linker Shc (Table 17.5-2) etc. In different tissues, the same agonist/receptor pair can cause different effects.

Contrary to other receptors of this group, the insulin receptor is permanently dimeric. The extracellular ligand binding domains of the insulin receptor are located on separate chains (Fig. 17.5-1). After agonist binding, the receptor undergoes a conformation change, which initiates autophosphorylation of up to 7 of the 13 tyrosine residues per β-chain. These Tyr-P residues are docking sites for the insulin receptor substrate (IRS-1, see below).

Phosphorylation of Ser and Thr residues of the insulin receptor (e.g. by protein kinase A) decreases its tyrosine kinase activity. In insulin resistant diabetes, a mutation in the tyrosine kinase domain takes place, which renders the enzyme inactive and interrupts the signal chain (see 17.1.3).

Figure 17.5-1. Structures of Some Receptor Tyrosine Kinases

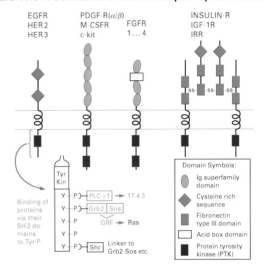

Two examples of receptor tyrosine kinase cascades are presented in detail:

EGF receptor cascade (Figure 17.5-2): The Grb2 protein binds via its SH2 domain to phosphorylated tyrosine residues of the EGF receptor and acts as a linker to the Sos1-protein (alternatively, Shc can be placed in between as an additional linker). This complex, acting as GDP/GTP exchange factor (GRF) releases GDP from Ras•GDP, thus permitting GTP association with this small G-protein. Ras•GTP binds and activates Raf by effecting a membrane contact, where Raf can be phosphorylated by a protein kinase. Raf is the first of a sequence of specific serine/threonine kinases, constituting a phosphorylation cascade (MAP kinase cascade; the MEK kinases additionally phosphorylate tyrosine). This finally leads to the phosphorylation of *nuclear* transcription factors (Elk1 etc.) and thus to modulation of gene expression (11.4.3).

Ras returns to the inactive state by dephosphorylation of the bound GTP to GDP. The intrinsic GTPase activity of Ras is accelerated by GTPase activating protein (GAP). The following members of the cascade are inactivated by membrane bound or soluble protein phosphatases, which are themselves regulated.

The protein serine/threonine phosphatases belong to the PPP family (Fe^{++}/Mn^{++} or Zn^{++} dependent) or to the PPM family (Mg^{++}/Mn^{++} dependent). The dual MAPKK activity is counteracted by the dual activity (Thr-P, Tyr-P) phosphatases of the PP2A or PP2C type.

Insulin receptor cascade (Figure 17.5-3): The insulin receptor substrate (IRS-1) binds to phosphorylated Tyr residues of the insulin receptor. This allows phosphorylation of specific Tyr residues of IRS-1, which in this way become docking sites for the SH2 domains of intracellular messengers, e.g., the adaptor protein GRF (Grb 2 + Sos 1). This activates the MAP kinase cascade

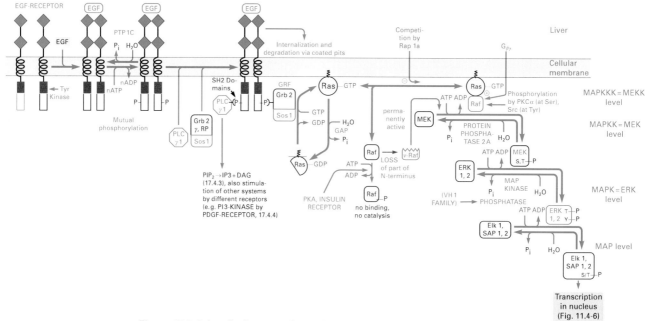

Figure 17.5-2. EGF Receptor Cascade (Terminology of the *human system*)

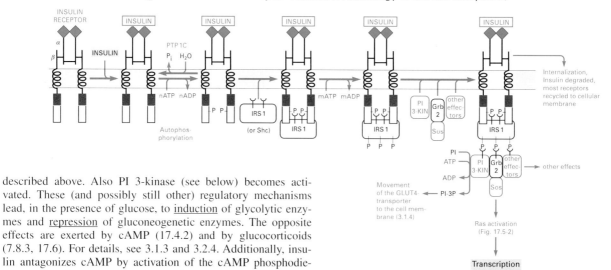

Figure 17.5-3. Insulin Receptor Cascade (Terminology of the *human system*)

described above. Also PI 3-kinase (see below) becomes activated. These (and possibly still other) regulatory mechanisms lead, in the presence of glucose, to <u>induction</u> of glycolytic enzymes and <u>repression</u> of gluconeogenetic enzymes. The opposite effects are exerted by cAMP (17.4.2) and by glucocorticoids (7.8.3, 17.6). For details, see 3.1.3 and 3.2.4. Additionally, insulin antagonizes cAMP by activation of the cAMP phosphodiesterase, thus decreasing its concentration.

Table 17.5-1. *Vertebrate* **Receptor Families With Integral Protein-Tyrosine Kinase Activity (RTK). For Structures, see Figure 17.5-1**

Important Receptor Families (14 known)	Examples for Members (R = receptor)	Ligands (GF = growth factor)	Targets of action (direct and indirect) – Effects of activated receptors	Oncogenes related to recept. or ligand genes
Epidermal growth factor (EGF) receptor family	EGFR (170 kDa), HER2 = Neu = ErbB2, HER3 = ErbB3; ErbB4	EGF, TGF-α, heparin-binding EGF-like GF	<u>EGFR</u>: Grb2 subunit of GRF (17.5.3) → Ras, Shc, Src, PLC-γ1 (17.4.3). – Mitogenic effects, influences Ca^{++} metabolism, activates transcription, cell division.	c-erbB, v-erbB encode truncated EGFR (without ligand bind. site)
Insulin receptor family	insulin R (IR), insulinlike growth factor-1 R (IGF-1R), insulin relat.R (IRR)	insulin IGF-1 ?	<u>IR</u>: IRS-1→ Raf, PI 3-kinase, possibly others. – Induces expression of glycolytic and represses expression of gluconeogenetic enzymes. Antagonizes cAMP and glucocorticoid effects. General anabolic activity. See 3.1.5.	
Platelet derived growth factor (PDGF)/macrophage colony stimulating factor (M-CSF) receptor family	PDGFR (α/β) M-CSFR, c-kit (Steel R)	PDGF M-CSF stem cell factor	<u>PDGFR</u>: Ras, Src, Shc, PLC-γ1, PI 3-kinase. – Mitogenic in smooth muscle cells, glial cells, fibroblasts. <u>MCSFR</u>.: Src, PI 3-kinase. – Differentiation of macrophages <u>c-kit</u>: PI 3-kinase. – Melanocyte and hematopoiesis	c-sis, v-sis encode homologue of PDGF c-fms, v-fms encode homol. of M-CSFR.
Fibroblast growth factor (FGF) receptor family	FGFR 1, 2, 3, 4 (4 genes + alternative splicing)	FGF (acidic, basic, keratinocyte GF, FGF4...9)	Raf-1 → ERK1+2, PLC-γ, S6 ribosomal kinase. – Growth, differentiation (mesoderm formation), survival of cells, pH↑, Ca^{++} ↑, PI turnover ↑	c-int 2 encodes homologue of FGF
Vascular endothelial cell growth factor (VEGF) receptor family	VEGFR/Flt 1 KDR/ Flk	VEGF	Mitogenic in endothelial cells	
Hepatocyte growth factor (HGF) receptor family	HGFR (extracellular + transmembr. subunits)	HGF/SF (scatter factor)	GAP, Src-related kinases, PLC-γ, PI 3-kinase. – Mitogenic in epithelial cells, stimulation of cell motility	
Neurotropin receptor family	TrkA, TrkB, TrkC LNGFR	neurotropins (NGF, NT-3, -4, -5, BDNF)	<u>Trk's</u>: Raf, PLC-γ, S6 ribosomal kinase, Tyr hydroxylase. – Growth, differentiation, survival of neurons. <u>LNGFR</u>: no enzymatic function. Role unknown	
Eph-like receptor tyrosine kinases (largest family)	Eph, Elk, Eck, Eek, Er, Cek4/HEK, Cek5	?	Cell adhesion?	
Axl receptor tyrosine kinase	Axl	?	Insulin receptor related	

Other cascades: There exist a number of parallel cascades, although many details of them are still unknown. The cascade sequences are schematically shown in Fig. 17.5-4. Ras appears to play a central role. The cascades interact in many ways, which are not presented in this Figure.

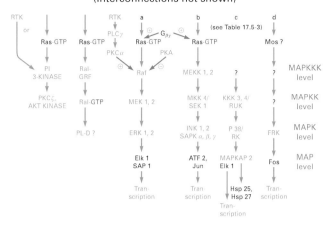

Fig. 17.5-4. Sequences in *Vertebrate* Ras Signalling Cascades
(Interconnections not shown)

Crosstalk with other pathways: Many receptor tyrosine kinases activate the phospholipase Cγ1 isoenzyme and thus connect with the inositol and the PKC regulated pathways (17.4.3, 17.4.4). Similarly, receptor tyrosine kinases activate the phosphatidylinositol 3-kinase (PI 3-K, 17.4.4), which yields mitogenic signals and is involved in vesicle trafficking, cell adherence etc. The downstream mechanism is still unclear. This activation apparently can also be performed by Ras•GTP. Furthermore, Ras•GTP can bind to and activate Ral-GRF, which promotes the formation of Ral•GTP and, in turn, the activation of phospholipase D. This initiates another sequence of lipid messenger signalling (Fig. 17.5-4).

Vice versa, Ras can be activated by $G_{\beta\gamma}$-proteins (17.4.1). Raf can be activated by protein kinase C catalyzed phosphorylation of multiple Ser and Thr residues.

Ras related small, monomeric G proteins (Table 17.5-2): The members of this superfamily resemble the heterotrimeric G-proteins (17.4), since hydrolysis of bound GTP inactivates these proteins, while replacement of GDP by GTP brings them into an active state. Individual families are Ras (containing Ras, Rap, Ral), Rac/Rho (cytoskeletal effects), Ran (transport into the *nucleus*, 14.3), Arf, Rab (vesicle transport, 14.2) etc.

17.5.4 Tyrosine Kinase-Associated Receptors (TKaR)

This group encompasses mostly cytokine receptors (19.1.6), B and T cell receptors, which are associated with tyrosine kinases.

Cytokine Receptors (Fig. 19.1-15): In many cases, these monomeric receptors undergo di- or oligomerization after ligand binding. Similarly to the RTK, this can occur by binding of dimeric ligands (e.g., CSF-1, SCF) or by binding of monomeric ligands with two binding sites (e.g., EPO). The dimerized receptors are able to bind *cytoplasmic* tyrosine kinases, which then phosphorylate themselves and the receptor. Many receptors can associate with tyrosine kinases of both the Src and the Jak families, which, however, leads to different pathways [a) and b), below].

Table 17.5-2. 'Small' G proteins, GDP/GTP Exchange Factors and Linker Proteins

Name	Properties	Action
Ras (p21, 3 isotypes)	Well conserved from *yeast* to *mammals*. Membrane-anchored by prenylation. Proto-oncogene. v-ras or c-ras oncogene products hydrolyze GTP at a slower rate.	Central role in RTK cascades. Activated by binding of GTP (effected by GRF). Activates Raf and other downstream effectors. GTPase function, enhanced by GAP.
Rap1a	Member of Ras family	GTPase, acting as antagonist to Ras in Raf binding
Rac, Rho, Ran, Arf, Rab	Other Ras-related small G-proteins (ca. 21 kDa)	
GRF (GEF, GNRP)	Guanine nucleotide releasing factor (GDP/GTP exchange factor). Subunits: Grb2 and Sos, associated by SH3 domains.	Activated by EGFR, PDGFR, CSF-1R. After binding to these receptors it exchanges GDP with GTP at Ras. Similar factors act on Ras-related proteins
GAP	GTPase activation protein	Enhances hydrolysis of Ras- bound GTP. Similar GAPs exist for other proteins
Shc	No catalytic domain. 52 or 46 kDa (different translation initiation).	Binds to and becomes phosphorylated by activated RTK or by tyrosine kinases associated with receptors or via a $G_{\beta\gamma}$ pathway. Linker function to GRF.
IRS1	No catalytic domain.	Gets highly phosphorylated by the activated insulin receptor, linker function

Table 17.5-3. 'Downstream' Protein Kinases, Counteracting Phosphatases and Transcription Factors

Level (K = kinase)	[1]	Abbreviation	Type, Properties	Phosphorylates and Activates	Counteracting Phosphatases
MAPKKK	a	Raf	Activated by direct interaction with Ras•GTP (RTK activated) and with PKC (phorbol ester activated). Inactivated by PKA-catalyzed phosphorylation	MEK1,2, Mos	PP1, PP2A ?
	b	MEKK1,2	MEK kinase. Activated by direct interaction with Ras•GTP (which was activated by stress, heat shock, UV irradiation, translational inhibition, cytokines, ceramide etc.). Also activated by anisomycin, okadaic acid (phosphatase inhibitor), phorbol esters.	MKK3,4	
(S/T)	c	?	Activated by osmolarity changes, heat shock, lipopolysaccharides (?)		
MAPKK	a	MEK1,2	ERK activating kinase MAPK kinase	ERK1,2	PP2A
	b	MKK4	MEK related kinase (= SEK 1, JNKK1)	JNK/SAPK; (p38/ RK?)	
(T/Y)	c	RKK/ MKK3,4	MEK related kinase	p38/RK	
MAPK	a	ERK1,2	Extracellular signal regulated kinase – mitogen activated protein kinase. Migrates into the *nucleus* after being phosphorylated at Thr-Glu-Tyr (= TEY).	TCF = Elk-1; SAP1,2; Ets1,2; S6K; PLA2	CL100, PAC-1 (nuclear[3])
	b	JNK1,2/p54/ SAPKα,β,γ	c-Jun N-terminal kinase/stress activated protein kinase, ERK related. Migrates into the *nucleus* after being phosphorylated at Thr-Pro-Tyr (= TPY).	c-Jun, Elk-1	CL100
	c	p38/RK/ Mpk2	Reactivating kinase. Migrates into the *nucleus* after being phosphorylated at Thr-Gly-Tyr (= TGY). Related to yeast kinase HOG-1.	Elk-1; MAPKAP kinase 2	CL100
(S/T)	d	FRK	Fos-regulating kinase	c-Fos	
MAP[2]	a	1) Elk-1; SAP-1,2; Ets-1,2; 2) S6K, 3) PLA₂	1) transcription factors: ternary complex factors (TCF, combine with SRF) = Elk-1, SAP-1,2, Ets-1,2 (Fig. 11.4-6). – 2) S6 kinase = Rsk, insulin stim. protein kinase, p90 – 3) PLA₂: formation of free arachidonic acid and lysophospholipids (17.4.8)	Rsk phosphorylates phosphatase PP1G and Fos	
	b	Jun, Elk-1, ATF2	Transcription factors. ATF2 = CREB1 (Fig. 11.4-6).		
	c	1) Elk-1, 2) MAP-KAPkinase 2	1) transcription factors Elk-1, – 2) MAPK activated protein kinase 2 (MAP-KAP kinase2), phosphorylates heat shock proteins and glycogen synthas (3.2.4)	MAPKAP kinase 2 phosphoryl. Hsp25, Hsp 27	
	d	Fos	Transcription factor (alternative phosphorylat. by S6 kinase or by ERK2 deactivates)		

[1] a, b, c, d designate different cascades, see Fig. 17.5-4. – S,T,Y indicate phosphorylation at serine, threonine, tyrosine residues, respectively.
[2] Mitogen activated proteins (transcription factors and kinases)
[3] Before phosphorylated MAPKs migrate to the *nucleus*, they can be dephosphorylated by PP2A and by a phosphotyrosine phosphatase.

Protein-tyrosine kinases: a) Receptors, which became phosphorylated by tyrosine kinases of the Src family are able to bind the adaptor molecule Shc via its SH2 domain. This acts as a linker to the Grb2/Sos protein, which is the starting point of the MAPK cascade (17.5.3, compare Fig. 17.5-2). Additionally, phosphorylation can provide binding sites for the SH2 domains of phosphatidylinositol 3-kinase (PI 3-K, 17.4.4), of hematopoietic cell phosphatase (which is a negative regulator of cytokine signalling, especially in hematopoietic pathways) and indirectly for PKC (17.4.3).

Receptor	Phosphorylated by	Receptor	Phosphorylated by
IL-2R	Fyn, Lck, Syk	G-CSFR	Lyn, Syk
IL-3R	Lyn, Tec?	GM-CSFR	Fes
IL-5R	Btk?	EPO-R	Fes
IL-6R	Btk, Hck, Tec	CD19 (B cell receptor)	Lck, Fyn, Lyn[1]
IL-7R	Fyn	CD4, CD8α (T cell receptor)	Lck[2]

[1] phosphorylate the ARAM region of the Igα/Igβ dimer (18.1.3)
[2] phosphorylates the ARAM region of CD3 (18.1.4)

b) After ligand binding and dimerization, the receptors for growth hormone, prolactin, erythropoietin, many interleukins and interferons associate at membrane-proximal sequences (boxes 1 and 2) with distinct protein-tyrosine kinases of the Jak family (Jak1...3, Tyk2). The kinases then phosphorylate both the receptor and themselves (Fig. 17.5-5). Thereafter, *cytoplasmic* STAT proteins (signal transducer and activator of transcription, so far 6 known) bind via their SH2 domains specifically to the phosphorylated receptors and are also phosphorylated. This effects their (homo- or hetero-) dimerization by mutual contact of Tyr-P and SH2 sites, release from the receptor, translocation into the *nucleus* and direct activation of transcription.

A number of two-chain cytokine receptors are composed of ligand specific α chains and common γ_c (or β_χ) chains, which associate with the same Jak after receptor dimerization (Fig. 19.1-15). This allows the combination of the intracellular messages from different external signals (redundancy).

Tyrosine phosphatases (PTP family) terminate the receptor action. They are structurally different from Ser/Thr phosphatases (17.5.3) and do not require metal ions. Catalysis is effected by Arg and Cys residues. Many of them are regulated by phosphorylation. Some tyrosine phosphatases are transmembrane proteins; their activity can be modulated by ligands (e.g. CD45, 17...5.5).

T cell receptors (Figure 17.5-6) recognize antigens, which are presented by MHC molecules (19.1.4). The receptors are associated with dimeric CD3 co-receptors accessory glycoproteins and either with the CD4 or the CD8 co-receptor. With their cytoplasmic section, CD4 and CD8 are associated with Lck (or Fyn) kinases of the Src family. In resting receptors, the kinase is kept inactivated by phosphorylation of the C-terminal tyrosine residue (catalyzed by the Csk kinase).

Upon contact of the T cell receptor with peptide-charged MHC molecules, the CD4 or CD8 co-receptors bind to monomorphic de-

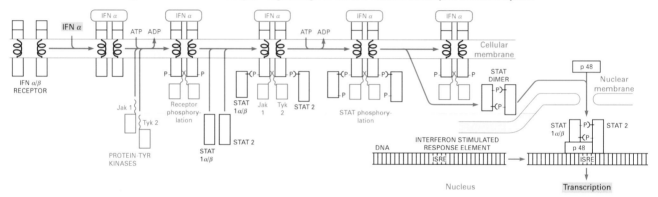

Figure 17.5-5. TKaR Receptor Signalling via STAT Proteins (Example: IFNα Receptor)

Table 17.5-4. Receptors Associated With Protein-Tyrosine Kinases (TKaR)

a) Cytokine Receptor Family. For Structures, see Figure 19.1-15

Type	Receptor Family	Group 1 Receptors	Group 2 Receptors	Group 3 Receptors
Ia	Hematopoietin rec. family	interleukin (IL)-2 R., IL-3 R., IL-4 R., IL-5 R., IL-7 R., IL-9 R.	granulocyte-macrophage colony stimulating factor (GM-CFS) R. (Fig.19.1-3)	erythropoetin R., growth hormone (GH) R., prolactin R. (Fig. 17.1-4)
Ib		IL-6 R., IL-12 R.	granulocyte colony stimulating factor (G-CSF) R. (Fig.19.1-3)	oncostatin M R., leukemia inhibitory factor (LIF) R.
II	Interferon rec. fam.	interferon R.; IL-10 R.		
III	TNF rec. family			tumor necrosis factor (TNF) R.; nerve growth factor (NGF) R.
IV	Ig superfamily rec.	IL-1 R.		

b) T and B Cell Receptors. For Structures, see Figure 19.1-9 and 19.1-8

Receptor	Co-receptor, associated with protein kinase	Co-receptor, associated with phosphatase, which activates this kinase	Target of phosphorylation	Ligands (kinases) docking at phosphorylated target
T cell R	CD4 + Lck (Lyn, Fys)	CD45	CD3ζ	ZAP-70, binds PLCγ1 and Grb2-Sos
T cell R	CD8 + Lck (Lyn, Fys)	CD45	CD3ζ	ZAP-70, binds PLCγ1 and Grb2-Sos
B cell R	CD19 + Src-type	CD45	Igα, Igβ	Syk, binds PLCγ1 and Grb2-Sos

Table 17.5-5. Protein-Tyrosine Kinases Associated With Receptors

Kinase Family	Members	Mol.Mass (kDa)	Properties
Src family	Blk, Csk, Fgr, Fyn, Hck, Lck, Lyn, Src, Yes, Yrk Oncogene variants of Fes (v-, c-), Fgr (c-), Src (v- in Rous sarcoma virus), Yes (c-) are known.	53...64	SH2 and SH3 domains, kinase domain. Membrane anchored via myristoylation. They are activated by Tyr-kinase associated receptors and also by receptor tyrosine kinases. Lck and Fyn kinase are deactivated by phosphorylation (Csk kinase) and activated by dephosphorylation (CD45 phosphatase) in T and B lymphocytes.
Jak family	Jak1...3, Tyk2, hopscotch (*Drosophila*)	ca. 130	Tyrosine kinase- and tyrosine kinase-like domain, no SH2 and SH3 domains. They are directly activated by IFN receptors and phosphorylate them. This allows binding of STAT proteins to the receptor and their phosphorylation.
Syk family	Syk, ZAP-70	72, 70	2 SH2 domains, not myristoylated, intermediates in B and T cell receptor cascades
Other	Abl, Arg, Btk, Fak, Fer, Fes, Tec, Tsk Oncogene variants of Abl (v-, bcr- in leukemia)	50...150	*Nuclear* Abl promotes the phosphorylation of the CTD domain of RNA Pol II (11.3.2). Its activity is regulated during the cell cycle by Rb protein (11.6.4)

terminants of the MHC molecules class II or I, respectively, and move the associated Src-type kinase closer to the receptor. The CD45 molecule (which is also associated with the T-cell receptor) contains phosphatase domains in its cytoplasmic section, which upon receptor stimulation dephosphorylate and thus activate the Src-type kinase. This kinase, in turn, phosphorylates tyrosine residues in the ARAM regions (antigen-recognition activation motif) of CD3 co-receptors. These phosphate groups are docking sites for the SH2 domains of downstream effectors:

- After binding, the Grb2-Sos complex (= GRF) starts the MAPK cascade (as described in 17.5.3) leading to the AP-1 transcription complex (11.4.3), which, in cooperation with NF-ATp (see below) and NF-κB, activates transcription of the IL-2 gene. The secreted IL-2 has growth promoting and proliferative properties (Fig. 19.1-16).

- After binding to the CD3ζ co-receptor, the protein kinase ZAP-70 becomes activated by phosphorylation and, in turn, phosphorylates and activates PLCγ1, which leads to PIP_2 cleavage, Ca^{++} release and formation of the Ca^{++}-calmodulin complex (17.4.3, 17.4.4). This complex attaches itself to the calcineurin dimer (PP2B) and enables it to dephosphorylate and activate the transcription factor NF-ATp (nuclear factor of activated T cells, preformed). This factor enters the *nucleus* and promotes transcription of the IL-2 gene (and other genes). Some immunosuppressors (e.g. cyclosporin A after binding to cyclophilin, a peptidyl-prolyl-*cis-trans* isomerase) interfere with this pathway by blocking the phosphatase activity of calcineurin.

Diacylglycerol (DAG), the other product of PLCγ1 catalysis, activates protein kinase C (PKC, 17.4.3), which, among other effects, activates Raf via phosphorylation. This also initiates the MEK cascade leading to transcription of the IL-2 gene (17.5.3).

Additional co-stimulatory signals are required to perform the transcription (19.1.4).

B cell receptors (Figure 19.1-8): The antigen binding receptor has likewise a complex structure. The receptor is a membrane-bound immunoglobulin. The co-receptor CD19 is in close contact with the glycoprotein CD21 (complement receptor 2, Table 19.2-2) and with Src-family tyrosine kinases (Lyn, Fyn, Lck). Upon contact of the receptors with antigen, activation of the tyrosine kinases via CD 45-catalyzed dephosphorylation takes place (analogously to the T cell receptor mechanism). The phosphorylation targets are the heterodimeric Igα-Igβ co-receptors, which act as signal transducers (instead of CD3 in the T cell receptor complex).

The further steps are analogous to the PLCγl pathway in T cell receptors (Table 17.5-4b, the role of Zap-70 is taken by Syk). As with the T cell receptors, costimulatory signals are required for transcription (Fig. 19.1-8).

Literature:
Cano, E., Mahadevan, L.C.: Trends in Biochem. Sci. 20 (1995) 117–122.
Cohen, G.B., Ren, R., Baltimore, D.: Cell 80 (1995) 237–248.
Fantl, W.J., *et al.*: Ann. Rev. Biochem 62 (1993) 453–481.
Howe, L.R., Weiss, A. : Trends in Biochem. Sci. 20 (1995) 59–64.
Ihle, J.N., *et al.*: Trends in Biochem. Sci. 19 (1994) 222–227.
Janeway, C.A. jr.: Ann. Rev. Immunol. 10 (1992) 645–674.
Karin M., Hunter, T.: Current Biology 5 (1995) 747–775
Macara, I.G., *et al.*: FASEB J. 10 (1996) 625–630.
Stahl, N., *et al.*: Science 267 (1995) 1949–1953.
Taniguchi, T.: Science 268 (1995) 251–255.
Treisman, R.: Current Opin. in Cell Biol. 8 (1996) 205–215.
van der Geer, P., *et al.*: Ann. Rev. Cell Biol. 10 (1994) 251–337.
(various authors): Trends in Biochem. Sci. 19 (1994) 439–513.
White, M.F., Kahn, C.R.: J. Biol. Chem. 269 (1994) 1–4.

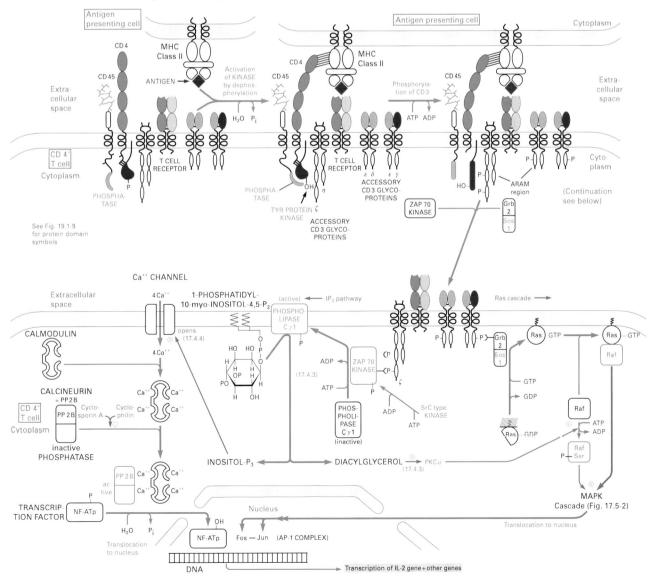

Figure 17.5-6. Signalling Pathways After Receptor Activation in CD4⁺ T Cells

Table 17.5-6. Some Specific Transcription Factors (see also Table 11.4-2 and Figure 11.4-6)

Name	Function	Characteristics	Oncogenes related to the encoding genes
Elk-1, SAP-1	Binds with SRF as ternary complex factor (TCF) to serum response element (Fig. 11.4-6)	member of Ets family	
Ets	Binds as a monomer to the Ets motif (C/AGGAA/T), binds to the TCRα enhancer, possibly involved in general transcription on TATA-less core promoters (11.4.1)	helix-loop-helix (Table 11.4-1)	v-ets (avian retroviral oncogene)
Fos	Binds heterodimeric with Jun to AP-1 binding site (Fig.11.4-6)	leucine zipper (Table 11.4-1)	c-fos, v-fos (retroviral oncogene)
Jun	Binds homodimeric (or heterodimeric with Fos) as AP-1 to the AP-1 binding site (Fig.11.4-6)	leucine zipper (Table 11.4-1)	c-jun, v-jun (retroviral oncogene)
Myc	Binds to Myc control site, induction of cell division	helix-loop-helix (Table 11.4-1)	c- and v-myc (lead to cell immortalization)
STAT (1...4)	Bind to phosphorylated IFN receptors, dimerize and translocate to the *nucleus* after being phosphorylated by Jak kinases	Mol. mass (kDa): 1a: 91; 1b: 81; 2: 113; 3 ?; 4 ?	

17.6 Receptors for Steroid and Thyroid Hormones, for Retinoids and Vitamin D

Although steroid (7.4) and thyroid (4.7.5) hormones, retinoids (7.3.2) and vitamin D (9.11) are structurally very different, they are all small, hydrophobic molecules.

For transport in blood, this type of hormones has to be bound to protein carriers (Fig.17.1-4). They remain in blood for extended periods (hours to days). Thus, their action is long-lasting.

The hydrophobicity of the hormones enables them to diffuse through the *cytoplasmatic membrane*. Contrary to other hormones, they bind and activate their cognate receptors intracellularly. The complexes, in turn, bind to the DNA hormone response elements (HRE), where they act directly as enhancers (or sometimes repressors) of transcription (11.4.3).

The receptors belong to the superfamily of nuclear (or steroid) receptors. They share a common structure principle (Fig. 17.6-1).

Generally, these receptors consist of a C-terminal hormone binding and dimerization domain (E) which is connected via a hinge region (D) with a DNA binding domain (C, containing 2 Cys$_2$-Cys$_2$ zinc fingers, Table 11.4-1) and a N-terminal transcription activating domain (A/B). The C (and a good portion of the E) domains are highly conserved, while the A/B domains differ greatly between receptors. Two receptor groups are to be distinguished:
- A group receptors (Table 17.6-1, Fig. 17.6-2): In absence of a hormone, an inhibitory protein (e.g. the heat shock protein Hsp90, possibly also Hsp70 and Hsp56) binds to the receptor and covers the DNA binding/dimerization domain. (Alternatively, intramolecular blocking of the active sites in absence of hormones has been proposed). Hormone binding causes a conformation change of the receptor, which effects the dissociation of the inhibitory protein and enables the formation of homodimers, which enter the *nucleus*. There, high affinity binding of the receptor to the palindromic DNA response element, interaction of its A/B site with TFIIB and accelerated formation of the pre-initiation complex of transcription (11.3.2) take place.
- B group receptors (Table 17.6-1, Fig. 17.6-2): They are not associated with Hsp's and bind to DNA even in absence of a ligand. In this state, the receptors interfere with the formation of a pre-initiation complex and cause transcriptional silencing. The conformation change occurring after hormone binding reverses the effect into gene activation. Frequently, the activated receptors have to cooperate with additional protein factors for modulation of transcription (e.g. AP-1, 11.4.3). Possibly, receptor phosphorylation also plays a regulatory role.

In many cases, the activated receptors induce only the transcription of so-called primary response genes. Some of the expressed proteins then initiate the transcription of secondary response genes, while others repress the further transcription of the primary response genes and thus limit their activity.

Table 17.6-1. Superfamily of *Nuclear* Receptors (There are Many Isoforms and Additionally > 40 'Orphan' Receptors With Unknown Ligands)

Group		Name (R = Receptor)	Binds to[1]	Effects, see Chapter
A (homo-dimeric)	GR	glucocorticoid R	AGAACA	7.8.3, 11.4.3, 17.1.5
	MR	mineralocorticoid R	AGAACA	7.8.3, 17.1.8
	AR	androgen R	AGAACA	7.6.3, 17.1.5
	PR	progesterone R	AGAACA	7.5.2, 17.1.5
	ER	estrogen R[2]	AGGTCA	7.7.3, 11.4.3, 17.1.5
B (hetero-dimeric)	TR	thyroid hormone R	AGGTCA	11.4.3, 17.1.5
	RAR	*all-trans*-retinoic acid R	AGGTCA	7.3.2
	RXR	9-*cis*-retinoic acid R	AGGTCA	7.3.2
	VDR	vitamin D R	AGGTCA	11.4.3, 17.1.7
	EcR	ecdysone R (*insects*)[3]	AGGTCA	7.2.3

[1] DNA half site. The receptor dimers bind to palindromic sequences or to direct repeats with various length of spacers
[2] Specific receptor, binds only estrogens (Other A group receptors are less selective)
[3] related receptor: ultraspiracle R, preferred dimerizing partner to EcR

Figure 17.6-1. General Structure of *Nuclear* Receptors

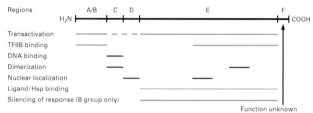

Figure 17.6-2. A Model for the Activation Mechanism of *Nuclear* Receptors (Structure is Schematized)

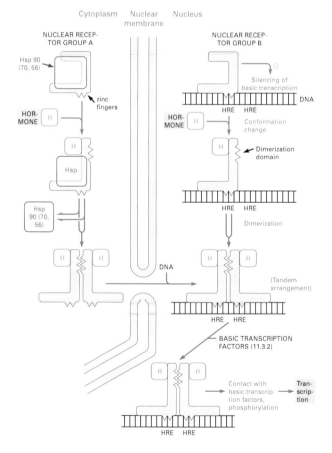

Literature:
Evans, R.M.: Science 240 (1988) 889–895.
Leid, M., *et al.*: Trends in Biochem. Sci. 17 (1992) 427–433.
Tsai, M.J., O'Malley, B.W.: Ann. Rev. of Biochem. 63 (1994) 451–486.
(various authors) In: Cooke, B.A., *et al.* (Eds.) *Hormones and Their Actions* Part I. *New Comprehensive Biochemistry* Vol. 18A. Elsevier (1988).

17.7 Cyclic GMP Dependent Pathways and Effects of Nitric Oxide (NO)

In several *vertebrate* organs, hormones activate guanylate cyclases (GC) to synthesize 3',5'-cyclic GMP (cGMP) according to the reaction

$$GTP = cGMP + PP_i.$$

Guanylate cyclase also plays a role in *lower animals* (e.g., chemoattraction of *amoebae*). There are indications of cGMP occurence in *plants*, *archaea* and *bacteria*.

cGMP, as a second messenger, acts directly on targets or activates protein kinase G (PKG), which, in turn, phosphorylates various proteins and thus modifies their actions.

17.7.1 Membrane Bound Guanylate Cyclases (Figure 17.7-1)

These enzymes are monomeric glycoproteins with different extracellular ligand-binding domains. They act as receptors.

- GC-A and -B in *brain, heart, smooth muscles* and *kidneys* can be stimulated by the atrial natriuretic factor (ANF, 17.1.8). Binding of this ligand causes phosphorylation and activation of the enzyme. The formed cGMP effects (via PKG) Na^+ and water secretion by the *kidneys* and vasodilation. ATP or ADP potentiate this effect.
- GC-C in the *intestinal mucosa* is stimulatable by the peptide guanylin (15 aa) and by structurally similar *bacterial* (*E.coli*) enterotoxins. This leads via cGMP and PKG to changes in ion transport mechanisms, e.g. resulting in enterotoxin-caused diarrhea.
- In *retinal rods*, GC sustains the cGMP level, which keeps the Na^+ channels open (17.4.6).

Since in the visual process, cGMP is hydrolyzed and the ion channels close, the intracellular Ca^{++} concentration decreases. The Ca^{++} binding protein recoverin (23 kDa), acting as a sensor, changes its conformation at low Ca^{++} levels and becomes able to activate GC. The cGMP level increases and opens the ion channels again (recovery phase).

17.7.2 Soluble Guanylate Cyclases and Their Activation by Nitric Oxide (NO)

NO metabolism: The NO• radical is a gas, which diffuses freely through membranes and is quickly oxidized ($t_{1/2} = 5...30$ sec). Therefore it acts only for short distances.

NO is synthesized in *vertebrates* by NO synthases from arginine in an unusual 5-electron reaction (Fig.17.7-2).

All NO synthases are homodimers and contain FAD, FMN, tetrahydrobiopterin and heme (protoporphyrin IX). 3 isoforms of NO synthase are known (Table 17.7-1). The NO formed by the constitutive type I and III enzymes mainly activates soluble guanylate cyclase. The major effect of the NO synthesized by the inducible type II enzyme is a cytotoxic activity.

The receptors, which activate the NO synthases type I and III are stimulated by neurotransmitters (via an increase of intracellular Ca^{++}). Even shear forces in the bloodstream are effective.

The synthesis of NO synthase type II is induced by bacterial lipopolysaccharides and cytokines (IFNγ). The resulting NO destroys bacteria and tumor cells and is thus, part of the cellular defense system (19.1.7).

Figure 17.7-2. Synthesis of Nitric Oxide

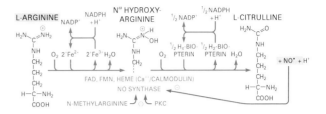

Nitroglycerol and other nitro drugs are metabolized to NO• and thus mimic its physiological effects. Similar effects are ascribed to carbon monoxide (CO) and to the hydroxyl radical (OH•, 4.5.8), which also activate guanylate cyclase.

Soluble guanylate cyclases (GC-S, Fig. 17.7-3): These enzymes are heterodimers from α_1 and β_1 or β_2 subunits. The enzymes contain heme as prosthetic group. Binding of NO activates the enzyme (contrary to most other heme containing enzymes).

cGMP formed this way activates protein kinase G (17.7.3). It also regulates some cAMP phosphodiesterases (17.4.2).

17.7.3 Protein Kinase G (PKG, Figure 17.7-1)

cGMP (formed by membrane bound or soluble GC) can activate protein kinase G.

This enzyme is a serine-threonine kinase (ca. 80 kDa). It contains a regulatory (R) and a catalytic (C) domain on each peptide chain of the homodimer. The structure of both domains resembles the 2 subunits of PKA (17.4.2, which, however, are located on different peptide chains). Apparently the binding of cGMP causes a conformation change, which removes the block from the active domain. High activities of PKG are found in the *smooth muscles* and the *brain* of *mammals*.

Phosphorylation by this enzyme
- stimulates the ion pumps which keep the intracellular Ca^{++} concentration low, causing *muscle* relaxation and vasodilatation
- keeps Na^+ channels open, resulting in Na^+ and water secretion and in neurological effects.

Literature:
Bredt, D.S., Snyder, S.H.: Ann. Rev. Biochem. 63 (1994) 175–195.
Chinkers, M., Garbers, D.L.: Ann. Rev. Biochem. 60 (1991) 553–575.
Ishii, K., *et al.*: J.Appl. Cardiology 4 (1989) 505–512.
Jeffrey, S.R., Snyder, S.H.: Ann. Rev. Cell Dev.Biol. 11 (1995) 417–440.
Tremblay, J., *et al.*: Adv. in Sec. Mess. Phosphoprot. Res. 22 (1988) 319–383.
Schmidt, H.H.H.W.: FEBS Letters 307 (1992) 102–107.

Table 17.7-1. Types of NO Synthase

NO synthase type	Regulation		Occurrence	Effects of formed NO
nNO synthase (Type I)	soluble	Activated by norepinephrine, acetylcholine, vasopressin, oxytocin, cytokines, glutamate via Ca^{++}/calmodulin	a) *CNS* (postsynaptic)	a) Retrograde diffusion → activation of presynaptic GC → modulation of signal transmission. Possibly involved in brain development and learning, but also in stroke damage and neurodegenerative disorders.
			b) *periph. neurons* (presynaptic)	b) activation of postsynaptic GC → muscle relaxation
eNO synthase (Type III)	membr. bound	Activated by hormones via Ca^{++}/calmodulin, shear forces	*endothelial cells* of *blood vessels*	Activation of GC in underlying muscle → relaxation and vasodilation. Diffusion to *platelets* → decrease of adhesion and aggregation.
iNO synthase (Type II)	soluble	Synthesis induced by bacterial lipopolysaccharides and by cytokines. Not stimulable by Ca^{++} (permanently saturated)	*macrophages, hepatocytes* etc.	Diffusion to neighboring tissues and cells → blocks Fe centers in the respiratory chain, aconitase, ribonucleotide reductase, causes DNA damage, acts bactericidal (19.1.7) and tumoricidal. Misregulation leads to autoimmune diseases, septic shock etc.

Figure 17.7-1. cGMP Formation and Protein Kinase G Activation by Membrane Bound Guanylate Cyclases

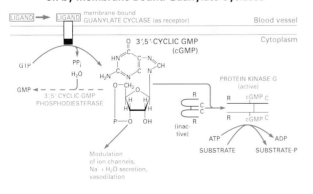

Figure 17.7-3. Vasodilatory and Antiaggregatory Effects of NO

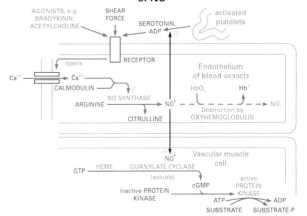

18 Eukaryotic Transport

18.1 Systems of Eukaryotic Membrane Passage

The *eukaryotic* cellular membrane is a permeability barrier to almost all molecules. The import of metabolites, nutrients, inorganic ions etc. into the cells and their export proceed through special structures. Many of the systems are described in detail in other contexts (quoted below). This section is a summary of the principles involved.

The important lipoprotein transport mechanisms are discussed separately in Section 18.2, while the different mechanisms of protein transport are the topic of Chapter 14.

Energy requirements (for details see 1.5.3): If an uncharged compound A is to be imported into a cell against a concentration gradient ($[A_{inside}] > [A_{outside}]$, Eq. [1.5-9] and [1.5-9a]), expenditure of free energy

$$\Delta G \,[kJ * mol^{-1}] = R * T * 2.303 * \log \frac{[A_{inside}]}{[A_{outside}]} = 5.706 * \log \frac{[A_{inside}]}{[A_{outside}]} \text{ (at 25 °C)}$$

is required. In the opposite direction, the same amount of free energy can be theoretically gained (it is lower under actual conditions). For the import of charged molecules the electric potential $\Delta\Psi$ also has to be considered (Eq. [1.5-12]):

$$\Delta G \,[kJ * mol^{-1}] = 5.706 * \log \frac{[A_{inside}]}{[A_{outside}]} + Z * 0.0965 * \Delta\Psi \,[mV] \text{ (at 25 °C)}$$

where Z = charge number of the ion. The membrane potential $\Delta\Psi$ of the cell at equilibrium can be calculated with the Goldman equation (17.2.1).

18.1.1 Channels and Transporters

1. Ion channels (Table 18.1-1) regulate the membrane permeability. They allow the flow of ions from higher to lower concentrations and therefore do not require any input of energy (passive transport).

   Many of them are used in a controlled way for conduction of electric signals or for initiation of metabolic functions (regulatory effects, Chapter 17), others for compensation of osmotic or charge dysequilibria.

2. Passive transport of uncharged molecules is best exemplified by the glucose transporter. There are known 5 different variants in *humans* (GLUT1...5, see 3.1.4, which differ in their tissue distribution and responsiveness to insulin.

   The protein contains 12 transmembrane domains. Glucose transport is effected by a conformation change. In *liver* and *pancreatic β-cells*, the intracellular glucose concentration is additionally kept low by phosphorylation (facilitated diffusion, compare 15.3). In presence of insulin, the variant GLUT4 (in *muscle* and *adipose tissue*, 3.1.4) is moved from *membraneous vesicles* to the cell surface and thus put into action. When the insulin concentration drops, endocytosis takes place.

   Unspecific pores are the gap junctions, which allow both the exchange of ions (17.2.4) and of uncharged molecules (e.g. in embryonal development). In *plants*, plasmodesmata (12.3.1) enable the exchange of many cellular contents.

Table 18.1-1. Passive Ion Channels (Selection). Classification according to Trends in Pharmacol. Sci. – Receptor and Ion Channels, Supplement 1996

Ion specificity	Channel types	Examples, Function	Occurrence	Structure §	Section
K^+ channels	Voltage gated	K_A: rapid, transient activation, opened during membrane depolarization, outward flow → repolarization of membranes	*neurons, secretory cells, muscles, heart*	Figs. 17.2-2, -3 4 * (6 TM + P)#	17.2.2
		K_V: delayed activation, opened during membrane depolarization, outward flow → repolarization of membranes, 'delayed rectifier'. Inactivated near the equilibrium potential.	*same*		
		K_{IR}: closed during membrane depolarization, opened near the K^+ equilibrium potential, inward flow → stabilization of resting potential, 'inward rectifier'. Often also regulated by $G_{\beta\gamma}$-proteins.	*same*	4 * (2 TM + P)#	
		K_{SR}: strong voltage dependency, low selectivity for K^+/Na^+	*muscles*		
	Ca^{++} activated, voltage gated	BK_{Ca}: high conductance, outward flow → repolarization of membranes	*neurons*	4 * (7 TM + P)#	
	Transmitter gated receptors type I	Neurotransmitter gated: Opened by ACh (nicotinic receptor), serotonin (5-HT$_3$-rec.), etc.: little specificity toward cations, not permeable for anions → fast postsynaptic signal transduction.	See Table 17.2-2	Figs. 17.2-4, -5: 5 * 4 TM	17.2.3
		K_{ATP}: High intracellular [ATP]/[ADP] ratio causes K^+ channel closing → depolarization (response to energy supply). In *pancreas* → insulin release. The channel closing can also be achieved by sulfonylurea binding (oral antidiabetic drugs!)	*neurons, heart, skeletal muscle, pancreas*		
	Coupled to receptors of type II	K_{ACh}: opened by muscarinic M_2 ACh receptor via $G_{i\alpha}$ protein	*atrium*		Tab. 17.4-1
		K_{5-HT}: regulated by serotonin receptor 5-HT$_1$ via G_o protein	*neurons*	5 * 4 TM	
Na^+ channels	Voltage gated	I, II, III, h1, u1 etc. Channels permeable for Na^+, less for K^+. Essential for signal transmission	*neurons, secretory cells, muscles, heart*	Figs. 17.2-2, -3#	17.2.2
	Transmitter gated receptors type I	See K^+ channels	See Table 17.2-2	Figs. 17.2-4, -5	17.2.3
		cGMP gated ion channel, slightly permeable for Ca^{++}	*retina*	4 * (6 TM + P)	17.4.6
Ca^{++} channels	Voltage gated	L: activated by high voltage, large conductance, modulated by phosphorylation (PKA), inhibited by dihydropyridines	*muscle, neurons, endocrine cells,* etc.	Figs. 17.2-2, -3#	17.2.2 17.4.4
		T: activated by low voltage, transient activation	*heart, neurons*		
		N: activated by high voltage, large conductance, inhibited by ω-conotoxin	*neurons*		
		P: activated by moderate voltage, moderate conductance, inhibited by ω-agatoxin	*neurons*		
	Transmitter gated receptors type I	See K^+ channels	See Table 17.2-2	Figs. 17.2-4, -5	17.2.3
		Ryanodine receptors	*sarcoplasmic/endoplasmic reticulum*	4 * (4...6 TM + P)	17.4.4,
		Inositol-P_3 receptor		4 * (6 TM + P)	17.4.5
	Coupled to receptors of type II	Coupled to GABA$_B$ receptors	*neurons*		
Cl^- channels (carry also other small anions)	Voltage gated	Opened by changes in membrane polarization	*neurons, epithel. cells*		
	Ca^{++} activated	Opened by increasing intracellular Ca^{++} concentration, also slightly voltage gated	*neurons, photoreceptors, secretory cells*		
	Secretory channel	CFTR channel, opened by phosphorylation via PKA (17.4.2). Defect → viscous mucus (cystic fibrosis / mucoviscidosis)	*epithelial cells (many organs)*	12 TM–ABC-Protein!	
	Background anion channels	Permanently open → charge compensation (of immobile anions and during depolarization), stabilizes membrane potential	*neurons, skeletal muscle* etc.		
	Transmitter gated receptors type I	Opened by glycine or GABA (GABA$_A$ receptor): general anion channels, anion influx → hyperpolarization, inhibitory action	*neurons*	5 * 4 TM	17.2.3

§ TM = transmembrane domains in the individual peptide chain, P = P domains, essential for pore structure   # Contains also additional subunits

3. <u>Primary active transport systems</u> (Table 18.1-2) for ions are driven by energy derived from biochemical reactions. Thus they are acting as <u>ion pumps</u>. They increase the concentration differences across membranes and are structurally and functionally different from ion channels.

The ions bind to specific sites of transport proteins with high affinity. <u>ATP hydrolysis</u> energizes the 'uphill' transport by effecting a configuration change, which allows the ion release on the side of higher concentration. Then the protein returns to its initial state, reversing the configuration change (Example: Na$^+$/K$^+$ exchanging ATPase, Fig. 18.1-1). However, an ion (H$^+$) flow in opposite direction can be used for generating ATP (<u>oxidative phosphorylation, photosynthesis</u>, 16.1.4, 16.2.1).

Related to the ion transport mechanism is the <u>binding protein dependent transport system</u> (BPDS) for nonionic compounds, employing coupled ATP hydrolysis. The transport proteins have an <u>ATP binding cassette</u> (<u>ABC</u>). BPDS is involved, e.g. in peptide fragment transport into the *ER* (<u>TAP1,2</u>; 18.1.5) and in multidrug resistance (export of antibiotics by the <u>P-glycoprotein</u>).

4. <u>Secondary active transport processes</u> (Table 18.1-3) couple two transmembrane movements. The flow of an ion from higher to lower concentrations even allows the 'uphill' transport of another ion or molecule from a lower to a higher concentration. Proteins which couple the transport of two or more compounds in the same direction are called <u>symporters</u>. If they do this in opposite direction, they are <u>antiporters</u>. These mechanisms are similar to those in *bacteria* (15.3.1).

A variant of antiport reactions is the exchange of acids, which are used, e.g. for ATP export from mitochondria (Fig.16.1-1) or in a cyclic fashion (as 'shuttles') for transport across membranes of H$^+$ (Fig.16.1-2), CO_2 (Fig. 16.2-10) etc.

Na$^+$/K$^+$-exchanging ATPase:

The enzyme keeps up the essential Na$^+$/K$^+$ gradients across the cellular membrane.

$$3\ Na^+_{inside} + 2\ K^+_{outside} + ATP = 3\ Na^+_{outside} + 2\ K^+_{inside} + ADP + P_i$$

It has to reconstitute the membrane polarization after the opening of ion channels (17.2), to drive many secondary active transport processes (above) and to counteract the slight leakage of ions through the cellular membranes. In resting cells, the enzyme consumes 17…52 % of the total energy turnover, the highest value is found in the *brain*.

Table 18.1-2. Primary Bulk Transport Mechanisms in *Eukarya*: ATP Energized Ion Pumps, ATP Synthesis (Selection)

Name (molar ratio cargo: ATP)	Function	Occurrence in membranes of	Structure (TM = transmem. segm.)	Section
P-Type ATPases (Type 2):				
Na$^+$/K$^+$-exchanging ATPase ($3_{out} : 2_{in}$: 1 ATP)	Upkeep of the Na$^+$ and K$^+$ concentration gradients → membrane potential, energizes Na$^+$ driven secondary transport mechanisms	*cells* (general)	Fig.18.1-1: 8 TM, tetrameric $\alpha_2\beta_2$	below + 17.2.1
H$^+$/K$^+$-exchanging ATPase ($2_{out} : 2_{in}$: 1 ATP)	Secretion of H$^+$ into the stomach	*gastric parietal cells*	homologous, 8 or 10 TM per α chain?	17.1.9
Ca^{++}-transporting ATPase (1_{out} : 1 ATP)	Upkeep of low cytosolic Ca^{++} concentration. Becomes activated by calmodulin/Ca^{++}. Possibly also export of H$^+$	*cells* (general)	130 kDa, oligomeric?	17.4.4
Ca^{++}-transporting ATPase ($2_{into\ ER/SR}$: 1 ATP)	Transport of Ca^{++} into ER/SR. Becomes activated by phospholamban in *slow skeletal* and in *cardiac muscle*.	*endoplasmic/sarcoplasmic reticulum*	100 kDa, 10 TM per chain, oligomeric?	17.4.4
V-Type ATPases (Type 3):				
H$^+$-transporting ATPase	Acidification of vacuoles (e.g. *lysosomes*). The H$^+$ gradient formed also serves to import other molecules via antiporters	*vacuoles (plants), acidic vesicles*	many subunits, some homologous to type 1	18.2.4
F-Type ATPases (Type 1), synthesizing ATP:				
H$^+$-transporting ATP synthase (3_{in} : 1 ATP)	Generation of ATP from a H$^+$ gradient, which is established by oxidation reactions (oxidative phosphorylation)	*mitochondria* (general)	Fig. 16.1-3	16.1.4
H$^+$-transporting ATP synthase (3_{in} : 1 ATP)	Generation of ATP from a H$^+$ gradient, which is established by photosynthetic reactions	*chloroplasts (green plants)*	similar	16.2.1
ABC-Type (ATP binding Cassette) ATPases Superfamily:				
P-glycoprotein	Multidrug resistance protein: export, e.g. of chemotherapeutic drugs. Mechanism not well known.	*carcinoma cells, Plasmodium falciparum*	single chain, 6 + 6 TM, 2 ABC	
TAP1, TAP2	Transport of peptide fragments to nascent MHC I in the *ER*	*endoplasm. reticulum*		19.1.5

Table 18.1-3. Secondary Bulk Transport Mechanisms in *Eukarya* (Selection)

Name (molar ratio of transported ions or compounds)	Function	Occurrence in membranes of	Section
Antiporters:			
Na$^+$/ Ca^{++} antiporter ($3_{in} : 1_{out}$)	Fast decrease of cytosolic Ca^{++} levels	*cells* (ubiquitous)	17.4.4
Na$^+$/ H$^+$ antiporter ($1_{in} : 1_{out}$)	H$^+$ export, activity controlled by pH. The Na$^+$ inflow is accompanied by water uptake.	*cells* (ubiquitous)	
H$^+$/amine antiporter (various)	e.g. in *adrenals*: concentration of norepinephrine and serotonin	*many cells*	17.1.5
Cl$^-$/ HCO$_3^-$ antiporter ($1_{in} : 1_{out}$)	Gas exchange, supply of Cl$^-$ for secretion of HCl	*intestine, kidney, erythrocytes*	17.1.9
Symporters:			
Na$^+$, K$^+$, Cl$^-$ symporter $(1:1:2)_{in}$	Import of Na$^+$, K$^+$, Cl$^-$; Na$^+$ is consecutively pumped out → KCl import. Regulated by second messengers (cAMP, cGMP)	*many cells*	
Na$^+$, glucose symporter $(2:1)_{in}$	Resorption of glucose, Na$^+$ driven	*intestine, kidney* *	3.1.4
Na$^+$, amino acid symporter (Na, amino acid)$_{in}$	Import of amino acids, Na$^+$ driven (several systems)	*liver, kidney, intestine etc.*	
Na$^+$, metabolite symporters (Na, metabolite)$_{in}$	Import of nucleosides and other metabolites (various systems)	*kidney, intestine etc.*	
Na$^+$, GABA$^-$, Cl$^-$ symporter $(2:1:1)_{in}$	Import of γ-aminobutyrate (GABA) into *neurons*	*mammalian brain*	17.2.3
Na$^+$, X, Cl$^-$ symporter (Na, X, Cl$^-$)$_{in}$	X = catecholamines, glycine etc.: recovery of neurotransmitters	*neuronal synapses*	17.2.3
Complex Transporters:			
Na$^+$, HCO$_3^-$/ H$^+$, Cl$^-$ complex transporter: (Na$^+$, HCO$_3^-$)$_{in}$, (H$^+$, Cl$^-$)$_{out}$	H$^+$ export, accumulation of bicarbonate, activity controlled by pH.	in *many vertebrates* and *invertebrates*	
Na$^+$, glutamate /K$^+$ complex transporter $(3:1)_{in} : (1)_{out}$	Import of glutamate into *neurons*	*neurons*	17.2.3
Na$^+$, serotonin, Cl$^-$/K$^+$ complex transporter (Na$^+$, serotonin, Cl$^-$)$_{in}$, (K$^+$)$_{out}$	Import of serotonin into *neurons*	*platelets, neurons*	17.2.3
Acid Exchange Mechanisms:			
ATP-ADP carrier ($1_{out} : 1_{in}$)	Export of ATP from *mitochondria*	*mitochondria*	16.1.2
Aspartate/glutamate + malate/oxoglutarate shuttle ($1_{out} : 1_{in} + 1_{in} : 1_{out}$)	Import of H$^+$ into *mitochondria*	*mitochondria*	16.1.2
Combination of shuttles	CO_2 pumping	C_4 *plants*	16.2.2

* In other body cells, glucose import takes place via facilitated diffusion (18.1.1)

The enzyme consists of 2 α (112 kDa, 8 transmembrane helices) and 2 β subunits (35 kDa, 1 transmembrane helix). It is a type 2 ATPase, acting via phosphorylation and dephosphorylation of an aspartyl residue. Phosphorylation occurs only in the presence of Na$^+$ (and Mg^{++}), dephosphorylation in presence of K$^+$. This results in a cyclic conformation change, which effects the transmembrane transport of the ions (Figure 18.1-1). Similar cyclic processes also occur with the H$^+$/K$^+$-exchanging ATPase and the Ca^{++}-transporting ATPase (Table 18.1-2).

Cardiac glycosides (7.2.2) inhibit the enzyme and thus diminish the transmembrane Na$^+$ gradient. This slows down the Na$^+$ driven Ca^{++} export from the cytosol; the increased Ca^{++} concentration intensifies the strength of cardiac contractions.

Figure 18.1-1. Structure and Postulated Mechanism of the Na$^+$/K$^+$-Exchanging ATPase

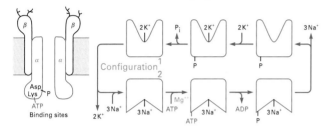

Literature:
Albers, R.W.: *Encyclopedia Human Biol.* Vol. 4. Academic Press (1991) 557–571.
Barnard, E.A.: Trends in Pharmacol. Sci. 17 (1996) 305–309.
Fasolato, C. *et al.*: Trends in Pharmacol. Sci. 15 (1994) 77–83.
Kuriyama, H., *et al.*: Pharmacol. Reviews 47 (1995) 387–515.
Maeda, M.: J. Biochem. 115 (1994) 6–14.
Mintz, E., Guillain, F.: Biochim. Biophys. Acta 1318 (1997) 52–70.
See also references in the quoted chapters.

18.1.2 Import by Endocytosis and Pinocytosis

Endocytosis is the import of macromolecules by formation of vesicles as invaginations of the plasma membrane. The intake can also take place with large cell fragments, microorganisms etc. (phagocytosis by specialized cells of *multicellular organisms* for defense purposes, 19.1.1) or with fluids (pinocytosis). A very important example is the lipoprotein metabolism, which shows many typical aspects of the import mechanism and the consecutive metabolization. It is described in the next section (18.2).

Most *animal* cells use a receptor-mediated system for endocytosis. The extracellular macromolecules bind specifically to their cognate receptors, which assemble in clathrin coated pits at the *plasma membrane*. Clathrin is an association of 3 heavy and 3 light protein subunits to a 'three-legged' triskelion. The units combine to form a polyhedral structure and promote the fast budding of coated vesicles from the internal surface of the membrane. After leaving the membrane, the clathrin coat is removed and returns to the membrane. The ingested molecules are thereafter metabolized, while frequently the receptor is recycled to the membrane. This mechanism results in a thousandfold enrichment of imported molecules.

Endocytosis is also a means to terminate the action of receptors involved in signal transduction. In a number of cases, not only the ligand, but also the receptor is thereafter degraded (Figures 17.5-2 and -3).

Pinocytosis is a constitutive procedure, which also starts at (receptor-less) coated pits. Pinocytotic vesicles are usually small.

Clathrin-coated vesicles are also used for the transport of enzymes from the *trans-Golgi network* to *endosomes,* while other intracellular transport steps use different coats. They are described in 14.2.

Literature:
Brodsky, F.M.: Science 242 (1988) 1396–1402.
Kirchhausen, T.: Curr. Opin. Struct. Biol. 3 (1993) 182–188.
Smythe, E.: Warren, G.: Eur. J. Biochem. 202 (1991) 689–699.

18.1.3 The Cytoskeleton as Means for Intracellular Transport and Cellular Movements in *Eukarya*

The cytoskeleton consists of
- microtubules (polymers of α- and β-tubulin, diameter 25 nm)
- actin filaments (polymerized β- and γ-actin, diameter 7 nm)
- intermediary filaments (diameter 8...15 nm).

In spite of its name, it contributes to cellular movements by providing tracks for molecular motors and by its own expansion and contraction.

Microtubules are present in all nucleated cells. They take part in cell division (11.6.5), intracellular transport and secretion, movement of flagella and cilia and are a member of the cytoskeleton.

Structure of microtubules (Fig. 18.1-2): Microtubules consist of heterodimeric, globular α/β tubulin subunits (53 and 55 kDa, respectively), which are longitudinally aligned. 13 of these arrangements form a hollow tube with structural polarity (+ and – ends). Each subunit binds GTP, only the β-unit bound GTP is exchangeable. Microtubule associated proteins (e.g. MAPs 1...4 with different forms) are bound to the surface and stimulate the microtubule assembly. This activity can be modified by PKA-catalyzed phosphorylation. Dynein (MAP1C), kinesin and kinesin-like proteins are molecular motors moving along the microtubuli for transport of cellular components. The assembly of microtubules is inhibited by, e.g. colchicine, vinblastine, griseofulvin, Ca^{++}/calmodulin, which bind to tubulin. Taxol, on the other hand, stimulates the assembly and stabilizes microtubuli.

Motion during mitosis: Microtubules perform several steps of chromatide and cell separation (11.6.5). Controlled polymerization and depolymerization reactions (dynamic instability) are used to generate force. When GTP-carrying tubulin heterodimers assemble in presence of Mg^{++} (preferably at the + end), the GTP is slowly hydrolyzed. Therefore, during fast assembly, several tubulin dimers at the + end still carry GTP (GTP-cap), which protects against disassembly. If the speed of assembly decreases, the GTP cap disappears and the end becomes prone to disassembly (Fig. 18.1-2). A number of molecular motors (dynein, kinesin-like proteins) are involved in these procedures, moving polar microtubuli along each other and chromatides along kinetochore microtubuli (Fig. 11.6-6). However, many details are still unknown.

Figure 18.1-2. Assembly, Disassembly and Structure of Microtubules

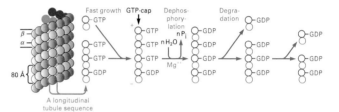

Intracellular movement along microtubules: Microtubules also exist during the interphase. They are more stable than those formed during mitosis. ($t_{1/2}$ > 300 sec vs. < 50 sec.) Posttranslational modification of tubulin (acetylation, removal of C-terminal tyrosine) may play a role.

In various cells (e.g. *neurons*, 17.2.5) motor proteins recognize the polarity of the microtubules and move cargoes (vesicles, proteins, neurotransmitters) either towards the plus end (kinesin and most kinesin-like proteins, KLPs, in *neurons* towards the nerve terminal = anterograde) or towards the minus end (dynein, in *neurons* = retrograde).

The motor protein kinesin consists of 2 heavy and 2 light subunits. The structure of the head portion resembles strongly the myosin structure. Also, the movements are energized by ATP hydrolysis and involve a cycle of conformation changes similar to myosin (Fig.17.4-7). The kinesin heads move along the microtubules, while the tail is attached to the cargo. Dynein has 2 or 3 heads.

Extracellular movements: Groups of cilia (hairlike organelles) transport fluids, mucus or small particles (dust) along the cell surface by wavelike movements, while the single, longer flagella propel, e.g., *sperm cells* and *flagellates*. The propelling units are similarly composed: Two central microtubules are surrounded by a ring of 9 fused microtubule doublets with attached dynein molecules. A major number of other proteins connect the structures. For movement, both dynein arms slide along the next microtubule, energized by ATP hydrolysis.

Actin filaments: The actin filaments, which form a network underneath the *plasma membrane,* are attached at focal adhesions. The structure of the filaments resembles those in *muscle* (17.4.5). Together with actin binding proteins, myosin etc., they modify the shape of the plasma membrane, forming spikes, invaginations etc.

Intermediary filaments: These vary greatly in size and structure. Generally, they consist of a lengthy α-helix structure, which associate with another one to form a coiled coil. Further associations lead to structures of higher order. The filaments carry a globular domain on each end. They mainly provide resilience. Examples are keratin (in outer *epidermal layer*, in *hair, bird feathers*), vimetin (in *fibroblasts, endothelial cells* etc.), nuclear laminin (underneath the inner *nuclear* membrane) and neuronal intermediary filaments (along the *axon*).

Literature:
Erickson, H.P., *et al.*: Ann. Rev. of Biophys. Biomol. Struct. 21 (1992) 145–166.
Klymowski, M.W.: Curr. Opin. in Cell Biol. 7 (1995) 46–54.
Walin, M.: *Encyclopedia Human Biol.* Vol. 5. Academic Press (1991) 11–29.

18.2 Plasma Lipoproteins

While here the situation in *humans* is described, a qualitatively similar situation exists in *higher vertebrates*.

Plasma lipoproteins are high molecular weight aggregates composed of lipids (mainly free or esterified cholesterol, triacylglycerols = triglycerides and phospholipids) and one or more specific apolipoproteins (Figure 18.2-1). Lipoproteins are the functional units of delivering water-insoluble lipids via the circulation to cells for utilization or storage. They play a decisive role in cholesterol homeostasis and in triglyceride transport. While they show a common basic structure (the polar groups of phospholipds, the OH-groups of cholesterol and the apolipoproteins at the outside, cholesterol esters and triacylglycerols in the core), they vary largely in size and composition.

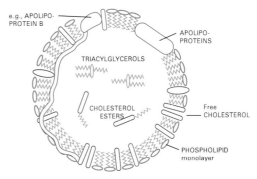

Figure 18.2-1. General Structure of Lipoproteins

Table 18.2-1. Classification, Properties and Composition of Plasma Lipoproteins (Data vary somewhat between authors)

	Chylomicrons	Remnants	Very low density lipoprotein (VLDL)	Intermediate density lipoprotein (IDL)	Low density lipoprotein (LDL)	High density lipoprotein (HDL$_2$)	High density lipoprotein (HDL$_3$)
Spec. mass (g/ml)	< 0.95	< 1.006	< 1.006	1.006…1.019	1.019…1.063	1.063…1.125	1.125…1.210
Diameter (nm)	80…500	> 30	30…80	25…35	18…28	9…12	5…9
Electroph. mobility*	origin	origin	pre-β	pre-β/β	β	α	α
Proteins (%)	1…2		5…10		20…24	**45…50**	
Cholesterol (%)	0.5…1		6…8		5…10	3…6	
Cholest. esters (%)	1…3		12…14		**35…40**	14…18	
Triacylglycerols (%)	**86…94**		55…65		8…12	3…6	
Phospholipids (%)	2…8		12…18		20…25	20…30	

* Migration speed during electrophoresis as compared with α and β globulin

18.2.1 Apolipoproteins (Apo)

There are also striking differences in the apolipoprotein composition of the lipoprotein classes. Tables 18.2-2 and 18.2-3 show the distribution and the properties of apolipoproteins.

18.2.2 Plasma Lipoprotein Metabolism (Figure 18.2-2)

There are four major pathways of lipoprotein metabolism. The first two involve chylomicrons and VLDL, which contain Apo B as an essential, not exchangeable membrane constituent.

Chylomicron metabolism: Chylomicrons are transport vehicles for dietary lipids (mainly triacylglycerols) from the *intestine* to *peripheral tissues* and to the *liver*. They are assembled in the *Golgi apparatus* of *intestinal mucosal cells* from biosynthesized Apo B-48 and absorbed lipids. Following secretion, chylomicrons acquire Apo E and Apo C-II primarily from HDL. Apo C-II activates lipoprotein lipase (LPL), which originates mainly from *adipose tissue* and *muscles*. The majority of it is bound to heparan sulfate (13.1.5) at the *cell membranes* of the *capillary vessel endothelium*. LPL also exists in a free form in the *circulation*. The fatty acids liberated from triacylglycerols by action of membrane bound LPL are directly transported into the *adipose* and *muscle cells*, while those generated by LPL in the bloodstream bind to serum albumin and are eventually taken up by *heart* and *muscle* cells. There, they can be degraded (6.1.5) or be utilized for synthetic reactions. As the size of the remnants decreases, a second (secreted) lipase, hepatic triacylglycerol lipase (HL), may play a role in triacylglycerol degradation. Eventually, the remnants are bound to remnant receptors of the *liver* (18.2.4) and internalized into liver cells similarly to the degradation of LDL (see below).

VLDL/LDL metabolism: VLDL transports lipids from the *liver* to *peripheral tissues* (such as *muscle*) and to *adipose tissue*. After assembly in the *Golgi apparatus* and secretion, triacylglycerol rich VLDL from the *liver* (containing Apo B-100) associates with Apo C-II and Apo-E which have been dissociated from HDL. VLDL then becomes converted to IDL by *endothelial* and *plasma* lipoprotein lipase (LPL) analogous to the conversion of chylomicrons to remnants; the fate of the liberated fatty acids is also the same. About one half of the IDL is removed from *plasma* by interaction of Apo E and Apo B-100 at the particle surface with LDL- or remnant receptors. The other half is converted into LDL. HL is necessary for this step. Lipid components (via lipid exchange proteins) and apolipoproteins (except Apo B-100) are exchanged with HDL.

Table 18.2-2. Distribution of Apolipoproteins in Plasma Lipoproteins
(% of Protein)

Apolipoprotein	Chylomicrons	VLDL	LDL	HDL$_2$	HDL$_3$
A-I	33	trace	trace	65	62
A-II	trace	trace	trace	10	23
A-IV	14	0	0	?	trace
B	5	25	> 95	trace	trace
C-I, C-II, C-III	32	55	0…2	13	0
D	?	?	?	2	4
E	10	15	0…3	3	1
Other	6	5	0…5	4	5

Apo (a) is omitted in this table. It is highly homologuous to plasminogen (20.5) and shows a high variation in its molecular mass (from 250 to > 500 kDa). The heterodimer Apo (a)-S-S-Apo B is the protein component of lipoprotein (a), a cholesterol-rich particle whose function is as yet unknown. The concentration of lipoprotein (a) varies considerably: from < 1 mg/100 ml up to ca. 200 mg/100 ml. Concentrations of > 30 mg/100 ml are considered to be atherogenic (18.2.5).

Table 18.2-3. Properties and Function of Plasma Apolipoproteins

Apolipoprotein	Mass (kDa) ca.	Conc.* (mg/ 100 ml)	Major occurrence (in parentheses site of synthesis)	Function
A-I	28	130	*liver, intestine*	structural, activator of LCAT, ligand for HDL receptor?
A-II (dimer)	18	40	*liver (intestine)*	structural, activator of HL, ligand for HDL receptor?
A-IV	45	?	*intestine (liver)*	activator for LCAT, ligand for HDL receptor?
B-100**	512	80	*liver*	structural, ligand for LDL receptor, secretion of VLDL
B-48*	250		*intestine*	structural, secretion of chylomicrons
C-I	7	6	*liver (intestine)*	activator of LCAT
C-II	9	3	*liver (intestine)*	activator of lipoprotein lipase
C-III	9	12	*liver (intestine)*	inhibition of premature removal of triacylglycerol-rich lipoprotein
D	20	10	*liver (intestine)*	exchange of cholesterol esters
E	34	5	*liver (macrophages)*	ligand for LDL receptor and putative chylomicron remnant receptor

* in normal fasting plasma

** Apo B-100 (4536 amino acids) and B-48 (2125 amino acids) are synthesized from a single Apo B gene. Apo B-100 is translated from the full length of mRNA in the *liver*, while Apo B-48 is synthesized from an Apo B post-transcriptional mRNA containing a premature translational stop codon introduced by mRNA modification in the *intestinal mucosa* (cf. 11.5.2).

LDL and IDL bind to LDL receptors of *liver, adrenal* and *peripheral cells* (including *fibroblasts* and *smooth muscle cells*), followed by endocytosis. For the intracellular metabolism, see 18.2.4.

In *birds*, oocytes express a special receptor for vitellogenin (an Apo B related protein at the surface of VLDL). This receptor directs VLDL to the ovaries in order to satisfy the lipid needs for egg production.

Scavenger pathway (not shown): Blood *monocytes*, which adhere to the intact epithelium, may migrate into the subendothelial space and differentiate to *macrophages* (Figure 19.1-3). Native LDL is not readily taken up by macrophages. However, oxidative modifications of LDL enhance the LDL uptake by the 'scavenger receptor' of the macrophages, which causes massive deposition of cholesterol ester droplets, thus converting macrophages into *foam cells*. Lipoprotein (a) and possibly other Apo B-containing, oxidatively modified lipoproteins may also convert macrophages into foam cells. These components accumulate in atherosclerotic lesions and therefore may be 'risk factors' for atherosclerosis, especially for coronary heart disease. As long as the macrophages are not too overloaded with cholesterol, excessive free cholesterol can be removed by HDL.

HDL metabolism: HDL seems to be involved in reverse cholesterol transport from *peripheral tissues* to the *liver*. Nascent HDL, composed primarily of Apo A-I, free cholesterol and phospholipid disks, appears to be assembled into stacked discoidal structures. These particles, synthesized by the *liver*, the *intestine* and *macrophages*, take up free cholesterol from *extrahepatic cells* including *foam cells* (The involvement of a specific HDL receptor is unclear). Cholesterol is esterified with fatty acid (from lecithin) by lecithin-cholesterol acyltransferase (LCAT). The esters enter the HDL core. Additional uptake of triacylglycerols occurs. The discoidal HDL changes into spherical HDL_3. Further uptake of triacylglycerols and cholesterol converts HDL_3 into a HDL_2 particle. (The reverse reaction is catalyzed by HL.)

The mechanism of HDL uptake and catabolism by liver cells is not completely known. In the *liver*, part of the cholesterol is converted into bile acids and thus removed from its interconversion cycles (7.9.2). This removal of body cholesterol from plasma explains the reversed relationship of HDL-cholesterol (especially HDL_2-cholesterol) and atherosclerosis risk.

18.2.3 Lipid Transport Proteins

The cholesterol ester transport protein (CETP) is mainly an *intracellular* enzyme in *humans*. In *plasma*, it apparently effects a net transfer of triacylglycerols from chylomicrons and VLDL to HDL and concomitantly of cholesterol esters (but not of free cholesterol) in the opposite direction. The latter activity is possibly a risk factor, the actual metabolic role of CETP remains unclear.

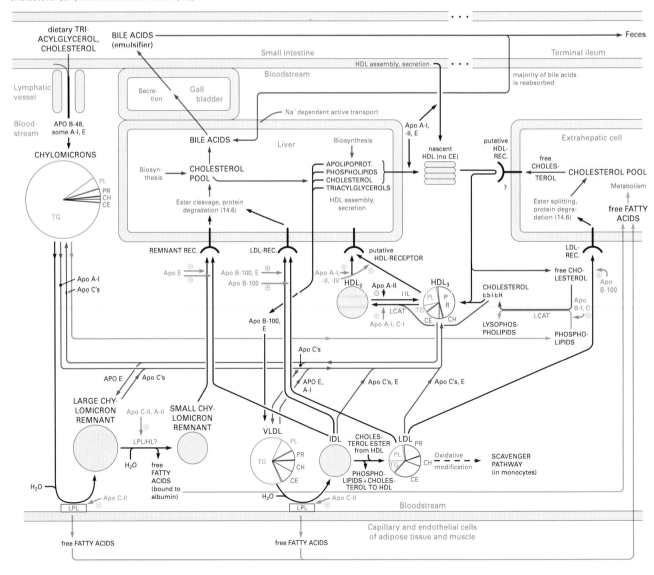

Figure 18.2-2. Metabolism of Chylomicrons, VLDL and HDL in *Mammals*
(in part after personal communications by Dr. A. van Tol, Rotterdam). Contrary to the arrow colors elsewhere, their meaning is here: (black) lipoprotein metabolism, (blue) transfer of proteins (PR), (green) transfer of triacylglycerols (TG), phospholipids (PL) and fatty acids, (red) transfer of cholesterol (CH) and cholesterol esters (CE)

18.2.4 Lipoprotein Receptors

The receptors effect the uptake of lipoproteins into the cells (18.1.2). The apolipoproteins at the lipoprotein surface act as ligands for specific receptors and thus have a 'targeting' effect.

LDL receptor (Figure 18.2-3): This glycoprotein of 160 kDa binds Apo B and E as ligands. This results in endocytosis and transport of LDL to *lysosomes* where the proteins (almost exclusively Apo B) are degraded; triacylglycerols and cholesterol esters are cleaved. Free cholesterol can react in various ways:

- It can enter the intracellular cholesterol pool (important for membrane synthesis and for regulatory effects).
- Excessive quantities are recycled by *cytosolic* acyl cholesterol acyltransferase (ACAT) to cholesterol esters, which are deposited as lipid droplets in the cell.
- In *liver*, it is converted to bile acids (7.9) and secreted via the gallbladder into the cystic ducts eventually serving in the *intestine* as a lipid emulgator. More than 80 % are later reabsorbed and returned to the liver, the rest is excreted in the feces.
- In specialized tissues (e.g. *adrenal glands*) cholesterol is converted to steroid hormones (7.4).

The LDL-receptor is synthesized at the ribosomes of the *ER* ①, transported to the *Golgi apparatus* ② and intercalated into the *cell membrane*. After clustering at the surface of coated pits ③ (18.1.2) and binding of LDL or IDL via Apo B-100 ④ the receptor-ligand complexes are internalized as coated vesicles. They are then converted into endosomes. Here, a more acidic pH exists, which leads to the dissociation of the ligand-receptor complex. The receptor is recycled to the surface of the cell ⑤. The endosomes combine with lysosomes. Subsequently, apolipoproteins, cholesterol esters and triacylglycerols are hydrolyzed. The resulting increase of the 'cholesterol pool' has several consequences:

- suppression of the transcription of the HMG-CoA reductase gene (7.1.4)
- acceleration of the degradation of the HMG-CoA synthase and reductase, geranyl transferase and squalene synthase (7.1.4)
- activation of acyl cholesterol acyltransferase (ACAT, 7.1.5)
- lowering of the concentration of mRNA coding for the LDL receptor (7.1.4)

The first three steps decrease the level of intracellular free cholesterol, while the fourth step leads to a lack of LDL-receptors on the cell surface and therefore to a dramatic decrease of cellular LDL uptake. As a consequence, plasma LDL rises. High plasma LDL leads to cholesterol deposits in skin and tendons (xanthoma), but especially in arteries (plaques). Thus, it is one of the main risk factors of atherosclerosis. In familial hypercholesterolemia, one of steps ① to ⑤ is defective (see 18.2.5).

Scavenger receptor: This trimeric integral membrane glycoprotein is composed of three 77 kDa subunits. It is located at the cell membrane of *macrophages* and mediates with broad specificity the uptake of chemically modified LDL (like oxidized LDL, glycated LDL, acetylated LDL etc.), which are poorly or not at all recognized by the LDL receptor.

Chylomicron remnant receptor: It is mainly responsible for the clearance of chylomicrons and recognizes Apo E as ligand. It has been presumed that the low-density lipoprotein receptor related protein (LRP) is this receptor. LRP appears to be a large sized relative of the LDL receptor; the extracellular domain of the LDL receptor is repeated about four times. LRP is definitely the receptor for α_2 macroglobulin. Whether or not it has a dual function is not known for certain at present.

HDL receptors: Their existence has not yet been completely proven. High affinity binding sites for HDL with specificity for Apo A-I, A-II and A-IV as ligands exist on many cell types. HDL receptor protein candidates have been described ranging from 58 to 120 kDa. Not much definite knowledge exists about the cholesterol removal mechanism from peripheral cells by HDL and its uptake by the liver ('reverse cholesterol transport').

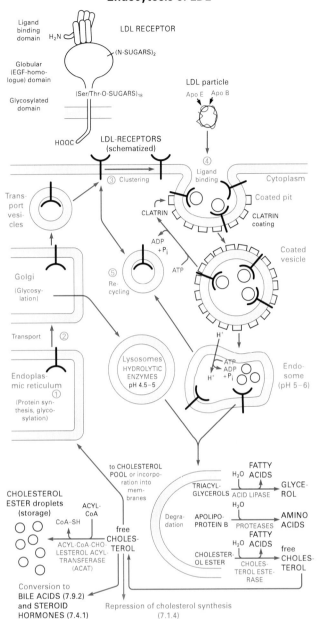

Figure 18.2-3. LDL Structure and Receptor Mediated Endocytosis of LDL

18.2.5 Lipid metabolic disorders

Some thirty genetic dyslipoproteinemias are known. In most cases, also the affected genes and their mutations are well characterized. Best known are disturbances in synthesis and function of the LDL receptor, numbered ① ... ⑤ in Figure 18.2-3. They cause familial hypercholesterolemia, a disease with extremely elevated levels of plasma cholesterol which usually leads to premature coronary heart disease.

Since atherosclerosis mainly seems to be caused by disturbances of lipoprotein metabolism (elevated LDL cholesterol, low HDL cholesterol, high triacylglycerols, foam cell formation etc.), lipid lowering therapy is the treatment of choice. If dietary means do not suffice, 3-hydroxy-3-methylglutaryl coenzyme A (HMG-CoA) reductase inhibitors, fibrates, nicotinic acid and derivatives, bile acid-binding ion exchangers, probucol (an antioxidant) and other drugs are applied. In extreme cases it may be necessary to remove LDL cholesterol by extracorporeal plasmapheresis.

Literature:
Assmann, G.: *Lipid Metabolism and Atherosclerosis.* Schattauer (1982).
Brown, M.S., Goldstein, J.L.: Science 232 (1985) 34–47.
Krieger, M., Herz, J.: Ann. Rev. of Biochem. 63 (1994) 601–637.
Schettler, G., Habenicht, A.J.R. (Eds.): *Principles and Treatment of Lipoprotein Disorders.* Springer (1994).
Vance, D.E., Vance, J. (Eds.): *Biochemistry of Lipids, Lipoproteins and Membranes.* New Comprehensive Biochemistry Vol. 31. Elsevier (1996).

19 Antimicrobial Defense Systems

19.1 Immune System

The immune system protects *higher vertebrate* organisms against pathological microorganisms and likely also plays a role in the recognition and elimination of malignantly transformed cells. In the following, the *human* system is described.

The immune system is composed of a multiplicity of components:
- Lymphoid tissue (lymph nodes, spleen, thymus, mucosal lymphoid tissue)
- epithelial cells of *skin* and *mucosa*, forming a barrier to the environment
- mobile and sessile nucleated cells of the hematopoetic system originating from the *bone marrow*
- numerous soluble components (e.g. antibodies and cytokines)

Functionally, one distinguishes between
- natural, innate, nonadaptive immune defense
- specific, acquired, adaptive immune defense

Both are composed of different cellular and soluble components. However, they cooperate closely in the formation of a specific immune response.

Table 19.1-1. Components of the Immune Defense Systems

	Nonadaptive defense	Adaptive defense
Cells	skin and mucosa cells phagocytes natural killer cells (NK cells) mast cells neutrophilic, basophilic, eosinophilic granulocytes	T lymphocytes B lymphocytes
Soluble components	complement factors (19.2) acute phase proteins collectins cytokines	antibodies cytokines

19.1.1 Cells of the Non Adaptive Immune Defense System

Monocytes / macrophages and neutrophilic granulocytes phagocytose foreign particles, senescent self cells or dead tissue. The recognition, membrane binding and endocytosis of material is facilitated either by lectin-type receptors or by Fc-receptors and complement receptors when the material is complexed with specific antibodies (19.1.2, opsonization). The substances are then intracellularly degraded by lysosomal enzymes. These cells also produce reactive oxygen species which contribute to the killing of microbes (4.5.8). Mast cells and basophilic granulocytes bind IgE via high affinity membrane Fcε-receptors and are the effector cells in the IgE-mediated allergic immune response (19.1.8). Eosinophilic granulocytes express also receptors for IgE and are involved in the defense against parasites and in the mucosal inflammation in (chronic) allergic asthma. All these cells recognize their targets via non clonally distributed surface receptors.

Natural killer cells (NK cells) are cytotoxic cells which lyse virus-infected cells and tumor cells. They express two types of surface receptors which recognize determinants of the MHC class I molecules: Killer cell inhibitory receptors (KIR) and CD94:NKG2 heterodimers. Their genes are polymorphic. These receptors can either trigger or inhibit the cytotoxic activity of NK cells.

19.1.2 Development and Maturation of the Cellular Components (Figure 19.1-3)

The mobile cells of the immune system develop from common precursor cells in the *bone marrow*, the pluripotent stem cells. Colony stimulating factors, CSF (growth and differentiation factors), influence the formation of colony forming units, CFU, which are precursors of the various differentiation lines. They are named by the initial letter of the final cells. After maturation, the hematopoetic cells migrate into the *circulatory system*. As an exception, the T lymphocytes leave the *bone marrow* as immature cells and migrate into the *thymus* for complete differentiation.

The various differentiation lines and steps of *leukocytes* are characterized by expression of surface molecules (antigens), which can be detected with monoclonal antibodies and are designated with CD-numbers (cluster of differentiation).

T lymphocytes are characterized phenotypically by the CD2 molecule, which appears early during their maturation in the *thymic cortex*. There, most T cells also express simultaneously the CD4 and CD8 molecules and the T cell receptor, which is associated with the CD3 polypeptide chains. During differentiation in the *thymic cortex*, the T lymphocytes temporarily carry the CD1 molecule, whose structure is similar to that of MHC class I molecules (19.1.5). Before leaving the *thymic cortex*, separation into two immunocompetent differentiation lines takes place. They carry on their surface either the CD4 molecule (T helper cells) or the CD8 molecule (mainly cytotoxic T cells).

B lymphocytes mature in the *bone marrow*. The various differentiation steps are characterized by the expressed isotypes of immunoglobulins (see 19.1.3). Mature B *lymphocytes* carry simultaneously IgM and IgD on their surface. After activation by antigen, a switch to expression of other Ig isotypes (e.g. IgG, IgA, IgE) can take place. Activated B lymphocytes differentiate into either short-lived plasma cells, which return to the *bone marrow* or into memory cells, which circulate as quiescent, inactive cells for extended periods in the body (mainly in *lymphoid tissue*).

From the *bloodstream*, both lymphocyte types enter the *lymphoid tissue* via *venules*, which are characterized by a high endothelium (high endothelial venules, HEV). For the mechanism of this passage, see 19.3.

Stimulation of lymphocytes by antigens: In the *lymphoid tissue*, lymphocytes are activated by antigens, which are presented to them by dendritic cells.

B-lymphocytes are stimulated in the *lymphoid follicles* (Fig. 19.1-1) by follicular dendritic cells (FDC), which carry antigens as antigen-antibody complexes on their surface.

T lymphocytes are activated in the *paracortex* of *lymph nodes* (Fig. 19.1-1) by peptide-MHC complexes on the surface of interdigitating cells.

These cells originate from *dendritic cells* in certain tissues (e.g., *Langerhans' cells* of the *epidermis*), where they pick up antigens, process them

Figure 19.1-1. Schematic Diagram of a Lymph Node

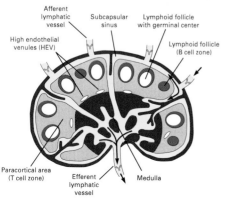

Figure 19.1-2. Structure of Immunoglobulin V and C Domains

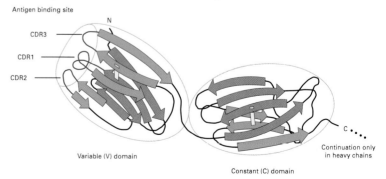

Figure 19.1-3. Development of Blood Cells

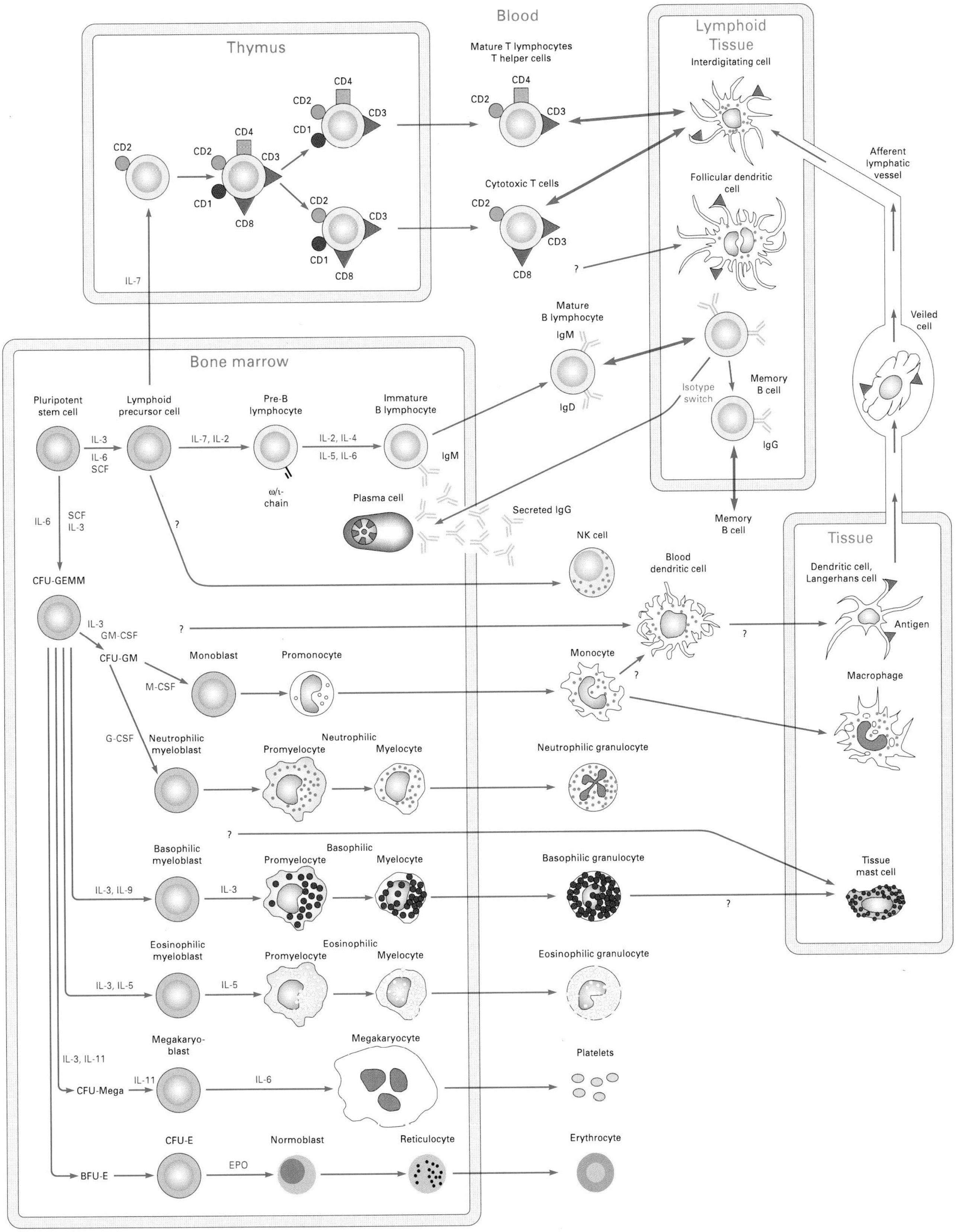

and transport them as veiled cells via the *lymphatic vessels* to the regional *lymph node*, where they settle in the *paracortical area*.

19.1.3 Antigen Recognition by B Lymphocytes
Structure of immunoglobulins (Ig, Figure 19.1-4): Immunoglobulins are the antigen recognition structures of B lymphocytes. They occur either as membrane receptors or as secreted soluble antibodies. Immunoglobulins are composed of heavy (H-, 53...75 kDa) and light (L-, 23 kDa) polypeptide chains, which are folded into so-called immunoglobulin domains (Figure 19.1-2). Disulfide bonds stabilize the domain structure or connect the polypeptide chains.

Each domain of ca. 100 amino acids forms two β-pleated sheets (Fig. 19.1-2) consisting of 3...5 antiparallel β-strands, which are stabilized by a disulfide bond. Polypeptide chains, which are folded according to this principle, are the basic structure of a large number of membrane molecules. Their coding genes compose the immunoglobulin supergene family.

Variable (V) domains: In immunoglobulins, the domain closest to the N-terminus of each polypeptide chain contains 3 hypervariable sequences, which are different in individual immunoglobulin molecules. These complementarity determining regions (CDR) of a H and a L chain are close to each other in the completed immunoglobulin molecule and form the specific binding site for an antigen.

Constant (C) domains: The other domains of the Ig polypeptide chains are constant ($C_H1...C_H3$ or C_H4, respectively, in heavy chains, C_L in light chains). They determine the biological properties of the Ig molecules and the membership in a particular immunoglobulin class (isotype). While IgG, IgD and IgE are bivalent (possess 2 binding sites), secreted IgM forms a decavalent pentamer and IgA can aggregate to a tetravalent dimer. IgM and dimeric IgA are stabilized by an extra covalently bound J chain (ca. 20 kDa). The notation of the isotype heavy chain domains is $C_\mu 1...4$ for IgM, $C_{\gamma 1} 1...3$ for IgG1 etc.

Both H polypeptide chains of IgG are connected by disulfide bridges, which are located between the C_H1 and C_H2 domains in the so-called hinge region. The size of this region is different in various Ig isotypes. In IgM and IgE heavy chains, the hinge region is replaced by an extra domain. During limited proteolysis of IgG with papain, the H-chains are split at the N-terminal side of the disulfide bridges. This yields Fab-molecules (fragment antigen binding), which contain the V_H and the C_H1 domain of a heavy chain and a complete light chain. Pepsin splits the IgG heavy chains at the C-terminal side of the disulfide bridges, resulting in an $F(ab')_2$ molecule, which is able to bind an antigen bivalently. It does not, however, show the biological properties of the native molecule, such as binding to Fc receptors (19.1.7) and activation of complement (19.2).

Structure and recombination of immunoglobulin genes: The huge diversity of V domains with different antigen-binding specificity originates
- in somatic recombination of gene segments
- in somatic point mutations in recombined genes

The genes for light chains (κ or λ chain) and for heavy chains are located on different chromosomes. In downstream direction (5'→3', Figure 19.1-5) they consist of
- a number of V segments (87 in genes for heavy chains in *humans* + 24 on 2 other chromosomes, about half of them functional), each preceded by a L (leader peptide) gene segment
- (only in H chain genes): a number of D (diversity) segments (28 in *humans*)
- a number of short J (joining) segments (9 in genes for heavy chains in *humans*, 6 of them functional)
   The V gene segments code for the major part of the V domain, while the (D and) J gene segments code for the CDR3 region of the V domain.
- (in κ light chain genes): a gene segment coding für the constant domain C_κ.
- (in λ light chain genes): several C_λ gene segments, each of them preceded by a single J gene segment
- (only in H chain genes): gene segments coding for
  - the constant domains of IgM ($C_\mu 1...4$), IgD ($C_\delta 1...3$), IgG3 ($C_{\gamma 3} 1...3$), IgG1 ($C_{\gamma 1} 1...3$), IgA1 ($C_{\alpha 1} 1...3$), IgG2 ($C_{\gamma 2} 1...3$), IgG4 ($C_{\gamma 4} 1...3$), IgE ($C_\epsilon 1...4$) and IgA2 ($C_{\alpha 2} 1...3$)
  - a sequence in secreted Ig's (S)
  - the transmembrane and cytoplasmic parts of Ig receptors (M_1, M_2)

Generation of diversity (Figure 19.1-5): The recombination of immunoglobulin genes proceeds in a fixed order of events. In a maturing B cell the gene segments coding for the H chain are rearranged on one chromosome at first. One of the D gene segments is combined with one of the J gene segments.

This rearrangement is catalyzed by specific recombinases and is dependent on signal sequences, which are located 5' and 3' to each D and 5' to each J gene segment (Figure 19.1-6). They consist of highly conserved 7 bp and 9 bp sequences, separated by spacers of 12 bp or 23 bp length, which are not conserved. The signal sequences combine in such a way, that the 12 bp spacer is located on one side, the 23 bp spacer on the other side of a loop (12 / 23 spacer rule). The recombinase system likely recognizes this palindromic structure, removes the DNA loop and ligates the D and J gene segments.

A sequence modification takes place by the introduction of a variable number of nucleotides between the cuts by the template-independent enzyme terminal deoxynucleotidyl transferase, TdT (so-called N region formation), by removal of nucleotides or by variation in the crossover points.

After the D-J recombination, the DJ segment is combined with a V segment by the same mechanism. This results in a functional heavy chain gene. This is transcribed into a pre-mRNA, which also contains downstream the genes for the constant domains of the µ chain (in case of IgM; for other isotypes see below). After splic-

Figure 19.1-4. Human Immunoglobulin Isotypes

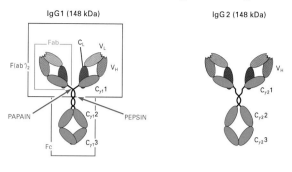

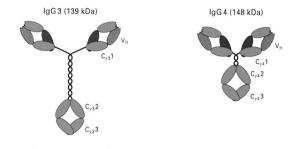

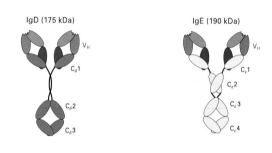

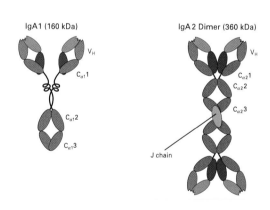

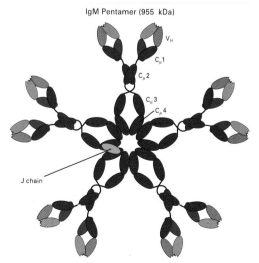

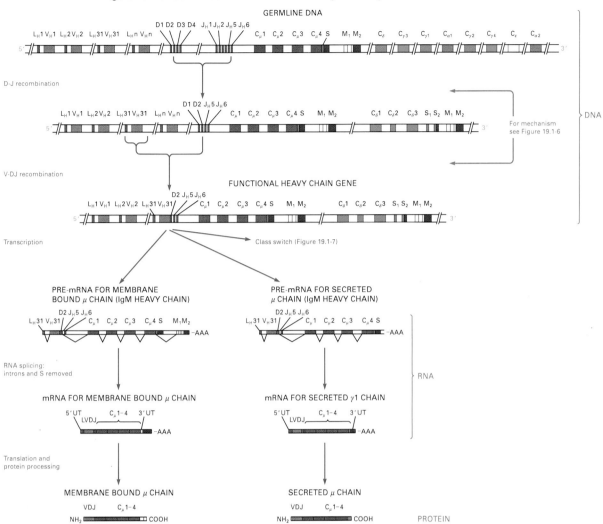

Figure 19.1.5. Generation of a Functional Heavy Chain by Somatic Recombination

ing, the mRNA is then translated into the μ polypeptide chain. The functional gene for the light chain is formed analogously.

If the recombination of the heavy chain gene segments results in a non-functioning gene, e.g. by generation of a stop codon, the recombination is repeated at the other chromosome. With the light chain gene, the V-J recombination takes place at first at the κ chain locus. If this recombination on either chromosome did not result in a functioning gene, then recombination of the V and J gene segments at the λ chain locus takes place. For heavy chains as well as for light chains, the information of only one chromosome is used for production of the immunoglobulin polypeptide chain ('allelic exclusion').

After activation of mature B lymphocytes by antigens, many point mutations may occur in the V segments of immunoglobulin genes (somatic hypermutation, several orders of magnitude more frequent as in other genes). This way, the affinity of antibodies for a certain antigen can increase (or decrease) during the immune response (maturation of affinity).

Thus, the huge diversity of binding specificity by the V domains of antibodies is achieved by the following mechanisms:
- selection of a single gene segment out of each of the V, D, J gene complexes yields many different VDJ combinations
- modification of the nucleotide sequence in the recombination area (junctional diversity)
- combination of various H and L chains. The V domains of both contribute to the structure of the antigen binding site
- somatic point mutations in the V segments of immunoglobulin genes occurring after activation of mature B-lymphocytes.

Membrane bound and soluble immunoglobulins: Immunoglobulins exist either as membrane-based receptors or as soluble antibody molecules. The DNA for the heavy chain of immunoglobulin M (Figure 19.1-5) contains:

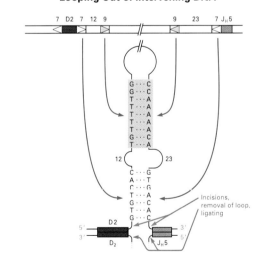

Figure 19.1-6. D-J Gene Recombination by Enzyme-Dependent Looping Out of Intervening DNA

- two gene segments (M_1, M_2) coding for the transmembrane region and the short cytoplasmatic portion of the H-chain
- one gene segment (S) coding for the hydrophilic C-terminal region of the soluble form of the H-chain

After transcription, the mRNAs for both types of expression are provided by different splicing mechanisms. Either the S segment or the M segments are removed, yielding after translation membrane bound or soluble proteins, respectively.

Figure 19.1-7. Class Switch by Different Splicing of Long Transcripts (Left) or by DNA Recombination (Right)

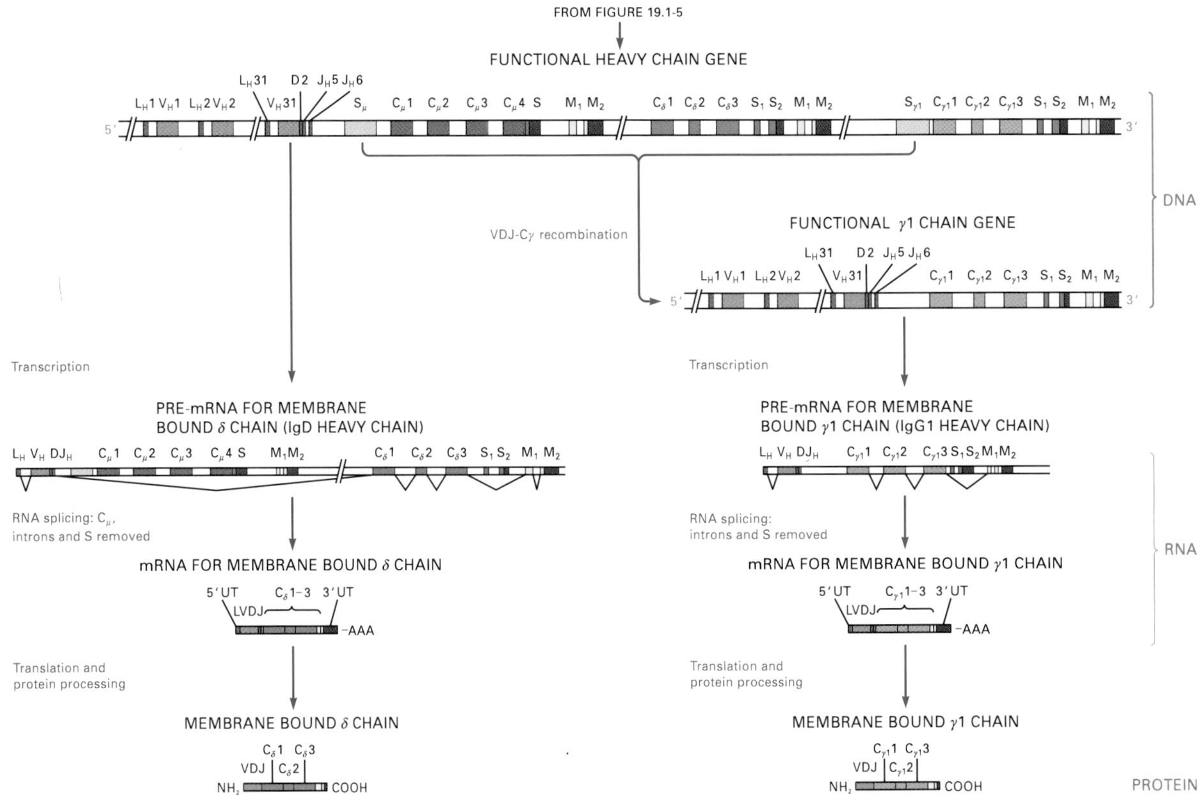

Classes of immunoglobulins: Immunocompetent B lymphocytes, which have not yet been specifically activated by antigen, simultaneously carry both IgM and IgD on their surface.

- **IgM:** The reactions leading to IgM molecules have been described above.
- **IgD** (Figure 19.1-7, left): The connection of the recombined VDJ sequence with the sequences for the constant domains of the δ heavy chain (C_δ) takes place post-transcriptionally by differential splicing of the long pre-mRNA. Hereby, the VDJ segment of the RNA becomes linked with the 3 RNA segments coding for the constant domains of the δ chain. The intermediate segments coding for the constant domains of the μ chain are eliminated.
- **IgG, IgA, IgE:** The genes for these immunoglobulin isotypes are generated by further DNA recombination procedures at the H-chain locus (Figure 19.1-7, right, shows the case of the $\gamma 1$ chain gene). The recombined VDJ gene segment for the V_H domain, which is at first connected with the gene segments for the constant domains of the μ chain ($C_\mu 1...4$, coding for IgM), gets connected with the gene segments coding for the constant domains of other H chain isotypes ($C_{\gamma 1...4}1...3$ for IgG's, $C_{\alpha 1...2}1...3$ for IgA's, $C_\varepsilon 1...4$ for IgE).

These switch recombinations are regulated by repetitive, GC-rich intron sequences (switch or S regions), which are located 5' of each H chain gene segment, except the δ gene segments. The DNA between S_μ (as donor sequence) and $S_{\gamma 1...4}$, $S_{\alpha 1...2}$ or S_ε (as acceptor sequences) forms a loop which is cut out (looping-out deletion). The functional heavy chain gene is transcribed into a primary RNA, which is processed into a mRNA and translated into a polypeptide chain (11.3; 11.5).

Antigen receptor complex on B lymphocytes (Figure 19.1-8): The antigen receptor complex on the surface of B lymphocytes contains the membrane form of an immunoglobulin and an heterodimer of the polypeptide chains Igα (CD79a, not to be confused with IgA!) and Igβ (CD79b), which are responsible for the signal transduction leading to gene activation after antigen binding (17.5.4). Closely associated with the B-cell receptor complex is the co-receptor molecule CD19. On its cytoplasmatic side, it is associated with a tyrosine kinase of the Src family (Tab. 17.5-5).

If the antigen exists as an immune complex with C3d molecules (19.2.3), its binding to the B-cell surface via the immunoglobulin receptor can be augmented by the complement receptor 2 (CD21).

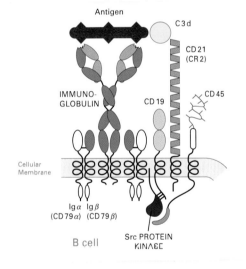

Figure 19.1-8. Antigen Receptor Complex of B Lymphocytes

19.1.4 Antigen Recognition by T Lymphocytes

While immunoglobulins recognize a wide range of different structures on the surface of soluble and insoluble molecules [e.g., linear peptide sequences, conformation determinants, small haptene molecules and carbohydrate structures (13.5)], T lymphocytes recognize with their antigen receptors only complexes of small peptides and MHC molecules (19.1.5) at the surface of *nucleated cells* (including *B lymphocytes*). These peptides originate from intracellular processing of intracellular or exogenous proteins (see below).

During their differentiation in the *thymus*, T lymphocytes undergo a complex selection procedure. As a consequence, mature T cells preferentially recognize peptides, which are bound to autologous MHC molecules. Concomitantly, the majority of T cells, which recognize peptide sequences of self proteins, are eliminated.

Structure of the T cell receptor (TCR, Figure 19.1-9): The T cell receptor is a covalently bound heterodimer of an α- (40...60 kDa) and a β-chain (40...50 kDa). A small subpopulation contains γ- and δ-chains, instead. Each chain consists of a variable domain and a constant domain next to the membrane. They are structured analogously to the immunoglobulin domains (19.1.3). The variable domains each contain 3 complementarity determining regions (CDR), which are responsible for the binding specificity of the T cell receptor.

The genes for the 4 different T cell receptor chains are located at different chromosomes. Analogously to immunoglobulins, the functional genes for the V domains are formed by somatic recombination of V, J and D segments (in case of β and δ chains) or of V and J segments (in case of α and γ chains). The same mechanisms as in immunoglobulins (19.1.3) are involved in the generation of the multiple receptor specificities. Somatic point mutations have generally not been observed with T cell receptors.

Like the B cell receptor, the T cell receptor is associated with polypeptide chains, which are primarily responsible for signal transduction (Fig. 17.5-6). These chains form the CD3 complex, which is composed of homo- or heterodimers of δ, ε, γ, ζ and η chains (Figure 19.1-9). In a wider sense, CD4 and CD8 molecules are also considered part of the T cell receptor complex. They contribute as coreceptors to the activation of T cells after recognition of antigens.

Interactions of T cell receptors, activation of T cells (Figure 19.1-9): While the T cell receptors interact mainly with the peptides presented by the MHC molecules (19.1.5) and with polymorphic determinants of the MHC molecules themselves, the coreceptors interact with monomorphic determinants on the MHC molecules. The CD4 coreceptor binds to the β2 domain of the MHC class II molecule and the CD8 coreceptor to the α3 domain of the MHC class I molecule. Since exclusively either CD4 or CD8 molecules are expressed on mature T lymphocytes, it follows that

- CD4⁺ T lymphocytes are activated by antigen peptides, which are presented by MHC class II molecules.
- CD8⁺ T lymphocytes recognize peptides, which are bound to MHC class I molecules.

The cytoplasmic section of the CD4 and CD8 coreceptors are associated with a tyrosine kinase of the Src family (17.5.4). This tyrosine kinase is activated by dephosphorylation, which is catalyzed by the cytoplasmic section of the CD45 molecule. CD45 is expressed in all *nucleated hematopoetic cells*, it occurs on various cells in different isoforms.

Also, other membrane molecules take part in the activation of T lymphocytes. For the primary antigen-specific activation of T cells, two signals are necessary. While the above mentioned binding is the first signal, the second one is contributed by interaction of the CD28 and CTLA4 molecules at the T lymphocyte surface with the B7.1 (CD80) and B7.2 (CD86) molecules, respectively, at the surface of the antigen-presenting cells (APC). Therefore, they are named co-stimulatory molecules.

Of importance is the enhancement of the cellular contact between T lymphocytes and antigen presenting cells. This is achieved by
- binding of the LFA-1 molecule (an integrin molecule, consisting of the CD11a and CD18 chains) to the ICAM-1 (CD54) molecule on the APC

Figure 19.1-9. Membrane Molecules Involved in the Activation of T Lymphocytes

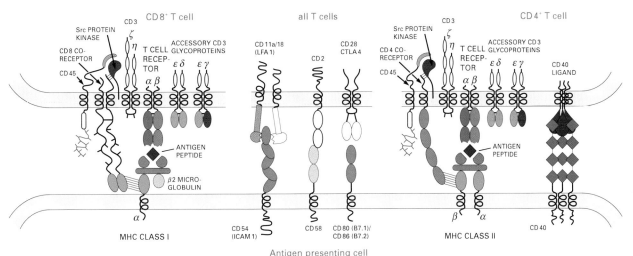

Table 19.1-2. Membrane Molecules Involved in the Activation of T Lymphocytes

Name	Structure	Molecular mass (kDa)	Family*	Function
T cell receptor	dimeric (αβ or γδ)	α: 50, β: 45	Ig	Binds to peptides presented by MHC class I or II molecules
CD2 (LFA-2)	monomeric	45...58	Ig	Cell adhesion molecule, binds to CD58 (LFA-3)
CD3	dimeric (γ, δ, ε, ζ, η)	γ: 25...28, δ: 20, ε: 20, ζ: 16, η: 22	Ig	Associated with T cell receptor, required for its expression, signal transduction
CD4	monomeric	55	Ig	Coreceptor for MHC class II molecules, associates with Lck (p56) protein kinase (17.5.4). Receptor for HIV-1 and -2
CD8	dimeric (αα or αβ)	α: 30, β: 32	Ig	Coreceptor for MHC class II molecules, associates with Lck (p56) protein kinase (17.5.4).
CD11a (LFA-1)	assoc. with CD18	180	integrin	Cell adhesion molecule, binds to CD50 (ICAM-3), CD54 (ICAM-1), CD102 (ICAM-3)
CD18	assoc. with CD11	95	integrin	Associates with CD11a (= LFA-1), CD11b and CD11c
CD28	homodimeric	44 * 2	Ig	Receptor for CD80 (B7.1) and for CD86 (B7.2)
CTLA-4	homodimeric		Ig	Receptor for CD80 (B7.1) and for CD86 (B7.2)
CD 40 ligand (CD154)	trimeric	39	TNF	Binds to CD40
CD40 (gp50)	mono-/trimeric	40	NGFR	Receptor for CD40 ligand (17.5.4), trimerizes after ligand binding, mediates costimulatory effects
CD45 (LCA)		180...240		Contains phosphotyrosine phosphatase at cytoplasmic side, activates Lck (p56) protein kinase (17.5.4).
CD54 (ICAM-1)		85...110	Ig	Cell adhesion molecule, binds to CD11a/CD18 (= LFA-1) etc.
CD58 (LFA-3)		55...70	Ig	Cell adhesion molecule, binds to CD2
CD80 (B7.1)		60	Ig	Ligand to CD28 and CTLA-4, provides costimulatory signals for T cells
CD86 (B7.2)		80	Ig	Ligand to CD28 and CTLA-4, provides costimulatory signals for T cells

* Ig = immunoglobulin superfamily, NGFR = nerve cell growth factor receptor superfamily, TNF = tumor necrosis factor superfamily

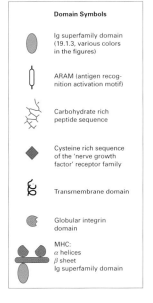

- interaction of the CD2 molecule with the CD58 molecule on the APC.

For B lymphocytes, the costimulatory signal is induced by binding of its CD40 molecule to the CD40 ligand (CD40L, CD154) which is expressed on activated CD4+ T lymphocytes.

19.1.5 Antigen Presentation by MHC Molecules

Besides the antigen receptors of B and T lymphocytes, the adaptive immune system comprises as third antigen binding structure the MHC molecules (major histocompatibility complex). In *humans*, they are named HLA molecules (human leukocyte antigen). Two types of MHC molecules are involved in antigen binding (Figure 19.1-10):

- MHC Class I molecules consist of an α-chain with 3 domains (ca. 44 kDa) and of β_2 microglobulin (β2m, 12 kDa), which is not covalently bound. They are expressed in every *nucleated cell* of the organism. The antigens presented by them are recognized by CD8+ T lymphocytes (cytotoxic T cells).
- MHC Class II molecules consist of an α- and a β-chain (33 and 28 kDa, with 2 domains each), which are not covalently bound. They occur primarily only on cells, which present extracellular protein antigens to the T lymphocytes, such as *dendritic cells, monocytes / macrophages* and *B lymphocytes* (see below). The antigens presented are recognized by CD4+ T lymphocytes (helper cells, 19.1.2).

The domains of the MHC molecules, which are adjacent to the membrane and the β_2 microglobulin domain show a structure similar to immunoglobulins (19.1.3). The sections distant to the membrane are forming a groove with a β-pleated sheet structure at the bottom and an α-helix at each wall. The antigen peptides are bound within the groove (Figures 19.1-11 and 19.1-12). The specificity of the binding is determined by the side chains of the amino acids composing the β-pleated sheet and the helices, which interact with the side chains of the peptides. The T-cell receptor binds both to the peptide and to the side chains of the α-helices.

The genes for the MHC chains are exceedingly polymorphic (Figure 19.1-13). About 200 different genes and pseudogenes are localized in the major histocompatibility complex. In *humans*, it is localized on the short arm of chromosome 6 and is subdivided into 3 regions. Closest to the centromere is the class II region, which contains the genes for the α and β chains of the HLA-DP, DQ and DR molecules. Then follows the class III region, which contains the genes for the complement factors C2, C4 and Bf (19.2), the cytokines tumor necrosis factor (TNF), lymphotoxin (LT) and others. Finally there is the class I region, which encompasses the genes for the α chains of the HLA-A, HLA-B and HLA-C molecules. For each of the MHC genes there exist a different number of alleles, which are codominantly transmitted. They vary mainly in the sequence coding for the peptide binding groove. Therefore

- MHC molecules derived from different alleles will bind a given peptide with different affinity
- antigen presenting cells from individuals differing in their HLA haplotype will present different sequences from a given protein to the T lymphocytes

Processing and presentation of antigens for T lymphocytes (Figure 19.1-14): Class I MHC molecules are charged with peptides immediately after synthesis of the α-chain and its complexing with β_2 microglobulin in the *endoplasmic reticulum*. Only peptides with a length of 9...10 amino acids fit into the groove. They originate mainly from intracellular proteins, which are cleaved into peptides by a protease complex (large multifunctional proteasome, LMP ①, cf. 14.6.7) present in the *cytoplasm*. (The genes for LMP and the TAP molecules are also present in the MHC, Fig. 19.1-13.) Then

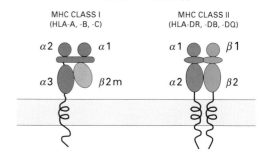

Figure 19.1-10. Structure of MHC Class I and Class II Molecules

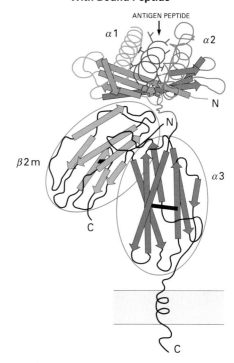

Figure 19.1-11. MHC Class I Molecule With Bound Peptide

Figure 19.1-12. MHC Class I Molecule With Bound Peptide, Top View

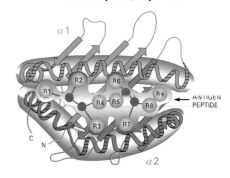

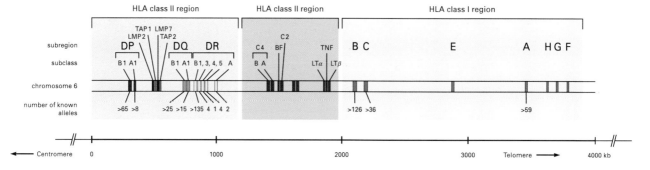

Figure 19.1-13. Major Histocompatibility Complex of *Humans*

Figure 19.1-14. Antigen Processing and Presentation

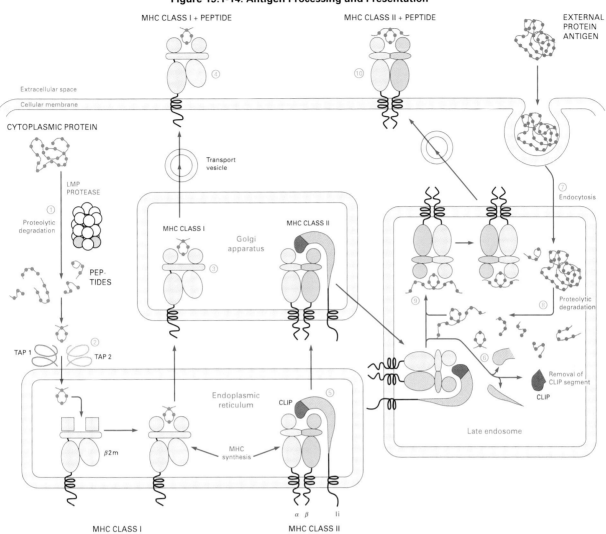

they are transported into the *endoplasmic reticulum* by special transport proteins (transporter associated with antigen processing, TAP1 and TAP2 ②), where they bind to MHC class I / β2 microglobulin and stabilize the complex, which is then transported via the *Golgi apparatus* ③ to the *cellular membrane* ④. Thus, class I MHC allows the recognition of all intracellularly synthesized proteins by T lymphocytes. This is of special importance for the recognition of virus infected cells by cytotoxic (CD8$^+$) T lymphocytes (19.1.7).

The primary task of Class II MHC molecules is the presentation of peptides derived from extracellular proteins. These enter the so-called antigen presenting cells (APC, e.g. *dendritic cells, monocytes / macrophages* and *B lymphocytes*) via phagocytosis or endocytosis (⑦, 18.1.2) and are split into peptides in *phagolysosomes / endosomes* (⑧, 14.6). The fragments bind to class II MHC molecules, which are present at the *endosome* membrane ⑨. The MHC-peptide complexes are transported to the *cellular membrane*, where they can be recognized by CD4$^+$ lymphocytes ⑩.

After synthesis in the *endoplasmic reticulum*, the class II MHC molecules form a complex with the invariant chain Ii ⑤, which prevents the premature binding of peptides by its CLIP segment (class II associated invariant chain peptide). The class II MHC-Ii complex reaches the *endosomes* via the *Golgi apparatus*. There, the CLIP segment is split off and the rest of the Ii chain dissociates ⑥. Thus, the groove is free for binding of external peptides. These peptides (15...24 amino acids) are longer than peptides bound by class I MHC molecules, since the class II groove is open on both sides.

19.1.6 Cytokines and Cytokine Receptors

The term 'cytokines' comprises various proteins and glycoproteins, which, being mainly soluble messengers, are responsible for the intercellular communication. They bind to specific receptors at the cellular surface and can induce or inhibit the activation of genes. This way, they influence growth, differentiation and activation of cells. Here only those cytokines are discussed, which are known to be important for the functions of the immune system.

Special properties of cytokines are
- a particular cytokine can act on different cells and hereby exert different functions (pleotropism)
- several cytokines can effect the same results (redundancy)

The cytokine effects can be autocrine (acting on the producing cell), paracrine (acting on cells in immediate contact to the producing cell) or endocrine (acting on target cells after reaching them via the circulatory system).

With respect to basic structure, cytokines are subdivided into 3 major groups, consisting of 4 short or 4 long α-helices or of antiparallel β-strands (Figure 19.1-15).

Likewise, the cytokine receptors are subdivided according to their structure (Figure 19.1-15):

- Type I or hematopoietin receptors, also called the cytokine receptor superfamily, are the largest group. They consist generally of 1 or 2 polypeptide chains (which are responsible for the specificity of cytokine binding) and of another chain (which effects the signal transduction). After binding of their ligand, they associate with the 'common' (c, since it is used by several receptors) signal transduction sequence. Frequently this chain starts with tyrosine kinases of the Jak type (17.5.4).
- Type II or interferon receptors also consist of 2 polypeptide chains (17.5.4, Fig. 17.5-5).
- Type III or TNF receptors contain the cystine rich sequences of the nerve growth factor receptor family. They form trimers after binding of cytokines.
- Type IV or Ig superfamily receptors contain cytoplasmic sequences with tyrosine kinase activity in a number of cases (17.5.3).

Figure 19.1-15. Cytokines and Cytokine Receptors

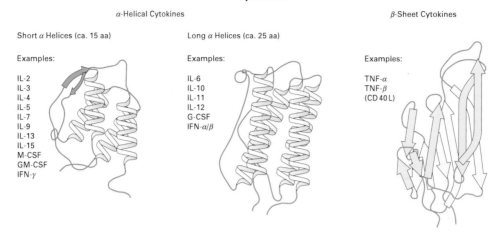

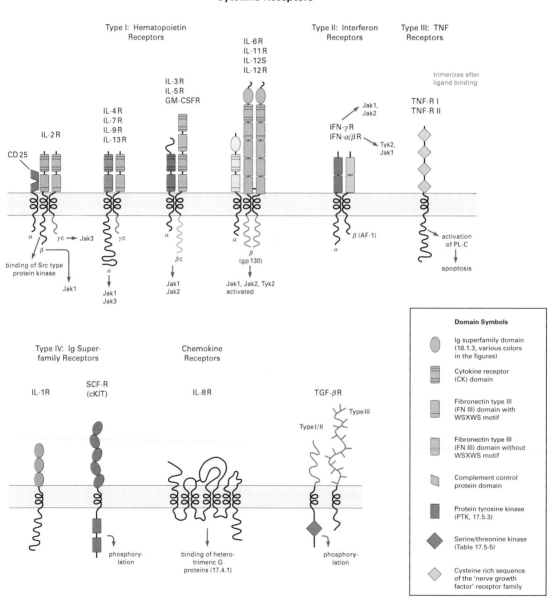

- Chemokine receptors possess a structure with 7 membrane passes, typical of G-protein coupled receptors (17.4.1).
- TGF (transforming growth factor) family receptors contain cytoplasmic sequences with serine-threonine kinase activity. They can associate with proteoglycan molecules for the enhancement of cytokine binding.

Immunologically important cytokines are listed in Figure 19.1-16, their effect on the differentiation of hematopoietic cells is shown in Figure 19.1-3.

Figure 19.1-16. Immunologically Important Cytokines, Their Sources and Main Effect on Target Cells
(The effect on the differentiation of hematopoietic cells is shown in Figure 19.1-3)

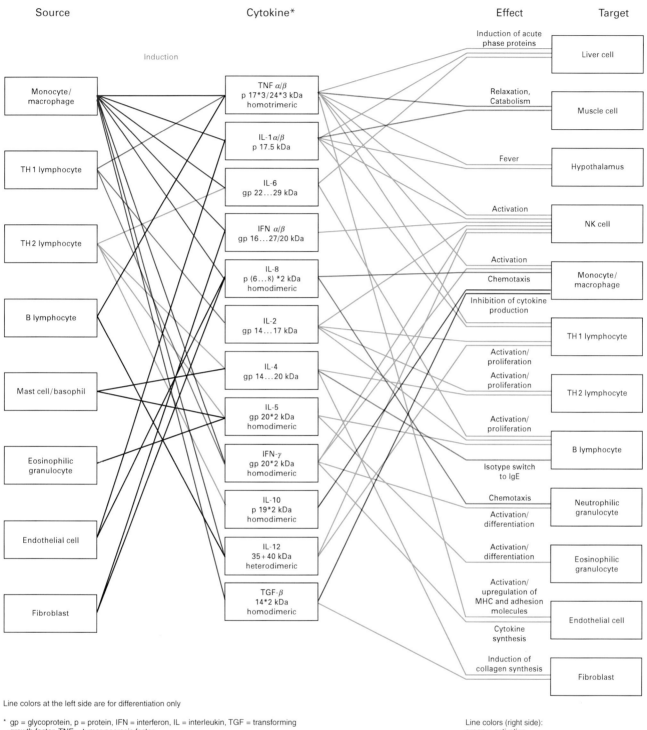

Line colors at the left side are for differentiation only

* gp = glycoprotein, p = protein, IFN = interferon, IL = interleukin, TGF = transforming growth factor, TNF = tumor necrosis factor

Line colors (right side):
green = activation
red = inhibition
blue = chemotaxis
orange = other

19.1.7 Regulation of the Immune Response

The essential steps during induction of cellular and humoral immune responses against *bacteria* and *viruses* are compiled in Figure 19.1-17. The CD4⁺ T effector cells (helper cells) are primarily responsible for coordination of the defense mechanisms. They activate as TH1 cells the cellular and as TH2 cells mainly the humoral arm of the immunological response.

Uptake and presentation of antigens: This action of the *antigen presenting cells* (APC, ①) is the first step for induction of the immune response.

The binding of antigens to the APC membrane can be augmented by various mechanisms. E.g., *bacteria* can be attached directly via lectin receptors, such as mannose receptors. Also, *bacteria* can be opsonized by antibodies, which are already present. Binding and uptake of opsonized *bacteria* occur either via Fcγ receptors or via complement receptors.

The internalized antigens are processed and those antigen peptides, which fit into the binding grooves, are bound by class II MHC molecules (compare Figure 19.1-14). After translocation to the *cellular membrane*, the complexes can be recognized by CD4⁺ T lymphocytes ②.

Activation and differentiation of T lymphocytes: Two signals are required for the primary antigen-specific activation:
- the binding of the T cell receptor to the peptide complex at the antigen presenting cell ②
- a costimulatory signal, which is generated by interaction between B7.1 (CD80) or B7.2 (CD86) and CD28. Interaction with CTLA4 may be inhibitory (③, see also 19.1.4 and Figure 19.1-9).

The activated CD4⁺ T lymphocytes differentiate into TH1 or TH2 effector cells ④, caused by signals, which are at present only partially known. Liberation of interleukin-12 (IL-12) by *macrophages* following a bacterial infection favors the formation of TH1 cells, while IL-4 guides towards TH2 cells ⑤. Autocrinic stimulation by interleukin-2 (IL-2) is the major cause of proliferation of T cells after stimulation ⑥.

Humoral arm of the immune response: These reactions are primarily promoted by TH 2 cells, which support the activation and the differentiation of B-lymphocytes primarily via the cytokines IL-4, IL-5 and IL-6 ⑦, whereby IL-6 is especially responsible for the differentiation into *plasma cells*. IL-10, which is formed by TH2 cells, inhibits the development of TH1 cells ⑧.

B lymphocytes can accumulate protein antigens specifically by their immunoglobulin receptors, even if the antigens are present only in small concentrations ⑨. After processing, the B cells present the antigen peptides bound to class II MHC molecules on their surface. *T lymphocytes,* which recognize these peptide-MHC complexes specifically, supply the B lymphocytes with the signals required for activation, proliferation and differentiation into plasma cells.

Analogously to *T lymphocytes,* usually two signals are also required for activation of *B lymphocytes*:
- binding of antigen to the Ig receptor on their surface
- interaction of the CD40 receptor with the CD40 ligand, which is expressed only in activated T-cells ⑩.

This coupled, dual recognition of protein antigen molecules by B and T lymphocytes ensures that *B lymphocytes* produce antibodies only against those antigens, which they recognize themselves, and whose peptide fragments are recognized by *T lymphocytes*. This is also necessary, because B lymphocytes can change the specificity of their receptors via somatic point mutations (19.1.3). The control of the B cell antibody production by T lymphocytes is very important for the orderly function of the immune system, since the secretion of antibodies against endogenous structures is restricted this way.

Under certain circumstances B lymphocytes are activated and produce antibodies without T cell help. The so-called T independent antigens (e.g. carbohydrates and repetitive determinants induce only IgM antibodies, since the isotype switch in B cells requires T cell assistance.

The *plasma cells* secrete specific antibodies (Ig's ⑪), which have various effects on infectious *microorganisms*, depending on the isotype. They agglutinate or opsonize extracellular *bacteria* or

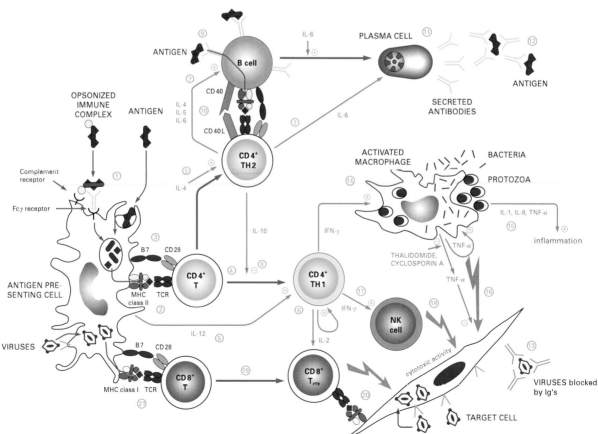

Figure 19.1-17. Regulation of the Antigen-Specific Immune Response

viruses (①, ⑫) and prevent the attachment of *viruses* to their specific receptors on the surface of target cells ⑬. They can induce the lysis of *bacteria* by activation of the complement system (19.2) and can also neutralize bacterial toxins.

Cellular arm of the immune response: Besides IL-2, IFN-γ is the most important cytokine, which is secreted from activated TH1 cells. One of its prevalent tasks is the activation of *macrophages* as a prerequisite for the killing of intracellularly growing microorganisms ⑭. *Activated macrophages* liberate the multiple active cytokines IL-1 and TNF-α. These cause local and systemic inflammatory reactions, the formation of cellular infiltrates being increased by chemokines ⑮. TNF-α can act cytotoxically on *tumor cells* ⑯. IFN-γ also stimulates the cytotoxic activity of natural killer cells (NK) ⑰ against virus infected cells or tumor cells ⑱.

The immune response against *viruses* is also started by *antigen presenting cells*. After infection, they present viral peptides (bound to HLA class I molecules) and stimulate CD8⁺ T lymphocytes to differentiate into *cytotoxic effector cells* ⑲. Second costimulatory signals required for activation are also provided by interaction between B7.1 (CD80) or B7.2 (CD86) and CD28 or CTLA4, respectively (㉑, see 19.1.4) or by IL-2, which is produced by CD4⁺ T cells. The major function of CD8⁺ T lymphocytes (cytotoxic T cells) is the recognition and killing of virus infected cells ⑳.

19.1.8 IgE Mediated Hypersensitivity of the Immediate Type

The allergic reaction of the immediate type is a pathological hypersensitivity proceeding in two phases (Figure 19.1-18):
- During the induction phase, the formation of IgE antibodies against a definite antigen is stimulated. The IgE antibodies bind to the high affinity IgE receptors (FcεR I) located on the membrane of *mast cells* and *basophilic granulocytes*.
- During the effector phase, the binding of multivalent antigens to the receptor-bound IgE antibodies leads to cross-linking of the Fcε receptors. This generates a signal, which causes the instant liberation of mediators by exocytosis, mostly histamine (4.8.2), tryptase, kallikrein and chemotactic factors for *neutrophilic* and *eosinophylic granulocytes*. Simultaneously, synthesis of prostaglandins, leukotrienes (17.4.8), platelet activating factor (6.3.3) and cytokines is induced; their secretion proceeds with some delay. The essential effects of the various mediators are shown in Figure 19.1-16.

The decisive step during the development of the immediate type allergy consists in the switch of the antibody production in activated *B-lymphocytes* to the isotype IgE. This isotype switch is induced by the cytokines IL-4 and IL-13 (from *TH2 cells*) and inhibited by the cytokine IFN-γ (from *TH1 cells*).

Literature:
Abbas, A.K. *et al.*: *Cellular and Molecular Immunology.* W.B. Saunders Co. (1991).
Brown, E. *et al.*: Curr. Opin. Immunol. 6 (1994) 43–145.
Callard, R.E., Gearing, A.J.H.: *The Cytokine Facts Book.* Academic Press (1994).
Germain, R.N.: Cell 76 (1994) 287–299.
Ikuta, K. *et al.*: Ann. Rev. Immunol. 10 (1992) 759–783.
Honjo, T., Alt, F.W. (Eds.): *Immunoglobin Genes,* 2nd ed. Academic Press (1995).
Janeway, C.A. jr., Travers, P.: *Immunobiology.* 2nd ed. Current Biology and Garland Publishing Inc. (1997).
Janeway, C.A. jr., Bottomly, K.: Cell 76 (1994) 275–285.
Kishimoto, T. *et al.*: Cell 76 (1994) 253–262.
Lichtenstein, L.M.: Scientific American 269 (1993) 117–124.
Schlossman, S.F. *et al.* (Eds.): *Leucocyte Typing V.* Oxford University Press (1995).
Paul, W.E.: *Fundamental Immunology.* 3rd ed. Raven Press (1993).
Rammensee, H.G., Monaco, J.: Curr. Opin. Immunol. 6 (1994) 1–71.
Reth, M.: Ann. Rev. Immunol. 10 (1992) 97–121.

Figure 19.1-18. Two Phases of the IgE Mediated Allergic Immune Response

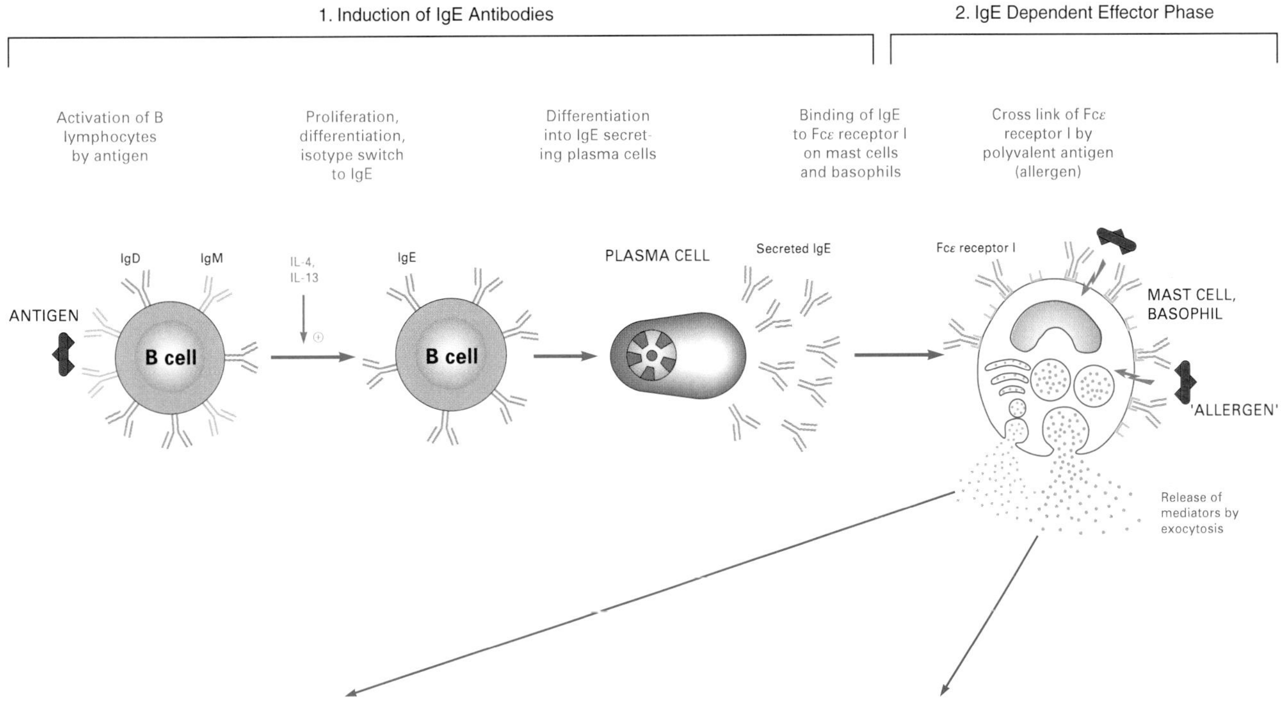

Newly synthesized mediators	Biological effect	See Section
Leukotriene C₄, Leukotriene D₄	bronchoconstriction, edema, hypersecretion, inflammation	17.4.8
Leukotriene B₄	chemotaxis	17.4.8
Thromboxane A₂, PDG₂	bronchoconstriction, edema, hypersecretion, vasodilatation	17.4.8
Platelet activating factor (PAF)	bronchoconstriction, vasodilatation, aggregation and activation of platelets	6.3.3, 20.4
Cytokines	activation of inflammatory cells	19.1.7

Preformed mediators	Biological effect	See Section
Histamine	vasodilatation, bronchoconstriction	4.8.2
Heparin	inhibition of blood coagulation	20.3
Tryptase	activation of complement factor C3	19.2
Kallikrein	generation of kinins, vasodilatation, edema	20.2.2
Eosinophilic chemotactic factor A (ECF-A)	chemotaxis of eosinophils	
Neutrophilic chemotactic factor (NCF) = IL-8	chemotaxis of neutrophils	

19.2 Complement System

Complement is a system of plasma proteins in *vertebrates* that received its name from its property to interact with bound antibodies and to 'complement' the antibacterial activity of antibodies. Two pathways can activate the effector systems of complement:
- the phylogenetically older alternative pathway (part of the innate immune system)
- the classical pathway (part of the adaptive humoral immune response, 19.1).

The main effects of complement are:
- lysis of pathogens and cells
- solubilization of immune complexes
- enhancement of phagocytosis of pathogens ('opsonization')
- induction of inflammation, activation of inflammatory cells.

The activation of the complement factors proceeds via a proteolytic cascade (similar to 20.1), each member cleaves and activates this way the next member of the sequence. The activated factors join the complex of their predecessors at the pathogen surface.

Except for the classical complement factors C4, C2 and C3, the numbering reflects the sequence of action. Attached small letters (e.g. C4a) indicate cleavage products. The nomenclature for the released fragment of C2 has recently been changed from C2b to C2a, the activated one is named C2b.

19.2.1 Activation of the Complement Pathways (Figure 19.2-1)

Classical pathway: Prerequisite for activation is the binding of an IgM pentamer molecule (19.1.3) to the surface of a pathogen or the aggregation of two IgG molecules at a cell surface or at an immune complex (within 30 ... 40 nm of each other). The activation is initiated by the binding of at least two of the six globular heads of the C1q molecule (Figure 19.2-3) to the Fc domains of the IgM pentamer or of both IgG monomers. This binding of C1q leads to autoactivation of C1r which cleaves and activates the associated C1s molecule. C1s is a serine protease (14.6.2), acting on C4 and C2.

Cleavage of C4 results in the release of C4a and the exposition of an internal thioester bond at C4b which is either hydrolyzed (inactivation) or mediates covalent binding of C4b to a nearby hydroxyl or amino group (at the surface of the pathogen or at the Ig's).

After attachment, C4b binds C2 which then becomes susceptible to cleavage by C1s, resulting in release of C2a and activation of C2b.

Binding of C3 to the C4b, 2b complex (classical pathway C3 convertase) leads to cleavage of C3, the release of the anaphylatoxin C3a and formation of the complex C4b, 2b, 3b (classical pathway C5 convertase), which then cleaves C5.

Alternative pathway: The alternative pathway is initiated by cleavage of the abundant protein C3 to C3b. This enzymatic reaction may be performed
- by the slow but continuous activation of C3 in the fluid phase to C3(H₂O), which is able to attach to B. B in the complex is then activated to C3 (H₂O) Bb or
- by the classical pathway C3 convertase C4b, 2b (coupling of both pathways) or
- by the alternative pathway C3 convertase C3b, Bb (amplification loop, see below).

Only if the resulting C3b succeeds in binding covalently with its thioester group to a hydroxyl or amino moiety at a pathogen surface, the following reactions take place. Otherwise, hydrolysis of the thioester group leads to inactivation.

Surface-attached C3b binds B which then becomes susceptible to cleavage by D, resulting in the release of Ba and formation of the C3b, Bb complex (alternative pathway C3 convertase). This convertase acts on another C3 molecule, leading to the release of the anaphylatoxin C3a and its association with the C3b formed. This results in the complex C3b, Bb, C3b (alternative pathway C5 convertase) which then cleaves C5.

On the other hand, the activity of the C3 convertase C3b, Bb generates extra C3b, which saturates rapidly the surface of the pathogen and is the seed of new C3b, Bb complexes. This mechanism is called 'amplification loop'.

Homology between classical and alternative pathway (Table 19.2-1): There is a striking homology between both pathways, starting from covalent binding of C4b and C3b, respectively, and leading to the formation of C5 convertase.

19.2.2 Formation of the Membrane-Attack Complex (MAC), Lysis of Pathogens and Cells (Figure 19.2-2)

The cleavage of C5 by the classical or alternative C5 convertase initiates the assembly of the terminal complement components to

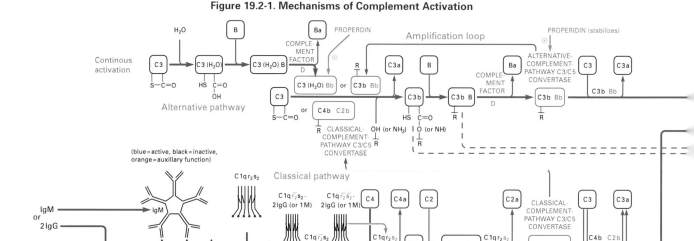

Figure 19.2-1. Mechanisms of Complement Activation

Table 19.2-1. Comparison of Pathway Functions

Steps	Classical Pathway	Alternative Pathway
Deposition of covalently bound complement protein	C4b	C3b
followed by binding of further complement protein	C4b, 2	C3b, B
This complex is cleaved by a serine protease	C1s	D
resulting in C3 convertase	C4b, 2b	C3b, Bb
Binding of C3 takes place	C4b, 2b, 3	C3b, Bb, 3
Activation, formation of C5 convertase	C4b, 2b, 3b	C3b, Bb, 3b

form the membrane-attack complex C5b...9. Binding of C6 to C5b enables the binding of C7. This exposes a hydrophobic site on C7 leading to insertion in the lipid bilayer. C8 binds to C5b...7 on membranes and induces the polymerization of approx. 10 to 16 C9 molecules which form a transmembrane ring-shaped structure with a hydrophilic inner channel (about 10 nm diameter). This channel allows free passage of water and other small molecules across the lipid bilayer, leading to cell lysis. It resembles the perforin pores generated by cytotoxic T cells and NK cells (19.1.7).

19.2.3 Other Effects of the Complement System

Solubilization of immune complexes (Figure 19.2.2, bottom): Aggregates of antigen and antibody can be solubilized by the classical complement pathway and covalent deposition of C4b and C3b. Immune complexes bind via deposited C4b and C3b to the complement receptor CR1 on *erythrocytes* and are removed from them by *macrophages* in liver and spleen.

Opsonization of pathogens (Figure 19.2.2, bottom): The deposition of covalently bound C3b and, to a much lesser extent, C4b at the surface of pathogens is very important. This plays a major role in their uptake and destruction by phagocytic cells. Pathogen-attached C3b (C4b) and some of their cleavage products bind specifically to the receptors present on various phagocytic cell types (opsonin activity, 19.1.7, Table 19.2-2). This interaction between opsonins and their receptors is a prerequisite for efficient engulfment and uptake of pathogens. However, additional signals are required to initiate phagocytosis.

In addition, binding of immune complexes via complement receptors to the surface of *follicular dendritic cells* in *lymphoid organs* results in long lasting antigen deposition for continuous B cell stimulation (19.1.2).

Induction of inflammation, activation of inflammatory cells: The C5a, C3a and C4a fragments, which are released during the complement activation steps induce directly a local inflammatory reaction, including smooth muscle contraction and increased vascular permeability. The increase in tissue fluid enhances the movement of pathogens to local lymph nodes. In addition, C5a (the most active one) acts on *mast cells, monocytes* and *neutrophils* and induces a multitude of activating effects.

Table 19.2-2. Complement Receptors

Receptor	CD Nr.	Specificity for	Present at	Function
CR1	35	C3b, C4b	*erythrocytes, macrophages, monocytes, neutrophils, B cells*	stimulation of phagocytosis, decay of C3b and C4b, transportation of immune complexes (erythrocytes)
CR2	21	C3d, C3dg, iC3b, EBV	*B cells*	EBV receptor, B cell stimulation
CR3 *	11b/18	iC3b	*macrophages, monocytes, neutrophils*	stimulation of phagocytosis
CR4 *	11c/18	iC3b	*macrophages, monocytes, neutrophils*	stimulation of phagocytosis
C1q Rec.		C1q	*macrophages, monocytes, B cells, platelets*	binding of immune complexes to phagocytic cells

* part of the integrin family (19.3)

Table 19.2-3. Complement Control Proteins

Factors	Function
Plasma Proteins	
C1 Inhibitor (C1INH)	binds to activated C1r and C1s, leading to displacement from C1q (limits duration of activation)
C4 binding protein (C4bp)	binds to C4b, displacement of C2b, cofactor for C4b cleavage by I
Factor H	binds to C3b, displacement of Bb, cofactor for C3b cleavage by I
Factor I	cleavage of C3b (cofactors H, MCP or CR1)
Membrane Proteins	
Complement-receptor 1 (CR1, CD 35, see above)	binds to C4b, displacement of C2b, binds to C3b, displacement of Bb, cofactor for I
Decay-accelerating factor (DAF, CD 55)	displacement of Bb from C3b, displacement of C2b from C4b
Membrane cofactor protein (MCP)	cofactor for C4b and C3b cleavage by I
CD 59	prevention of MAC formation on homologous cells

19.2.4 Control Mechanism of the Complement System

Inherent amplification mechanisms of the complement system require a tight control in solution and on host cells. The activated components are rapidly inactivated unless they bind to a surface.

Additionally, complement control proteins exist in *plasma* and as *membrane* proteins. If, e.g., the C3/C5 convertases (C4b, C2b; C4b, C2b, C3b or C3b, Bb; C3b, Bb, C3b, respectively) are attached to host cells instead of pathogens, they become inactivated by the control proteins listed in Table 19.2-3.

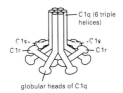

Figure 19.2-3. Structure of the C1 Complex consisting of C1q (C1r)$_2$ (C1s)$_2$

19.2.5 Medical Aspects

Deficiency of C1INH causes hereditary angioneurotic edema. A glycophosphoinositol tail attaches DAF and CD59 to the cellular membrane. If this cellular fixation is defective, paroxysmal nocturnal hemoglobinuria occurs due to erythrocyte lysis. Epstein-Barr virus binds to CR2 expressed at B cells as part of the infection cycle, resulting in infectious mononucleosis.

Literature:
Janeway, C.A., Travers, P.: *Immunbiology.* Current Biology and Garland Publishing (1997) p. 8:35–8:58, 9:8–9:10.
Law, S.K.A., Reid, K.B.M.: *Complement.* IRL Press (1988).

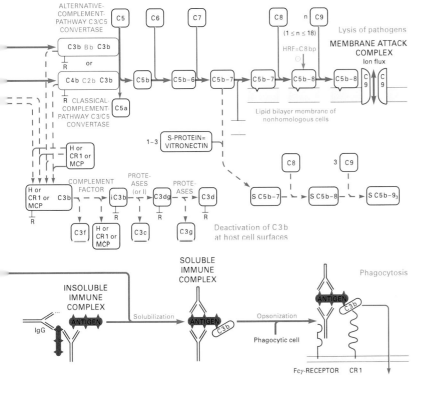

Figure 19.2-2. Actions of Complement

19.3 Adhesion of Leukocytes

Cell adhesion is a fundamental feature of multicellular organisms including their defense mechanisms. In the latter case in *mammals*, leukocytes play a central role. They bind *bacteria, parasites, viruses, tumor cells* etc (19.1.1). Furthermore, their interactions with the *endothelium* are of special importance. Two situations are relevant in this respect:
- During an inflammation or immune reaction (19.1.7), *specialized leukocytes (neutrophilic and eosinophilic granulocytes, monocytes)* adhere to and pass through the *endothelium* of the *blood vessels* and the *underlying matrix*.
- Generally, *lymphocyte* adhesion and passage from the *bloodstream* to the *lymphatic system* occurs in the *high endothelial venules (HEV)* of the *lymph nodes* (19.1.2). This way, lymphocytes can communicate with each other in the lymphatic system and search for foreign compounds after their recirculation to the bloodstream, thus fulfilling their role in the immune system (19.1).

In both cases, the reaction passes through the following steps (Figures 19.3-2 and 19.3-3):

Rolling: The flow of cells is slowed down by first making contacts to the *endothelium*. The interaction takes place between P-, E- and L-selectins and their receptors (Table 19.3-1, Figure 19.3-1); it is weak and reversible. Involved are:
- P-Selectin. It is released from granules of *endothelial cells* to their surface within minutes after irritation. Mediators for release are, e.g., thrombin, histamine or complement proteins. It recognizes the sialyl-Lewis^x- structure (13.5) at the mucin PSGL-1 of *leukocytes*.
- E-Selectin. It is expressed between 3 to 6 hours after irritation of the *endothelium* by *de novo* synthesis. Mediators of the expression are, e.g., interleukin-1, tumor necrosis factor-α (TNF-α) or bacterial lipopolysaccharides. It recognizes sialyl-di-Lewis^x structures and their sulfated derivatives at the *leukocyte* surface.
- L-Selectin. It is constitutively expressed on *leukocytes* and is shedded into the serum. It recognizes sialyl-Lewis^x structures or sulfated derivatives (13.5). The receptor for L-selectin at the *endothelial cells* has not yet been characterized. In the *lymphatic system*, L-selectin binds to MadCAM and to the mucins CD34 and GlyCAM.

Adhesion: After activation of *leukocyte* integrins, firm contacts are established between them and *endothelium* molecules of the immunoglobulin superfamily. These interactions are irreversible. Involved are the integrins LFA-1 (binds to ICAM-1 and 2), Mac-1 (binds to ICAM-1) and VLA-4 (binds to VCAM-1 or MadCAM-1). Additionally, Mac-1 also binds to fibrinogen (forming network bridges) and to the inactivated complement factor iC3b (therefore, Mac-1 is also named complement receptor 3, CR-3, 19.2.3). VLA-4 also binds to fibronectin expressed on the *endothelium*.

Both *leukocyte* and *endothelial* factors are regulated. Activation of integrins is effected by cytokines (e.g., GM-CSF), chemokines (e.g., IL-8) and chemoattractants (e.g., complement factor C5a). Increase in transcription of the endothelial ligands ICAM-1 and VCAM-1 is caused by the same agents as for E-selectin.

Flattening of the cells and diapedesis: Adhering *leukocytes* crawl to an intercellular junction of the *endothelium* and then transmigrate to or even through the *intercellular matrix*. This is mediated by a homophilic interaction of PECAM (platelet-endothelial cell adhesion molecule 1; CD31). It is expressed both on *endothelial cells* (concentrated at the junctions) and on *granulocytes* and *monoytes*. During inflammation, *granulocytes* follow an IL-8 gradient in the tissue towards the focus of infection.

The transmigration of *lymphocytes* to the *lymphatic system* is mediated by interaction of L-selectin or $β_7α_4$-integrin (at the *lymphocytes*) with MadCAM (at the cells of the *high endothelial venules*).

Other adhesion reactions: Similar cell-cell interactions as described for *leukocytes* and their modification play also a role in:
- Embryonic development for segregation of different groups of cells and dispersion/aggregation of migratory cells, including the formation of the *nerve system*. Besides similar molecules as listed in Table 19.3-1 for *leukocytes*, also cadherines are involved. They occur in most cells and mediate Ca^{++} dependent homophilic adhesion (that is, between the same cell type). For dispersion of cells, their expression is downregulated.
- Maintaining correct contacts between cells or with the *extracellular matrix* during postnatal life
- Aggregation of *platelets* during blood coagulation
- Tumor cell migration. Extravasation of migrating tumor cells into certain tissues is meditad by a similar cascade of events as described here for *leukocytes*. Additionally, in several invasive tumor cell types, the levels of E-cadherin or specific integrins are reduced (apparently causing separation from the tissue of origin), while the levels of other adhesion molecules are elevated (needed for invasion?).

Table 19.3-1. Molecules Involved in *Leukocyte* Binding (Examples)

Name	CD No.	Mol. mass kDa	Present in	Notes	Ligand Name	Family	Mol. mass kDa	Present in
Selectins								
P-selectin = PADGEM	62P	140	*endothelial cells, platelets*	released within minutes from α-granula	PSGL-1 (sialyl-Le^x-structure)	sialomucin	220 dimer	*leukocytes*
E-selectin = ELAM-1	62E	105...115	*endothelial cells*	expressed between 3–6 hours	sialyl-di-Le^x- structures = CLA	glycoprotein		*granulocytes, memory T- cells*
L-selectin	62L	75...110	*neutrophils, monocytes, eosinophils etc.*	constitutive, shedded from cells after leukocyte activation	sialyl-Le^x-structures, also sulfated	glycoprotein		*endothelial cells*
same			*lymphocytes*		CD 34, MAdCAM, GlyCAM (secreted)	sialomucin mucinous immuno- globulins	105...120 58–66 50	*periph. lymph nodes lymphoid high endothelial venule*
Integrins *								
LFA-1 ($β_2α_L$)	11a	180 (α) + 95 (β)	*leukocytes*	activated by cytokines, chemokines etc.	ICAM-1 (CD 54) ICAM-2 (CD 102)	immunoglob. immunoglob.	90...115 55–65	*endothelial cells endothelial cells*
Mac-1 = CR3 ($β_2α_M$)	11b/18	170 (α) + 95 (β)	*leukocytes*	activated by cytokines, chemokines etc.	ICAM-1 (CD 54) fibrinogen (soluble)	immunoglob. fibrous prot.	90...115	*endothelial cells bridge to endothelial cells*
					iC3b	inactiv. complem. factor		*attached to endothelial cells*
p150,95 ($β_2α_X$)	11c/18		*leukocytes*		fibrinogen (soluble)	fibrous prot.		*bridge to endothelial cells*
					iC3b	inactiv. complem. factor		*attached to endothelial cells*
VLA-4 ($β_1α_4$)	49d/29	150 (α) + 130 (β)	*lymphocytes, monocytes etc. not neutrophils*	activated by cytokines, chemokines etc.	VCAM-1 (CD106) fibronectin	immunoglob. fibrous prot.	90–110	*endothelial cells bridge to endothelial cells*
$β_7α_4$ = MLA	49d	150 (α) + 120 (β)	*lymphocytes*		MAdCAM-1	mucinous immunogl.	58–66	*lymphoid high endothelial venule*
Immunogl.								
PECAM-1	31	120...130	*platelets, monocytes, granulocytes, T cells*	constitutively expressed	PECAM-1 (see left)	immunoglob.	120...130	*endothelial cells*

* The suffixes indicate the types of the subunits, not their numbers

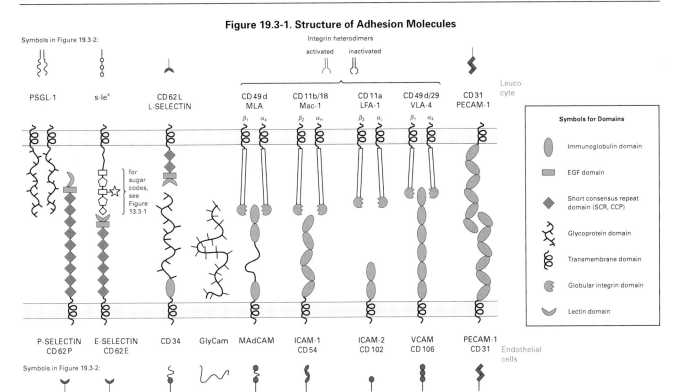

Figure 19.3-1. Structure of Adhesion Molecules

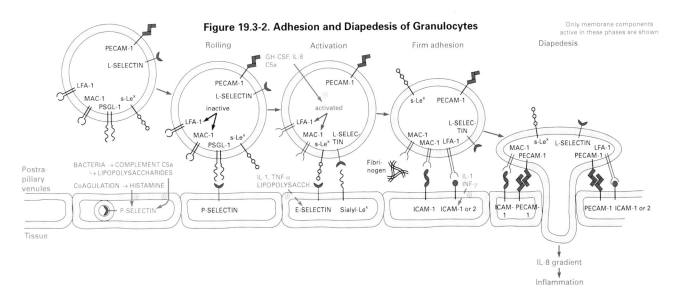

Figure 19.3-2. Adhesion and Diapedesis of Granulocytes

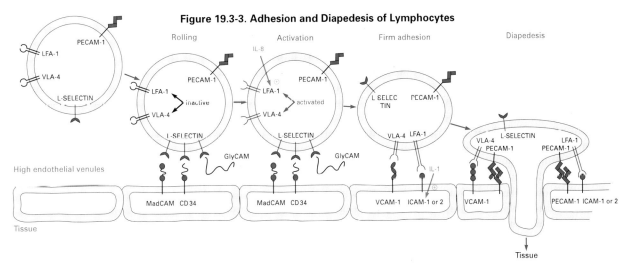

Figure 19.3-3. Adhesion and Diapedesis of Lymphocytes

Literature:
Carlos, T.M., Harlan, J.M.: Blood 84 (1994) 2068–2101.
Lasky, L.A.: Ann. Rev. of Biochem. 64 (1995) 113–139.
Patel, T.P. et al.: Biochemistry 33 (1994) 14815–14824.
Rosen. S.D., Bertozzi, C.R.: Current Opinion in Cell Biology 6 (1994) 663–673.
van der Merwe, P.A., Barclay, A.N., Trends in Biol. Sci. 19 (1994) 354–358.

20 Blood Coagulation and Fibrinolysis

20.1 Hemostasis

Hemostasis is the balanced equilibrium between coagulation and fibrinolysis in the *blood vessels* of all *vertebrates*. Its purpose is
- to avoid blood loss in injured vessels
- to seal intravasal (non-bleeding) wounds
- to initiate wound healing processes
- to dissolve the wound sealing clot after healing of the lesion.

In order to fulfill these tasks, the hemostasis system has to react fast and extensively, but in a precisely regulated way: The coagulation initiating event must be neutralized completely, but accidental coagulation due to shear forces and surface effects within the blood vessels has to be avoided. Because of the dynamics of blood flow, these reactions must be localized.

Fast action is realized by a concerted action of different mechanisms, which will activate each other. Since most partners of these mechanisms are always present, they need only a minute trigger for activation. Involved are:
- *endothelial cells, plasma cells* and *platelets*,
- membrane bound proteins and receptors
- non physiological surfaces
- plasma factors and secreted soluble protein factors
- liberated low molecular weight factors affecting the *vessel wall musculature* and the blood pressure.

Extensive action (especially in case of bleeding) is achieved by a cascade of proteolytic activation steps (20.2, 20.3). Each step specifically activates the next one, thereby multiplying the effect. This way, a huge amount of the last member of the cascade becomes activated: either the fibrin forming thrombin (for coagulation) or the fibrin degrading plasmin (for fibrinolysis).

Proteolytic activation leads to the splitting of the peptide chain of the target protein, but in case of disulfide bonds the residues remain connected. Depending on the target, a peptide domain or small activation peptides are then released. The molecular basis of protease activation is not only to allow access to the active center, but also a structural arrangement: The newly created N-termini often play a central role in forming the active center.

Regulation takes place at different levels:
- Most reactions require the presence of special surfaces (non physiological surfaces, liberated and constitutive receptors, binding sites caused by the lesion, surface of blood clots) and are thereby localized.
- The dimension of the initial event (surface) may determine the amount of the reaction.
- Permanently active inhibitors in plasma stop unlocalized or incorrectly started reactions.
- Intermediates or end products of both systems (coagulation and fibrinolysis) cause feedback inhibitions and activations.
- Both systems act antagonistically to each other and also may stimulate each other.

After an injury, the following reactions take place:
- Platelets and the injured vessel wall secrete vasoconstrictive compounds (20.4). The smooth muscle cells of the *vessel wall* contract for about 1 minute and shut down the blood flow, which assists initiation of coagulation. Thereafter, vasodilatoric compounds (e.g. bradykinin) are liberated by contact activation. This allows a slow, clot feeding blood flow and also activation of inflammation processes by permeation of humoral factors (e.g. proteases) and cells (e.g. *granulocytes*, 19.3) through the vessel wall.
- Platelets adhere to the collagen layers of the *subendothelium*, which become exposed after injury of the endothelial cell layer. They form aggregates among each other (20.4), with other cells and with proteins, especially fibrinogen.
- Starting at the injured surface and the adhering platelets, interconnected coagulation cascades (employing plasmatic factors) cause activation of thrombin, which converts fibrinogen into insoluble fibrin. This and platelet reactions lead to contraction of the clot, effecting a tight sealing of the injury (20.3).
- Formation of the clot is controlled by coagulation inhibitors. Later on, fibrinolysis dissolves the clots (20.5).

A schematic survey on coagulation and fibrinolysis is given in Fig. 20.1-1 (surfaces and inhibitors are not shown). Most coagulation factors involved in hemostasis are denoted by Roman numerals, the activated forms with an additional 'a'. The numbering does not follow the biological sequence of action for historical reasons.

Literature on Blood Coagulation and Fibrinolysis

Bombeli, T. *et al.*: Thrombosis Hemostasis 77 (1997) 408–423.
Binder, B.R.: Fibrinolysis 9 (1995) 3–8.
Blockmans, D. *et al.*: Blood Reviews 9 (1995) 143–156.
Collen, D.: Thrombosis Hemostasis 43 (1980) 77–89.
Colman, R.W. *et al.*: Blood 90 (1997) 3819–3843.
Esmon, C.T.: Current Biology 5 (1995) 743–746.
Kalafatis, M. *et al.*: Biochimica et Biophysica Acta 1227 (1994) 113–129.
Lane, D.A. *et al.*: Thrombosis Hemsostasis 76 (1996) 651–662.
Lijnen, H.R. *et al.*: J. Biol. Chem. 261 (1986) 1253–1258.
Lorand, L., Mann, K.G. (Eds.): *Methods in Enzymology* Vol. 222, Academic Press (1993).
Petersen, L.C. *et al.*: Thrombosis Research 79 (1995) 1–47.
Zwaal, R.F.A. *et al.*: Blood 89 (1997) 1121–1132.

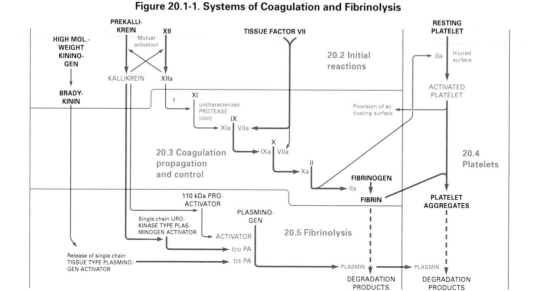

Figure 20.1-1. Systems of Coagulation and Fibrinolysis

20.2 Initial Reactions

Usually, two principal events are known to lead to coagulation:
- exposure of tissue factor to blood ('extrinsic pathway')
- contact activation ('intrinsic pathway').

This well-known scheme was created more than 30 years ago, based mainly on *in-vitro* research. However, new results obtained during the last 15 years require to modify this scheme considerably (e.g. influences of blood flow and surface reactions *in vivo*). Many details, however, remain to be elucidated. The main components involved are listed in Table 20.3.2.

20.2.1 Reactions Initiated at the Tissue Factor (Figure 20.2-1, Lower Right)

The tissue factor (TF, thromboplastin, factor III) is constitutively expressed as an integral membrane protein on the surface of most cells surrounding the *blood vessels* underneath the *endothelial cell layer*.

Therefore, TF gets into contact with plasmatic factors only in case of a mechanical lesion (including, e.g., myocardial infarction). Other pathological conditions (e.g. inflammation, sepsis or as consequence of arteriosclerosis) lead to the expression of TF at the surface of *endothelial cells* and *monocytes* and thereby cause systemic coagulation without the prerequisite of a mechanical lesion.

After binding of the plasmatic factor VII to TF, **factor VIIa** can be generated by autoactivation (noticeable proteolytic activity in contrast to XII and PK) or by attack from other coagulation factors (e.g. thrombin, Xa or hepsin, a membrane-associated protease). TF does not only localize the reaction, but also triggers structural changes in VII as a prerequisite for proteolytic attack. For full activity, Ca^{++} and a special phospholipid composition surrounding TF is required. Formation of the TF-VIIa complex seems to be the main cause of starting the coagulation cascade (20.3.3).

20.2.2 Contact Activation (Figure 20.2-1, Top)

Active factors (α-XIIa or KK) docking at a lesion site (LS) activate reciprocally inactive prekallikrein (PK) to kallikrein (KK, with HK as cofactor) and factor XII to α-XIIa. Kallikrein, in turn, splits α-XIIa to non-binding β-XIIa and also high molecular weight kininogen (HK) to bradykinin, releasing peptide fragments.

Although contact activation is well documented, details of the *in vivo* activation of the initiating factors and of the contact activating surface (at the lesion site, see below) are still unclear.

The major function of contact activation is the activation of fibrinolysis (20.5), involving the following reactions:
- Bradykinin stimulates strongly the secretion of sct-plasminogen activator (t-PA) from internal storage pools in *endothelial cells*.
- Plasma kallikrein, in collaboration with plasmin, converts the weakly active scu-plasminogen activator into the highly active tcu-plasminogen activator.
- Plasma kallikrein activates an 110 kDa proactivator, which consecutively converts plasminogen into plasmin.

The *in vivo* importance of contact activation for coagulation (as derived from *in vitro* experiments) must be relativized. A defect in XII apparently does not cause bleeding. Patients with this defect rather suffer from thrombotic episodes (e.g., in case of extracorporeal circulation) or are free from symptoms.

Although XIIa is able to activate XI to XIa (a binding epitope for XIIa exists on XI), it is not proven, that XIIa is the main trigger for activation of XI and initiates the consecutive coagulation cascade (20.3) *in vivo*. Also, activation of XI by thrombin is only documented *in vitro*. Receptor dependent binding of XI to *platelets* or binding of XI to *growing clots* seems to be the essential prerequisite for its activation. The physiological activator of XI has not been characterized yet.

Contact activation may also start the complement system by attack of β-XIIa on complement component C1qr$_2$s$_2$ (19.2.1).

20.2.3 Generation of Binding Surfaces (BS)

Localized coagulation requires the presence of factor specific binding surfaces (BS) for activation of the various factors. They are generated by damage to or by activation of *endothelial cells, platelets* or other *plasmatic cells* (20.2.3, lesion sites). These surfaces are composed of:
- Accumulated negatively charged phospholipids. Their main component is phosphatidylserine (6.3.2), that is actively accumulated on the inner surface of all living cells by a specific, ATP driven flip-flop mechanism. It reaches the outer surface by inhibition of this flip-flop mechanism

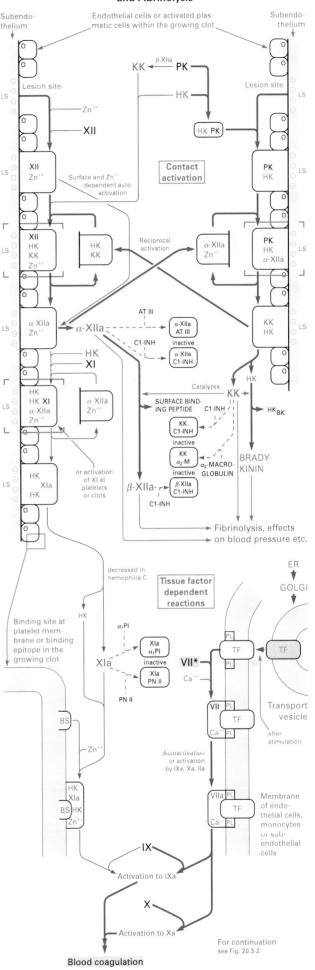

Figure 20.2-1. Initial Reactions of Coagulation and Fibrinolysis

Blood coagulation

– in *cells* in case of effector signalling, injury or death
– in *microparticles* (inside-out particles) discharged from, e.g., stimulated *platelets*
* Proteins (Table 20.3.2), which form complexes with the phospholipids:
  – TF, an integral membrane protein (20.2.1) and/or
  – Cofactors VIII/VIIIa and V/Va (20.3) and/or
  – other integral membrane proteins, e.g. thrombomodulin or 'receptors' for XI/XIa, IXa, protein C and protein S

Calcium enables interactions between binding sites and coagulation factors.

In case of contact activation, the nature of this surface and the mechanism of its generation are under discussion. The binding surface is possibly an exposed collagen layer. There is some evidence, that the contact activating surface may also be composed of glycosaminoglycans (13.1.2) present in the membrane of plasmatic cells, which are activated by inclusion into the growing clot. *In vitro*, coagulation can be started at e.g. sulfated glycoproteins or kaolin.

20.3 Coagulation Propagation and Control

The main purpose of these reactions is to provide thrombin for coagulation and other effects and to regulate its formation. The proteolytic activation mechanism (20.1) occurs in definite cascade complexes, which allow localization, activation and limited protection of the cascade reactions from inhibition.

20.3.1 Requirements for Protease Activity

All of the cascade complexes assemble at binding surfaces (20.2.3). In the complex, aided by cofactors, the conformation of the protease and the substrate orientation are optimized (Table 20.3-1):

Table 20.3-1. Activity Increase by Coagulation Complex Formation

Protease	Substrate	Cofactors	Effectivity*
VIIa	X	TF	3.0×10^5
IXa	X	VIIIa + platelets	1.7×10^7
Xa	II	Va + platelets	6.5×10^6

* Ratio of activities (k_2/K_M values) with and without cofactors (1.5.4)

Many factors involved in the coagulation cascade (VII, IX, X, prothrombin, as well as the coagulation control factors PC and PS, indicated in Figure 20.3-2 with *) contain domains with 9...12 Ca^{++}-binding amino acids γ-carboxyglutamate (Gla). This compound is formed posttranslationally by carboxylation of glutamate in presence of vitamin K (9.13). K-avitaminosis or presence of K antagonists (e.g. dicoumarol) prevents the reaction. Lack of carboxyglutamate seriously impairs binding of the factors to the binding site and their activation. This causes severe bleeding.

20.3.2 Pathways Leading to Thrombin

Several interconnected and regulated pathways are leading to thrombin activation (a survey is given in Figure 20.3-1, next page).

20.3.3 Key Events (Figure 20.3-2)

Factor X is activated to Xa by the TF-VIIa complex ①. (At the same time, TF-VIIa activates IX ②, see below). The consecutive thrombin formation from prothrombin (II to IIa by Xa) requires the presence of cofactor Va, which has to be activated from V by thrombin. Thus, thrombin exerts a feedback activation mechanism. Therefore the 'first' thrombin molecule is generated in a suboptimal way.

TFPI (tissue factor pathway inhibitor) quickly inactivates Xa by forming a complex, which, in turn is a feedback inhibitor of the activation of X and IX by TF-VIIa ①,②. This avoids starting the coagulation cascade, if the triggering cause is too weak.

Although the concentration of free TFPI in plasma is very low, TFPI is almost unlimitedly available (liberation of a membrane-bound form by heparin, secretion from *endothelial cells*). Therefore most – if not all – of the VIIa-TF complexes are blocked this way.

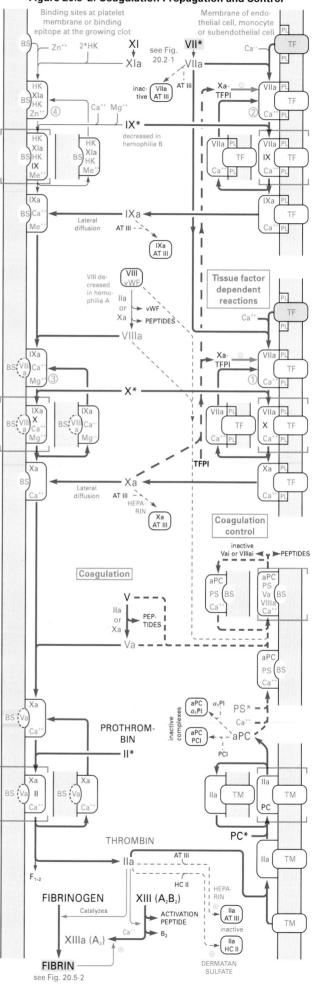

Figure 20.3–2. Coagulation Propagation and Control

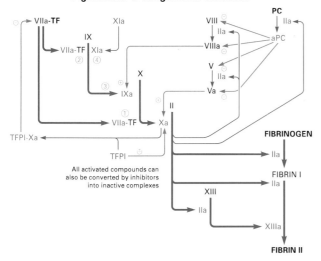

Figure 20.3-1. Coagulation Cascades

These inhibitory reactions of coagulation are also controlled: The quantity of available aPC is limited by complex formation with PC-inhibitor, α_1-proteinase inhibitor or α_2-macroglobulin. PS can be proteolytically inactivated by thrombin.

A newly discovered mutant of cofactor V (Factor V Leiden) cannot be inactivated by aPC. In these patients, thrombin generation is less restricted. About 30% of all thrombotic events are caused by this mutant. High levels of homocysteine (4.5.4, causing damage to the endothelial cell layer) are another source of thrombotic events.

Proteins C, S and others also play a role in, e.g., inflammation, cell growth.

Thrombin can be inactivated by complex formation with heparin cofactor II (HC-II) or AT III. These reactions are mediated by dermatan sulfate or heparan sulfate, or by heparin, respectively.

The presence of TM and PS receptor at the surface of intact *endothelial cells* is the cause of their anticoagulant properties. If lesions or noxious agents disturb these cells, procoagulant effects come into action: establishment of binding sites, secretion of TF (20.2), von Willebrand factor and PAI-1 (20.5), loss of thrombomodulin and PS-receptor.

20.3.4 Controlled Propagation (Figure 20.3-2)

The progress of the coagulation cascades depends essentially on the activation of sufficient IX to IXa by TF-VIIa ②), before the inhibition of this reaction by Xa-TFPI becomes effective. IXa, in turn, causes activation of X to Xa ③ and thus circumvents the TFPI inhibition of X activation. Insufficient activity of this pathway causes bleeding: In haemophilia A, cofactor VIII and in hemophilia B, factor IX is lacking.

In later stages of coagulation, generation of more thrombin is essential for proper clot formation. This is accomplished by a supplementary pathway for IX activation by XIa ④. The physiological activator of XI is unknown (20.2).

Another important feedback control system of coagulation is initiated by thrombin itself: Binding to thrombomodulin (TM, present at the membrane of intact *endothelial cells*) converts thrombin into an anticoagulant. It proteolytically activates protein C (PC) to aPC, which in a complex with protein S (PS) inactivates the essential cofactors of the coagulation cascade VIIIa and Va as well as their unactivated precursors.

20.3.5 Generation of Fibrin (Fig. 20.3-3 and 20.3-4)

Fibrinogen consists of 2 sets of triple protein helices $[(A\alpha)(B\beta)\gamma]_2$, which are symmetrically joined in the center by disulfide bonds. To effect coagulation, thrombin has to release four strongly negatively charged peptides ($2*A$ and $2*B$) from this central area. This way, electric repulsion is removed and (with the new N-termini) high-affinity ligands for the binding epitopes at both ends of the fibrinogen molecule are generated. This allows self-assembly

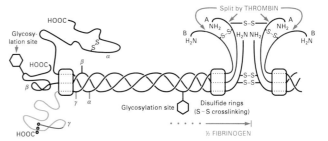

Figure 20.3-3. Fibrinogen Structure

Table 20.3-2. Properties of Coagulation Factors

Abbreviation	Factor	Trivial Name	Mol.Wt. (kDa)	Conc. (µmol/l)	Acts (in activated form) on	Pathology
Proteases and Precursors						
II/IIa	factor II/IIa	(pro-)thrombin	72/37	1.4	V, VIII, Fibrinogen, XIII, PC	hypoprothrombinemia
VII/VIIa	factor VII/VIIa	(pro-)convertin	50/50	0.01	IX, X	hypoproconvertinemia
IX/IXa	factor IX/IXa	Christmas factor	57/46	0.09	X; VII	hemophilia B
X/Xa	factor X/Xa	Stuart-Prower-factor	59/46	0.17	II; VII, V, VIII	Stuart disease
XI/XIa	factor XI/XIa	plasma thromboplastin	160/160	0.04	IX	hemophilia C
XII/α-XIIa/β-XIIa	factor XII/α-XIIa/β-XIIa	Hageman factor	80/80/30	0.4	PK, 110 kDa proactivator; XI	Hageman trait
PK/KK	plasma (pre)kallikrein	Fletcher factor	86/86	0.6	XII, HK, scu-PA	Fletcher trait
PC/aPC	protein C		62/56	0.07	Va, VIIIa	thromboembolism
Other Enzymes and Precursors						
XIII/XIIIa	factor XIII/XIIIa	fibrin stabilizing factor	320/160	0.1	fibrin	hemorrhagic diathesis
Protease Inhibitors						
α_1PI	α_1 proteinase inhibitor	α_1-antitrypsin	55	50	XIa, aPC	
α_2M	α_2-macroglobulin		720	0.003	IIa; KK; aPC	
AT III	antithrombin III		58	2...3	IIa, Xa; IXa, α-XIIa, VIIa-TF	thromboembolism
C1-INH	complement factor C1 inhib.		110	1.8	XIIa(α+β), KK; XIa	hereditary angioedema
HC II	heparin cofactor II	antithrombin BM	66	1.2	IIa	thromboembolism (rare)
PCI	protein C inhibitor (= PAI-3)		56	0.1	aPC	
PN II	protease nexin II		51	(platelets)	XIa; IXa?	
TFPI	tissue factor pathway inhibitor	lipid associated coagulation inhibitor	43/41/34	0.003	Xa, VIIa	disseminated intravascular coagulation
Cofactors and Receptors						
V/Va/Vai	factor V/Va/Vai	proaccelerin	330/180	0.02	Xa	APC-resistance
VIII/VIIIa/VIIIai	factor VIII/VIIIa/VIIIai	antihaemophil. globulin A	285/165	0.0007	IXa	hemophilia A
BS	'binding site'		–	–		
Fg/Fb	fibrinogen/fibrin, factor I/Ia		340/333	7	thrombospondin, XIII, scu-/tcu-PA, plasminogen	dysfibrinogenemias
HK/HKi	high mol.-weight kininogen	Fitzgerald factor	110	0.7	PK/KK, XI/XIa	Fitzgerald trait
LS	lesion site					
PS	protein S		75	0.3	aPC	thromboembolism
TF	tissue factor	thromboplastin	45	(membrane)	VII/VIIa	
TM	thrombomodulin		100	(membrane)	IIa	[lethal genetic defect]
vWF	von-Willebrand factor	Ristocetin cofactor	(2...100)*250		platelets/endothelial cells	vW's disease

of fibrin and formation of polymers (fibrin I_n and thereafter II_n). Then cross-linking of specific lysine and glutamate residues at the C terminals of the γ-strands is effected by thrombin-activated factor XIIIa (which is not a protease, but a transglutaminase). This converts the clot into an insoluble form and causes contraction. Fibrin and fibrinogen also have binding epitopes for other proteins to effect crosslinking with platelets (20.4) and other cells. This cross-linking is the central mechanism for the fast stopping of bleeding.

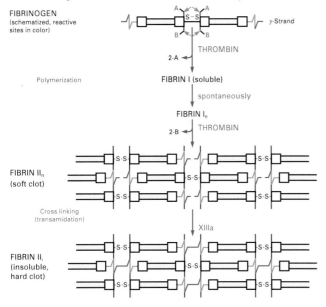

Figure 20.3-4. Formation of Fibrin Polymers

20.4 Platelets (Thrombocytes, Fig. 20.4-1)

Platelets play a central part in the hemostatic process. When injuries to the vessel wall cause a tearing of the *endothelial layer*, the *subendothelial* collagen gets exposed. Platelets adhere to it

- directly via integral membrane glycoproteins (GP) acting as receptors, e.g., GPIa–IIa, GPIIIb and GPVI (Table 20.4-2)
- indirectly via linker proteins present in *blood plasma* (e.g., von Willebrand factor, fibronectin) or being released from the *endothelium* (e.g., von Willebrand factor, fibronectin) or from *platelets* after activation (e.g., thrombospondin, TSP, Table 20.4-1).

The contact to the *subendothelium* activates the platelets. The granule membranes fuse with the platelet membrane, causing release of the contents of the secretory organelles *(dense bodies, α-granules)* and integration of α-granule membrane proteins (e.g., P-selectin) into the plasma membrane. This has multiple consequences (Figure 20.4-1):

- **Vascular effects:** ADP, serotonin (from *dense bodies*) and epinephrine quickly cause vasoconstriction, slowing down the bloodflow and assisting in this way the thrombus formation.

  They also induce the production of NO (17.7.2, formerly called endothelium-derived relaxing factor, EDRF), which has strong vasodilatory effects, thus limiting the primary actions.

- **Change of platelet shape:** The microtubular bundles (which keep up the normal platelet form) depolymerize. Binding of ADP to its receptors effects release of Ca^{++} ions from *dense bodies* into the *cytosol,* where they interact with calmodulin (IP_3 pathway, 17.4.3). This causes activation of the myosin light chain kinase (17.4.5), which, in turn, activates myosin. The activated myosin interacts with the actin filaments of the cytoskeleton, changing the platelet shape from a discoidal to a globular form with long pseudopods.

- **Platelet aggregation and clot formation:** The released factors (e.g., ADP, thromboxane A_2 and platelet activating factor, PAF,

6.3.3) enhance the activation of surrounding platelets. By formation of fibrinogen or thrombospondin bridges to other platelets via the activated GPIIb–IIIa receptor complex, platelets aggregate and provide the primary hemostatic plug.

On formation of the first traces of thrombin by the coagulation cascade (20.3), these fibrinogen-platelet aggregates are converted into a stable fibrin-platelet network that also includes other blood cells. Even small amounts of thrombin activate factor XIII (from plasma and from α-granules), which catalyzes the covalent crosslinking inside of the fibrin clot.

- **Microparticle formation:** Platelet activation leads to the disintegration of the platelet membrane and to the formation of *microparticles*. Already in intact platelets, phosphatidylserine appears at the outer surface due to inhibition of the flip-flop mechanism (20.2.3) following activation. Factor V (from *plasma* and from *α-granules*) as well as vitamin K dependent coagulation factors (factors II, IX, X from *plasma*) bind at the procoagulant surface (20.2.3) of these particles. This provides effective progress of the coagulation and assures quick sealing of the vessel wall after injury.

- **Control of coagulation:** Not only the coagulation progress, but also its feedback inhibition via protein C and protein S (20.3.4) depend on the modified phospholipid bilayer structures (20.2.3) present in, e.g., *activated platelets* and *microparticles*. In addi-

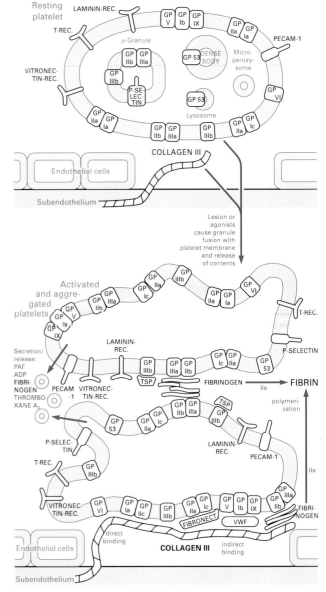

Figure 20.4-1. Role of Platelets in Coagulation

tion, the intact *endothelium* synthesizes substances that inhibit platelet function and prevent unbalanced platelet activation. The most important inhibitory factors are prostacyclin I_2 and D_2 (17.4.8), ADP-decomposing nucleotidase and nitric oxide (NO, 17.7.2). All these inhibitors cause an increase of the intracellular cAMP concentration, which effects the sequestration of Ca^{++} ions into intracellular storages areas (17.4.4). This way, all processes based on calcium ions are inhibited.

The major contents of the platelet vesicles and the receptors at the membrane are listed in Table 20.4-1 and -2.

Table 20.4-1. Compounds Stored in Vesicles of Thrombocytes

Storage Vesicles	Compounds Involved in Aggregation	Compounds Causing Vasoconstriction	Other Compounds
α-Granules	fibrinogen, fibronectin, thrombospondin, von Willebrand factor (vWF), factors V XIII, multimerin (stabilizes factor V)		platelet factor 4 (PF4, suppresses anticoagulation), β-thromboglobulin (β-TG), α$_2$-antiplasmin
Dense bodies	Ca^{++}, ADP	ADP, serotonin, epinephrine	
Lysosomes			hydrolyzing enzymes (involved in inflammation)
Microperoxisomes			catalase (involved in inflammation)

Table 20.4-2. Receptors on the Platelet Membrane

The glycoproteins usually have large extracellular domains. These are not shown in detail in Fig. 20.4-1.

Platelet Glycoprotein (Complexes)	CD-No.	Family	Mol. Mass kDa	Copies (resting platelet)	Receptor for	Physiological Function
GP Ib/IX/V	42b,c/a/d	Leu-rich glycoproteins (LRG)	143 (Ibα), 22 (Ibβ)/ 22/82	Ib: 25 000 IX: 25 000 V: 11 000	vW factor (by GP Ib$_a$), thrombin (by GP Ib$_a$)	shear stress induced adhesion of unactivated platelets to the *subendothelium* of injured vessels, modulation of thrombin induced activation, activation of GP IIb/IIIa for binding of vWF
GP Ia/IIa	49b/29	α$_2$β$_1$ integrin	153/130	Ia: 2000 IIa: 7000	collagen, laminin in fibroblasts and endothelial cells, fibronectin	collagen receptor, collagen induced adhesion (Mg^{++} dependent), activation of tyrosine kinase adhesion of platelets to fibronectin
GP Ic/IIa	49e/29	α$_5$β$_1$ integrin	160/130	Ic: ? IIa: 7000		
GP Ic*/IIa		α$_6$β$_1$ integrin	?/7000		laminin	adhesion of platelets to laminin
GP IIb/IIIa	41/61	α$_{IIb}$β$_3$/α$_V$β$_3$ integrin	125(IIba), 23 (IIbb)/ 95	40000... 80000	fibrinogen, vW factor, fibronectin, thrombospondin, vitronectin	after activation: irreversible attachment of platelets to the subendothelium via vWF and fibronectin, aggregation of platelets, signal transmission
GP IIIb	36	scavenger receptor (18.2,4)	88	12000... 19000	collagen type I, thrombospondin, oxidized LDL	binding between platelets, endothelial cells and monocytes, signal transmission, receptor for oxidized LDL
GP VI			62	?	collagen types I, III, V, VI, VIII (fibrillar collagens)	collagen receptor, collagen-induced signal transduction
GP 53	63					?
Laminin receptor			67		laminin	adhesion of platelets to laminin
Thrombin receptor		G-protein (17.3)	47	1800	thrombin	initiation of platelet aggregation and secretion, activation of PL C, inhibition of adenylate cyclase
Vitronectin receptor		α$_\omega$β$_3$ integrin		500	vW factor, fibronectin, vitronectin, thrombospondin	adhesion of cells to the extracellular matrix, mediates cell spreading, phagocytosis and intercellular interactions (entrance of virusus into platelets and endothelial cells?)
PECAM-1	31	immunoglobulin (19.3)	130	5000	PECAM-1 (autoassociation)	
P-Selectin	62 P	selectin (19.3)	140		neutrophile granulocytes, monocytes	mediates contact of activated platelets and endothelial cells, to other cells (incl. tumor cells)
ADP receptor	not yet identified		43 ?		ADP	$G_{i\alpha}$ inhibition of adenylate cyclase
Thromboxane receptor(s)		possibly 2 receptors	37		thromboxane A_2	G protein activation of PLC and Ca^{++} channel
Receptors for adenosine, PGI$_2$, PGE$_2$, PGD$_2$					adenosine, PGI$_2$ PGE$_2$, PGD$_2$	G protein activation of adenylate cyclase

* different from GP 1c

Table 20.5-1. Properties of Fibrinolytic Factors

Abbreviation	Factor	Mol.Wt. (kDa)	Acts (in activated form) on
Precursors and Proteases			
scet-PA/tct-PA	single/two chain tissue type plasminogen activ.	70	plasminogen
scu-PA/tcu-PA	single/two chain urokinase type plasminogen activ.	55/34	plasminogen
Plg/Pl	plasminogen/plasmin	90/78	fibrin (plasmin dissolves)
Protease Inhibitors			
α2AP	α2-antiplasmin	ca. 70	plasmin
α2M	α2-macroglobulin	720	plasmin
PAI-1	plasminogen activator inhibitors	52	sct-PA, tct-PA
PAI-2		60 (glycosyl.)	tcu-PA
PAI-3		56	tcu-PA
PN II	protease nexin II	51	sct-PA, tct-PA
HRGP	histidine rich glycoprotein	60	plasminogen
Cofactors and Receptors			
BE	'binding epitope' at fibrin surface		

20.5 Fibrinolysis

The fibrinolytic system has the task of removing fibrin deposits and thrombo-embolic fibrin plugs in the blood vessels. It fulfills important control functions and forms a dynamic equilibrium with the coagulation system.

Fibrinolysis is controlled by activation and inhibition processes. Specific activators convert the inactive proenzyme plasminogen into the active serine protease plasmin, which degrades the fibrin clot (Figure 20.5-1).

Figure 20.5-1. Scheme of Fibrinolytic Activation

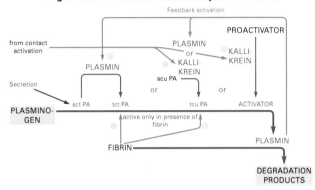

20.5.1 Pathways of Plasminogen Activation (Figure 20.5-2, Table 20.5-1)

Plasminogen can be activated via several pathways:
- Small quantities of tissue plasminogen activator (t-PA) are permanently liberated from the *endothelial layer* of the blood vessels. Furthermore, acute stimuli (e.g., vascular congestion due to thrombosis) cause secretion of t-PA from *intracellular depots*. t-PA requires fibrin for effective conversion of plasminogen into plasmin, therefore it is inoperative in the absence of a fibrin clot.

  The physiological concentration of plasminogen is about 1.5 µmol/l. The K_M of t-PA for plasminogen is 65 µmol/l in the absence and 0.15 µmol/l in the presence of fibrin. Thus, the presence of fibrin ensures quick activation.

  There is strong evidence that t-PA is essential for starting the fibrinolytic system. At first, small quantities of plasminogen are converted into plasmin. This causes cleavage of some fibrin chains and exposition of lysine residues, which are attachment sites for more plasminogen and t-PA. This potentiates the formation of plasmin. Although t-PA is already fully active in its single chain form (sct-PA), it is converted by plasmin into the two-chain form (tct-PA).

- Urokinase type plasminogen activator (u-PA) occurs in the circulation only in a weakly active single-chain form (scu-PA). Its physiological source is unknown. In cell cultures it was found in the supernatant of *endothelial, epithelial* and *tumor* cells. Plasmin converts scu-PA into the much more active 2-chain form tcu-PA. Both scu-PA and tcu-PA act only on plasminogen, which is bound to fibrin.

  Presumably, the intravascular u-PA pathway becomes operative after some plasmin has been obtained by the t-PA pathway. It presumably acts as an amplification system for the essential t-PA pathway.

  Additionally, urokinase is involved in many extravascular processes, e.g., wound healing (removal of the fibrin layer, neovascularization, development of organs, tumor invasion/metastasis and degradation of the extracellular matrix by plasmin).

  Streptokinase (a bacterial toxin from *Streptococcus haemolyticus*, which is therapeutically used) forms a complex with plasminogen, which also acts as plasminogen activator.

- The contact activation of coagulation (20.2) represents another system of initiating fibrinolysis. It generates kallikrein, which converts a (still not completely characterized) 110 kDa proactivator into an active plasminogen activator, resulting in plasmin formation. Additionally, kallikrein activates scu-PA to tcu-PA, while bradykinin (20.2.2) stimulates the liberation of tPA from intracellular depots in the *endothelium*. Thus this system acts as another amplification pathway of fibrinolysis.

20.5.2 Control of Fibrinolysis (Figure 20.5-2)

Although plasmin has no definite substrate specificity (similar to trypsin), under physiological conditions its fibrinolytic activities are stricty confined to fibrin. Likewise, tPA requires the presence of fibrin in order to generate the fibrin-degrading plasmin.

Circulating free plasmin and free activators (except scu-PA) are effectively controlled by the respective inhibitors. These inhibitors (PAI-1, -2, -3; α_2-antiplasmin, α_2-macroglobulin) form inactive complexes with unbound fibrinolytic proteases.

Its in probable, that in plasma only PAI-1 is responsible for the control of t-PA. PAI-2 occurs in plasma only in special situations, such as pregnancy or certain malignancies. The active form of PAI-1 is conserved for extended periods when bound to vitronectin (Tab. 20.4-2). Otherwise, it is quickly converted into the latent form. This form can be reactivated by binding to anionic phospholipids, e.g. from microparticles.

Fibrin is not only the target of the fibrinolysis, but also the central cofactor, which determines initiation, location and limits of this event. Binding of proteases to fibrin protects the enzymes from inhibition.

Figure 20.5-2. System of Fibrinolysis

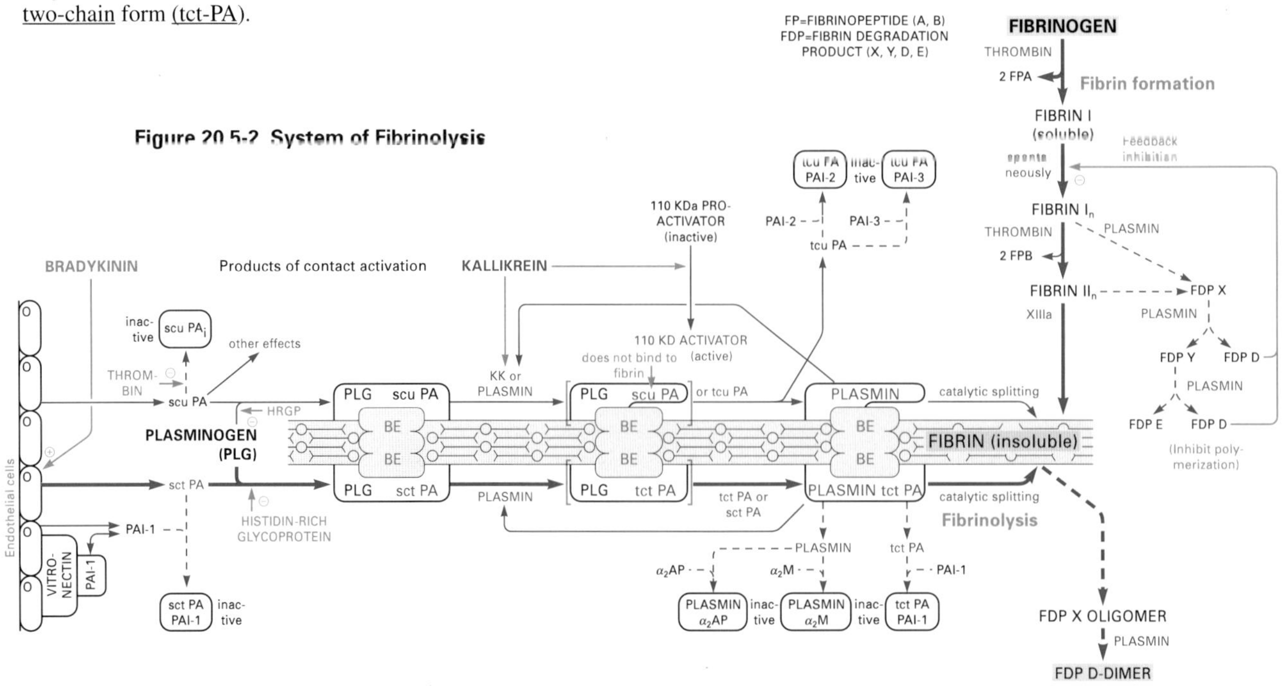

21 Further Information

21.1 Electronic Storage of Biochemical Information

This book can only provide a general survey of biochemistry. A more detailed and expanded information on these topics can be obtained from various data banks available through the Internet. They are also very helpful in locating original papers. A selection of major sources existing at the time of publication is given in Table 21-1. Almost all of these databases are interconnected via hyperlinks.

Table 21-1. Selection of Data Banks on Biochemical Topics

Nr.	Name	Compiled by	Internet address Capitalized words have to be observed!	Species	Characteristics (see below)				
1	KEGG	Kyoto Univ., JP	http://www.genome.ad.jp/kegg/kegg.html	G	PA	EC		PQ	DQ
1a	LIGAND	Kyoto Univ., JP	http://www.genome.ad.jp/dbget/ligand.html	G		EC			
2	BIOCHEMICAL PATHWAYS	Boehringer Mannheim, DE	http://expasy.hcuge.ch/cgi-bin/search-biochem-index	G	PA	EC			
3	PUMA	Argonne National Laboratories, US	http://www.mcs.anl.gov/home/compbio/PUMA/Production/puma.html	G	PA				
4	WIT	Mich. St. U., US	http://www.cme.msu.edu/wit	G	PA				
5	ENTREZ	National Center for Bioinformatics, US	http://www.ncbi.nlm.nih.gov/Entrez	G	PA	EC			
6	EcoCyc	PangeaSystems, US	http://ecocyc.PangeaSystems.com/ecocyc/ecocyc.html	B	PA	EC	PS		DQ
7	ENZYME	Univ. Geneva, CH	http://expasy.hcuge.ch/sprot/enzyme.html	G		EC			
8	PDB	Brookhaven National Laboratories, US	http://www.pdb.bnl.gov	G			PS		
9	CATH	Uni. Coll. London, UK	http://www.biochem.ucl.ac.uk/bsm/cath	G			PS		
10	SCOP	Univ. Cambridge, UK	http://scop.mrc-lmb.cam.ac.uk/scop	G			PS		
11	SWISS-PROT	EBI, UK Univ. Geneva, CH	http://www.ebi.ac.uk/ebi_docs/swissprot_db/swisshome.html http://expasy.hcuge.ch/sprot/sprot-top.html	G				PQ	
11a	TREMBL	EBI, UK Univ. Geneva, CH	http://www.ebi.ac.uk/ebi_docs/swissprot_db/swisshome.html or ftp://ftp.expasy.ch/databases/trembl						
12	TRANSFAC	GBF, DE	http://transfac.gbf-braunschweig.de	E			PS	PQ	DQ
13	PEDANT	MIPS, DE	http://pedant.mips.biochem.mpg.de/frishman/pedant.html	G				PQ	
14	PDD	NCI, US	http://www-lecb.ncifcrf.gov/PDD	hu		EC		PQ	DQ
15	REBASE	Johns Hopkins U., US	http://www.gdb.org/Dan/rebase/rebase.html	B		EC			
16	Nucleotide sequ.	EBI, UK	http://www.ebi.ac.uk/ebi_docs/embl_db/ebitopembl.html	G					DQ
17	GENBANK	NCBI, US	http://www.ncbi.nlm.nih.gov/web/genbank	G					DQ
17a	dbEST	NCBI, US	http://www.ncbi.nlm.nih.gov/dbEST/index.html						DQ
18	GDB	Johns Hopkins U., US	http://gdb.www.gdb.org	hu		EC		PQ	DQ
19	OMIM	Johns Hopkins U., US	http://www3.ncbi.nlm.nih.gov/Omim/	hu		EC	PS	PQ	DQ
20	MGD	Jackson Lab., US	http://www.informatics.jax.org	mi		EC		PQ	DQ
21	FLYBASE	Harvard Univ., US	http://flybase.harvard.edu:7081 or http://embl-ebi.ac.uk/flybase	dr		EC		PQ	DQ
22	ARABIDOPSIS	Stanford U., US	http://genome-www3.stanford.edu/Arabidopsis	ar					DQ
23	YEAST	MIPS, DE	http://www.mips.biochem.mpg.de/mips/yeast	ye					DQ
24	MAGPIE	Argonne National Laboratories, US	http://www-c.mcs.anl.gov/home/gaasterl/magpie.html	B					DQ

Contents of the Data Bank

Characteristics	PA Pathways	EC Enzyme Characteristics	PS Protein Structure	PQ Protein Sequence	DQ DNA Sequence
Kingdoms	G General	E Eukarya	B Bacteria	F Fungi	
Species	dr Drosophila	hu Human	mi Mice	ar Arabidopsis	ye Yeast

Details:

1 KEGG/LIGAND (Kyoto University, JP)
The databases established by the KEGG project (Kyoto Encyclopedia of Genes and Genomes) contain information on
• enzyme reactions (ENZYME),
• chemical compounds (COMPOUND)
• metabolic pathways (PATHWAYS), with links to gene catalogues (GENES)
KEGG release 6.1+/06–23, June 98: 1754 entries for PATHWAY; 9042 entries for GENES. – LIGAND release 16.0+/06–23, June 98: 3603 entries for ENZYME; 5439 entries for COMPOUND.

2 BIOCHEMICAL PATHWAYS (Boehringer Mannheim, DE)
The Internet version of the wall chart (edited by G. Michal), showing pathways in animals, plants and bacteria. It provides graphic information on more than 2000 metabolic reactions, including their interrelationships and regulation. The enzymes are linked to ENZYME (7) and SWISS-PROT (11).

3 PUMA (Argonne National Laboratory, IL, US)
Contains pathways diagrams and supporting information. Predecessor to WIT (4). It encompasses a wider range of species than WIT. The updates were discontinued in early 1996.

4 WIT (Michigan State University, US and Argonne National Laboratory, US)
The WIT system contains
• almost 2000 pathway diagrams based on the 'Database of Enzymes and Metabolic Pathways' (E. Selkov, Puschino, RU)

- functional overviews of the metabolism (assignment of genes to function) of several bacteria and yeast.
- descriptions of the functional role and structural alignments of proteins (Baylor College of Medicine, US).

5 ENTREZ (National Center for Bioinformatics, US)
This search system connects (among others) to
- PubMed/MEDLINE, a collection of more than 9 million journal articles including summaries
- Pre-MEDLINE in medicine and related fields (biochemistry, biology etc.). In many cases, the full text can be obtained by links to the publishers.
- PROTEIN DB   ⎱ compiled frome a variety of sources
- NUCLEOTIDE DB ⎰ (see other numbers of this list)
- NM3D: Data from crystallographic and NMR structural determinations.

6 EcoCyc (PangeaSystems, US)
Description of published pathways in Escherichia coli, signal transduction proteins, genes, tRNAs, nucleotide sequences. Jan. 1998: 988 reactions,. 4909 genes, 1303 compounds (79 tRNAs, 60 signal transduction proteins).

7 ENZYME (Med. Biochem. Dept., University of Geneva, CH)
The Enzyme Database presents information on the nomenclature of enzymes, primarily based on the recommendations of the Enzyme Committee (IUMB). It describes the enzymes for which an EC number has been provided. July 1997: 3651 entries.

8 PDB (Brookhaven National Laboratory, US)
The Protein Data Bank contains 3D coordinates of macromolecular structures, crystallographic and NMR structure models. June 1998: 7306 proteins, 556 nucleic acids, 12 carbohydrates.

9 CATH (University College of London, UK)
The CATH database is a hierarchical domain classification of protein structures, present in PDB (8). The hierarchy levels are: class, architecture, topology (fold family) and homologous superfamily. 1998: 5605 entries, 13,338 domains.

10 SCOP (University of Cambridge, UK)
A structural classification of protein domains. June 1998: 13,073 domains.

11 SWISS-PROT (European Bioinformatics Institute, UK/University of Geneva, CH)
A collection of protein sequences directly obtained or translated from nucleotide sequences. It includes descriptions of protein functions, domain structure, posttranslational modifications, variants etc. July 1998: 73,459 entries.

11a TREMBL (European Bioinformatics Institute, UK/University of Geneva, CH)
Translations of coding sequences present in the Nucleotide Sequence database (16) not yet integrated into SWISS-PROT (11) are found in TREMBL. June 1998: 150,329 entries.

12 TRANSFAC (Gesellschaft für biologische Forschung, DE)
Contains transcription factors and their genomic binding sites, consensus binding sequences and binding profiles. Release 3.4 (June 1998): 2376 entries.

13 PEDANT [Max-Planck Institute of Biochemistry/Inst. of Protein Sequences and Protein Identification Resource (PIR), DE]
A tool for analysis of complete or partial genome sequences, which enables form based protein sequence database searches.

14 PDD (National Cancer Institute, US)
The Protein-Disease Database provides information on the correlation of diseases with proteins and protein patterns observable in human body fluids.

15 REBASE (Johns Hopkins University, US)
The Restriction Enzyme Database is a collection of information about restriction enzymes and methylases.

16 Nucleotide Sequence (European Bioinformatics Institute, UK)
The nucleotide sequence database is a comprehensive database of DNA and RNA sequences. Data collection is done in cooperation with GenBank (US, 17) and DNA DataBank of Japan (DDBJ).

17 GenBank (National Center for Biotechnological Information, National Library of Medicine, US)
The NIH genetic sequence database is an automated collection of all publicly available DNA sequences. For details see Nucleotide Sequence (16). June 1998: 2,356,000 entries, 1,622,000,000 bases.

17a dbEST (National Center for Biotechnological Information, National Library of Medicine, US)
dbEST is a division of Gen Bank that contains sequence data and other information on 'single pass' cDNA sequences or expressed sequence tags from a number of organisms. Entries: 1,080,934.

18 GDB (Oak Ridge National Laboratories, US)
The Human Genome Database contains information on the location of human genes, DNA fragments, fragile sites and breakpoints, as well as on populations, polymorphisms, maps and mutations. It is linked to the OMIM database (19).

19 OMIM (National Center for Bioinformatics and Johns Hopkins University, US)
The Online Mendelian Inheritance in Man Database is a catalog of human genes and genetic disorders.

20 MGD (The Jackson Laboratory, US)
The Mouse Genome Database provides information on the genetics of the laboratory mouse, e.g., genetic markers, homologies, probes and clones and experimental marker mapping data. Expression data are also included.

21 FLYBASE (Harvard University, US)
FlyBase is a comprehensive database on the genetics and molecular biology of *Drosophila*.

22 ARABIDOPSIS (Stanford University, US)
Genomic and related data, physical and genetic maps of *Arabidopsis thaliana*, a plant widely used in research.

23 YEAST [Max-Planck Institute of Biochemistry/Munich Inst. of Protein Sequences (MIPS), GE]
The Yeast Genome Database provides the annotated complete DNA sequence of *Saccharomyces cerevisiae*: 16 chromosomes, amino acid sequences of > 6000 open reading frames.

24 MAGPIE (Argonne National Laboratory and University of Chicago, US)
Compilation of bacterial genome sequencing projects.

Links and possible updates for the URL's listed in this section can be called from the Spektrum internet homepage (http://www.spektrum-verlag.com).

21.2 Printed Sources

This is a selection of major review journals, review books and textbooks. They have proven helpful during the writing of this book.

Review Journals and Books
Advances in Enzymology
Advances in Protein Chemistry
Advances in Microbial Physiology
Annual Review of Biochemistry
Annual Review of Cell Biology
Annual Review of Genetics
Annual Review of Physiology
Annual Review of Plant Physiology and Plant Molecular Biology
Current Opinion in Cell Biology
Current Opinion in Structural Biology
Current Topics in Cellular Regulation
Trends in Biochemical Sciences
Trends in Cell Biology
Trends in Genetics

Book Series
Methods in Enzymology. Academic Press
New Comprehensive Biochemistry. Elsevier
The Enzymes. Third Edition. Academic Press

Textbooks
Alberts, B. *et al.*: *Molecular Biology of the Cell*. Third Edition. Garland Publ. Co (1994).
Campbell, N.A.: *Biology*. Benjamin/Cummings Publ. Co (1996).
Freifelder, M.: *Physical Biochemistry*. Freeman (1982).
Heldt, H.W.: *Plant Biochemistry and Molecular Biology*. Oxford University Press (1998).
Janeway, C.A., Travers, P.: *Immunology*. Current Biology/Garland Publishing (1997)
Lehninger, A.L. *et al.*: *Principles of Biochemistry*. Second Edition. Worth Publishers (1993).
Lewin B.: *Genes* VI. Oxford University Press (1997).
Lodish, H. *et al.*: *Molecular Cell Biology*. Third Edition. Freeman (1995).
Stryer, L.: *Biochemistry*. Fourth Edition. Freeman (1995).
Voet, D., Voet, J.G.: *Biochemistry*. Second Edition. Wiley (1995).

Index

Keywords in this index are listed with section numbers (not page numbers)

A

ABC system (DNA repair) 10.3.3
ABC transporters 15.3.1
abietic acid 7.3.1
ABO-system (blood groups) 13.5.1
acceptor helix 10.6.1
acetaldehyde 3.3.5, 15.4
acetaldehyde dehydrogenase 15.4
acetaminophen 17.4.8
acetate 3.8.1, 15.4
acetate CoA-transferase 15.4
acetate kinase 15.4
acetoacetate 6.1.6, 15.4
acetoacetate decarboxylase 15.4
acetoacetyl thiolase 6.1.5
acetogenesis 9.5.2, 15.5.3
acetoin 15.4
acetolactate 4.6.1, 15.4
acetolactate decarboxylase 15.4
acetolactate synthase 15.4
acetone 6.1.6, 15.4
acetylcholine 17.1.4, 17.1.9
– neurotransmitter action 6.3.5, 17.2.3 f
acetylcholine receptors 17.2.5
acetylcholinesterase 6.3.5, 13.3.4
acetyl-CoA 3.8.1, 15.4, 15.7
acetyl-CoA acetyltransferase (thiolase) 6.1.5, 15.4
acetyl-CoA C-acyltransferase 6.1.5
acetyl-CoA carboxylase 6.1.1, 6.1.2, 9.8.2
– regulation 7.1.4
acetyl-CoA carboxylase type 2 6.1.2
acetyl-CoA
– carboxylation 6.1.1
– export to the cytoplasm 3.8.1, 6.1.1
– role in metabolism 3.3.3
acetyl-P 15.4
α-N-acetylhexosamidase 13.1.5
β-N-acetylhexosaminidase 13.4.3
N-acetyl muramic acid 3.7.1, 13.1.3
N-acetyl-D-mannosamine 3.7.1
N-acetyl-galactosamine 13.4.1
N-acetyl-glucosamine 13.1.3
N-acetyl-glutamate 4.2.2
N-acetyl-neuraminic acid 3.7.1, 13.1.1, 13.2.1
acetylsalicylic acid 17.4.8
acid exchange mechanisms 18.1.1
aconitase 3.8.1
aconitate 3.8.1
aconitate hydratase 3.8.1
acridinium derivates 6.3.5
acromegalin 17.1.5
acrylate pathway 15.4
actin filaments 18.1.3
actin, structure 17.4.5
actinomycin D 10.4.5, 11.3.10
action potential 17.2.1, 17.2.2, 17.2.3
activation energy 1.5.4, 2.4.1
activator protein 10.5.1
active site (enzymes) 2.4.1
active transport 15.3.1
acute phase proteins 19.1
acyl adenylate 6.1.4
acyl carrier protein (ACP) 6.1.1, 9.7.1
acyl group carriers 9.7
acyl transferase 6.1.1
acylation, proteins 13.3.5
acyl-CoA cholesterol acyl transferase (ACAT) 7.1.5
acyl-CoA dehydrogenase 6.1.5
acyl-CoA oxidase 6.1.5
acyl-CoA synthetases 6.1.4
acyl-CoA thioester 6.1.4
acylglycerol 1.4.2, 13.4.3
acylglycerol 3-P 6.2.1
1-acylglycerol-3-phosphate O-acyltransferase 6.2.1
β-N-acyl-hexosamidase 13.2.1
β-N-acylhexosamidase B 13.2.1
acylsulfatase 13.2.1
adaptor molecule Shc 17.5.4
Addison's disease 7.8.4
adenine 8.1
adenine P-ribosyl transferase 8.1.2
adenine, deamination of cytosine 11.2.1
adenosine 8.1
adenosine deaminase, defective 8.1.7
adenosine phosphosulfate 15.5.1, 15.6.1
adenosine phosphosulfate reductase 15.6.1

adenosine-3′,5′-diphosphate 4.5.6
adenosine-5′-diphosphate (ADP) 8.1.2
adenosine-5′-triphosphate (ATP) 8.1.2
adenosine-5′-monophosphate (AMP) 8.1.2
adenovirus 12.2
adenylate cyclase 17.1.3, 17.4.1
– activation 17.4.2
– in olfactory process 17.4.7
– isoforms 17.4.2
adenylate kinase 15.6.1
adenylsuccinate 8.1.2
adenylsulfate 4.5.6
adenylsulfate reductase 15.5.1, 15.6.1
adhesion belts 17.4.5
adhesion molecules 17.1.6, 19.3
adhesion reactions 19.3
ADP receptor 20.4
ADP ribosylating factor 14.2.2, 17.4.1
ADP/ATP-carrier 14.4
ADP 8.1.2
ADP-decomposing nucleotidase 20.4
ADP-D-glucose 16.2.2
adrenal cortex 17.1.5
adrenaline 4.7.3, 4.7.4
β adrenergic receptor kinase 17.4.1
adrenergic receptors 17.2.4, 17.2.5
adrenocorticoidal carcinoma 7.8.4
adrenodoxin 7.5.1
adrenosteron 7.6.1
aerobic respiration 15.4
aglycons 1.2.2, 3.5.1
agonists 17.1.2, 17.3
A group receptors 17.6
AICAR 4.8.1, 8.1.1
AIDS 12.1.1, 12.4
AIRC gene 8.1.1
alanine 3.1.1
– metabolism 4.2.3
alanine dehydrogenase 3.1.1
alanine fermentation 15.4
alanine-oxo acid transactinase 15.7
L-alanine-2-oxoglutarate transaminase (ALAT, GPT) 4.2.4, 15.4
alanine transaminase 3.1.1, 4.2.4
albinism 4.7.3
albumin 7.4.3, 7.5.2
alcohol dehydrogenase 3.3.5, 15.4, 15.7
alcohol dehydrogenase III 14.4
alcoholic fermentation 3.3.6, 15.4
alcohols, oxidation 9.3.2
aldehyde dehydrogenase 15.7
aldehyde ferredoxin oxidoreductase family 9.6.4
aldehyde reductase 3.1.1
aldehydes, oxidation 9.3.2
aldonic acids 3.5.2
aldose 1.2.1
aldosterone 7.8.1, 17.1.8
– hormone action 7.8.3
aldosteronism 7.8.4
alginate 3.5.1, 3.5.6
alkaline environment 15.3.1, 16.1.3
alkaloids, blocking of axonal transport 17.2.6
alkane oxidation 15.7
alkane-1-monooxygenase 15.7
alkylacylglycerol 6.3.3, 13.2.2
alkylation compounds 10.3.1
alkyltransferases 11.2.2
allantoate 8.1.6
allelic exclusion 19.1.3
allergic immune response 19.1.8
allergic reactions 4.8.2, 19.1.8
allolactose 10.5.1
allopurinol 8.1.7
allosteric regulation 2.5.2
allostery 5.2.3
alternative pathway (complement) 19.2.1
Alzheimer's disease 6.3.5
α-Amanitin 11.3.10
amido-P ribosyltransferase 8.1.1
amino acid oxidases 4.2.2
amino acids 4.2
– aromatic 4.7.1, 4.7.3
– branched-chain 4.6.2
– essential 4.1
– fermentations 15.4
– glucogenic 4.2.2
– ketogenic 4.2.2
– non-standard 1.3.1, 11.5.3

– standard 1.3.1
L-2-aminoadipate 15.8.1
aminoacyl-(A-)site (ribosome) 10.6.1, 11.5.2
aminoacyl-tRNA ligases 10.6.2, 11.5.1
4-aminobutyrate (GABA) 4.2.2, 17.2.4
5-aminolevulinate 5.1
5-aminolevulinate dehydratase 5.1
5-aminolevulinate synthase 5.1
aminomethyltransferase 4.4.2
2-amino-3-oxiadipate 3.8.1
8-amino-7-oxo-pelargonate 9.8.1
6-aminopenicillanic acid 15.8.1
aminopterin 8.2.5
amino sugars 13.1.2
– biosynthesis 3.7
ammonia monooxygenase 15.6.1
ammonia oxidation 15.6.1
ammonotelic animals 4.9
AMP 8.1.2
amplification loop 19.2.1
amylase 3.2.3
amylo-1,6-glucosidase 3.2.3
amylopectin 3.2.1, 3.2.2
amylose 3.2.1, 3.2.2
anaerobic lithotrophy 15.5
anaerobic respiration 15.4, 15.5
analgetic effects 17.1.5
anaphase 11.6.5
anaphase-promoting complex (APC) 11.6.5, 14.6.7
anaplerotic reactions 3.3.4, 3.8.2
androgens 7.4.1, 7.6
– biological function 7.6.3
– biosynthesis 7.6.1
– degradation 7.6.2
– synthetic analogues 7.6.4
– transport 7.6.2
4-androstene-3,17-dione 7.6.1, 7.7.1
aneurin 9.2
angiotensin II 17.4.8
anhydrotetracycline 15.8.4
animal cells 2.2.4
anomers 1.2.1
antenna complexes 16.2.1
anthranilate 4.7.1, 4.7.3
antibacterial drugs 8.2.5
antibiotics
– classification 15.8
– resistance 15.8.1, 15.8.2, 15.8.3, 15.8.4
antibodies 19.1
– secreted soluble 19.1.3
– specific 19.1.7
anticancer drugs 8.2.5
anticoagulants 9.13
anticodon helix 10.6.1
antidiabetic drugs 18.1.1
antiestrogens 7.7.4
antigen peptides 19.1.4
antigen presentation 19.1.5
antigen-presenting cells (APC) 19.1.4, 19.1.5, 19.1.7
antigen recognition 19.1.3
antigenic surface components 15.1
antigen-recognition activation motifs 17.5.4
antigens 19.1.2
– processing and presentation 19.1.5
– T independent 19.1.7
– uptake and presentation 19.1.7
antiglucocorticoids 7.8.4
antiinflammatory drugs 17.4.8
antimineralocorticoids 7.8.4
antimycin 16.1.2
antioncogenes 17.5.1
α₂-antiplasmin 20.5.2
antiporters 15.3.1, 18.1.1
antitermination 10.5.3
antitermination factors 10.4.2
antiviral defense 12.1.1
antizymes 4.9.3
aorta 13.1.5
AP-endonucleases 10.3.3, 11.2.3
Apo B-48 18.2.2
Apo C-II 18.2.2
Apo E 18.2.2
Apo E 18.2.4
apoenzyme 2.4.1
apolipoproteins 18.2.1
– hydrolyzation 18.2.4

apoptosis 11.6.6, 14.6, 14.6.7
apoptotic bodies 11.6.6
aprotinin 14.6.6
APS reductase pathway 15.6.1
apurinic site 10.3.3, 11.2.3
apyrimidinic site 10.3.3, 11.2.3
arabinan-endo-arabinosidase 3.6.3
arabinogalactan 3.6.3
arabinogalactan-endo-galactosidases 3.6.3
L-arabinose 3.6.2, 3.6.4
arachidonic acid 6.3.2
– hormone-like effects 17.4.8
– release of 17.4.8
ARAM regions 17.5.4
ArcAB system 10.5.4
archaea 2.2.1, 13.2.2, 15.5.2
Arf 17.4.1
arginase 4.9.1
arginine 4.2.2, 4.9.1
– biosynthesis 4.9.1
– degradation 4.9.1
argininosuccinate 4.2.4, 4.9.1
arogenate 4.7.1
arogenate pathway 4.7.1
aromatase inhibitors 7.7.4
aromatase reaction 7.7.1
aromatase system 7.7.1
aromatic amino acid monooxygenases 9.6.3
arrestin 17.4.1
arylsulfatase A 13.4.3
ascorbate 3.5.2, 3.6.2, 4.5.8
– biochemical function 9.10.2
ascorbate decarboxylase 4.2.4
ascorbate metabolism 9.10
ascorbate peroxidase 4.5.8
asialoglycoprotein receptor 13.1.1
A-site (ribosome) 11.5.2
asparagine 15.8.4
– metabolism 4.2.4
asparagine-ammonia ligase 4.2.4
aspartate 3.8.1
– metabolism 4.2.4
aspartate carbamoyl transferase 8.2.1
aspartate cycle 4.9.1
aspartate kinases 4.5.1
aspartate oxoglutarate carrier 16.1.2
aspartate peptidases, reaction mechanism 14.6.4
aspartate shuttle 3.3.5
aspartate transaminase (ALAT, GOT) 4.2.4
aspartate-glutamate shuttle 16.1.2
aspartylglucosaminuria 13.1.4
Asp-N-metallo-endoproteinase 14.6.1
aspulvinone 7.3
atherosclerosis 18.2.2, 18.2.5
– risk factors 18.2.4
ATM kinase 11.6.6
ATP
– conservation of energy 8.1.3
– hydrolysis 18.1.1
– metabolic role 8.1.3
– synthesis 16.1.4
ATP/ADP carrier 16.1.4
ATP binding cassette 15.3.1, 18.1.1
ATP-glucose-1-P-adenyltransferase 16.2.2
ATPases 14.4
– Ca^{++} transporting 17.4.4
– types 18.1.1
– type-2 18.1.1
ATP synthase
– H^+ transporting 8.1.3, 15.6.1, 16.1.4, 16.2.1
– H^+-dependent 15.6.1
– Na^+-transporting 16.1.3
– structure 16.1.4
atrazine 16.2.1
atrial natriuretic factor 17.7.1, 17.1.8
attenuation 10.5.3
attenuator sequences 10.5.3
autocrine 17.1.1, 19.1.6
autonomic nervous system 17.4.5
autonomously replicating DNA sequences 11.6.3
autotrophy 2.1, 15.6.1
auxin 4.7.3
avidin 9.8.1
axerophthol 9.1
axonal transport 17.2.6
azotobacter 4.1

B

B7.1 19.1.4, 19.1.7
B7.2 19.1.4, 19.1.7
Bacillus subtilis 4.7.1

bacitracin 13.3
bacteria
– anaerobic 16.1.5
– chemolithotrophic 15.6.1
– electron transport 16.1.3
– endospore forming 10.5.1
– Gram-negative 15.1
– Gram-positive 15.1
– homoacetogenic 15.5.1
– lysis 19.1.7
– methanogenic 9.3.2
– nitrosofying 15.6.1
– obligate anaerobic 16.1.3
– opsonization 19.1.7
– oxidative phosphorylation 16.1.2
– photosynthetic 16.2.1
– sulfidogenic 15.5.1
– viral infection 12.2.1
bacterial cell walls 13.1, 13.1.3
bacterial envelope 15.1
bacterial taxonomy 15.1
bacteriochlorin ring systems 5.4
bacteriochlorophylls 5.4, 16.2.1
bacteriophage λ 12.2.1
bacteriopheophytine 5.4, 16.2.1
bacteriorhodopsin 16.2.1, 17.4.6
balsams 7.3.1
barrels (protein structure) 2.3.1
base excision repair 10.3.3, 11.2.3
Basedow's disease 17.1.5
basic helix-loop-helix 11.4.2
basic transcription factors 11.3.2
Bax 11.6.6
B cell receptors 17.5.1, 17.5.4
Beggiatoa 15.6.1
beriberi 9.1.2
betaine aldehyde dehydrogenase 6.3.5
betaine metabolism 6.3.5
B group receptors 17.6
bi-bi reactions 1.5.4
bile acid conjugates 7.9.1
bile acids 18.2.4
– biosynthesis 7.9.2
– configuration 7.9.2
– from cholesterol 7.1.1
– occurence 7.9
– primary 7.9.2
– secondary 7.9.2
– secretion 7.9.2
bile alcohol 7.9.1
bile pigments 5.3.1
bilins 5.3.2
bilirubin 5.3.1
biliverdin 5.3.1
binding protein dependent transport system 15.3.1, 18.1.1
binding proteins 13.3.2
binding surfaces, factor-specific 20.2.3
biogenic amines 17.2.4, 17.4, 4.7.3
biotin
– biochemical function 9.8.2
– biosynthesis 9.8
biotin carboxylase 6.1.1, 9.8.2
biotin carrier protein 9.8.2
biotin decarboxylase 9.8.2
biotinidase 9.8.1
BiP/Grp78 14.1.3
2,3 biphosphoglycerate 3.1.1, 3.2.3
biphosphoglycerate mutase 3.1.1
bisphosphoglycerate phosphatase 3.1.1
bitter tastes 17.4.7
blastocyte state 17.1.6
blood 19.1.2
blood agglutination 13.5.1
blood cells, development 19.1.2
blood coagulation 13.1.5, 19.3, 20.1
blood group specificity 13.2.1
blood groups 13.5.1
blood lipoproteins 7.1.2
blood pressure, control 17.1.8
blunt end 10.7.2
B lymphocytes 19.1.2, 19.1.5
– activation 19.1.7
– antigen receptor complex 19.1.3
body fluids 13.5.1
body temperature regulation 17.1.5
Bohr effect 5.2.3
bone marrow 19.1.2
A boxes 11.4.1
B boxes 11.4.1
bradykinin 17.4, 17.4.8, 20.1, 20.2.2, 20.5.1
brain 18.1.1

branched-chain α-keto acid dehydrogenase 4.6.2
brassinosteroids 7.2.2
brown adipose tissue 6.2.2, 16.1.4
Brownian ratchet model 14.4.2
bundle-sheath cells 16.2.2
butane diol 15.4
butanol 15.4
butanol acid fermentation 15.4
butanol dehydrogenase 15.4
butyraldehyde 15.4
butyrate 15.4
butyrate kinase 15.4
butyric acid fermentation 15.4
butyryl-CoA 15.4
butyryl-CoA dehydrogenase 15.4
butyryl-P 15.4

C

C2 19.2.1
C3 19.1.8, 19.2.1
C4 19.2.1
C5 19.2.1, 19.2.2
Ca^{++} channels 17.4.5, 18.1.1
– types 17.4.4
Ca^{++} influx 17.4.4
Ca^{++}/Na^+ antiport mechanism 17.4.4
Ca^{++} sensors 17.2.3
Ca^{++} vesicular transport 14.2.2
Ca^{++} waves 17.4.4
cadherines 19.3
cadinene 7.3.1
CAK protein 8.2.1
calbindin 9.11.2
calciferol
– biochemical function 9.11.2
– biosynthesis 9.11
calcineurin 17.5.4
calcitonin 17.1.7
calcitriol 9.11.1
calcium binding proteins 9.11.2
calcium metabolism 9.11.2
calcium
– biochemical functions 17.1.7, 17.4.4
– concentration, in cells 17.4.4
– in blood coagulation 20.2.3
– in muscle contraction 17.4.5
– inositol phosphates 17.4.4
– metabolism 17.1.7
– regulation 17.1.7
see also Ca^{++}
caldesmon 17.4.5
calmodulin 17.4.2, 17.4.5, 18.1.3, 20.4
– role of 17.4.4
calmodulin-dependent protein kinase II (CaM kinase) 17.4.5
calmodulin-Ca^{++}, targets 17.4.4
calnexin 14.1.3
calponin 17.4.5
calreticulin 17.4.4
calsequestrin 17.4.4
Calvin cycle 16.2.2
CaM kinase II 17.4.4
cAMP 17.5.3
– biochemical functions 17.4.2
– level 10.5.1
cAMP phosphodiesterases 17.7.2
campestrol 7.2.2
camphor 7.3.1
canalicular bile 18.2.4
CAP (catabolite genes activator protein) 10.5.1, 10.5.4
capillary vessel endothelium 18.2.2
capping 11.3.3
capsid 12.1.1
capsidiol 7.3.1–7.3.4
carbamoyl phosphate 4.9.1, 8.2.1
carbamoyl synthase 4.9.1
carbamoyl synthase II 8.2.1
carbamoyl-P synthase 8.2.1
carboanhydrase 4.9.1, 5.2.3, 16.2.2
carbohydrate structures, antigenic 13.5
carbohydrates
– classification 1.2.1
– microheterogenity 13.4.1
– nomenclature 1.2.1
– storage forms 3.2
– transport form 3.4.1
carbon monoxide 17.7.2
carbon monoxide dehydrogenase 15.5.3, 15.6.1
carbon sources, non-sugar 15.4
carbonate dehydratase 4.9.1, 5.2.3

carbonic anhydrase 4.9.1, 5.2.3, 16.2.2
carbon-monoxide: acceptor oxidoreductase 15.6.1
carboxidobacteria 15.6.1
γ-carboxyglutamate 1.3.1, 9.13, 20.3.1
carboxyl carrier protein 6.1.1
carboxylation reactions, vitamin K dependent 9.13
carboxylations 9.8.2
carboxyltransferases 3.3.4
carboxypeptidase A, reaction mechanism 14.6.5
carcinoma cells 18.1.1
cardiac glycosides 7.2.2
cardiolipin 6.3.1, 6.3.2
carnitine 4.5.2
carnitine O-octanoyltransferase 6.1.4
carnitine O-palmitoyltransferase 6.1.4
carnitine shuttle 6.1.4
carnitine-acylcarnitine translocase 6.1.4
carnosine 4.2.4, 4.8.2
β-carotene 4.5.8, 7.3.2, 9.1.1, 16.2.1
carotenoids 7.3.2
– absorption spectrum 16.2.1
– metabolism 7.3.1–7.3.4
cartilage 13.1.5
cartilage proteoglycan 13.1.2
cascade mechanism 2.5.2
caspases 11.6.6
catabolite repression 10.5.1
catalase 4.5.8, 5.2.4, 6.1.5
catalytic efficiency 1.5.4
catalytic facilitation 4.7.1
catechol 4.7.3
catechol estrogens 7.7.1
catecholamines 3.1.2, 4.7.4, 17.2.3
– biosynthesis 17.1.4
– metabolic effects 17.1.4
catechol-O-methyltransferase 4.7.4
C_4 cycle 16.2.2
CD molecules 19.1.4
CD1 19.1.2
CD2 19.1.2
CD3 17.5.4, 19.1.2
CD3 complex 19.1.4
CD3ζ co-receptor 17.5.4
C3d 19.1.3
CD4 17.5.4, 19.1.2, 19.1.4
CD4 (Co)receptor 12.4.1, 19.1.4
$CD4^+$ T effector cells 19.1.7
$CD4^+$ T lymphocytes 19.1.4, 19.1.5, 19.1.7
CD8 17.5.4, 19.1.2, 19.1.4
CD8 receptor 19.1.4
$CD8^+$ T lymphocytes 19.1.4, 19.1.5, 19.1.7
CD19 17.5.4, 19.1.3
CD21 17.5.4, 19.1.3
CD28 19.1.4, 19.1.7
CD31 19.3
CD34 19.3
CD40 receptor 19.1.7
CD45 17.5.4
CD59 19.2.5
CD79a 19.1.3
CD79b 19.1.3
CD80 19.1.7
CD86 19.1.7
Cdc6 11.2.1, 11.6.3
Cdc28 11.6.2
Cdc55 11.6.6
CDK activating kinase 11.6.4
CDK4 11.6.4
CDK6 11.6.4
CD numbers 19.1.2
CDP-ethanolamine 6.3.2
cedrene 7.3.1
cell adhesion 13.3.4, 19.3
cell-cell interactions 13.1.1, 13.2.1, 19.3
cell-cell-recognition 13.5
cell cycle
   checkpoints 11.6.6
– control mechanism 11.6.1
– eukaryotic 11.1.1, 11.6
– phases 11.1.1
cell differentiation 14.6.7
cell division 14.6.7, 18.1.3
cell envelope 2.2.1
cell growth, regulation 17.5.1
cell lysis 13.5.1
cell membrane 2.2.4
– components 13.1
cell surface, antigenic properties 13.5
cell transformation 12.2
cell wall 2.2.1
cell wall synthesis 3.6.3, 15.1

cellobiose 3.2.3
cells, veiled 19.1.2
cellular immune response 19.1.7
cellular membranes 1.4.9
cellular movements 18.1.3
cellulase 3.2.3, 3.6.3
cellulose 3.1.1, 3.2.1, 3.6.3
– degradation 3.2.3
– synthesis 3.2.2
cellulose fibers 3.6.3
cellulose synthases 3.2.2
centromere DNA section 11.6.5
centromeres 2.6.4
cephalosporin 15.8.1
cephamycin C 15.8.1
ceramide 1.4.6, 6.3.4, 13.2.1
– biosynthesis 6.3.4
ceramide galactosyltransferase 13.4.3
ceramide glucosyltransferase 13.4.3
cerebrosides 1.4.6
cerebroside sulfatase 13.4.3
cervix cancer 12.1.1
c-Fos 11.2.5, 11.4.3, 17.5.4
cGMP phophodiesterase 17.4.6
cGMP, in the visual process 17.7.1
chalcone isomerase 4.7.6
chalcone synthase 4.7.6
channeling effect 4.7.1
chaperone homologues 14.1.2
chaperones 14.1.1, 14.4.2, 14.5.2
– Hsp70-like 14.5.2
chemical information 17.3
chemiosmotic theory 15.5.1
chemiosmotic coupling 15.5.2, 15.6.1, 16.1
chemoattractants 19.3
chemokine receptors 19.1.6
chemokines 19.1.7, 19.3
chemolithotrophy 2.1, 4.5.6, 15.6
chemoorganotrophy 2.1, 15.6
chemotaxis 19.1.6
chemotrophy 2.1
chenodeoxycholyl-CoA 7.9.1
chicken pox 12.1.1
chi-site 10.3.5
chitin 3.6.3, 3.7.1
chitin synthase 3.7.1
chloramphenicol 15.8
chlorin ring 5.4
chlorophyll 16.2.1
– biosynthesis 5.4
chloroplast proteins 14.5
chloroplasts 2.2.3
– components for import 14.5.1
chlorotetracycline 15.8.4
5β-cholanic acid 7.9.1
cholecalciferol 9.11.1, 17.1.7
cholecystokinin 17.1.9
cholera toxin 17.4.1
cholesterol
– as a precursor 7.1, 7.4
– biosynthesis 7.1.1, 7.1.4, 7.12
– daily demand 7.1.2
– de novo synthesis 17.1.5
– degradation 7.1.1
– esterification 7.1.1
– homeostasis 7.1.5, 18.2
– in membranes 7.1.3
– removal by HDL 18.2.2
– transport 18.2.2
– turnover 7.1.2
cholesterol-7α-hydroxylase 7.1.5
– regulation 7.9.3
cholesterol acyltransferase 18.2.4
cholesterol deposits 18.2.4
cholesterol ester droplets 18.2.2
cholesterol ester transport protein 18.2.3
cholesterol esters
– hydrolysis 17.1.5
– hydrolyzation 18.2.4
cholesterol monooxygenase 7.5.1
cholesterol pool 7.1.2, 18.2.4
cholesterol transport, reverse 18.2.4
choline 6.3.2
choline acetyltransferase 6.3.5
choline dehydrogenase 6.3.5
choline metabolism 6.3.5
choline monooxygenase 6.3.5
chondroblasts 13.1.2
chondroitin 4-sulfate 13.1.5
choriongonadotropin (CG) 17.1.6
chorionsomato(mammo)tropin (CS) 17.1.6
chorionthyrotropin (CT) 17.1.6

chorismate 4.7.1
chromatide separation 11.6.5
chromatin 2.6.4
– replication competent state 11.1.1
chromatin filaments 2.6.4
chromosome segregation 11.6
chromosomes
– bacterial 10.2
– eukaryotic 11.1.2
– organisation level 2.6.4
chylomicrons 6.2.1
– clearance 18.2.4
– metabolism 18.2.2
chymotrypsin, reaction mechanism 14.6.2
cilia 18.1.3
cinnamic acid 4.7.6
cis-platin 10.3.1
citramalate 15.4
citrate 3.3.5, 3.8.1, 15.4
– high concentrations 3.1.2
citrate cycle
– energy balance 3.8.3
– reaction sequence 3.8.1
– reductive 3.8
– regulation 3.8.2
citrate lyase 3.8.1
citrate synthase 3.3.5, 3.8.1
citrate-pyruvate shuttle 6.1.1
citrulline 1.3.1, 4.9.1, 4.9.2
c-Jun 11.2.5, 11.4.3, 17.5.4
class II MHC molecules 19.1.7
class switch 19.1.3
classical pathway (complement) 19.2.1
clathrin 18.1.2
Cl⁻ channels 18.1.1
cleavage and polyadenylation specificity factor 11.3.3
cleavage stimulating factor 11.3.3
Cleland nomenclature 1.5.4
CLIP segment 19.1.5
Clostridium thermoaceticum 15.5.3
cluster of differentiation (CD) 19.1.2
C_1 metabolism 9.6.2
c-Myc 11.2.5, 11.4.3, 17.5.4
CO_2 fixation 15.5.3, 16.2.2
CO_2 reduction 15.5.1, 15.5.2, 15.5.3
coagulation 14.6, 20.2.2
– contact activation 20.5.1
– propagation 20.3.4
coagulation cascades 20.1
coagulation control 20.3.1
coagulation factor Xa 14.6.1
coagulation factors 9.13, 20.3
coagulation inhibitors 20.1
coated pits 7.1.3, 18.1.2, 18.2.4
coated vesicles 18.1.2, 18.2.4
cobalamin
– biochemical function 9.5.2
– biosynthesis 9.5
Cockayne's syndrome 11.2.6
coding region 10.4.2
coding strand 10.1.1
codon 10.1.2
coenzyme A, synthesis 9.7.1
coenzyme B (HS-HTP) 15.5.2
coenzyme B_{12}, see cobalamin
coenzyme F_{420} 9.3.2, 15.5.2
coenzyme F_{430} 9.5.3, 15.5.2
coenzyme M 15.5.2
coenzymes 2.4.1
coeruloplasmin 5.2.1
cofactor Va 20.3.3
coiled-coil 2.3.1, 18.1.3
colchicine 17.2.6, 18.1.3
colipase 6.2.2
collagen 2.3.1, 4.3, 13.4.1
collagen synthesis 19.1.6
collecting tubules 17.1.8
collectins 19.1
colony forming units 19.1.2
colony stimulating factors 19.1.2
Com70 14.5.2
commitment point 11.1.1
committed steps 6.1.2
compactin 7.1.4
competence factors 17.5.1
competitive inhibition 1.5.4, 2.5.2
complement activation 14.6, 19.2.1
complement control receptors 19.2.4
complement factors 19.1, 19.2
complement pathways 19.2.1
complement receptor-2 19.1.3

complement receptors 19.1.1, 19.1.7, 19.2.3
complement system 19.2
– activation 19.1.7
– control mechanism 19.2.4
complement, main effects 19.2
complementarity determining regions 19.1.3, 19.1.4
complex mannose structures 13.4.1
concatamers 12.2.1
cone cells 17.4.6
congenital adrenal hyperplasia 7.8.4
conjugation 2.2.1
conjugation reactions 4.5.7
Conn syndrome 7.8.4
connective tissues 13.1.5
connexins 17.2.4
consensus sequences 11.3.3
constant domains 19.1.3
constitutive pathways 14.2.1
contact activation 20.1, 20.2, 20.2.2
contact inhibition 11.6.4
contractile ring 17.4.5
contraction cycle 17.4.5
C5 convertase 19.2.1, 19.2.2
cooperative feedback control 4.5.1
copper heme 16.1.3
COP proteins 14.2.2
coproporphyrinogen III 5.1
coprostanol 7.1.1
core polysaccharide 15.1
core promotor elements 11.3.2, 11.4.1
core protein 13.1.2
core region 11.3.5, 11.4.1
corrin ring 9.5.1
cortexone 7.8.1
corticosteroid-binding globulin 7.5.2
corticosteroids 7.8
– degradation 7.8.2
corticosterone 7.8.1
corticotropin 7.5.1, 17.1.5
corticotropin releasing hormones (CRH) 17.1.5
cortisol 7.8.1, 17.1.5
– action 17.1.5
– secretion 17.1.5
cortisone 7.8.1
cos-sequence 12.2.1
coumarin derivatives 9.13, 14.6.6
coumarol 4.7.6
cow pox 12.1.1
C_4 plants 16.2.2, 18.1.1
C-protein 17.4.5
CPSF 11.3.3
C1q 19.2.1
CR2 19.2.5
creatine 4.9.2, 6.3.5
creatine kinases, isoenzymes 2.4.2
creatinine 4.9.2
cretinism 4.7.5, 17.1.5
Creutzfeld-Jacob encephalopathy 13.2.1
critical micelle concentration 1.4.9
Cro-protein 12.2.1
crotonyl-CoA 15.4
C1s 19.2.1
CSA/ERCC6 11.2.3
CSA/ERCC8 11.2.3
Csk kinase 17.5.4
CTLA4 19.1.4
CTLA4 19.1.7
CTP 8.2.2
curcumene 7.3.1
Cushing's syndrome 7.8.4
CutI 11.6.5
cyanobacteria 4.1, 16.2.1
3′5′-cyclic nucleotide phosphodiesterase 17.4.2
cyclin A-CDK2 11.6.4
cyclin box 11.6.1
cyclin D 11.6.4
cyclin D-CDK4,6 11.6.4
cyclin E 11.6.4
cyclin E-CDK2 11.6.4
cyclin-CDK complex 11.6.1
cyclins 11.1.1, 11.6.1
cycloartenol 7.1.1
cycloheximide 11.5.2
cyclooxygenase pathway 17.4.8
cyclopentanoperhydrophenanthrene 1.4.7
cyclophilin 17.5.4
cyclosome 11.6.5
cyclosporin A 17.5.4
cystathionine 4.5.4
cysteine 15.8.1
– biological role 4.5.5
– metabolism 4.5.5

cysteine peptidases, reaction mechanism 14.6.3
cystic fibrosis 18.1.1
cystine 4.5.5
Cyt aa_3 16.1.2
Cyt b_{560} 16.1.2
Cyt b_6 16.2.1
Cyt b_H 16.1.2
Cyt bc_1 16.1.2
Cyt bd 16.1.3
Cyt bo 16.1.3
Cyt d 16.1.3
cytidine 8.1
cytidine-5′-triphosphate 8.2.2
cytochrome c 16.1.2
cytochrome c oxidase complex IV 16.1.2
cytochrome oxidase 15.6.1
cytochrome P-450 5.2.1
cytochrome P-450 monooxygenases 5.2.1, 7.4.1, 17.4.8
cytochrome-b_5 reductase 6.1.3
cytochromes 16.2.1
– biological function 5.2.4
cytokine receptor superfamily 19.1.6
cytokine receptors 17.5.4, 19.1.6
cytokines 7.3.4, 12.4.1, 17.1.6, 19.1, 19.1.6, 19.1.8, 19.3
– anti-proliferative 11.6.4
cytokinesis 11.6.5
cytokinines 7.3
cytoplasm 2.2.1
cytosine 8.1
– degradation 8.2.4
cytoskeleton 2.2.2
– components 18.1.3
cytosol 2.2.2
cytotoxic effector cells 19.1.7

D

DAF 19.2.5
Dam-methylase 10.3.4
dark reactions 16.2.2
data banks 21.1
5′-deazaflavin coenzyme F_{420} 15.5.2
decarboxylations 9.8.2
decatenation 10.2.3, 11.1.4
2-dehydro-3-deoxy-6-P-gluconate 15.4
7-dehydrocholesterol 7.1.2
11-dehydrocorticosterone 7.8.1
dehydroepiandrosterone 7.6.1, 7.7.1
dendritic cells 19.1.2, 19.1.5
denitrification 4.1, 15.5.1
deoxy hexoses 3.5.6
deoxyadenosylcobalamin 9.5
11-deoxycortisol 7.8.1
deoxyribodipyrimidine photo-lyase 10.3.2, 11.2.2
deoxyribonucleotides, from ribonucleotides 8.1.4
deoxyribophosphodiesterase 11.2.3
3-deoxyribose 8.1
deoxythymidine 8.1
deoxythymidine-5′-monophosphate 8.2.3
deoxyuridine pyrophosphorylase 8.2.3
deoxyuridine-5′-monophosphate 8.2.3
Deoxy-D-xylulose 7.3, 9.2.1, 9.4.1
Dep-1 enzyme 11.5.5
depolarization 17.2.2, 17.2.3
dermatan sulfate 13.1.5, 20.3.4
dermatoses 4.5.8
desensitization 17.4.1
desmin 17.4.5
desmosterol 7.1.1, 7.1.2
D-desosamine 15.8.3
destruction box 11.6.6
dexamethasone 7.8.4
dextrane 3.2.1, 3.2.2, 3.4.1
α-dextrin 3.2.3
DHPR Ca^{++} channels 17.4.5
diabetes 6.1.2
– insulin resistant 17.1.3
– ketone body level 6.1.6
diacetyl 15.4
diacetyl fermentation 15.4
diacetyl synthase 15.4
diacylglycerol 17.4.3, 17.5.4
diacylglycerol O-acyltransferase 6.2.1
1,2-diacylglycerol-3-P 6.2.1
1,2-diacylglycerols 6.2.1
7,8-diaminopelargonate 9.8.1
diaminopimelate 4.5.2
diapedesis 19.3
dicoumarol 9.13, 20.3.1
differentiation 19.1.6

– regulation 17.5.1
diffusion
– faciliated 15.3.1
– passive 15.3
digestion 14.6
digestive fluids 17.1.9
digoxin 7.2.2
DIHETRE 17.4.8
dihydrodopicolinate synthase 4.5.1
dihydrofolate 9.6.1
dihydrofolate reductase 4.7.1, 8.2.3, 8.2.5
dihydrolipoamide dehydrogenase 4.5.7
dihydrolipoyl acetyltransferase 3.3.1
dihydrolipoyl dehydrogenase 3.3.1, 4.4.2
dihydropteroate reductase 9.6.3
dihydropyridine 17.4.4
dihydropyridine receptor 17.4.4
dihydrosphingosine 1.4.6
dihydrostreptomycin 15.8.2
dihydrostreptose 15.8.2
5α-dihydrotestosterone 7.6.1, 17.1.5
dihydroxyacetone-P (glycerone-P) 3.1.1, 15.7
1α, 25-dihydroxycalciferol 9.11.1, 17.1.7
dihydroxyphenylalanine (dopa) 4.7.3
dihydroxypyruvate 16.2.2
dimethylallyl pyrophosphate 7.1.1, 7.3.4
dimethylglycine dehydrogenase 6.3.5
dinucleotides 8
diosgenin 7.2.2
dioxygenases 9.10.2
diphtheria toxin 11.5.2
diploptene 7.2.1
diplopterol 7.2.1
dipyrromethane cofactor 5.1
discs (retina) 17.4.6
diseases, viral 12.1.1
dissociation constant 1.5.4
disulfide bonds, reversible reduction 14.1.1
diterpenes 7.3.1
DJ segment 19.1.3
DMSO reductase familiy 9.6.4
DNA
– apurinic sites 10.3.1
– bending 11.4.3
– B-type 2.6.3
– compaction levels 2.6.4
– damage 10.3.1, 11.2.1, 11.6.6
– deamination 10.3.1
– defects 11.2.6
– double-strand breaks 11.2.5
– double-strand repair 10.3.5, 11.2.5
– elements, cis-acting 11.4
– initiator motif 11.3.2
– methylation, non-enzymatic 11.2.1
– mismatches 11.2.4
– photoreactivation 10.3.2
– photoproducts 11.2.3
– recombination 11.2.5
– repair systems 10.3.1, 11.2
– structure 2.6.3
– supercoiling 2.6.3
– viral 12.2
DNA alkylation 10.3.1
DNA glycosylases 10.3.3, 11.2.3
DNA dependent protein kinase 11.2.5
DNA helicase II 10.3.4
DNA helicases 11.1.4
DNA ligase 10.2.3, 10.3.3, 10.3.4, 11.1.4, 11.2.3
DNA methyltransferase 10.7.2
DNA polymerase I 10.2.3, 10.7.1
DNA polymerase III 10.2.2, 10.3.4
DNA polymerases
– α 11.1.4
– δ 11.1.4
– ε 11.2.3
– DNA-dependent 12.4.1
– eukarotic 11.1.3, 11.1.4
– proofreading ability 11.1.6
– RNA-dependent 12.4.1
– yeast 11.1.3
DNA primase 10.2.3
DNA repair methyltransferases 10.3.2
DNA replication 10.1.1, 14.6.7
– bacterial 10.2
– errors 10.3.1
– fidelity 10.2.4, 11.1.6
– heterochromatin 11.1.1
– in yeast 11.6.2
– initiation 10.2.2, 11.1.1
– mechanism 11.1.4
– proteins involved 10.2.2
– reaction mechanism 10.2

– telomeres 11.1.1
– termination 10.2.3
DNA response elements 11.4, 11.4.3
DNA topoisomerase 11.3.2
DNA topoisomerase II 11.1.4
DNA topoisomerase IB 11.1.4
DNA viruses 12.2
DnaA 10.2.1, 10.2.2
DnaB 10.2.2, 10.2.3
DnaC 10.2.2
DnaG 10.2.2, 10.2.3
DnaJ 14.1.1
DnaK-ATP 14.1.1
DnaKJ/GrpE 14.1.1
DNA-PK 11.2.5
DNases 10.7
dolichol anchors 13.3.3
dolichol esters 7.3.3
dolichol kinase 13.3
dolichol-P 7.3.3, 13.2.3, 13.3.3
dolichol-P-mannose 13.3.3
dolichol sugars 7.3.1–7.3.4
dopamine 4.7.3, 4.7.4, 17.2.4
dopamine receptors 17.2.5
dopamine β-monooxygenase 4.7.4, 17.1.4
dopaquinone 4.7.3, 15.7
double-strand breaks 10.3.5, 11.2.5
double-strand repair 11.2.5
downstream elements 11.4.1
Dsb proteins 14.1.1
DsbA 14.1.3
DsbE 14.1.3
D segments 19.1.3
dwarfism 17.1.5
dynein 11.6.5, 17.2.6, 18.1.3
dynorphine 17.1.5, 17.2.4
dyslipoproteinemia 18.2.5

E

early transcription phase 12.2.1
EC list 2.4.5
ecdysone 7.2.3
EF hand 17.4.4
eicosanoids
– degradation pathways 17.4.8
– hormone-like effects 17.4.8
Einstein (unit) 16.2.1
Eisenthal and Cornish-Bowden plot 1.5.4
electrical membrane potential 1.5.3, 15.3.1
electrochemical gradient 15.2, 15.6.1
electrolytes 17.1.9
electron acceptors 15.4, 16.1
– inorganic 15.5
– typical 15.5.1
electron donors, typical 15.5.1
electron flow 16.2.1
electron transferring flavoprotein 6.1.5
electron transport 15.5.1
– bacterial 15.5.1, 16.1.3
– mitochondrial 16.1.2
elicitors 4.7.6
elongation
– complex 10.4.2
– factors 10.6.1, 11.3.2
– regulation 10.5.2
– translation 10.6.3
embryogenesis 17.1.5
embryonal development 18.1.1
emulsifiers 1.4.2
endergonic reactions 8.1.3
endocrine 17.1.1, 19.1.6
endocytosis 14.2.1, 18.1.1, 18.1.2, 19.1.5
endodeoxyribonucleases 10.7, 10.7.2
endometrium 17.1.5
endonucleases 10.7, 10.7.2
endopeptidases 14.6.1
endoplasmic reticulum (ER) 2.2.2, 14.2.1
– functions 13.3
endoproteinase Glu-C 14.6.1
β-endorphin 17.2.4
energy-rich bond 8.1.3, 9.7.2
endorphins 17.1.5
energy-rich bond 8.1.3, 9.7.2
endosomes 18.2.4, 19.1.5
endothelium 19.3
endothelium-derived relaxing factor, EDRF 20.4
end-plate potential 17.2.3
enhancers 11.4, 11.4.3
enkephalins 17.2.4
enolase 15.7
enoyl reductase 6.1.1

enterohepatic circulation 7.9.2
enterokinase 14.6.1
enteropeptidase 14.6.1
enterotoxins 17.4.1, 17.7.1
enthalpy 1.5.1
Entner-Doudoroff pathway 3.5.3, 15.4
entropy 1.5.1
enzyme activity 1.5.4
– regulation 2.5
enzyme cofactors 2.4.1
enzyme kinetics 1.5.4
enzymes
– biotin-dependent 9.8.1
– catabolic 17.1.5
– catalytic mechanisms 2.4.1
– classification 2.4.5
– covalent modification 2.5.2
– induction 3.1.2
– inhibition 1.5.4
– kinetics 1.5.4
– multifunctional 8.1.1
– site of regulation 2.5.3
enzyme-substrate complex 2.4.1
eosinophilic chemotactic factor 19.1.8
EPETREs 17.4.8
epimers 1.2.1
epinephrine 3.1.5, 4.7.3, 4.7.4, 17.1.3, 17.1.4, 17.4, 20.4
– regulation of fatty acid synthesis 6.1.2
episomes 2.2.1, 10.2
epithelial cells 19.1
epoxygenase P-450 pathway 17.4.8
cis-epoxy-eicosatetranoic acids 17.4.8
Epstein-Barr virus 19.2.5
equilibrium constant 1.5.1, 2.4.1
equilibrium potential 1.5.3, 17.3
ergocalciferol 7.2.2, 9.11.1
ergosterol 7.2.2
ergot alkaloids 7.3
ergotamin 7.3.1–7.3.4
ergotoxin 7.3.1–7.3.4
erythrocytes 13.5.1
– lifetime 5.3.1
erythromycin 15.8.3
erythronolide 15.8.3
erythropoietin 17.5.4
erythrose-4-P 16.2.2
E-selectin 19.3
E-site 11.5.2
estradiol 7.7.1, 17.1.5
– biosynthesis 7.7, 17.1.5
– hydroxylation 7.7.2
estriol 7.7.1
estrogen production 7.7.1
estrogen receptor 7.7.3
estrogens effects 17.1.5
estrogens 7.4.1, 17.1.6
– biological function 7.7.3
– degradation 7.7.2
– metabolism 7.7.1
– substitution therapy 7.7.4
– synthesis 7.7.1, 17.1.5
– transport 7.7.2
estrogen-progesterone interaction 7.5.2
estrone 7.7.1
– biosynthesis 7.7, 17.1.5
– hydroxylation 7.7.2
ethane 4.5.4
ethanol 3.3.5, 15.4
ethanolamine 6.3.2
ethanolamine-phosphate 13.3.4
ether lipids 6.3.3
etheric oils 7.3.1
N-ethylmaleinimide sensitive protein 14.2.2
euchromatin 2.6.4
eukaryotic cells 2.2.2
excision exonucleases 10.3.3
excision nuclease complex 11.2.3
excitatory signals 17.2.3
exciton 16.2.1
exciton transfer 5.4
exergonic reactions 8.1.3
exit(E-)site (ribosomes) 10.6.1
exo-α-sialidase 13.2.1, 13.4.3
exodeoxyribonucleases 10.7
exoglycosidases 13.2.1
exons 11.3.3
exonuclease I 10.3.4
exonuclease III 10.7.1
exonuclease MFI 11.1.4
exonuclease V 10.3.5
exonuclease VII 10.3.4

exonucleases 10.7
exopeptidases 14.6.1
exoskeleton 3.7.1
extracellular matrix 2.2.4, 13.1, 13.1.2
extracellular movements 18.1.3
extrahepatic cells 18.2.2
extrinsic pathway (blood coagulation) 20.2

F

Fab-molecules 19.1.3
Fabry's disease 13.2.1
facilitated diffusion 18.1.1
F-actin 17.4.5
factor F_{420} 9.3.2
factor III 20.2.1
factor V Leiden 20.3.4
factor VII 20.2.1, 20.3.3
factor IX 20.3.3
factor X 20.3.3
facort XI 20.2.2, 20.3.3
factor XII 20.2.2
σ-factor 10.4.1, 10.5.1
facultative anaerobes 15.5.1
facultative methyltrophs 15.7
FAD 9.3
familial hypercholesterolemia 7.1.5, 18.2.4, 18.2.5
farnesol 7.3.1
farnesyl-PP 7.1.1
fasting 6.1.2
fatty acid binding proteins 6.1.4
fatty acid substituents, remodelling 6.3.2
fatty acid synthase 6.1.1
fatty acids 1.4.1
– activation 6.1.4
– biosynthesis 6.1.1, 6.1.2, 7.1.4
– branched-chain 4.6.2
– chain elongation 6.1.3
– desaturation 6.1.3
– degradation 6.1.5
– essential 6.1.3, 9.14.2
– liberated 18.2.2
– methyl-branched 6.1.1
– odd-numbered 6.1.1, 6.1.5
– oxidation 6.1.5, 6.1.6
– polyunsaturated 6.1.3
– transport 6.1.4
– unsaturated 6.1.3
fatty-acid-CoA ligases 6.1.4, 6.2.1
Fcε-receptors 19.1.1, 19.1.8
Fcγ receptors 19.1.7
Fc-receptors 19.1.1
feedback inhibition 2.5.3, 4.2.1, 4.5.1, 4.6
feed-forward activation 3.1.2
Fenton reaction 4.5.8
fermentation 15.4
– alcoholic 3.3.6
– general principle 15.4
– non-glycolytic 15.4
– of heterocyclic compounds 15.4
– of sugars 15.4
ferredoxin 4.5.8, 16.2.1, 16.2.2
ferredoxin NADP$^+$ reductase 16.2.1
ferredoxin
– nitrogenase reaction 4.1
– reduced 4.5.6
ferredoxin-thioredoxin reductase 16.2.2
ferric reductase 15.5.1
ferrioxidase 5.2.1
ferritin 5.2.1
ferritin mRNA 11.5.4
ferritin reductase 5.2.1
ferrochelatase 5.2.1
fertilization 14.6
ferulol 4.7.6
FeS clusters 16.1.2
fever 19.1.6
fibrin 20.5.1, 20.3.5, 20.5.2
fibrin deposits 20.5
fibrinogen 19.3, 20.3.5
fibrinolysis 14.6, 20.1, 20.5
– activation 20.2.2
– amplification pathway 20.5.1
– control 20.5.2
fibrinolytic factors 20.5
fibroblasts 13.1.2, 18.2.2
fibronectin 19.2
10 nm filament 2.6.4
FIS protein 10.2.2
Fischer convention 1.2.1
flagella 2.2.1, 18.1.3
flagellates 18.1.3

FLAP 17.4.8
flavin mononucleotide, see FMN
flavin-adenine dinucleotide, see FAD
flavonoids 4.7.6
flavoproteins, reactions with O_2 9.3.2
fluid mosaic modelc1.4.9
5-fluorouracil 8.2.5
FMN 9.3
foam cells 18.2.2
focal adhesions 18.1.3
follicle-stimulating hormone (FSH) 7.5.1, 17.1.5
follicular dendritic cells 19.1.2, 19.2.3
folylpolyglutamate 9.6.1
formaldehyde 15.7
formaldehyde dehydrogenase 15.7
formate 15.4
formate dehydrogenase 10.6.5, 15.5.1, 15.5.3, 15.7, 16.1.3
formate : hydrogen lyase complex 15.4
formimino-THF 9.6.2
formyl-methionine 4.5.4, 10.6.3, 11.5.2
10-formyl-THF 9.6.2
forskolin 17.4.2
frameshift 10.1.2
free energy 1.5.1
fructans 3.2.1, 3.2.2, 3.4.1
β-fructofuranosidase 3.4.1
fructokinase 3.1.1
D-fructose 3.1.1
fructose bisphosphatase 3.1.1, 3.1.2, 16.2.2
fructose bisphosphate aldolase 3.6.1, 15.7, 16.2.2
fructose bisphosphate aldolase B 3.1.1
fructose metabolism 3.1.1
fructose-1,6-bisphosphatase 3.1.1, 16.2.2
fructose-1,6-P_2 3.1.1
fructose-1-P 3.1.1
fructose-2,6-bisphosphatase 3.1.1, 3.1.2
fructose-2,6-P_2 3.1.1, 3.1.2
fructose-6-P 3.1.1, 3.1.2, 15.7
Fts Y 15.2
fucose 13.5.1
L-fucose 3.5.6
fucosidosis 13.1.4
fucosyl transferases 13.5.1
fucosylation 13.5.1
fumagallin 15.8
fumarate 3.8.1, 4.9.1, 15.4
– as terminal electron acceptor 16.1.3
fumarate hydratase 15.4
fumarate reductase 15.4, 15.5.1, 16.1.3
fumarate respiration 15.5.1
furanoses 1.2.1
fusidic acid 15.8

G

GABA receptors 17.2.5
GADD45 11.6.6
galactinol 3.4.1
galactose catabolism 3.1.1
α-galactosidase 13.4.3
α-galactosidase A 13.2.1
β-galactosidase 3.4.2
galactosyl transferase 3.4.2
galactosyl ceramidase 13.2.1, 13.4.3
galacturonate 3.5.3, 3.5.6
Gallionella 15.6.1
gallstones 7.9.4
ganglioside β-galactosidase 13.2.1, 13.4.3
ganglioside galactosyl transferase 13.4.3
gangliosides 13.2.1
– functions 13.2.1
G_{M1} gangliosidosis 13.1.4, 13.2.1
GAP 17.1.5, 17.5.3
gap junctions 2.2.4, 17.2.4, 18.1.1
GART gene 8.1.1
gastric inhibitory peptide 17.1.3
gastric parietal cells 18.1.1
gastrin 17.1.9
gastrin releasing peptide 17.1.9
Gaucher's disease 13.2.1
GDP 8.1.2
GDP/GTP exchange factors 17.5.2, 17.5.3
gene activators 12.4.1
gene product cII 12.2.1
genetic code 10.1.2
– mitochondria 10.1.2
genital herpes 12.1.1
genomes
– bacterial 2.6.4
– eukaryotic 2.6.4

geraniol 7.3.1
geranyl transferase 18.2.4
geranylgeranyl-PP 7.3.4
geranyl-PP 7.1.1
gestagens 7.5
gibberelic acid 7.3.1
gigantism 17.1.5
globin gene 11.3.3
globin synthesis 5.2.2
GLP-1 17.1.3
GLP-2 17.1.3
glucagon 3.1.2, 17.1.3
– regulation of fatty acid synthesis 6.1.2
glucagon receptor 17.1.3
glucagon-like peptides 17.1.3
glucan branching enzyme 3.2.2, 3.2.3
4α-glucanotransferase 3.2.3
glucaric acid 3.5.1
glucocorticoids 3.1.2, 7.4.1, 17.1.4, 17.5.3
– biological function 7.8.3
– biosynthesis 7.8.1
– substitution therapy 7.8.4
glucokinase 3.1.1, 3.1.2, 3.1.5
gluconate 15.4
D-gluconate 3.5.2
gluconeogenesis 3.1.3, 3.1.5, 4.2.2
– initiation 3.3.5
gluconeogenetic enzymes 17.1.5
– repression 17.5.3
glucoreceptors 17.1.5
D-glucosamine-6-P 3.7.1
glucose 3.1.1, 15.4
– dephosphorylation 3.1.2
– high level 3.1.5
– low level 3.1.5
– phosphorylation 3.1.2
– resorption 3.1.4
glucose level 3.1.5
glucose shortage 3.1.5, 15.3.1
glucose transport 3.1.4, 18.1.1
glucose transporter 18.1.1
glucose-1-P-phosphodismutase 3.1.1
glucose-1,6-bisphosphate 3.1.1
glucose-6-P 3.1.5, 3.6.1
glucose-6-P-dehydrogenase 3.6.1, 4.5.7
glucose-6-P-isomerase 3.1.1, 16.2.2
glucose-6-phosphatase 3.1.1, 3.1.2, 3.1.5
glucose-alanine cycle 4.2.3
glucosidase 3.2.3, 14.1.3
α-glucosidase 3.4.1
glucosyl-ceramidase 13.2.1, 13.4.3
D-glucuronate 3.5.2
glucuronides 3.5.1
glutamate 3.8.1, 4.2.1
– carboxylation 11.5.3
– neurotransmitter action 17.2.4
glutamate-ammonia ligase 4.1, 4.2.1
glutamate dehydrogenase 4.1, 4.2.2
glutamate fermentation 15.4
glutamate metabolism 4.2.2
glutamate pathway 5.1
glutamate receptors 17.2.5
glutamate synthase 4.1, 4.2.2
glutaminase 4.2.1
glutamine metabolism 4.2.1
glutamine synthase 4.2.1
γ-glutamyl cycle 4.5.7
γ-glutamyltranspeptidase 4.5.7
glutathione 4.5.7
glutathione peroxidase 4.5.8
glutathione reductase 4.5.7
GlyCAM 19.3
D-glyceraldehyde 1.2.1, 3.1.1
D-glyceraldehyde-3-P 15.4
glyceraldehyde-3-P-dehydrogenase 3.1.1, 16.2.2
glycerate 3.1.1, 15.7, 16.2.2
glycerate kinase 3.1.1, 15.7
glycerol 3.1.1
– phosphorylation 6.2.1
glycerol dehydratase 3.1.1
glycerol kinase 3.1.1, 15.3.1
glycerol-3-P dehydrogenase 3.1.1, 15.5.1
glycerol-3-phosphate O-acyltransferases 6.2.1
glycerol-3-phosphate shuttle 16.1.2
glycerone-P (dihydroxyacetone-P) 3.1.1
glycerophospholipids 1.4.4, 4.4.1, 6.3.2
glyceryl-ether hydroxylase 9.6.3
glycine 15.7, 16.2.2
– neurotransmitter action 17.2.4
glycine amidinotransferase 4.9.2
glycine betaine 6.3.5
glycine cleavage enzyme 3.9.2, 4.4.2

glycine conjugates 7.9.1
glycine dehydrogenase 4.4.2
glycine fermentation 15.4
glycine hydroxymethyltransferase 4.4.2, 6.3.5, 9.6.2
glycine metabolism 4.4.2
glycine N-methyltransferase 6.3.5
glycine receptor 17.2.5
glycine reductase 10.6.5, 15.4
glycocholic acid 7.9
glycoconjugates 13.5
glycogen 3.1.1, 3.2.1
glycogen degradation 3.2.3
glycogen metabolism, regulation 3.2.4
glycogen phosphorylase 3.1.5, 3.2.4
glycogen storage diseases 3.2.5
glycogen synthase 3.1.5, 3.2.2, 3.2.3, 3.2.4
glycogenin 3.2.2, 3.2.3
glycoglycerolipids 1.4.2, 13.2.2
glycolate 16.2.2
glycolipid anchors 3.5.4
glycolipids 13.4
– degradation 13.2.1
– synthesis 13.4.3
– types 13.2
glycolysis 3.1.1, 3.1.5, 15.4
– committed step 3.1.2
– regulation 3.1.2
glycolytic enzymes, induction 17.5.3
glycophorin 13.5.1
glycoprotein degradation diseases 13.1.4
glycoproteins 13.4
– carbohydrate content 13.1.1
– classification 13.1
– folding mechanism 13.3.3, 14.1.3
– in plant cell walls 3.6.3
– synthesis 13.4.1
– targeting 13.1
glycosaminoglycans 13.1, 13.1.2
glycosidases 13.4
glycosides 3.4.3
– cardiac 18.1.1
glycosidic bonds 1.2.2
glycosphingolipids 1.4.6, 6.3.4, 13.2.1
– biosynthesis 13.4.3
– degradation 13.2.f
glycosyl groups, carriers 13.2.3
N-glycosylation 13.1, 13.3
O-glycosylation 13.1, 13.3
O-glycosylation core 13.4
glycosylation reactions 13.3, 13.4
glycosylation
– cotranslational 13.3
– in the Golgi apparatus 13.4
glycosylphosphatidylinositol anchors 13.3.3
glycosylphosphatylinositol 13.3.4
glycosylphosphopolyprenol 13.2.3
glycosyltransferases 13.4
glyoxylate 3.9.1, 3.9.2, 15.7, 16.2.2
– oxidation 3.9.2
– reactions 3.9.2
– transamination 4.4.2
glyoxylate cycle 3.1.3, 3.9.1
glyphosate 4.7.1
GMP 8.1.2
goiter 4.7.5, 17.1.5
Goldman equation 1.5.3, 17.2.1, 18.1
Golgi apparatus 2.2.2, 14.2.1
– functions 13.4
cis-Golgi network 13.4
gonadotropin releasing hormone (GnRH, LHRH) 17.1.5
gonadotropins 17.1.5
gout 8.1.7
gp41 12.4.1
gp120 12.4.1
gp160 12.4.1
gpB 12.2.1
gpC 12.2.1
gpN 10.5.3
gpNu3 12.2.1
gpR 12.2.1
gpS 12.2.1
G_1 phase 11.1.1, 11.6
G_2 phase 11.6
G-protein system 17.4.2
G-proteins 17.1.3, 17.2.3
– as second messengers 7.5.1, 17.4.1
– small 17.5.2
Gram staining 15.1
grana 2.2.3
granulocytes 19.1, 19.3

– basophilic 19.1.1
– eosinophilic 19.1.1
– neutrophilic 19.1.1
granulosa cells 7.7.1
Grb2 protein 17.5.3
Grb2/Sos protein 17.5.4
Greek key 2.3.1
GRF 17.5.3, 17.5.4
griseofulvin 15.8, 18.1.3
GroEL 12.2.1
GroEL/ES complex 14.1.1
group transfer reactions 11.5.3
growth factor genes 11.5.4
growth factors 17.1.6
growth hormone (GH) 6.2.2, 17.1.5, 17.5.4
growth hormone release inhibiting hormone (somatostatin) 17.1.5
growth hormone releasing hormone (GRH) 17.1.5
growth inhibitory factors 11.6.4
Grp94 14.1.3
GrpE 14.1.1
G subunits 17.4.1
GTP 8.1.2
GTP-binding proteins 14.5.2, 17.4
GTP-cap 18.1.3
GTPase activating proteins 14.2.2, 17.5.3
GTPase interacting factor 14.3
guanidino compounds 4.9.1
guanine 8.1
guanine nucleotide exchange factor 14.2.2
guanisino acetate 4.9.2
guanosine 8.1
guanosine nucleotide releasing factors 17.5.2
guanosine tetraphosphate 10.5.1
guanosine-5′-diphosphate (GDP) 8.1.2
guanosine-5′-monophosphate (GMP) 8.1.2
guanosine-5′-triphosphate (GTP) 8.1.2
guanylate cyclases 17.3, 17.4.6, 17.7
– membrane bound 17.7.1
– soluble 17.7.2
guanylate kinase 8.1.2
guanylin 17.7.1
L-gulonate 3.5.2
gulonolactone oxidase 9.10.1
L-guluronate 3.5.1
gustatory perception 17.4.7
gustducin 17.4.7
gyrase 10.2.2, 10.2.3

H

haemophilia 20.3.4
halobacteria 9.1.2, 17.4.6
– photosynthesis 16.2.1
Hanes plot 1.5.4
H chain genes 19.1.3
HDL receptors 18.2.4
heat-shock conditions 10.5.1
heat-shock promotors 10.4.2
heat-shock proteins 14.1.1
heavy chain gene 19.1.3
helicase 10.2.3
α-helix 2.3.1
helix-turn-helix 11.3.2, 11.4.2
helper cells 19.1.5, 19.1.7
hematopoietic cell phosphatase 17.5.4
hematopoietic cells 19.1.2
hematopoietic system 19.1
hematopoietin receptors 19.1.6
heme
– biosynthesis 5.2.1
– derivatives 5.2.4
– reduction 9.10.2
heme-controlled repressor 5.2.2, 11.5.4
heme oxygenase 5.3.1
heme synthase 5.2.1
hemerythrin 5.2.2
hemiacetal 1.2.1
hemicelluloses 3.6.3
hemiketal 1.2.1
hemocyanin 5.2.2
hemoglobin
– oxidation 5.3.1
– oxygen binding 5.2.3
– properties 5.2.2
– structure 5.2.2
– synthesis 11.5.4
hemolytic anemia 3.6.1, 4.5.7
hemosiderin 5.2.1
hemostasis 20.1
heparan sulfate 13.1.5, 18.2.2, 20.3.4
heparin 13.1.5, 19.1.8, 20.3.4

heparin cofactor II 20.3.4
hepatic triacylglycerol lipase 18.2.2
hepatitis B 12.1.1
hepsin 20.2.1
hereditary angioneurotic edema 19.2.5
hereditary fructose intolerance 3.1.1
hereditary galactosemia 3.4.2
heterodisulfide 15.5.2
heterodisulfide reductase 15.5.2
heterolactate fermentation 15.4
heterotrimeric G-proteins 17.3
– action 17.4.1
– structural features 17.4
heterotrophy 2.1
heterotropic regulation 2.5.2
HETEs 17.4.8
Hevea brasiliensis 7.3.3
hexameric helicase 10.2.2
hexitols 3.5.5
hexokinase 3.1.1, 3.1.2
hexoses 1.2.1
hexulose-6-P isomerase 15.7
hexulose-6-P synthase 15.7
Hh system 13.5.1
high density lipoproteins (HDL) 18.2
– metabolism 18.2.2
– receptors 18.2.4
high endothelial venules (HEV) 19.1.2, 19.3
high mannose structures 13.4.1
Hill coefficient 5.2.3
hinge region 19.1.3
histamine 4.8.2, 17.4, 19.1.8
histidine
– biosynthesis 4.8
– decarboxylation 4.8.2
– degradation 4.8.2
histidinemia 4.8.2
histones 2.6.4, 11.1.4, 11.3.3
histone mRNAs 11.5.5
HIV
– genome structure 12.4.1
– proviral state 12.4.1
– replication cycle 12.4.1
HIV protease 12.4.1
HLA molecules 19.1.5
HMG-CoA 18.2.5
HMG-CoA reductase 7.1.4, 18.2.4
HMG-CoA synthase 18.2.4
Holliday junction 10.3.5
holoenzyme 2.4.1
holoenzyme assembly model 11.3.2
holoenzyme synthetases 9.8.1
homoacetate fermentation 15.4, 15.5.3
homocysteine 4.5.4, 6.3.5
– high levels 20.3.4
homocysteine methyltransferases 9.5.2
homogentisate 4.7.3
homolactate fermentation 15.4
homologue recombination 11.2.5
homophilic adhesion 19.3
homoserine 4.5.4
homoserine dehydrogenase 4.5.1
homoserine kinase 4.5.1
homoserine succinyltransferase 4.5.1
homotropic regulation 2.5.2
Hoogsteen pairs 2.6.1
hopanoids 7.1.1, 7.2.1
hormone action 17.1.1
– deactivating mechanism 17.1.2
hormone binding domain 11.4.2
hormone binding kinetics 17.1.2
hormone cascade 17.1.1
hormone concentrations 17.1.1
hormone response elements 17.6
hormones
– anabolic 17.1.3
– carrier proteins 17.1.1
– catabolic 17.1.3
– gastric 17.1.9
– general characteristics 17.1.1
– hypothalamus 17.1.5
– inhibitory 17.1.5
– of the gastrointestinal tract 17.1.9
– pancreatic 17.1.9
– pituitary 17.1.5
– placental 17.1.6
– releasing 17.1.5
– trophic 17.1.5
housekeeping genes 11.3.2, 11.4.3
Hpr 15.3.1
Hsc70 14.5.2
Hsp56 17.6

Hsp70 14.4.2, 17.6
Hsp70 IAP 14.5.2
Hsp90 14.3.1, 17.6
H^+ transporting ATP synthase 15.5.1
HU protein 10.2.2
human immunodeficiency virus, see HIV
human leukocyte antigen 19.1.5
humoral immune response 19.1.7
Hunter's syndrome 13.1.5
Hurler's syndrome 13.1.5
hyaluronic acid 13.1.2, 13.1.5
hyaluronidase 13.1.5
hybrid mannose structures 13.4.1
hydrogen shuttles 16.1.2
hydrogen, as an electron donor 15.5
hydrogenase 10.6.5, 15.5.1, 15.5.2, 15.6.1
hydrogen peroxide 4.5.7
hydrolases 2.4.5
hydroperoxide radical 4.5.8
hydroperoxy-eicosatetraenoic acid 17.4.8
2-hydroxy acid dehydrogenase 15.7
3-hydroxyacyl-CoA dehydrogenase 6.1.5
2-hydroxybutyrate 6.1.6
hydroxybutyryl-CoA 15.4
hydroxybutyryl-CoA dehydrogenase 15.4
25-hydroxycholesterol 7.1.5
hydroxycobalamin 9.5.1
hydroxycytosine 10.3.1
20-hydroxyecdisone 7.2.3
hydroxyeicosatetraenoic acid (5-HETE) 17.4.8
2-hydroxyglutarate 3.8.1
hydroxyl radical 4.5.8
– activation of guanylate cyclase 17.7.2
hydroxylamine:Cyt c oxidoreductase 15.6.1
hydroxylysine 1.3.1, 4.3, 13.4.1
hydroxymethylbilane 5.1
hydroxymethylbilane synthase 5.1
hydroxymethylglutaryl-CoA 7.1.1
hydroxymethylglutaryl-CoA reductase, regulation 7.1.4, 7.9.3
hydroxyproline 1.3.1
– biosynthesis 4.3
– degradation 4.3
– racemization 4.3
2-hydroxypropionaldehyde 3.1.1
hydroxypyruvate 3.1.1, 15.7
hydroxypyruvate reductase 3.1.1
hyperammonemia 4.9.1
hyperpolarization 17.2.2, 17.2.3
hyperthyroidism 17.1.5
D-hypervitaminosis 9.11.2
hyperuricemia 8.1.7
hypoaldosteronism 7.8.4
hypothalamo-pituitary-adrenal axis 7.4.2, 17.1.5
hypothalamo-pituitary-liver/bone axis 17.1.5
hypothalamo-pituitary-ovary/uterus axis 17.1.5
hypothalamo-pituitary-testis axis 17.1.5
hypothalamo-pituitary-thyroid axis 17.1.5
hypothalamus 17.1.5
– feedback inhibition 7.8.3
hypothalamus-anterior pituitary hormone system 17.1.5
hypothyroidism 17.1.5
hypoxanthine 8.1
hypoxanthine P-ribosyl transferase 8.1.2, 8.1.7

I

ICE family 11.6.6
I-cell disease 13.1
iduronate 2-sulfatase 13.1.5
α-L-iduronidase 13.1.5
IFN-γ 19.1.7
Ig isotypes 19.1.2
Ig superfamily receptors 19.1.6
Igα 19.1.3
Igα-Igβ co-receptors 17.5.4
Igβ 19.1.3
IgA 19.1.3
IgD 19.1.2, 19.1.3
IgE 19.1.3, 19.1.8
IgE mediated hypersensitivity 19.1.8
IgE receptors 19.1.8
IgG 19.1.3, 19.2.1
IgM 19.1.2, 19.1.3, 19.2.1
IHF protein 10.2.2
IIA 15.3.1
I/i antigens 13.5.1
IIB 15.3.1
IIC 15.3.1
IL-1 19.1.7
IL-2 12.4.1, 17.5.4, 19.1.7

IL-4 19.1.7, 19.1.8
IL-5 19.1.7
IL-6 19.1.7
IL-10 19.1.7
IL-12 19.1.7
IL-13 19.1.8
immune complexes, solubilization 19.2.3
immune reaction 19.3
immune response, regulation 19.1.7
immune system, components 19.1
immunoglobulin domains 19.1.3
immunoglobulin genes 19.1.3
immunoglobulin receptors 19.1.7
immunoglobulin superfamily 19.3
immunoglobulin supergene family 19.1.3
immunoglobulins 11.3.3, 19.3
– classes of 19.1.3
– generation of diversity 19.1.3
– isotypes 19.1.2, 19.1.3
– membrane bound 19.1.3
– soluble 19.1.3
– structure 19.1.3
immunological defense 19.1
immunosuppression effect 7.8.3
immunosuppressors 17.5.4
IMP 8.1.1
importins 14.3.1
import mechanisms 18.1.2
IMPS gene 8.1.1
indole acetate 4.7.3
indomethacin 17.4.8
induced fit 2.5.2
infectious mononucleosis 19.2.5
inflammation 14.6, 19.2.3, 19.3
– induction 19.2
– suppression 7.8.3, 17.1.5
inflammatory cells 19.2, 19.2.3
inflammatory reactions 19.1.7
influenza A 12.1.1
inhibitory signals 17.2.3
initial reaction 1.5.4
initiation complex 10.4.2, 10.6.3, 11.3.2, 11.3.5, 11.3.7
80 S initiation complex 11.5.2
initiation factors 10.6.1, 11.5.1
– phosphorylation 11.5.4
initiation
– abortive 10.4.2
– translation 11.5.2
initiator DNA 11.1.4
initiator element 11.4.1
initiator tRNA 10.6.3
initiator tRNA$_{Met}$ 11.5.2
inner mitochondrial membrane 16.1.2
inorganic compounds, reduced 15.6
inosine 8.1
inosine-5′-phosphate, biosynthesis 4.2.4, 8.1.1
inositol (1,4,5)-P$_3$ 17.4.3
inositol (1,4,5)-P$_3$ receptor 17.4.4
inositol phosphatases, higher 17.4.4
inositol phosphates 3.5.4
– metabolism 17.4.4
inositol polyphosphate 1-phosphatase 17.4.4
inositolmonophosphate phosphatase 17.4.4
inositol-P$_3$-receptor 17.4.5
insect hormones 7.3.1
insulin 3.1.2, 3.1.5, 18.1.1
– metabolic effects 17.1.3
– mode of action 3.1.2
– regulation of fatty acid synthesis 6.1.2
– regulation of glycogen synthesis 3.2.4
– structure 17.1.3
insulin receptor 17.1.3, 17.1.5, 17.5.3
insulin receptor cascade 17.5.3
insulin resistant diabetes 17.5.3
insulin-like growth factors 17.1.5
integrase, viral 12.4.1
integration host factor 12.2.1
integrins 17.1.6
intercellular communication 19.1.6
intercellular matrix 19.3
interdigitating cells 19.1.3
interferon receptors 19.1.6
interferon-γ 6.3.4
interferons 17.5.1, 17.5.4
interleukin-1 19.3
interleukins 17.5.4
intermediary filaments 18.1.3
intermediate density lipoproteins 18.2
International Union of Biochemistry and Molecular Biology 2.4.5
interphase 11.6, 11.6.5, 18.1.3

interphase nucleus 2.6.4
intestinal movements 17.1.9
intestinal mucosa 18.2.2
intracellular communication 13.1.1, 17.1, 19.1.6, 19.3, 19.5
intracellular movement 18.1.3
intracellular transport 18.1.3
intrinsic factor 9.5.1
intrinsic pathway 20.2
introns 11.3.3
– definition 11.3.3
invertase 3.4.1
ion channels 18.1.1
– ligand gated 17.1.2
– transmitter gated 17.2.2, 17.3
– voltage gated 17.2.2
ion pumps 18.1.1
ion transport mechanism 15.4
iron metabolism 5.2.1
iron response element binding protein 11.5.4
iron sulfur proteins 16.1.2
iron-sensing protein 5.2.1
ischemia-reperfusion injury 4.5.8
isoacceptor-tRNA 10.6.2
isoalloxazine ring 9.3.2
isocitrate 3.8.1
isocitrate dehydrogenase 3.8.1, 3.9.1
isocitrate lyase 3.9.1
isoenzymes 2.4.2, 2.5.3
isoleucine
– biosynthesis 4.6
– degradation 4.6.2
isomerases 2.4.5
isopenicillin 15.8.1
isopenicillin N epimerase 15.8.1
isopentenyl pyrophosphate 7.1.1
isopentyl-adenosine 7.3
isopeptide 14.6.7
isoprenoid anchors 7.3.4
isoprenoid side chains 7.3.4
isotype switch 19.1.6, 19.1.8
itaconate 3.8.1

J

Jak families 17.5.4
jaundice 5.3.1
J chain 19.1.3
J segments 19.1.3
junctional diversity 19.1.3
juvenile hormones 7.3.1
juxtacrine 17.1.1

K

kainate receptor 17.2.5
kallikrein 19.1.8, 20.2.2, 20.5.1
katal 1.5.4
kauric acid 7.3.1
K$^+$ channels 17.2.2, 18.1.1
K$^+$, concentration 17.1.8
– equilibrium potential 17.2.1
keratan sulfate 13.1.5
keratin 2.3.1, 18.1.3
kestoses 3.2.1
keto… see also at oxo…
ketohexokinase 3.1.1
ketone bodies 4.2.2, 6.1.6
ketose 1.2.1, 6.1.7
ketosis 6.1.6
kexin 14.6.1
kinases
– CDK-activating 11.6.1
– cyclin-dependent 11.1.1, 11.6.1, 11.6.5
– DNA-dependent protein 11.2.5
kinesin 17.2.6, 18.1.3
kinesin-like proteins 11.6.5
kinetic proofreading 10.6.3
kinetochore microtubuli 18.1.3
kinetochores 11.6.5
kininogen 20.2.2
KKXX motif 13.3.2
Knallgas bacteria 15.6.1
Koshland-Nemethy-Filmer sequential model 2.5.2
Krebs-Henseleit cycle 4.9.1
kynurenate 4.7.3
kynureninase 4.7.3
kynurenine 4.7.3

L

lac operon 10.5.1

lac promotor 10.5.4
lac repressor 10.5.1
α-lactalbumin 3.4.2
lactase 3.4.2
lactate 3.1.1, 15.4
lactate dehydrogenase 3.1.1, 15.4, 15.5.1
– isoenzymes 2.4.2
lactate racemase 3.1.1, 15.4
lactose intolerance 3.4.2
lactose metabolism 3.4.2
lactose permease 15.3.1
lactose synthase 3.4.2
lactosyl ceramide 13.2.1
lagging strand (DNA) 10.2.3, 11.1.4, 11.1.5
laminin receptor 20.4
Lancefield groups 15.1
Langerhans cells 19.1.3
Langerhans islets 17.1.3
lanosterol 7.1.1
large multifunctional proteasome (LMP) 19.1.5
lariat 11.3.3
Lassa fever 12.1.1
lathosterol 7.1.1, 7.1.2
LDL, high plasma level 18.2.4
LDL receptor gene 7.1.5
LDL receptors 18.2.2
– mRNA 18.2.4
leader peptide 10.5.3
leading strand (DNA) 10.2.3, 11.1.4
lecithin 6.3.2
lecithin-cholesterol acyltransferase 18.2.2
lectin receptors 19.1.7
leghemoglobin 4.1
leptin 6.2.1
Lesch-Nyhan syndrome 8.1.7
leucine
– biosynthesis 4.6
– degradation 4.6.2
leucine zipper 11.4.2
leucocyte binding 19.3
leucocyte immigration, inhibition 7.8.3
leukocytes 19.1.2
– adhesion 19.3
– differentiation 13.2.1
leukotrienes 6.3.2, 17.4.8, 19.1.8
– biosynthesis 17.4.8
Lewis antigens 13.5.1
LexA repressor protein 10.3.6
Leydig cells 7.6.1, 17.1.5
LFA-1 19.1.4, 19.3
liberins 17.1.5, 7.4.2
licensing factors 11.1.1, 11.2.1, 11.6.3
ligases 2.4.5
light chain genes 19.1.3
light harvesting complexes 16.2.1
lignin 3.6.3, 4.7.6
lignin degrading enzyme 3.6.3
limonene 7.3.1
Lineweaver-Burk plot 1.5.4
linker DNA 2.6.4
linoleic acid 1.4.1, 6.1.3, 9.14.2
linolenic acid 1.4.1, 6.1.3, 9.14.2
lipase, hormone-sensitive 6.2.2
lipases, bacterial 6.2.2
lipid A 15.1
lipid aggregates 1.4.9
lipid anchor 13.4.3
lipid bilayers 1.4.9
lipid lowering therapy 18.2.5
lipid metabolism disorders 18.2.5
lipid transport proteins 18.2.3
lipoate, biosynthesis 9.14.1
lipocortin 7.8.3
lipogenesis 6.2.1
lipopolysaccharides 13.2.3, 15.1
lipoprotein lipase 6.1.4, 6.2.1, 18.2.2
lipoprotein receptors 18.2.4
lipoproteins 1.4.8, 6.1.4
– function 18.2
– metabolism 18.1.2, 18.2.2
– peroxidation 4.5.8
– scavenger pathway 18.2.2
– transport 18.1
lipoteichoic acids 13.2.2, 15.1
5-lipoxygenase 17.4.8
5-lipoxygenase-activating protein 17.4.8
lipoxygenase pathway 17.4.8
lithium ions 17.4.4
local mediators 17.4.8
loop structures 2.3.1
looping-out deletion 19.1.3
low-density lipoproteins 18.2

low-density lipoprotein receptor related protein 18.2.4
L-selectin 19.3
LTA$_4$ 17.4.8
LTA$_4$ hydroxylase 17.4.8
LTB$_4$-LTA$_4$ glutathione transferase 17.4.8
LTC$_4$ 17.4.8
LTC$_4$ synthase 17.4.8
LTD$_4$ 17.4.8
lumen targeting domain 14.5.2
lumisterol 9.11.1
luteinizing hormone (LH) 7.5.1, 7.6.1, 17.1.5, 17.4
luteinizing hormone releasing hormone (LHRH) 17.1.5
luteolysis 17.1.5
lyases 2.4.5
lymph node 19.1.1
lymphatic system 19.3
lymphocytes 19.1, 19.3
– adhesion 19.3
– diapedesis 19.3
– stimulation 19.1.2
lymphoid organs 19.2.3
lymphoid tissue 19.1, 19.1.2
Lynch syndrome II 11.2.6
lysine
– biological role 4.5.2
– biosynthesis 4.5.1
– hydroxylation 11.5.3
– metabolism 4.5.2
– methylation 11.5.3
lysogenic cycle 12.2.1
lysophosphatidate 6.2.1
lysophosphoglycerides 1.4.4
lysophospholipases 6.3.2
lysophospholipids 6.3.2
lysosomal storage diseases 13.1.4
lysosomes 2.2.2, 13.2.1, 18.2.4
lytic cycle 12.2.1

M

Mac-1 19.3
α_2-macroglobulin 20.5.2
α-macroglobulins 14.6.6
macrophages 18.2.2, 19.1.1, 19.1.5
– activation 19.1.7
Mad1 11.6.6
Mad2 11.6.6
MadCAM 19.3
magnesium chelatase 5.4
magnesium, biochemical functions 17.1.7
maize 16.2.2
major histocompatibility complex 11.3.3, 19.1.5
malate 3.3.5, 3.8.1, 15.4, 16.2.2
malate dehydrogenase 3.3.5, 15.4, 15.7, 16.2.2
malate shuttle 3.3.5
malate synthase 3.9.1
malate-CoA ligase 15.7
malate-glutamate carrier 16.1.2
malic enzyme 3.3.4
malonate 16.1.2
malonate pathway 4.7.6
α-maltose 3.2.3
maltotriose 3.2.3
malyl-CoA 15.7
malyl-CoA lyase 15.7
manic-depressive conditions 17.4.4
mannose 3.5.6, 13.4.1
mannose isomerase 3.1.1
mannose-6-P 3.5.6
mannose-6-P isomerase 3.1.1
mannuronates 3.5.1, 3.5.6
MAP kinase cascade 17.5.3
MAPs 18.1.3
Maroteaux-Lamy's syndrome 13.1.5
mast cells 19.1, 19.1.1, 19.2.3
matrix associated regions 2.6.4
maturation of affinity 19.1.3
maximum reaction rate 1.5.4
Mbf 11.6.2
Mcm proteins 11.2.1, 11.6.3
Mdj1 14.4.2
σ-mechanism 10.2
MEK-kinases 17.5.2
melanin synthesis 17.1.5
melanins 4.7.3
melanocyte stimulating hormones 17.1.5
melanotropins 4.7.3, 17.1.5
melatonin 4.7.3
membrane anchors 6.3.2, 13.2

membrane-attack complex 19.2.2
membrane fusion 6.3.2
membrane permeability 18.1.1
membrane potential 1.5.3, 17.2.1, 18.1
membrane proteins, integration 15.2
membrane signalling events 13.3.4
membrane translocation complex 13.3
membranes, characteristics 1.4.9, 18.1
memory cells 19.1.2
menaquinone 4.7.2, 7.3.4, 9.13, 16.1.3
menopause 17.1.5
menstrual cycle 17.1.5
2-mercaptoethanesulfonic acid 15.5.2
7-mercaptoheptanoylthreonine phosphate 15.5.2
mesaconate 15.4
mesophyll cells 16.2.2
metabolon 4.9.1
metachromatic leukodystrophia 13.2.2
metal chelators 14.6.6
metal ions, reduction 15.5.1
metal oxides, reduction 15.5.1
metallopeptidases, reaction mechanism 14.6.5
metaphase 11.6.5
metaphase chromosome 2.6.4
metaphase plate 11.6.5
metarhodopsin 17.4.6
Met-enkephalin 17.1.5
methane monooxygenase 15.7
methane oxidation 15.7
Methanobacterium thermoautotrophicum 15.5.2
methanofuran 15.5.2
methanogenesis 15.5.2, 9.5.3, 15.5.2
methanogenic habitats 15.5.2
methanogens 15.5.2
methanol 15.7
methanopterin 9.6.5
methemoglobin 5.2.3
methemoglobin reductase 5.2.3
methemoglobinemia 5.2.3
methenyl-THF 15.5.3
methionine 11.5.2
– biological roles 4.5.4
– biosynthesis 4.5.1
– formylation 10.6.3
– metabolism 4.5.4
– synthesis 6.3.5
methotrexate 8.2.5
methyl cobalamin 9.5.2
methyl transferases 9.5.2
3-methyl-2-oxobutanoate dehydrogenase 4.6.2
methyladenine 10.3.1
3-methyladenine DNA-glycosylase 11.2.3
methylaspartate 4.6.1, 15.4
methylcobalamin 9.5
methylcytosine 10.3.1
methyl-directed pathways 10.3.4
α-methyldopa 4.7.4
5-methyl-THF 9.6.2
methylene-tetrahydrofolate reductase 15.5.3
methylene-THF 9.6.2, 15.5.3
N-methyl-L-glucosamine 15.8.2
methylguanine 10.3.1
O^6-methylguanine 11.2.3
O^6-methylguanine-DNA-protein-cysteine S-methyl-transferase 11.2.2
7-methylguanosine 11.3.3
3-methylhistidine 4.8.2
methyllysine 1.3.1
methylmalonate 4.5.4, 4.6.2
methylmalonate semialdehyde 4.6.2
methylmalonyl-CoA 4.5.4, 6.1.1, 15.4
methylmalonyl-CoA carboxyltransferase 9.8.2, 15.4
methylmalonyl-CoA mutase 15.4
methylmalonyl-CoA-epimerase 15.4
methylotrophs 15.7
6-methylpretetramide 15.8.4
methyl-THF 15.5.3
5'-methylthioadenosine 4.9.3
α-methyltyrosine 4.7.4
mevalonate 7.1.1
mevinolin 7.1.4
MF1 11.1.4
Mgel 14.4.2
MHC class I 19.1.1, 19.1.2, 19.1.5
MHC class II 19.1.5
MHC genes 19.1.5
MHC molecules 19.1.4, 19.1.5
micelles 1.4.9
Michaelis constant K_M 1.5.4, 2.4.4
Michaelis-Menten equation 1.5.4, 2.4.4, 2.5.2
β_2-microglobulin 19.1.5

microtubules 11.6.5, 18.1.3
mineralocorticoids 7.4.1
– biological function 7.8.3
– biosynthesis 7.8.1
miniband 2.6.4
mismatch repair 10.3.4, 11.2.4
mitochondria 2.2.2
– components for import 14.4.2
– electron transport system 16.1.2
– genetic code 10.1.2
mitochondrial matrix 16.1.2
mitogenic stimulation 11.6.4
mitogens 11.6.4
mitosis 2.6.4, 11.6.5, 18.1.3
mitotic cell division 11.6.5
mitotic spindle 11.6.5
mixed acid fermentation 15.4
mixed function steroid monooxygenases 7.4.1
mixed inhibition 1.5.4, 2.5.2
MN antigens 13.5.1
mobilferrin 5.2.1
Moeller-Barlow disease 9.10.2
MoFe protein 4.1
molecular chaperones 14.5.2
molecular motors 11.6.5
molecular scavengers 4.5.8
molten globules 14.1
molybdenum cofactor 9.6.4
molybdoenzymes 9.6.4
molybdopterin 9.6.4
monensin 15.8
monoacylglycerols 6.2.1
2-monoacylglycerol lipase 6.2.2
monoamine oxidase 4.7.3, 4.7.4
monocytes 18.2.2, 19.1.1, 19.1.5, 19.2.3
Monod-Changeux-Wyman symmetry model 2.5.2
monodehydroascorbate 9.10.2
monohydroxy-eicosatetraenoic acids 17.4.8
monooxygenase reactions 9.3.2, 9.10.2
monoterpenes 7.3.1
Morquio syndrome 13.1.4
motor proteins 18.1.3
movement protein 12.3.1
M-protein 15.1, 17.4.5
Mps1 11.6.6
MPT synthase 9.6.4
mRNA 10.1.1, 10.4, 10.4.3, 10.6.1
– capping 11.3.3
– degradation 11.5.5
– half-life 11.5.5
– mature 11.3.3
– polyadenylation 11.3.3
– polycistronic 10.4.3
– processing 11.3.3
– splicing 11.3.3
– stability 11.3.3
– temporary masking 11.5.4
MSF 14.4.2
MSH 17.1.5
mtGrpE 14.4.2
mtHsp70 14.4.2
mucin 5.2.1, 13.1.1, 17.1.9
mucopolysaccharides 13.1, 13.1.2
mucopolysaccharidoses 13.1.4
mucosa cells 19.1
Muir-Torre syndrome 11.2.6
MukB 10.2.3
Müllerian inhibitory substance 17.1.5
multidrug resistance 18.1.1
multienzyme complexes 2.4.3
multiple enzyme control 4.5.1, 4.6.1, 4.7.1
multiple inhibition 2.5.3
mumps 12.1.1
murein 3.7.1, 4.2.3, 13.1.3, 13.2.3
murein synthesis 15.1
muscarinic receptor 17.2.4, 17.2.5
muscle contraction 17.4.5
muscle fibers 17.4.5
muscles
– cardiac 17.4.5
– smooth 17.4.5
– striated 17.4.5
muscular dystrophy 7.6.4
MutH 10.3.4
MutL 10.3.4
MutS 10.3.4
mycarose 15.8.3
mycarosylerythronolide 15.8.3
mycoplasma 2.2.1
mycosterols 7.1.1, 7.2.2
myelin sheath 13.2.2
myoadenylate deaminase insufficiency 8.1.2

myofibril bundles 17.4.5
myoglobin
– oxygen binding 5.2.3
– properties 5.2.2
– structure 5.2.2
myo-inositol 3.5.4, 15.8.2
myosin, structure 17.4.5
myosin light chain 17.4.5
myosin light chain kinase 17.4.5, 20.4
myosin light chain phosphatase 17.4.5
myristic acid 1.4.1
myristoylation 13.3.5
myxothiazol 16.1.2

N

Na^+
– equilibrium potentials 17.2.1
– concentration 17.1.8
– turnover 17.1.8
Na^+ channels 17.1.8, 17.2.2, 18.1.1
– increased expression 7.8.3
– rod cells 17.4.6
Na^+-glucose symporter 3.1.4
Na^+ gradient 15.3.1
Na^+/K^+ exchanging ATPase 17.1.8, 17.2.1, 18.1.1
Na^+/K^+ channel, increased expression 7.8.3
Na^+-rich environments 16.1.3
NAD(H) biosynthesis 4.7.5
NAD(H) metabolism 9.9.1
$NAD(P)^+$ reduction 16.2.1
$NAD(P)^+$ transhydrogenase 16.2.1
NADH dehydrogenase 6.1.5, 16.1.2, 16.1.3, 16.2.1
NADPH oxidase 4.5.8
NADPH, lack of 4.5.7
naphtoquinones 9.13
natural killer cells 19.1, 19.1.1, 19.1.7
natural rubber 7.3.3
nebulin 17.4.5
neoendorphins 17.1.5
neopterin 9.6.3
Nernst equation 1.5.3
nerve conduction 17.1.2
nerve growth factor 6.3.4
nerve-muscular synapse 17.2.3
neuraminidase 13.4.3
neurodegenerative defects 13.2.1
neurohypophysis 17.1.5
neuronal integration 17.2.3
neuronal intermediary filaments 18.1.3
neurons 18.1.1, 18.1.3
neuropeptide receptors 17.2.5
neuropeptide Y 6.2.1
neuropeptides 17.2.4
Neurospora crassa 4.7.1
neurotransmitters 17.2.3, 17.2.4
neutral cholesterol esterase 7.1.5
neutrophilic chemotactic factor 19.1.8
neutrophils 19.2.3
NF-κB 17.5.4
NF-ATp 17.5.4
niacin 9.9
nicotinamide 9.9
nicotinamide coenzymes, biochemical importance 9.9.3
nicotinamide-adenine dinucleotide phosphate, see NADP(H)
nicotinamide-adenine dinucleotide, see NAD(H)
nicotinate 9.9
nicotinic receptor 17.2.4
Niemann-Pick patients 6.3.4
nikkomycin 15.8
nitrate ammonification 4.1, 15.5.1
nitrate reductase 15.5.1
nitric oxide (NO) 17.7.2, 20.4
nitric oxide reductase 15.5.1
nitric oxide synthase 7.8.3, 9.6.3
nitride reductase 15.5.1
nitrification 4.1
nitrite, oxidation 15.6.1
nitrite: acceptor oxidoreductase 15.6.1
nitro drugs 17.7.2
Nitrobacter 15.6.1
nitrogen assimilation, regulation 10.5.4
nitrogen, circulation 4.1
nitrogen fixation 4.1
nitrogenase 4.1
nitrogenous compounds, fermentations 15.4
nitrogenous oxides, reduction 15.5.1
nitroglycerol 17.7.2
Nitrosomonas 15.6.1

nitrous oxide reductase 15.5.1
NO, see nitric oxide
nocturnal hemoglobinuria 19.2.5
noncompetitive inhibition 1.5.4, 2.5.2
nonpolyposis colorectal cancer 11.2.6
noradrenaline 4.7.3, 4.7.4
norepinephrine 4.7.3, 4.7.4, 17.1.4, 17.2.4
N protein 12.2.1
N region 19.1.3
NSF attachment protein 14.2.2
NTF-2 14.3.2
NtrB 10.5.4
NtrC 10.5.4
nuclear lamina 2.2.2, 2.6.4, 18.1.3
nuclear localization sequence 4.3.1
nuclear matrix 2.2.2, 2.6.4
nuclear pore complex 14.3, 14.3.2
nuclear pores 11.3.3
nuclear receptors 17.6
nuclear scaffold proteins 2.6.4
nucleases 10.7
– eukaryotic 11.2.3
– see also endo- or exonucleases
nucleated hematopoetic cells 19.1.4
nucleic acids
– components 2.6.1
– degradation 10.7
– see also DNA and RNA
nucleocapsid proteins 12.4.1
nucleolar organizer region 11.3.5
nucleolus 11.3.5
nucleoplasm 14.3.2
nucleoplasmin 2.6.4
nucleoporins 14.3, 14.3.1
nucleoprotein 2.6.4
nucleoside diphosphate glycosides 13.3.3
nucleosides 8
– phosphorylation 8.1.2
nucleosome 2.6.4
nucleotide excision repair 10.3.3, 11.2.3
nucleotide sugars 3.4.3
nucleotides 2.2.1, 2.6.4
– degradation 8.1.6
– methylation 10.3.2
– modified 11.3.8
– post-transcriptional modification 11.3.6
nucleus 2.2.2

O

O-antigen 15.1
obesity 6.2.1
ob gene product 6.2.1
octanoic acid 9.14.1
OEP-complex 14.5.2
oils 1.4.2
Okazaki fragments 10.2.3, 11.1.4
oleyl-CoA 6.1.3
oleyl-corticosterone 7.8.1
olfactory process 17.4.7
olfactory receptors 17.4.7
oligosaccharides, dolichol-bound 13.3.3
oligosaccharyltransferase complex 13.3.3
oncogenes 11.5.4, 12.4, 17.5.1
oncoproteins 11.6.4
oncorna viruses 12.4
one-electron reactions 9.3.2
one-substrate reactions 1.5.4
operator region 10.4.2, 10.5.1
operons 10.4.2
– CAP controlled 10.5.1
– monocistronic 8.1.1
– NtrBC controlled 10.5.4
– polycistronic 10.4.3
opioid receptors 17.1.5
opsin 17.4.6
opsonins 19.2.3
opsonization 19.1.1, 19.2.1, 19.2.3
Orc-proteins 11.6.3
ordered multistep model 11.3.2
ordered sequential mechanism 2.4.1
organelles, secretory 20.4
organisms, classification 2.1
organophosphates 6.3.5, 14.6.6
oriC 10.2.1, 10.2.2
origin of replication 10.2.2
origin recognition complex (ORC) 11.2.1
ornithine 1.3.1, 4.2.2, 4.9.1
ornithine carbamoyltransferase 14.4
orotate 8.2.1
orotate reductase 8.2.1
orsomucoid 7.4.3, 7.5.2

osmoprotection 6.3.5
osmoregulation 17.1.8
osmotic situation 1.4.2
O-specific side chains 15.1
osteoblasts 13.1.2
osteocalcin 9.11.2, 9.13
osteomalacia 9.11.2
osteoporosis 9.11.2
ouabain 7.2.2
ovicalcin 9.13
ovulation 17.1.5
oxaloacetate 3.3.5, 3.8.1, 15.4, 15.7, 16.2.2
oxalosuccinate 3.8.1
β-oxidation, fatty acids 6.1.5
oxidative damages 4.5.8
oxidative deaminations 9.3.2
oxidative phosphorylation
– energy balance 16.1.1
– free energy 16.1.1
– photosynthesis 18.1.1
– systems 16.1
oxidoreductases 2.4.5
3-oxo-6-P-hexulose 15.7
oxoacyl reductase 6.1.1
3-oxoacyl synthase 6.1.1
3-oxoacyl-CoA thiolase 6.1.5
2-oxobutyrate 4.5.4
2-oxoglutarate 3.8.1
2-oxoglutarate dehydrogenase 3.8.1
oxygen evolving complex 16.2.1
oxygen-free habitats 15.5
oxytetracycline 15.8.3, 15.8.4
oxytocin 17.1.5, 17.1.8

P

p11 12.4.1
p21 11.6.6
p53 11.6.4, 11.6.6
p107 11.6.4
p130 11.6.4
PAB II 11.3.3
palindromic sequences 10.7.2
palmitate 6.1.3
palmitoyl-CoA 6.1.2
palmitoyl thioesterase 6.1.1
palmitoyl transferase 6.1.1
palmitoylation 13.3.5
pancreatic enzyme secretion 17.1.9
pancreozymin 17.1.9
panthothenate 4.2.4
– biochemical function 9.7.2
– biosynthesis 9.7
PAP 4.5.6
papain, reaction mechanism 14.6.3
PAPS 4.5.6
PAPS reductase 4.5.6
Par proteins 10.2.3
paracortex 19.1.3
paracrine 17.1.1, 19.1.6
parathyroid hormone 17.1.7
Parkinson's syndrome 4.7.3
PARS 11.2.5
passive transport 18.1.1
Pasteur effect 3.1.2
pathogens, opsonization 19.2.3
pathological hypersensitivity 19.1.8
PCNA 11.1.4
PdsI 11.6.5
PECAM 19.3
PECAM-1 20.4
pectate 3.5.6
pellagra 9.9.1
penicillin 15.8.1
– semisynthetic 15.8.1
penicilloic acid 15.8.1
pentitols 3.5.5
pentose metabolism, in humans 3.6.4
pentose phosphate cycle 3.5.2, 3.6.1
pentoses 1.2.1, 3.6.1
pepsin 17.1.9
– reaction mechanism 14.6.4
pepsinogen 17.1.9
pepstatin 14.6.6
peptidases, functions 14.6
– inhibitors 14.6.6
peptide aldehydes 14.6.6
peptide bond, degradation 14.6
– formation 10.6.3
peptide bonds 1.3.2
peptide hormones 7.4.2, 17.1.1
peptide synthesis, non-ribosomal 15.8.1

peptides viral 19.1.7
peptidoglycans 13.1, 13.1.3, 15.1
peptidyl-(P-)site (ribosome) 10.6.1
peptidyl-prolyl *cis-trans* isomerase 4.3, 14.1.1
peptidyltransferase 11.5.2
periplasm 15.1, 15.3
pernicious anemia 9.5.1
peroxidases 4.5.8, 5.2.4
peroxide radicals 5.3.1
peroxyl radicals 9.10.2
peroxysomes 2.2.2
pertussis toxin 17.4.1
PEST sequences 11.6.6
pH gradient 15.3.1
phages, temperent 12.2.1
– proteins 12.2.1
phagocytes 19.1, 19.2.3
phagocytosis 4.5.8, 18.1.2, 19.1.5, 19.2
phagolysosomes 19.1.5
M phase 11.1.1, 11.6
phenol oxidases 4.7.3
phenylalanine
– biosynthesis 4.7.1
– degradation 4.7.3
phenylalanine-4-monooxygenase 4.7.1
phenylalanine-ammonia lyase 4.7.6
phenylbutazone 17.4.8
phenylethanolamine N-methyltransferase 4.7.4, 17.1.4
phenylketonuria 4.7.1, 4.7.3
phenylpropanoids 4.7.6
pheophytine 5.4, 16.2.1
phorbol 17.4.3
phosphagene 4.9.2
phosphate acetyltransferase 15.4
phosphate carrier 16.1.4
phosphate metabolism 9.11.2
phosphate transfer 15.3.1
phosphatidate 6.2.1, 6.3.2
phosphatidate cytidyltransferase 6.3.2
phosphatidate phosphatase 6.2.1
phosphatides 6.2.1
phosphatidic acid 1.4.4
phosphatidyl glycerol 6.3.1
phosphatidylcholine 6.3.2, 6.3.5
phosphatidylethanolamine 6.3.1, 6.3.2, 13.3.4
phosphatidylglycerol 6.3.2
phosphatidylinositol 6.3.2, 13.3.4
phosphatidylinositol-3-kinase 17.5.3, 17.5.4
phosphatidylinositol-3-kinase/protein kinase 17.4.4
phosphatidylinositol membrane anchors 13.3.2
phosphatidylinositol phosphates, reconstitution 17.4.4
phosphatidylserine 6.3.2, 20.2.3
phosphatidylserine synthase 6.3.2
3'-phosphoadenylylsulfate 4.5.6
phosphoarginine 4.9.2
phosphocreatine 4.9.2, 17.4.5
phosphoenolpyruvate 3.1.1, 15.4, 15.7
phosphoenolpyruvate carboxykinase 3.3.4f, 16.2.2
phosphoenolpyruvate carboxylase 3.3.4, 15.4
6-phosphofructo-1-kinase 3.1.1, 3.1.2
6-phosphofructo-2-kinase 3.1.1, 3.1.2
phosphofructokinase 3.1.2, 16.2.2
phosphoglucokinase 3.1.1
phosphoglucomutase 3.1.1, 15.8.2, 16.2.2
phosphogluconate dehydrogenase 15.4
3-phosphoglycerate 3.9.2, 16.2.2
phosphoglycerate kinase 3.1.1, 16.2.2
phosphoglycerate mutase 3.1.1, 15.7
phosphoglycerides see glycerophospholipids
2-phosphoglycolate 16.2.2
phosphoglycolate phosphatase 16.2.2
phosphoglycoproteins 18.1.1
3-phosphohydroxypyruvate 3.1.1
phosphoketolase 3.6.1, 15.4
phospholamban 17.4.4
phospholipase A_2 6.3.2, 17.4.1, 17.4.8
phospholipase C 6.3.2, 17.4.3, 17.4.7, 17.4.8
phospholipase Cβ 17.4.1
phospholipase Cγ1 17.5.3
phospholipase D 6.3.2, 17.4.8, 17.5.3
phospholipases 6.3.2
phospholipids 6.3, 20.2.3
– occurrence 6.3.1
phosphoprotein phosphatase-1 17.4.2
5'-phosphoribosyl pyrophosphate 8.1.1, 8.2.1
5'-phosphoribosylamine 8.1.1
5'-phospho-ribosyl-5-amino-4-imidazole carboxamide 4.8.1

phosphoribosyl-PP 4.8.1
phosphoribulokinase 16.2.2
phosphorylase a 3.2.4
phosphorylase b 3.2.4
phosphorylase kinase 3.2.4, 17.4.4
phosphorylation cascades 17.1.2
phosphorylation potential 16.1.1
phosphoserine phosphatase 4.4.1
phosphotransferase system 15.3.1
photoautotrophy 16.2.1
photon energy 16.2.1
photoreceptor nerve cells 17.4.6
photorespiration 4.2.1, 4.4.1, 16.2.2
photorespiration products, recycling 3.9.2
photosensitisation 4.5.8
photosynthesis
– bacteria 16.2.1
– dark reactions 16.2.2
– electron flow 16.2
– green plants 16.2.1
– halobacteria 16.2.1
– light reaction 16.2
– redox potentials 16.2.1
photosynthetic reaction centers 16.2.1
photosystems 16.2.1
phototaxis 17.4.6
phototrophy 2.1
phycocyanin, absorption spectrum 16.2.1
phycocyanin 5.3.1
phycocyanobilin 5.3.2, 16.2.1
phycoerythrin, absorption spectrum 16.2.1
phycoerythrin 5.3.1
phycoerythrobilin 5.3.2, 16.2.1
phylloquinone 4.2.2, 4.7.2, 7.3.4, 9.13
phylloquinone B 16.2.1
physostigmin 6.3.5
phytate 17.4.4
phytoalexins 4.7.6, 7.3.1–7.3.4
phytochelatins 4.5.7
phytochrome 5.3.1
phytochrome system 17.4.6
phytoene 7.3.2
phytol 7.3.1
phytosterol, conversion to cholesterol 7.2.3
phytosterols 7.1.1, 7.2.2
phytyl-PP 7.3.4
picolinate 4.7.5
piercidine 16.1.2
pili 2.2.1
pimeloyl-CoA 9.8.1
α-pinene 7.3.1
ping-pong Bi-Bi reaction 1.5.4
ping-pong-mechanism 2.4.1
pinocytosis 4.7.5, 18.1.2
placenta 17.1.6
Planck's constant 16.2.1
plant cell walls 2.2.3, 3.6.3, 13.1.1
plant hormones 7.2.2, 7.3.4
plant respiration 7.3.4
plant viruses 12.3.1
plants, viral infection 12.3.1
plasma apolipoproteins, properties and function 18.2.1
plasma cells 19.1.2, 19.1.7
plasma lipoproteins 18.2
– distribution of apolipoproteins 18.2.1
plasma membranes 2.2.1, 7.1.2
plasma protein carriers 7.4.3
plasmalogens 1.4.5, 6.3.3
plasmanylcholine 6.3.3
plasmanylethanolamine 6.3.3
plasmanylethanolamine desaturase 6.3.3
plasmids 2.2.1, 10.2
plasmin 20.2.2, 20.5
– substrate specificity 20.5.2
plasminogen 20.2.2, 20.5
plasminogen activation 20.5.1
plasminogen activators 20.2.2, 20.5.1
plasmodesmata 2.2.3, 3.6.3, 12.3.1, 18.1.1
Plasmodium falciparum 18.1.1
plastocyanine 16.2.1
plastoquinol-plastocyanine reductase 16.2.1
plastoquinone 4.7.3, 7.3.4
platelet activating factor (PAF) 6.3.3, 19.1.8, 20.4
platelet aggregation 20.4
platelet membrane, receptors 20.4
platelet-endothelial cell adhesion molecule 19.3
platelets 18.1.1, 20.1
cis-platin 10.3.1
β pleated sheet 2.3.1
pleotropsin 19.1.6
pluripotent stem cells 19.1.2

point desmosomes 2.2.4
poliomyelitis 12.1.1
poly (ADP-ribose) synthase 11.2.5
poly(A) binding protein-1 11.5.5
poly(A)osome 11.3.3
polyadenylation 11.3.3
polyadenylation signal 11.3.3
polyamine oxidase 4.9.3
polyamines 4.9.1, 4.9.3
polyA-polymerase 11.3.3
polycistronic 10.6.1
polygalacturonase 3.6.3
polyglutanylation 9.6.1
polyketide pathway 15.8.4
polypeptide hormones 17.5.1
polyprenyl-*cis*-transferase 7.3.3
polysaccharides, biosynthesis 3.2.2
polysome 10.6.1
POMC 17.1.5
P/Q quotient 16.1.1
porins 15.3
porphobilinogen 5.1
porphobilinogen deaminase 5.1
porphobilinogen synthase 5.1
porphyria 5.1
portal blood 7.9.2
postsynaptic membrane 17.2.2
postsynaptic reactions 17.2.3
potassium, see K^+
potato spindle tuber viroid 12.1
poxvirus 12.2
PP2B 17.5.4
precursor proteins 14.6
pregnancy 17.1.6
pregnenolone 7.5.1, 7.6.1
– precursor of glucocorticoids 7.8.1
pre-initiation complex 10.2.2, 11.2.1, 11.3.2, 11.3.5, 11.3.7
pre-mRNA 11.3.3
prephenate 4.7.1
prepriming complex 10.2.2
prepro-hormones 17.1.5
prepro-insulin 17.1.3
preproteins 14.6, 15.2
pre-replication complex 11.1.1, 11.6.3
– control of formation 11.6.3
presynaptic membrane 14.2.2, 17.2.3
pre-tRNA 10.4.3
previtamin D 9.11.1
Pri proteins 10.2.2
primary acceptor 16.2.1
primary response genes 17.6
primary structure 2.3.1
primary transcripts 10.4.3
primase 10.2.2, 11.1.4
primer t-RNA, viral 12.4.1
primosome 10.2.2
probucol 18.2.5
procollagen 4.5.2
procollagen-lysine-5-dioxygenase 4.3, 4.5.2
procollagen-proline-4-dioxygenase 4.3
profilin 17.4.3
progesterone 7.4.1, 7.5.2, 7.6.1, 17.1.6
– biosynthesis 7.5.1f
– degradation 7.8.2
– physiological effects 7.5.2
– precursor of mineralocorticoids 7.8.1
progesterone receptors 7.5.2
progression factors 17.5.1
prohormones 17.1.1, 17.1.5
proinsulin 17.1.3
prokaryotic cells 2.2.1
prolactin 17.1.5, 17.5.4
proliferating cell nuclear antigen (PCNA) 11.1.4
proline 4.2.2
– biosynthesis 4.3
– degradation 4.3
– hydroxylation 11.5.3
– isomerization 14.1.1
– racemization 4.3
proline-4-hydroxylase 4.3
prometaphase 11.6.5
promoter binding 10.5.1
promoter recognition 10.5.1
promoter region 10.4.2
promoters 11.4
– bacterial 10.4.2
promyelocytic leukemia 7.3.2
proofreading 10.6.2
proopiomelanocortin 17.1.5
propanol 15.4
propanol dehydrogenase 15.4

prophase 11.6.5
propionate 4.5.4, 15.4
propionate fermentation 4.5.4, 15.4
propionyl-CoA 4.5.4, 6.1.1, 15.4, 15.8.3
propionyl-CoA carboxylase 4.5.4
proproteins 14.6
prostacyclin I$_2$ 17.4.8, 20.4
prostacyclin synthase 17.4.8
prostacyclins 17.4.8
prostaglandin H$_2$ synthase 17.4.8
prostaglandins 6.3.2, 17.4.8, 19.1.8
prostanoids, biosynthesis 17.4.8
prosthetic groups 2.4.1
protease inhibitors 20.3
proteases 14.6, 20.3
– viral 12.4.1
proteasome 2.4.3, 11.6.6, 14.6.7
protein antigens, dual recognition 19.1.7
protein C 20.3.4
protein channels 15.3
protein degradation 11.6.6, 14.6
– by the ubiquitin system 14.6.7
protein disulfide isomerase 13.3.2, 14.1.1, 14.1.3
protein factors, trans-acting 11.4
protein folding
– in bacteria 14.1.1
– in the cytoplasm 14.1.2
– in the ER 14.1.3
protein glycosylation 13.3
– pathways 13.4.1
protein kinase A 3.2.4, 6.1.2, 6.2.2, 17.4.5
– substrates 17.4.2
protein kinase C 6.1.3, 17.4.1, 17.4.3, 17.4.5, 17.5.3
protein kinase G 17.7, 17.7.2, 17.7.3
protein kinases
– AMP-dependent 6.1.2
– cyclin-dependent 11.1.1
– substrates 17.4.3
protein phosphatase-2A 6.1.2, 11.6.6
protein phosphatases 11.6.6, 17.5.2, 17.5.3
protein S 20.3.4
protein secretion
– bacterial 15.2
– cotranslational 15.2
protein serine/threonine kinases 17.5.2
protein serine/threonine phosphatases 17.5.3
protein synthesis
– bacterial 10.6
– eukaryotic 11.5
protein targeting 14.3.1, 14.4.1, 14.5.1
protein transport 18.1
– into chloroplasts 14.5
– into mitochondria 14.4
– into the nucleus 14.3
protein turnover 14.6
protein tyrosine kinase receptors 17.4.4
protein tyrosine kinases 17.5.2, 17.5.3, 17.5.4
protein tyrosine phosphatases 17.5.3
protein-disulfide reductase 4.5.7
proteins
– acylation 13.3.5
– biological function 2.3.2
– contractile 11.6.5
– dietary 14.6
– exported 13.1
– farnesylated 7.3
– fibrous 2.3.1
  folding 14.1
– geranylgeranylated 7.3
– globular 2.3.1
– glycosylation 11.5.3
– kinesin-like 18.1.3
– lipid-anchored 13.3.4
– maturation 14.6
– membrane attachment 11.5.3
– membrane-associated 1.4.9, 13.3.2
– microtubule-associated 18.1.3
– myristoylated 13.3.5
– N-glycosylated 13.4.1
– nuclear 13.3
– nuclear scaffold 2.6.4
– O-glycosylated 13.4.1
– phosphorylation 11.5.3
– post-ribosomal modifications 2.3
– post-translational processing 2.3, 11.5.3, 13.3
– post-translational translocation 15.2
– primary structure 2.3.1
– proteolytic cleavage 13.4
– quaternary structure 2.3.1
– recycling 13.4
– regulatory 10.5.1

– ribosomal 11.3.5
– secondary structure 2.3.1
– secretion 13.5, 14.2
– secretory 13.3, 13.3.2
– soluble ER 13.3.2
– spliceosome-associated 11.3.3
– telomere binding 11.1.5
– tertiary structure 2.3.1, 14.1
– transmembrane 1.4.9, 13.3.2
– type I membrane 13.3.2
– type II membrane 13.3.2
– viral 12.4.1
protein-tyrosine phosphatases 17.5.2
proteoglycans 3.5.1, 13.1, 13.1.2, 13.4
– components 13.1.5
– synthesis 13.4.2
proteolytic activation steps 20.1
prothrombin 20.3.3
protoheme 5.2.1
protoheme ferro-lyase 5.2.1
proton gradient 15.3.1
proton motive force 15.3.1
proton translocation 15.5.1
proto-oncogenes 17.5.1
protopectin 3.6.3
protoporphyrin IX 5.1
protoporphyrinogen IX 5.1
3′ protruding tails 10.7.2
5′ protruding termini 10.7.2
proviruses 12.4.1
proximal sequence element 11.4.1
P-selectin 19.3, 20.4
pseudo-uridine 10.4.3, 11.3.6, 11.3.8
P-site (ribosomes) 11.5.2
psoralene 4.7.6
psoriasis 7.6.4
P system (blood groups) 13.5.1
pteridine ring 9.6
pterol-poly-γ-glutamate hydrolase 9.6.1
puberty 17.1.5
purine nucleotides
– biosynthesis 8.1.2
– interconversions 8.1.2
purine repressor 8.1.2
purine ring atoms 8.1.1
puromycin 11.5.2, 15.8
purple bacteria 16.2.1
– electron transfer 16.2.1
putrescine 4.9.3
pyranoses 1.2.1
pyridine nucleotide cycles 9.9.1
pyridoxal 9.4.1
pyridoxal phosphate, biochemical function 9.4.2
pyridoxamine 9.4.1
pyridoxine, biosynthesis 9.4.1
pyrimidine deoxyribonucleotides, interconversions 8.2.3
pyrimidine ribonucleotides, interconversions 8.2.2
pyrimidines
– dimerization 10.3.1
– synthesis 8.2.1
pyrrolo-quinoline quinone 15.7
pyruvate 3.1.1, 15.4, 16.2.2
– carboxylation 3.3.4f
– oxidation 3.3.1
– reactions 3.3.1
– transamination 4.2.3
pyruvate carboxylase 3.3.4f
pyruvate decarboxylase 3.3.6, 15.4
pyruvate dehydrogenase 3.3.1, 3.3.5, 15.4
– regulation 3.3.2
pyruvate formate lyase 15.4
pyruvate kinase 3.1.2
pyruvate: ferredoxin oxidoreductase 15.4
pyruvate-P dikinase 16.2.2

Q

Q cycle 16.2.1
quinoenzymes 15.7
quinolinate 4.7.5, 9.9.1
quinone cofactors 4.7.2
quinone pool 4.7.2, 16.1.2, 16.1.3, 16.2.1
quisqualate receptor 17.2.5

R

Rab proteins 7.3.4
rabies 12.1.1
RAD51 11.2.5
RAD52 11.2.5

Raf 17.4.1
raffinose 3.4.1
Ramachandran diagram 1.3.2
Ran 14.3, 14.3.2
RanBPI 14.3.2
random sequential mechanism 2.4.1
RanGAP1 14.3.2
RAR receptor 9.1.2
Ras 17.5.3
rate constant 1.5.4
Rb protein 11.6.4
reaction rate 1.5.1, 1.5.4, 2.4.4
reaction rate constants 2.4.1
reading frames 10.1.2
RecA 10.3.5, 10.3.6, 12.2.1
RecBCD 10.3.5
receptor antagonists 17.3
receptor families 17.5.1
receptor tyrosine kinases 17.4.3, 17.5.3
receptor-mediated endocytosis 18.1.2
receptors
– general characteristics 17.1.2, 17.2.3
– G-protein coupled 17.4
– ionotropic 17.2.3
– lectin-type 19.1.1
– metabotropic 17.2.3
– postsynaptic 17.2.3, 17.2.5
– transmitter-gated 18.1.1
receptor-tyrosine kinases, structure 17.5.3
RecJ 10.3.4
recombinases 19.1.3
– recombination, somatic 19.1.3
recoverin 17.7.1
redox potentials 1.5.2, 15.5.1, 16.1.5, 16.2.1
– respiratory chain 16.1.5
redox reactions 1.5.2, 15.5.1, 15.6.1
reducing equivalents 3.1.1, 3.6.1
reductive citrate cycle 3.8
regulated pathway 14.2.1
regulon 4.9.1
release factors 10.6.1, 10.6.3
releasing hormones 7.4.2
remnant receptors 18.2.2
renin-angiotensin system 17.1.8
replication enhancers 11.1.2
replication forks 10.2, 11.1.4
replication origins 11.1.2
replication, see DNA replication
replicon 10.2
replisome 10.2.2
λ-repressor 12.2.1
repressor establishment 12.2.1
repressor proteins 10.5.1, 11.5.4
repressors 11.4.3
resilience 18.1.3
resins 7.3.1
resonance interaction 16.2.1
respiratory chain
– ATP yield 16.1.1
– components 16.1.2
– inhibitors 16.1.2
– redox potentials 16.1.5
– supply of hydrogen 16.1.2
resting phase 11.6.4
restriction endonucleases 10.7.2
restriction endoproteinases 16.6.1
restriction point 11.1.1, 11.6.2, 11.6.4
restriction-modification system 10.7.2
retina 17.4.6
retinal 9.1, 17.4.6
– isomerization 16.2.1
– light-induced interconversion 9.1.2
all-trans retinal 7.3.2
retinoblastoma tumor suppressor protein (Rb) 11.6.4
retinoic acid 9.1
– biochemical function 9.1.2
– hormone-like activity 7.3.2
retinoic acid binding protein 7.3.2
retinoids
– metabolism 7.3.1–7.3.4
– receptors for 17.6
all-trans retinol 7.3.2
retinol 7.3.2
– daily requirement 9.1.1
retinol binding proteins 9.1.1
retinol-protein complexes 7.3.2
retinyl palmitate 9.1.1
retinyl phosphate 9.1.2
retroviruses 12.1, 12.3, 12.4
– mutation rate 12.4.1
– regulatory genes 12.4.1

reverse transcriptases, retroviral 12.4.1
reverse transcription, mechanism 12.4.1
reversed electron pumping 15.6.2
RFA 11.1.4
RFC 11.1.4
rhamnogalacturonan 3.5.6, 3.6.3
rhamnolipids 15.7
L-rhamnose 3.5.6
rhamnosidases 3.6.3
rhesus antigens 13.5.1
Rhodobacter 16.2.1
Rhodopseudomonas sphaeroides 4.5.1, 16.2.1
Rhodopseudomonas viridis 16.2.1
rhodopsin, absorption spectra 17.4.6
rhodopsin kinase 17.4.6
α-ribazole 9.5.1
ribitol 3.6.1
riboflavin
– biochemical function 9.3.2
– biosynthesis 9.3.1
ribonuclease H 10.2.3, 11.1.4, 12.4.1
ribonuclease P 10.4.3
ribonucleases 10.4.3, 10.7, 10.7.3
ribonucleoparticles, small nuclear 11.3.4
ribonucleoproteins, small nuclear 11.3.3, 11.3.6
ribonucleoside-diphosphate reductase 8.1.4, 8.1.7
ribonucleotides, reduction 4.5.7, 8.1.4, 9.5.2
D-ribose 3.6.1
ribose-5-P 3.6.1, 16.2.2
ribose-5-P isomerase 16.2.2
ribose-P pyrophosphokinase 8.1.1
ribosomal recycling factor 10.6.1
ribosome receptor 13.3.1
ribosomes
– bacterial 10.6.1, 11.5.1
– biogenesis 11.3.5
– composition 11.5.1
– eukaryotic 11.5.1
– subunits 10.4.3, 11.3.5
ribothymidine 10.4.3
ribozymes 2.6.1, 10.4.3, 10.7.3
ribulose 16.2.2
ribulose bisphosphate carboxylase/oxygenase 3.9.2, 16.2.2
ribulose bisphosphate pathway 15.6.1
ribulose monophosphate pathway 15.7
ribulose-5-P 3.6.1, 15.4, 16.2.2
ribulose-P-3 epimerase 16.2.2
ribulose-P-4 epimerase 15.4
rickets 9.11.2
Rieske protein 16.1.2, 16.2.1
rifamycin B 10.4.5
RNA
– export 14.3
– functions 2.6.2
– nuclear 2.6.4
– polyadenylation 11.3.3
– small nuclear 11.3.4, 11.4.1
– structure 2.6.2
– synthesis 10.4, 11.3
– viral 12.3
RNA polymerase I
– core promotors 11.4.1
– transcription factors 11.3.5
RNA polymerase II 11.2.5
– core promotors 11.3.7, 11.4.1
RNA polymerase III
– core promotors 11.4.1
– transcription factors 11.3.7
RNA polymerases
– distinguish 11.3.10
– DNA-directed 10.4.1, 11.3.1, 11.6.4
– RNA-directed 12.3
– types of 11.3.1
RNA primers 10.2.3, 11.1.4
RNA replication, error rate (virus) 12.3
RNA viruses 12.3
RNase, see ribonuclease
rolling circle mechanism 10.2, 12.2
root structure 13.2.1, 13.4.3
rotenone 16.1.2
Rous sarcoma virus 12.4
RPA 11.1.4
rRNA 10.4
– genes 11.3.5
– processing 11.3.6
5S 11.3.9
– transcription 11.3.5
R state 2.5.2
RU-486 7.5.2
rubisco 3.9.2, 16.2.2

rubisco binding protein 14.5.2
rubredoxin 15.7
rubredoxin-NAD reductase 15.7
ruminants 6.1.3
rustiycyanin 15.6.1
RuvC 10.3.5
RXR receptor 9.1.2, 17.6
ryanodine 17.4.4
ryanodine receptor 17.4.4, 17.4.5

S

saccharides 3.4.3
saccharopine 4.5.2
S-adenosyl-L-methionine 4.5.4
S-adenosylhomocysteine 4.5.4
S-adenosylmethioninamide 4.5.4
sal-box 11.3.5
salicylic acid 4.7.6
salty tastes 17.4.7
salvage compartment 13.3.2
salvage pathway 8.1.2
Sandhoff disease 13.2.1
Sanfilippo's syndrome 13.1.5
sapogenins 7.2.2
sarcoplasm 17.4.5
sarcoplasmic reticulum 2.2.2, 17.4.4
sarcosine dehydrogenase 6.3.5
sarcosine metabolism 6.3.5
sarcosine oxidase 6.3.5
Sbf 11.6.2
scaffold-attached regions (SAR) 2.6.4
scanning model 11.5.2
Scatchard plot 17.1.2
scavenger receptor 18.2.4
SCID cells (severe combined immunodeficiency) 8.1.7, 11.2.5
scurvy 9.10.2
Sec61 complex 13.3.1
SecA 14.5.2
SecA dependent pathway 15.2
SecB 15.2
second messenger 17.1.2, 17.7
secondary aldosteronism 7.8.4
secondary response genes 17.6
secondary structure 2.3.1, 14.1
secretion 14.2.1, 18.1.3
secretory ATPase 15.2
secretory channel 18.1.1
secretory chaperone 15.2
secretory granula 17.1.1
secretory vesicles 14.2.1
Sec-type kinase 17.5.4
SecYEG 15.2
sedoheptulose bisphosphatase 16.2.2
sedoheptulose-bisphosphatase-P 16.2.2
sedoheptulose-1,7-P_2 16.2.2
sedoheptulose-7-P 16.2.2
SelB 10.6.5
selectins 19.3
selenocysteine 1.3.1, 6.3.5, 10.6.5
selenocysteine codon 10.1.2
selenophosphate 10.6.5
selenophosphate synthetase 10.6.5
self-splicing 10.7.3, 11.3.3
semiquinones 4.5.8
sense strand 10.1.1
sensor protein 10.5.4
sequential control 4.7.1
sequential inhibition 2.5.3
sequential model (allostery) 2.5.2
serine 3.1.1, 15.7, 16.2.2
– glycosylation 13.3
– metabolism 4.4.1
serine dehydratase 4.4.1
serine hydroxymethyltransferase 3.9.2, 4.4.2
serine-isocitrate lyase pathway 15.7
serine peptidases, reaction mechanism 14.6.2
serine/threonine kinases 17.4.3, 17.4.4, 17.5.3
serotonin 4.7.3, 17.2.4, 20.1
serotonin receptors 17.2.5
serotyping 15.1
serpentine receptors 17.4.6
serpins 14.6.6
Sertoli cells 17.1.5
serumalbumin 6.1.4
sesquiterpenes 7.3.1
severe combined immunodeficiency disease 8.1.7, 11.2.5
sex-hormone binding globulin 7.4.3, 7.7.2
sexual differentiation 17.1.5
sexual dimorphism 11.3.3

Shc 17.5.3
shikimate pathway 4.7.1
Shine-Dalgarno sequence 10.6.3, 11.5.2
sialate 3.7.1
sialic acid 13.1.1, 13.2.1
sialidase 13.2.1, 13.4.3
sialyation, terminal 13.5.1
sickle cell anemia 5.2.3
Sic1 11.6.2
Sid system 13.5.1
signal amplification 17.4.1
signal anchor sequence 13.3.2
signal cascades, components 17.5.2
signal peptidase 13.3.2, 15.2
signal peptidase complex 13.3
signal peptide 15.2
– for GPI addition 13.3.4
signal recognition particle 11.5.2, 13.3.1, 15.2
signal sequence receptor complex 13.3
signal sequences 11.5.2, 13.3.1, 19.1.3
signal transduction 6.3.2, 17.1.2, 18.1.2, 19.1.4
signalling proteins 17.5.3
signals, co-stimulatory 17.5.4
silencers 11.4, 11.4.3
sinapol 4.7.6
single-stranded binding protein 10.2.2–10.2.4
single-substrate reaction 2.4.1
singlet oxygen 4.5.8, 9.10.2
siroheme 4.5.6, 9.5.3, 15.5.1, 15.6.1
sister chromatids 11.6.5
sitosterol 7.2.2
skin 13.1.5
smooth muscle 18.2.2
snake venoms 6.3.2
SNAP receptor 14.2.2
snRNA 11.3.4, 11.4.1
snRNA transcription 11.3.4, 11.4.1
snRNP 11.3.4
soaps 1.4.9
sodium, see Na+
somatic hypermutation 19.1.3
somatic recombination 19.1.3, 19.1.4
somatomedins 17.1.5
somatostatin 17.1.3, 17.1.5, 17.1.6, 17.1.9
somatotropin 17.1.5
sorbitol 3.1.1
SOS boxes 10.3.6
SOS response 10.3.1, 10.3.6
Sos1 protein 17.5.3
sour tastes 17.4.7
soybean trypsin inhibitor 14.6.6
12/23 spacer rule 19.1.3
special pair 16.2.1
specific activity 1.5.4
sperm cells 18.1.3
sperm nucleus 2.6.4
spermatogenesis 17.1.5
spermatozoa 13.2.2, 18.1.3
spermidine 4.9.3
spermine 4.9.3
S-phase 11.1.1, 11.6
sphinganine 6.3.4
sphingolipids 1.4.6
sphingomyelin 1.4.6, 6.3.4
sphingophospholipids 1.4.6, 6.3.4
sphingosine 1.4.6, 13.2.1
spindle assembly checkpoint 11.6.5, 11.6.6
spironolactone 7.8.4
spliceosome 11.3.3
splicing 11.3.3
– alternative 11.3.3
sporulation 14.6
squalene, cyclization 7.1.1
(S-)squalene-2,3 epoxide 7.1.1
squalene synthase 18.2.4
Src familiy 17.5.4
Src homology domains-2 17.5.3
SRE-1 (sterol regulatory element-1) 7.1.5
SREBP-1 (sterol regulatory element binding protein) 7.1.5
S-Region 19.1.3
SRF 11.2.5, 14.5.2
SRP receptor 13.3.1, 15.2
SRP 14.5.2
SRY gene 17.1.5
SSB 10.2.2, 10.2.3, 10.3.4
starch 3.1.1, 3.2.1
– degradation 3.2.3
starch phosphorylase 3.2.3
starch synthase 3.2.2, 3.2.3
starch synthesis 3.2.2
START 11.1.1

START signal 11.6.2
starvation 3.1.5, 6.1.6
STAT proteins 17.5.4
statins 7.4.2, 17.1.5
stearoyl-ACP desaturase 6.1.3
stearoyl-CoA desaturase 6.1.3
stem loops 2.6.2
stercobilin 5.3.1
stercobilinogen 5.3.1
steroid alcaloids 7.2.2
steroid hormones 17.1.2
– activation 7.4.2
– biosynthesis 7.4.1
– secretation 7.4.2
– transport 7.4.3
steroid receptors 11.4.3, 17.6
steroids 1.4.7
– degradation 7.4.4
sterol carrier protein 7.1.1, 7.5.1
sterols 1.4.3
sterol regulatory element 7.1.5
sticky end 10.7.2
stigmasterol 7.2.2
stilbene derivatives 7.7.4
stilbenes 4.7.6
stimulons 10.5.4
stop codons 11.5.2
stop transfer effector sequences 13.3.2
streptidine-6-P 15.8.2
streptokinase 20.5.1
streptolydigin 10.4.5
streptomycin 15.8.2
stress situations 4.7.4, 17.1.4, 17.1.5
stringent control 10.5.1
stringent factor 10.5.1
stroma 2.2.3
stromal peptidase 14.5.2
structural motifs 2.3.1
structure, fibrinogen 20.3.5
suberine 4.7.6
substrate channeling 3.1.1, 4.9.1
substrate level phosphorylation 8.1.3, 15.4, 15.6.1
succinate 3.8.1, 15.4,
succinate dehydrogenase 3.8.1, 15.5.1, 16.1.2, 16.1.3, 16.2.1
succinate pathway 15.4
succinate-propionate CoA transferase 15.4
succinyl-CoA 3.8.1, 4.5.4, 15.4
succinyl-CoA ligase 3.8.1
succinyl-CoA pathway 5.1
sucrose 3.2.2, 16.2.2
– role in metabolism 3.4.1
sucrose synthase 3.2.2
sucrose-6-P 3.4.1
sucrose-P phosphatase 3.4.1
sucrose-P synthase 3.4.1
sudden energy demand 3.1.5, 6.1.2
sugar cane 16.2.2
sugars
– biosynthesis 13.2.2
– transfer 13.2.2
sulfate, transfer 13.2.2
sulfatases 13.2.2
sulfate adenylyl transferase (ATP sulfurylase) 15.5.1, 15.6.1
sulfate respiration 15.5.1
sulfide dehydrogenase 15.6.1
sulfide oxidase 15.6.1
sulfides, metabolism 4.5.6
sulfidogenesis 15.5.1
sulfite dehydrogenase 15.6.1
sulfite oxidase family 9.6.4
sulfite oxidation 15.6.1
sulfite reductase 4.5.6, 15.5.1, 15.6.1
sulfite:cytochrome C oxidoreductase 15.6.1
Sulfolobus 15.6.1
sulfonamides 9.6.1
sulfur dioxygenase 15.6.1
sulfur metabolism 4.5.6
sulfur respiration 15.5.1
superoxide dismutases 4.5.8
superoxide radical 4.5.8, 9.10.2
supersecondary structure 2.3.1
suprasterol 9.11.1
surface receptors 19.1.1
sweet tastes 17.4.7
switch recombinations 19.1.3
Syk 17.5.4
symmetry model 2.5.2
symporters 15.3.1, 18.1.1
synapsin 17.2.3

synapsin I 14.2.2
synaptic cleft 17.2.3
synaptic transmission, types 17.2.3
synaptic vesicles 14.2.2
synaptobrevin 14.2.2
synaptotagmin 14.2.2
synovial fluid 13.1.5
synpatic transmission 17.2.3
syntaxin 14.2.2

T

tachysterol 9.11.1
tagetitoxin 11.3.10
tamoxiphen 7.7.4
tannin 4.7.6
TAP1 19.1.5
TAP2 19.1.5
targeting presequence 14.4.1, 14.5.1
targeting sequences 11.5.3
TATA box 11.4.1
taurine 4.5.5
taurine conjugates 7.9.1
taurocholic acid 7.9
taxol 18.1.3
Tay-Sachs disease 13.2.1
TBP 11.3.2
T-cell activation 13.3.4
T-cell mitogens 12.4.1
T cell receptors 17.5.1, 17.5.4, 19.1.2
– interactions 19.1.4f
T cells
– activation 19.1.4
– cytotoxic 19.1.2, 19.1.5
teichoic acids 3.6.1, 13.2.3, 15.1
teichuronic acids 15.1
telangiectasia 11.2.6
telomerase 11.1.4, 11.1.5
telomeres 2.6.4, 11.1.5
telophase 11.6.5
teresantol 7.3.1
terminal deoxynucleotidyl transferase 19.1.3
terminal oxidase 15.6.1
terminal reductase 15.5.1
termination hairpin 10.5.3
termination sequence 10.4.2
termination sites 10.2.3, 10.4.2
terpenes
– biosynthesis 7.3.1
– metabolism 7.3.1–7.3.4
tertiary structure 2.3.1, 14.1
testis determining factor 17.1.5
testosterone 7.6.1, 17.1.5
tetracycline 15.8.4
tetrahydrobiopterin 4.7.1, 9.6.3
tetrahydrofolate (THF) 15.5.3
– biosynthesis 9.6.1
tetrahydrofolylpolyglutamate synthase 9.6.1
tetrahydromethanopterin 9.6.5, 15.5.2
tetrahymenol 7.2.1
tetroses 3.6.1
TFIID 11.2.5
TFIIH 11.2.3
TFIIIB 11.6.4
TFIIIC 11.3.7
TFIIJ 11.2.3
TFPI 20.3.3
TF-VIIa complex 20.2.1
TGF factors 11.6.4
TGF family receptor 19.1.6
TH1 effector cells 19.1.7
TH2 effector cells 19.1.7
thalassemia 5.2.3, 11.3.3
theca interna cells 17.1.5
T helper cells 19.1.2
thermogenesis 6.2.2
thiamin
– biochemical function 9.2.2
– biosynthesis 9.2.1
thiamine pyrophosphate 3.6.1
thick filaments 17.4.5
thin filaments 17.4.5
Thiobacillus 15.6.1
Thiobacillus ferroxidans 15.5.1
thioesterase 6.1.1
thioredoxin 4.5.7, 14.1.1, 16.2.2
threonine
– biosynthesis 4.5.1
– cleavage 4.5.3
– metabolism 4.5.3
threonine aldolase 4.4.2
threonine dehydratase 4.5.1, 4.5.3, 4.6.1

threonine peptidase 14.6.1, 14.6.7
thrombin 14.6.1, 20.2.1
– activation 20.3.2
– inactivation 20.3.4
thrombin receptor 20.4
thrombocytes, see platelets
thrombomodulin 20.3.4
thromboplastin (tissue factor, TF) 20.2.1
thromboxane A_2 17.4.8, 20.4
thromboxane receptor 20.4
thromboxane synthase 17.4.8
thromboxanes 17.4.8, 19.1.8
thylakoid, import of proteins 14.5.2
thylakoid membrane 13.2.2
thylakoid space 2.2.3
thymic cortex 19.1.2
thymidylate synthase 8.2.3, 8.2.5
thymine 8.1
– degradation 8.2.4
thymine glycol 10.3.1
thymine-DNA glycosylase 11.2.4
thymineless death 8.2.5
thymus 19.1.2
thyreoglobulin 4.7.5
thyreoperoxidase 4.7.5
thyroid follicles 17.1.5
thyroid hormones 4.7.3, 4.7.5, 17.1.2, 17.1.5
– receptors for 17.6
thyroid stimulating hormone (TSH) 17.1.5
thyrotropin (TSH) 17.1.5
thyroxine 17.1.5
thyroxine binding globulin 4.7.5, 17.1.5
thyroxine deiodinase 4.7.5
tight junctions 2.2.4
Tim44 14.4.2
TIM-complex 14.4.2
tissue disintegration 14.6
tissue factor 20.2.1
tissue factor pathway inhibitor 20.3.3
tissue plasminogen activator 20.5.1
T lymphocytes 19.1.2
– activation and differentiation 19.1.7
– antigen recognition 19.1.4
– selection procedure 19.1.4
TNF receptors 19.1.6
TNFα 12.4.1, 19.1.7
tobacco mosaic virus 12.3.1
α-tocopherol 4.5.8, 4.7.3, 7.3.4, 9.12
TOM complex 14.4.2
tonoplast 2.2.3
topoisomerase I 11.3.2, 11.3.5
topoisomerase II 11.1.4
topoisomerase type IB 11.1.4
topoisomerases 10.2.3
torsion angles 2.3.1
TRAM protein 13.3.1
transaldolase 3.6.1
transamination reactions 4.2.1, 4.2.2, 4.2.5
transamination, oxaloacetate 4.9.1
transcarboxylase 6.1.1, 9.8.2
transcarboxylations 9.8.2
transcobalamin 9.5.1
transcortin 7.4.3, 7.5.2, 7.8.2, 17.1.5
transcription
– accuracy 10.4.4
– bacterial 10.4
– elongation 11.3.2
– eukaryotic 11.3
– inhibitors 10.4.5, 11.3.10
– initiation 10.4.2, 11.3.2
– modulation by hormones 17.1.2
– multiple copies 12.3
– regulation 10.5,11.4
– repression 11.6.4
– rRNA 11.3.5
– snRNA 11.3.4
– termination 11.3.2, 11.3.5
– tRNA 11.3.7
transcription anti-terminator 12.2.1
transcription bubble 10.4.2
transcription errors 10.1.3
transcription factors 11.3.2, 11.4, 17.3, 17.5.2, 17.5.3
– activation 11.1.1, 11.4.3, 11.6.2
– specific 11.4.2, 11.4.3
– structure 11.4.1
transcription rate 11.4.3
transcription start site 11.3.2
transcription termination sequence 11.3.2
transcription termination, modification 10.5.3
transcription-repair coupling factor (TRCF) 10.3.3

transducin 17.4.6
transesterification reactions 11.3.3
transferases 2.4.5
transferrin 5.2.1
transferrin reductase 5.2.1
transforming growth factor 11.6.4, 19.1.6
transforming growth factor-β 17.5.1
transglycosylase 15.1
trans-Golgi network 13.4
transhydrogenase 9.9.3, 16.1.2
transit peptides 14.5.1
transition complex 1.5.4
transition state 2.4.1
transketolase 3.6.1, 8.2.1, 16.2.2
translation
– bacterial 10.6.1
– control point 11.5.2
– elongation 11.5.2
– fidelity 10.6.4
– inhibitors 10.6.3
– initiation 10.6.3
– reinitiation 11.5.4
– regulation 11.5.4
– termination 10.6.3, 11.5.2
– translocation 10.6.3
translation factors 10.6.1, 11.5.1
translation initiation codon 10.6.3
translocase complex 15.2
translocation contact sites 14.4.2, 14.5.2
translocation motor model 14.4.2
translocon 13.3.1
transmembrane receptor proteins 17.3
transmitter gated ion channels 17.2.2
transmitter gated signalling 17.2.3
transmitter gating
– direct 17.2.3
– indirect 17.2.3
transpeptidase 15.1
transport
– primary active 18.1.1
– secondary active 18.1.1
– through membranes 1.5.3
transport proteins, associated with antigen processing 19.1.5
transport mechanisms
– axonal 17.2.6
– bacterial 15.3
– binding proteins dependent 15.3.1
– passive 18.1.1
– secondary active 15.3.1, 18.1.1
transport vesicles 14.2.2
trans-splicing 11.3.3
trehalolipids 15.7
TRH 17.1.5
triacylglycerols 1.4.2, 3.1.5
– biosynthesis 6.2.1
– hydrolyziation 18.2.4
– mobilization 6.2.2
– transport 18.2
trichothiodystrophy 11.2.6
triglycerides, see triacylglycerols
trihydroxyphenylalanine quinone 15.7
triiodothyronine 4.7.5, 17.1.5
trimethoprim 8.2.5
trimethylammonium compounds 6.3.5
triolein 1.4.2
triose-P isomerase 15.7
triskelion 18.1.2
tRNA 10.1.2, 10.4
– aminoacyl-tRNA ligases 10.6.2
– eukaryotic 11.3.8
– mature 11.3.7, 11.3.8
– modification 10.4.3, 11.3.6
– processing 11.3.8
– structure 10.6.1, 11.5.1
– transcription 11.3.7
tRNA genes 11.3.7
trombone model 11.1.4
trophic hormones 7.4.2
tropins 7.4.2, 17.1.5
tropomyosin 17.4.5
troponin 17.4.5
trp operon 10.5.1
– attenuation 10.5.3
trp repressor 10.5.1
trypsin 14.6.1
tryptamine 4.7.3
tryptase 19.1.8
tryptophan
– biosynthesis 4.7.1, 10.5.1
– degradation 4.7.3
– derivatives 4.7.3

tryptophan synthase 4.7.1
tryptophantryptophyl quinone 15.7
TSH releasing hormone (TRH) 17.1.5
T state 2.5.2
TTF-I protein 11.3.5
tubulin 18.1.3
– mitotic spindles 11.6.5
– posttranslational modification 18.1.3
tumor cell invasion 14.6
tumor cell migration 19.3
tumor necrosis factor 6.3.4
tumor necrosis factor-α 17.5.1, 19.3
tumor suppressor proteins 11.6.4
tumor viruses 12.4
tumorigenesis 11.6.4, 12.2, 17.5.1
tungsten enzymes 9.6.4
tunicamycin 13.3
turgor 2.2.3
turnover number 1.5.4
Tus proteins 10.2.3
two-substrate reactions 1.5.4, 2.4.1, 2.4.4
tyramine 4.7.3
tyrosine
– biosynthesis 4.7.1
– degradation 4.7.3
tyrosine derivatives 4.7.3
tyrosine kinase 17.5.1, 17.5.4
– Jak type 19.1.6
– of the Src family 19.1.4
tyrosine kinase cascacdes 17.3
tyrosine kinase domain 17.5.2
tyrosine kinase pathways 17.4.1
tyrosine kinase-associated receptors 17.5.4
tyrosine phosphatases 17.5.4
tyrosine-3-monooxygenase 17.1.4

U

UBF 11.6.4
ubihydroquinone 16.1.2
ubiquinol-cytochrome c reductase complex 16.1.2, 16.2.1
ubiquinone 7.3.4, 16.1.2, 16.1.3
– biosynthesis 4.7.2
ubiquitin 11.6.5, 11.6.6, 14.6.7
ubiquitin-protein ligase 11.1.1, 11.6.2, 11.6.5
UDP-D-galacturonate 3.5.1, 3.6.2
UDP-D-glucuronate 3.5.1, 3.6.2
UDP-L-iduronate 3.5.1
UDP-D-glucose 3.2.3, 16.2.2
UDP-glucose: glycoprotein glucosyltransferase 14.1.3
UDP-N-acetyl-D-glucosamine 3.7.1
UDP-N-acetylgalactosamine 3.7.1
umbelliferone 4.7.6
UMP 8.2.1
uncompetitive inhibition 1.5.4, 2.5.2
uncoupling protein 6.2.2, 16.1.4
undecaprenyl-P 7.3.3, 13.2.3
undecaprenyl sugars 7.3.1–7.3.4
uniport 15.3.1
upstream binding factor 11.3.5, 11.4.1
upstream control element 11.3.5, 11.4.1
upstream regulatory sequence 10.4.2, 10.5.4
uracil 8.1
– degradation 8.2.4
uracil DNA-glycosylase 11.2.3
urate 8.1.6
– fermentation 15.4
urea cycle 4.9.1
uric acid 4.5.8
uricotelic animals 4.9, 8.1.6
uridine 8.1
uridine-5′-phosphate (UMP), biosynthesis 8.2.1
uridine-5′-triphosphate (UTP) 3.2.2, 8.2.2
uridylyltransferase 16.2.2
urobilin 5.3.1
urobilinogen 5.3.1
urocanate 4.8.2
urocortisol 7.8.1
urocortisone 7.8.1
urokinase 20.5.1
uronic acids 3.5.1
– UDP-derivatives 3.6.2
uroporphyrinogen-III (co)synthase 5.1
uroporphyrinogen III 5.1
urotelic animals 4.9
urothione 9.6.4
UTP 3.2.2, 8.2.2
UvrA 10.3.3, 10.3.6
UvrB 10.3.3, 10.3.6
UvrC 10.3.3, 10.3.6

V

cis-vaccenate 6.1.3
vacuoles 2.2.3, 18.1.1
valine 15.8.1
– biosynthesis 4.6
– degradation 4.6.2
variable domains 19.1.3
vasoactive intestinal peptide 17.1.9
vasoconstriction 20.4
vasoconstrictive compounds 20.1
vasodilatoric compounds 20.1
vasopressin 17.1.5, 17.1.8
very low density lipoproteins (VLDL) 18.2
vesicles, presynaptic 17.2.3
vesicular transport 14.2
vimetin 17.4.5,18.1.3
vinblastin 17.2.6, 18.1.3
violaxanthin 7.3.1–7.3.4
viral envelope glycoprotein 12.4.1
viral glycoprotein precursor 12.4.1
virion 12.1
viroids 12.1
viruses
– genomic characteristics 12.1.1
– immune response against 19.1.7
– reproduction 12.1.1
– self-assembly 12.2.1, 12.3.1, 12.4.1
– structure 12.1.1
(+)RNA viruses 12.3
(–)RNA viruses 12.3
virus families 12.1.1
visual process 9.1.2, 17.4.6
vitamin A 7.3.2
– see also retinol
vitamin B, see thiamin
vitamin B_{12}, see cobalamin
vitamin B_2, see riboflavin
vitamin B_6 group 9.4
vitamin B_6, see pyridoxin
vitamin C 3.5.2, 3.6.2, 9.10
– see also ascorbate
vitamin D bindung protein 9.11.1
vitamin D group 9.11.1
vitamin D
– receptors for 9.11.2, 17.6
– see also calciferol
vitamin D_2 7.2.2
vitamin E 4.5.8, 7.3.2, 7.3.4
– see also tocopherol
vitamin F 9.14.2
vitamin K 4.2.2, 7.3.4, 20.3.1
– see also menaquinone
vitamin K_1 4.7.2
vitamin K_2 4.7.2, 7.3.4
vitamin K epoxide reductase 9.13
vitamin K hydroquinone 9.13
vitellogenin 18.2.2
vitronectin 20.5.2
vitronectin receptor 20.4
VLA-4 19.3
VLDL 6.2.1
voltage gated 18.1.1
voltage gated ion channels 17.2.2
voltage gated transmission 17.2.4
V segments 19.1.3
Vsr protein 10.3.4

W

warfarin 9.13
warts 12.1.1
water turnover 17.1.8
Waterhouse-Friederichsen syndrome 7.8.4
Watson-Crick pairing rules 2.6.1, 10.1.1, 10.4
wax alcohools 1.4.3
waxes 1.4.3
wobble hypothesis 2.6.2, 10.1.2, 11.5.2
Wolinella succinogenes 15.5.1, 15.6.1, 16.1, 16.1.3, 16.1.5

X

xanthine 8.1, 11.2.1
xanthine oxidase 8.1.6
xanthine oxidase family 9.6.4
xanthoma 18.2.4
xanthophylls 7.3.2, 16.2.1
xanthosine 8.1
xanthosine-5′-phosphate 8.1.2
xanthurenate 4.7.3
XDP-sugars 13.4.1

xenobiotics 4.5.7
Xeroderma pigmentosum 11.2.6
xylan xylosidase 3.6.3
xyloglucan 3.6.3
D-xylose 3.6.2, 3.6.4
xylose isomerase 3.1.1
D-xylulose-5-phosphate 3.6.4, 15.4, 16.2.215

Y

yeast
– DNA polymerase 11.1.3
– initiation of DNA replication 11.1.1
Yersinia outer proteins 15.2

Z

ZAP-70 17.5.4
zeaxanthin 7.3.1–7.3.4
Zellweger's syndrome 7.9.4
zinc fingers 11.3.2, 11.4.2
Zollinger-Ellison's syndrome 7.9.4
zona fascilata 7.8.1
zona glomerulosa 7.8.1
zymogen activation 2.5.1
zymogens 14.6
zymosterol 7.1.1